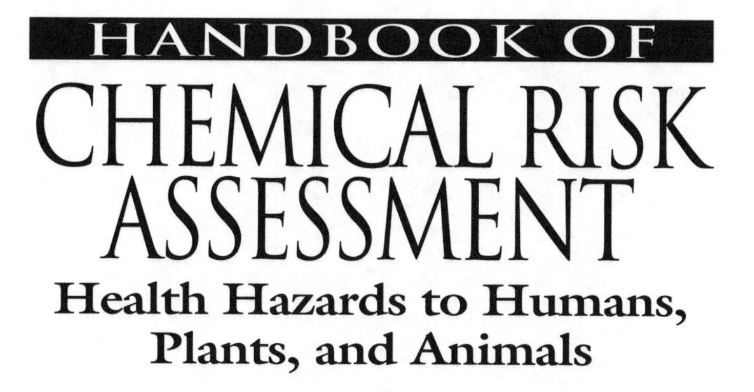

Health Hazards to Humans, Plants, and Animals

VOLUME 3

Metalloids, Radiation, Cumulative Index to Chemicals and Species

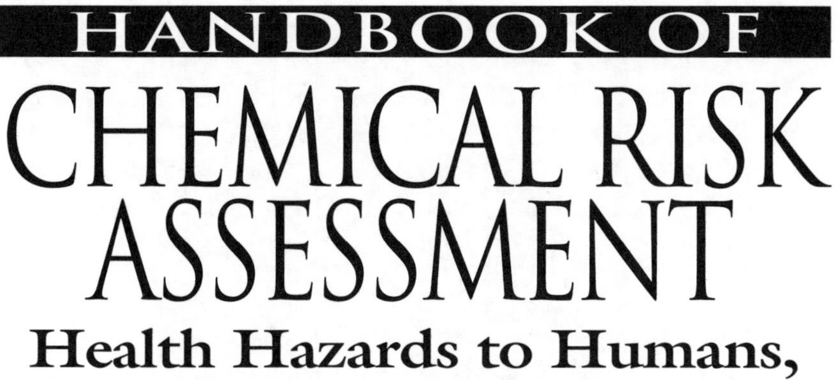

Health Hazards to Humans, Plants, and Animals

VOLUME 3

Metalloids, Radiation, Cumulative Index to Chemicals and Species

Ronald Eisler, Ph.D.

U.S. Geological Survey
Patuxent Wildlife Research Center
Laurel, Maryland

Library of Congress Cataloging-in-Publication Data

Eisler, Ronald, 1932–
Handbook of chemical risk assessment: health hazards to humans, plants, and animals / by Ronald Eisler
p. cm.
Includes bibliographical references and index.
Contents: v. 1. Metals — v. 2. Organics — v. 3. Metalloids, radiation, cumulative index to chemicals and species.
ISBN 1-56670-506-1 (alk. paper)
1. Pollutants — Handbooks, manuals, etc. 2. Hazardous substances — Risk assessment — Handbooks, manuals, etc. 3. Environmental chemistry — Handbooks, manuals, etc. 4. Environmental risk assessment — Handbooks, manuals, etc. I. Title.

TD196.C45 E38 2000
363.17¢9 — dc21

99-053394
CIP

Direct all inquiries to CRC Press LLC, 2000 N. W. Corporate Blvd., Boca Raton, Florida 33431.

First CRC Press LLC Printing, 2000
Lewis Publishers is an imprint of CRC Press LLC

International Standard Book Number 1-56670-506-1
Library of Congress Card Number 99-053394
Printed in the United States of America 1 2 3 4 5 6 7 8 9 0
Printed on acid-free paper

About the Author

Ronald Eisler received his A.B. degree from New York University in biology and chemistry, and M.S. and Ph.D. degrees from the University of Washington in aquatic sciences and radioecology, respectively. He attended the University of Miami Marine Laboratory prior to military service. He is a member of the American Chemical Society, the American Fisheries Society, the Marine Biological Association of the United Kingdom, and the Society for Environmental Geochemistry and Health. Since 1955 he has authored more than 125 technical articles and books on contaminant hazards to living organisms.

From 1984 to the present, Dr. Eisler has been a senior research biologist at the U.S. Geological Survey, Patuxent Wildlife Research Center, located in Laurel, Maryland. Prior to 1984, he held the following positions, in order: bioscience advisor, U.S. Fish and Wildlife Service, Washington, D.C.; research aquatic toxicologist, U.S. Environmental Protection Agency, Narrangansett, Rhode Island; fishery research biologist, U.S. Fish and Wildlife Service, Highlands, New Jersey; radiochemist, University of Washington Laboratory of Radiation Ecology, Seattle, Washington; aquatic biologist, New York State Department of Environmental Conservation, Raybrook, New York; biochemist, U.S. Army Medical Nutrition Laboratory, Medical Service Corps, Denver, Colorado.

He has held several adjunct professor appointments and was especially active at the Graduate School of Oceanography, University of Rhode Island; the Department of Oceanography and Marine Biology, Hebrew University of Jerusalem; and the Department of Biology, American University, Washington, D.C.

Dr. Eisler is currently on the editorial board of *Marine Ecology Progress Series* and other technical journals.

About the Author

Contents

Chapter 28
Arsenic
28.1 Introduction 1501
28.2 Sources, Fate, and Uses 1502
28.3 Chemical and Biochemical Properties 1505
28.4 Essentiality, Synergism, and Antagonism 1508
28.5 Concentrations in Field Collections 1509
28.5.1 General 1509
28.5.2 Nonbiological Samples 1510
28.5.3 Biological Samples 1514
28.6 Lethal and Sublethal Effects 1528
28.6.1 General 1528
28.6.2 Carcinogenesis, Mutagenesis, and Teratogenesis 1529
28.6.3 Terrestrial Plants and Invertebrates 1530
28.6.4 Aquatic Biota 1533
28.6.5 Birds 1542
28.6.6 Mammals 1544
28.7 Recommendations 1551
28.8 Summary 1556
28.9 Literature Cited 1557

Chapter 29
Boron
29.1 Introduction 1567
29.2 Environmental Chemistry 1568
29.2.1 General 1568
29.2.2 Sources and Uses 1568
29.2.3 Chemical Properties 1570
29.2.4 Mode of Action 1571
29.3 Concentrations in Field Collections 1573
29.3.1 General 1573
29.3.2 Nonbiological Materials 1573
29.3.3 Plants and Animals 1576
29.4 Effects 1581
29.4.1 General 1581
29.4.2 Terrestrial Plants 1581
29.4.3 Terrestrial Invertebrates 1584
29.4.4 Aquatic Organisms 1585
29.4.5 Birds 1591
29.4.6 Mammals 1593
29.5 Recommendations 1600
29.6 Summary 1604
29.7 Literature Cited 1605

Chapter 30
Molybdenum
30.1 Introduction 1613
30.2 Environmental Chemistry 1614

30.2.1 General 1614
30.2.2 Sources and Uses 1614
30.2.3 Chemical Properties 1615
30.2.4 Mode of Action 1615
30.3 Concentrations in Field Collections....................1617
30.3.1 General 1617
30.3.2 Nonbiological Samples 1617
30.3.3 Biological Samples....................1619
30.4 Effects....................1625
30.4.1 General 1625
30.4.2 Terrestrial Plants....................1626
30.4.3 Terrestrial Invertebrates....................1626
30.4.4 Aquatic Organisms....................1626
30.4.5 Birds 1630
30.4.6 Mammals....................1631
30.5 Recommendations 1637
30.6 Summary 1641
30.7 Literature Cited 1641

Chapter 31

Selenium

31.1 Introduction 1649
31.2 Environmental Chemistry....................1650
31.3 Concentrations in Field Collections....................1653
31.4 Deficiency and Protective Effects 1672
31.5 Lethal Effects 1675
31.5.1 Aquatic Organisms....................1675
31.5.2 Mammals and Birds 1678
31.6 Sublethal and Latent Effects....................1682
31.6.1 Aquatic Organisms....................1682
31.6.2 Terrestrial Invertebrates....................1685
31.6.3 Birds 1685
31.6.4 Mammals....................1685
31.7 Recommendations 1687
31.8 Summary 1692
31.9 Literature Cited 1693

Chapter 32

Radiation

32.1 Introduction 1707
32.2 Physical Properties of Radiation....................1708
32.2.1 General 1708
32.2.2 Electromagnetic Spectrum 1709
32.2.3 Radionuclides 1710
32.2.4 Linear Energy Transfer 1716
32.2.5 New Units of Measurement....................1716
32.3 Sources and Uses 1717
32.3.1 General 1717
32.3.2 Natural Radioactivity 1717
32.3.3 Anthropogenic Radioactivity 1718
32.3.4 Dispersion....................1723

32.4 Radionuclide Concentrations in Field Collections 1725
32.4.1 General 1725
32.4.2 Abiotic Materials 1725
32.4.3 Aquatic Ecosystems 1729
32.4.4 Birds 1731
32.4.5 Mammals 1732
32.5 Case Histories 1749
32.5.1 Pacific Proving Grounds 1749
32.5.2 Chernobyl 1752
32.6 Effects: Nonionizing Radiations 1769
32.7 Effects: Ionizing Radiations 1771
32.7.1 General 1771
32.7.2 Terrestrial Plants and Invertebrates 1774
32.7.3 Aquatic Organisms 1777
32.7.4 Amphibians and Reptiles 1784
32.7.5 Birds 1786
32.7.6 Mammals 1788
32.8 Proposed Criteria and Recommendations 1800
32.9 Summary 1805
32.10 Literature Cited 1807
32.11 Glossary 1824

Chapter 33

Cumulative Index to Chemicals and Species

33.1 Introduction 1829
33.2 Index to Chemicals 1829
33.3 Index to Species 1829
33.4 Summary 1830
33.5 Literature Cited 1830

Index 1893

LIST OF TABLES

28.1 Total arsenic concentrations in selected nonbiological materials 1511
28.2 Arsenic concentrations in field collections of selected species of plants and animals 1516
28.3 Lethal and sublethal effects of various arsenic compounds on selected species of terrestrial plants and invertebrates 1532
28.4 Lethal and sublethal effects of various arsenic compounds on selected species of aquatic biota 1536
28.5 Lethal and sublethal effects of various arsenicals on selected species of birds 1543
28.6 Lethal and sublethal effects of various arsenicals on selected species of mammals 1547
28.7 Proposed arsenic criteria for the protection of selected natural resources and human health 1553

29.1 Environmental sources of domestic boron 1569
29.2 Boron concentrations in selected nonbiological materials 1574
29.3 Boron concentrations in field collections of selected species of plants and animals 1577
29.4 Boron toxicity to some terrestrial plants 1583
29.5 Boron concentrations in soil water associated with optimal growth and plant injury 1584
29.6 Lethal and sublethal effects of boron on terrestrial invertebrates 1585
29.7 Lethal and sublethal effects of boron on aquatic organisms 1586
29.8 Lethal and sublethal effects of boron on birds 1592
29.9 Lethal and sublethal effects of boron on mammals 1595
29.10 Proposed boron criteria for the protection of natural resources and human health 1602

30.1 Molybdenum concentrations in selected nonbiological materials 1618
30.2 Molybdenum concentrations in field collections of selected species of animals and plants 1621
30.3 Molybdenum effects on selected species of aquatic organisms 1628
30.4 Molybdenum effects on selected species of mammals 1634
30.5 Proposed molybdenum criteria for the protection of living resources and human health 1638

31.1 Selenium concentrations in nonbiological materials 1653
31.2 Selenium concentrations in field populations of selected species of flora and fauna 1657
31.3 Toxicity of selenium salts to aquatic biota 1675
31.4 Selenium effects on birds 1680
31.5 Proposed criteria for prevention of selenium deficiency and for protection against selenosis 1687

32.1 Selected radionuclides: symbol, mass number, atomic number, half-life, and decay mode 1710
32.2 Radiation weighting factors for various types of ionizing radiations 1716
32.3 New units for use with radiation and radioactivity 1716
32.4 Annual effective dose equivalent to humans from natural sources of ionizing radiation 1717

32.5 Annual whole-body radiation doses to humans from various sources 1719
32.6 Sources and applications of atomic energy 1719
32.7 Annual effective dose equivalent from nuclear weapons testing to humans in the north temperate zone 1720
32.8 Estimated fallout of ^{90}Sr and ^{137}Cs over the Great Lakes, 1954–1983, in cumulative millions of Bq/km^2 1720
32.9 Fission products per kg ^{235}U reactor charge at 100 days cooling 1722
32.10 Radioactive waste disposal at sea 1722
32.11 Theoretical peak dose, in microsieverts per year, received from plutonium and americium by three human populations 1723
32.12 Time required to transport selected radionuclides added into marine waters at surface from the upper mixed layer by biological transport 1724
32.13 Radionuclide concentrations in field collections of selected abiotic materials 1726
32.14 Radionuclide concentrations in field collections of selected living organisms 1734
32.15 Radionuclide concentrations in selected samples from the Pacific Proving Grounds 1750
32.16 Selected fission products in the Chernobyl reactor core and their estimated escape into the environment 1753
32.17 Regional total effective human dose-equivalent commitment from the Chernobyl accident 1754
32.18 Radionuclide concentrations in biotic and abiotic materials from various geographic locales before or after the Chernobyl nuclear accident on April 26, 1986 1761
32.19 Radiation effects on selected terrestrial plants 1775
32.20 Radiation effects on selected terrestrial invertebrates 1777
32.21 Radiation effects on selected aquatic organisms 1778
32.22 Concentration factors for cesium-137 and strontium-90 in aquatic organisms 1782
32.23 Approximate maximum concentration factors for selected transuranics in marine sediments, macroalgae, and fish 1782
32.24 Maximum concentration factors reported for selected elements in marine organisms at various trophic levels 1783
32.25 Radiation effects on selected amphibians and reptiles 1785
32.26 Radiation effects on selected birds 1787
32.27 Radiation effects on selected mammals 1788
32.28 Recommended radiological criteria for the protection of human health 1802

33.1 Chemical and trade names of substances listed in the *Handbook of Chemical Risk Assessment* series 1831
33.2 Common and scientific names of plants and animals listed in the *Handbook of Chemical Risk Assessment* series 1847

LIST OF FIGURES

32.1 The spectrum of electromagnetic waves, showing the relationship between wavelength, frequency, and energy 1709
32.2 The principal uranium-238 decay series, indicating major decay mode and physical half-time of persistence 1714
32.3 The three still-existing natural decay series 1715
32.4 Natural radiations in selected radiological domains 1718
32.5 Plutonium-239+240 in environmental samples at Thule, Greenland, between 1970 and 1984, after a military accident in 1968 1730
32.6 Chernobyl air plume behavior and reported initial arrival times of detectable radioactivity 1754
32.7 Acute radiation dose range fatal to 50% (30 days postexposure) of various taxonomic groups 1772
32.8 Relation between diet, metabolism, and body weight with half-time retention of longest-lived component of cesium-137 1774
32.9 Survival time and associated mode of death of selected mammals after whole-body doses of gamma radiation 1795
32.10 Relation between body weight and radiation-induced LD50 (30 days postexposure) for selected mammals 1796

Contents of Volumes 1 and 2

Volume 1

Metals

1. Cadmium 1
2. Chromium 45
3. Copper 93
4. Lead 201
5. Mercury 313
6. Nickel 411
7. Silver 499
8. Tin 551
9. Zinc 605

Volume 2

Organics

10. Acrolein 739
11. Atrazine 767
12. Carbofuran 799
13. Chlordane 823
14. Chlorpyrifos 883
15. Cyanide 903
16. Diazinon 961
17. Diflubenzuron 983
18. Dioxins 1021
19. Famphur 1067
20. Fenvalerate 1089
21. Mirex 1133
22. Paraquat 1159
23. Pentachlorophenol 1193
24. Polychlorinated Biphenyls 1237
25. Polycyclic Aromatic Hydrocarbons 1343
26. Sodium Monofluoroacetate 1413
27. Toxaphene 1459

Preface

Risk assessment is an inexact science. Successful risk assessment practitioners rely heavily on extensive and well-documented databases. In the case of chemicals entering the environment as a result of human activities, these databases generally include the chemical's source and use; its physical, chemical, and metabolic properties; concentrations in field collections of abiotic materials and living organisms; deficiency effects, in the case of chemicals essential for life processes; lethal and sublethal effects, including effects on survival, growth, reproduction, metabolism, mutagenicity, teratogenicity, and carcinogenicity; proposed regulatory criteria for the protection of human health and sensitive natural resources; and recommendations for additional research when databases are incomplete. Of the hundreds of thousands of chemicals discharged into the environment yearly from agricultural, domestic, industrial, mining, manufacturing, municipal, and military operations, there is sufficient information on only a small number to attempt preliminary risk assessments.

The chemicals selected for inclusion in this handbook series were recommended by environmental specialists of the U.S. Fish and Wildlife Service and other resource managers responsible for protecting our fish, wildlife, and biological diversity. Their choices were based on real or potential impact of each contaminant on populations of free-living natural resources — including threatened and endangered species — and on insufficient knowledge on how best to mitigate damage effects. Each chapter selectively reviews and synthesizes the technical literature on a specific priority contaminant in the environment and its effects on notably terrestrial plants and invertebrates, aquatic plants and animals, avian and mammalian wildlife, and other natural resources. Early versions of individual chapters were published between 1985 and 1999 in the *Contaminant Hazard Reviews* series of the U.S. Department of the Interior. This series rapidly became an important reference tool for contaminant specialists, scientists, educational institutions, state and local governments, natural resources agencies, business and industry, and the general public. More than 105,000 copies of individual reports were distributed in response to specific requests before supplies became exhausted. The current offering is updated to reflect the burgeoning literature in this subject area. I authored the original versions and the updates while stationed at the Patuxent Wildlife Research Center; however, all interpretations are my own and do not necessarily reflect those of the U.S. government. Moreover, mention of trade names or commercial products is not an endorsement or recommendation for use by the U.S. government.

Ronald Eisler, Ph.D.
Senior Research Biologist
U.S. Geological Survey
Patuxent Wildlife Research Center
Laurel, Maryland

Acknowledgments

This project was completed under the aegis of James A. Kushlan, director of the U.S. Geological Survey Patuxent Wildlife Research Center (PWRC), and Richard L. Jachowski, chief scientist at PWRC. I am grateful for their support and encouragement. Other research managers and scientists at PWRC who contributed significantly to the project over the 17 years that it was operative — and to whom I am indebted — include Peter H. Albers, Andre A. Belisle, W. Nelson Beyer, Lawrence J. Blus, Donald R. Clark, Jr., Nancy C. Coon, Harry N. Coulombe, Thomas W. Custer, Lawrence R. DeWeese, W. James Fleming, John B. French, Jr., Martha L. Gay, Christopher E. Grue, Russell J. Hall, Gary H. Heinz, Charles J. Henny, Paula F. P. Henry, Elwood F. Hill, David J. Hoffman, Kirk A. King, Alex J. Krynitsky, Jerry R. Longcore, T. Peter Lowe, Mark J. Melancon, John F. Moore, Harold J. O'Conner, Harry M. Ohlendorf, Oliver H. Pattee, H. Randolph Perry, Jr., Matthew C. Perry, Barnett Rattner, Gregory J. Smith, Donald W. Sparling, Charles J. Stafford, Kenneth L. Stromborg, Douglas M. Swineford, Michael W. Tome, David L. Trauger, Donald H. White, and Stanley N. Wiemeyer.

Computer and graphics assistance was kindly provided by PWRC specialists Kinard Boone, Henry C. Bourne, Lois M. Loges, and Robert E. Munro. Library services — a major component of this project — were expertly provided by Lynda J. Garrett, chief librarian at PWRC, and a number of assistant librarians, most notably P. Wanda Manning.

I also acknowledge the technical assistance provided in early phases of this project by numerous scientists and editors from other U.S. Department of the Interior installations, especially researchers from the U.S. Geological Survey Columbia Environmental Research Center, and by scientists from the U.S. Environmental Protection Agency, the U.S. Department of Agriculture, various universities, the International Atomic Energy Agency, the Canadian Wildlife Service, and the U.S. Department of Energy.

CHAPTER 28

Arsenic

28.1 INTRODUCTION

Anxiety over arsenic (As) is understandable and frequently justifiable. Arsenic compounds were the preferred homicidal and suicidal agents during the Middle Ages, and arsenicals have been regarded largely in terms of their poisonous characteristics in the nonscientific literature (National Academy of Sciences [NAS] 1977). Acute arsenic poisoning was reported in the 1800s in horses, cows, deer, and foxes as a result of emissions from metal smelters (Ismail and Roberts 1992). In 1885, arsenic accounted for nearly one third of the homicide poisonings in France (North et al. 1997). Data collected on animals, including humans, indicated that inorganic arsenic can cross the placenta and produce mutagenic, teratogenic, and carcinogenic effects in offspring (Nagymajtenyi et al. 1985; Agency for Toxic Substances and Disease Registry [ATSDR] 1992). Correlations between elevated atmospheric arsenic levels and mortalities from cancer, bronchitis, and pneumonia were established in an epidemiological study in England and Wales, where deaths from respiratory cancer were increased at air concentrations >3 μg As/m^3 (National Research Council of Canada [NRCC] 1978). Chronic arsenical poisoning, including skin cancer and a gangrenous condition of the hands and feet called Blackfoot's disease, has occurred in people from several communities in Europe, South America, and Taiwan who were exposed to elevated concentrations of arsenic in drinking water (U.S. Environmental Protection Agency [USEPA] 1980; Lin et al. 1998). Inorganic arsenic in dietary staples such as yams and rice may have contributed substantially to exposure and adverse health effects observed in an endemic Taiwanese population historically exposed to arsenic in drinking water (Schoof et al. 1998). Severe health effects, including cancer and death, have recently been documented in Mongolia (Luo et al. 1997; Ma et al. 1998), Bangladesh (Biswas et al. 1998), Chile (Smith et al. 1998), and India (Chowdhury et al. 1997) from ingestion of naturally elevated levels of arsenic in the drinking water, and in Thailand from the use of traditional arsenic-containing medicines and arsenical wastes from mining activities (Choprapawon and Rodcline 1997). In Japan, about 12,000 infants were poisoned (128 deaths) after consuming dry milk containing 15 to 24 mg inorganic As/kg, which originated from contaminated sodium phosphate used as a milk stabilizer. Fifteen years after exposure, the survivors sustained an elevated frequency of severe hearing loss and brain-wave abnormalities (Pershagen and Vahter 1979).

Many reviews on ecotoxicological aspects of arsenic in the environment are available, including those by Woolson (1975), NAS (1977), NRCC (1978), Pershagen and Vahter (1979), USEPA (1980, 1985), Hood (1985), Andreae (1986), Sorensen (1987), Eisler (1988, 1994), Phillips (1990), ATSDR (1992), Abernathy et al. (1997), Society for Environmental Geochemistry and Health [SEGH] (1998) and the U.S. Public Health Service [USPHS] (1998). These authorities agree on six points:

1. Arsenic is a relatively common element and is present in air, water, soil, plants, and in all living tissues.
2. Arsenicals have been used in medicine as chemotherapeutics since 400 BCE, and organoarsenicals were used extensively for this purpose until about 1945, with no serious effects when judiciously administered.
3. Large quantities of arsenicals are released into the environment as a result of industrial and especially agricultural activities, and these may pose potent ecological dangers.
4. Exposure of humans and wildlife to arsenic may occur through air (emissions from smelters, coal-fired power plants, herbicide sprays), water (mine tailings runoff, smelter wastes, natural mineralization), and food (especially seafoods).
5. Chronic exposure to arsenicals by way of the air, diet, and other routes has been associated with liver, kidney, and heart damage, hearing loss, brain-wave abnormalities, and impaired resistance to viral infections.
6. Exposure to arsenic has been associated with different types of human cancers such as respiratory cancers and epidermoid carcinomas of the skin, as well as precancerous dermal keratosis. The epidemiological evidence of human carcinogenicity is supported by carcinogenesis in experimental animals (Deknudt et al. 1986).

28.2 SOURCES, FATE, AND USES

Global production of arsenic is estimated to be 75,000 to 100,000 tons annually, of which the United States produces about 21,000 tons and uses about 44,000 tons. Major quantities are imported from Sweden, the world's leading producer (NAS 1977; USEPA 1980). Imports of arsenic trioxide (As_2O_3) to the United States have increased from about 16 million kg in 1985 to about 30 million kg in 1989; imports of elemental arsenic have increased from about 0.45 million kg in 1985/88 to 1.2 million kg in 1989 (ATSDR 1992). The United States exports about 0.3 million kg arsenic compounds annually, mainly as arsenic acid, sodium arsenate, and lead arsenate (ATSDR 1992). Almost all (97%) of the arsenic made worldwide enters end-product manufacture in the form of arsenic trioxide, and the rest is used as additives in producing special lead and copper alloys (NAS 1977). About 74% of the total arsenic trioxide is in products used for wood preservation (ATSDR 1992). Most of the rest (about 19%) is used in production of agricultural chemicals, such as insecticides, herbicides, fungicides, algicides, and growth stimulants for plants and animals (ATSDR 1992). Smaller amounts are used in the production of glass, in the electronic industry, dyestuffs, and in veterinary and human medicines, including medicines for the eradication of tapeworm in sheep and cattle (NAS 1977; ATSDR 1992; Eisler 1994). The sole producer and refiner of As_2O_3 in the United States is a copper smelter in Tacoma, Washington (NAS 1977).

Arsenic naturally occurs as sulfides and as complex sulfides of iron, nickel, and cobalt (Woolson 1975). In one form or another, arsenic is present in rocks, soils, water, and living organisms at concentrations of parts per billion to parts per million (NAS 1977). Soil arsenic levels are normally elevated near arseniferous deposits and in mineralized zones containing gold, silver, and sulfides of lead and zinc (Dudas 1984; Thornton and Farago 1997). Secondary iron oxides formed from the weathering of pyrite act as scavengers of arsenic (Dudas 1984). Pyrite is a known carrier of arsenic and may contain up to 5600 mg/kg. For example, total arsenic is 10 times above normal background levels in soils derived from pyritic shale (Dudas 1984). Natural weathering of rocks and soils adds about 40,000 tons of arsenic to the oceans yearly, accounting for <0.01 mg/L input to water on a global basis (NRCC 1978). Many species of marine plants and animals often contain naturally high concentrations of arsenic (NAS 1977), but it is usually present in a harmless organic form (Woolson 1975). Anthropogenic input of arsenic to the environment is substantial and exceeds that contributed by natural weathering processes by a factor of about 3 (NRCC 1978).

The most important concept with respect to arsenic cycling in the environment is constant change. Arsenic is ubiquitous in living tissue and is constantly being oxidized, reduced, or otherwise metabolized. In soils, insoluble or slightly soluble arsenic compounds are constantly being resolubilized, and the arsenic is being presented for plant uptake or reduction by organisms and chemical processes. Humans reportedly modify the arsenic cycle only by causing localized high concentrations (NAS 1977). The speciation of arsenic in the environment is affected partly by indiscriminate biological uptake, which consumes about 20% of the dissolved arsenate pool and results in measurable concentrations of reduced and methylated arsenic species. The overall arsenic cycle is similar to the phosphate cycle; however, regeneration time for arsenic is much slower — on the order of several months (Sanders 1980). The ubiquity of arsenic in the environment is evidence of the redistribution processes that have been operating since early geologic time (Woolson 1975). A prehuman steady-state solution to the global arsenic cycle (Austin and Millward 1984) indicates that major reservoirs of arsenic (in kilotons) are magma (50 billion), sediments (25 billion), oceanic deep waters (1.56 million), land (1.4 million), and ocean mixed layers (270,000); minor amounts occur in ocean particulates (100), and in continental (2.5) and marine tropospheres (0.069). Arsenic is significantly mobilized from the land to the troposphere by both natural and anthropogenic processes. Industrial emissions account for about 30% of the present-day burden of arsenic in the troposphere (Austin and Millward 1984). Agronomic ecosystems, for example, may receive arsenic from agricultural sources such as organic herbicides, irrigation waters, and fertilizers, and from such nonagricultural sources as fossil fuels and industrial and municipal wastes (Woolson 1975). Arsenic is mobile and nonaccumulative in air, plant, and water phases of agronomic ecosystems; arsenicals sometimes accumulate in soils, but redistribution mechanisms usually preclude hazardous accumulations (Woolson 1975).

Arsenic compounds have been used in medicine since the time of Hippocrates, ca. 400 BCE (Woolson 1975). Inorganic arsenicals have been used for centuries, and organoarsenicals for at least a century in the treatment of syphilis, yaws, amoebic dysentery, and trypanosomiasis (NAS 1977). During the period 1200 to 1650, however, arsenic was used extensively in homicides (NRCC 1978). In 1815, the first accidental death was reported from arsine (AsH_3) poisoning; and between 1900 and 1903, accidental poisonings from consumption of arsenic-contaminated beer were widely reported (NRCC 1978). In 1938, it was established that arsenic can counteract selenium toxicity (NRCC 1978; ATSDR 1992). The introduction of arsphenamine, an organoarsenical, to control venereal disease earlier this century gave rise to intensive research by organic chemists, which resulted in the synthesis of at least 32,000 arsenic compounds. But the advent of penicillin and other newer drugs nearly eliminated the use of organic arsenicals as human therapeutic agents (USEPA 1980). Arsenical drugs are still used in treating certain tropical diseases, such as African sleeping sickness and amoebic dysentery, and are used in veterinary medicine to treat parasitic diseases, including filariasis in dogs (*Canis familiaris*) and blackhead in turkeys (*Meleagris gallopavo*) and chickens, *Gallus* spp. (NAS 1977). Today, abnormal sources of arsenic that can enter the food chain from plants or animals include arsenical pesticides such as lead arsenate; arsenic acid, $HAsO_3$; sodium arsenite, $NaAsO_2$; sodium arsenate, Na_2AsO_4; and cacodylic acid, $(CH_3)_2As(OH)$ (NAS 1977).

The major uses of arsenic are in the production of herbicides, insecticides, desiccants, growth stimulants for plants and animals, and especially wood preservatives. Much smaller amounts are used in the manufacture of glass (nearly all of which contains 0.2% to 1.0% arsenic as an additive — primarily as a decolorizing agent) and textiles, and in medical and veterinary applications (NAS 1977; USEPA 1980; Hamasaki et al. 1995). Arsenic is also an ingredient in lewisite, a blistering poison gas developed (but not used) during World War I, and in various police riot-control agents (NAS 1977). The availability of arsenic in certain local areas has been increased by various human activities:

- Smelting and refining of gold, silver, copper, zinc, uranium, and lead ores
- Combustion of fossil fuels, such as coal and gasoline
- Burning of vegetation from cotton gins treated with arsenical pesticides
- Careless or extensive use of arsenical herbicides, pesticides, and defoliants

- Dumping of land wastes and sewage sludge (1.1 mg/L) in areas that allow leaching into groundwater
- Use of domestic detergents in wash water (2.5 to 1000 mg As/L)
- Manufacture of glass
- Sinking of drinking water wells into naturally arseniferous rock (NRCC 1978; USEPA 1980).

There are several major anthropogenic sources of environmental arsenic contamination: industrial smelters — the effluent from a copper smelter in Tacoma, Washington, contained up to 70 tons arsenic discharged yearly into nearby Puget Sound (NRCC 1978); coal-fired power plants, which collectively emit about 3000 tons arsenic annually in the United States (USEPA 1980); and production and use of arsenical pesticides, coupled with careless disposal of used pesticide containers (NAS 1977). Elevated levels of arsenic have been reported in soils near smelters, in acid mine spoils, and in orchards receiving heavy applications of lead arsenate (NAS 1977; Dudas 1984). Air concentrations of arsenic are elevated near metal smelters, near sources of coal burning, and wherever arsenical pesticides are applied (NAS 1977). Atmospheric deposition of arsenic has steadily increased for at least 30 years, as judged by sedimentary evidence from lakes in upstate New York (Smith et al. 1987). Arsenic is introduced into the aquatic environment through atmospheric deposition of combustion products and through runoff from flyash storage areas near power plants and nonferrous smelters (Smith et al. 1987). Elevated arsenic concentrations in water were recorded near mining operations and from mineral springs and other natural waters — usually alkaline and with high sodium and bicarbonate contents (NAS 1977). In the United States, the most widespread and frequent increases in dissolved arsenic concentrations in river waters were in the northern Midwest. All evidence suggests that increased atmospheric deposition of fossil fuel combustion products was the predominant cause of the trend (Smith et al. 1987).

Agricultural applications provide the largest anthropogenic source of arsenic in the environment (Woolson 1975). Inorganic arsenicals (arsenic trioxide; arsenic acid; arsenates of calcium, copper, lead, and sodium; and arsenites of sodium and potassium) have been used widely for centuries as insecticides, herbicides, algicides, and desiccants. Paris green (cuprous arsenite) was successfully used in 1867 to control the Colorado potato beetle (*Leptinotarsa decemlineata*) in the eastern United States. Arsenic trioxide has been applied widely as a soil sterilant. Sodium arsenite has been used for aquatic weed control, as a defoliant to kill potato vines before tuber harvest, as a weed killer along roadsides and railroad rights-of-way, and for control of crabgrass (*Digitaria sanguinalis*). Calcium arsenates have been applied to cotton and tobacco fields to protect against the boll weevil (*Anthonomus grandis*) and other insects. Lead arsenate has been used to control insect pests of fruit trees, and for many years was the only insecticide that controlled the codling moth (*Carpocapsa pomonella*) in apple orchards and the horn worm larva (Sphyngidae) on tobacco. Much smaller quantities of lead arsenate are now used in orchards because fruit growers rely primarily on carbamate and organophosphorus compounds to control insect pests; however, lead arsenate is still being used by some growers to protect orchards from certain chewing insects. The use of inorganic arsenicals has decreased due to the banning of sodium arsenite and some other arsenicals for most purposes, although they continue to be used on golf greens and fairways in certain areas to control annual bluegrass (*Poa annua*). In recent decades, inorganic arsenicals have been replaced by organoarsenicals for herbicidal application, and by carbamate and organophosphorus compounds for insect control (Woolson 1975). By the mid-1950s, organoarsenicals were used extensively as desiccants, defoliants, and herbicides (NRCC 1978). Organoarsenicals marketed in agriculture today, which are used primarily for herbicidal application, include cacodylic acid (also known as dimethylarsinic acid) and its salts — monosodium and disodium methanearsonate (Woolson 1975; NAS 1977). Organoarsenicals are used as selective herbicides for weedy grasses in turf, and around cotton and noncrop areas for weed control. At least 1.8 million ha (4.4 million acres) have been treated with more than 8000 tons of organoarsenicals (NAS 1977). In 1945, it was discovered that one organoarsenical (3-nitro-4-hydroxyphenyl arsonic acid) controlled coccidiosis and promoted growth in domestic chicken (Woolson 1975). Since that time, other substituted phenylarsonic acids have been shown to have both therapeutic and growth-promoting properties as feed additives for poultry and swine (*Sus* spp.), and are used for this purpose today under existing regulations (Woolson 1975; NAS 1977) — although the use of arsenicals in poultry food was banned in France in 1959 (NRCC 1978).

28.3 CHEMICAL AND BIOCHEMICAL PROPERTIES

Elemental arsenic is a gray, crystalline material characterized by atomic number 33, atomic weight of 74.92, density of 5.727, melting point of 817°C, sublimation at 613°C, and chemical properties similar to those of phosphorus (Woolson 1975; NAS 1977; NRCC 1978; USEPA 1980, 1985; ATSDR 1992). Arsenic has four valence states: –3, 0, +3, and +5. Arsines and methylarsines, which are characteristic of arsenic in the –3 oxidation state, are generally unstable in air. Elemental arsenic, As^0 is formed by the reduction of arsenic oxides. Arsenic trioxide (As^{+3}) is a product of smelting operations and is the material used in synthesizing most arsenicals. It is oxidized catalytically or by bacteria to arsenic pentoxide (As^{+5}) or orthoarsenic acid (H_3AsO_4). Arsenic in nature is rarely in its free state. Usually, it is a component of sulfidic ores, occurring as arsenides and arsenates, along with arsenic trioxide, which is a weathering product of arsenides. Most arsenicals degrade or weather to form arsenate, although arsenite may form under anaerobic conditions. Biotransformations may occur, resulting in volatile arsenicals that normally are returned to the land where soil adsorption, plant uptake, erosion, leaching, reduction to arsines, and other processes occur. This natural arsenic cycle reflects a constant shifting of arsenic between environmental compartments. Atomic absorption spectrometry is the most common procedure for measuring arsenic in biological materials, although other methods are used, including neutron activation (ATSDR 1992). A variety of sensitive techniques have been used to obtain speciation data for the forms of arsenic at trace levels (Hamasaki et al. 1995; Crecelius et al. 1998).

Arsenic species in flooded soils and water are subject to chemically and microbiologically mediated oxidation or reduction and methylation reactions (Tamaki and Frankenberger 1992; Hamasaki et al. 1995). At high Eh values (i.e., high oxidation–reduction potential) typical of those encountered in oxygenated waters, pentavalent As^{+5} tends to exist as H_3AsO_4, H_2AsO_4, $HAsO_2$, and AsO_4^{-3}. At lower Eh, the corresponding trivalent arsenic species can be present, as well as AsS_2 (Thanabalasingam and Pickering 1986). In aerobic soils, the dominant arsenic species is As^{+5}, and small quantities of arsenite and monomethylarsonic acid are present in mineralized areas. In anaerobic soils, As^{+3} is the major soluble species (Haswell et al. 1985). Inorganic arsenic is more mobile than organic arsenic, and thus poses greater problems by leaching into surface waters and groundwater (NRCC 1978). The trivalent arsenic species are generally considered to be more toxic, more soluble, and more mobile than As^{+5} species (Thanabalasingam and Pickering 1986). Soil microorganisms metabolize arsenic into volatile arsine derivatives. Depending on conditions, 17% to 60% of the total arsenic present in soil may be volatilized (NRCC 1978). Estimates of the half-life of arsenic in soil vary from 6.5 years for arsenic trioxide to 16 years for lead arsenate (NRCC 1978).

In water, arsenic occurs in both inorganic and organic forms, and in dissolved and gaseous states (USEPA 1980). The form of arsenic in water depends on Eh, pH, organic content, suspended solids, dissolved oxygen, and other variables (USEPA 1985). Arsenic in water exists primarily as a dissolved ionic species; particulates account for less than 1% of the total measurable arsenic (Maher 1985a). Arsenic is rarely found in water in the elemental state (0) and is found in the –3 state only at extremely low Eh values (Lima et al. 1984). Common forms of arsenic encountered in water are arsenate, arsenite, methanearsonic acid, and dimethylarsinic acid (USEPA 1985). The formation of inorganic pentavalent arsenic, the most common species in water, is favored under conditions of high dissolved oxygen, basic pH, high Eh, and reduced content of organic material; reverse conditions usually favor the formation of arsenites and arsenic sulfides (NRCC 1978;

Pershagen and Vahter 1979; USEPA 1980), although some arsenite is attributed to biological activity (Maher 1985a). Water temperature seems to affect arsenic species composition in the estuary of the River Beaulieu in the United Kingdom, where reduced and methylated species predominate during warmer months and inorganic As^{+5} predominates during the colder months. The appearance of methylated arsenicals during the warmer months is attributed both to bacterial and abiotic release from decaying plankton and to grazing by zooplankton (Howard et al. 1984). Also contributing to higher water or mobile levels are the natural levels of polyvalent anions, especially phosphate species. Phosphate, for example, displaces arsenic held by humic acids, and it sorbs strongly to the hydrous oxides of arsenates (Thanabalasingam and Pickering 1986).

Marine algae transform arsenate into nonvolatile methylated arsenic compounds such as methanearsonic and dimethylarsinic acids (Tamaki and Frankenberger 1992). Freshwater algae and macrophytes, like marine algae, synthesize lipid-soluble arsenic compounds and do not produce volatile methylarsines. Terrestrial plants preferentially accumulate arsenate over arsenite by a factor of about 4. Phosphate inhibits arsenate uptake by plants, but not the reverse. The mode of toxicity of arsenate in plants is to partially block protein synthesis and interfere with protein phosphorylation — a process that is prevented by phosphate (Tamaki and Frankenberger 1992).

Physical processes play a key role in governing arsenic bioavailability in aquatic environments. For example, arsenates are readily sorbed by colloidal humic material under conditions of high organic content, low pH, low phosphate, and low mineral content (USEPA 1980; Thanabalasingam and Pickering 1986). Arsenates also coprecipitate with, or adsorb on, hydrous iron oxides and form insoluble precipitates with calcium, sulfur, aluminum, and barium compounds (USEPA 1980). Removal of arsenic from seawater by iron hydroxide scavenging seems to be a predominant factor in certain estuaries. The process involves both As^{+3} and As^{+5} and results in a measurable increase in arsenic levels in particulate matter, especially at low salinities (Sloot et al. 1985; Tremblay and Gobeil 1990). Arsenic sulfides are comparatively insoluble under conditions prevalent in anaerobic aqueous and sedimentary media containing hydrogen sulfide. Accordingly, these compounds may accumulate as precipitates and thus remove arsenic from the aqueous environment. In the absence of hydrogen sulfide, these sulfides decompose within several days to form arsenic oxides, sulfur, and hydrogen sulfide (NAS 1977).

In reduced environments, such as sediments, arsenate is reduced to arsenite and methylated to methylarsinic acid or dimethylarsenic acids: these compounds may be further methylated to trimethylarsine or reduced to dimethylarsine, and may volatilize to the atmosphere where oxidation reactions result in the formation of dimethylarsinic acid (Woolson 1975). Arsenates are more strongly adsorbed to sediments than are other arsenic forms, the adsorption processes depending strongly on arsenic concentration, sediment characteristics, pH, and ionic concentration of other compounds (USEPA 1980). An important mechanism of arsenic adsorption onto lake sediments involves the interaction of anionic arsenates and hydrous iron oxides. Evidence suggests that arsenic is incorporated into sediments at the time of hydrous oxide formation, rather than by adsorption onto existing surfaces (Aggett and Roberts 1986). Arsenic concentrations in lake sediments are also correlated with manganese; hydrous manganese oxides — positively charged for the adsorption of Mn^{+2} ions — play a significant role in arsenic adsorption onto the surface of lake sediments (Takamatsu et al. 1985). The mobility of arsenic in lake sediments and its release to the overlying water is related partly to seasonal changes. In areas that become stratified in summer, arsenic released from sediments accumulates in the hypolimnion until turnover, when it is mixed with epilimnetic waters. This mixing may result in a 10 to 20% increase in arsenic concentration (Aggett and O'Brien 1985). Microorganisms (including four species of fungi) in lake sediments oxidized inorganic As^{+3} to As^{+5} and reduced inorganic As^{+5} to As^{+3} under aerobic conditions; under anaerobic conditions, only reduction was observed (Freeman et al. 1986). Inorganic arsenic can be converted to organic alkyl arsenic acids and methylated arsines under anaerobic conditions by fungi, yeasts, and bacteria — although biomethylation may also occur under aerobic conditions (USEPA 1980).

Most arsenic investigators now agree on the following points:

1. Arsenic may be absorbed by ingestion, inhalation, or through permeation of the skin or mucous membranes.
2. Cells accumulate arsenic using an active transport system normally used in phosphate transport.
3. Arsenicals are readily absorbed after ingestion, most being rapidly excreted in the urine during the first few days, or at most a week (the effects seen after long-term exposure are probably a result of continuous daily exposure, rather than of accumulation).
4. The toxicity of arsenicals conforms to the following order, from greatest to least toxicity: arsines > inorganic arsenites > organic trivalent compounds (arsenoxides) > inorganic arsenates > organic pentavalent compounds > arsonium compounds > elemental arsenic.
5. Solubility in water and body fluids appears to be directly related to toxicity (the low toxicity of elemental arsenic is attributed to its virtual insolubility in water and body fluids, whereas the highly toxic arsenic trioxide, for example, is soluble in water to 12.0 g/L at 0°C, 21.0 g/L at 25°C, and 56.0 g/L at 75°C).
6. The mechanisms of arsenical toxicity differ considerably among arsenic species, although signs of poisoning appear similar for all arsenicals (Woolson 1975; NRCC 1978; Pershagen and Vahter 1979; Eisler 1988, 1994; ATSDR 1992; Abernathy et al. 1997; SEGH 1998).

The primary toxicity mode of inorganic As^{+3} is through reaction with sulfhydryl groups of proteins and subsequent enzyme inhibition; inorganic pentavalent arsenate does not react as readily as As^{+3} with sulfhydryl groups, but may uncouple oxidative phosphorylation (Howard et al. 1984; USEPA 1985). Inorganic As^{+3} interrupts oxidative metabolic pathways and sometimes causes morphological changes in liver mitochondria. Arsenite *in vitro* reacts with protein-SH groups to inactivate enzymes such as dihydrolipoyl dehydrogenase and thiolase, producing inhibited oxidation of pyruvate and beta-oxidation of fatty acids (Belton et al. 1985). Inorganic As^{+3} may also exert toxic effects by the reaction of arsenous acid (HAsO) with the sulfhydryl (SH) groups of enzymes. In the first reaction, arsenous acid is reduced to arsonous acid ($AsOH_2$), which then condenses to either monothiols or dithiols to yield dithioesters of arsonous acid. Arsonous acid may then condense with enzyme SH groups to form a binary complex (Knowles and Benson 1984a, 1984b).

Methylation to methylarsonic acid [$(CH_3)_2AsO_3H_2$] and dimethylarsinic acid [$(CH_3)_2AsO_2H$] is usually the major detoxification mechanism for inorganic pentavalent arsenates and trivalent arsenites in mammals. Methylated arsenicals rapidly clear from all tissues, except perhaps the thyroid (Marafante et al. 1985; Vahter and Marafante 1985; Yamauchi et al. 1986). Methylated arsenicals are probably common in nature. Methylation of arsenic (unlike methylation of mercury) greatly reduces toxicity and is a true detoxification process (Woolson 1975; Hamasaki et al. 1995; Aposhian et al. 1997). Before methylation (which occurs largely in the liver), As^{+5} is reduced to As^{+3} — the kidney being an important site for this transformation (Belton et al. 1985). Arsenate reduction and subsequent methylation are rapid: both arsenite and dimethylarsinate were present in hamster (*Cricetus* sp.) plasma only 12 min postinjection of inorganic As^{+5} (Hanlon and Ferm 1986c). Demethylation of methylated arsenicals formed *in vivo* has not yet been reported (USEPA 1980). Although terrestrial biota usually contain much lower total concentrations of arsenic than marine biota, metabolism of arsenic is similar in marine and terrestrial systems (Irgolic et al. 1998).

Toxic effects of organoarsenicals are exerted by initial metabolism to the trivalent arsenoxide form, and then by reaction with sulfhydryl groups of tissue proteins and enzymes to form an arylbis (organylthio) arsine (NAS 1977). This form, in turn, inhibits oxidative degradation of carbohydrates and decreases cellular ATP, the energy-storage molecule of the cell (NRCC 1978). Among the organoarsenicals, those physiologically most injurious are methylarsonous acid [$CH_3As(OH)_2$] and dimethylarsinous acid [$(CH_3)_2AsOH$] (Knowles and Benson 1984b). The enzyme inhibitory forms of organoarsenicals (arsonous acid) are formed from arsenous acid and the corresponding arsonic acids by a wide variety of enzymes and subcellular particles (Knowles and Benson 1984a). Organoarsenicals used as growth promoters and drugs are converted to more easily excretable (and sometimes more toxic) substances, although most organoarsenicals are eliminated without being converted to inorganic arsenic or to demethylarsinic acids (Pershagen and Vahter 1979; Edmonds et al. 1993).

28.4 ESSENTIALITY, SYNERGISM, AND ANTAGONISM

Limited data are available on the beneficial, protective, and essential properties of arsenic and on its interactions with other chemicals. Arsenic apparently behaves more like an environmental contaminant than as a nutritionally essential mineral (NAS 1977). Nevertheless, low doses (<2 μg/day) of arsenic stimulated growth and metamorphosis in tadpoles and increased viability and cocoon yield in silkworm caterpillars (NAS 1977). Arsenic deficiency has been observed in rats: signs include rough haircoat, low growth rate, decreased hematocrit, increased fragility of red cells, and enlarged spleen (NAS 1977). Similar results have been documented in goats and pigs fed diets containing less than 0.05 mg As/kg (NAS 1977). In these animals, reproductive performance was impaired, neonatal mortality was increased, birth weight was lower, and weight gains in second-generation animals were decreased. These effects were not evident in animals fed diets containing 0.35 mg As/kg (NAS 1977; ATSDR 1992).

The use of phenylarsonic feed additives to promote growth in poultry and swine and to treat specific diseases does not seem to constitute a hazard to the animal or to its consumers. Animal deaths and elevated tissue arsenic residues occur only when the arsenicals are fed at excessive dosages for long periods (NAS 1977). Arsenic can be detected at low levels in tissues of animals fed organoarsenicals, but it is rapidly eliminated when the arsenicals are removed from the feed for the required 5-day period before marketing (Woolson 1975).

Selenium and arsenic are antagonists in several animal species. Dietary arsenic, as arsenate, alleviates the toxic effects of selenium, as seleno-DL-methionine, on mallard (*Anas platyrhynchos*) reproduction and duckling growth and survival (Stanley et al. 1994). Mallard ducklings fed arsenate in the diet at 200 mg As/kg ration were protected against the toxic effects of 60 mg Se/kg ration as selenomethionine, including selenium-induced mortality, impaired growth, hepatic lesions, and enzyme disruptions (Hoffman et al. 1992). In rats, dogs, swine, cattle, and poultry, the arsenic protects against selenium poisoning if arsenic is administered in the drinking water and selenium through the diet (NAS 1977; NRCC 1978; Pershagen and Vahter 1979). Inorganic arsenic compounds decrease the toxicity of inorganic selenium compounds by increasing biliary excretion (NRCC 1978). However, in contrast to antagonism shown by inorganic arsenic/inorganic selenium mixtures, the toxic effects of naturally methylated selenium compounds (trimethylselenomium chloride and dimethyl selenide) are markedly enhanced by inorganic arsenicals (NRCC 1978).

The toxic effects of arsenic can be counteracted with:

1. Saline purgatives
2. Various demulcents that coat irritated gastrointestinal mucous membranes
3. Sodium thiosulfate (NAS 1977)
4. Mono- and dithiol-containing compounds and 2,3-dimercaptopropanol (Pershagen and Vahter 1979)

Arsenic uptake in rabbit intestine is inhibited by phosphate, casein, and various metal-chelating agents (USEPA 1980). Mice and rabbits are significantly protected against sodium arsenite intoxication by *N*-(2,3-dimercaptopropyl)phthalamidic acid (Stine et al. 1984). Conversely, the toxic effects of arsenite are potentiated by excess dithiols, cadmium, and lead, as evidenced by reduced food efficiency and disrupted blood chemistry in rodents (Pershagen and Vahter 1979).

Arsenic effectively controls filariasis in cattle; new protective uses are under investigation. The control of parasitic nematodes (*Parafilaria bovicola*) in cattle was successful after 30 weekly treatments in plungement dips containing 1600 mg As_2O_3/L. However, the muscle of treated cattle contained up to 1.3 mg As/kg, or 12 times the amount in controls (Nevill 1985). Existing anionic organic arsenicals used to control tropical nematode infections in humans have sporadic and unacceptable lethal side effects. Cationic derivatives have been synthesized in an attempt to avoid the side effects and have been examined for effects on adult nematodes (*Brugia pahangi*) in gerbils (*Meriones unguiculatus*). All arsenicals were potent filaricides; the most effective compounds tested killed 95% of adult *B. pahangi* after five daily subcutaneous injections of 3.1 mg As/kg body weight (Denham et al. 1986).

Animals previously exposed to sublethal levels of arsenic may develop tolerance to arsenic on reexposure. Although the mechanism of this process is not fully understood, it probably includes the efficiency of *in vivo* methylation processes (USEPA 1980). For example, resistance to toxic doses of As^{+3} or As^{+5} increases in mouse fibroblast cells pretreated with a low As^{+3} concentration (Fischer et al. 1985). Also, growth is better in arsenic-conditioned mouse cells in the presence of arsenic than in previously unexposed cells, and inorganic arsenic is more efficiently methylated. *In vivo* biotransformation and excretion of inorganic arsenic as monomethylarsonic acid (MMA) and dimethylarsinic acid (DMA) have been demonstrated in a number of mammalian species, including humans. It seems that cells may adapt to arsenic by increasing the biotransformation rate of the element to methylated forms, such as MMA and DMA (Fischer et al. 1985). Pretreatment of Chinese hamster (*Cricetus* spp.) ovary cells with sodium arsenite provided partial protection against adverse effects of methyl methanesulfonate (MMS) and may even benefit the MMS-treated cells. However, posttreatment dramatically increases the cytotoxic, clastogenic, and mitotic effects induced by MMS (Lee et al. 1986b).

Although arsenic is not an essential plant nutrient, small yield increases have sometimes been observed at low soil arsenic levels, especially for tolerant crops such as potatoes, corn, rye, and wheat (Woolson 1975). Arsenic phytotoxicity of soils is reduced with increasing lime, organic matter, iron, zinc, and phosphates (NRCC 1978). In most soil systems, the chemistry of As becomes the chemistry of arsenate; the estimated half-time of arsenic in soils is about 6.5 years, although losses of 60% in 3 years and 67% in 7 years have been reported (Woolson 1975). Additional research is warranted on the role of arsenic in crop production, and in nutrition, with special reference to essentiality for aquatic and terrestrial wildlife.

28.5 CONCENTRATIONS IN FIELD COLLECTIONS

28.5.1 General

In abundance, arsenic ranks 20th among the elements in the earth's crust (1.5 to 2 mg/kg), 14th in seawater, and 12th in the human body (Woolson 1975). It occurs in various forms, including inorganic and organic compounds and trivalent and pentavalent states (Pershagen and Vahter 1979). In aquatic environments, higher arsenic concentrations are reported in hot springs, in groundwaters from areas of thermal activity or in areas containing rocks with high arsenic content, and in some waters with high dissolved salt content (NAS 1977). Most of the other elevated values reported in lakes, rivers, and sediments are probably due to anthropogenic sources, which include smelting and mining operations; combustion of fossil fuel; arsenical grasshopper baits; synthetic detergent and sewage sludge wastes; and arsenical defoliants, herbicides, and pesticides (NAS 1977). Most living organisms normally contain measurable concentrations of arsenic, but except for marine biota, these are usually less than 1 mg/kg fresh weight. Marine organisms, especially crustaceans, may contain more than 100 mg As/kg dry weight, usually as arsenobetaine, a water-soluble organoarsenical that poses little risk to the organism or its consumer. Plants and animals collected from naturally arseniferous areas or near anthropogenic sources may contain significantly elevated tissue residues of arsenic. Additional and more detailed information on background concentrations of arsenic in abiotic and living resources has been given by NAS (1977), Hall et al. (1978), NRCC (1978), USEPA (1980), Jenkins (1980), Eisler (1981, 1994), and Phillips (1990).

28.5.2 Nonbiological Samples

Arsenic is a major constituent of at least 245 mineral species, of which arsenopyrite is the most common (NAS 1977). In general, background concentrations of arsenic are 0.2 to 15 mg/kg in the lithosphere, 0.005 to 0.1 μg/m^3 in air, <10 μg/L in water, and <15 mg/kg in soil (NRCC 1978; ATSDR 1992). The commercial use and production of arsenic compounds have raised local concentrations in the environment far above the natural background concentrations (Table 28.1).

Weathering of rocks and soils adds about 45,000 tons of arsenic to the oceans annually, accounting for less than 0.01 mg/L on a global basis (NRCC 1978). However, arsenic inputs to oceans have increased during the past century, both from natural sources and as a result of industrial use, agricultural and deforestation activities, emissions from coal and oil combustion, and loss during mining of metal ores. If present activities continue, arsenic concentrations in oceanic surface waters will have increased overall by about 2% by the year 2000, with most of the increased burden in estuaries and coastal oceans (e.g., Puget Sound, Washington; the Tamar, England; and the Tejo, Portugal) (Sanders 1985). Estimates of the residence times of arsenic are 60,000 years in the ocean and 45 years in a freshwater lake (NRCC 1978). In the hydrosphere, inorganic arsenic occurs predominantly as As^{+5} in surface water, and significantly as As^{+3} in groundwater containing high levels of total arsenic. The main organic species in freshwater are methylarsonic acid and dimethylarsinic acid, and these are usually present in lower concentrations than inorganic arsenites and arsenates (Pershagen and Vahter 1979). Total arsenic concentrations in surface water and groundwater are usually <10 μg/L. In certain areas, however, levels above 1 mg/L have been recorded (Pershagen and Vahter 1979).

In air, most arsenic particulates consist of inorganic arsenic compounds, often as As^{+3}. Burning of coal and arsenic-treated wood, and smelting of metals are major sources of atmospheric arsenic contamination (i.e., >1 μg/m^3). In general, atmospheric arsenic levels are higher in winter, due to increased use of coal for heating (Pershagen and Vahter 1979).

The main carrier of arsenic in rocks and in most types of mineral deposits is iron pyrite (FeS_2), which may contain >2000 mg As/kg (NRCC 1978). In localized areas, soils are contaminated by arsenic oxide fallout from smelting ores (especially sulfide ores) and combustion of arsenic-rich coal (Woolson 1975).

Arsenic in lacustrine sediment columns is subject to control by diagenetic processes and adsorption mechanisms, as well as anthropogenic influences (Farmer and Lovell 1986; Farag et al. 1995). For example, elevated levels of arsenic in surface or near-surface sediments may have several causes (Farmer and Lovell 1986), including natural processes (Loch Lomond, Scotland) and human activities such as smelting (Lake Washington, Washington; Kelly Lake, Ontario, Canada), manufacture of arsenical herbicides (Brown's Lake, Wisconsin), and mining operations (Northwest Territories, Canada; Clark Fork River, Montana). Elevated levels of arsenic in sediments of the Wailoa River, Hawaii, are the result of As_2O_3 applied as an anti-termite agent between 1932 and 1963. These elevated levels are found mainly in anaerobic sediment regions where the chemical has been relatively undisturbed by activity. Low levels of arsenic in the biota of the Wailoa River estuary suggest that arsenic is trapped in the anaerobic sediment layers (Hallacher et al. 1985).

Arsenic geochemistry in Chesapeake Bay, Maryland, depends on anthropogenic inputs and phytoplankton species composition (Sanders 1985). Inputs of anthropogenic arsenic into Chesapeake Bay are estimated at 100 kg daily, or 39 tons/year — probably from sources such as unreported industrial discharges, use of arsenical herbicides, and from wood preservatives (Sanders 1985). The chemical form of the arsenic in solution varies seasonally and along the axis of the bay. Arsenic is present only as arsenate in winter, but substantial quantities of reduced and methylated forms are present in summer in different areas. The forms and distribution patterns of arsenic during the summer suggest that separate formation processes exist. Arsenite, present in low salinity regions, may have been formed by chemical reduction in anoxic, subsurface waters and then mixed into the surface layer. Methylated arsenicals are highly correlated with standing crops of algae. One particular form, methylarsonate, is significantly correlated with the dominant alga, *Chroomonas*. Since both arsenic reactivity and toxicity are altered by transformation of chemical form, the observed variations in arsenic speciation have considerable geochemical and ecological significance (Sanders 1985).

Table 28.1 Total Arsenic Concentrations in Selected Nonbiological Materials

Material and Units (in parentheses)	Concentration[a]	Reference[b]
AIR (μg/m³)		
Remote areas	<0.02	1
Urban areas	(0.0–0.16)	1
Near smelters		
U.S.S.R.	(0.5–1.9)	2
Texas	Max. 1.4	2
Tacoma, Washington	Max. 1.5	2
Romania	Max. 1.6	2
Germany	(0.9–1.5)	2
Coal-fired power plant, Czechoslovakia	(19–69)	3
Orchard spraying of Pb arsenate	Max. 260,000	3
Near U.S. cotton gin burning vegetation treated with arsenic	Max. 400	3
DRINKING WATER (mg/L)		
Nationwide, U.S.	2.4 (0.5–214)	4
Fairbanks, Alaska	224 (1–2450)	4
Bakersfield, California	(6–393)	4
Nevada, 3 communities	(51–123)	4
Mexico, from plant producing As_2O_3	(4000–6000)	2
Japan, near factory producing arsenic sulfide	3000	2
Ghana, near gold mine	1400	2
Minnesota, contaminated by residual arsenical grasshopper bait	(11,800–21,000)	1
Methylated arsenicals use areas, U.S.	Usually <0.3 (0.01–1.0)	5
DUST (mg/kg)		
Tacoma, Washington		
Near smelter	1300	1
Remote from smelter	70	1
FOSSIL FUELS (mg/kg)		
Coal		
Canada	4 (0.3–100)	3
United States	5	2
Czechoslovakia	Max. 1500	2
Worldwide	13 (0.0–2000)	1
Coal ash	(<20–8000)	3
Flyash	(2.8–200)	3
Petroleum	0.2	3
Petroleum ash	Max. 100,000	3
Automobile particulates	298	3

Table 28.1 (continued) Total Arsenic Concentrations in Selected Nonbiological Materials

Material and Units (in parentheses)	Concentration[a]	Reference[b]
GROUNDWATER (µg/L)		
Near polymetallic sulfide deposits	Max. 400,000	3
Near gold mining activities	>50	2
U.S., Atlantic coastal plain	Usually <10	2, 20
U.S., Southwest and New England	17.9 (0.01–800)	3, 20
LAKE WATER (µg/L)		
Dissolved solids		
<2000 mg/L	(0.0–100)	6
>2000 mg/L	(0.1–2000)	6
Lake Superior	(0.1–1.6)	6
Japan, various	(0.2–1.9)	6
Germany, Elbe River	(20–25)	6
Searles Lake, California	(198,000–243,000)	1, 4
California, other lakes	(0.0–100)	1, 4
Michigan	Max. 2.4	1, 4
Wisconsin	(4–117)	1, 4
Florida	Max. 3.6	1, 4
Lake Chatauqua, New York	(3.5–35.6)	1. 4
Lake Ohakuri, New Zealand	(30–60)	7
Finfeather Lake, Texas	Max. 240,000	8
Thermal waters, worldwide	Usually 20–3800; Max. 276,000	1, 2, 3, 9
RAIN (µg/L)		
Canada	(0.01–5)	3
Rhode Island	0.8	1
Seattle, Washington	17	1
RIVER WATER (µg/L)		
Polluted, U.S.	Max. 6000	4
Nonpolluted, U.S.	Usually <5	4
Nationwide, U.S., 1974–1981		
25th percentile	<1	10
50th percentile	1	10
75th percentile	3	10
ROCK (mg/kg)		
Limestones	1.7 (0.1–20)	1
Sandstones	2 (0.6–120)	1
Shales and clays	14.5 (0.3–490)	1
Phosphates	22.6 (0.4–188)	1
Igneous, various	1.5–3 (0.06–113)	1
SEAWATER (µg/L)		
Worldwide	2 (0.15–6)	6
Pacific Ocean	(1.4–1.8)	11
Atlantic Ocean	(1.0–1.5)	11
South Australia		
Total dissolved As	1.3 (1.1–1.6)	12
As^{+5}	1.29	12
As^{+3}	0.03	12
Particulate As	<0.0006	12

Table 28.1 (continued) Total Arsenic Concentrations in Selected Nonbiological Materials

Material and Units (in parentheses)	Concentration[a]	Reference[b]
U.K., Beaulieu estuary		
Water temperature <12°C		
Inorganic arsenic	(0.4–0.9)	13
Suspended arsenic	(0.02–0.24)	13
Organoarsenicals	(0.19–0.75)	13
Water temperature >12°C		
Inorganic arsenic	(0.6–1.1)	13
Suspended arsenic	(0.2–0.6)	13
Organoarsenicals	ND	13
SEDIMENTS (mg/kg dry weight)		
Near sewer outfall	35	3
From areas contaminated by smelteries, arsenical herbicides, or mine tailings		
Surface	(198–3500)	1, 7, 9, 14, 15
Subsurface	(12–25)	1, 7, 9, 14, 15
Upper Mississippi River	2.6 (0.6–6.2)	16
Lake Michigan	(5–30)	1
Naturally elevated	>500	1, 9
Oceanic	33.7 (<0.4–455)	3
Lacustrine	Usually 5–26.9; Max. 13,000	3
SNOW (mg/kg)		
Near smelter	>1000	3
SOIL PORE WATERS (µg/L)		
Mineralized areas		
Arsenate	(79–210)	17
Arsenite	(2–11)	17
Monomethyl arsonic acid (MMAA)	(4–22)	17
Total arsenic	(93–240)	17
Unmineralized areas		
Arsenate	(18–49)	17
Arsenite	(1–7)	17
MMAA	<1	17
Total arsenic	(13–59)	17
SOILS (mg/kg dry weight)		
U.S., uncontaminated	7.4	18
Worldwide, uncontaminated	7.2	18
Canada		
Near gold mine		
Air levels 3.9 mg As/m^3	21,213	3
80 km distant	(10–25)	3
England		
Near arsenic refinery	156	19
Reference site	5	19
Near smelter		
Japan	Max. 2470	2
Tacoma, Washington	Max. 380	2
Treated with arsenical pesticides		
U.S.	165 (1–2554)	6
Canada	121	6

Table 28.1 (continued) Total Arsenic Concentrations in Selected Nonbiological Materials

Material and Units (in parentheses)	Concentration[a]	Reference[b]
SYNTHETIC DETERGENTS (mg/kg)		
Household, heavy duty	(1–73)	1, 2

[a] Concentrations are listed as mean, minimum–maximum (in parentheses), and maximum (Max.).

[b] **1,** NAS 1977; **2,** Pershagen and Vahter 1979; **3,** NRCC 1978; **4,** USEPA 1989; **5,** Hood 1985; **6,** Woolson 1975; **7,** Freeman et al. 1986; **8,** Sorensen et al. 1985; **9,** Farmer and Lovell 1986; **10,** Smith et al. 1987; **11,** Sanders 1980; **12,** Maher 1985a; **13,** Howard et al. 1984; **14,** Hallacher et al. 1985; **15,** Takamatsu 1985; **16,** Wiener et al. 1984; **17,** Haswell et al. 1985; **18,** Dudas 1984; **19,** Ismail and Roberts 1992; **20,** Welch et al. 1998.

28.5.3 Biological Samples

Background arsenic concentrations in living organisms are usually <1 mg/kg fresh weight in terrestrial flora and fauna, birds, and freshwater biota. These levels are higher, sometimes markedly so, in biota collected from mine waste sites, arsenic-treated areas, near smelters and mining areas, near areas with high geothermal activity, and near manufacturers of arsenical defoliants and pesticides (Table 28.2). For example, bloaters (*Coregonus hoyi*) collected in Lake Michigan near a facility that produced arsenical herbicides consistently had the highest (1.5 to 2.9 mg As/kg fresh weight whole body) arsenic concentrations measured in freshwater fishes in the United States between 1976 and 1984 (Schmitt and Brumbaugh 1990). Marine organisms normally contain arsenic residues of several to more than 100 mg/kg dry weight (Lunde 1977). However, as will be discussed later, these concentrations present little hazard to the organism or its consumers.

Arsenic concentrations in tissues of marine biota show a wide range of values, being highest in lipids, liver, and muscle tissues, and varying with the age of the organism, geographic locale, and proximity to anthropogenic activities (Table 28.2). In general, tissues with high lipid content contained high levels of arsenic. Crustacean tissues sold for human consumption and collected in U.S. coastal waters usually contained 3 to 10 mg As/kg fresh weight (Hall et al. 1978), or 1 to 100 mg/kg dry weight (Fowler and Unlu 1978), and were somewhat higher than those reported for finfish and molluscan tissues. Marine finfish tissues usually contained 2 to 5 mg As/kg fresh weight (Table 28.2). However, postmortem reduction of As^{+5} to As^{+3} occurs rapidly in fish tissues (Reinke et al. 1975), suggesting a need for additional research in this area. Maximum arsenic values recorded in elasmobranchs (mg/kg fresh weight) were 30 in the muscle of a shark (*Mustelus antarcticus*) and 16.2 in the muscle of a ray (*Raja* sp.) (Eisler 1981). The highest arsenic concentration recorded in a marine mammal, 2.8 mg As/kg fresh weight lipid, was from a whale (Eisler 1981).

Arsenic appears to be elevated in marine biota because of their ability to accumulate arsenic from seawater or food sources — and not because of localized pollution (Maher 1985b). The great majority of arsenic in marine organisms exists as water-soluble and lipid-soluble organoarsenicals that include arsenolipids, arsenosugars, arsenocholine, arsenobetaine [$(CH_3)_3AsCH_2COOH$], monomethylarsonate [$CH_3AsO(OH)_2$], and demethylarsinate [$(CH_3)_2AsO(OH)$], as well as other forms (Edmonds et al. 1993). There is no convincing hypothesis to account for the existence of all the various forms of organoarsenicals found in marine organisms. One suggested hypothesis is that each form involves a single anabolic/catabolic pathway concerned with the synthesis and turnover of phosphatidylcholine (Phillips and Depledge 1986). Arsenosugars (arsenobetaine precursors) are the dominant arsenic species in brown kelp (*Ecklonia radiata*), giant clam (*Tridacna maxima*), shrimp (*Pandalus borealis*), and ivory shell (*Buccinum striatissimum*) (Shiomi et al. 1984a, 1984b; Francesconi et al. 1985; Matsuto et al. 1986; Phillips and Depledge 1986). For most marine species, however, there is general agreement that arsenic exists primarily as arsenobetaine, a water-soluble organoarsenical that has been identified in tissues of western rock lobster (*Panulirus cygnus*),

American lobster (*Homarus americanus*), octopus (*Paroctopus* sp.), sea cucumber (*Stichopus japonicus*), blue shark (*Prionace glauca*), sole (*Limanda* sp.), squid (*Sepioteuthis australis*), prawn (*Penaeus latisulcatus*), scallop (*Pecten alba*), and many other species (including teleosts, molluscs, tunicates, and crustaceans) (Shiomi et al. 1984b; Francesconi et al. 1985; Hanaoka and Tagawa 1985a, 1985b; Maher 1985b; Norin et al. 1985; Matsuto et al. 1986; Ozretic et al. 1990; Phillips 1990; Kaise and Fukui 1992; Edmonds et al. 1993). Degradation of arsenobetaine in muscle and liver of the star spotted shark (*Mustelus manazo*) to inorganic arsenic occurs in a natural environment and suggests that arsenobetaine bioconverted from inorganic arsenic in seawater is degraded to original inorganic arsenic. About 13% of the arsenobetaine in shark muscle and 4% in liver was degraded to inorganic arsenic within 40 days (Hanaoka et al. 1993). The potential risks associated with consumption of seafoods containing arsenobetaine seem to be minor. The chemical was not mutagenic in the bacterial *Salmonella typhimurium* assay (Ames test), had no effect on metabolic inhibition of Chinese hamster ovary cells at 10,000 mg/L, and showed no synergism or antagonism on the action of other contaminants (Jongen et al. 1985). Arsenobetaine was not toxic to mice at oral doses of 10,000 mg/kg body weight during a 7-day observation period and was rapidly absorbed from the gastrointestinal tract and rapidly excreted in urine without metabolism, owing to its high polar and hydrophilic characteristics (Kaise et al. 1985; Kaise and Fukui 1992).

Shorebirds (seven species) wintering in the Corpus Christi, Texas, area contained an average of 0.3 mg As/kg fresh weight in livers (maximum of 1.5 mg/kg), despite the presence of smelters and the heavy use of arsenical herbicides and defoliants. These values probably reflect normal background concentrations (White et al. 1980). Similar arsenic levels were reported in livers of brown pelicans (*Pelecanus occidentalis*) collected from South Carolina (Blus et al. 1977). Bone arsenic concentrations in 23 species of birds collected in southwestern Russia during 1993 to 1995 ranged from 0.1 to 1.7 mg As/kg DW; arsenic concentrations were similar for terrestrial and aquatic birds, and for urban and rural environments (Lebedeva 1997). The highest arsenic concentration recorded in seemingly unstressed coastal birds was 13.2 mg/kg fresh weight lipids (Table 28.2). This tends to corroborate the findings of others that arsenic concentrates in lipid fractions of marine plants, invertebrates, and higher organisms. An abnormal concentration of 16.7 mg As/kg fresh weight was recorded in the liver of an osprey (*Pandion haliaetus*) from the Chesapeake Bay region (Wiemeyer et al. 1980). This bird was alive but weak, with serious histopathology including the absence of subcutaneous fat, presence of serous fluid in the pericardial sac, and disorders of the lung and kidney. The bird died shortly after collection. Arsenic concentrations in the livers of other ospreys collected in the same area were usually <1.5 mg As/kg fresh weight. Chicks of the avocet (*Recurvirostra americana*) reared from eggs taken near arsenic-contaminated ponds in California (127 to 1100 μg As/L) had reduced hatch, as well as impaired growth and immune function, when compared to chicks from a reference site (29 μg As/L). Other contaminants present included boron and selenium, although the authors concluded that arsenic played a significant role in observed effects (Fairbrother et al. 1994).

Arsenic concentrations in tissues of small mammals (field mouse, *Apodemus sylvaticus*; bank vole, *Clethrionomys glareolus*; field vole, *Microtus agrestis*; common shrew, *Sorex araneus*) from the vicinity of an English arsenic refinery were usually less than 1 mg/kg FW and did not reflect arsenic levels of the surrounding soil or vegetation (Ismail and Roberts 1992). In general, mean arsenic concentrations were highest in spleen, followed — in descending order — by bone, heart, kidney, brain, muscle, and liver (Ismail and Roberts 1992). For human adults, seafood contributes 74% to 96% of the total daily arsenic intake, and rice and rice cereals most of the remainder. For infants, 41% of the estimated total arsenic intakes arise from seafood and 34% from rice and rice cereals (Tao and Bolger 1998). Effective biomarkers of arsenic exposure in humans include elevated arsenic concentrations in hair, fingernails, and especially urine (Lin et al. 1998). In Taiwan, urinary levels of inorganic and organic arsenic metabolites are associated with previous exposure to high-arsenic artesian well water (Hsueh et al. 1998). Humans who had previously been exposed to high-arsenic drinking water and had switched to tap water containing <50 μg total As/L had — after 30 years — elevated levels of arsenic in urine, especially MMA and DMA (Hsueh et al. 1998).

Table 28.2 Arsenic Concentrations in Field Collections of Selected Species of Plants and Animals (Values listed are in mg As/kg fresh weight [FW] or dry weight [DW].)

Ecosystem, Species, and Other Variables	Concentration (mg/kg[a])	Reference[b]
TERRESTRIAL PLANTS		
Colonial bentgrass, *Agrostis tenuis*		
On mine waste site	1480 DW; Max. 3470 DW	1
On low arsenic soil	(0.3–3) DW	1
Scotch heather, *Calluna vulgaris*		
On mine waste site	1260 DW	1
On low arsenic soil	0.3 DW	1
Coontail, *Ceratophyllum demersum*		
From geothermal area, New Zealand	(20–1060) DW	1
Cereal grains		
From arsenic treated areas	Usually <3 DW; Max. 252 DW	2
Nontreated areas	Usually <0.5 DW; Max. 5 DW	2
Grasses		
From arsenic treated areas	(0.5–60,000) DW	2
Nontreated areas	(0.1–0.9) DW	2
Apple, *Malus sylvestris*		
Fruit	<0.1 FW; <1.8 DW	1
Alfalfa, *Medicago sativa*		
U.S.	1.6 FW	1
Montana, smelter area	(0.4–5.7) FW	1
Mushrooms; Austria; 1995; near arsenic roasting facility in operation for about 500 years and closed for about 100 years; soil had 730 mg total As/kg DW		
Collybia maculata		
Total	30.0 DW	60
Arsenobetaine	>27.0 DW	60
Collybia butyracea		
Total	10.9 DW	60
Arsenobetaine	8.8 DW	60
Dimethylarsinic acid	1.9 DW	60
Amantia muscaria		
Total	21.9–22.0 DW	60, 61
Arsenate	Trace–0.3 DW	60, 61
Arsenite	Trace–0.4 DW	60, 61
Arsenobetaine	Max. 15.1 DW	60
Arsenocholine	2.5–2.7 DW	60, 61
Dimethylarsinic acid	Trace–0.7 DW	60, 61
Tetramethylarsonium salt	0.5–0.8 DW	60, 61
White spruce, *Picea alba*		
Arsenic-contaminated soil		
Branch	(2.8–14.3) DW	1
Leaf	(2.1–9.5)	1
Trunk	(0.3–55) DW	1
Root	(45–130) DW	1
Control site		
All samples	<2.4 DW	1
Pine, *Pinus silvestris*, needles		
Near U.S.S.R. metals smelter; soil levels 120.0 mg As/kg	22 FW	3
Peppers, *Piper* spp.; China		
Fresh	<1 FW	56
Dried over arsenic-rich (35 g As/kg) coal fires	500 DW	56
Trees, nontreated areas	Usually <1 DW	2
Lowbush blueberry, *Vaccinium angustifolium*		
Maine, leaf		
Arsenic-treated soil	(6.8–15) DW	1

Table 28.2 (continued) Arsenic Concentrations in Field Collections of Selected Species of Plants and Animals (Values listed are in mg As/kg fresh weight [FW] or dry weight [DW].)

Ecosystem, Species, and Other Variables	Concentration (mg/kg[a])	Reference[b]
Control	0.8 DW	1
Various species		
From uncontaminated soils	(<0.01–5) DW	2
From arsenic-impacted (80 mg/kg) soils	1.2 (<0.2–5.8) DW	4
Vegetables		
From arsenic-treated areas	Usually <3 DW; Max. 145 DW	2
Nontreated areas	Usually <1 DW; Max. 8 DW	2
Vegetation		
Near gold mine, Canada		
Air levels up to 3.9 mg As/m^3	Max. 11,438 DW	5
80 km distant	(12–20) DW	5
Near arsenic refinery (vs. control site), England	37 DW vs. 0.2 DW	54
FRESHWATER PLANTS		
Aquatic Plants		
Arsenic-treated areas	(20–1450) DW	2
Untreated areas	(1.4–13) DW	2
Irish moss, *Chondrus crispus*		
Whole	(5–12) DW	1
Pondweeds, *Potamogeton* spp.		
Whole		
Near geothermal area	(11–436) DW	1
Control site	<6 DW	1
Widgeongrass, *Ruppia maritima*		
From Kern National Wildlife Refuge, California, contaminated by agricultural drainwater of 12–190 μg As/L	Max. 430 DW	32
FRESHWATER FISHES		
Alewife, *Alosa pseudoharengus*		
Whole, Michigan	0.02 FW	1
Muscle, Wisconsin	0 FW	1
California; San Joaquin River; September–November 1986; whole fish		
Common carp, *Cyprinus carpio*	0.2–0.5 (0.1–0.6) DW	50
Mosquitofish, *Gambusia affinis*	0.5–1.1 (0.4–1.6) DW	50
Bluegill, *Lepomis macrochirus*	0.3–1.0 (0.2–1.3) DW	50
Largemouth bass, *Micropterus salmoides*	0.2–0.3 DW	50
Sacramento blackfish, *Orthodon microlepidotus*	0.7 DW	50
White sucker, *Catostomus commersoni*		
Muscle	(0.03–0.13) FW	1
Whole	(0.05–0.16) FW	1
Common carp, *Cyprinus carpio*		
Upper Mississippi River, 1979		
Whole	0.4 (0.2–0.6) DW	6
Liver	0.4 (0.3–1) DW	6
Nationwide		
Whole	0.05 FW	1
Muscle	(0.0–0.2) FW	1
Northern pike, *Esox lucius*		
Muscle		
Canada	(0.05–0.09) FW	1
Great Lakes	<0.05 FW	1
Sweden	0.03 FW	1
New York	<0.1 FW	1
Wisconsin	<0.01 FW	1

Table 28.2 (continued) Arsenic Concentrations in Field Collections of Selected Species of Plants and Animals (Values listed are in mg As/kg fresh weight [FW] or dry weight [DW].)

Ecosystem, Species, and Other Variables	Concentration (mg/kg[a])	Reference[b]
Fish, various species		
Whole	Max. 1.9 FW	2
Whole	(0.04–0.2) FW	7
Netherlands, 1977–1984		
Muscle	(0.04–0.15) FW	8
Nationwide, U.S., whole fish		
1976–1977	0.27 FW; Max. 2.9 FW	9, 33
1978–1979	0.16 FW; Max. 2.1 FW	33
1980–81	0.15 FW; Max. 1.7 FW	33
1984	0.14 FW; Max. 1.5 FW	33
Near smelter (water arsenic 2.3–2.9 mg/L)		
Muscle, 3 species		
Total arsenic	0.05–0.24) FW	10
Inorganic arsenic	(0.01–0.02) FW	10
Liver, 2 species		
Total arsenic	0.15 FW	10
Inorganic arsenic	0.01 FW	10
Control location (water arsenic <0.5 mg/L)		
Muscle		
Total arsenic	(0.06–0.09) FW	10
Inorganic arsenic	<0.03 FW	10
Liver		
Total arsenic	0.09 FW	10
Inorganic arsenic	<0.01 FW	10
Channel catfish, *Ictalurus punctatus*		
Muscle		
Native	(0.0–0.3) FW	1
Cultured	(0.2–3.1) FW	1
Whole, nationwide	(<0.05–0.3) FW	1
Green sunfish, *Lepomis cyanellus;* liver		
Polluted waters (from manufacturer of arsenical defoliants and pesticides), Texas. Mean water concentration 13.5 mg As/L; sediment content of 4700 mg/kg		
Age 1–2 years	(19.7–64.2) DW	11
Age 3	15 DW	11
Age 4	(6.1–11.5) DW	11
Bluegill, *Lepomis macrochirus*		
From pools treated with arsenic		
Muscle	1.3 FW	1
Skin and scales	2.4 FW	1
Gills and GI tract	17.6 FW	1
Liver	11.6 FW	1
Kidney	5.9 FW	1
Ovary	8.4 FW	1
Control locations		
All tissues	<0.2 FW	1
Whole		
Nationwide	(<0.05–0.15) FW	1
Upper Mississippi River, 1979	0.3 (0.2–0.4) DW	6
Smallmouth bass, *Micropterus dolomieui*		
Muscle		
Wisconsin	<0.13 FW	1
Lake Erie	0.22 FW	1
New York	(0.03–0.51) FW	1
Whole, nationwide	(<0.05–0.28) FW	1

Table 28.2 (continued) Arsenic Concentrations in Field Collections of Selected Species of Plants and Animals (Values listed are in mg As/kg fresh weight [FW] or dry weight [DW].)

Ecosystem, Species, and Other Variables	Concentration (mg/kg[a])	Reference[b]
Largemouth bass, *Micropterus salmoides*		
Whole, nationwide	(<0.05–0.22) FW	1
Muscle		
Wisconsin	(0.0–0.12) FW	1
New York	(0.03–0.16) FW	1
Striped bass, *Morone saxatilis*		
Muscle	(0.2–0.7) FW	1
Coho salmon, *Oncorhynchus kisutch*		
Muscle		
Wisconsin	<0.15 FW	1
Lake Erie	(0.07–0.17) FW	1
New York	<0.5 FW	1
U.S.	0.09 FW	1
Yellow perch, *Perca flavescens*		
All tissues	<0.16 FW	1
Rainbow trout, *Oncorhynchus mykiss*		
All tissues	<0.4 FW	1
Atlantic salmon, *Salmo salar*		
Oil		
Liver	6.7 FW	1
Muscle	(0.8–3.1) FW	1
Brown trout, *Salmo trutta*		
Kidney; Clark Fork River, Montana vs. reference site	11.3 DW vs. 1.8 DW	37
Lake trout, *Salvelinus namaycush*		
Whole, nationwide	(0.06–0.68) FW	1
MARINE PLANTS		
Algae		
Green	(0.5–5) DW	2
Brown	Max. 30 DW	2
11 species	(2–58) DW	12
Various species	(10–100) DW	13
Seaweed, *Chondrus crispus*	5.2 DW	2
Alga, *Fucus* spp.		
Oil	(6–27) FW	2
Fatty acid	(5–6) FW	2
Brown alga, *Fucus vesiculosus*		
Whole	(35.2–80) DW	1
Brown alga, *Laminaria digitata*		
Whole	94 DW	2
Whole	(42–50) DW	1
Oil	(155–221) DW	2
Fatty acid	(8–36) DW	2
Alga, *Laminaria hyperborea*		
Total arsenic	142 DW	12
Organic arsenic	139 DW	12
Sargassum weed, *Sargassum fluvitans*		
Total arsenic	19.5 FW	7
As^{+3}	1.8 FW	7
As^{+5}	17.7 FW	7
Organoarsenicals	0.2 FW	7
Seaweed, *Sargassum* sp.		
Total arsenic	(4.1–8.7) FW	5
As^{+3}	(0.14–0.35) FW	5

Table 28.2 (continued) Arsenic Concentrations in Field Collections of Selected Species of Plants and Animals (Values listed are in mg As/kg fresh weight [FW] or dry weight [DW].)

Ecosystem, Species, and Other Variables	Concentration (mg/kg[a])	Reference[b]
As^{+5}	(1.9–7.3) FW	5
Organoarsenicals	Max. 0.1 FW	5
Seaweeds		
Whole	(3.8–93.8) DW	2
Whole	(10–109) DW	12
Oil fraction	(5.7–221) FW	12
MARINE MOLLUSCS		
Bivalves, California, 1984–1986, soft parts		
Clam, *Corbicula* sp.	5.4–11.5 DW	34
Clam, *Macoma balthica*	7.6–12.1	34
Ivory shell, *Buccinium striatissimum*		
Muscle		
Total arsenic	38 FW	14
Arsenobetaine	24.2 FW	14
Midgut gland		
Total arsenic	18 FW	14
Arsenobetaine	10.8 FW	14
Oysters, *Crassostrea* spp.		
Soft parts	(1.3–10) DW; (0.3–3.4) FW	12
American oyster, *Crassostrea virginica*		
Soft parts	2.9 FW	1
Soft parts	10.3 DW	15
Spindle shells, *Hemifusus* spp.		
Hong Kong, 1984, Muscle		
Total arsenic	Max. 500 FW	16
Inorganic arsenic	<0.5 FW	16
Limpet, *Littorina littorea*		
Soft parts		
Near arsenic source	11.5 DW	12
Offshore	4 DW	12
Squid, *Loligo vulgaris*		
Soft parts	(0.8–7.5) FW	1
Hardshell clam, *Mercenaria mercenaria*		
Soft parts		
Age 3 years	3.8 DW	12
Age 4 years	4.7 DW	12
Age 10 years	9.3 DW	12
Age 15 years	8.4 DW	12
Molluscs, edible tissues		
Hong Kong, 1976–1978		
Bivalves	(3.2–39.6) FW	17
Gastropods	(19–176) FW	17
Cephalopods	(0.7–5.5)	17
United States.		
6 species	(2–3) FW	18
8 species	(3–4) FW	18
3 species	(4–5) FW	18
4 species	(7–20) FW	18
Yugoslavia, northern Adriatic Sea, summer 1986, 6 species	21–31 FW	35
Mussel, *Mytilus edulis*		
Soft parts	2.5 (1.4–4.6) FW	8
Soft parts	(1.6–16) DW	12
Scallop, *Placopecten magellanicus*		
Soft parts	1.6 (1.3–2.4) FW	1

Table 28.2 (continued) Arsenic Concentrations in Field Collections of Selected Species of Plants and Animals (Values listed are in mg As/kg fresh weight [FW] or dry weight [DW].)

Ecosystem, Species, and Other Variables	Concentration (mg/kg[a])	Reference[b]
MARINE CRUSTACEANS		
Blue crab, *Callinectes sapidus*		
Florida, whole	7.7 FW	1
Maryland, soft parts	(0.5–1.8) FW	1
Dungeness crab, *Cancer magister*		
Muscle	6.5 (2.2–37.8) FW	1
Muscle	4 FW	19
Alaskan snow crab, *Chinocetes bairdii*		
Muscle	7.4 FW	19
Copepods		
Whole	(2–8.2) DW; (0.4–1.3) FW	12
Shrimp, *Crangon crangon*		
Netherlands, 1977–1984		
Muscle	3 (2–6.8) FW	8
Crustaceans, edible tissues		
Hong Kong, 1976–1978		
Crabs	(5.4–19.1) FW	17
Lobsters	(26.7–52.8) FW	17
Prawns and shrimps	(1.2–44) FW	17
United States		
6 species	(3–5) FW	18
3 species	(5–10) FW	18
4 species	(10–20) FW	18
2 species	(20–30) FW	18
1 species	(40–50) FW	18
American lobster, *Homarus americanus*		
Muscle	(3.8–7.6) DW; Max. 40.5 FW	1
Hepatopancreas	22.5 FW	1
Whole	(3.8–16) DW; (1–3) FW	12
Stone crab, *Menippe mercenaria*		
Whole	(9–11.8) FW	1
Deep sea prawn, *Pandalus borealis*		
Head and shell	68.3 DW	1
Muscle	61.6 DW	1
Oil	42 DW; 10.1 FW	1
Egg	3.7–14 FW	1
Prawns, *Pandalus* spp.		
Whole	(7.3–11.5) FW	12
Alaskan king crab, *Paralithodes camtschatica*		
Muscle	8.6 FW	19
Brown shrimp, *Penaeus aztecus*		
Muscle	(3.1–5.2) FW	1
Whole	0.6 DW	1
White shrimp, *Penaeus setiferus*		
Muscle		
Mississippi	(1.7–4.4) FW	1
Florida	(2.8–7.7) FW	1
Shrimp, *Sergestes lucens*		
Muscle		
Total arsenic	5.5 FW	20
Arsenobetaine	4.5 FW	20
Shrimps		
Exoskeleton	15.3 FW	7
Muscle, 2 species	(18.8–41.6) FW; (3.8–128) DW	2

Table 28.2 (continued) Arsenic Concentrations in Field Collections of Selected Species of Plants and Animals (Values listed are in mg As/kg fresh weight [FW] or dry weight [DW].)

Ecosystem, Species, and Other Variables	Concentration (mg/kg[a])	Reference[b]
MARINE FISHES AND ELASMOBRANCHS		
Whitetip shark, *Carcharhinus longimanus*		
Muscle	3.1 FW	21
Black sea bass, *Centropristis striata*		
Muscle	6.4 DW	1
Elasmobranchs		
Muscle		
Sharks	Max. 30 FW	12
Rays	Max. 16.2 FW	12
Roundnose flounder, *Eopsetta grigorjewi*		
Muscle	20.1 FW	22
Finfishes		
Near metal smelter, water concentration 2.3–2.9 mg As/L		
Muscle, 6 species		
Total arsenic	(0.2–2.6) FW	10
Inorganic arsenic	(0.02–0.1) FW	10
Liver, 4 species		
Total arsenic	(0.4–1.8) FW	10
Inorganic arsenic	(0.02–0.07) FW	10
Control location, water concentration <2.0 mg As/L		
Muscle, 5 species		
Total arsenic	(0.1–1.2) FW	10
Inorganic arsenic	(0.02–0.15) FW	10
Liver, 4 species		
Total arsenic	(0.2–1.5) FW	10
Inorganic arsenic	(0.02–0.05) FW	10
Finfish, Hong Kong, 1976–1978		
Edible tissues	Max. 21.1 FW	17
Finfish, Netherlands, 1977–1984		
Muscle, 4 species	(2.8–10.9) FW	8
Finfish, North America		
Liver		
49 species	(0.7–5) FW	18
26 species	(5–20) FW	18
6 species	(20–50) FW	18
Muscle		
91 species	(0.6–4) FW	18
41 species	(4–8) FW	18
27 species	(8–30) FW	18
6 species	(0.18–0.30) DW	36
4 species		
Total arsenic	(1.4–10) FW	23
Inorganic arsenic	<0.5 FW	23
Whole		
16 species	(1–8) FW	18
Finfish, worldwide		
Various tissues		
Total arsenic	(ND–142) FW	2
Inorganic arsenic	(0.7–3.2) FW	2
Organic arsenic	(3.4–139) FW	2
Atlantic cod, *Gadus morhua*		
Muscle	2.2 FW	2
Liver	9.8 FW	2

Table 28.2 (continued) Arsenic Concentrations in Field Collections of Selected Species of Plants and Animals (Values listed are in mg As/kg fresh weight [FW] or dry weight [DW].)

Ecosystem, Species, and Other Variables	Concentration (mg/kg[a])	Reference[b]
North East Irish Sea; total As (<1% inorganic); muscle; sludge disposal site vs. reference site		
Plaice, *Pleuronectes platessa*	20.5 FW vs. <7.5 FW	39
Whiting, *Merlangius merlangus*	6.2 FW vs. <4.0 FW	39
Blue pointer, *Isurus oxyrhinchus*		
Muscle	9.5 FW	21
Striped bass, *Morone saxatilis*		
Muscle	(0.3–0.5) FW; 1.8 DW	12
Liver	0.7 FW	12
Striped mullet, *Mugil cephalus*		
Viscera	Max. 1.3 FW	24
Norway; Glomma estuary; March–December 1988; muscle		
Atlantic cod	4.1 FW	38
Flounder, *Platichthys flesus*	5.2 FW	38
English sole, *Pleuronectes vetulus*		
Muscle	1.1 (0.6–11.5) FW	1
Skate, *Raja* sp.		
Muscle	16.2 FW	1
Windowpane flounder, *Scophthalmus aquosus*		
Muscle	(1.4–2.8) FW	1
Spiny dogfish, *Squalus acanthias*		
Muscle	10 DW	25
Liver	5.7 DW	25
Spleen	9.8 DW	25
Yolk sac	9.1 DW	25
Embryo	2.6 DW	25
Muscle		
Arsenate	<0.03 DW	58
Arsenite	<0.03 DW	58
Arsenobetaine	15.6–16.0 DW	58
Arsenocholine	0.02–0.04 DW	58
Dimethylarsinic acid	0.28–0.49 DW	58
Methylarsonic acid	<0.03 DW	58
Tetramethylarsonium	0.23–0.38 DW	58
Trimethylarsine oxide	<0.03 DW	58
Unknown	0.16–0.32 DW	58
AMPHIBIANS AND REPTILES		
Alligator, *Alligator mississippiensis*		
Egg	(0.05–0.2) FW	1
Southern toad, *Bufo terrestris*; adults; whole body		
From coal ash settling basins (sediments with 39.6 mg As/kg DW) vs. reference site (0.3 mg As/kg DW sediment)	1.6 DW vs. 0.2 DW	63
From reference site to settling basin for 7–12 weeks	0.2 DW vs. 0.8–1.2 DW	63
Crocodile, *Crocodylus acutus*		
Egg	0.2 FW	26
Bullfrog, *Rana catesbeiana*; tadpoles; South Carolina; 1997		
With digestive tract		
Body	0.8 FW; 3.9 DW	65
Tail	0.3 FW; 1.8 DW	65
Whole	0.6 FW; 3.1 DW	65
Without digestive tract		
Body without gut	3.3 DW	65
Tail	1.9 DW	65

Table 28.2 (continued) Arsenic Concentrations in Field Collections of Selected Species of Plants and Animals (Values listed are in mg As/kg fresh weight [FW] or dry weight [DW].)

Ecosystem, Species, and Other Variables	Concentration (mg/kg[a])	Reference[b]
Digestive tract	17.3 DW	65
Whole	3.1 DW	65
Frogs, *Rana* spp.		
All tissues	<0.4 FW	1
Toads, 2 species		
All tissues	<0.05 FW	1
BIRDS		
American black duck, *Anas rubripes*		
Egg	0.2 FW	12
Ducks, *Anas* spp.		
All tissues	<0.4 FW	1
Scaup, *Aythya* spp.		
All tissues	<0.4 FW	1
California; 1986–87; eggs; Merced County vs. Fresno County; mallard, *Anas platyrhynchos* and gadwall, *Anas strepera*		
1986	<0.3 FW in 97% of eggs; Max. 0.84 FW	40
1987	All <0.3 FW	40
Canada; Vancouver Island, British Columbia; near copper mine; 1976 vs. 1981–82		
Western grebe, *Aechmophorus occidentalis*; liver	1.1 FW vs. 0.08 FW	43
Glaucous-winged gull, *Larus glaucescens*; liver	1.6 FW vs. 0.1 FW	43
Marbeled murrelet, *Synthliboramphus antiguis*		
Liver	3.2 FW vs. 0.8 FW	43
Diet, all locations	0.03–0.15 FW	43
Willet, *Catoptrophorus semipalmatus*		
Southwest Texas; summer 1986; 4 locations; sediments had 2.1–11.0 mg As/kg DW		
Liver	1.3–7.5 (ND–15.0) DW	41
Stomach contents	1.2–2.3 (ND–4.7) DW	41
Gulls, 3 species		
Oil	(0.6–13.2) FW	12
Kenya, Lake Nakaru; 1990 vs. 1970		
White pelican, *Pelecanus onocratalus*		
Kidney	0.05 (0.03–0.8) FW vs. 0.009 FW	53
Liver	0.04 FW vs. 0.01 FW	53
White-necked cormorant, *Phalacrocorax carbo*		
Kidney	0.05 (0.04–0.11) FW vs. no data	53
Liver	0.04 (0.03–0.06) FW vs. no data	53
Lesser flamingo, *Phoeniconaias minor*		
Kidney	0.06 FW vs. 0.10 FW	53
Liver	0.04 FW vs. 0.07 FW	53
Osprey, *Pandion haliaetus*		
Liver	Max. 16.7 FW	27
Brown pelican, *Pelecanus occidentalis*		
Egg		
South Carolina, 1971–72	0.3 (0.08–0.8) FW	28
Florida, 1969–70	0.1 (0.07–0.2) FW	28
Liver, 1971–72, GA, FL, SC		
Found dead	(0.2–1) FW	28
Shot	(0.3–0.9) FW	28
Russia, southwestern region, 1993–95, bone		
Herons, *Ardea* spp.	0.12 DW	52
21 species	0.15–1.6 DW	52
Hawks, *Buteo* spp.	1.7 DW	52

Table 28.2 (continued) Arsenic Concentrations in Field Collections of Selected Species of Plants and Animals (Values listed are in mg As/kg fresh weight [FW] or dry weight [DW].)

Ecosystem, Species, and Other Variables	Concentration (mg/kg[a])	Reference[b]
Shorebirds		
Corpus Cristi, Texas; 1976–1977		
Liver, 7 species	(0.05–1.5) FW	29
New Zealand, 5 species		
Feather	<1 FW	12
Liver	Max. 2.6 FW	12
Spain; Ebro Delta bird sanctuary; livers; January–April 1989; found dead		
7 species	ND	42
5 species	0.2–0.5 FW	42
2 species	1.0–1.2 FW	42
Starling, *Sturnus vulgaris*		
Whole, nationwide, U.S., 1971	(<0.01–0.21) FW	2
Icelandic redshank, *Tringa totanus robusta*		
Netherlands, 1979–1982		
Feather		
Juveniles	Max. 0.8 FW	30
Adults	(0.5–3.2) FW	30
MAMMALS		
Fin whale, *Balaenoptera physalis*		
Blubber oil	1.8 FW	1
Cow, *Bos bovis*		
Downwind from copper smelter		
16–21 km		
Hair	8.9 FW	1
Milk	0.013 FW	1
Blood	0.026 FW	1
60 km		
Hair	0.46 FW	1
Milk	0.002 FW	1
Blood	0.009 FW	1
Controls		
Milk	<0.001 FW	31
Muscle	0.005 FW	31
Liver	(0.008–0.012) FW	31
Kidney	(0.017–0.053) FW	31
Red deer, *Cervus elaphus* vs. roe deer, *Capreolus capreolus*; Slovakia, Europe		
Hair	0.1–0.3 (Max. 1.0) DW vs. 0.1 (Max. 25.2) DW	51
Internal tissues	<0.1 DW vs.<0.1 DW	51
Domestic animals		
All tissues	<0.3 FW	2
Humans, *Homo sapiens*		
U.S.; 1991–96; total daily intake, in µg		
Age 6–11 months	2	57
Age 2 years	22	57
Age 6 years	19	57
Age 10 years	13	57
14–16-year-old males	15	57
14–16-year-old females	21	57
25–30-year-old males	54	57
25–30-year-old females	26	57
40–45-year-old males	44	57
40–45-year-old females	35	57

Table 28.2 (continued) Arsenic Concentrations in Field Collections of Selected Species of Plants and Animals (Values listed are in mg As/kg fresh weight [FW] or dry weight [DW].)

Ecosystem, Species, and Other Variables	Concentration (mg/kg[a])	Reference[b]
60–65-year-old males	87	57
60–65-year-old females	68	57
70-year-old male	66	57
70-year-old female	43	57
Finland; urine; drinking water contains 17–980 μg As/L		
Current users	0.058 FW	66
Ex-users (ceased drinking from As-contaminated source 2–4 months previously)	0.017 FW	66
Controls (<1 μg As/L)	0.005 FW	66
Taiwan; urine; more than 20 years after consumption of high As-contaminated well water		
Total	0.267 FW	67
Inorganic	0.086 FW	67
Livestock		
All tissues	<0.6 FW	2
Marine mammals		
Four species; liver		
Total	0.17–2.40 FW	59
Arsenobetaine	0.05–1.7 FW	59
Arsenocholine	0.005–0.044 FW	59
Arsenic acid	<0.001 FW	59
Arsenous acid	<0.001 FW	59
Dimethylarsinic acid	<0.001–0.11 FW	59
Methylarsonic acid	<0.001–0.025 FW	59
Tetramethylarsonium cation		
Pinnipeds	<0.009–0.043 FW	59
Cetaceans	ND	59
Unidentified compound	0.002–0.027 FW	59
Pinnipeds		
All tissues	Max. 1.7 FW	12
Cetaceans		
Muscle	0.4 DW	12
Oil	(0.6–2.8) FW	12
White-tailed deer, *Odocoileus virginianus*		
New York; found dead		
Kidney	56.0 FW	44
Liver	102.0 FW	44
Tennessee, killed from arsenic herbicide		
Liver	19 FW	1
Kidney	17.8 FW	1
Rumen contents	22.5 FW	1
Harbor seal, *Phoca vitulina*		
UK, all tissues	<0.3 FW	1
Mammals, (mice, voles, shrews); England; near arsenic refinery		
Spleen	Max. 33.9 FW	54
Brain	Max. 9.7 FW	54
Bone	Max. 9.5 FW	54
Heart	Max. 7.0 FW	54
Kidney	Max. 6.2 FW	54
Liver	Max. 5.2 FW	54
Muscle	Max. 4.4 FW	54
Whole body	Max. 3.2 FW	54
Mexican free-tailed bat, *Tadarida brasiliensis*; New Mexico and Oklahoma; May–August 1991; livers	Usually ND; Max. 0.35 FW	45

Table 28.2 (continued) Arsenic Concentrations in Field Collections of Selected Species of Plants and Animals (Values listed are in mg As/kg fresh weight [FW] or dry weight [DW].)

Ecosystem, Species, and Other Variables	Concentration (mg/kg[a])	Reference[b]
Fox, *Vulpes* sp.		
All tissues	<0.7 FW	1
Wildlife		
All tissues	<1 FW	2
INTEGRATED STUDIES		
Alaska; Cook Inlet; summer 1997; edible portions of seafood products		
Total arsenic	0.6 FW (salmon)–5.0 FW (clams, cod, flounder)	55
Dimethylarsinic acid	0.02–0.6 FW	55
As^{+3} plus As^{+5}	0.001–0.01 FW	55
Monomethylarsinic acid	Trace	55
Germany; near former arsenic roasting facility; total As		
Ants, *Formica* sp.; whole	12.6 DW	62
Ant-hill material	5420.0 DW	62
Spruce, *Picea* sp.; needles	1.2 DW	62
Larch, *Larix* sp.; needles	3.7 DW	62
Korea; near silver and gold mine closed in 1970; surveyed October/November 1997		
Tailings	5000 DW	68
Downstream		
Sediments	Max. 4400 DW	68
Irrigation water	Max. 1.89 FW	68
Rice; mine area vs. reference site		
Grain	Max. 0.6 DW vs. Max. 0.3 DW	68
Stalks and leaves	Max. 9.1 DW vs. Max. 3.6 DW	68
South Texas; Lower Laguna Madre; 1986–87		
Sediments	1.9 (0.4–8.2) DW	46
Shoalgrass, *Halodule wrightii*; rhizomes	12.2 (2–25) DW	46
Grass shrimp, *Palaemonetes* sp.; whole	26.9 (9.7–55.0) DW	46
Brown shrimp, *Penaeus aztecus*; whole	17.9 (8.1–50.0) DW	46
Blue crab, *Callinectes sapidus*; whole minus legs, carapace and abdomen	18.4 (2.7–50.0) DW	46
Pinfish, *Lagodon rhomboides*; whole	9.4 (1.7–20.0) DW	46
Montana; mining waste-contaminated wetland vs. reference site; 1990–92		
Sediments	45.3 DW vs. 5.8 DW	47, 48
Soil	52.5–67.1 DW vs. 7.7 DW	47, 48
Surface water	0.014 FW vs. no data	47, 48
Vegetation		
Above ground		
Forbs	1.1–1.9 DW vs. 0.4–0.5 DW	47, 48
Grasses	6.7 DW vs. 2.1 DW	47, 48
Floating macrophytes	18.9–35.2 DW vs. <0.02 DW	47
Below ground		
Forbs	9.3 DW vs. 2.1 DW	47, 48
Grasses	100.1 DW vs. 6.4 DW	47, 48
Macrophytes	52.0 DW vs. <0.2 DW	47
Invertebrates		
Aquatic		
Arthropods	7.7 DW vs. <0.02 DW	47
Snails	<0.02 DW vs. <0.02 DW	47
Terrestrial		
Grasshoppers	3.9 DW vs. <0.02 DW	47
Earthworms	10.9 DW vs. 1.7 DW	47

Table 28.2 (continued) Arsenic Concentrations in Field Collections of Selected Species of Plants and Animals (Values listed are in mg As/kg fresh weight [FW] or dry weight [DW].)

Ecosystem, Species, and Other Variables	Concentration (mg/kg[a])	Reference[b]
Fishes, whole	0.15 FW vs. 0.05–0.12 FW	47
Bird eggs	<0.1 DW vs. no data	47
Deer mice, *Peromyscus maniculatus*		
Carcass	<0.02 FW vs. no data	48
Liver, kidney, gonads	ND vs. no data	48
Meadow vole, *Microtus pennsylvanicus*		
Carcass	0.25 FW vs. no data	48
Liver, kidney, gonads	ND vs. no data	48
Ontario, Canada; Moira Lake (recipient of mining wastes — including arsenic — since the 1830s); June 1991		
Water	Max. 0.106 FW	49
Sediments	Max. 1000 DW	49
Fishes; 13 species; whole	0.03–0.34 FW	49
Creek chub, *Semotilus atromaculatus*; whole	2.4 FW	49
Rock bass, *Ambloplites rupestris*		
Bone	0.5 FW	49
Intestine	0.6 FW	49
Liver	0.04 FW	49
Muscle	0.08 FW	49
Whole	0.13 (0.07–1.9) FW	49
Taiwan, 1995–96		
Fish; 8 species; muscle	Max. 5.3 DW	64
Shrimp; 2 species; muscle	Max. 5.1 DW	64
Clam, *Meretrix lusoria*; soft parts	13.7 (9.5–20.1) DW	64
Pacific oyster, *Crassostrea gigas*; soft parts	11.8 (7.2–16.3) DW	64

[a] Concentrations are listed as mean, minimum-maximum (in parentheses), nondetectable (ND), and maximum (Max.).

[b] **1,** Jenkins 1980; **2,** NAS 1977; **3,** Mankovska 1986; **4,** Merry et al. 1986; **5,** NRCC 1978; **6,** Wiener et al. 1984; **7,** Woolson 1975; **8,** Vos and Hovens 1986; **9,** Lima et al. 1984; **10,** Norin et al. 1985; **11,** Sorensen et al. 1985; **12,** Eisler 1981; **13,** Pershagen and Vahter 1979; **14,** Shiomi et al. 1984a; **15,** Zaroogian and Hoffman 1982; **16,** Phillips and Depledge 1986; **17,** Phillips et al. 1982; **18,** Hall et al. 1978; **19,** Francesconi et al. 1985; **20,** Shiomi et al. 1984b; **21,** Hanaoka and Tagawa 1985a; **22,** Hanaoka and Tagawa 1985b; **23,** Reinke et al. 1975; **24,** Hallacher et al. 1985; **25,** Windom et al. 1973; **26,** Hall 1980; **27,** Wiemeyer et al. 1980; **28,** Blus et al. 1977; **29,** White et al. 1980; **30,** Goede 1985; **31,** Vreman et al. 1986; **32,** Camardese et al. 1990; **33,** Schmitt and Brumbaugh 1990; **34,** Johns and Luoma 1990; **35,** Ozretic et al. 1990; **36,** Ramelow et al. 1989; **37,** Farag et al. 1995; **38,** Stavelind et al. 1993; **39,** Leah et al. 1992; **40,** Hothem and Welsh 1994; **41,** Custer and Mitchell 1991; **42,** Guitart et al. 1994; **43,** Vermeer and Thompson 1992; **44,** Mathews and Porter 1989; **45,** Thies and Gregory 1994; **46,** Custer and Mitchell 1993; **47,** Pascoe et al. 1996; **48,** Pascoe et al. 1994; **49,** Azcue and Dixon 1994; **50,** Saiki et al. 1992; **51,** Findo et al. 1993; **52,** Lebedeva 1997; **53,** Kairu 1996; **54,** Ismail and Roberts 1992; **55,** Bigler and Crecelius 1998; **56,** Finkelman et al. 1998; **57,** Tao and Bolger 1998; **58,** Goessler et al. 1998a; **59,** Goessler et al. 1998b; **60,** Kuehnelt et al. 1997a; **61,** Kuehnelt et al. 1997b; **62,** Kuehnelt et al. 1997c; **63,** Hopkins et al. 1998; **64,** Han et al. 1998; **65,** Burger and Snodgrass 1998; **66,** Kurttio et al. 1998; **67,** Hsueh et al. 1998; **68,** Ahn et al. 1999.

28.6 LETHAL AND SUBLETHAL EFFECTS

28.6.1 General

As will be discussed later, most authorities agree on ten points:

1. Inorganic arsenicals are more toxic than organic arsenicals, and trivalent forms are more toxic than pentavalent forms.
2. Episodes of arsenic poisoning are either acute or subacute; cases of chronic arsenosis are rarely encountered, except in humans.

3. Sensitivity to arsenic is greatest during the early developmental stages.
4. Arsenic can traverse placental barriers: as little as 1.7 mg As^{+5}/kg body weight at critical stages of hamster embryogenesis, for example, can produce fetal death and malformation.
5. Biomethylation is the preferred detoxification mechanism for inorganic arsenicals.
6. Arsenic is bioconcentrated by organisms, but not biomagnified in the food chain.
7. In soils, depressed crop yields were recorded at 3 to 28 mg of water-soluble As/L, or about 25 to 85 mg total As/kg soil; adverse effects on vegetation were recorded at concentrations in air of >3.9 μg As/m^3.
8. Some aquatic species were adversely affected at water concentrations of 19 to 48 μg As/L, or 120 mg As/kg in the diet, or tissue residue of 1.3 to 5 mg As/kg fresh weight.
9. Sensitive species of birds died following single oral doses of 17.4 to 47.6 mg As/kg body weight.
10. Adverse effects were noted in mammals at single oral doses of 2.5 to 33 mg As/kg body weight, at chronic oral doses of 1 to 10 mg As/kg body weight, and at feeding levels of 50 mg — and sometimes only 5 mg — As/kg in the diet.

The literature emphasizes that arsenic metabolism and toxicity vary greatly between species and that its effects are significantly altered by numerous physical, chemical, and biological modifiers. Adverse health effects, for example, may involve respiratory, gastrointestinal, cardiovascular, and hematopoietic systems, and may range from reversible effects to cancer and death, depending partly on the physical and chemical forms of arsenic tested, the route of administration, and the dose.

28.6.2 Carcinogenesis, Mutagenesis, and Teratogenesis

Inorganic arsenic is a known human carcinogen and has been assigned Group A classification by the U.S. Environmental Protection Agency, Group 1 classification by the International Agency for Research on Cancer, and been designated Category A1 by the American Conference of Governmental Industrial Hygienists (ATSDR 1992; USPHS 1998). Epidemiological studies show that an increased risk of cancers in the skin, lung, liver, and lymph and hematopoietic systems of humans is associated with exposure to inorganic arsenicals. These increased cancer risks are especially prevalent among smelter workers and in those engaged in the production and use of arsenical pesticides where atmospheric levels exceed 54.6 μg As/m^3 (NRCC 1978; Belton et al. 1985; Pershagen and Bjorklund 1985). Skin tumors, mainly of low malignancy, have been reported after consumption of arsenic-rich drinking waters; a total dose of several grams, probably as As^{+3}, is usually required for the development of skin tumors (Pershagen and Vahter 1979). High incidences of skin cancer and hyperpigmentation were noted among several population groups, especially Taiwanese and Chileans, consuming water containing more than 0.6 mg As/L. The frequency of cancer was highest among people over age 60 who demonstrated symptoms of chronic arsenic poisoning (NRCC 1978). In some areas of India, however, as many as 60% of the children between age 4 and 10 years had arsenical melanosis because of the exceptionally high levels of arsenic in the drinking water (Chowdhury et al. 1997). Elimination of arsenic in drinking water decreased the mortality incidence of arsenic-related cancers of the liver, lung, kidney, and skin in communities where blackfoot disease is endemic (Tsai et al. 1998).

Arsenic reportedly inhibits cancer formation in species having a high incidence of spontaneous cancers (NRCC 1978). In fact, arsenic may be the only chemical for which there is sufficient evidence for carcinogenicity in humans but not in other animals (Woolson 1975; Belton et al. 1985; Lee et al. 1985). In general, animal carcinogenicity tests with inorganic and organic arsenicals have been negative (Hood 1985), even when the chemicals were administered at or near the highest tolerated dosages for long periods (NAS 1977). Most studies of arsenic carcinogenesis in animals were presumably of insufficient duration to simulate conditions in long-lived species such as humans (NRCC 1978). However, mice developed leukemia and lymphoma after 20 subcutaneous injections of 0.5 mg As^{+5}/kg body weight: 46% of the experimental group developed these signs vs. none of the controls (NRCC 1978). And mice given 500 μg As/L (as sodium arsenate) in the drinking water for lifetime exposures developed tumors in the lung, liver, and GI tract (Ng et al. 1998). Pulmonary tumorigenicity has been demonstrated in hamsters administered calcium arsenate intratracheally (Pershagen and Bjorklund 1985). Inorganic arsenic interacts with benzo[a]pyrene in the induction of lung adenocarcinomas in hamsters (ATSDR 1992). Cacodylic acid and other organoarsenicals are not carcinogenic, but may be mutagenic at very high doses (Hood 1985). Monomethylarsonic acid (MMA) and dimethlarsinic acid (DMA) are primary metabolites of inorganic arsenic, a known human carcinogen. However, the rapid elimination (Tb 1/2 = 2 h) and low retention (<2%) of MMA and DMA explain, in part, their low acute toxicity and cancer risk (Hughes and Kenyon 1998).

Several inorganic arsenic compounds are weak inducers of chromosomal aberrations, sister chromatid exchange, and *in vitro* transformation of mammalian and piscine cells. However, there is no conclusive evidence that arsenic causes point mutations in any cellular system (Pershagen and Vahter 1979; Belton et al. 1985; Lee et al. 1985; Deknudt et al. 1986; Manna and Mukherjee 1989). Studies with bacteria suggest that arsenite is a comutagen, or may inhibit DNA repair (Belton et al. 1985).

Arsenic is a known teratogen in several classes of vertebrates and has been implicated as a cause of birth defects in humans (Domingo 1994). Specific developmental malformations have been produced experimentally in mammals using inorganic As^{+3} or As^{+5}, either through a single dose or a continuous dose during embryogenesis (Hanlon and Ferm 1986b). Teratogenic effects are initiated no later than 4 h after administration of arsenic; fetal abnormalities are primarily neural tube defects (Hanlon and Ferm 1985c) but might also include protruding eyes, incomplete development of the skull, abnormally small jaws, and other skeletal anomalies (NRCC 1978). Inorganic As^{+3} and As^{+5}, but not organoarsenicals, cross placental barriers in many species of mammals and result in fetal deaths and malformations (NRCC 1978; USEPA 1980; Domingo 1994). Studies with hamsters, for example, showed that sodium arsenite can induce chromatid breaks and chromatid exchanges in Chinese hamster ovary cells in a dose-dependent manner (Lee et al. 1986b). In an earlier study (Lee et al. 1985), As^{+3} was about 10 times more potent than As^{+5} in causing transformations. The birth defects were most pronounced in golden hamsters exposed to As^{+5} during the 24-hr period of critical embryogenesis — day 8 of gestation (Ferm and Hanlon 1985) — when 1.7 mg As^{+5}/kg body weight induced neural tube defects in about 90% of the fetuses. Hanlon and Ferm (1986a) showed that hamsters exposed to As^{+5} and heat stress (39°C for 50 min) on day 8 of gestation produced a greater percentage of malformed offspring (18% to 39%) than did hamsters exposed to As^{+5} alone (4% to 8%).

28.6.3 Terrestrial Plants and Invertebrates

In general, arsenic availability to plants is highest in coarse-textured soils having little colloidal material and little ion exchange capacity, and lowest in fine-textured soils high in clay, organic material, iron, calcium, and phosphate (NRCC 1978). To be absorbed by plants, arsenic compounds must be in a mobile form in the soil solution. Except for locations where arsenic content is high (e.g., around smelters), the accumulated arsenic is distributed throughout the plant body in nontoxic amounts (NAS 1977). For most plants, a significant depression in crop yields was evident at soil arsenic concentrations of 3 to 28 mg/L of water-soluble arsenic and 25 to 85 mg/kg of total arsenic (NRCC 1978). Yields of peas (*Pisum sativum*), a sensitive species, were decreased at 1 mg/L of water-soluble arsenic or 25 mg/kg of total soil arsenic. Rice (*Oryza sativum*) yields were decreased 75% at 50 mg/L of disodium methylarsonate in silty loam, and soybeans (*Glycine max*) grew poorly when residues exceeded 1 mg As/kg (Table 28.3) (NRCC 1978). Forage plants grown in soils contaminated with up to 80 mg total As/kg from arsenical orchard sprays contained up to 5.8 mg As/kg dry weight; however, these plants were considered nonhazardous to grazing ruminants (Merry et al. 1986).

Attention focused on inorganic arsenical pesticides after accumulations of arsenic in soils eventually became toxic to several agricultural crops, especially in former orchards and cotton fields. Once toxicity is observed, it persists for several years even if no additional arsenic treatment is made (Woolson 1975). Poor crop growth was associated with bioavailability of arsenic in soils. For example, alfalfa (*Medicago sativa*) and barley (*Hordeum vulgare*) grew poorly in soils containing only 3.4 to 9.5 mg As/kg, provided the soils were acidic, lightly textured, low in phosphorus and aluminum, high in iron and calcium, and contained excess moisture (Woolson 1975). Use of inorganic arsenical herbicides, such as calcium arsenate, to golf course turfs for control of fungal blight sometimes exacerbates the disease. The use of arsenicals on Kentucky bluegrass (*Poa pratensis*) is discouraged under conditions of high moisture and root stress induced by previous arsenical applications (Smiley et al. 1985).

Methylated arsenicals, whether herbicides or defoliants, are sprayed on plant surfaces. They can reach the soil during application or can be washed from the plants. Additional arsenic enters soils by exchange from the roots or when dead plant materials decay (Hood 1985). Cacodylic acid and sodium cacodylate are nonselective herbicides used in at least 82 products to eliminate weeds and grasses around trees and shrubs, and to eradicate vegetation from rights-of-way and other noncrop areas (Hood 1985). Normal application rates of various organoarsenicals for crop and noncrop purposes rarely exceed 5 kg/ha (Woolson 1975). At recommended treatment levels, organoarsenical soil residues are not toxic to crops, and those tested (soybean, beet, wheat) were more resistant to organoarsenicals than to comparable levels of inorganic arsenicals (Woolson 1975).

Air concentrations up to 3.9 μg As/m^3 near gold mining operations were associated with adverse effects on vegetation. Higher concentrations of 19 to 69 μg As/m^3, near a coal-fired power plant in Czechoslovakia, produced measurable contamination in soils and vegetation in a 6-km radius (NRCC 1978).

The phytotoxic actions of inorganic and organic arsenicals are different, and each is significantly modified by physical processes. The primary mode of action of arsenite in plants is inhibition of light activation, probably through interference with the pentose phosphate pathway (Marques and Anderson 1986). Arsenites penetrate the plant cuticle to a greater degree than arsenates (NAS 1977). One of the first indications of plant injury by sodium arsenite is wilting caused by loss of turgor, whereas stress due to sodium arsenate does not involve rapid loss of turgor (NAS 1977). Organoarsenicals, such as cacodylic acid, enter plants mostly by absorption of sprays; uptake from the soil contributes only a minor fraction (Hood 1985). The phytotoxicity of organoarsenical herbicides is characterized by chlorosis, cessation of growth, gradual browning, dehydration, and death (NAS 1977). In general, plants cease to grow and develop after the roots have absorbed much arsenic (NRCC 1978). Plants can absorb arsenic through the roots and foliage, although translocation is species dependent. Concentrations of arsenic in plants correlate highly and consistently with water-extractable soil arsenic, and usually poorly with total soil arsenic (NRCC 1978). For example, concentrations of arsenic in corn (*Zea mays*) grown in calcareous soils for 25 days were significantly correlated with the soil water-extractable arsenic fraction, but not other fractions. Extractable phosphorus was correlated positively to both arsenic in corn and to the water-soluble arsenic fraction (Sadiq 1986). In the moss *Hylocomium splendens*, arsenate accumulation from solution was through living shoots, optimum uptake being between pH 3 and 5 (Wells and Richardson 1985). Some plants, such as beets (*Beta vulgaris*), accumulated arsenic more readily at elevated temperatures, but the addition of phosphate fertilizers markedly depressed uptake (Merry et al. 1986).

Soils amended with arsenic-contaminated plant tissues were not measurably affected in CO_2 evolution and nitrification, suggesting that the effects of adding arsenic to soils does not influence the decomposition rate of plant tissues by soil microorganisms (Wang et al. 1984). The half-life of cacodylic acid is about 20 days in untreated soils and 31 days in arsenic-amended soils (Hood 1985). Estimates of the half-time of inorganic arsenicals in soils are much longer, ranging from 6.5 years for arsenic trioxide to 16 years for lead arsenate (NRCC 1978).

Data on arsenic effects to soil biota and insects are limited. In general, soil microorganisms are capable of tolerating and metabolizing relatively high concentrations of arsenic (Wang et al. 1984). This adaptation seems usually to be due to decreased permeability of the microorganism to arsenic (NAS 1977). Tolerant soil microbiota can withstand concentrations up to 1600 mg/kg; however, growth and metabolism were reduced in sensitive species at 375 mg As/kg and, at 150 to 165 mg As/kg, soils were devoid of earthworms and showed diminished quantities of bacteria and protozoans (NRCC 1978). Earthworms (*Lumbricus terrestris*) held in soils containing 40 to 100 mg As^{+5}/kg DW soil for 8 to 23 days had significantly reduced survival, especially among worms held in soils less than 70 mm in depth when compared to worms held at 500 to 700 mm. Survivors had negligible arsenic residues. Dead worms had higher concentrations, suggesting that arsenic homeostasis breaks down after death (Meharg et al. 1998). Honeybees (*Apis mellifera*) killed accidentally by sprayed As^{+3} contained 4 to 5 µg As per bee (NAS 1977), equivalent to 21 to 31 mg/kg body weight (Table 28.3). Larvae of the western spruce budworm (*Choristoneura occidentalis*) continued to feed on As^{+3}-contaminated vegetation until a threshold level of about 2300 to 3300 mg As/kg dry weight whole larvae was reached; death then sometimes occurred (Table 28.3) (Robertson and McLean 1985). Larvae that had accumulated sufficient energy reserves completed the first stage of metamorphosis but developed into pupae of subnormal weight. Larvae containing <2600 mg As^{+3}/kg ultimately developed into adults of less than normal weight, and some containing >2600 mg/kg dry weight died as pupae (Robertson and McLean 1985).

Table 28.3 Lethal and Sublethal Effects of Various Arsenic Compounds on Selected Species of Terrestrial Plants and Invertebrates

Ecosystem, Organism, and Other Variables	Arsenic Concentration and Effects	Reference[a]
TERRESTRIAL PLANTS		
Crops		
Total water soluble As in soils	Depressed crop yields at 3–28 mg/L	1
Total soil As concentrations	Depressed crop yields at 25–85 mg/kg	1
Common bermudagrass, *Cynodon dactylon*		
Arsenite	Plants grown on As-amended soils (up to 90 mg As^{+3}/kg) contained up to 17 mg As/kg dry weight in stems, 20 in leaves, and 304 in roots	2
Fruit orchards		
Inorganic arsenites and arsenates	Soils contain 31–94 mg/kg dry weight (vs. 2.4 in untreated orchards); whole rodents contain <0.002 mg As/kg fresh weight (vs. ND in untreated orchards)	3
Soybean, *Glycine max*		
Total As	Toxic signs at plant residues >1 mg total As/kg	1
Grasslands		
Cacodylic acid	Kill of 75 to 90% of all species at 17 kg/ha; recovery modest	3
Rice, *Oryza sativum*		
Disodium methylarsonate	75% decrease in yield at soil (silty loam) concentrations of 50 mg/kg	1
Scots pine, *Pinus sylvestris*		
Inorganic As^{+5}	Seedlings die when soil (sandy) concentrations exceed 250 mg/kg DW. Maximum BCF factors low: 0.6 for roots; 0.1 for shoots. Residues >62 mg As/kg DW in shoots are toxic, and 3300 mg/kg DW is usually fatal	4
Pea, *Pisum sativum*		
Sodium arsenite	15 mg/L inhibits light activation and photosynthetic CO_2 fixation in chloroplasts	5
Sandhill plant communities		
Cacodylic acid	No lasting effect at 2.25 kg/ha. Some species defoliated at 6.8 kg/ha. Significant effect, including 75% defoliation of oaks and death of all pine trees, at 34 kg/ha	3
Cowpea, *Vigna* sp.		
Total water soluble As in soils	Decreased yields at 1 mg/L	1
Total soil As concentrations (loamy sand)	Toxic at 25 mg/kg	1
Yeast		
Arsenate	At 75 mg/L, 60% reduction in phosphate transport and glucose metabolism in 30 min; at 375 mg/L, 100% reduction	1

Table 28.3 (continued) Lethal and Sublethal Effects of Various Arsenic Compounds on Selected Species of Terrestrial Plants and Invertebrates

Ecosystem, Organism, and Other Variables	Arsenic Concentration and Effects	Reference[a]
TERRESTRIAL INVERTEBRATES		
Honeybee, *Apis mellifera*		
Inorganic arsenite	Following arsenic spray dusting, dead bees contained 20.8–31.2 mg/kg FW (adults) or 5–13 mg/kg FW (larvae)	6
Beetles		
Cacodylic acid	Dietary levels of 100–1000 mg/kg fatal to certain pestiferous species	3
Western spruce budworm, *Christoneura occidentalis*, sixth instar stage		
Arsenic trioxide	Dietary levels of 99.5 mg/kg FW killed 10%, 2250 mg/kg killed 50%, and 65,300 mg/kg was fatal to 90%. Newly molted pupae and adults of As-exposed larvae had reduced weight. Regardless of dietary levels, concentrations of As ranged up to 2640 mg/kg DW in dead pupae, and 1708 mg/kg DW in adults	7
Earthworm, *Lumbricus terrestris*		
As^{+5}	Earthworms held in soils containing 40 mg As/kg DW for 23 days had no accumulations in first 12 days, with BCF of 3 by day 23	8
As^{+5}	100 mg As/kg DW soil was fatal to 50% in 8 days	8
As^{+5}	400 mg As/kg DW soil was fatal to 50% in 2 days	8

[a] **1,** NRCC 1978; **2,** Wang et al. 1984; **3,** Hood 1985 **4,** Sheppard et al. 1985; **5,** Marques and Anderson 1986; **6,** Jenkins 1980; **7,** Robertson and McLean 1985; **8,** Meharg et al. 1998.

28.6.4 Aquatic Biota

Adverse effects of arsenicals on aquatic organisms have been reported at concentrations of 19 to 48 µg/L in water, 33 mg/kg in diets, and 1.3 to 5 mg/kg fresh weight in tissues (Table 28.4). The most sensitive of the aquatic species tested that showed adverse effects were three species of marine algae, which showed reduced growth in the range of 19 to 22 µg As^{+3}/L; developing embryos of the narrow-mouthed toad (*Gastrophryne carolinensis*), of which 50% were dead or malformed in 7 days at 40 µg As^{+3}/L; and a freshwater alga (*Scenedesmus obliquis*), in which growth was inhibited 50% in 14 days at 48 µg As^{+5}/L (Table 28.4). Chronic studies with mass cultures of natural phytoplankton communities exposed to low levels of arsenate (1.0 to 15.2 µg/L) showed that As^{+5} differentially inhibits certain plants, causing a marked change in species composition, succession, and predator–prey relations. The significance of these changes on carbon transfer between trophic levels is unknown (Sanders and Cibik 1985; Sanders 1986). Adverse biological effects have also been documented at water concentrations of 75 to 100 µg As/L:

- At 75 µg As^{+5}/L, growth and biomass in freshwater and marine algae was reduced.
- At 85 to 88 µg/L of As^{+5} or various methylated arsenicals, mortality was 10% to 32% in amphipods (*Gammarus pseudolimnaeus*) in 28 days.
- At 95 µg As^{+3}/L, marine red alga failed to reproduce sexually.
- At 100 µg As^{+5}/L, marine copepods died and goldfish behavior was impaired (Table 28.4).

Rainbow trout (*Oncorhynchus mykiss*) fed diets containing up to 90 mg As^{+5}/kg were slightly affected, but those given diets containing >120 mg As/kg (as As^{+3} or As^{+5}) grew poorly, avoided food, and failed to metabolize food efficiently. No toxic effects were reported over 8 weeks of exposure to diets containing 1600 mg/kg, as methylated arsenicals (Table 28.4). Dietary disodium heptahydrate (DSA) is more toxic to juvenile rainbow trout than dietary arsenic trioxide, dimethylarsinic acid, or arsanilic acid (Cockell et al. 1991; Cockell and Bettger 1993). Diets containing 55 to 60 mg As^{+5}/kg ration as DSA were associated with changes in the hepatobiliary system of juvenile rainbow trout after 12 weeks of feeding (Cockell et al. 1992). The most sensitive indicator of DSA insult in juvenile rainbow trout was chronic inflammation of the gallbladder wall, as found in 71% of trout exposed to 33 mg As/kg ration for 24 weeks and 100% of those exposed to 65 mg As/kg ration for 24 weeks. There was no damage at 13 mg/kg ration and lower in the 24-week study (Cockell et al. 1991). The whole-body arsenic concentrations in moribund rainbow trout poisoned by arsenate compounds in 11-week exposures ranged between 4 and 6 mg As/kg fresh weight (vs. 2.0 mg/kg fresh weight in controls) — and dead whole trout had 8 to 12 mg As/kg fresh weight — suggesting that a critical arsenic body concentration is reached before death or toxicant insult occurs (McGeachy and Dixon 1990). In bluegills (*Lepomis macrochirus*), tissue residues of 1.35 mg As/kg fresh weight in juveniles and 5 mg/kg in adults are considered elevated and potentially hazardous (NRCC 1978).

Toxic and other effects of arsenicals to aquatic life are significantly modified by numerous biological and abiotic factors (Woolson 1975; NAS 1977; NRCC 1978; USEPA 1980, 1985; Howard et al. 1984; Michnowicz and Weaks 1984; Bryant et al. 1985; Sanders 1986; Hamilton and Buhl 1990; McGeachy and Dixon 1990). The LC50 values, for example, are markedly affected by water temperature, pH, Eh, organic content, phosphate concentration, suspended solids, and presence of other substances and toxicants, as well as arsenic speciation, and duration of exposure. In general, inorganic arsenicals are more toxic than organoarsenicals to aquatic biota, and trivalent species are more toxic than pentavalent species. Early life stages are most sensitive, and large interspecies differences are recorded, even among species closely related taxonomically.

Arsenic is accumulated from the water by a variety of organisms. However, there is no evidence of magnification along the aquatic food chain (Woolson 1975; NAS 1977; NRCC 1978; Hallacher et al. 1985; Hood 1985; Lindsay and Sanders 1990; Woodward et al. 1995). In a marine ecosystem based on the alga *Fucus vesiculosus*, arsenate (7.5 μg As^{+5}/L) was accumulated by all biota. After 3 months, arsenic was concentrated most efficiently by *Fucus* (120 mg/kg dry weight in apical fronds) and filamentous algal species (30 mg/kg dry weight); little or no bioaccumulation occurred in invertebrates, although arsenic seemed to be retained by gastropods and mussels (Rosemarin et al. 1985). In a simplified estuarine food chain, there was no significant increase in arsenic content of grass shrimp (*Palaemonetes pugio*) exposed to arsenate-contaminated food or to elevated water concentrations (Lindsay and Sanders 1990). In a freshwater food chain composed of algae, daphnids, and fish, water concentrations of 0.1 mg cacodylic acid/L produced residues (mg As/kg dry weight), after 48 h, of 4.5 in algae and 3.9 in daphnids, but only 0.09 in fish (NAS 1977). Microcosms of a Delaware cordgrass (*Spartina alterniflora*) salt marsh exposed to elevated levels of As^{+5} showed that virtually all arsenic was incorporated into plant tissue or strongly sorbed to cell surfaces (Sanders and Osman 1985). Studies with radioarsenic and mussels (*Mytilus galloprovincialis*) showed that accumulation varied with nominal arsenic concentrations, tissue, age of the mussel, and temperature and salinity of the medium (Unlu and Fowler 1979). Arsenate uptake increased with increasing arsenic concentration in the medium, but the response was not linear, accumulation being suppressed at higher external arsenic concentrations. Smaller mussels took up more arsenic than larger ones. In both size groups, arsenic was concentrated in the byssus and digestive gland. In general, arsenic uptake and loss increased at increasing temperatures. Uptake was significantly higher at 1.9% salinity than at 3.8%, but loss rate was about the same at both salinities. Radioarsenic loss followed a biphasic pattern; the biological half-life was 3 and 32 days for the fast and slow compartments, respectively; secretion of the byssal thread played a key role in elimination (Unlu and Fowler 1979). Factors known to modify rates of arsenic accumulation and retention in a marine shrimp (*Lysmata seticaudata*) include water temperature and salinity, arsenic concentration, age, and especially frequency of molting (Fowler and Unlu 1978).

Bioconcentration factors (BCF) experimentally determined for arsenic in aquatic organisms are, except for algae, relatively low. The BCF values for inorganic As^{+3} in most aquatic invertebrates and fish exposed for 21 to 30 days did not exceed 17 times; the maxima were 6 times for As^{+5}, and 9 times for organoarsenicals (USEPA 1980, 1985). Significantly higher BCF values were recorded in other aquatic organisms (NRCC 1978), but they were based on mean arsenic concentrations in natural waters that seemed artificially high. A BCF of 350 times was reported for the American oyster (*Crassostrea virginica*) held in 5 μg As^{+5}/L for 112 days (Zaroogian and Hoffman 1982). There was no relation between oyster body burdens of arsenic and exposure concentrations. However, diet seemed to contribute more to arsenic uptake than did seawater concentrations (Zaroogian and Hoffman 1982). An arsenic-tolerant strain of freshwater alga (*Chlorella vulgaris*) from an arsenic-polluted environment showed increasing growth up to 2000 mg As^{+5}/L, and it could survive at 10,000 mg As^{+5}/L (Maeda et al. 1985). Accumulations up to 50,000 mg As/kg dry weight were recorded (Maeda et al. 1985), suggesting a need for additional research on the extent of this phenomenon and its implications on food-web dynamics.

Some investigators have suggested that arsenic in the form of arsenite is preferentially utilized by marine algae and bacteria (Johnson 1972; Bottino et al. 1978; Johnson and Burke 1978). Arsenate reduction to arsenite in seawater depends on phosphorus in solution and available algal biomass (Johnson and Burke 1978). During algal growth, as phosphate is depleted and the P^{+5}:As^{+5} ratio drops, the rate of As^{+5} reduction increases. The resultant As^{+3}, after an initial peak, is rapidly oxidized to As^{+5}, indicating the possibility of biological catalysis of oxidation as well as mediation of As^{+5} reduction. Researchers generally agree that As^{+3} is more toxic than arsenates to higher organisms; however, As^{+5} had a more profound effect on growth and morphology of marine algae than As^{+3}. It is possible that marine algae erect a barrier against the absorption of As^{+3}, but not against As^{+5}. Within the cell, As^{+5} can then be reduced to the possibly more toxic As^{+3}. For example, the culture of two species of marine algae (*Tetraselmis chui* and *Hymenomonas carterae*) in media containing various concentrations of As^{+5} or As^{+3} showed that arsenic effects varied with oxidation state, concentration, and light intensity. Arsenate was incorporated and later partly released by both species. Differences between rates of uptake and release suggest that arsenic undergoes chemical changes after incorporation into algal cells (Bottino et al. 1978). When bacterial cultures from the Sargasso Sea and from marine waters of Rhode Island were grown in As^{+3}-enriched media, the bacteria reduced all available As^{+5} and utilized As^{+3} during the exponential growth phase, presumably as an essential trace nutrient. The arsenate reduction rate per cell was estimated to be 75×10^{-11} mg As/minute (Johnson 1972).

The ability of marine phytoplankton to accumulate high concentrations of inorganic arsenicals and transform them to methylated arsenicals that are later efficiently transferred in the food chain is well documented (Irgolic et al. 1977; Benson 1984; Matsuto et al. 1984; Freeman 1985; Froelich et al. 1985; Maeda et al. 1985; Norin et al. 1985; Sanders 1985; Yamaoka and Takimura 1986). Algae constitute an important source of organoarsenic compounds in marine food webs. In the food chain composed of the alga *Dunaliella marina*, the grazing shrimp *Artemia salina*, and the carnivorous shrimp *Lysmata seticaudata*, organic forms of arsenic were derived from *in vivo* synthesis by *Dunaliella* and efficiently transferred, without magnification, along the food chain (Wrench et al. 1979). Laboratory studies with five species of euryhaline algae grown in freshwater or seawater showed that all species synthesized fat-soluble and water-soluble arseno-organic compounds from inorganic As^{+3} and As^{+5}. The BCF values in the five species examined ranged from 200 times to about 3000 times — accumulations being highest in lipid phases (Lunde 1973). In Charlotte Harbor, Florida, a region that has become phosphate enriched due to agricultural activity, virtually all of the arsenic taken up by phytoplankton was biomethylated and returned to the estuary, usually as monomethylarsonic and dimethylarsenic acids (Froelich et al. 1985). The ability of marine phytoplankton to methylate arsenic and release the products to a surrounding environment varies between species and even within a particular species in relation to their possession of necessary methylating enzymes (Sanders 1985). The processes involved in detoxifying arsenate after its absorption by phytoplankton are not firmly established, but seem to be nearly identical in all plants, suggesting a similar evolutionary development. Like phosphates and sulfates, arsenate may be fixed with ADP, reduced to the arsonous level, and successfully methylated and adenosylated, ultimately producing the 5-dimethylarsenosoribosyl derivatives accumulating in algae (Benson 1984).

Sodium arsenite has been used extensively as an herbicide for control of mixed submerged aquatic vegetation in freshwater ponds and lakes. Concentrations of 1.5 to 3.8 mg As^{+3}/L have usually been effective and are considered safe for fish (NAS 1977). However, As^{+3} concentrations considered effective for aquatic weed control may be harmful to several species of freshwater teleosts, including bluegills, flagfish (*Jordenella floridae*), fathead minnows (*Pimephales promelas*), and rainbow trout (*Oncorhynchus mykiss*) (Table 28.4). Fish exposed to 1 to 2 mg total As/L for 2 to 3 days may show one or more of several signs: hemorrhagic spheres on gills; fatty infiltration of liver; and necrosis of heart, liver, and ovarian tissues (NRCC 1978). In green sunfish (*Lepomis cyanellus*), hepatocyte changes parallel arsenic accumulations in the liver (Sorensen et al. 1985). Organoarsenicals are usually eliminated rapidly by fish and other aquatic fauna. Rainbow trout, for example, fed a marine diet containing 15 mg organic arsenic/kg had only negligible tissue residues 6 to 10 days later, although some enrichment was noted in the eyes, throat, gills, and pyloric caeca (Pershagen and Vahter 1979). Oral administration of sodium arsenate to estuary catfish (*Cnidoglanis macrocephalus*) and school whiting (*Sillago bassensis*) resulted in tissue accumulations of trimethylarsine oxide. Arsenobetaine levels, which occur naturally in these teleosts, were not affected by As^{+5} dosing. The toxicity of trimethylarsine oxide is unknown, but the ease with which it can be reduced to the highly toxic trimethylarsine is cause for concern (Edmonds and Francesconi 1987). Recent studies, however, suggest that humans are capable of metabolizing trimethylarsine to the comparatively innocuous arsenobetaine (Goessler et al. 1997).

Table 28.4 Lethal and Sublethal Effects of Various Arsenic Compounds on Selected Species of Aquatic Biota

Ecosystem, Species, Arsenic Compound[a], and Other Variables	Arsenic Concentration (mg/L)	Effect	Reference[b]
FRESHWATER PLANTS			
Algae, various species			
As^{+3}	1.7	Toxic	1
As^{+3}	4	Decomposition	1
As^{+3}	2.3	95%–100% kill in 2–4 weeks of 4 species	2, 3
As^{+5}	0.075	Decreased growth	3
Alga, *Ankistrodesmus falcatus*			
As^{+5}	0.26	EC50 (14 days)	3
Alga, *Scenedesmus obliquus*			
As^{+5}	0.048	EC50 (14 days)	3
Alga, *Selenastrum capricornutum*			
As^{+5}	0.69	EC50 (4 days)	3
FRESHWATER INVERTEBRATES			
Cladoceran, *Bosmina longirostris*			
As^{+5}	0.85	50% immobilization in 96 h	4
Cladoceran, *Daphnia magna*			
As^{+3}	0.63–1.32	MATC[c]	3

Table 28.4 (continued) Lethal and Sublethal Effects of Various Arsenic Compounds on Selected Species of Aquatic Biota

Ecosystem, Species, Arsenic Compound[a], and Other Variables	Arsenic Concentration (mg/L)	Effect	Reference[b]
As^{+3}	0.96	LC5 (28 days)	5
As^{+3}			
Starved	1.5	50% immobilization (96 h)	6
Fed	4.3	50% immobilization (96 h)	6
As^{+5}	0.52	Reproductive impairment of 16% in 3 weeks	3
As^{+5}	0.93	LC5 (28 days); maximum bioconcentration factor (BCF) of 219	5
As^{+5}	7.4	LC50 (96 h)	2
DSMA	0.83	LC0 (28 days)	5
SDMA	1.1	LC0 (28 days)	5
Total As	1	18% decrease in body weight in 3 weeks	1
Total As	1.4	50% reproductive impairment in 3 weeks	1
Total As	2.8	LC50 (21 days)	1
Total As	4.3–7.5	Immobilization (21 days)	1
Cladoceran, *Daphnia pulex*			
As^{+5}	49.6	50% immobilization (48 h)	4
As^{+3}	1.3	LC50 (96 h)	2, 3
As^{+3}	3	EC50 (48 h)	7
Amphipod, *Gammarus pseudolimnaeus*			
As^{+3}	0.87	50% immobilization (96 h)	6
As^{+3}	0.088	LC20 (28 days)	5
As^{+3}	0.96	LC100 (28 days)	5
As^{+5}	0.97	LC20 (28 days); no accumulations	5
DSMA	0.086	LC10 (28 days)	5
DSMA	0.97	LC40 (28 days)	5
SDMA	0.85	LC0 (28 days)	5
Snail, *Helisoma campanulata*			
As^{+3}	0.96	LC10 (28 days)	5
As^{+5}	0.97	LC0 (28 days); maximum BCF of 99	5
DSMA	0.97	LC0 (28 days)	5
SDMA	0.085	LC0 (28 days)	5
SDMA	0.085	LC32 (28 days)	5
Red crayfish, *Procambarus clarkii*			
MSMA	Nominal concentration of 0.5 mg/L, equivalent to 0.23 mg As/L	Whole-body As concentration after 8-week exposure plus 8-week depuration was 0.3 mg/kg DW whole body vs. 0.4 for controls	15
MSMA	Nominal exposure of 5 mg/L, equivalent to 2.3 mg As/L	Exposure and depuration as above; maximum As concentration was 4.3 mg/kg DW whole body during exposure, 0.6 at end of depuration	15
MSMA	Nominal concentration of 50 mg/L, equivalent to 23.1 mg As/L	Exposure and depuration as above; maximum As concentration during exposure was 9 mg/kg DW whole animal and 2.1 at end of depuration	15
MSMA	Nominal exposure of 100 mg/L, equivalent to 46.3 mg As/L	No effect on growth or survival during 24-week exposure. but hatching success reduced to 17% vs. 78% for controls	16
MSMA	1019	LC50 (96 h)	16
Stonefly, *Pteronarcys californica*			
As^{+3}	38	LC50 (96 h)	7

Table 28.4 (continued) Lethal and Sublethal Effects of Various Arsenic Compounds on Selected Species of Aquatic Biota

Ecosystem, Species, Arsenic Compound[a], and Other Variables	Arsenic Concentration (mg/L)	Effect	Reference[b]
Stonefly, *Pteronarcys dorsata*			
As^{+3}	0.96	LC0 (28 days)	5
As^{+5}	0.97	LC20 (28 days); maximum BCF of 131	5
DSMA	0.97	LC0 (28 days)	5
SDMA	0.85	LC0 (28 days)	5
Cladoceran, *Simocephalus serrulatus*			
As^{+3}	0.81	LC50 (96 h)	3
Zooplankton			
As^{+3}	0.4	No effect	1
As^{+3}	1.2	Population reduction	1
FRESHWATER VERTEBRATES			
Marbled salamander, *Ambystoma opacum*			
As^{+3}	4.5	EC50 (8 days) concentration producing death and malformations in developing embryos	3
Goldfish, *Carassius auratus*			
As^{+5}	0.1	15% behavioral impairment in 24 h; 30% impairment in 48 h	1
As^{+5}	24.6–41.6	LC50 (7 days)	1
As^{+3}	0.49	EC50 (7 days)	3
MSMA	5	LC50 (96 h)	3
Narrow-mouthed toad, *Gastrophryne carolinensis*			
As^{+3}	0.04	50% death or malformations in developing embryos in 7 days	3
Channel catfish, *Ictalurus punctatus*			
As^{+3}	25.9	LC50 (96 h)	8
Flagfish, *Jordanella floridae*			
As^{+3}	14.4	LC50 (96 h)	6
As^{+3}	2.1–4.1	MATC[c]	3
Bluegill, *Lepomis macrochirus*			
As^{+3}			
Juveniles	0.69	Reduced survival 16 weeks after a single treatment	2, 3
Adults	0.69	Histopathology after 16 weekly treatments	2
As^{+3}	4	Population reduction of 42% after several monthly applications	8
As^{+3}	30–35	LC50 (96 h)	7, 8
MSMA	1.9	LC50 (96 h)	3
Total As	Tissue levels of 1.35 mg/kg fresh weight (juveniles) and 5 mg/kg (adults)	Threshold acute toxic value	1
Spottail shiner, *Notropis hudsonius*			
As^{+3}	45	LC50 (25 h)	8
As^{+3}	29	LC50 (48 h); survivors with fin and scale damage	8

Table 28.4 (continued) Lethal and Sublethal Effects of Various Arsenic Compounds on Selected Species of Aquatic Biota

Ecosystem, Species, Arsenic Compound[a], and Other Variables	Arsenic Concentration (mg/L)	Effect	Reference[b]
Chum salmon, *Oncorhynchus keta*			
As^{+3}	11	LC50 (48 h)	8
Chinook salmon, *Oncorhynchus tshawytscha*			
Mean weight 0.5 g; fresh water			
As^{+3}	25.1	LC50 (96 h)	21
As^{+5}	90.0	LC50 (96 h)	21
Mean wight 2.0 g; brackish water			
As^{+3}	21.4	LC50 (96 h)	21
As^{+5}	66.5	LC50 (96 h)	21
Minnow, *Phoxinus phoxinus*			
As^{+3}	20	Equilibrium loss in 36 h	8
As^{+5}	234–250	Lethal	8
Fathead minnow, *Pimephales promelas*			
As^{+5}	0.53–1.50	MATC[c]	3
As^{+3}	2.1–4.8	MATC[c]	6
As^{+3}	6.0	LC30 (96 h)	17
As^{+3}	14.1	LC50 (96 h)	6
As^{+3}	17.7	LC76 (96 h)	17
As^{+5}	25.6	LC50 (96 h)	3
Rainbow trout, *Oncorhynchus mykiss*			
As^{+3}	0.13	EC10 (28 days)	3
As^{+3}			
Embryos	0.54	LC50 (28 days)	2
Adults	23–26.6	LC50 (96 h)	5
As^{+3}	0.96	LC50 (28 days)	7, 8
As^{+5}	60 at 15°C	50% loss of equilibrium (LOE) in 210 h; whole-body arsenic concentration (WBC) of 8.1 mg/kg FW	18
As^{+5}	120 at 5°C	50% LOE in 57 h; WBC of 8.6 FW	18
As^{+5}	120 at 15°C	50% LOE in 35 h; WBC of 8.6 FW	18
As^{+5}	240 at 5°C	50% LOE in 32 h; WBC of 13.5 FW	18
As^{+3} or As^{+5}	Fed diets of 120–1600 mg As/kg for 8 weeks	Growth depression, food avoidance, and impaired feed efficiency at all levels	9
As^{+5}	Fed diets of 10–90 mg As/kg for 16 weeks	No effect level at about 10 mg/kg diet. Some adaptation to dietary As observed in trout fed 90 mg/kg diet, as initial negative growth gave way to slow positive growth over time	9
As^{+5}	0.97	LC0 (28 days); no accumulations	5
DSA	13–33 mg As/kg ration for 12–24 weeks or 0.281–0.525 mg As/kg BW daily	MATC[c]	19
DSA	55–60 mg As^{+5} per kg ration for 8–12 weeks	Elevated accumulations in tissues; adverse effects on hepatobiliary system	20

Table 28.4 (continued) Lethal and Sublethal Effects of Various Arsenic Compounds on Selected Species of Aquatic Biota

Ecosystem, Species, Arsenic Compound[a], and Other Variables	Arsenic Concentration (mg/L)	Effect	Reference[b]
DSA	58 mg As^{+5}/kg ration for 12 weeks	Diet avoidance and reduced growth. Gallbladder lesions within 24 h of exposure developing to chronic inflammation with fibrosis of the gallbladder wall. Tissue arsenic concentrations at day 84, in mg As/kg DW, were 21 in liver, 39 in gallbladder, and 60 (as As^{+3}) in bile vs. <0.5 for all control tissues	22
DSMA	0.97	LC0 (28 days)	5
SDMA	0.85	LC0 (28 days)	5
SC	1000	LC0 (28 days)	14
DMA or ABA	Fed diet containing 120–1600 mg/kg for 8 weeks	No toxic response at any level tested	9
Brook trout, *Salvelinus fontinalis*			
As^{+3}	15	LC50 (96 h)	3
MARINE PLANTS			
Algae, 2 species			
As^{+3} or As^{+5}	1	No effect	10
As^{+5}	1000	No deaths	10
Algae, 3 species			
As^{+3}	0.019–0.022	Reduced growth	3
Red alga, *Champia parvula*			
As^{+3}	0.065	Normal sexual reproduction	10
As^{+3}	0.095	No sexual reproduction	10
As^{+3}	0.30	Death	10
As^{+5}	10	Normal growth, but no sexual reproduction	10
Phytoplankton			
As^{+5}	0.075	Reduced biomass of populations in 4 days	3
Red alga, *Plumaria elegans*			
As^{+3}	0.58	Arrested sporeling development 7 days posttreatment after exposure for 18 h	12
Alga, *Skeletonema costatum*			
As^{+5}	0.13	Growth inhibition	3
Alga, *Thalassiosira aestivalis*			
As^{+5}	0.075	Reduced chlorophyll *a*	3
MARINE INVERTEBRATES			
Copepod, *Acartia clausi*			
As^{+3}	0.51	LC50 (96 h)	3
Dungeness crab, *Cancer magister*			
As^{+3}	0.23	LC50 (96 h), zoea	3
Amphipod, *Corophium volutator*			
As^{+5}			
Water temperature, °C			
5	8	LC50 (230 h)	11
10	8	LC50 (150 h)	11

Table 28.4 (continued) Lethal and Sublethal Effects of Various Arsenic Compounds on Selected Species of Aquatic Biota

Ecosystem, Species, Arsenic Compound[a], and Other Variables	Arsenic Concentration (mg/L)	Effect	Reference[b]
15	8	LC50 (74 h)	11
15	4	LC50 (140 h)	11
15	2	LC50 (192 h)	11
Pacific oyster, *Crassostrea gigas*			
As^{+3}	0.33	LC50 (96 h) for embryos	3
American oyster, *Crassostrea virginica*			
As^{+3} (eggs)	7.5	LC50 (48 h)	8
Copepod, *Eurytemora affinis*			
As^{+5}	0.025	No effect	12
As^{+5}	0.1	Reduced juvenile survival	12
As^{+5}	1	Reduced adult survival	12
Clam, *Macoma balthica*			
As^{+5}			
Water temperature, °C			
5	220	LC50 (192 h)	11
10	60	LC50 (192 h)	11
15	15	LC50 (192 h)	11
Mysid, *Mysidoposis bahia*			
As^{+3}	0.63–1.27	MATC[c]	3
As^{+5}	2.3	LC50 (96 h)	3
Blue mussel, *Mytilus edulis*			
As^{+3}	16	Lethal in 3–16 days	8
Mud snail, *Nassarius obsoletus*			
As^{+3}	2	Depressed oxygen consumption in 72 h	8
Oligochaete annelid, *Tubifex costatus*			
As^{+5}			
Water temperature, °C			
5	500	LC50 (130 h)	11
10	500	LC50 (115 h)	11
15	500	LC50 (85 h)	11
MARINE VERTEBRATES			
Grey mullet, *Chelon labrosus*			
As^{+3}	27.3	LC50 (96 h); some skin discoloration	13
Dab, *Limanda limanda*			
As^{+3}	28.5	LC50 (96 h); respiratory problems	13
Pink salmon, *Oncorhynchus gorbuscha*			
As^{+3}	2.5	LC0 (10 days)	8
As^{+3}	3.8	LC54 (10 days)	3
As^{+3}	7.2	LC100 (7 days)	3
Winter flounder, *Pleuronectes americanus*			
As^{+3}	28, 56, or 112 mg As/kg BW by sc injection	All doses induced liver metallothionein protein within 24 h	23
Teleosts, 3 species			
As^{+3}	12.7–16	LC50 (96 h)	3

Table 28.4 (continued) Lethal and Sublethal Effects of Various Arsenic Compounds on Selected Species of Aquatic Biota

[a] As^{+3}, inorganic trivalent arsenite; As^{+5}, inorganic pentavalent arsenate; DMA, dimethylarsinic acid; ABA, *p*-aminobenzenearsonic acid; DSA, disodium arsenate heptahydrate; DMSA, disodium methylarsenate [$CH_3AsO(ONa)_2$]; SDMA, sodium dimethylarsenate [$(CH_3)_2AsO(ONa)$]; MSMA, monosodium methanearsonate; SC, sodium cacodylate.

[b] **1,** NRCC 1978; **2,** USEPA 1980; **3,** USEPA 1985; **4,** Passino and Novak 1984; **5,** Spehar et al. 1980; **6,** Lima et al. 1984; **7,** Johnson and Finley 1980; **8,** NAS 1977; **9,** Cockell and Hilton 1985; **10,** Thursby and Steele 1984; **11,** Bryant et al. 1985; **12,** Sanders 1986; **13,** Taylor et al. 1985; **14,** Hood 1985; **15,** Naqvi et al. 1990; **16,** Naqvi and Flagge 1990; **17,** Dyer et al. 1993; **18,** McGeachy and Dixon 1992; **19,** Cockell et al. 1991; **20,** Cockell et al. 1992; **21,** Hamilton and Buhl 1990; **22,** Cockell and Bettger 1992; **23,** Jensen-Eller and Crivello 1998.

[c] Maximum acceptable toxicant concentration. Lower value in each pair indicates highest concentration tested producing no measurable effect on growth, survival, reproduction, or metabolism during chronic exposure; higher value indicates lowest concentration tested producing a measurable effect.

28.6.5 Birds

Signs of inorganic trivalent arsenite poisoning in birds (muscular incoordination, debility, slowness, jerkiness, falling, hyperactivity, fluffed feathers, drooped eyelid, huddled position, unkempt appearance, loss of righting reflex, immobility, seizures) were similar to those induced by many other toxicants and did not seem to be specific for arsenosis. Signs occurred within 1 h and deaths within 1 to 6 days postadministration; remission took up to 1 month (Hudson et al. 1984). Internal examination suggested that lethal effects of acute inorganic arsenic poisoning were due to the destruction of blood vessels lining the gut, which resulted in decreased blood pressure and subsequent shock (Nystrom 1984). Coturnix (*Coturnix coturnix*), for example, exposed to acute oral doses of As^{+3} showed hepatocyte damage (i.e., swelling of granular endoplasmic reticulum); these effects were attributed to osmotic imbalance, possibly induced by direct inhibition of the sodium pump by arsenic (Nystrom 1984).

Arsenic, as arsenate, in aquatic plants (up to 430 mg As/kg plant dry weight) from agricultural drainwater areas can impair normal development of mallard ducklings that consume these plants (Camardese et al. 1990) (Table 28.5). Pen studies with ducklings showed that a diet of 30 mg As/kg ration adversely affects growth and physiology, and a 300 mg As/kg diet alters brain biochemistry and nesting behavior. Decreased energy levels and altered behavior can further decrease duckling survival in a natural environment (Camardese et al. 1990).

Western grasshoppers (*Melanophis* spp.) poisoned by arsenic trioxide were fed, with essentially no deleterious effects, to nestling northern bobwhites (*Colinus virginianus*), mockingbirds (*Mimus polyglottos*), American robins (*Turdus migratorius*), and other songbirds (NAS 1977). Up to 134 poisoned grasshoppers, containing a total of about 40 mg As, were fed to individual nestlings with no apparent toxic effect. Species tested that were most sensitive to various arsenicals were the brown-headed cowbird (*Molothrus ater*) with an LD50 (11-day) value of 99.8 mg copper acetoarsenite/kg diet; California quail (*Callipepla californica*) with an LD50 single oral dose value of 47.6 mg sodium arsenite/kg body weight; and chicken with 33 and turkey with 17.4 mg/kg body weight of 3-nitro-4-hydroxy phenylarsonic acid as a single oral dose (Table 28.5).

Chickens rapidly excrete arsenicals; only 2% of dietary sodium arsenite remained after 60 h (NAS 1977), and arsanilic acid was excreted largely unchanged (Woolson 1975). Excretion of arsanilic acid by chickens was affected by uptake route: excretion was more rapid if administration was by intramuscular injection than if it was oral (NRCC 1978). Studies with inorganic As^{+5} and chickens indicated that (Fullmer and Wasserman 1986):

1. Arsenates rapidly penetrated mucosal and serosal surfaces of epithelial membranes.
2. As^{+5} intestinal absorption was essentially complete within 1 h at 370 mg As^{+5}/kg BW, but only 50% complete at 3700 mg/kg BW.
3. Vitamin D_3 was effective in enhancing duodenal As^{+5} absorption in rachitic chicks.
4. As^{+5} and phosphate did not appear to share a common transport pathway in the avian duodenum.

Table 28.5 Lethal and Sublethal Effects of Various Arsenicals on Selected Species of Birds

Species and Arsenic Compound	Effect	Reference[a]
CHUKAR, *Alectoris chukar*		
Silvisar-510 (mixture of cacodylic acid and triethanolamine cacodylate)	Single oral LD50 dose of 2000 mg/kg body weight (BW); signs of poisoning evident within 10 min and mortalities within 1–2 days postadministration. Remission took up to 1 month	1
MALLARD, *Anas platyrhynchos*		
Sodium arsenate	Adult breeding pairs fed diets with 0, 25, 100, or 400 mg As/kg ration for up to 173 days. Ducklings produced were fed the same diet as their parents for 14 days. Dose-dependent increase in liver arsenic from 0.23 mg As/kg DW in controls to 6.6 in the 400-mg/kg group and in eggs from 0.23 in controls to 3.6 mg/kg DW in the 400-mg/kg group. Dose-dependent adverse effects on growth, onset of egg laying, and eggshell thinning. In ducklings, arsenic accumulated in the liver from 0.2 mg As/kg DW in controls to 33.0 in the 400-mg/kg group and caused dose-dependent decrease in whole-body and liver growth rate	9
Sodium arsenate	Ducklings were fed 30, 100, or 300 mg As/kg diet for 10 weeks. All treatment levels produced elevated hepatic glutathione and ATP concentrations and decreased overall weight gain and rate of growth in females. Arsenic concentrations were elevated in brain and liver of ducklings fed 100 or 300 mg/kg diets; at 300 mg/kg, all ducklings had altered behavior, i.e., increased resting time; males had reduced growth	6
Sodium arsenate	Day-old ducklings fed diets containing 200 mg As/kg ration for 4 weeks. When protein was adequate (22%), some growth reduction resulted. With only 7% protein in diet, growth and survival was reduced and frequency of liver histopathology increased	7
Sodium arsenate	Adult males fed diets containing 300 mg As/kg ration. Equilibrium reached in 10–30 days; 50% loss from liver in 1–3 days on transfer to uncontaminated diet	8
Sodium arsenite	323 mg/kg BW is LD50 acute oral value	1–3
Sodium arsenite	500 mg/kg diet is fatal to 50% in 32 days; 1000 mg/kg diet fatal to 50% in 6 days	2
Sodium cacodylate	1740–5000 mg/kg diet fatal to 50% in 5 days	4
Silvisar 510	Single oral LD50 >2400 mg/kg BW; regurgitation and excessive drinking noted	1
Lead arsenate	5000 mg/kg diet not fatal in 11 days	2
Copper acetoarsenite	5000 mg/kg diet fatal to 20% in 11 days	2
CALIFORNIA QUAIL, *Callipepla californica*		
Sodium arsenite	LD50 single oral dose of 47.6 mg/kg BW	1
COMMON BOBWHITE, *Colinus virginianus*		
Copper acetoarsenite	480 mg/kg in diet fatal to 50% in 11 days	2
Sodium cacodylate	1740 mg/kg in diet for 5 days produced no effect on behavior, no signs of intoxication, and negative necropsy	4
Monosodium methanearsonate, CH_4AsNaO_3	Single oral LD50 dose of 3300 mg/kg BW	4
CHICKEN, *Gallus gallus*		
Inorganic trivalent arsenite	Up to 34% dead embryos at dose range of 0.01–1 mg As^{+3}/embryo; threshold for malformations at dose range 0.03–0.3 mg/embryo	3
Inorganic pentavalent arsenate	Up to 8% dead at dose range 0.01–1 mg As^{+5}/embryo; threshold for malformations at dose range 0.3–3 mg/embryo	3
Disodium methyl arsenate	Teratogenic to embryos when injected at 1–2 mg/egg	3, 4
Sodium cacodylate	Developmental abnormalities at embryonic injected doses of 1–2 mg/egg	4
Dodecyclamine *p*-chlorophenylarsonate	At dietary levels of 23.3 mg/kg, liver residues were 2.9 mg/kg FW at 9 weeks. No ill effects noted	5

Table 28.5 (continued) Lethal and Sublethal Effects of Various Arsenicals on Selected Species of Birds

Species and Arsenic Compound	Effect	Reference[a]
3-Nitro-4-hydroxyphenyl-arsonic acid	At 18.7 mg/kg diet for 9 weeks, liver residues of 2.4 mg/kg FW. Those fed diets containing 187 mg/kg for 9 weeks had no ill effects; liver content of 7.5 mg/kg FW	5
3-Nitro-4-hydroxyphenyl-arsonic acid	LC50 dose of 33 mg/kg BW (single oral) or 9.7 mg/kg BW (intraperitoneal injection)	2
Arsanilic acid	Fed diets containing 45 mg/kg for 9 weeks; no effect except slightly elevated liver content of 1.2 mg/kg fresh weight. At dietary levels of 455 mg/kg, liver residues were 6.4 mg/kg FW after 9 weeks; no other effects evident	5
Cacodylic acid	Dosed orally without effect at 100 mg/kg BW daily for 10 days	4
CHICKENS, *Gallus* spp.		
Arsanilic acid	50% excreted in 36–38 h	3
Arsenate	50% excreted in 60–63 h	3
TURKEY, *Meleagris gallopavo*		
3-Nitro-4-hydroxyphenyl-arsonic acid	Single oral LD50 dose of 17.4 mg/kg BW	2
BROWN-HEADED COWBIRD, *Molothrus ater*		
Copper acetoarsenite	All survived 11 mg/kg diet for 6 months; maximum whole-body residue of 1.7 mg As/kg dry weight	2
Copper acetoarsenite	All survived 33 mg/kg diet for 6 months (whole-body content of 6.6 mg As/kg dry weight) or 7 months (8.6 DW)	2
Copper acetoarsenite	99.8 mg/kg in diet fatal to 50% in 11 days	2
Copper acetoarsenite	100 mg/kg in diet for 3 months fatal to 100%; tissue residues of 6.1 dry weight in brain, 40.6 in liver	2
GRAY PARTRIDGE, *Perdix perdix*		
Lead arsenate	300 mg/kg BW fatal in 52 h	2
RING-NECKED PHEASANT, *Phasianus colchicus*		
Sodium arsenite	Single oral dose of 386 mg/kg BW is LD50 value	1
Copper acetoarsenite	Single oral dose of 1403 mg/kg BW is LD50 value	3
Lead arsenate	4989 mg/kg in diet fatal	2

[a] **1,** Hudson et al. 1984; **2,** NAS 1977; **3,** NRCC 1978; **4,** Hood 1985; **5,** Woolson 1975; **6,** Camardese et al. 1990; **7,** Hoffman et al. 1992; **8,** Pendleton et al. 1995; **9,** Stanley et al. 1994.

28.6.6 Mammals

Mammals are exposed to arsenic primarily through the ingestion of naturally contaminated vegetation and water, or through human activity. In addition, feed additives containing arsonic acid derivatives are often fed to domestic livestock to promote growth and retard disease. Some commercial pet foods contain up to 2.3 mg As/kg dry weight (NRCC 1978). Uptake may occur by ingestion (the most likely route), inhalation, and absorption through skin and mucous membranes. Soluble arsenicals are absorbed more rapidly and completely than are the sparingly soluble arsenicals, regardless of the route of administration (NRCC 1978). In humans, inorganic arsenic at high concentrations is associated with adverse reproductive outcomes, including increased rates of spontaneous abortion, low birth weight, congenital malformations, and death (Hopenhayn-Rich et al. 1998). However, at environmentally relevant levels and routes of exposure, humans are not at risk for birth defects due to arsenic (Holson et al. 1998). *In vitro* tests with human erythrocytes demonstrate that inorganic As^{+5} as sodium arsenate was up to 1000 times more effective than inorganic As^{+3} as sodium arsenite after exposure for 5 h to 750 mg As/L in causing death, morphologic changes, and ATP depletion (Winski and Carter 1998).

Acute episodes of poisoning in warm-blooded organisms by inorganic and organic arsenicals are usually characterized by high mortality and morbidity over a period of 2 to 3 days (NAS 1977;

Selby et al. 1977). General signs of arsenic toxicosis include intense abdominal pain, staggering gait, extreme weakness, trembling, salivation, vomiting, diarrhea, fast and feeble pulse, prostration, collapse, and death. Gross necropsy shows a reddening of gastric mucosa and intestinal mucosa, a soft yellow liver, and red edematous lungs. Histopathological findings show edema of gastrointestinal mucosa and submucosa, necrosis and sloughing of mucosal epithelium, renal tubular degeneration, hepatic fatty changes and necrosis, and capillary degeneration in the gastrointestinal tract, vascular beds, skin, and other organs. In subacute episodes, where animals live for several days, signs of arsenosis include depression, anorexia, increased urination, dehydration, thirst, partial paralysis of rear limbs, trembling, stupor, coldness of extremities, and subnormal body temperatures (NAS 1977; Selby et al. 1977; ATSDR 1992). In cases involving cutaneous exposure to arsenicals, a dry, cracked, leathery, and peeling skin may be a prominent feature (Selby et al. 1977). Nasal discharges and eye irritation were documented in rodents exposed to organoarsenicals in inhalation toxicity tests (Hood 1985). Subacute effects in humans and laboratory animals include peripheral nervous disturbances, melanosis, anemia, leukopenia, cardiac abnormalities, and liver changes. Most adverse signs rapidly disappear after exposure ceases (Pershagen and Vahter 1979).

Arsenic poisoning in most animals is usually manifested by acute or subacute signs; chronic poisoning is infrequently seen (NAS 1977). The probability of chronic arsenic poisoning from continuous ingestion of small doses is rare because detoxification and excretion are rapid (Woolson 1975). Chronic toxicity of inorganic arsenicals is associated with weakness, paralysis, conjunctivitis, dermatitis, decreased growth, and liver damage (NRCC 1978). Arsenosis, produced as a result of chronic exposure to organic arsenicals, was associated with demyelination of the optic and sciatic nerves, depressed growth, and decreased resistance to infection (NRCC 1978).

Research results on arsenic poisoning in mammals (Table 28.6) show general agreement on eight points:

1. Arsenic metabolism and effects are significantly influenced by the organism tested, the route of administration, the physical and chemical form of the arsenical, and the dose.
2. Inorganic arsenic compounds are more toxic than organic arsenic compounds, and trivalent species are more toxic than pentavalent species.
3. Inorganic arsenicals can cross the placenta in most species of mammals.
4. Early developmental stages are the most sensitive, and humans appear to be one of the more susceptible species.
5. Animal tissues usually contain low levels (<0.3 mg As/kg fresh weight) of arsenic. After the administration of arsenicals, these levels are elevated, especially in liver, kidney, spleen, and lung; and several weeks later, arsenic is translocated to ectodermal tissues (hair, nails) because of the high concentration of sulfur-containing proteins in these tissues.
6. Inorganic arsenicals are oxidized *in vivo*, biomethylated, and usually excreted rapidly in the urine, but organoarsenicals are usually not subject to similar transformations.
7. Acute or subacute arsenic exposure can lead to elevated tissue residues, appetite loss, reduced growth, loss of hearing, dermatitis, blindness, degenerative changes in liver and kidney, cancer, chromosomal damage, birth defects, and death.
8. Death or malformations have been documented at single oral doses of 2.5 to 33 mg As/kg body weight, at chronic doses of 1 to 10 mg As/kg body weight, and at dietary levels >5 and <50 mg As/kg diet.

Episodes of wildlife poisoning by arsenic are infrequent. White-tailed deer (*Odocoileus virginianus*) consumed, by licking, fatal amounts of sodium arsenite used to debark trees. The practice of debarking trees with arsenicals for commercial use has been almost completely replaced by mechanical debarking equipment (NAS 1977). In another incident, white-tailed deer were found dead of arsenic poisoning in a northern New York forest and had 102 mg As/kg fresh weight in liver and 56 mg As/kg FW in kidney. These tissue concentrations are 2 to 3 times higher than those in cattle that died of arsenic poisoning — estimated at 241 to 337 mg As/kg BW (Mathews and

Porter 1989). It is speculated that these deer licked trees injected with Silvisar 550, which contains monosodium methanearsonate, probably because of its salty taste (Mathews and Porter 1989). Snowshoe hares (*Lepus* sp.) appear to be especially sensitive to methylated arsenicals; hares died after consuming plants heavily contaminated with monosodium methanearsonate as a result of careless silviculture practices (Hood 1985).

Unlike wildlife, reports of arsenosis in domestic animals are common in bovines and felines, less common in ovines and equines, and rare in porcines and poultry (NAS 1977). In practice, the most dangerous arsenic preparations are dips, herbicides, and defoliants in which the arsenical is in a highly soluble trivalent form, usually as trioxide or arsenite (Selby et al. 1977). Accidental poisoning of cattle with arsenicals, for example, is well documented. In one instance, more than 100 cattle died after accidental overdosing with arsenic trioxide applied topically to control lice. On necropsy, there were subcutaneous edematous swellings and petechial hemorrhages in the area of application, and histopathology of the intestine, mucosa, kidney, and epidermis (Robertson et al. 1984). In Bangladesh, poisoned cattle showed depression, trembling, bloody diarrhea, restlessness, unsteady gait, stumbling, convulsions, groaning, shallow labored breathing, teeth grinding, and salivation (Samad and Chowdhury 1984). Cattle usually died 12 to 36 h after the onset of signs; necropsy showed extensive submucosal hemorrhages of the gastrointestinal tract (Samad and Chowdhury 1984) and tissue residues >10 mg/kg fresh weight in liver and kidney (Thatcher et al. 1985). It sometimes appears that animals, especially cattle, develop an increased preference for weeds sprayed with an arsenic weed killer, not because of a change in the palatability of the plant, but probably because arsenic compounds are salty and thus attractive to animals (Selby et al. 1977).

When extrapolating animal data from one species to another, the species tested must be considered. For example, the metabolism of arsenic in the rat (*Rattus* sp.) is unique and very different from that in man and other animals. Rats store arsenic in blood hemoglobin, excreting it very slowly — unlike most mammals which rapidly excrete ingested inorganic arsenic in the urine as methylated derivatives (NAS 1977). Blood arsenic, whether given as As^{+3} or As^{+5}, rapidly clears from humans, mice, rabbits, dogs, and primates; the half-life is 6 h for the fast phase and about 60 h for the slow phase (USEPA 1980). In the rat, however, blood arsenic is mostly retained in erythrocytes and clears slowly; the half-life is 60 to 90 days (USEPA 1980). In rats, the excretion of arsenic into bile is 40 times faster than in rabbits and up to 800 times faster than in dogs (Pershagen and Vahter 1979). Most researchers now agree that the rat is unsatisfactory for use in arsenic research (NAS 1977; NRCC 1978; Pershagen and Vahter 1979; USEPA 1980; Webb et al. 1986).

Dimethylarsinic acid is the major metabolite of orally administered arsenic trioxide, and is excreted rapidly in the urine (Yamauchi and Yamamura 1985). The methylation process is true detoxification, since methanearsonates and cacodylates are about 200 times less toxic than sodium arsenite (NAS 1977). The marmoset monkey (*Callithrix jacchus*), unlike all other animal species studied to date, was not able (for unknown reasons) to metabolize administered As^{+5} to demethylarsinic acid; most was reduced to As^{+3}. Only 20% of the total dose was excreted in urine as unchanged As^{+5}, and another 20% as As^{+3}. The rest was bound to tissues, giving distribution patterns similar to arsenite (Vahter and Marafante 1985). Accordingly, the marmoset, like the rat, may be unsuitable for research with arsenicals.

Arsenicals were ineffective in controlling certain bacterial and viral infections. Mice experimentally infected with bacteria (*Klebsiella pneumonias*) or viruses (pseudorabies, encephalitis, encephalmyocarditis) showed a significant increase in mortality when treated with large doses of arsenicals compared to nonarsenic-treated groups (NAS 1977; Aranyi et al. 1985).

It has been suggested, but not yet verified, that many small mammals avoid arsenic-supplemented feeds and consume other foods if given the choice (NAS 1977), and that cacodylic acid, which has negligible effects on wildlife, reduces species diversity due to selective destruction of vegetation (Hood 1985). Both topics merit more research.

Table 28.6 Lethal and Sublethal Effects of Various Arsenicals on Selected Species of Mammals

Organism and Arsenical	Effect	Reference[a]
COW, *Bos bovis*		
Arsenate	Cows fed 33 mg As^{+5} daily per animal for 3 months had slightly elevated levels in muscle (0.02 mg/kg fresh weight vs. 0.005 in controls) and liver (0.03 vs. 0.012), but normal levels in milk and kidney	1
Arsenite	Cows fed 33 mg As^{+3} daily per animal for 15–28 months had tissue levels, in mg/kg fresh weight, of 0.002 for milk (vs. <0.001 for controls), 0.03 for muscle (vs. 0.005), 0.1 for liver (vs. 0.012), and 0.16 for kidney (vs. 0.053)	1
CATTLE, *Bos* spp.		
Arsenic pentoxide (wood ashes treated with arsenic preservative)	Several deaths after eating ashes (780 mg/kg dry weight); tissue residues, in mg As/kg fresh weight, of 13.9 in liver, 23.7 in kidney, and 25.8 in rumen contents (vs. normal values of <0.5)	2
Arsenic trioxide	Single oral dose of 15–45 grams per animal is fatal	3
Arsenic trioxide	Toxic dose is 33–55 mg/kg body weight (BW), or 13.2–22 g for a 400-kg animal. Animals accidentally poisoned topically contained up to 15 mg As/kg fresh weight liver, 23 in kidney, and 45 in urine (vs. <1 for all normal tissues)	4
Cacodylic acid, $(CH_3)_2AsO(OH)$	Calves were anorexic in 3–6 days when fed diets containing 4700 mg/kg. Adult oral dose of 10 mg/kg BW daily for 3 weeks, followed by 20 mg/kg BW daily for 5–6 weeks was lethal. Adverse effects at 25 mg/kg BW daily for 10 days	5
Methanearsonic acid	Calves were anorexic in 3–6 days when fed diets containing 4000 mg/kg	5
Monosodium methanearsonate	10 mg/kg BW daily for 10 days was fatal	3
Sodium arsenite	Single oral dose of 1–4 g fatal	3
DOG, *Canis familiaris*		
Cacodylic acid	Single oral LD50 value of 1000 mg/kg BW; diets containing 30 mg/kg for 90 days had no ill effects	5
Methanearsonic acid	Fed diets containing 30 mg/kg for 90 days with no ill effects	5
Sodium arsenite	50–150 mg fatal	3
DOMESTIC GOAT, *Capra* sp.		
Arsenic acid	Single oral dose of 2.5–7.5 mg/kg BW (50–150 mg) was acutely toxic	3
GUINEA PIG, *Cavia* sp.		
Arsenic acid	Dietary levels of 350 mg/kg resulted in blindness and optic disc atrophy in 25–30 days	6
Arsenic trioxide	Fed diets containing 50 mg/kg for 21 days; elevated As residues, in mg/kg fresh weight, of 4 in blood, 15 in heart (vs. <1 for all control tissues)	7
Sodium arsanilate	Subcutaneous injection of 70 mg/kg BW caused degeneration of sensory walls of inner ear; elevated As residues in cochlea	6
Sodium arsenate	Intraperitoneal injection of 0.2 mg/kg BW at age 2 months causes deafness	6
HAMSTER, *Cricetus* sp.		
Arsenate	Maternal dose of 5 mg As^{+5}/kg BW caused some fetal mortality, but no malformations; higher dose of 20 mg/kg BW caused 54% fetal deaths and malformations	3
Calcium arsenate	Pulmonary tumorigenicity demonstrated 70 weeks after 15 intratracheal weekly injections of 3 mg/kg BW	8

Table 28.6 (continued) Lethal and Sublethal Effects of Various Arsenicals on Selected Species of Mammals

Organism and Arsenical	Effect	Reference[a]
Dimethylarsinate	50% growth reduction in Chinese hamster ovary cells (CHOC) at 90–112 mg/L	9
Gallium arsenide	Single oral dose of 100 mg/kg BW mostly (85%) eliminated in 5 days, usually in form of organoarsenicals; all tissue levels <0.25 mg/kg	10
Sodium arsenate	Dosed intravenously on day 8 of gestation: 2 mg/kg BW had no measurable effect; 8 mg/kg produced increased incidence of malformation and resorption; 16 mg/kg BW killed all embryos	6
Sodium arsenate	50% growth reduction in CHOC at 2.25 mg/L	9
Sodium arsenite	Chinese hamster ovary cells (CHOC) show 50% growth reduction at 0.3 mg/L	9
Sodium cacodylate	Single intraperitoneal injection of 900–1000 mg/kg during midgestation results in some maternal deaths and increased incidences of fetal malformations	5
HORSE, *Equus caballus*		
Sodium arsenite	Daily doses of 2–6 mg/kg BW (1–3 grams) for 14 weeks is fatal	3
CAT, *Felis domesticus*		
Inorganic arsenate or arsenite	Chronic oral toxicity at 1.5 mg/kg BW	6
HUMAN, *Homo sapiens*		
Arsenic trioxide	Fatal at 70–189 mg, equivalent to about 1–2.6 mg As/kg BW	6
Arsenic trioxide	LD50 dose of 7 mg/kg BW	3
Cacodylic acid	LD50 of 1350 mg/kg BW	3
Lead arsenate	Some deaths at 7 mg/kg BW	3
Total arsenic	Accumulations of 1 mg/kg BW daily for 3 months in children, or 80 mg/kg BW daily for 3 years in adults produced symptoms of chronic arsenic poisoning	3
Total arsenic, daily oral dose	Prolonged dosages of 3–4 mg daily produced clinical symptoms of chronic arsenic intoxication	3
Total arsenic in drinking and cooking water	Prolonged use produced symptoms of chronic arsenic intoxication (0.6 mg/L) or skin cancer (0.29 mg/L)	3
Total arsenic, probably arsenate	12,000 Japanese infants poisoned (128 deaths) from consumption of dry milk contaminated with arsenic; average exposure of 3.5 mg As daily for 1 month. Severe hearing loss, brain wave abnormalities, and other CNS disturbances noted 15 years after exposure	6
Total inorganic arsenic	Daily dose of 3 mg for 2 weeks may cause severe poisoning in infants, and symptoms of toxicity in adults	6
CYNOMOLGUS MONKEY, *Macaca* sp.		
Fish-arsenic meal (witch flounder, *Glyptocephalus cynoglossus*) containing 77 mg total As/kg	Given a single meal at 1 mg/kg BW; tissue residues normal after 14 days	11
As above, except arsenic trioxide substituted for total As	As above	11
MAMMALS, many species		
Calcium arsenate	Single oral LD50 range of 35–100 mg/kg BW	3
Lead arsenate	Single oral LD50 range of 10–50 mg/kg BW	3
MAMMALS, most species		
Arsenic trioxide	3–250 mg/kg BW lethal	12
Sodium arsenite	1–25 mg/kg BW lethal	12

Table 28.6 (continued) Lethal and Sublethal Effects of Various Arsenicals on Selected Species of Mammals

Organism and Arsenical	Effect	Reference[a]
MOUSE, *Mus* spp.		
Arsenate	Maternal dose of 10 mg As^{+5}/kg BW results in some fetal deaths and malformations	3
Arsenic trioxide	Single oral LD50 (96 h) value of 39.4 mg/kg BW; LD0 (96 h) of 10.4 mg/kg BW	12
Arsenic trioxide	"Adapted" group (50 mg As/L in drinking water for 3 months) had subcutaneous LD50 value of 14 mg/kg BW vs. 11 for nonadapted group	12
Arsenic trioxide	Air concentrations of 28.5 mg/m^3 for 4 h daily on days 9–12 of gestation caused fetotoxic effects and chromosomal damage to liver cells by day 18; effects included reduced survival, impaired growth, retarded limb ossification, and bone abnormalities. At 2.9 mg/m^3, a 9.9% decrease in fetal weight was recorded; at 0.26 mg/m^3, a 3.1% decrease was measured	13
Cacodylic acid	Oral dosages of 400–600 mg/kg BW on days 7–16 of gestation produces fetal malformations (cleft palate), delayed skeletal ossification, and fetal weight reduction	5
Dimethylarsinic acid	200–600 mg/kg BW daily for 10 days (DMA) produced fetal and maternal toxicity	25
Single oral dose		
Arsenous oxide	34 mg As/kg BW fatal to 50%	21
Tetramethylarsonium iodide	890 mg As/kg BW fatal to 50%	21
Dimethylarsinic acid	1200 mg As/kg BW fatal to 50%	21
Dimethylarsonic acid	1800 mg As/kg BW fatal to 50%	21
Arsenocholine	6500 mg As/kg BW fatal to 50%	21
Trimethylarsinoxide	10,600 mg As/kg BW fatal to 50%	21
Arsenobetaine	>100,000 mg As/kg BW fatal to 50%	21
Sodium arsenate	Maximum tolerated doses in terms of abortion or maternal death over 24 h in 18-day pregnant mice were 20 mg As^{+5}/kg BW, intraperitoneal route, and 50 mg/kg BW when administered orally. Residue half-life was about 10 h, regardless of route of administration	14
Sodium arsenate	Given 500 µg As/L in drinking water for as long as 26 months, equivalent to 0.07–0.08 mg As/kg BW daily. No tumors in controls vs. 41.1% of mice in treated groups with 1 or more tumors — mostly of the lung, liver, and GI tract	22
Sodium arsenite	Fed 5 mg/kg diet for 3 generations: reduced litter size, but outwardly normal	6
Sodium arsenite	LD50 of 9.6 mg/kg BW, sc route; LD90 (7 days postadministration) of 11.3 mg/kg BW, sc route	15
Sodium arsenite	LD50 of 12 mg/kg BW intraperitoneal route. At 10 mg/kg BW, damage to bone marrow and sperm	16
Sodium cacodylate	Single intraperitoneal injection of 1200 mg/kg BW during midgestation results in increased rates of fetal skeletal malformations	5
MULE DEER, *Odocoileus hemionus hemionus*		
Silvisar-510 (mixture of cacodylic acid and triethanolamine cacodylate)	Single oral LD50 dose >320 mg/kg BW; appetite loss	17
WHITE-TAILED DEER, *Odocoileus virginianus*		
Sodium arsenite (used to debark trees)	Lethal dose of 923–2770 mg equivalent to about 34 mg/kg BW; liver residues of 40 mg/kg fresh weight	12
Arsenic acid (to control Johnson grass)	23 deer killed from apparent misuse. Arsenic levels, in mg/kg fresh weight, in deer found dead were 19 in liver, 18 in kidney, and 22.5 in rumen. Soils from area contained ~2.4 mg As/kg, and water 0.42 mg As/L	12

Table 28.6 (continued) Lethal and Sublethal Effects of Various Arsenicals on Selected Species of Mammals

Organism and Arsenical	Effect	Reference[a]
RABBIT, *Oryctolagus* sp.		
Monomethylarsonic acid (MMA)	50 mg/kg ration for 7–12 weeks produces hepatotoxicity	25
DOMESTIC SHEEP, *Ovis aries*		
Arsanilic acid	One-year-old castrates fed diets with 273 mg As/kg for 28 days had 0.54 mg As/L in blood, 29 mg/kg dry weight in liver, 24 in kidney, and 1.2 in muscle (vs. <0.01 in all control tissues). After 6 days on an As-free diet, liver residues were <5 mg/kg DW. Maximum tissue levels in sheep fed diets containing 27 mg As/kg for 28 days were 3.2 mg/kg DW kidney; for a 144 mg/kg DW diet, the maximum tissue level was 27 mg/kg DW liver	7
Sodium arsenite	Single oral dose of 5–12 mg/kg BW (0.2–0.5 g) was acutely toxic	3
Soluble arsenic	Lambs fed supplemental arsenic for 3 months at 2 mg As/kg DW diet contained maximum concentrations of 2 mg/kg FW brain (vs. 1 in controls), 14 in muscle (2), 24 in liver (4), and 57 in kidney (10)	18
Total arsenic	Sheep fed diets containing lakeweed (*Lagarosiphon major*) (288 mg As/kg DW) at 58 mg total As/kg diet for 3 weeks without ill effect. Tissue residues increased during feeding, but rapidly declined when lakeweed was removed from diet	7
RAT, *Rattus* spp.		
Arsanilic acid	No teratogenesis observed in 7 generations at dietary level of 17.5 mg/kg; positive effect on litter size and survival	6
Arsenate	Fed diets containing 50 mg/kg for 10 weeks with no effect on serum uric acid levels	19
Arsenic trioxide	Single oral LD50 (96 h) value of 15.1 mg/kg BW	12
Arsenic trioxide	Single dose of 17 mg/kg BW administered intratracheally is maximally tolerated nonlethal dose; 2 weeks later, blood As elevated (36 mg/L) and lung histopathology evident	20
Arsenic trioxide	After 21 days on diet containing 50 mg/kg, tissue arsenic levels were elevated in blood (125 mg/L vs. 15 in controls), heart (43 mg/kg FW vs. 3.3), spleen (60 vs. <0.7), and kidney (25 vs. 1.5)	7
Arsenite	Oral administration of 1.2 mg/kg BW daily for 6 weeks reduced uric acid levels in plasma by 67%	19
Arsenite	10 mg As/L in drinking water for 7 months; urinary metabolites were mainly methylated arsenic metabolites with about 6% in inorganic form	24
Arsenobetaine	100 mg As/L in drinking water for 7 months; eliminated in urine unchanged without transformation	24
Cacodylic acid	Fetal and maternal deaths noted when pregnant rats dosed by gavage at 50–60 mg/kg BW daily during gestation days 6–13. Fetal abnormalities observed when dams given oral dosages of 40–60 mg/kg BW on days 7–16 of gestation	5
Dimethylarsinic acid (DMA)	Main metabolites in urine after 7 months of exposure to 100 mg As/L drinking water were DMA and trimethylarsin oxide (TMAO) with minute amounts of tetramethylarsonium (TMA)	24
Dimethylarsinic acid	40–60 mg/kg BW daily for 10 days associated with fetal and maternal toxicity	25
Monomethylarsonic acid (MMA)	100 mg As/L in drinking water for 7 months produced main products in urine of unchanged MMA and DMA, plus small amounts of TMA and TMAO	24
3-Nitro-4-hydroxyphenylarsonic acid	Single oral LD50 value of 44 mg/kg BW	12
Sodium arsenate	LD75 (48 h) value of 14–18 mg/kg BW (intraperitoneal route)	12

Table 28.6 (continued) Lethal and Sublethal Effects of Various Arsenicals on Selected Species of Mammals

Organism and Arsenical	Effect	Reference[a]
Sodium arsenate	Single intraperitoneal injection of 5–12 mg/kg on days 7–12 of gestation produced eye defects, exencephaly, and faulty development of kidney and gonads	6
Sodium arsenite	LD75 (48 h) value of 4.5 mg/kg BW (intraperitoneal injection)	12
Trimethylarsin oxide (TMAO)	100 mg As/L in drinking water for 7 months was excreted in urine mostly unchanged with some TMA	24
RODENTS, various species		
Cacodylic acid	LD50 (various routes) values range from 470–830 mg/kg BW	5
Sodium cacodylate	LD50 (various routes) values range from 600–2600 mg/kg BW	5
COTTON RAT, *Sigmodon hispidus*		
Sodium arsenite	Adult males given 0, 5, or 10 mg As^{+3}/L drinking water for 6 weeks had dose-dependent decrease in daily food intake. Minimal effects on immune function, tissue weights, and blood chemistry	23
PIG, *Sus* sp.		
Sodium arsenite	Drinking water containing 500 mg/L lethal at 100–200 mg/kg BW	12
3-Nitro-4-hydroxyphenylarsonic acid	Arsenosis documented after 2 months on diets containing 100 mg/kg, or after 3–10 days on diets containing 250 mg/kg	12
RABBIT, *Sylvilagus* sp.		
Cacodylic acid	Adverse effects at dermal dosages equivalent to 4–6 g/kg BW	5
Calcium arsenate	Single oral dose of 23 mg/kg BW fatal in 3 days	12
Copper acetoarsenite	Single oral dose of 10.5 mg/kg BW fatal in 50 h	12
Inorganic arsenate	Single oral LD50 value of 8 mg/kg BW	3
Lead arsenate	Single oral dose of 40.4 mg/kg BW fatal in 52 h	12

[a] **1,** Vreman et al. 1986; **2,** Thatcher et al. 1985; **3,** NRCC 1978; **4,** Robertson et al. 1884; **5,** Hood 1985; **6,** Pershagen and Vahter 1979; **7,** Woolson 1975; **8,** Pershagen and Bjorklund 1985; **9,** Belton et al. 1985; **10,** Yamauchi et al. 1986; **11,** Charbonneau et al. 1978; **12,** NAS 1977; **13,** Nagymajtenyi et al. 1985; **14,** Hood et al. 1987; **15,** Stine et al. 1984; **16,** Deknudt et al. 1986; **17,** Hudson et al. 1984; **18,** Veen and Vreman 1986; **19,** Jauge and Del-Razo 1985; **20,** Webb et al. 1986; **21,** Hamasaki et al. 1995; **22,** Ng et al. 1998; **23,** Savabieasfahani et al. 1998; **24,** Yoshida et al. 1998; **25,** Hughes and Kenyon 1998.

28.7 RECOMMENDATIONS

Numerous criteria for arsenic have been proposed to protect natural resources and human health (Table 28.7). But many authorities recognize that these criteria are not sufficient for adequate or (in some cases) reasonable protection, and that many additional data are required if meaningful standards are to be promulgated (NAS 1977; NRCC 1978; Pershagen and Vahter 1979; USEPA 1980, 1985; Eisler 1994; Abernathy et al. 1997; SEGH 1998). Specifically, data are needed on the following subjects:

1. Cancer incidence and other abnormalities in natural resources from areas with elevated arsenic levels, and the relation to potential carcinogenicity of arsenic compounds
2. Interaction effects of arsenic with other carcinogens, cocarcinogens, promoting agents, inhibitors, and common environmental contaminants
3. Controlled studies with aquatic and terrestrial indicator organisms on physiological and biochemical effects of long-term, low-dose exposures to inorganic and organic arsenicals, including effects on reproduction and genetic makeup

4. Methodologies for establishing maximum permissible tissue concentrations for arsenic
5. Effects of arsenic in combination with infectious agents
6. Mechanisms of arsenical growth-promoting agents
7. Role of arsenic in nutrition
8. Extent of animal adaptation to arsenicals and the mechanisms of action
9. Identification and quantification of mineral and chemical forms of arsenic in rocks, soils, and sediments that constitute the natural forms of arsenic entering water and the food chain
10. Physicochemical processes influencing arsenic cycling.

In addition, the following techniques should be developed and implemented: (1) more sophisticated measurements of the chemical forms of arsenic in plant and animal tissues; (2) correlation of biologically observable effects with particular chemical forms of arsenic; and (3) management of arsenical wastes that accommodates recycling, reuse, and long-term storage.

Some proposed arsenic criteria merit additional comment, such as those on aquatic life protection, levels in seafoods and drinking water, and use in food-producing animals as growth stimulants or for disease prevention and treatment.

For saltwater life protection, the current water quality criterion of 36 μg As^{+3}/L (USEPA 1985; Table 28.7) seems to offer a reasonable degree of safety. Only a few species of algae show adverse effects at <36 μg/L (e.g., reduced growth at 19 to 22 μg/L). In 1980, this criterion was 508 μg/L (USEPA 1980), about 14 times higher than the current criterion. The downward modification seems to be indicative of the increasingly stringent arsenic criteria formulated by regulatory agencies. The current criterion for freshwater-life protection of 190 μg As^{+3}/L (USEPA 1985; Table 28.7), however, which is down from 440 μg As^{+3}/L in 1980 (USEPA 1980), is unsatisfactory. Many species of freshwater biota are adversely affected at <190 μg/L of As^{+3}, As^{+5}, or various organoarsenicals (Table 28.4). These adverse effects include death and malformations of toad embryos at 40 μg/L, growth inhibition of algae at 48 to 75 μg/L, mortality of amphipods and gastropods at 85 to 88 μg/L, and behavioral impairment of goldfish (*Carassius auratus*) at 100 μg/L. A downward adjustment in the current freshwater aquatic-life protection criterion seems warranted.

In Hong Kong, permissible concentrations of arsenic in seafood destined for human consumption range from 6 to 10 mg/kg fresh weight (Table 28.7). However, these values are routinely exceeded in 22% of finfish, 20% of bivalve molluscs, 67% of gastropods, 29% of crabs, 21% of shrimp and prawns, and 100% of lobsters (Phillips et al. 1982). The highest arsenic concentrations recorded in Hong Kong seafood products were in gastropods (*Hemifusus* spp.), in which the concentrations of 152 to 176 mg/kg FW were among the highest recorded in any species to date (Phillips et al. 1982). A similar situation exists in Yugoslavia, where almost all seafoods exceed the upper limit prescribed by food quality regulations (Ozretic et al. 1990). Most of the arsenic in seafood products is usually arsenobetaine or some other comparatively harmless form. In effect, arsenic criteria for seafoods are neither enforced nor enforceable. Some toxicologists from the U.S. Food and Drug Administration believe that the average daily intake of arsenic in the different food commodities does not pose a hazard to the consumer (Jelinek and Corneliussen 1977). It is now clear that formulation of maximum permissible concentrations of arsenic in seafoods for health regulation purposes should recognize the chemical nature of arsenic (McGeachy and Dixon 1990; Table 28.7).

For maximum protection of human health from the potential carcinogenic effects of exposure to arsenic through drinking water or contaminated aquatic organisms, the ambient water concentration should be zero, based on the nonthreshold assumption for arsenic. But a zero level may not be attainable. Accordingly, the levels established are those that are estimated to increase cancer risk over a lifetime to only one additional case per 100,000 population. These values are estimated at 0.022 μg As/L for drinking water and 0.175 μg As/L for water containing edible aquatic resources (USEPA 1980; Table 28.7).

Various phenylarsonic acids — especially arsanilic acid, sodium arsanilate, and 3-nitro-4-hydroxyphenylarsonic acid — have been used as feed additives for disease control and for improvement of weight gain in swine and poultry for almost 40 years (NAS 1977). The arsenic is present as As^{+5}

and is rapidly excreted; present regulations require withdrawal of arsenical feed additives 5 days before slaughter for satisfactory depuration (NAS 1977). Under these conditions, total arsenic residues in edible tissues do not exceed the maximum permissible limit of 2 mg/kg fresh weight (Jelinek and Corneliussen 1977). Organoarsenicals probably will continue to be used as feed additives unless new evidence indicates they should not be.

Table 28.7 Proposed Arsenic Criteria for the Protection of Selected Natural Resources and Human Health

Resource and Other Variables	Criterion or Effective Arsenic Concentration (Reference)
AQUATIC LIFE	
Freshwater biota: medium concentrations	4-day mean water concentration not to exceed 190 µg total recoverable inorganic As^{+3}/L more than once every 3 years; 1-h mean not to exceed 360 µg inorganic As^{+3}/L more than once every 3 years. Insufficient data for criteria formulation for inorganic As^{+5}, or for any organoarsenical (USEPA 1985)
Freshwater biota: tissue residues	Diminished growth and survival reported in immature bluegills (*Lepomis macrochirus*) when total arsenic residues in muscle are >1.3 mg/kg fresh weight (FW) or >5 mg/kg in adults (NRCC 1978)
Saltwater biota: medium concentration	4-day average water concentration not to exceed 36 µg As^{+3}/L more than once every 3 years; 1-h mean not to exceed 69 µg As^{+3}/L more than once every 3 years. Insufficient data for criteria formulation for inorganic As^{+5}, or for any organoarsenical (USEPA 1985)
Saltwater biota: tissue residues	Depending on chemical form of arsenic, certain marine teleosts may be unaffected at muscle total arsenic residues of 40 mg/kg FW (NRCC 1978)
BIRDS	
Tissue residues	Residues, in mg total As/kg FW, liver or kidney in 2–10 range are considered elevated; residues >10 mg/kg are indicative of arsenic poisoning (Goede 1985)
Mallard, *Anas platyrhynchos*	
Sodium arsenate in diet	Reduced growth in ducklings fed more than 30 mg As/kg diet (Camardese et al. 1990)
Turkey, *Meleagris gallopavo*	
Arsanilic acid in diet	Maximum dietary concentration for turkeys less than 28 days old is 300–400 mg/kg feed (NAS 1977)
Phenylarsonic feed additives for disease control and improvement of weight gain in domestic poultry; safe dietary levels	Maximum levels in diets, in mg/kg feed, are 50–100 for arsanilic acid, 25–188 for 3-nitro-4-hydroxy-phenylarsonic acid (for chickens and turkeys, not recommended for ducks and geese), and 180–370 for others (NAS 1977)
DOMESTIC LIVESTOCK	
Prescribed limits for arsenic in feedstuffs	
Straight feedstuffs, except those listed below	<2 mg total As/kg FW (Vreman et al. 1986)
Meals from grass, dried lucerne, or dried clover	<4 mg total As/kg FW (Vreman et al. 1986)
Phosphate mealstuffs	<10 mg total As/kg FW (Vreman et al. 1986)
Fish meals	<10 mg total As/kg FW (Vreman et al. 1986)
Tissue residues	
Poisoned	
Liver, kidney	5–>10 total As/kg FW (Thatcher et al. 1985; Vreman et al. 1986)
Normal, muscle	<0.3 mg total As/kg FW (Veen and Vreman 1986)
VEGETATION	
No observable effects	<1 mg total water-soluble soil As/L; <25 mg total As/kg soil; <3.9 µg As/m^3 air (NRCC 1978)

Table 28.7 (continued) Proposed Arsenic Criteria for the Protection of Selected Natural Resources and Human Health

Resource and Other Variables	Criterion or Effective Arsenic Concentration (Reference)
HUMAN HEALTH	
Diet	
Permissible levels	
Total diet	<0.5 mg As/kg dry weight diet (Sorensen et al. 1985); 0.0003–0.0008 mg/kg body weight (BW) daily (ATSDR 1992)
Fruits, vegetables	The tolerance for arsenic residues as As_2O_3, resulting from pesticidal use of copper, magnesium, and sodium arsenates is 3.5 mg/kg (Jelinek and Corneliussen 1977)
Muscle of poultry and swine, eggs, swine edible by-products	<2 mg total As/kg FW (Jelinek and Corneliussen 1977)
Edible by-products of chickens and turkey, liver and kidney of swine	<2 mg total As/kg FW (Jelinek and Corneliussen 1977)
Seafood products	In Hong Kong, limited to <6 mg As^{+3}/kg FW for edible tissues of finfish and <10 mg As $^{+3}$/kg for molluscs and crustaceans (Phillips et al. 1982; Edmonds and Francesconi 1993); in Yugoslavia, these values are 2 for fish and 4 for molluscs and crustaceans (Ozretic et al. 1990); in Australia, <1 mg inorganic As/kg FW and in New Zealand <2 mg inorganic As/kg FW — there is no limit on organoarsenicals (Edmonds and Francesconi 1993). In the UK, seafood products should contain <1 mg As/kg FW contributed as a result of contamination (Edmonds and Francesconi 1993)
Adverse effects	
Consumption of aquatic organisms living in As-contaminated waters	
Cancer risk of	
10^{-5}	0.175 µg As/L (USEPA 1980)
10^{-6}	0.0175 µg As/L (USEPA 1980)[a]
10^{-7}	0.00175 µg As/L (USEPA 1980)
Drinking water	
Allowable concentrations	
Total arsenic	<10 µg/L (NAS 1977; Hering et al. 1997; Kurttio et al. 1998)
Total arsenic	<50 µg/L (Pershagen and Vahter 1979; USEPA 1980; Norin et al. 1985; ATSDR 1992; USPHS 1998)
Total arsenic, Maine	<30 µg/L (ATSDR 1992)
Adverse effects	
Cancer risk of	
10^{-5}	0.022 µg As/L (USEPA 1980)
10^{-6}	0.0022 µg As/L (USEPA 1980)[a]
10^{-7}	0.00022 µg As/L (USEPA 1980)
Symptoms of arsenic toxicity observed	9% incidence at 50 µg As/L, 16% at 50–100 µg/L, and 44% at >100 µg As/L (NRCC 1978)
Harmful after prolonged consumption	>50–960 µg As/L (NRCC 1978)
"Cancer"	In Chile, cancer rate estimated at 0.01% at 82 µg As/L, 0.17% at 600 µg As/L (NRCC 1978)
Skin cancer	0.26% frequency at 290 µg/L and 2.14% at 600 µg/L (USEPA 1980)
Total intake	
No observable effect	
North America	0.007–0.06 mg As daily (Pershagen and Vahter 1979); 2 µg/kg BW daily (ATSDR 1992)
Japan	0.07–0.17 mg As daily (Pershagen and Vahter 1979)
United States	
1960s	0.05–0.1 mg As daily (Pershagen and Vahter 1979)
1974	0.015 mg As daily (Pershagen and Vahter 1979)
1998	<0.021 mg As daily, based on <0.0003 mg/kg BW daily for 70-kg adult (USPHS 1998)
Canada	0.03 mg As daily (NRCC 1978)

Table 28.7 (continued) Proposed Arsenic Criteria for the Protection of Selected Natural Resources and Human Health

Resource and Other Variables	Criterion or Effective Arsenic Concentration (Reference)
Netherlands	
Acceptable	2 µg total inorganic As/kg body weight (BW) (about 0.14 mg daily for 70 kg adult); 0.094 mg daily through fishery products (Vos and Hovens 1986)
Adverse effects (prolonged exposure)	
Subclinical symptoms	0.15–0.6 mg As daily (NRCC 1978)
Intoxication	3–4 mg As daily (NRCC 1978)
Blackfoot disease	Total dose of 20 g over several years increases prevalence of disease by 3% Pershagen and Vahter 1979)
Mild chronic poisoning	0.15 mg As daily or about 2 µg/kg BW daily (NRCC 1978)
Chronic arsenic poisoning	Lifetime cumulative absorption of 1 g As, or intake of 0.7–2.6 g/year for several years (in medications) can produce symptoms after latent period of 4–24 years (NRCC 1978)
Tissue residues	
No observed effect levels	
Urine	<0.05 mg As/L (NRCC 1978)
Liver, kidney	<0.5 mg As/L (NRCC 1978)
Blood	<0.7 As/kg (NRCC 1978)
Hair[b]	<2 mg As/kg (NRCC 1978)
Fingernail	<5 mg As/kg (NRCC 1978)
Arsenic-poisoned	
Liver, kidney	2–100 mg As/kg FW; confirmatory tests >10 mg As/kg FW; residues in survivors several days later were 2–4 mg/kg FW (NAS 1977)
Whole body	In children, symptoms of chronic arsenicism evident at 1 mg As/kg BW, equivalent to intake of about 10 mg/month for 3 months; for adults, these values were 80 mg/kg BW, equivalent to about 2 g/year for 3 years (NRCC 1978)
Air	
Allowable concentrations	
Arsine	<200 µg/m³ for U.S. industrial workers; proposed mean arsine limit of <4 µg/m³ in 8-h period and <10 µg/m³ maximum in 15 min (NAS 1977)
Arsine	<4 µg/m³ (NRCC 1978)
Inorganic arsenic	
Occupational	<2 µg/m³ (ATSDR 1992)
Residential	<10 µg/m³ (ATSDR 1992; USPHS 1998)
Organic arsenic	<500 µg/m³ (ATSDR 1992)
Total As	<3 µg/m³ in U.S.S.R. and Czechoslovakia, <500 µg/m³ for U.S. industrial workers (NAS 1977)
Total As (threshold limit value-time weighted mean:8 h/day, 40-h work week)	Proposed limit of <50 µg/m³, maximum of 2 µg/m³ in 15 min, <10 µg airborne inorganic As/m³ (USEPA 1980)
Arsenic trioxide	<0.3 µg/m³ in U.S.S.R., <0.1 µg/m³ in U.S. (Nagymajtenyi et al. 1985)
Adverse effects	
Increased mortality	Associated with daily time-weighted average arsenic exposure of >3 µg/m³ for 1 year (NRCC 1978)
Respiratory cancer (increased risk)	Associated with chronic exposure >3 µg/m³, or occupational exposure (lifetime) of >54.6 µg As/m³ (NRCC 1978)
Respiratory cancer (increased risk)	Exposure to 50 µg As/m³ for more than 25 years associated with 3-fold increase (Pershagen and Vahter 1979)
Skin diseases	Associated with ambient air concentrations of 60–13,000 µg As/m³ (NRCC 1978)
Dermatitis	Associated with ambient air concentrations of 300–81,500 µg As/m³ (NRCC 1978)

[a] One excess cancer per million population (10^{-6}) is estimated during lifetime exposure to 0.0022 µg arsenic per liter of drinking water, or to lifetime consumption of aquatic organisms residing in waters containing 0.0175 µg As/L (USEPA 1980).
[b] Thai children, age 6–9 years from the Ronpiboon district with >5 mg As/kg DW hair had abnormally low IQs compared to those with 2.01–5 mg As/kg DW. Both groups had significantly lower IQs than controls (<1 mg As/kg DW hair), as measured by the Wechsler Intelligence Scale Test for Children (Unchalee et al. 1999). This study needs verification.

28.8 SUMMARY

Arsenic (As) is a relatively common element that occurs in air, water, soil, and all living tissues. It ranks 20th in abundance in Earth's crust, 14th in seawater, and 12th in the human body. Arsenic is a teratogen and carcinogen that can traverse placental barriers and produce fetal death and malformations in many species of mammals. Although it is carcinogenic in humans, evidence for arsenic-induced carcinogenicity in other mammals is scarce. Paradoxically, evidence is accumulating that arsenic is nutritionally essential or beneficial. Arsenic deficiency effects, such as poor growth, reduced survival, and inhibited reproduction, have been recorded in mammals fed diets containing <0.05 mg As/kg, but not in those fed diets with 0.35 mg As/kg. At comparatively low doses, arsenic stimulates growth and development in various species of plants and animals.

Most arsenic produced domestically is used in the manufacture of agricultural products such as insecticides, herbicides, fungicides, algicides, wood preservatives, and growth stimulants for plants and animals. Living resources are exposed to arsenic by way of atmospheric emissions from smelters, coal-fired power plants, and arsenical herbicide sprays; from water contaminated by mine tailings, smelter wastes, and natural mineralization; and from diet, especially from consumption of marine biota. Arsenic concentrations are usually low (<1.0 mg/kg fresh weight) in most living organisms, but they are elevated in marine biota (in which arsenic occurs as arsenobetaine and poses little risk to organisms or their consumers) and in plants and animals from areas that are naturally arseniferous or are near industrial manufacturers and agricultural users of arsenicals. Arsenic is bioconcentrated by organisms but is not biomagnified in the food chain.

Arsenic exists in four oxidation states, as inorganic or organic forms. Its bioavailability and toxic properties are significantly modified by numerous biological and abiotic factors, including the physical and chemical forms of arsenic tested, the route of administration, the dose, and the species of animal. In general, inorganic arsenic compounds are more toxic than organic compounds, and trivalent species are more toxic than pentavalent species. Arsenic may be absorbed by ingestion, inhalation, or through permeation of the skin or mucous membranes; cells take up arsenic through an active transport system normally used in phosphate transport. The mechanisms of arsenic toxicity differ greatly among chemical species, although all appear to cause similar signs of poisoning. Biomethylation is the preferred detoxification mechanism for absorbed inorganic arsenicals; methylated arsenicals usually clear from tissues within a few days.

Episodes of arsenic poisoning are either acute or subacute. Chronic cases of arsenosis are seldom encountered in any species except human beings. Single oral doses of arsenicals fatal to 50% of sensitive species tested ranged from 17 to 48 mg/kg body weight (BW) in birds and from 2.5 to 33 mg/kg BW in mammals. Susceptible species of mammals were adversely affected at chronic doses of 1 to 10 mg As/kg BW, or 50 mg As/kg diet. Sensitive aquatic species were damaged at water concentrations of 19 to 48 μg As/L (the U.S. Environmental Protection Agency drinking water criterion for human health protection is 50 μg/L), 120 mg As/kg diet, or (in the case of freshwater fish) tissue residues >1.3 mg/kg fresh weight. Adverse effects to crops and vegetation were recorded at 3 to 28 mg of water-soluble As/L (equivalent to about 25 to 85 mg total As/kg soil) and at atmospheric concentrations >3.9 μg As/m^3.

Numerous and disparate arsenic criteria have been proposed for the protection of sensitive natural resources. However, the consensus is that many of these criteria are inadequate and that additional information is needed in at least five categories:

1. Developing standardized procedures to permit correlation of biologically observable effects with suitable chemical forms of arsenic
2. Conducting studies under controlled conditions with appropriate aquatic and terrestrial indicator organisms to determine the effects of chronic exposure to low doses of inorganic and organic arsenicals on reproduction, genetic makeup, adaptation, disease resistance, growth, and other variables

3. Measuring interaction effects of arsenic with other common environmental contaminants, including carcinogens, cocarcinogens, and promoting agents
4. Monitoring the incidence of cancer and other abnormalities in the natural resources of areas with relatively high arsenic levels, and correlating these with the possible carcinogenicity of arsenic compounds
5. Developing appropriate models of arsenic cycling and budgets in natural ecosystems.

28.9 LITERATURE CITED

Abernathy, C.O., R.L. Calderon, and W.R. Chappell (eds.). 1997. *Arsenic. Exposure and Health Effects.* Chapman & Hall, London. 429 pp.

Agency for Toxic Substances and Disease Registry (ATSDR). 1992. Toxicological Profile for Arsenic. U.S. Dept. Health Human Serv., Publ. Health Serv., ATSDR, TP-92/02. 186 pp.

Aggett, J. and G.A. O'Brien. 1985. Detailed model for the mobility of arsenic in lacustrine sediments based on measurements in Lake Ohakuri. *Environ. Sci. Technol.* 19:231-238.

Aggett, J. and L.S. Roberts. 1986. Insight into the mechanism of accumulation of arsenate and phosphate in hydro lake sediments by measuring the rate of dissolution with ethylenediaminetetracetic acid. *Environ. Sci. Technol.* 20:183-186.

Ahn, J.S., H.T. Chon, and K.W. Kim. 1999. Arsenic and heavy metal contamination and uptake by rice crops around an abandoned Au–Ag mine in Korea. Pages 932-933 in W.W. Wenzel, D.C. Adriano, B. Alloway, H.E. Doner, C. Keller, N.W. Lepp, M. Mench, R. Naidu, and G.M. Pierzynski (eds.). Proceedings of the Fifth International Conference on the Biogeochemistry of Trace Elements, July 11-15, 1999, Vienna, Austria.

Andreas, M.O. 1986. Organoarsenic compounds in the environment. Pages 198-228 in *P.J.* Craig (ed.). *Organometallic Compounds in the Environment. Principles and Reactions.* John Wiley, New York.

Aposhian, H.V., R. Zakharyan, Y. Wu, S. Healy, and M.M. Aposhian. 1997. Enzymatic methylation of arsenic compounds. II. An overview. Pages 296-321 in C.O. Abernathy, R.L. Calderon, and W.R. Chappel (eds.). *Arsenic. Exposure and Health Effects.* Chapman & Hall, London.

Aranyi, C., J.N. Bradof, W.J. O'Shea, J.A. Graham, and F.J. Miller. 1985. Effects of arsenic trioxide inhalation exposure on pulmonary antibacterial defenses in mice. *Jour. Toxicol. Environ. Health* 15:163-172.

Austin, L.S. and G.E. Millward. 1984. Modelling temporal variations in the global tropospheric arsenic burden. *Atmosph. Environ.* 18:1909-1919.

Azcue, J.M. and D.G. Dixon. 1994. Effects of past mining activities on the arsenic concentration in fish from Moira Lake, Ontario. *Jour. Great Lakes Res.* 20:717-724.

Belton, J.C., N.C. Benson, M.L. Hanna, and R.T. Taylor. 1985. Growth inhibitory and cytotoxic effects of three arsenic compounds on cultured Chinese hamster ovary cells. *Jour. Environ. Sci. Health* 20A:37-72.

Benson, A.A. 1984. Phytoplankton solved the arsenate–phosphate problem. Pages 55-59 in O. Holm-Hansen, L. Bolis and R. Gilles (eds.). *Lecture Notes on Coastal and Estuarine Ecology. 8. Marine Phytoplankton and Productivity.* Springer-Verlag, Berlin.

Bigler, J. and E. Crecelius. 1998. Methods for the analysis of arsenic speciation in seafood from Cook Inlet, Alaska. *SEGH 3rd Inter. Conf. Arsenic Expos. Health Effects: 9.*

Biswas, B.K., R.K. Dhar, G. Samanta, B.K. Mandal, T.R. Chowdhury, D. Chakraborti, S. Kabir, and S. Roy. 1998. Arsenic groundwater contamination and sufferings of people in Bangladesh. *SEGH 3rd Inter. Conf. Arsenic Expos. Health Effects: 22.*

Blus, L.J., B.S. Neely, Jr., T.G. Lamont, and B. Mulhern. 1977. Residues of organochlorines and heavy metals in tissues and eggs of brown pelicans 1969–73. *Pestic. Monitor. Jour.* 11:40-53.

Bottino, N.R., R.D. Newman, E.R. Cox, R. Stockton, M. Hoban, R.A. Zingaro, and K.J. Irgolic. 1978. The effects of arsenate and arsenite on the growth and morphology of the marine unicellular algae *Tetraselmis chui* (Chlorophyta) and *Hymenomonas carterae* (Chrysophyta). *Jour. Exp. Mar. Biol. Ecol.* 33:153-168.

Bryant, V., D.M. Newbery, D.S. McLusky, and R. Campbell. 1985. Effect of temperature and salinity on the toxicity of arsenic to three estuarine invertebrates (*Corophium volutator*, *Macoma balthica*, *Tubifex costatus*). *Mar. Ecol. Prog. Ser.* 24:129-137.

Burger, J. and J. Snodgrass. 1998. Heavy metals in bullfrog (*Rana catesbeiana*) tadpoles: effects of depuration before analysis. *Environ. Toxicol. Chem.* 17:2203-2209.

Camardese, M.B., D.J. Hoffman, L.J. LeCaptain, and G.W. Pendleton. 1990. Effects of arsenate on growth and physiology in mallard ducklings. *Environ. Toxicol. Chem.* 9:785-795.

Charbonneau, S.M., K. Spencer, F. Bryce, and E. Sandi. 1978. Arsenic excretion by monkeys dosed with arsenic-containing fish or with inorganic arsenic. *Bull. Environ. Contam. Toxicol.* 20:470-477.

Choprapawon, C. and A. Rodcline. 1997. Chronic arsenic poisoning in Ronpibool Nakhon Sri Thammarat, the southern province of Thailand. Pages 69-77 in C.O. Abernathy, R. Calderon, and W.R. Chappell (eds.). *Arsenic. Exposure and Health Effects.* Chapman & Hall, London.

Chowdhury, T.R., B.K. Mandal, G. Samanta, G.K. Basu, P.P. Chowdhury, C.R. Chanda, N.K. Karan, D. Lodh, R.K. Dhar, D. Das, K.C. Saha, and D. Chakraborti. 1997. Arsenic in groundwater in six districts of West Bengal, India: the biggest arsenic calamity in the world: the status report up to August, 1995. Pages 93-111 in C.O. Abernathy, R. Calderon, and W.R. Chappel (eds.). *Arsenic. Exposure and Health Effects.* Chapman & Hall, London.

Cockell, K.A. and W.J. Bettger. 1993. Investigations of the gallbladder pathology associated with dietary exposure to disodium arsenate heptahydrate in juvenile rainbow trout (*Oncorhynchus mykiss*). *Toxicology* 77:233-248.

Cockell, K.A. and J.W. Hilton. 1985. Chronic toxicity of dietary inorganic and organic arsenicals to rainbow trout (*Salmo gairdneri* R.). *Feder. Proc.* 44(4):938.

Cockell, K.A., J.W. Hilton, and W.J. Bettger. 1991. Chronic toxicity of dietary disodium arsenate heptahydrate to juvenile rainbow trout (*Oncorhynchus mykiss*). *Arch. Environ. Contamin. Toxicol.* 21:518-527.

Cockell, K.A., J.W. Hilton, and W.J. Bettger. 1992. Hepatobiliary and hematological effects of dietary disodium arsenate heptahydrate in juvenile rainbow trout (*Oncorhynchus mykiss*). *Comp. Biochem. Physiol.* 103C:453-458.

Crecelius, E., R. Schoof, L.J. Yost, D. Menzel, J. Eichkoff, and D. Cragin. 1998. Arsenic speciation methods for food. *SEGH 3rd Inter. Conf. Arsenic Expos. Health Effects: 62.*

Custer, T.W. and C.A. Mitchell. 1991. Contaminant exposure of willets feeding in agricultural drainages of the lower Rio Grande valley of south Texas. *Environ. Monitor. Assess.* 16:189-200.

Custer, T.W. and C.A. Mitchell. 1993. Trace elements and organochlorines in the shoalgrass community of the lower Laguna Madre, Texas. *Environ. Monitor. Assess.* 25:235-246.

Deknudt, G., A. Leonard, J. Arany, G.J. Du Buisson, and E. Delavignette. 1986. *In vivo* studies in male mice on the mutagenic effects of inorganic arsenic. *Mutagenesis* 1:33-34.

Denham, D.A., S.L. Oxenham, I. Midwinter, and E.A.H. Friedheim. 1986. The antifilarial activity of a novel group of organic arsenicals upon *Brugia pahangi. Jour. Helminthol.* 60:169-172.

Domingo, J.L. 1994. Metal-induced developmental toxicity in mammals: a review. *Jour. Toxicol. Environ. Health* 42:123-141.

Dudas, M.J. 1984. Enriched levels of arsenic in post-active acid sulfate soils in Alberta. *Soil Sci. Soc. Amer. Jour.* 48:1451-1452.

Dyer, S.D., G.L. Brooks, K.L. Dickson, B.M. Sanders, and E.G. Zimmerman. 1993. Synthesis and accumulation of stress proteins in tissues of arsenite-exposed fathead minnows (*Pimephales promelas*). *Environ. Toxicol. Chem.* 12:913-924.

Edmonds, J.S. and K.A. Francesconi. 1987. Trimethylarsine oxide in estuary catfish (*Cnidoglanis macrocephalus*) and school whiting (*Sillago bassensis*) after oral administration of sodium arsenate; and as a natural component of estuary catfish. *Sci. Total Environ.* 64:317-323.

Edmonds, J.S. and K.A. Francesconi. 1993. Arsenic in seafoods: human health aspects and regulations. *Mar. Pollut. Bull.* 26:665-674.

Edmonds, J.S., K.A. Francesconi, and R.V. Stick. 1993. Arsenic compounds from marine organisms. *Natur. Prod. Rep.* 10:421-428.

Eisler, R. 1981. *Trace Metal Concentrations in Marine Organisms.* Pergamon Press, New York. 687 pp.

Eisler, R. 1988. Arsenic Hazards to Fish, Wildlife, and Invertebrates: A Synoptic Review. U.S. Fish. Wildl. Serv. Biol. Rep. 85(1.12). 92 pp.

Eisler, R. 1994. A review of arsenic hazards to plants and animals with emphasis on fishery and wildlife resources. Pages 185-259 in J.O. Nriagu (ed.). *Arsenic in the Environment. Part II. Human Health and Ecosystem Effects.* John Wiley, New York.

Fairbrother, A., M. Fix, T. O'Hara, and C.A. Ribic. 1994. Impairment of growth and immune function of avocet chicks from sites with elevated selenium, arsenic, and boron. *Jour. Wildl. Dis.* 30:222-233.

Farag, A.M., M.A. Stansbury, C. Hogstrand, E. MacConnell, and H.L. Bergman. 1995. The physiological impairment of free-ranging brown trout exposed to metals in the Clark Fork River, Montana. *Canad. Jour. Fish. Aquat. Sci.* 52:2038-2050.

Farmer, J.G. and M.A. Lovell. 1986. Natural enrichment of arsenic in Loch Lomond sediments. *Geochim. Cosmochim. Acta* 50:2059-2067.

Ferm, V.H. and D.P. Hanlon. 1985. Constant rate exposure of pregnant hamsters to arsenate during early gestation. *Environ. Res.* 37:425-432.

Ferm, V.H. and D.P. Hanlon. 1986. Arsenate-induced neural tube defects not influenced by constant rate administration of folic acid. *Pediatric Res.* 20:761-762.

Findo, V.S., P. Hell, J. Farkas, B. Mankovska, M. Zilinec, and M. Stanovsky. 1993. Accumulation of selected heavy metals in red and roe deer in the central west Carpathian mountains (central Slovakia). *Z. Jagdwiss.* 39:181-189 (in German with English summary).

Finkelman, R.B., J.A. Centano, and B. Zheng. 1998. Etiology of arsenism in Guizhou Province, southwest China. *SEGH 3rd Inter. Conf. Arsenic Expos. Health Effects: 100.*

Fischer, A.B., J.P. Buchet, and R.R. Lauwerys. 1985. Arsenic uptake, cytotoxicity and detoxification studied in mammalian cells in culture. *Arch. Toxicol.* 57:168-172.

Fowler, S.W. and M.Y. Unlu. 1978. Factors affecting bioaccumulation and elimination of arsenic in the shrimp *Lysmata seticaudata. Chemosphere* 9:711-720.

Francesconi, K.A., P. Micks, R.A. Stockton, and K.J. Irgolic. 1985. Quantitative determination of arsenobetaine, the major water-soluble arsenical in three species of crab, using high pressure liquid chromatography and an inductively coupled argon plasma emission spectrometer as the arsenic-specific detector. *Chemosphere* 14:1443-1453.

Freeman, M.C. 1985. The reduction of arsenate to arsenite by an *Anabaena* bacteria assemblage isolated from the Waikato River. *N.Z. Jour. Mar. Freshwater Res.* 19:277-282.

Freeman, M.C., J. Aggett, and G. O'Brien. 1986. Microbial transformations of arsenic in Lake Ohakuri, New Zealand. *Water Res.* 20:283-294.

Froelich, P.N., L.W. Kaul, J.T. Byrd, M.O. Andreas, and K.K. Roe. 1985. Arsenic, barium, germanium, tin, dimethylsulfide and nutrient biogeochemistry in Charlotte Harbor, Florida, a phosphorus-enriched estuary. *Estuar. Coastal Shelf Sci.* 20:239-264.

Fullmer, C.S. and R.H. Wasserman. 1985. Intestinal absorption of arsenate in the chick. *Environ. Res.* 36:206-217.

Goede, A.A. 1985. Mercury, selenium, arsenic and zinc in waders from the Dutch Wadden Sea. *Environ. Pollut.* 37A:287-309.

Goessler, W., D. Kuehnelt, C. Schlagenhaufen, Z. Slejkovec, and K.J. Irgolic. 1998a. Arsenobetaine and other arsenic compounds in the National Research Council of Canada certified reference materials DORM 1 and DORM 2. *Jour. Anal. Atom. Spectrom.* 13:183-187.

Goessler, W., C. Schlagenhaufen, D. Kuehnelt, H. Greschonig, and K.J. Irgolic. 1997. Can humans metabolize arsenic compounds to arsenobetaine? *Appl. Organometall. Chem.* 11:327-335.

Guitart, R., M. Torra, S. Cerradelo, P. Puig-Casado, R. Mateo, and J. To-Figueras. 1994. Pb, Cd, As, and Se concentrations in livers of dead wild birds from the Ebro Delta, Spain. *Bull. Environ. Contam. Toxicol.* 52:523-529.

Hall, R.A., E.G. Zook, and G.M. Meaburn. 1978. National Marine Fisheries Service Survey of Trace Elements in the Fishery Resources. U.S. Dep. Commerce NOAA Tech. Rep. NMFS SSRF-721. 313 pp.

Hall, R.J. 1980. Effects of Environmental Contaminants on Reptiles: A Review. U.S. Fish Wildl. Serv. Spec. Sci. Rep. — Wildl. 228. 12 pp.

Hallacher, L.E., E.B. Kho, N.D. Bernard, A.M. Orcutt, W.C. Dudley, Jr., and T.M. Hammond. 1985. Distribution of arsenic in the sediments and biota of Hilo Bay, Hawaii. *Pac. Sci.* 39:266-273.

Hamasaki, T., H. Nagase, Y. Yoshioka, and T. Sato. 1995. Formation, distribution, and ecotoxicity of methylmetals of tin, mercury, and arsenic in the environment. *Crit. Rev. Environ. Sci. Technol.* 25:45-91.

Hamilton, S.J. and K.J. Buhl. 1990. Safety assessment of selected inorganic elements to fry of chinook salmon (*Oncorhynchus tshawytscha*). *Ecotoxicol. Environ. Safety* 20:307-324.

Han, B.C., W.L. Jeng, R.Y. Chen, G.T. Fang, T.C. Hung, and R.J. Tseng. 1998. Estimation of target hazard quotients and potential health risks for metals by consumption of seafood in Taiwan. *Arch. Environ. Contam. Toxicol.* 35:711-720.

Hanaoka, K., T. Kogure, Y. Miura, S. Tagawa, and T. Kaise. 1993. Post-mortem formation of inorganic arsenic from arsenobetaine in a shark under natural conditions. *Chemosphere* 27:2163-2167.

Hanaoka, K. and S. Tagawa. 1985a. Isolation and identification of arsenobetaine as a major water soluble arsenic compound from muscle of blue pointer *Isurus oxyrhincus* and whitetip shark *Carcharhinus longimanus*. *Bull. Japan. Soc. Sci. Fish.* 51:681-685.

Hanaoka, K. and S. Tagawa. 1985b. Identification of arsenobetaine in muscle of roundnose flounder *Eopsetta grigorjewi*. *Bull. Japan. Soc. Sci. Fish.* 51:1203.

Hanlon, D.P. and V.H. Ferm. 1986a. Teratogen concentration changes as the basis of the heat stress enhancement of arsenate teratogenesis in hamsters. *Teratology* 34:189-193.

Hanlon, D.P. and V.H. Ferm. 1986b. Concentration and chemical status of arsenic in the blood of pregnant hamsters during critical embryogenesis. 1. Subchronic exposure to arsenate using constant rate administration. *Environ. Res.* 40:372-379.

Hanlon, D.P. and V.H. Ferm. 1986c. Concentration and chemical status of arsenic in the blood of pregnant hamsters during critical embryogenesis. 2. Acute exposure. *Environ. Res.* 40:380-390.

Haswell, S.J., P. O'Neill, and K. C Bancroft. 1985. Arsenic speciation in soil-pore waters from mineralized and unmineralized areas of south-west England. *Talanta* 32:69-72.

Hering, J.G., P.Y. Chen, and J.A. Wilkie. 1997. Arsenic removal from drinking-water by coagulation: the role of adsorption and effects of source water composition. Pages 369-381 in C.O. Abernathy, R.L. Calderon, and W.R. Chappell (eds.). *Arsenic. Exposure and Health Effects.* Chapman & Hall, London.

Hoffman, D.J., C.J. Sanderson, L.J. LeCaptain, E. Cromartie, and G.W. Pendleton. 1992. Interactive effects of arsenate, selenium, and dietary protein on survival, growth, and physiology in mallard ducklings. *Arch. Environ. Contam. Toxicol.* 22:55-62.

Holson, J.F., J.M. DeSesso, A.R. Scialli, and C.F. Farr. 1998. Inorganic arsenic and prenatal development: a comprehensive evaluation for human risk assessment. *SEGH 3rd Inter. Conf. Arsenic Expos. Health Effects: 23.*

Hood, R.D. 1985. Cacodylic Acid: Agricultural Uses, Biologic Effects, and Environmental Fate. VA Monograph. 171 pp. Avail. from Sup. Documents, U.S. Govt. Printing Off., Washington, D.C. 20402.

Hood, R.D., G.C. Vedel-Macrander, M.J. Zaworotko, F.M. Tatum, and R.G. Meeks. 1987. Distribution, metabolism, and fetal uptake of pentavalent arsenic in pregnant mice following oral or intraperitoneal administration. *Teratology* 35:19-25.

Hopenhayn-Rich, C., K.D. Johnson, and I. Hertz-Picciotto. 1998. Reproductive and developmental effects associated with chronic arsenic exposure. *SEGH 3rd Inter. Conf. Arsenic Expos. Health Effects: 21.*

Hopkins, W.A., M.T. Mendonca, C.L. Rowe, and J.D. Congdon. 1998. Elevated trace metal concentrations in southern toads, *Bufo terrestris*, exposed to coal combustion waste. *Arch. Environ. Contam. Toxicol.* 35:325-329.

Hothem, R.L. and D. Welsh. 1994. Contaminants in eggs of aquatic birds from the grasslands of central California. *Arch. Environ. Contam. Toxicol.* 27:180-185.

Howard, A.G., M.H. Arbab-Zavar, and S. Apte. 1984. The behaviour of dissolved arsenic in the estuary of the River Beaulieu. *Estuar. Coastal Shelf Sci.* 19:493-504.

Hsueh, Y.M., Y.L. Huang, C.C. Huang, W.L. Wu, H.M. Chen, M.H. Yang, L.C. Lue, and C.J. Chen. 1998. Urinary levels of inorganic and organic arsenic metabolites among residents in an arseniasis-hyperendemic area in Taiwan. *Jour. Toxicol. Environ. Health* 54A:431-444.

Hudson, R.H., R.K. Tucker, and M A. Haegele. 1984. Handbook of Toxicity of Pesticides to Wildlife. U.S. Fish Wildl. Serv. Resour. Publ. 153. 90 pp.

Hughes, M.F. and E.M. Kenyon. 1998. Dose-dependent effects on the disposition of monomethylarsonic acid and dimethylarsinic acid in the mouse after intravenous administration. *Jour. Toxicol. Environ. Health* 53A:95-112.

Irgolic, K.J., W. Goessler, and D. Kuehnelt. 1998. Arsenic compounds in terrestrial biota. *SEGH 3rd Inter. Conf. Arsenic Expos. Health Effects: 11.*

Irgolic, K.J., E.A. Woolson, R.A. Stockton, R.D. Newman, N.R. Bottino, R.A. Zingaro, P.C. Kearney, R.A. Pyles, S. Maeda, W.J. McShane, and E.R. Cox. 1977. Characterization of arsenic compounds formed by *Daphnia magna* and *Tetraselmis chuii* from inorganic arsenate. *Environ. Health Perspec.* 19:61-66.

Ismail, A. and R.D. Roberts. 1992. Arsenic in small mammals. *Environ. Technol.* 13:1091-1095.

Jauge, P. and L.M. Del-Razo. 1985 Uric acid levels in plasma and urine in rats chronically exposed to inorganic As(III) and As(V). *Toxicol. Lett.* 26:31-35.

Jelinek, C.F. and P.E. Corneliussen. 1977. Levels of arsenic in the United States food supply. *Environ. Health Perspec.* 19:83-87.

Jenkins, D.W. 1980. Biological Monitoring of Toxic Trace Metals. Vol. 2. Toxic Trace Metals in Plants and Animals of the World. Part 1. U.S. Environ. Protection Agen. Rep. 600/3-80-090:30-138.

Jessen-Eller, K. and J.F. Crivello. 1998. Changes in metallothionein mRNA and protein after sublethal exposure to arsenite and cadmium chlorine in juvenile winter flounder. *Environ. Toxicol. Chem.* 17:891-896.

Johns, C. and S.N. Luoma. 1990. Arsenic in benthic bivalves of San Francisco Bay and the Sacramento/San Joaquin River delta. *Sci. Total Environ.* 97/98:673-684.

Johnson, D.L. 1972. Bacterial reduction of arsenate in seawater. *Nature (Lond.)* 240:44-45.

Johnson, D.L. and R.M. Burke. 1978. Biological mediation of chemical speciation. II. Arsenate reduction during marine phytoplankton blooms. *Chemosphere* 8:645-648.

Johnson, W.W. and M.T. Finley. 1980. Handbook of Acute Toxicity of Chemicals to Fish and Aquatic Invertebrates. U.S. Fish Wildl. Serv. Resour. Publ. 137. 98 pp.

Jongen, W.M.F., J.M. Cardinaals, and P.M.J. Bos. 1985. Genotoxicity testing of arsenobetaine, the predominant form of arsenic in marine fishery products. *Food Chem. Toxicol.* 23:669-673.

Kairu, J.K. 1996. Heavy metal residues in birds of Lake Nakuru, Kenya. *African Jour. Ecol.* 34:397-400.

Kaise, T. and S. Fukui. 1992. The chemical form and acute toxicity of arsenic compounds in marine organisms. *Appl. Organometall. Chem.* 6:155-160.

Kaise, T., S. Watanabe, and K. Itoh. 1985. The acute toxicity of arsenobetaine. *Chemosphere* 14:1327-1332.

Knowles, F.C. and A.A. Benson. 1984a. The mode of action of arsenical herbicides and drugs. *Z. ges. Hyg.* (Berlin) 30:407-408.

Knowles, F.C. and A.A. Benson. 1984b. The enzyme inhibitory form of inorganic arsenic. *Z. ges. Hyg. (Berlin)* 30:625-626.

Kuehnelt, D., W. Goessler, and K.J. Irgolic. 1997a. Arsenic compounds in terrestrial organisms. I. *Collybia maculata*, *Collybia butyracea* and *Amantia muscaria* from arsenic smelter sites in Austria. *Appl. Organometall. Chem.* 11:289-296.

Kuehnelt, D., W. Goessler, and K.J. Irgolic. 1997b. Arsenic compounds in terrestrial organisms II: arsenocholine in the mushroom *Amantia muscaria*. *Appl. Organometall. Chem.* 11:459-470.

Kuehnelt, D., W. Goessler, C. Schlagenhaufen, and K.J. Irgolic. 1997c. Arsenic compounds in terrestrial organisms. III. Arsenic compounds in *Formica* sp. from an old arsenic smelter site. *Appl. Organometall. Chem.* 11:859-867.

Kurttio, P., H. Komulainen, E. Hakala, and J. Pekkanen. 1998. Urinary excretion of arsenic species after exposure to arsenic present in drinking water. *Arch. Environ. Contam. Toxicol.* 34:297-305.

Leah, R.T., S.J. Evans, and M.S. Johnson. 1992. Arsenic in plaice (*Pleuronectes platessa*) and whiting (*Merlangius merlangus*) from the North East Irish Sea. *Mar. Pollut. Bull.* 24:544-549.

Lebedeva, N.V. 1997. Accumulation of heavy metals by birds in the southwest of Russia. *Russian Jour. Ecol.* 28:41-46.

Lee, T.C., K.C.C. Lee, C. Chang, and W.L. Jwo. 1986a. Cell-cycle dependence of the cytotoxicity and clastogenicity of sodium arsenate in Chinese hamster ovary cells. *Bull. Inst. Zool. Acad. Sinica* 25:91-97.

Lee, T.C., M. Oshimura, and J.C. Barrett. 1985. Comparison of arsenic-induced cell transformation, cytotoxicity, mutation and cytogenetic effects in Syrian hamster embryo cells in culture. *Carcinogenesis* 6:1421-1426.

Lee, T.C., S. Wang-Wuu, R.Y. Huang, K.C.C. Lee, and K.Y. Jan. 1986b. Differential effects of pre-and posttreatment of sodium arsenite on the genotoxicity of methyl methanesulfonate in Chinese hamster ovary cells. *Cancer Res.* 46:1854-1857.

Lima, A.R., C. Curtis, D.E. Hammermeister, T.P. Markee, C.E. Northcott, and L.T. Brooke. 1984. Acute and chronic toxicities of arsenic (III) to fathead minnows, flagfish, daphnids, and an amphipod. *Arch. Environ. Contam. Toxicol.* 13:595-601.

Lin, T.H. and Y.L. Huang. 1998. Arsenic species in drinking water, hair, fingernails, and urine of patients with blackfoot disease. *Jour. Toxicol. Environ. Health* 53A:85-93.

Lindsay, D.M. and J.G. Sanders. 1990. Arsenic uptake and transfer in a simplified estuarine food chain. *Environ. Toxicol. Chem.* 9:391-395.

Lunde, G. 1973. The synthesis of fat and water soluble arseno organic compounds in marine and limnetic algae. *Acta Chem. Scand.* 27:1586-1594.

Lunde, G. 1977. Occurrence and transformation of arsenic in the marine environment. *Environ. Health Perspec.* 19:47-52.

Luo, Z.D., Y.M. Zhang, L. Ma, G.Y. Zhang, X. He, R. Wilson, D.M. Byrd, J.G. Griffiths, S. Lai, L. He, K. Grumski, and S.H. Lamm. 1997. Chronic arsenicism and cancer in Inner Mongolia — consequences of well-water arsenic levels greater than 50 µg/L. Pages 55-68 in C.O. Abernathy, R.L. Calderon, and W.R. Chappell (eds.). *Arsenic. Exposure and Health Effects.* Chapman & Hall, London.

Ma, H.Z., Y.J. Xia, K.G. Wu, and T.Z. Sun. 1998. Arsenic exposure and health effects in inner Mongolia, China. *SEGH 3rd Inter. Conf. Arsenic Expos. Health Effects: 18.*

Maeda, S., S. Nakashima, T. Takeshita, and S. Higashi. 1985. Bioaccumulation of arsenic by freshwater algae and the application to the removal of inorganic arsenic from an aqueous phase. II. By *Chlorella vulgaris* isolated from arsenic-polluted environment. *Separation Sci. Technol.* 20:153-161.

Maher, W.A. 1985a. Arsenic in coastal waters of South Australia. *Water Res.* 19:933-934.

Maher, W.A. 1985b. The presence of arsenobetaine in marine animals. *Biochem. Physiol.* 80C:199-201.

Mankovska, B. 1986. Accumulation of As, Sb, S, and Pb in soil and pine forest. *Ekologia (CSSR)* 5:71-79.

Manna, G.K. and P.K. Mukherjee. 1989. A study of the genotoxic potentiality of the inorganic weedicide, sodium arsenite in the experimentally treated tilapia fish. *Jour. Freshwater Biol.* 1:147-159.

Marafante, E., M. Vahter, and J. Envall. 1985. The role of the methylation in the detoxication of arsenate in the rabbit. *Chem.-Biol. Interact.* 56:225-238.

Marques, I.A. and L.E. Anderson. 1986. Effects of arsenite, sulfite, and sulfate on photosynthetic carbon metabolism in isolated pea (*Pisum sativum* L., cv Little Marvel) chloroplasts. *Plant Physiol.* 82:488-493.

Mathews, N.E. and W.F. Porter. 1989. Acute arsenic intoxication of a free-ranging white-tailed deer in New York. *Jour. Wildl. Dis.* 25:132-135.

Matsuto, S., H. Kasuga, H. Okumoto, and A. Takahashi. 1984. Accumulation of arsenic in blue-green alga, *Phormidium* sp. *Comp. Biochem. Physiol.* 78C:377-382.

Matsuto, S., R.A. Stockton, and K.J. Irgolic. 1986. Arsenobetaine in the red crab, *Chionoecetes opilio. Sci. Total Environ.* 48:133-140.

McGeachy, S.M. and D.G. Dixon. 1990. Effect of temperature on the chronic toxicity of arsenate to rainbow trout (*Oncorhynchus mykiss*). *Canad. Jour. Fish. Aquat. Sci.* 47:2228-2234.

McGeachy, S.M. and D.G. Dixon. 1992. Whole-body arsenic concentrations in rainbow trout during acute exposure to arsenate. *Ecotoxicol. Environ. Safety* 24:301-308.

Meharg, A.A., R.F. Shore, and K. Broadgate. 1998. Edaphic factors affecting the toxicity and accumulation of arsenate in the earthworm *Lumbricus terrestris. Environ. Toxicol. Chem.* 17:1124-1131.

Merry, R.H., K.G. Tiller, and A.M. Alston. 1986. The effects of contamination of soil with copper, lead and arsenic on the growth and composition of plants. I. Effects of season, genotype, soil temperature and fertilizers. *Plant Soil* 91:115-128.

Michnowicz, C.J. and T.E. Weaks. 1984. Effects of pH on toxicity of As, Cr, Cu, Ni, and Zn to *Selenastrum capricornutum* Printz. *Hydrobiologia* 118:299-305.

Nagymajtenyi, L., A. Selypes, and G. Berencsi. 1985. Chromosomal aberrations and fetotoxic effects of atmospheric arsenic exposure in mice. *Jour. Appl. Toxicol.* 5:61-63.

Naqvi, S.M. and C.T. Flagge. 1990. Chronic effects of arsenic on American red crayfish, *Procambarus clarkii*, exposed to monosodium methanearsonate (MSMA) herbicide. *Bull. Environ. Contam. Toxicol.* 45:101-106.

Naqvi, S.M., C.T. Flagge, and R.L. Hawkins. 1990. Arsenic uptake and depuration by red crayfish, *Procambarus clarkii*, exposed to various concentrations of monosodium methanearsonate (MSMA) herbicide. *Bull. Environ. Contam. Toxicol.* 45:94-100.

National Academy of Sciences (NAS). 1977. *Arsenic*. Natl. Acad. Sci., Washington, D.C. 332 pp.

National Research Council of Canada (NRCC). 1978. Effects of Arsenic in the Canadian Environment. Natl. Res. Coun. Canada Publ. No. NRCC 15391. 349 pp.

Nevill, E.M. 1985. The effect of arsenical dips on *Parafilaria bovicola* in artificially infected cattle in South Africa. *Onderstepoort Jour. Vet. Res.* 52:221-225.

Ng, J.C., A.A. Seawright, L. Qi, C.M. Garnett, M.R. Moore, and B. Chiswell. 1998. Tumours in mice induced by chronic exposure of high arsenic concentrations in drinking water. *SEGH 3rd Inter. Conf. Arsenic Expos. Health Effects: 28.*

Norin, H., M. Vahter, A. Christakopoulos, and M. Sandstrom. 1985. Concentration of inorganic and total arsenic in fish from industrially polluted water. *Chemosphere* 14:1125-334.

North, D.W., H.J. Gibb, and C.O. Abernathy. 1997. Arsenic: past, present and future considerations. Pages 406-423 in C.O. Abernathy, R.L. Calderon, and W.R. Chappell (eds.). *Arsenic. Exposure and Health Effects.* Chapman & Hall, London.

Nystrom, R.R. 1984. Cytological changes occurring in the liver of coturnix quail with an acute arsenic exposure. *Drug Chem. Toxicol.* 7:587-594.

Ozretic, B., M. Krajinovic-Ozretic, J. Santin, B. Medjugorac, and M. Kras. 1990. As, Cd, Pd, and Hg in benthic animals from the Kvarner-Rijeka Bay region, Yugoslavia. *Mar. Pollut. Bull.* 21:595-597.

Pascoe, G.A., R.J. Blanchet, and G. Linder. 1994. Bioavailability of metals and arsenic to small mammals at a mining waste-contaminated wetland. *Arch. Environ. Contam. Toxicol.* 27:44-50.

Pascoe, G.A., R.J. Blanchet, and G. Linder. 1996. Food chain analysis of exposures and risks to wildlife at a metals-contaminated wetland. *Arch. Environ. Contam. Toxicol.* 30:306-318.

Passino, D.R.M. and A.J. Novak. 1984. Toxicity of arsenate and DDT to the cladoceran *Bosmina longirostris*. *Bull. Environ. Contam. Toxicol.* 33:325-329.

Pendleton, G.W., M.R. Whitworth, and G.H. Olsen. 1995. Accumulation and loss of arsenic and boron, alone and in combination, in mallard ducks. *Environ. Toxicol. Chem.* 14:1357-1364.

Pershagen, G. and N.E. Bjorklund. 1985. On the pulmonary tumorigenicity of arsenic trisulfide and calcium arsenate in hamsters. *Cancer Lett.* 27:99-104.

Pershagen, G. and M. Vahter. 1979. Arsenic — A Toxicological and Epidemiological Appraisal. Naturvardsverket Rapp. SNV PM 1128, Liber Tryck, Stockholm. 265 pp.

Phillips, D.J.H. 1990. Arsenic in aquatic organisms: a review, emphasizing chemical speciation. *Aquat. Toxicol.* 16:151-186.

Phillips, D.J.H. and M.H. Depledge. 1986. Chemical forms of arsenic in marine organisms, with emphasis on *Hemifusus* species. *Water Sci. Technol.* 18:213-222.

Phillips, D.J.H., G.B. Thompson, K.M. Gabuji, and C.T. Ho. 1982. Trace metals of toxicological significance to man in Hong Kong seafood. *Environ. Pollut.* 3B:27-45.

Ramelow, G.J., C.L. Webre, C.S. Mueller, J.N. Beck, J.C. Young, and M.P. Langley. 1989. Variations of heavy metals and arsenic in fish and other organisms from the Calcasieu River and Lake, Louisiana. *Arch. Environ. Contam. Toxicol.* 18:804-818.

Reinke, J., J.F. Uthe, H.C. Freeman, and J.R. Johnston. 1975. The determination of arsenite and arsenate ions in fish and shellfish by selective extraction and polarography. *Environ. Lett.* 8:371-380.

Robertson, I.D., W.E. Harms, and P.J. Ketterer. 1984. Accidental arsenical toxicity of cattle. *Aust. Veterin. Jour.* 61:366-367.

Robertson, J.L. and J.A. McLean. 1985. Correspondence of the LC 50 for arsenic trioxide in a diet-incorporation experiment with the quantity of arsenic ingested as measured by X-ray, energy-dispersive spectrometry. *Jour. Econ. Entomol.* 78:1035-1036.

Rosemarin, A., M. Notini, and K. Holmgren. 1985. The fate of arsenic in the Baltic Sea *Fucus vesiculosus* ecosystem. *Ambio* 14:342-345.

Sadiq, M. 1986. Solubility relationships of arsenic in calcareous soils and its uptake by corn. *Plant Soil* 91:241-248.

Saiki, M.K., M.R. Jennings, and T.W. May. 1992. Selenium and other elements in freshwater fishes from the irrigated San Joaquin valley, California. *Sci. Total Environ.* 126:109-137.

Samad, M.A. and A. Chowdhury. 1984. Clinical cases of arsenic poisoning in cattle. *Indian Jour. Vet. Med.* 4:107-108.

Sanders, J.G. 1980. Arsenic cycling in marine systems. *Mar. Environ. Res.* 3:257-266.

Sanders, J.G. 1985. Arsenic geochemistry in Chesapeake Bay: dependence upon anthropogenic inputs and phytoplankton species composition. *Mar. Chem.* 17:329-340.

Sanders, J.G. 1986. Direct and indirect effects of arsenic on the survival and fecundity of estuarine zooplankton. *Canad. Jour. Fish. Aquat. Sci.* 43:694-699.

Sanders, J.G. and S.J. Cibik.1985. Adaptive behavior of euryhaline phytoplankton communities to arsenic stress. *Mar. Ecol. Prog. Ser.* 22:199-205.

Sanders, J.G. and R.W. Osman. 1985. Arsenic incorporation in a salt marsh ecosystem. *Estuar. Coastal Shelf Sci.* 20:387-392.

Savabieasfahani, M., R.L. Lochmiller, D.P. Rafferty, and J.A. Sinclair. 1998. Sensitivity of wild cotton rats (*Sigmodon hispidus*) to the immunotoxic effects of low-level arsenic exposure. *Arch. Environ. Contam. Toxicol.* 34:289-296.

Schmitt, C.J. and W.G. Brumbaugh. 1990. National Contaminant Biomonitoring Program: concentrations of arsenic, cadmium, copper, lead, mercury, selenium, and zinc in U.S. freshwater fish, 1976–1984. *Arch. Environ. Contam. Toxicol.* 19:731-747.

Schoof, R.A., L.J. Yost, E. Crecelius, K. Irgolic, W. Goessler, H.R. Guo, and H. Greene. 1998. Dietary arsenic intake in Taiwanese districts with elevated arsenic in drinking water. *Human Ecol. Risk Assess.* 4:117-135.

Selby, L.A., A.A. Case, G.D. Osweiler, and H.M. Hages, Jr. 1977. Epidemiology and toxicology of arsenic poisoning in domestic animals. *Environ. Health Perspec.* 19:183-189.

Sheppard, M.I., D.H. Thibault, and S.C. Sheppard. 1985. Concentrations and concentration ratios of U, As and Co in Scots pine grown in a waste-site soil and an experimentally contaminated soil. *Water Air Soil Pollut.* 26:85-94.

Shiomi, K., A. Shinagawa, K. Hirota, H. Yamanaka, and T. Kikuchi. 1984a. Identification of arsenobetaine as a major arsenic compound in the ivory shell *Buccinum striatissimum*. *Agric. Biol. Chem.* 48:2863-2864.

Shiomi, K., A. Shinagawa, T. Igarashi, H. Yamanaka, and T. Kikuchi. 1984b. Evidence for the presence of arsenobetaine as a major arsenic compound in the shrimp *Sergestes lucens*. *Experientia* 40:1247-1248.

Sloot, H.A. van der, D. Hoede, J. Wijkstra, J.C. Duinker, and R.F. Nolting. 1985. Anionic species of V, As, Se, Mo, Sb, Te and W in the Scheldt and Rhine estuaries and the southern bight (North Sea). *Estuar. Coastal Shelf Sci.* 21:633-651.

Smiley, R.W., M.C. Fowler, and R.C. O'Knefski. 1985. Arsenate herbicide stress and incidence of summer patch on Kentucky bluegrass turfs. *Plant Dis.* 69:44-48.

Smith, A.H., M.L. Biggs, L. Moore, R. Haque, and C. Steinmaus. 1998. Extrapolating cancer risks for arsenic in drinking water. *SEGH 3rd Inter. Conf. Arsenic Expos. Health Effects: 25.*

Smith, R.A., R.B. Alexander, and M.G. Wolman. 1987. Water-quality trends in the nation's rivers. *Science* 235:1607-1615.

Society for Environmental Geochemistry and Health (SEGH). 1998. Third Int. Conf. on Arsenic Expos. and Health Effects. San Diego, CA, July 12-15, 1998. Book of Abstracts. 173 pp.

Sorensen, E.M.B. 1987. The effects of arsenic on freshwater teleosts. *Rev. Environ. Toxicol.* 3:1-53.

Sorensen, E.M.B., R.R. Mitchell, A. Pradzynski, T.L. Bayer, and L.L. Wenz. 1985. Stereological analyses of hepatocyte changes parallel arsenic accumulation in the livers of green sunfish. *Jour. Environ. Pathol. Toxicol. Oncol.* 6:195-210.

Spehar, R.L., J.T. Fiandt, R.L. Anderson, and D.L. DeFoe. 1980. Comparative toxicity of arsenic compounds and their accumulation in invertebrates and fish. *Arch. Environ. Contam. Toxicol.* 9:53-63.

Stanley, T.R., Jr. J.W. Spann, G.J. Smith, and R. Rosscoe. 1994. Main and interactive effects of arsenic and selenium on mallard reproduction and duckling growth and survival. *Arch. Environ. Contam. Toxicol.* 26:444-451.

Stavelind, G., I. Marthinsen, G. Norheim, and K. Julshamn. 1993. Levels of environmental pollutants in flounder (*Platichthys flesus* L.) and cod (*Gadus morhua* L.) caught in the waterway of Glomma, Norway. II. Mercury and arsenic. *Arch. Environ. Contam. Toxicol.* 24:187-193.

Stine, E.R., C.A. Hsu, T.D. Hoovers, H.V. Aposhian, and D.E. Carter. 1984. N-(2,3-dimercaptopropyl) phthalamidic acid: protection, *in vivo* and *in vitro*, against arsenic intoxication. *Toxicol. Appl. Pharmacol.* 75:329-336.

Takamatsu, T., M. Kawashima, and M. Koyama. 1985. The role of Mn^{2+}-rich hydrous manganese oxide in the accumulation of arsenic in lake sediments. *Water Res.* 19:1029-1032.

Tamaki, S. and W.T. Frankenberger, Jr. 1992. Environmental biochemistry of arsenic. *Rev. Environ. Contam. Toxicol.* 124:79-110.

Tao, S.S.H. and P.M. Bolger. 1998. Dietary intakes of arsenic in the United States. *SEGH 3rd Inter. Conf. Arsenic Expos. Health Effects: 85.*

Taylor, D., B.G. Maddock, and G. Mance. 1985. The acute toxicity of nine 'grey list' metals (arsenic, boron, chromium, copper, lead, nickel, tin, vanadium and zinc) to two marine fish species: dab (*Limanda limanda*) and grey mullet (*Chelon labrosus*). *Aquat. Toxicol.* 7:135-144.

Thanabalasingam, P. and W.F. Pickering. 1986. Arsenic sorption by humic acids. *Environ. Pollut.* 12B:233-246.

Thatcher, C.D., J.B. Meldrum, S.E. Wikse, and W.D. Whittier. 1985. Arsenic toxicosis and suspected chromium toxicosis in a herd of cattle. *Jour. Amer. Vet. Med. Assoc.* 187:179-182.

Thies, M. and D. Gregory. 1994. Residues of lead, cadmium, and arsenic in livers of Mexican free-tailed bats. *Bull. Environ. Toxicol. Chem.* 52:641-648.

Thornton, I. and M. Farago. 1997. The geochemistry of arsenic. Pages 1-16 in C.O. Abernathy, R.L. Calderon, and W.R. Chappell (eds.). *Arsenic. Exposure and Health Effects.* Chapman & Hall, London.

Thursby, G.B. and R.L. Steele. 1984. Toxicity of arsenite and arsenate to the marine macroalgae *Champia parvula* (Rhodophyta). *Environ. Toxicol. Chem.* 3:391-397.

Tremblay, G.H. and C. Gobeil. 1990. Dissolved arsenic in the St. Lawrence estuary and the Saguenay Fjord, Canada. *Mar. Pollut. Bull.* 21:465-468.

Tsai, S.M., T.N. Wang, and Y.C. Ko. 1998. Cancer mortality trends in a blackfoot disease endemic community of Taiwan following water source replacement. *Jour. Toxicol. Environ. Health* 55A:389-404.

Unchalee, S., V. Pongsakdi, P. Mandhana, and V. Thatavatchai. 1999. Association between chronic arsenic exposure and children's intelligence in Thailand. Pages 1140-1141 in W.W. Wenzel, D.C. Adriano, B. Alloway, H.E. Doner, C. Keller, N.W. Lepp, M. Mench, R. Naidu, and G.M. Pierzynski (eds.). Proceedings of the Fifth International Conference on the Biogeochemistry of Trace Elements, July 11-15, 1999, Vienna, Austria.

U.S. Environmental Protection Agency (USEPA). 1980. Ambient Water Quality Criteria for Arsenic. U.S. Environ. Protection Agen. Rep. 440/5-80-021. 205 pp.

U.S. Environmental Protection Agency (USEPA). 1985. Ambient Water Quality Criteria for Arsenic — 1984. U.S. Environ. Protection Agen. Rep. 440/5-84-033. 66 pp.

U.S. Public Health Service (USPHS). 1998. Toxicological Profile for Arsenic (update). Draft for public comment. U.S. Dept. Health Human Serv., Agen. Toxic Subst. Dis. Regist. 365 pp.

Unlu, M.Y. and S.W. Fowler. 1979. Factors affecting the flux of arsenic through the mussel *Mytilus galloprovincialis*. *Mar. Biol.* 51:209-219.

Vahter, M. and E. Marafante. 1985. Reduction and binding of arsenate in marmoset monkeys. *Arch. Toxicol.* 57:119-124.

Veen, N.G. van der and K. Vreman. 1985. Transfer of cadmium, lead, mercury and arsenic from feed into various organs and tissues of fattening lambs. *Neth. Jour. Agric. Sci.* 34:145-153.

Vermeer, K. and J.A.J. Thompson. 1992. Arsenic and copper residues in waterbirds and their food down inlet from the island copper mill. *Bull. Environ. Contam. Toxicol.* 48:733-738.

Vos, G. and J.P.C. Hovens. 1986. Chromium, nickel, copper, zinc, arsenic, selenium, cadmium, mercury and lead in Dutch fishery products 1977-1984. *Sci. Total Environ.* 52:25-40.

Vreman, K., N.G. van der Veen, E.J. van der Molen, and W.G. de Ruig. 1986. Transfer of cadmium, lead, mercury and arsenic from feed into milk and various tissues of dairy cows: chemical and pathological data. *Neth. Jour. Agric. Sci.* 34:129-144.

Wang, D.S., R.W. Weaver, and J.R. Melton. 1984. Microbial decomposition of plant tissue contaminated with arsenic and mercury. *Environ. Pollut.* 34A:275-282.

Webb, D.R., S.E. Wilson, and D.E. Carter. 1986. Comparative pulmonary toxicity of gallium arsenide, gallium (III) oxide, or arsenic (III) oxide intratracheally instilled into rats. *Toxicol. Appl. Pharmacol.* 82:405-416.

Welch, A.H., D.B. Westjohn, D. Helsel, and R. Wanty. 1998. Arsenic in ground water of the United States. *SEGH 3rd Inter. Conf. Arsenic Expos. Health Effects: 4.*

Wells, J.M. and D.H.S. Richardson. 1985. Anion accumulation by the moss *Hylocomium splendens*: uptake and competition studies involving arsenate, selenate, selenite, phosphate, sulphate and sulphite. *New Phytol.* 101:571-583.

White, D.H., K.A. King, and F.M. Prouty. 1980. Significance of organochlorine and heavy metal residues in wintering shorebirds at Corpus Christi, Texas, 1976–77. *Pestic. Monitor. Jour.* 14:58-63.

Wiemeyer, S.N., T.G. Lamont, and L.N. Locke. 1980. Residues of environmental pollutants and necropsy data for eastern United States ospreys, 1964–1973. *Estuaries* 3:55-167.

Wiener, J.G., G.A. Jackson, T.W. May, and B.P. Cole. 1984. Longitudinal distribution of trace elements (As, Cd, Cr, Hg, Pb, and Se) in fishes and sediments in the upper Mississippi River. Pages 139-170 in J.G. Wiener, R.V. Anderson, and D.R. McConville (eds.). *Contaminants in Upper Mississippi River.* Butterworth Publ., Stoneham, MA.

Windom, H., R. Stickney, R. Smith, D. White, and F. Taylor. 1973. Arsenic, cadmium, copper, mercury, and zinc in some species of North Atlantic finfish. *Jour. Fish. Res. Board Canada* 30:275-279.

Winski, S.L. and D.E. Carter. 1998. Arsenate toxicity in human erythrocytes: characterization of morphologic changes and determination of the mechanism of damage. *Jour. Toxicol. Environ. Health* 53A:345-355.

Woodward, D.F., A.M. Farag, H.L. Bergman, A.J. DeLonay, E.E. Little, C.E. Smith, and F.T. Barrows. 1995. Metals-contaminated benthic invertebrates in the Clark Fork River, Montana: effects on age-0 brown trout and rainbow trout. *Canad. Jour. Fish. Aquat. Sci.* 52:1994-2004.

Woolson, E.A. (ed.). 1975. *Arsenical Pesticides.* Amer. Chem. Soc. Symp. Ser. 7. 176 pp.

Wrench, J., S.W. Fowler, and M.Y. Unlu. 1979. Arsenic metabolism in a marine food chain. *Mar. Pollut. Bull.* 10:18-20.

Yamaoka, Y. and O. Takimura. 1986. Marine algae resistant to inorganic arsenic. *Agric. Biol. Chem.* 50:185-186.

Yamauchi, H., K. Takahashi, and Y. Yamamura. 1986. Metabolism and excretion of orally and intraperitoneally administered gallium arsenide in the hamster. *Toxicology* 40:237-246.

Yamauchi, H. and Y. Yamamura. 1985. Metabolism and excretion of orally administered arsenic trioxide in the hamster. *Toxicology* 34:113-121.

Yoshida, K., Y. Inoue, K. Kuroda, H. Chen, H. Wanibuchi, S. Fukushima, an G. Endo. 1998. Urinary excretion of arsenic metabolites after long-term oral administration of various arsenic compounds to rats. *Jour. Toxicol. Environ. Health* 54A:179-192.

Zaroogian, G.E. and G.L. Hoffman. 1982. Arsenic uptake and loss in the American oyster, *Crassostrea virginica. Environ. Monitor. Assess.* 1:345-358.

CHAPTER 29

Boron

29.1 INTRODUCTION

Borax ($Na_2B_4O_7 \cdot 10H_2O$) was the first of the boron (B) minerals to be traded by the Babylonians more than 4000 years ago for use in the working and welding of gold (Greenwood and Thomas 1973). Borax has been known as a cleaning agent since the days of the ancient Greek and Roman empires and was used as a food preservative in Europe and America, although its use for the latter purpose has been discontinued (Weir and Fisher 1972). Boron and its compounds were used in the Egyptian and Roman eras to prepare borosilicate glass. Borax glazes were known from about the year 200; by 1556, borax was widely used throughout Europe as a flux (Greenwood and Thomas 1973). Boric acid (H_3BO_3) was first synthesized in 1707 (Greenwood and Thomas 1973). Boric acid and borates are the main boron compounds of ecological significance; other boron compounds usually degrade or are transformed to borates or boric acid (Sprague 1972).

Boron is an essential trace element for the growth and development of higher plants, although the range between insufficiency and excess is generally narrow, varying with the plant; boron is not required in fungi and animals (Sprague 1972; Weir and Fisher 1972; Birge and Black 1977; Goldbach and Amberger 1986; Stone 1990). In the southwestern United States, naturally elevated boron concentrations in surface waters used for irrigation may be sufficiently high to cause toxicity to plants of commercial importance (Benson et al. 1984). Another major source of boron entering ground and surface waters results from the use of borax-containing laundry products, coupled with ineffective removal of boron by conventional sewage processes (Benson et al. 1984). Agricultural drainwaters contaminated with boron are considered potentially hazardous to waterfowl and other wildlife populations throughout areas of the western United States (Smith and Anders 1989).

Medical and household uses of boric acid solutions as antiseptics have led to numerous accidental poisonings by ingestion or absorption through abraded skin, particularly in infants (U.S. Environmental Protection Agency [USEPA] 1975; Dixon et al. 1976; Landolph 1985; Siegel and Wason 1986). Poisonings have been reported in English children consuming milk containing 0.7 g boric acid/L, and in burn patients treated topically with saturated boric acid solutions (NAS 1980). In the 1940s, topical preparations of boric acid became a popular remedy for diaper rash in England. By 1953, at least 60 fatal cases of boric acid poisoning had been reported in English infants (O'Sullivan and Taylor 1983). Inhalation of boranes, especially diborane (B_2H_6), pentaborane (B_5H_9), and decaborane ($B_{10}H_{14}$) — which is used as a rocket propellant — is toxic to exposed workers (Dixon et al. 1976; NAS 1980). Boron compounds, especially boric acid, can also accumulate in animal tissues and produce a reduction in fertility, an increase in developmental abnormalities — especially those involving the skeletal system — stillbirth, and death (Weir and Fisher 1972; Lee et al. 1978; Landolph 1985). There seems to be a reasonable margin between a toxic dose in man and other vertebrates and in boron levels that may occur as incidental residues from the use of borax and boric acid in agriculture and industry (Weir and Fisher 1972). Additional, and more extensive, information on ecological and toxicological aspects of boron in the environment is presented in reviews by Sprague (1972), USEPA (1975), NAS (1980), Anonymous (1983), Klasing and Pilch (1988), Butterwick et al. (1989), Eisler (1990), Stone (1990), U.S. Public Health Service [USPHS] (1991), and Culver et al. (1998).

29.2 ENVIRONMENTAL CHEMISTRY

29.2.1 General

The United States supplies about 70% of the global boron demand, and Turkey supplies 18%. Of the total annual U.S. production of about 500,000 tons, 45% is used in the manufacture of glass and glassware, 15% in laundry products, 10% in enamels and glazes, and 8% in agricultural chemicals. It is estimated that boron compounds enter the North American environment at a rate of 32,000 tons annually as a result of human activities, primarily from laundry products, irrigation drainwater, agricultural chemicals, coal combustion, and mining and processing. Boron compounds tend to accumulate in aquatic ecosystems because of the relatively high water solubility of these compounds.

The chemistry of boron is exceedingly complex and rivals that of carbon in its diversity. Most boron compounds, however, enter or degrade in the environment to borates (B-O compounds), such as borax and boric acid, and these are considered to be the most significant ecologically.

Toxicosis in animals has resulted from ingestion of boric acid or borax solutions, from topical applications of boric acid solutions to damaged skin, and from inhalation of boranes; the exact mechanisms of action are not understood. Boron and its compounds are potent teratogens when applied directly to the embryo, but there is no evidence of mutagenicity or carcinogenicity. Boron's unique affinity for cancerous tissues has been exploited in neutron capture radiation therapy of malignant human brain tumors.

29.2.2 Sources and Uses

Boron is a dark brown element that is widespread in the environment but occurs naturally only in combined form, usually as borax, colemanite ($Ca_2B_6O_{11} \cdot 5H_2O$), boronatrocalcite ($CaB_4O_7NaBO_2 \cdot 8H_2O$), and boracite ($Mg_7Cl_2B_{16}O_{30}$) (USEPA 1975; NAS 1980). In the United States, boron deposits in the form of borax are concentrated in the desert areas of southern California, especially near Boron, California (USEPA 1975; USPHS 1991). Proven deposits of sodium tetraborates — from which borax is prepared and from which boron can be isolated — also exist in Nevada, Oregon, Turkey, Russia, and China (Sprague 1972; NAS 1980). About 300,000 metric tons of boron are removed from mined ore each year (Argust 1998). The United States supplies about 70% of the world boron demand, and Turkey supplies 18%; the most common commercial compounds are boric acid and borax (Sprague 1972; Butterwick et al. 1989; USPHS 1991).

In 1988, the United States produced 566,093 metric tons of boric oxide, imported an additional 60,000 metric tons of boron-containing minerals, and exported 589,680 metric tons of boric acid and borates (USPHS 1991). The majority of the boron produced annually at facilities in Oklahoma, New Jersey, Nevada, and Pennsylvania is in the form of sodium tetraborate compounds (USPHS 1991). Of the total production, about 42% occurs as anhydrous borax ($Na_2B_4O_7$), 29% as borax pentahydrate ($Na_2B_4O_7 \cdot 5H_2O$), 10% as borax decahydrate or borax, and 16% as boric acid or boric oxide (B_2O_3) (Sprague 1972; USEPA 1975). Boron and its compounds are used in the manufacture of glassware (40% to 45%); soaps and cleansers (15%); enamels, frits, and glazes (10%); fertilizers

(5%); and herbicides (2% to 3%); 22% to 28% goes for other uses including cosmetics, insecticides, antifreeze, as neutron absorbers in atomic reactors, and leather tanning (Sprague 1972; USEPA 1975; NAS 1980; USPHS 1991; Woods 1994; Argust 1998). Borates have some toxicity to insects and, in relatively high concentrations, can control cockroaches, woodboring insects, gypsy moths, and larvae of flies in manure piles and in dog runs (Sprague 1972; USEPA 1975). Some organoboron compounds are used to sterilize fuel distribution and storage systems against fungi and bacteria (Sprague 1972). Radioboron-10 is widely used in radiation therapy against brain tumors, especially in Japan (Hatanaka 1986). In medicine, certain amine-carboxyborane derivatives show promise in reducing serum cholesterol and triglyceride concentrations, in alleviating some forms of chronic arthritis, and as antineoplastic agents (Hall et al. 1994; Newnham 1994). Other boron compounds are used widely as thermal protection materials in space probes, in fireproofing of fabrics and wood, in leather manufacture, in numerous pharmaceuticals and hygiene products, in steel hardening, in deoxidation of bronze, as a high-energy fuel, as neutron-absorbing shielding near atomic reactors, and as water softeners, pH adjusters, emulsifiers, neutralizers, stabilizers, buffers, and viscosifiers (NAS 1980; Parry and Kodama 1980; Schillinger et al. 1982; Siegel and Wason 1986; USPHS 1991).

The major global environmental reservoirs of boron (metric tons) include continental and oceanic crusts (10^{15}), oceans (10^{12}), groundwater (10^8), ice (10^8), coal deposits (10^7), commercial borate deposits (10^7), biomass (10^7), and surface waters (10^5) (Argust 1998). The largest flows of boron in the environment arise from the movement of boron into the atmosphere from the oceans at 1.3 to 4.5×10^6 tons annually. Drainage from soil systems into groundwaters and surface waters accounts for 1.3×10^5 to 1.3×10^6 tons yearly. And boron mining and volcanic eruptions account for 4×10^5 and 2×10^5 tons of boron per annum, respectively (Argust 1998).

Boron enters the environment at about 32,000 metric tons annually in the United States (Table 29.1). Most ends up in the aquatic environment because of the relatively high water solubility of all boron compounds, especially boron-containing laundry products and sewage (USEPA 1975). Conventional sewage treatment removes little or no boron (USEPA 1975; Vengosh 1998). Studies of domestic wastewaters in California and Israel, using boron isotopic composition techniques, show that boron in sewage is derived from sodium borate components used in household detergents (Vengosh 1998). Of the total boron in coal, as much as 71% may be lost to the atmosphere upon combustion. More than 50% of the boron found in coal ash is readily water soluble (Pagenkopf and Connolly 1982). The release of boron from coal flyash to leachate water is dependent on the ash to water ratio: at 1 g ash/L, up to 90% of the boron is soluble; at 50 g/L, only 40% is released; at 100 g/L, less than 30% is released. Coating of coal ash with aluminum solution reduces boron solubility by about 90% due to the formation of an insoluble aluminum–borate complex (Pagenkopf and Connolly 1982).

Table 29.1 Environmental Sources of Domestic Boron

Source	Metric Tons, Annually
Laundry products	14,000
Agricultural chemicals and fertilizers	7000
Coal combustion	4000
Mining and processing	3000
Glass and ceramics	1500
Miscellaneous	2500
Total	32,000

Data from U.S. Environmental Protection Agency (USEPA). 1975. Preliminary Investigation of Effects on the Environment of Boron, Indium, Nickel, Selenium, Tin, Vanadium and Their Compounds. Vol. 1. Boron. U.S. Environ. Protection Agen. Rep. 56/2-75-005A. 111 pp.

Boron compounds listed in the "Commodity List of Explosives and Other Dangerous Articles" are boron trichloride (BCl_3), boron trifluoride (BF_3), decaborone ($B_{10}H_{14}$), and pentaborane (B_5H_9) (USEPA 1975). Boron trichloride is a corrosive liquid; the maximum quantity allowed in containers by rail is 1 L, and by air only one container is permitted per aircraft. Boron trifluoride is a nonflammable gas restricted to 140 kg in one outside container by rail, and to 140 kg in cargo planes only. Decaborane is a flammable solid, and transport by rail or air is limited to 12 kg. Pentaborane is a flammable liquid and is prohibited for transport by air or rail. Diborane (B_2H_6) and higher boranes are unstable and are classified as dangerous articles in transport; no more than 0.1 kg can be shipped in a cylinder (USEPA 1975). Organic boron–oxygen compounds readily hydrolyze and should be stored and transferred in an inert atmosphere; usually, glass containers are used for shipping small quantities, and steel containers or tank cars are used for bulk items. Hazardous atmospheric conditions resulting from high concentrations of boron compounds are localized and are not considered a serious environmental problem (USEPA 1975).

29.2.3 Chemical Properties

The element boron has an atomic number of 5, a molecular weight of 10.811, an oxidation state of 3 for simple compounds (but other oxidation states for carboranes and other polyhedral cage boron compounds), a specific gravity of 2.34, a melting point of 2300°C, sublimation at 2550°C, and is almost insoluble in water. Boron exists as B-10 (19.78%) and B-11 (80.22%) isotopes, and it contributes about 0.001% to Earth's crust, although it does not occur free in nature (USEPA 1975; Smith 1985). The chemistry of boron is exceedingly complex and rivals that of carbon in diversity. Reviews on boron's chemistry are especially abundant and include those by Steinberg and McCloskey (1964), Brotherton and Steinberg (1970a, 1970b), Greenwood and Thomas (1973), Grimes (1982), Evans and Sparks (1983), Smith (1985), Emin et al. (1986), Heller (1986), Niedenzu and Trofimenko (1986), Hermanek (1987), and USPHS (1991).

Most boron compounds degrade in the environment to B–O (borate) compounds, and these are the boron compounds of ecological significance — especially borax and boric acid (Sprague 1972; Antia and Cheng 1975; Thompson et al. 1976). Sodium tetraborate decahydrate (borax) has a melting point of 75°C, a boiling point of 320°C, and is soluble in water to 20 g/L at 0°C and to 1700 g/L at 100°C (USEPA 1975). Boric acid has a melting point of 169°C, a boiling point of 300°C and, like borax, is exceedingly soluble in water: 63.5 g/L at 30°C and 276 g/L at 100°C (USEPA 1975; USPHS 1991).

Boron exists in several forms in the soil (USEPA 1975); in soil solution, it exists largely as the undissociated weak monobasic acid that accepts hydroxyl groups (Gupta and Macleod 1982). Most plant-available boron in soils is associated with soil organic matter (Gupta and Macleod 1982), with the hot-water soluble boron fraction (Hingston 1986), and with soil solution pH ranges of 5.5 to 8.5 and 10 to 11.5 (Goldberg and Glaubig 1986). It is assumed that boron adsorbs to soil particles and aluminum and iron oxide minerals (Goldberg and Glaubig 1986). Boron mobility in soils is reduced under conditions of pH 7.5 to 9.0, and with high abundance of amorphous aluminum oxide, iron oxide, and organic content (USPHS 1991).

In water, boron readily hydrolyzes to form the electrically neutral, weak monobasic acid H_3BO_3 and the monovalent ion $B(OH)_4^-$. Waterborne boron may be adsorbed by soils and sediments (USPHS 1991). The predominant boron species in seawater is boric acid (Thompson et al. 1976); concentrations are higher at higher salinities and in proximity to industrial waste discharges (Liddicoat et al. 1983; Narvekar et al. 1983). In seawater, borate or boric acid occurs naturally at 4.5 to 5.5 mg/L. About 76% of the total inorganic boron in seawater occurs as undissociated boric acid [$B(OH)_3$], and the remainder is identified as the borate ion [$B(OH)_4^-$]. Of the total borate ion, 44% appears to be complexed with sodium, magnesium, and calcium (Antia and Cheng 1975).

Other evidence suggests additional complexation of borate with ferric ions and polyhydroxylated organic compounds (Antia and Cheng 1975).

Atmospheric boron is in the form of particulates or aerosols of borides, boron oxide, borates, boranes, organoboron compounds, trihalide boron compounds, or borazines. The half-time persistence of airborne boron particles is short, usually on the order of days (USPHS 1991).

Despite the development of sophisticated instrumentation and techniques, the accurate determination of boron in biological materials is difficult at concentrations less than 1 mg B/kg (Downing et al. 1998). Problems associated with analysis of boron from biological sources include contamination from Teflon® vessels during microwave digestion; losses due to freeze drying; variations in boron isotope ratios, standards preparation, and reagent backgrounds; and instrumental interference (Downing et al. 1998). Inductively coupled plasma-mass spectrometry now allows quantitation of percutaneous absorption of ^{10}B in ^{10}B-enriched boric acid, borax, and disodium octaborate tetrahydrate in biological materials (Wester et al. 1998a), although absorption through intact human skin is significantly less than the mean daily dietary intake (Wester et al. 1998b).

29.2.4 Mode of Action

A proposed essential role for boron is as a regulator of enzymatic pathways closely involved with energy substrate metabolism, insulin release, and the immune system. Boron influences the activities of at least 26 enzymes — including reductases, transferases, hydrolases, and isomerases — examined in various biological systems by acting on the enzyme directly and binding to cofactors or substrates (Hunt 1998). The complexing ability of the boron atom is considered to be the key explanation of why it is essential to higher plants (USEPA 1975), although the exact mechanism of action is still unknown. In biological systems, boron probably is complexed with hydroxylated species, and inhibition or stimulation of enzymes and coenzymes is pivotal in its mode of action (Woods 1994). Boron interacts with substances of biological interest, including polysaccharides, pyridoxine, riboflavin, dehydroascorbic acid, and pyridine nucleotides (Samman et al. 1998). Boron's complexing ability is thought to beneficially influence transport of sugars and other organic compounds, production of plant growth regulators, biosynthesis of nucleic acids and phenolic acid, carbohydrate metabolism, respiration, and pollen germination (USEPA 1975; Nielson 1986).

Boron poisoning in animals is primarily an experimental phenomenon, although livestock in certain regions may be exposed to high concentrations in drinking water — up to 80 mg B/L — that have not been shown to be toxic (NAS 1980). Toxicosis in humans has resulted from ingestion of boric acid or borax solutions, topical applications of boric acid solutions to burn-damaged skin, and inhalation of boranes (NAS 1980). In mammals, boron is thought to regulate parathyroid function through metabolism of phosphorus, magnesium, and especially calcium. Boron has a close relationship with calcium metabolism, most likely at the cell membrane level (Nielsen 1986). The toxicological effects of boric acid and borax are similar for different species. Other inorganic borates that dissociate to boric acid display similar toxicity, whereas those that do not dissociate to boric acid may display a different toxicological profile (Hubbard 1998).

Dietary boron at nontoxic concentrations, as sodium borate or boric acid, is rapidly and almost completely absorbed from the gastrointestinal tract, does not seem to accumulate in healthy tissues, and is excreted largely in urine, usually within hours, but sometimes as long as 23 days. Similar patterns are evident for humans, dogs, cows, rabbits, rodents, and guinea pigs (NAS 1980; Benson et al. 1984; Nielsen 1986; Siegel and Wason 1986; USPHS 1991; Murray 1998). Urinary boron excretion changes rapidly with changes in boron intake, suggesting that the kidney is the site of homeostatic regulation (Sutherland et al. 1998). Boron does not seem to accumulate in soft tissues of animals, but does accumulate in bone; cessation of exposure to dietary boron resulted in a rapid drop in bone boron, usually within 24 h (Moseman 1994). Boric acid poisoning in animals, regardless of route of administration, is characterized by the following signs: generalized erythema (boiled lobster appearance) starting in the axillary, inguinal, and face regions, eventually covering the entire body with conjunctival redness, followed by massive desquamation 2 to 3 days later; acute gastroenteritis, including nausea and vomiting; diarrhea; anorexia; cardiac weakness; excessive urinary excretion of riboflavin; decreased oxygen uptake by the brain; hypoacidity; altered enzyme activity levels; impaired growth and reproduction; and death from circulatory collapse and shock, usually within 5 days (Dani et al. 1971; Sprague 1972; USEPA 1975; NAS 1980; Schillinger et al. 1982; Settimi et al. 1982; Siegel and Wason 1986; USPHS 1991).

Boron hydrides or boranes, such as B_2H_6, B_4H_{10}, and B_5H_9, from chemical processes produce acute central nervous system (CNS) pulmonary damage and lung disease through inhalation (NAS 1980; Klaassen et al. 1986). Boranes produce toxic effects by creating embolisms of hydrogen gas as they react with tissue, and by depleting biogenic amines of the CNS and inhibiting aminotransferases and other pyridoxol-dependent enzymes (Korty and Scott 1970; USEPA 1975). Boranes produce similar effects in humans and animals, and these are generally ascribable to CNS depression and excitation (Naeger and Leibman 1972; Smith 1985). Symptoms of borane intoxication include pulmonary irritation, headache, chills, fatigue, muscular weakness and pain, cramps, dizziness, chest tightness, and pneumonia (NAS 1980). Boranes may adversely affect male reproductive capacity (Klaassen et al. 1986), but this requires verification. Decaborane ($B_{10}H_{14}$), as one example, is a highly lipid-soluble compound that can enter the body through inhalation, ingestion, or the skin. In water, decaborane is rapidly transformed into intermediate products that are eventually degraded to boric acid. The intermediate products, but not decaborane or boric acid, reduce phosphomolybdic acid and inhibit glutamic-oxaloacetic transaminase; treatment of intermediates with pyridoxol phosphase tends to reverse the inhibitory activity (Naeger and Leibman 1972). Low decaborane doses cause behavioral effects such as depression, catatonia, and convulsions (USEPA 1975).

Inorganic borates are comparatively toxic, apparently complexing hydroxy compounds and interfering with protein synthesis (USEPA 1975). Organoborate compounds exert physiological effects on the CNS and peripheral nervous system, acting as spasmolytics, sedatives, and convulsants, depending on their structure (USEPA 1975). Boron trihalides such as BBr_3, BCl_3, and BF_3 are corrosive to the eyes, skin, and mucous membranes, and will cause burns on the skin — apparently due to the hydrolysis of the trihalides to their halogen acids, and not to boron (USEPA 1975; Smith 1985).

Boron is a potent teratogen when applied directly to the embryo. Boric acid injected into chicken and amphibian embryos produced abnormal development of the neural tube, notochord, tail, and limbs, perhaps through complexing polyhydroxy compounds and interfering with riboflavin metabolism (Landauer 1952, 1953a, 1953b, 1953c; Landauer and Clark 1964; USEPA 1975; Settimi et al. 1982). Boron and its compounds, however, are neither mutagenic nor carcinogenic (Landolph 1985; Dieter 1994). Nonmutagenicity is based on results of the *Salmonella typhimurium*-mammalian microsome mutagenicity assay; boron neither enhances nor inhibits the activity of benzo[*a*]pyrene, a known mutagen (Anonymous 1983; Benson et al. 1984). There is no evidence that boron is a possible carcinogen, although long-term, selective uptake of boron by tumors has been reported (USEPA 1975).

Boron seems to have an affinity for cancerous tumors, and this property has been exploited in radiation therapy (Hamada et al. 1983; Hatanaka 1986). Boron-10 has been used in neutron capture therapy to cure malignant sarcomas implanted in the hind legs of mice, as well as spontaneous malignant melanomas in pigs (Slatkin et al. 1986). The sulfhydral borane monomer $(B_{12}H_{11}SH)^{2-}$ is used as a B-10 carrier in neutron therapy of malignant human brain tumors and seems to be most effective at 30 µg B-10/kg tissue (Hatanaka 1986). Polyhedral boranes attached to monoclonal antibodies that are tumor specific may become useful in tumor therapy by neutron irradiation (Parry and Kodama 1980). It is possible, however, that uptake of boron may be a nonspecific attribute of tumors and of a variety of normal tissues that lack a blood–brain barrier. Thus, the potential usefulness of selected B-10 carriers for treating extracranial neoplasms seems questionable at this time (Slatkin et al. 1986).

29.3 CONCENTRATIONS IN FIELD COLLECTIONS

29.3.1 General

Terrestrial plants are normally rich sources of boron. Levels in meat and fish are usually low. But these generalizations are based on limited data. Boron is ubiquitous in the environment as a result of natural weathering processes (Woods 1994). However, human activities such as mining, coal burning, and use of borax laundry detergents have resulted in elevated boron loadings in air, water, and soils (USPHS 1991). Comparatively high levels of boron occur in fish, aquatic plants, and insects at Kesterson National Wildlife Refuge that was contaminated by agricultural drainwater (Hothem and Ohlendorf 1989; Eisler 1990). The availability of inductively coupled plasma-mass spectrometry (ICP-MS) technology enables measurement of boron concentrations and isotope ratios in a large number of biological samples with minimal sample preparation at detection limits of 0.11 μg/L (Vanderpool et al. 1994).

29.3.2 Nonbiological Materials

Boron is distributed widely in the environment (Ahl and Jonsson 1972; USEPA 1975). Naturally elevated boron levels are usually associated with marine sediments, thermal springs, large deposits of boron minerals, seawater, and certain groundwaters (Table 29.2). Human activities, however, have resulted in elevated boron concentrations near coal-fired plants, in mine drainage waters, in municipal wastes, and in agricultural drainage waters. In one case, agricultural drainwater practices in western California produced boron concentrations in local rivers, groundwaters, and surface waters that exceeded the established limits for the protection of crops and aquatic life (Schuler 1987; Klasing and Pilch 1988; Table 29.2). Contamination of pristine groundwaters (<0.05 mg B/L) by domestic wastewater and agriculture-return flows (0.5 to 1.0 mg B/L) is documented by the isotopically distinguished signature of borate compounds. For example, in areas where calcium borates are applied as fertilizers, the B^{11}/B^{10} ratios of the soil water and leachates are expected to be low and can be used as diagnostic tools for tracing agriculture-return flows (Vengosh 1998).

Coal-fired power plants are major sources of atmospheric boron contamination. At least 30% of boron in coal is lost in this manner (Cox et al. 1978; Gladney et al. 1978). The apparently large amounts of boron lost to the environment through stack emissions may be directly related to the organic content of coal (Table 29.2) (Gladney et al. 1978). Also, disposal of B-laden drainage waters from boron mines is a major problem in certain geographic areas. In Turkey, for example, which possesses about 60% of the world's boron reserves — localized in a rectangular area about 100 × 200 km near the Simav River — drainage waters discharged from the mines as a result of borate production have elevated boron concentrations in the Simav River to levels unsuitable for crop irrigation purposes. About 68,000 ha of agricultural land irrigated by the Simav River is now threatened by boron pollution (Okay et al. 1985). In the United States, laundry detergents originating from household use may contribute as much as 50% of the boron loadings in effluents discharged into aquatic environments; lesser amounts are contributed by soil minerals, rainfall, and industry and sewage effluents (USEPA 1975).

Table 29.2 Boron Concentrations in Selected Nonbiological Materials (Concentrations are in mg B/kg fresh weight [FW], dry weight [DW], or ash weight [AW], except where noted.)

Material	Concentration, (mg/kg or mg/L)	Reference[a]
AIR		
Workplace (borax mining and refining plants, sites where boric acid is manufactured)	1–14 mg/m^3	18
Non-workplace locations	0.02 (<0.0005–0.08) $\mu g/m^3$	19
AFFECTED BY AGRICULTURAL DRAINAGE WATERS		
Western San Joaquin Valley, California		
River waters	Median 1.1 FW	16
Surface waters	Median 3.1 (Max. 83.0) FW	16
Groundwaters	Median 7.4 (Max. 120.0) FW; frequently >100 FW	16, 18
Kesterson National Wildlife Refuge, 1984		
Subsurface waters	20 (12–41) FW	17
Sediments	20 (10–71) DW	17
COAL-FIRED POWER PLANTS		
Chalk Point Power Plant, Maryland		
Coal	13 AW	1
Bottom slag	19 AW	1
Flyash	33 AW	1
Four Corners Power Plant, New Mexico		
Coal	92 AW	1
Bottom slag	120 AW	1
Flyash	240 AW	1
Coal ash		
Anthracite	90 AW	1
Volatile bituminous		
Low	123 AW	1
Medium	218 AW	1
High	770 AW	1
Lignite	1020 AW	1
Coal ash	5–200 DW	2, 15
DRINKING WATER		
Worldwide	Usually <0.4 FW, range 0.0005–>2.0 FW	18
California		
50th percentile	0.1 FW	20
90th percentile	0.4 FW	20
Northern Chile	0.31–15.2 FW	20
Bottled water; U.S. and Europe	0.75 (<0.005–4.35) FW	20
GROUNDWATERS		
Pristine	Usually <0.05 FW	21
Worldwide	Usually <0.5 FW	9
Greece	2.3–5.4 FW	4
United States	Max. 5.0 FW	12
MINE DRAINAGE WATERS		
Turkish boron mines		
Avsar mine	16 FW	4
Simav mine	260 FW	4
Yenikoy mine	390 FW	4

Table 29.2 (continued) Boron Concentrations in Selected Nonbiological Materials (Concentrations are in mg B/kg fresh weight [FW], dry weight [DW], or ash weight [AW], except where noted.)

Material	Concentration, (mg/kg or mg/L)	Reference[a]
RAIN		
Sweden	0.002 FW	7
France	0.002–0.004 FW	7
U.S.		
Mississippi	Usually <0.01 FW	11
Florida	~0.01 FW	7, 8
India	0.03 (0.002–0.007) FW	11
England	0.08 FW	7
Japan	0.1 FW	7
RIVER WATER		
Germany	0.02 FW; Max. 0.18 FW	20
U.K., northern Italy	0.002–0.87 FW	20
SEAWATER		
British Columbia		
Surface	3.5 (0.2–4.7) FW	13
Depth 5 m	3.9 FW	13
Open ocean	4.5 FW	8
Coastal	4.6 FW	8, 18, 20
Total inorganic	4.5–5.5 FW	14
As undissociated boric acid	3.4–4.2 FW	14
As borate ion, $B(OH)_4^-$	1.1–1.3 FW	14
As complex with sodium, magnesium, and calcium	0.5–0.6 FW	14
SEDIMENTS		
Nonmarine clays	<10 DW	3
Postglacial marine	Max. 500 DW	3
SOILS		
Worldwide	Usually 45–124 DW, range 4–200, mostly as biologically unavailable tourmaline	5, 6
United States	30 (10–300) DW	7, 18, 19
SEWAGE WATERS		
Scandinavia	0.4 (Max. 0.7) FW	3
SURFACE FRESHWATERS		
Canada, 1988	0.16 FW	20
Worldwide	0.0001–<0.5 FW	8–10, 16
Norway, 1970	Usually <0.004 FW, median 0.013, range 0.001–1.05 FW	3
Sweden, 1970	0.12 (0.001–1.0) FW	3, 7
Southeastern U.S., 1969–1970		
Streams, swamps, ponds	Usually <0.1 FW	11
Reservoirs	0.007 (<0.001–0.09) FW	11
In regions where marine deposits are common	>0.06 FW	3
United States	Generally <0.1 FW; frequently 0.1–0.3 FW; rarely 360 FW in areas of boron-rich deposits	8, 12, 18, 19

Table 29.2 (continued) Boron Concentrations in Selected Nonbiological Materials (Concentrations are in mg B/kg fresh weight [FW], dry weight [DW], or ash weight [AW], except where noted.)

Material	Concentration, (mg/kg or mg/L)	Reference[a]
United States		
10th percentile	0.01 FW	20
50th percentile	0.076 FW	20
90th percentile	0.387 FW	20
Nevada		
Humboldt River	0.2 FW	6
Borax Flat	Up to 80 FW	6
Turkey		
Uncontaminated	<0.5 FW	4
Contaminated with boron mine wastes	4 (Max. 7) FW	4
Western U.S.	Sometimes 5–15 FW	10, 12
Japan	1–15 FW	7
THERMAL SPRINGS		
Greece	43 FW	4
WELL WATER		
India	0.08–0.5 FW	7

[a] **1,** Gladney et al. 1978; **2,** Pagenkopf and Connolly 1982; **3,** Ahl and Jonsson 1972; **4,** Okay et al. 1985; **5,** Gupta and Macleod 1982; **6,** NAS 1980; **7,** Sprague 1972; **8,** USEPA 1975; **9,** Benson et al. 1984; **10,** Lewis and Valentine 1981; **11,** Boyd and Walley 1972; **12,** Birge and Black 1977; **13,** Thompson et al. 1976; **14,** Antia and Cheng 1975; **15,** Cox et al. 1978; **16,** Klasing and Pilch 1988; **17,** Schuler 1987; **18,** USPHS 1991; **19,** Howe 1998; **20,** Coughlin 1998; **21,** Vengosh 1998.

29.3.3 Plants and Animals

Boron accumulates in both aquatic and terrestrial plants but it does not seem to biomagnify in the food chain (Howe 1998). Boron does not biomagnify in aquatic food chains and has low potential to accumulate in aquatic organisms, as judged by studies in the San Joaquin River, California, and its tributaries (Saiki et al. 1993). Marine and freshwater plants, fishes, and invertebrates concentrated boron from the medium by factors of less than 100, suggesting that biota is not a significant removal mechanism of boron from water (USPHS 1991). Boron concentrations in livers of birds collected from Baja California, Mexico, in 1986 were highest in the seed-eating mourning dove *Zenaida macroura* (maximum 28.5 mg B/kg FW) and lowest (maximum 8.7 mg B/kg FW) in fish-eating and omnivorous species (Mora and Anderson 1995).

Boron occurred at high concentrations in plants, insects, and fish at Kesterson National Wildlife Refuge in California — the recipient of contaminated agricultural drainwater — when compared to a nearby control area (Table 29.3) (Ohlendorf et al. 1986; Schuler 1987). The authors indicated that little is known about the effect of boron ingestion on bird reproduction, although both boric acid and borax produced mortality and teratogenic development when injected into eggs. Studies on the effects of boron on waterfowl growth, physiology (Hoffman et al. 1990), and reproduction (Smith and Anders 1989) are discussed later.

Terrestrial plants, especially nuts and some fruits and vegetables, are rich sources of boron (Table 29.3). Honey is another good source of boron, and concentrations up to 7.2 mg/kg dry weight have been reported (Nielsen 1986). Boron concentrations are also elevated in marine plants, zooplankton, and corals, but are low in fish and certain marine invertebrates (Table 29.3). No data were found on boron levels in terrestrial mammalian wildlife. The average daily intake of boron in humans ranges between 1 and 25 mg; however, populations residing in areas of the western

United States with natural boron-rich deposits may be exposed to higher-than-average levels of boron (USPHS 1991). Boron intake from drinking water is highly variable and dependent on the geographic source, the quantity of water consumed, and the water sources used to bottle other beverages (Coughlin 1998). Current estimates of boron in domestic diets for normal human adults is about 1 mg daily; however, toddlers age 2 years consumed 3.7 times more boron than mature males when adjusted for body weight (Meacham and Hunt 1998). There is great variability in the human diet, and people from different countries have different sources of dietary boron. In the United States, for example, major sources of dietary boron include coffee, milk, orange juice, peanut butter, wine, pinto beans, and other juices and fruits. By contrast, in Mexico and Kenya, major sources of dietary boron include corn, kidney beans, maguey (an alcoholic drink from fermented bananas), cactus, and mangoes (Rainey and Nyquist 1998). In rats, increasing concentrations of boron in drinking water were associated with increasing tissue boron concentrations, plasma testosterone, and Vitamin D, and a decrease in HDL cholesterol (Samman et al. 1998). Data for humans and domestic animals indicate that boron levels are elevated in bony tissues, but are always less than 0.6 mg B/kg fresh weight or 1.5 mg/kg dry weight in other tissues examined (Table 29.3).

Table 29.3 Boron Concentrations in Field Collections of Selected Species of Plants and Animals (Values shown are in mg B/kg fresh weight [FW], dry weight [DW], or ash weight [AW].)

Ecosystem, Organism, and Other Variables	Concentration (mg/kg)	Reference[a]
TERRESTRIAL PLANTS		
Cereal grains	1–5 DW	1, 2, 19
Box thorn, *Lycium andersonii*		
Stem	7–26 DW	8
Leaf	26–163 DW	8
Root	25–74 DW	8
Prunes, raisins, dates	9–27 DW	2
Tropical fruits	Usually <10 FW	1
Nuts	16–23 DW	2
Vegetables	Usually >13 DW	2
Angiosperms	50 DW	8
Gymnosperms	63 DW	8
Pteridophytes	77 DW	8
Beet, *Beta vulgaris*	76 DW	19
Dandelion, *Taraxacum* sp.	80 DW	19
Sagebrush, *Artemisia tridentata*		
On high B soil		
Whole	Max. 250 DW	8
Leaf	Max. 156 DW	8
Stem	Max. 54 DW	8
Quince, *Cydonia* sp.	160 FW	19
Apple, pear, tomato, red pepper	440–1250 DW	2
FRESHWATER ORGANISMS		
Lake trout, *Salvelinus namaycush*, muscle	0.2–0.6 FW	8
Cattail, *Typha latifolia*, whole	15–30 DW	8
Aquatic macrophytes		
22 species	Usually <20 DW; mean 11.3 (1.2–100) DW	4
Various	2–19 DW	5
Waterweed, *Elodea* spp., whole	18–44 DW	8
Pondweed, *Potamogeton* spp., whole	18–170 DW	8
Yellow pond lily, *Nuphar* spp., whole	23–31 DW	8
Watermilfoil, *Myriophyllum* spp., whole	25–54 DW	8

Table 29.3 (continued) Boron Concentrations in Field Collections of Selected Species of Plants and Animals (Values shown are in mg B/kg fresh weight [FW], dry weight [DW], or ash weight [AW].)

Ecosystem, Organism, and Other Variables	Concentration, (mg/kg)	Reference[a]
Kesterson National Wildlife Refuge, California, contaminated with irrigation drainwater		
1983		
Aquatic plants	382 (270–510) DW	3
Aquatic insects	45 (36–54) DW	3
Mosquitofish, *Gambusia affinis*, whole	11 (8–20) DW	3
1984		
Wigeongrass, *Ruppia maritima*		
Whole	371 (120–780) DW	9, 13
Seeds	1860 (450–3500) DW	9, 13
Filamentous algae	501 (390–787) DW	9, 13
Aquatic insects	43–186 (22–340) DW	9
1985		
Water	20.0 FW	16
Aquatic plants, whole	340–1800 DW	16
Aquatic insects, whole	Max. 280 DW	16
Mosquitofish, whole	20.6 DW; Max. 32.0 DW	16
Volta Wildlife area, California (control area)		
1983		
Aquatic plants	34 DW	3
Aquatic insects	13 (6–35) DW	3
Mosquitofish, whole	2.8 (Max. 3.6) DW	3
1984		
Widgeongrass		
Whole	100 (37–540) DW	9
Seeds	36 (32–43) DW	9
Filamentous algae	85 (64–140) DW	9
Aquatic insects	12–32 (7–47) DW	9
1985		
Water	1.8 FW	16
Aquatic plants, whole	220–520 DW	16
Aquatic insects, whole	Max. 60 DW	16
Mosquitofish, whole	4.9 DW; Max. 7.4 DW	16
Western San Joaquin Valley, California, contaminated with irrigation drainwater		
Vegetation and seeds	Max. 3390 DW	10
Clams, 2 species, muscle	Max. 9.3 FW	10
Bluegill, *Lepomis macrochirus*, whole	<0.8–1.9 FW; Max. 3.9 FW	10
Common carp, *Cyprinus carpio*, whole	0.5–5.7 FW; Max. 6.2 FW	10
San Joaquin River, California; 1987; maximum concentrations		
Water	2.9 FW	17
Sediments	6.9 DW	17
Detritus	190.0 DW	17
Filamentous algae, whole	280.0 DW	17
Plankton	47.0 DW	17
Chironomid larvae, whole	27.0 DW	17
Amphipods, whole	22.0 DW	17
Crayfish, whole	23.0 DW	17
Mosquitofish, whole	8.4 DW	17
Bluegill, whole	7.9 DW	17
Largemouth bass, whole	2.0 DW	17
MARINE ORGANISMS		
Seaweeds, whole, Japan, 41 species	106 (16–319) DW; 762 (231–3038) AW	6
Marine algae	4–120 DW	8

Table 29.3 (continued) Boron Concentrations in Field Collections of Selected Species of Plants and Animals (Values shown are in mg B/kg fresh weight [FW], dry weight [DW], or ash weight [AW].)

Ecosystem, Organism, and Other Variables	Concentration, (mg/kg)	Reference[a]
Zooplankton	18–216 DW	7
Ctenophore, *Beroe cucumis*	115 AW	6
Corals, 34 species		
Deep open ocean	50–85 DW	6
Shallow open ocean	65–100 DW	6
Shallow coastal zone	40–110 DW	6
Tunicate, *Salpa fusiformis*, whole	50 AW	6
Chaetognath, *Sagitta elega*	130 AW	6
Dungeness crab, *Cancer magister*, whole	1.8 (0.9–3.3) FW	6, 7
Molluscs, bivalves		
Soft parts, 11 species	1.6–4.5 FW	6
Soft parts, British Columbia		
Clams, 8 species	0.9–5.3 FW	7
Oysters, 2 species	3.1–4.0 FW	7
Mussels, 2 species	2.0–5.5 FW	7
Octopus, *Polypus bimaculatus*, whole	1.3 FW	7
Sockeye salmon, *Oncorhynchus nerka*		
Soft tissues	0.5–0.7 FW	6, 7
Bone	1.5 (1.1–4.4) FW	6, 7
Anchovetta, *Cetengraulis mysticetus*, whole	3.3–3.8 AW	8
Yellowfin tuna, *Thunnus albacares*		
Muscle	39.0 AW	8
Whole	9.0 AW	8
Eyeball	5.6 AW	8
Spleen	3.3 AW	8
Gill	1.8 AW	8
Heart	1.5 AW	8
Harbor seal, *Phoca vitulina*		
Blood	2.0 FW	8
Spleen	0.5 FW	8
Muscle	0.3 FW	8
Liver	0.2 FW	8
Heart	0.1 FW	8
Kidney	0.01 FW	8
BIRDS		
Aquatic birds; central California, 1985–88; livers; South Grasslands (contaminated) vs. North Grasslands; freshwater substituted for irrigation drainwater in fall 1985		
Cinnamon teal, *Anas cyanoptera*		
1987	4.1 DW vs. not detected (ND)	11
1988	3.0 DW vs. ND	11
Mallard, *Anas platyrhynchos*		
1985	7.4 DW vs. 6.6 DW	11
1988	29.0 DW vs. 1.8 DW	11
Gadwall, *Anas strepera*		
1987	6.9 DW vs. ND	11
1988	4.3 DW vs. 2.5 DW	11
American coot, *Fulica americana*		
1985	27.0 DW vs. 11.0 DW	11
1987	9.1 DW vs. 5.3 DW	11
1988	8.5 DW vs. 3.8 DW	11
Mexicali Valley, Baja California; 1986; livers; 5 species	1.2–28.5 FW	12
Willet, *Catoptrophorus semipalmatus*; 1994; San Diego Bay; sediments vs. stomach contents		
Naval Air Station	<10 DW vs. 3.9 DW	20
Tijuana Slough National Wildlife Refuge	21 DW vs. 6.5 DW	20

Table 29.3 (continued) Boron Concentrations in Field Collections of Selected Species of Plants and Animals (Values shown are in mg B/kg fresh weight [FW], dry weight [DW], or ash weight [AW].)

Ecosystem, Organism, and Other Variables	Concentration, (mg/kg)	Reference[a]
Sandhill crane, *Grus canadensis*; Nebraska; 1989–90; liver; found dead after collision with powerline	1.2 (1.0–2.4) DW	13
Whooping crane, *Grus americana*		
Male; 4-months old; Colorado; 1985		
Muscle	4.9 FW	14
Liver	10.0 FW	14
Egg; NWT, Canada		
1986	4.7–5.0 FW	14
1987	1.0 FW	14
1989	1.0 FW	14
Aransas/Wood Buffalo National Park, Canada; 1989		
Male; 1.5 years old		
Muscle	<0.05 FW	14
Liver	<1.0 FW	14
Female; 4-years old		
Muscle, liver	Not detected	14
Avocet, *Recurvirostra americana*; south Central Valley of California; 1991		
Contaminated ponds (29–109 mg B/L)		
Egg	8.3–10.5 DW	15
Kidney	1.9 DW	15
Liver	4.5–10.3 DW	15
Reference ponds (<1.0 mg B/L)		
Egg	8.5 DW	15
Kidney	Not detected	15
Liver	9.6 DW	15
MAMMALS, TERRESTRIAL		
Animal tissues, blood, urine		
Normal	Usually <10 FW	19
Poisoned	Max. 2000 FW	19
Human, *Homo sapiens*		
Teeth	18.2 (0.5–69) DW	1
Nails	16 (7.5–83) FW	19
Rib	6.2–10.2 AW	1
Hair	4.3 (0.8–10.2) FW	19
Kidney, lung, lymph nodes	0.6 FW	1, 2
Urine	0.75 (0.2–2.9) FW	19
Blood	0.06–0.4 FW	1, 2, 19
Serum	0.02–0.2 FW	2, 19
Muscle	0.1 FW	1, 2
Testes	0.09 FW	2
Milk	0.06–0.08 FW	2
Brain	0.06 FW	1, 2
Diet (foods with the highest boron concentrations)		
Avocado	11.1–14.3 FW	21
Peanut butter	5.9–14.5 FW	21
Prune juice	5.2–5.6 FW	21
Chocolate powder	4.3 FW	21
Wine	3.6 FW	21
Grape juice	3.4–3.7 FW	21
Pecans	2.6–6.6 FW	21
Animal meat for human consumption	0.2 DW	2
Animal muscle and organs	0.5–1.5 DW	1
Milk, cow	0.2–1.0 FW	1, 19
Dairy products	1.1 DW	2

Table 29.3 (continued) Boron Concentrations in Field Collections of Selected Species of Plants and Animals (Values shown are in mg B/kg fresh weight [FW], dry weight [DW], or ash weight [AW].)

[a] **1,** NAS 1980; **2,** Nielsen 1986; **3,** Ohlendorf et al. 1986; **4,** Boyd and Walley 1972; **5,** Ahl and Jonsson 1972; **6,** Eisler 1981; **7,** Thompson et al. 1976; **8,** Jenkins 1980; **9,** Schuler 1987; **10,** Klasing and Pilch 1988; **11,** Paveglio et al. 1992; **12,** Mora and Anderson 1995; **13,** Fannin 1991; **14,** Lewis et al. 1991; **15,** Fairbrother et al. 1994; **16,** Hothem and Ohlendorf 1989; **17,** Saiki et al. 1993; **18,** Whitworth et al. 1991; **19,** Moseman 1994; **20,** Hui and Beyer 1998; **21,** Meacham and Hunt 1998.

29.4 EFFECTS

29.4.1 General

Boron is essential for the growth of higher plants and has been applied to boron-deficient soils for at least 50 years to improve yields of many crops. Phytotoxic levels of boron usually occur as a result of human activities, such as boron-contaminated irrigation waters and excess applications of boron-rich fertilizers, sewage sludges, and flyashes. Boron compounds at comparatively high concentrations are used to control pestiferous insects through direct biocidal action, enhancement of disease sensitivity, or use as a chemosterilant.

Representative species of aquatic plants, invertebrates, fishes, and amphibians can usually tolerate up to 10 mg B/L medium for extended periods without adverse effects, although it has been suggested that concentrations greater than 0.1 mg B/L may ultimately affect reproduction in rainbow trout (*Oncorhynchus mykiss*), and greater than 0.2 mg/L may impair survival of other fish species. In waterfowl, diets that contain 30 or 100 mg B/kg fresh weight adversely affect growth rate. Elevated tissue residues were recorded in ducks fed diets containing between 100 and 300 mg B/kg, and reduced survival occurred at dietary levels of 1000 mg B/kg. Boron is a potent avian teratogen when injected directly into embryos during the first 96 h of development. In mammals, the lethal dose of boron, as boric acid, varies according to species, and usually ranges between 210 and 603 mg B/kg body weight (BW); early development stages are especially sensitive. Excessive boron consumption adversely affects growth and reproduction in sensitive species of mammals (i.e., >1000 mg B/kg diet, >15 mg B/kg BW daily, >1.0 mg B/L drinking water, or >3 g B/kg BW single dose on the first day of pregnancy). Boron is not considered essential for mammalian growth but does protect against fluorosis and bone demineralization.

29.4.2 Terrestrial Plants

Boron was accepted as being an essential micronutrient for higher plants in 1923, with toxicity owing to excess boron much less common in the environment than boron deficiency (Howe 1998; Samman et al. 1998). The role of boron in nutrition and toxicity of terrestrial crops and forest trees has been reviewed extensively by Eaton (1944), USEPA (1975), Gupta (1979, 1983), Gupta and Macleod (1982), Pilbeam and Kirkby (1983), Gupta et al. (1985), Eisler (1990), and Stone (1990). It is generally agreed that boron is essential for the growth of higher plants and some species of fungi, bacteria, and algae, and that excess boron is phytotoxic. It is also agreed that plants vary greatly in their sensitivity to B toxicity (Boyd and Walley 1972; USEPA 1975; Birge and Black 1977; Goldberg and Glaubig 1986; Dear and Lipsett 1987; USPHS 1991). The exact mode of action of boron is unknown; however, its complexing ability facilitates the movement of sugars and other materials, and it is involved in cell wall bonding, conversion of glucose-1-phosphate to starch, and metabolism of nucleic acids (Sprague 1972; Gupta et al. 1985; Goldbach and Amberger 1986). Blevins and Lukaszewski (1994) suggest that boron toxicity to plants is attributed, in part, to interactions between borates and divalent cations like manganese, resulting in altered metabolic pathways of allantoate aminohydrolase in the case of manganese. The boron level in plants depends on the content and availability of soil boron, season, disease state, inherent species or variety differences, and interactions with other substances (USEPA 1975; Gestring and Soltanpour 1987). Most of the plant-available boron comes from the decomposition of soil organic matter and from boron adsorbed and precipitated onto soil surface particles. However, soil solution boron is the most important form, and plants take it up directly from this source (Gupta et al. 1985). Boron availability to plants is strongly associated with the hot-water-soluble fraction. This usually ranges from 0.4 to 4.7% of the total boron. The highest percentage occurs in fine-textured soils, and the lowest occurs in coarse-textured soils (Gupta and Macleod 1982). Uptake of boron by plants is about 4 times higher at pH 4 than at pH 9, highest in the temperature range 10 to 30°C, and higher with increased light intensity (Sprague 1972).

For the past 60 years, boron has been applied to B-deficient soils to improve crop yields of grains, fruits, vegetables, legumes, pine trees, tobacco, cotton, sunflowers, and peanuts (USEPA 1975; Gupta 1979; Lipsett et al. 1979; Shorrocks and Nicholson 1980; Hopmans and Flinn 1984; Gupta and Cutcliffe 1985; Willett et al. 1985; Combrink and Davies 1987; Dear and Lipsett 1987; Mozafar 1987; Nuttall et al. 1987; Rerkasem et al. 1988; Stone 1990). Boron is unique among the essential micronutrients because it is the only element normally present in soil solution as a non-ionized molecule over the pH range suitable for plant growth (Gupta 1979). Boron deficiency in plants is widespread and has been reported in one or more crops in at least 43 states, almost all Canadian provinces, and many other countries (Gupta 1979). Boron deficiency in crops is more widespread than that of any other micronutrient (Gupta et al. 1985). It is more likely to occur in light-textured acid soils in humid regions because of boron's tendency to leach. However, deficiency may also occur in heavy-textured soils with high pH because boron is readily adsorbed under these conditions (Gupta et al. 1985). Deficiency signs include browning and spotting of leaves, chlorosis, abnormal thickening of cell walls, increased production of indoleacetic acid, accumulation of polyphenolic compounds, changes in membrane permeability, necrosis, and finally death (USEPA 1975; Gupta 1979). Visible signs of deficiency in corn are accentuated by calcium deficiency, and are least evident when calcium is added to excess. Under conditions of boron and calcium deficiency combined, yields are low, and starch phosphorylase activity in corn leaves increases markedly, as does that of ribonuclease and polyphenol oxidase (Chatterjee et al. 1987). Interaction effects were also measured between boron and potassium in alfalfa (Walker et al. 1987). Boron deficiency is usually corrected by application of 0.5 to 3 kg B/ha, depending on crop and formulation (Gupta 1979). Adding boron promotes translocation rate of photosynthetic products and increases CO_2 incorporation into free amino acids (Gupta 1979).

Boron toxicity has been reported in many species of grasses, fruits, vegetables, grains, trees, and other terrestrial plants (Gupta and Macleod 1982; Dye et al. 1983; Glaubig and Bingham 1985; Francois 1986; Nicholaichuk et al. 1988; Stone 1990) (Table 29.4). Toxic levels generally do not occur on agricultural lands unless boron compounds have been added in excessive quantities, such as with fertilizer materials, irrigation water, sewage sludge, or coal ash (Gupta and Macleod 1982; Gestring and Soltanpour 1987). Boron-contaminated irrigation water is one of the main causes of boron toxicity to plants. The continued use and concentration of boron in the soil due to evapotranspiration is the reason for eventual toxicity problems (Gupta et al. 1985). Borates have also been used as herbicides for complete kill of vegetation at application rates of 2244 kg/ha (equivalent to 2000 pounds/acre) (Sprague 1972). Borates are frequently applied at elevated concentrations (i.e., >2 g/kg soil) in combination with organic pesticides in order to produce bacteriostatic effects. The resultant B-produced reduction in microbial degradation of the pesticide effectively extends the pesticide's biocidal properties (Sprague 1972). In some cases, cooling tower drift from geothermal steam containing boron may cause foliar boron toxicity in the vicinity of generating units (Glaubig and Bingham 1985; Sage et al. 1989).

Boron poisoning in plants is characterized by stunted growth, leaf malformation, browning and yellowing, chlorosis, necrosis, increased sensitivity to mildew, wilting, and inhibition of pollen germination and pollen tube growth (USEPA 1975; Glaubig and Bingham 1985; Mitchell et al. 1987).

In barley (*Hordeum vulgare*), for example, excess boron caused decreased growth and grain yield, elevated residues in leaves, and increased rate of leaf senescence (Riley 1987; Table 29.4). Barley grown on zinc-deficient soils tended to accumulate boron up to 2.5 times within 7 days; a similar pattern was evident for excess phosphorus (Graham et al. 1987). Thus, under conditions of marginally high boron in the rooting zone, low zinc, and high phosphorus, boron may accumulate to toxic levels in plants (Graham et al. 1987). Toxic effects in plants — including leaf injury — were observed in 26% of plants at or below substrate concentrations that resulted in greatest growth, indicating considerable overlap between injurious and beneficial effects of boron in plants (Eaton 1944). In general, deficiency effects in plants were evident when boron concentrations in soil solution were <2 mg/L; optimal growth occurred at 2 to 5 mg/L; and toxic effects were evident at 5 to 12 mg B/L (Gupta et al. 1985). However, there is considerable variation in resistance to boron between species (Table 29.5). Sensitive species are known to include citrus, stone fruits, and nut trees; semitolerant species include cotton, tubers, cereals, grains, and olives; tolerant species usually include most vegetables (Gupta et al. 1985).

Table 29.4 Boron Toxicity to Some Terrestrial Plants

Species, Dose, and Other Variables	Effect (Reference[a])
BIGLEAF MAPLE, *Acer macrophyllum*	
0.9–5.4 mg B/L, in saturated soil extracts	Reduced growth; >25% foliar damage; leaf residues of 76–324 mg B/kg ash weight (1)
MADRONE, *Arbutus menziesii*	
2.2–5.4 mg B/L, in saturated soil extracts	Growth inhibition; >25% foliar damage; leaf residues of 216–540 mg B/kg ash weight (1)
BEET, *Beta vulgaris*	
Soil B solutions	
5 mg/L	Optimal growth (2)
15 mg/L	Injury evident (2)
BROCCOLI, *Brassica oleracea italica*	
Grown in nutrient solutions containing	
0.08 mg B/L	Chlorophyll levels and net photosynthetic rates were significantly lower than those grown in 0.41–0.81 mg B/L solutions (3)
4.1 and 8.1 mg B/L	Leaf damage evident; lower chlorophyll levels and lower net photosynthetic rate than 0.4 and 0.8 mg B/L groups (3)
RHODES GRASS, *Chloris gayana*	
Grown in flyash containing 3 mg hot-water-soluble B/L	Toxic. Residues >149 mg/kg DW (4)
LEMON, *Citrus limonia osbeck*	
Soil B concentrations	
0.03–0.04 mg/L	Optimal growth (2)
1 mg/L	Injury evident (2)
SOYBEAN, *Glycine max*	
Grown in soils amended with scrubber sludge residues (4.1 g B/kg) from coal-fired power plant for 2–3 years	Higher sludge B levels of 2 mg B/kg soil surface at year 1, and 1.2 mg B/kg at year 2 produced signs of B toxicity, including decreased growth and elevated residues (>83 mg/kg DW) in leaf and (>47 mg/kg DW) in seeds (5)
SUNFLOWER, *Helianthus annuus*	
50 mg B/L growth medium	Adversely affects phospholipid composition and synthesis in roots and microsomes from seedlings by inhibition of choline phosphotransferase (6)
10 mg B/L growth medium	Tolerated level (6)
BARLEY, *Hordeum vulgare*	
Residues, in mg B/kg DW	
0.5–1.0 in soil	Residues of 46–100 mg/kg DW in leaves (7)
30 in shoots	Damage to older leaves (8)
50–70 in shoots	Reduction of 10% in dry weight of shoots (7)
60–80 in leaf	Toxicity evident (8)
80–120 in shoots	Toxic signs, but no yield reductions (8)
120–130 in shoots	Grain yield reduced 10% (8)

Table 29.4 (continued) Boron Toxicity to Some Terrestrial Plants

Species, Dose, and Other Variables	Effect (Reference[a])
ALFALFA, *Medicago sativa*	
850–975 mg B/kg dry weight plant	Reduced yield (9)
RICE, *Oryza sativa*	
Whole plant B residues	
38 mg/kg DW	No signs of toxicity (10)
43–55 mg/kg DW	Signs of toxicity evident (10)
Soil waters	
2.5–5 mg B/L	Toxic (10)
FRENCH BEAN, *Phaseolus vulgaris*	
Grown in flyash containing 3 mg hot-water-soluble B/L	Toxic. Residues >209 mg/kg DW (4)
Residues in whole plant, in mg B/kg dry weight	
9–12	Slow flowering and pod formation; general yellowing of tips (11)
>125	Reduced growth; burned older leaves dark brown (11)
DIGGER PINE, *Pinus sabiniana,* seedlings	
13–17 mg B/L in saturated soil extracts	Growth reduction; foliar damage >25%; needle residues 1242–1512 mg B/kg ash weight (1)
PEAR, *Pyrus communis*	
82–164 kg B/ha applied to soil around pear trees in a nonirrigated orchard over a 6-year period	Toxicity observed during application and during 4 years postapplication. Toxicity was associated with residues, in mg B/kg DW, of 90–115 in blossom clusters and 45–55 in fruit. Within 5 years postapplication, soil B levels were <2 mg/kg, and all visible signs of toxicity had disappeared (12)
VEGETATION, various species	
2244 kg borates/ha (2000 lbs/acre)	Total kill of most species (2)
CORN, *Zea mays*	
Soil B concentrations	
1 mg/L	Optimal growth (2)
5 mg/L	Injury evident (2)
Plant residues	
>98 mg B/kg DW	Marginal burning and dark brown tips of older leaves (11)

[a] **1,** Glaubig and Bingham 1985; **2,** Sprague 1972; **3,** Petracek and Sams 1987; **4,** Aitken and Bell 1985; **5,** Ransome and Dowdy 1987; **6,** Belver and Donaire 1987; **7,** Riley 1987; **8,** Kluge and Podlesak 1985; **9,** Gestring and Soltanpour 1987; **10,** Cayton 1985; **11,** Gupta 1983; **12,** Crandall et al. 1981.

Table 29.5 Boron Concentrations in Soil Water Associated with Optimal Growth and Plant Injury

	Boron Concentration in Soil Water (mg/L)	
Plant Category	**Optimal Growth**	**Plant Injury**
Sensitive species	Trace–1	Usually 1–5
Semitolerant species	Usually 1–5	Usually 5–15
Tolerant species	Usually 5–10	Usually 5–25

Data from Sprague, R.W. 1972. The ecological significance of boron. United States Borax and Chemical Corp., Los Angeles, 58 pp.

29.4.3 Terrestrial Invertebrates

Relatively high concentrations of boron compounds are used to control fruitflies, cockroaches, gypsy moth larvae, houseflies, and woodboring insects (Sprague 1972; USEPA 1975; Table 29.6). Boric acid is an effective stomach poison for several insect species, including German cockroaches (*Blattella germanica*), that are unable to detect the presence of boric acid (USEPA 1975). Insect infestation of wood and other substrates can be prevented by pretreatment with boric acid or borax at

Table 29.6 Lethal and Sublethal Effects of Boron on Terrestrial Invertebrates

Organism, Dose, and Other Variables	Effect (Reference[a])
FRUITFLY, *Anastrepha ludens*	
Baits containing cottonseed hydrolysate and borax	Reduced infestation in oranges by 68%, and in mangoes by 98% (1)
HONEY BEE, *Apis mellifera*	
8.7 mg B/L syrup (50 mg boric acid/L)	No effect on survival (2)
17.5 mg B/L syrup (100 mg boric acid/L)	Fatal to about 50% (2)
GERMAN COCKROACH, *Blattella germanica*	
Baits containing 25% boric acid plus honey	Population reduction of 50% in about 5 days, 80% in 4 weeks, and 98% in 6–9 months (3)
Sugar diet containing	
11% boric acid	44% dead in 72 hours (1)
25% boric acid	79% dead in 72 hours (1)
50% boric acid	80% dead in 72 hours (1)
100% boric acid	91% dead in 72 hours (1)
Baits containing 20% boric acid	88% population reduction in 2 weeks; 92 to 95% reduction in 4–12 weeks (4)
GYPSY MOTH, *Lymantria dispar*, larvae	
0.25% boric acid solution (436 mg B/L)	No effect on gypsy moth nucleopolyhedrosis virus (NPV) (5)
0.5% boric acid	Enhanced NPV activity by 2-fold (5)
1% boric acid	Enhanced NPV activity by 11-fold (5)
HOUSEFLIES, *Musca domestica*	
250–5000 mg B/kg diet, as boric acid	Inhibits reproduction (2)
Isobornyl thiocyanoacetate	
27.3 µg/fly	LD50 (1)
Aerosols, >2%	50% knockdown in 6 minutes (1)
AMERICAN COCKROACH, *Periplaneta americana*	
Baits containing 1.5% boric acid	All dead in 6 days (6)
WOODBORING INSECTS	
Common houseborer	
430 mg boric acid/m³ wood	Adequate wood protection (2)
Termites, 3 species	
>10,000 mg boric acid/m³ wood	Required for wood protection (2)

[a] **1,** USEPA 1975; **2,** Sprague 1972; **3,** Gupta and Parrish 1984; **4,** Wright and Dupree 1982; **5,** Shapiro and Bell 1982; **6,** Lizzio 1986.

doses of 0.25 to 0.55 kg/m^3 of wood (USEPA 1975). Boric acid and other boron compounds are effective chemosterilants of the cotton boll weevil (*Anthonomus grandis*) and houseflies (USEPA 1975).

29.4.4 Aquatic Organisms

Boron effects on aquatic plants are highly species specific (Glandon and McNabb 1978; Rao 1981) (Table 29.7). Borate, like silicate, is an essential micronutrient for the growth of aquatic plants, such as diatoms, and it seems that a chemical combination of both nutrients in the form of silicoborate may be required by certain diatoms (Antia and Cheng 1975). In aquatic plants, boron affects nucleic acid metabolism, carbohydrate biosynthesis and transport, and membrane integrity, and it interacts with growth substances (Frick 1985). Diatoms (*Cylindrotheca fusiformis*) cultured under B-deficient conditions stop dividing and swell in size despite increased photosynthetic rates. Boron-deficient diatoms accumulate rubidium, phenolic compounds, nitrates, and phosphates, and they show increased activity of various enzymes, especially glucose-6-phosphate dehydrogenase. However, respiratory adjustment is negligible until nutrient stress becomes irreversible in about 48 h (Smyth and Dugger 1980, 1981). Boron, under conditions of excess, alleviates nutrient deficiency in some phytoplankters and can cause temporal variations of phytoplankton composition in coastal waters (Rao 1981). Phytoplankton can tolerate up to 10 mg inorganic B/L in the absence of stress from pH adversity and nutrient deficiency, although higher borate concentrations up to 100 mg/L are expected to cause species redistribution by favoring the growth of some species and suppressing that of others (Antia and Cheng 1975; Table 29.7).

Available data for aquatic invertebrates and boron suggest that the no-observable-effect levels were 13.6 mg B/L for freshwater organisms and 37 mg B/L for marine biota (Table 29.7). Juvenile Pacific oysters (*Crassostrea gigas*) accumulated boron in relation to availability, but showed no prolonged retention following cessation of exposure (Thompson et al. 1976). At industrial discharge levels of about 1.0 mg B/L, no hazard is apparent to oysters and aquatic vertebrates (Thompson et al. 1976).

Boron may be an essential nutrient in several species of aquatic vertebrates. Insufficient boron (<3 μg B/L; 62 μg B/kg ration) interfered with the normal development of the South African clawed frog (*Xenopus laevis*) during organogenesis, and substantially impaired normal reproductive function in adult frogs (Fort et al. 1998). Impaired growth of rainbow trout (*Oncorhynchus mykiss*) embryos was documented at <90 μg B/L, and death of zebrafish (*Brachydanio rerio*) embryos at <2 μg B/L (Rowe et al. 1998).

The most sensitive aquatic vertebrates tested for which data are available were coho salmon (*Oncorhynchus kisutch*), with an LC50 (16-day) value of 12 mg B/L in seawater, and sockeye salmon (*O. nerka*), showing elevated tissue residues after exposure for 3 weeks in seawater containing 10 mg B/L (Table 29.7). Boron concentrations between 0.001 and 0.1 mg/L had little effect on survival of rainbow trout embryos after exposure for 28 days (Table 29.7). These low levels may represent a reduction in reproductive potential of rainbow trout, and concentrations more than 0.2 mg B/L may impair survival of other fish species, according to Birge and Black (1977). However, additional data are needed to verify these speculations. Birge and Black (1977) reported that concentrations of 100 to 300 mg B/L killed all species of aquatic vertebrates tested, that embryonic mortality and teratogenesis were greater in hard water than in soft water, but that larval mortality of fish and amphibians was higher in soft water than in hard water, and that boron compounds were more toxic to embryos and larvae than to adults. Moreover, they found no measurable effect on boron toxicity to aquatic vertebrates in water temperature in the range of 13 to 29°C, dissolved oxygen between 6.4 and 10.3 mg/L, and pH between 7.5 and 8.5. Elevated boron concentrations of 50 to 100 mg B/L adversely affects the development of amphibian embryos (Laposata and Dunson 1998). In central Pennsylvania ponds, embryos from two species of salamanders (spotted salamander, *Ambystoma maculatum*; Jefferson salamander, *Ambystoma jeffersonianum*), the wood frog (*Rana sylvatica*), and the American toad (*Bufo americanus*) were exposed to wastewater effluents of 0, 50, or 100 mg B/L. At 50 and 100 mg B/L, there were significant increases in the frequency of deformed larvae and reduced hatching success (Laposata and Dunson 1998).

Table 29.7 Lethal and Sublethal Effects of Boron on Aquatic Organisms

Taxonomic Group, Organism, Compound, Dose, and Other Variables	Effect	Reference[a]
AQUATIC PLANTS		
Blue-green alga, *Anacystis nidulans*, boric acid, H_3BO_3		
0.01–4.0 mg B/L	Grows well in B-deficient media; growth neither stimulated nor inhibited at higher levels	1, 15
50 mg B/L	No effect on growth or organic constituents	2
75–100 mg B/L	Growth and chlorophyll content reduced; at 72 hours, photosynthetic pigments depleted	2
100 mg B/L	Decrease in protein content causing inhibition in nitrate uptake and nitrate reductase activity. Decreased chlorophyll content and photosynthesis inhibition within 72 h	2, 15

Table 29.7 (continued) Lethal and Sublethal Effects of Boron on Aquatic Organisms

Taxonomic Group, Organism, Compound, Dose, and Other Variables	Effect	Reference[a]
Green alga, *Chlorella pyrenoidosa*, boric acid		
10 mg B/L	No effect on growth or cell composition. Bioconcentration factor (BCF) of 4 after 7 days	3
50 mg B/L	BCF of 5 after 7 days	3
50–100 mg B/L	Altered cell division and amino acid activity after 72 h; reversible photosynthesis inhibition. Giant cells formed with increased nitrate and protein	4
100 mg B/L	BCF of 4.8 after 7 days	3
>100 mg B/L	100% inhibitory for cell division and biomass synthesis in 72 h	4
Duckweed, *Lemna minor*, boric acid		
Control media, 10–20 mg B/L, pH 5.0	Normal growth	5
100 mg B/L, pH 5.0	Growth inhibited; recovery on transfer to control media	5
20 mg B/L, pH 4.0	Residues of 93 mg B/kg FW vs. 63 in controls	5
20 mg B/L, pH 7.0	Growth inhibited. Residues of 257 mg/kg FW vs. 49 in controls	5
Marine algae, 19 species, boric acid		
5–10 mg B/L	No inhibitory effect on growth rate in 60 days; stimulatory to some species	6
10–50 mg B/L	Prolonged survival of peak populations of certain diatoms after growth cessation: *Bellerochea polymorpha* at 10 mg B/L, *Skeletonema costatum* at 50 mg B/L	6
50 mg B/L	Growth inhibition in 26% of species tested; adaptation and recovery by most species	6
100 mg B/L	Growth inhibition in 12 of 19 species tested; 8 species did not recover and died	6
Marine phytoplankton		
30 mg B/L, high nitrates, phosphates, silicates, and low temperatures	Increased primary production and carbon assimilation	7
30 mg B/L, low nutrients, high temperatures	Photosynthesis inhibited up to 62%	7
30 mg B/L, unialgal cultures, 5-days-old	Photosynthesis inhibition	7
As above, 14-days-old	Enhanced photosynthesis in certain species	7
INVERTEBRATES		
Sea urchin, *Anthocidaris crassispina*, embryos, boric acid		
37 mg B/L	Normal development	8
75 mg B/L	Fatal	8
Chironomid, *Chironomus decorus*, fourth instar		
20 mg/L	Growth rate reduced in 96 h	21
1376 mg/L	LC50 (48 h)	21
Cladoceran, *Daphnia magna*		
6.4 mg B/L	Highest concentration tested in 21-day exposure producing no measurable effect	9, 10
13.6 mg B/L	Lowest concentration tested in 21-day exposure causing reduction in number of broods, total young produced, mean brood size, and mean size	9, 10
27 mg B/L	LC14 (21 days)	10
53 mg B/L	LC50 (21 days)	10
54–200 mg B/L	No deaths in 48 h	9, 10
106 mg B/L	LC100 (21 days)	10

Table 29.7 (continued) Lethal and Sublethal Effects of Boron on Aquatic Organisms

Taxonomic Group, Organism, Compound, Dose, and Other Variables	Effect	Reference[a]
115–246 mg B/L	LC50 (48 h)	9, 10, 21
420 mg B/L	LC100 (48 h)	9
Mosquito larvae, 3 species, boric acid, mg/L		
250 (43.7 mg B/L)	LC97–LC99 through hatching	11
4000 (700 mg B/L)	LC100 (48 h), freshly-hatched	11
3000 (524 mg B/L)	LC100 (48 h), second instar	11
10,000 (1748 mg B/L)	LC100 (48 h), third instar	11
16,000 (2797 mg B/L)	LC100 (48 h), pupae	11
VERTEBRATES		
Amphibians; 3 species; eggs; exposed to 0, 50 or 100 mg B/L		
Jefferson salamander, *Ambystoma jeffersonianum*; spotted salamander, *Ambystoma maculatum*; wood frog, *Rana sylvatica*	At 50 and 100 mg B/L, there was a significant increase in frequency of deformed larvae	23
American toad, *Bufo americanus*	Reduced hatching success at 100 mg B/L	23
Zebrafish, *Brachydanio rerio*; exposed for 6 months from embryos to adults		
0.002 mg/L	Embryonic death	25
>9.2 mg/L	Adult death	25
Fowler's toad, *Bufo fowleri*, embryos, through day 4 posthatch		
Boric acid		
Soft water, 50 mg $CaCO_3$/L		
25 mg B/L	LC1 (7.5 days)	12
145 mg B/L	LC50 (7.5 days)	12
Hard water, 200 mg $CaCO_3$/L		
5 mg B/L	LC1 (7.5 days)	12
123 mg B/L	LC50 (7.5 days)	12
Toad, *Bufo vulgaris*, embryos		
874 mg B/L, as boric acid. Exposure for 24 h from 2-cell stage to tailbud stage	Malformations included edema, microcephalia, short tail, and suppressed forebrain development	11
Goldfish, *Carassius auratus*, embryos, through day 4 posthatch		
Boric acid		
Soft water		
0.6 mg B/L	LC1 (7 days)	12
46 mg B/L	LC50 (7 days)	12
Hard water		
0.2 mg B/L	LC1 (7 days)	12
75 mg B/L	LC50 (7 days)	12
Borax, $Na_2B_4O_7 \cdot 10H_2O$		
Soft water		
1.4 mg B/L	LC1 (7 days)	12
65 mg B/L	LC50 (7 days)	12
Hard water		
0.9 mg B/L	LC1 (7 days)	12
59 mg B/L	LC50 (7 days)	12
Endangered fishes, three species, Green River, Utah; boron tested as boric acid		
Bonytail, *Gila elegans*		
280 (226–347) mg B/L	LC50 (96 h), fry	17
552 (452–707) mg B/L	LC50 (96 h), juveniles	17
Colorado squawfish, *Ptychocheilus lucius*		
279 (216–360) mg B/L	LC50 (96h), fry	17
527 (430–667) mg B/L	LC50 (96 h), juveniles	17

Table 29.7 (continued) Lethal and Sublethal Effects of Boron on Aquatic Organisms

Taxonomic Group, Organism, Compound, Dose, and Other Variables	Effect	Reference[a]
Razorback sucker, *Xyrauchen texanus*		
233 (172–293) mg B/L	LC50 (96 h), fry	17
279 (216–360) mg B/L	LC50 (96 h), juveniles	17
Mosquitofish, *Gambusia affinis*, adults		
Boric acid		
5600 mg/L (979 mg B/L)	LC50 (96 h)	12
Sodium borate		
3600 mg/L	LC50 (96 h)	12
Channel catfish, *Ictalurus punctatus*, embryos, through day 4 posthatch		
Boric acid		
Soft water		
0.5 mg B/L	LC1 (9 days)	12
155 mg B/L	LC50 (9 days)	12
Hard water		
0.2 mg B/L	LC1 (9 days)	12
22 mg B/L	LC50 (9 days)	12
Borax		
Soft water		
5.5 mg B/L	LC1 (9 days)	12
155 mg B/L	LC50 (9 days)	12
Hard water		
1.7 mg B/L	LC1 (9 days)	12
71 mg B/L	LC50 (9 days)	12
Bluegill, *Lepomis macrochirus*		
Boron trifluoride, BF_3		
15,000 mg B/L	LC50 (24 h)	12
Dab, *Limanda limanda*		
74.0 mg B/L	LC50 (96 h)	13
88.3 mg B/L	LC50 (24 h)	13
Largemouth bass, *Micropterus salmoides*; embryos exposed 2–4 h after fertilization through 8 days posthatch (about 11 days after fertilization)		
1.39 mg B/L	No observable effect	20
12.2 mg B/L	Lowest observable effect concentration (reduced survival, increased developmental abnormalities)	20
92 (84–100) mg B/L	LC50	20
Striped bass, *Morone saxatilis*		
Juveniles exposed continuously to full strength agricultural drainwater containing 48.8 mg B/L	All dead in 23 days	22
Coho salmon, *Oncorhynchus kisutch*		
Fry		
447 (356–561) mg B/L	LC50 (96 h), freshwater	18
600 (511–705) mg B/L	LC50 (96 h), brackish water	18
Underyearlings		
12 mg B/L	LC50 (283–384 h), seawater	14
113 mg B/L	LC50 (283–552 h), freshwater	14
Rainbow trout, *Oncorhynchus mykiss*		
Embryos exposed until hatch to various concentrations of borates		
<0.09 mg B/L	Impaired growth	25
>0.09–5.0 mg B/L	Dose-dependent increase in embryonic growth	25
>100 mg B/L	Lethal	25

Table 29.7 (continued) Lethal and Sublethal Effects of Boron on Aquatic Organisms

Taxonomic Group, Organism, Compound, Dose, and Other Variables	Effect	Reference[a]
Embryos, through day 4 posthatch		
Boric acid		
Soft water		
0.1 mg B/L	LC1 (28 days)	12
100 mg B/L	LC50 (28 days)	12
Hard water		
0.001 mg B/L	LC1 (28 days)	12
79 mg B/L	LC50 (28 days)	12
Borax		
Soft water		
0.07 mg B/L	LC1 (28 days)	12
27 mg B/L	LC50 (28 days)	12
Hard water		
0.07 mg B/L	LC1 (28 days)	12
54 mg B/L	LC50 (28 days)	12
Adults		
339 mg B/L	LC50 (48 h)	10,12, 16
350 mg B/L	No effect in 30 min	16
3500 mg B/L	All alive after 30 min, but in obvious distress	16
14,000 mg B/L	After exposure for 30 min, all recovered if placed in flowing boron-free water	16
Sockeye salmon, *Oncorhynchus nerka*		
10 mg B/L, exposure in seawater for 3 weeks	Maximum residues, in mg/kg FW, were 17 in bone, 12 in kidney, 10 in gill, 9 in liver, and 8 in muscle. Max. control values were always <1.0, except bone, which was 4.4 mg/kg FW	14
Chinook salmon, *Oncorhynchus tshawytscha*		
Fry		
600 mg B/L	LC50 (96 h), brackish water	18
725 mg B/L	LC50 (96 h), freshwater	18
Juveniles exposed to boron concentrations as high as 6.05 mg/L as boric acid for 90 days in freshwater	No increase in whole-body B concentrations	19
Juveniles exposed to full-strength agricultural drainwater from San Joaquin Valley, California, for 28 days. Drainwater had 48.8 (44–53) mg B/L	77% dead; survivors had reduced growth and 192 (190–200) mg B/kg DW whole body vs. 3.1 mg B/kg DW in controls	22
Leopard frog, *Rana pipiens*, embryos, through day 4 posthatch		
Boric acid		
Soft water		
13 mg B/L	LC1 (7.5 days)	12
130 mg B/L	LC50 (7.5 days)	12
Hard water		
22 mg B/L	LC1 (7.5 days)	12
135 mg B/L	LC50 (7.5 days)	12
Borax		
Soft water		
5 mg B/L	LC1 (7.5 days)	12
47 mg B/L	LC50 (7.5 days)	12
Hard water		
3 mg B/L	LC1 (7.5 days)	12
54 mg B/L	LC50 (7.5 days)	12
South African clawed frog, *Xenopus laevis*		
Embryos allowed to develop in media containing from <1 to 5000 μg/L	Developmental malformations at <3 μg B/L, but not at higher concentrations. Malformations included abnormal	24

Table 29.7 (continued) Lethal and Sublethal Effects of Boron on Aquatic Organisms

Taxonomic Group, Organism, Compound, Dose, and Other Variables	Effect	Reference[a]
	development of the gut, craniofacial region and eye, visceral edema, myotomes, and notochord	
Adults fed diets containing 62 or 1851 μg B/kg ration for 28 days, then mated and offspring cultured in media with various levels of B	Frogs fed diets containing 62 μg B/kg produced a greater number of necrotic eggs and abnormal embryos than those given 1851 μg B/kg ration. Embryos cultured in media with less than 4 μg B/L had a high incidence of malformations when compared to those raised in media of 4 μg B/L and higher	24

[a] **1,** Martinez et al. 1986b; **2,** Martinez et al. 1986a; **3,** Fernandez et al. 1984; **4,** Maeso et al. 1985; **5,** Frick 1985; **6,** Antia and Cheng 1975; **7,** Rao 1981; **8,** Kobayashi 1971; **9,** Gersich 1984; **10,** Lewis and Valentine 1981; **11,** USEPA 1975; **12,** Birge and Black 1977; **13,** Taylor et al. 1985; **14,** Thompson et al. 1976; **15,** Mateo et al. 1987; **16,** Sprague 1972; **17,** Hamilton 1995; **18,** Hamilton and Buhl 1990; **19,** Hamilton and Wiedmeyer 1990; **20,** Black et al. 1993; **21,** Maier and Knight 1991; **22,** Saiki et al. 1992; **23,** Laposata and Dunson 1998; **24,** Fort et al. 1998; **25,** Rowe et al. 1998.

29.4.5 Birds

Boron stimulated growth in Vitamin D_3-deficient chicks. Supplemental dietary boron alleviated or corrected cholecalciferol deficiency-induced elevations in plasma glucose concentrations in chicks (Hunt 1994). There is no need to supplement the diets of laying hens with boron, provided that basal diets contained about 11 mg B/kg ration (Qin and Klandorf 1991).

Boron is a potent teratogen to domestic chicken embryos when injected into eggs. Injection of boron into the yolk sac of chicken embryos during the first 96 h of development with 1.0 to 2.5 mg of boric acid — equivalent to 3.2 to 8.0 mg B/kg fresh weight egg (55 g egg) — produced a wide range of developmental abnormalities (Table 29.8). Several compounds are known to counteract B-induced avian developmental abnormalities, or to reduce the frequency of malformations, although the mode of action is unclear. These compounds include sodium pyruvate, to counteract rumplessness (Landauer 1952); nicotinamide, to decrease frequency of facial defects (Landauer 1952) and melanin formation (Landauer 1953c); and riboflavin, which greatly reduced the teratogenic effects of boric acid (Landauer 1952, 1953a, 1953b; Landauer and Clark 1964). Other polyhydroxy compounds, such as D-ribose, pyridoxine hydrochloride, and D-sorbitol hydrate, also reduced or abolished boric acid-induced teratogenicity in chick embryos (Landauer 1953b).

High concentrations of boron have been found in the San Joaquin Valley of California in irrigation drainwater and in aquatic plants consumed by waterfowl. Measured boron concentrations in that locale exceeded 20 mg/L in subsurface agricultural drainage waters, 400 mg/kg dry weight in widgeongrass (*Ruppia maritima*) and algae, 150 mg/kg dry weight in aquatic insects, 1860 mg/kg dry weight in some aquatic plants, and up to 3390 mg/kg dry weight in seeds consumed by waterfowl (Schuler 1987; Klasing and Pilch 1988; Smith and Anders 1989; Hoffman et al. 1990). At present, only selenium has been implicated as the cause of abnormal development among waterfowl in western areas impacted by irrigation drainwaters (Ohlendorf et al. 1986; Hoffman et al. 1988, 1990). However, studies by Smith and Anders (1989) and Hoffman et al. (1990) with mallards demonstrate that dietary boron concentrations well below levels that can occur in the environment represent a toxicological hazard that has not been considered in the management of agricultural drainwater. For example, dietary concentrations of 300 to 400 mg B/kg of feed on a fresh weight basis — substantially lower than boron levels reported in the vicinity of some western wildlife refuges contaminated by agricultural drainwater — adversely affect mallard growth, behavior, and brain biochemistry and are often associated with elevated tissue boron levels (Table 29.8). Dietary levels of 100 mg B/kg fresh weight result in reduced growth of female mallard ducklings (Hoffman et al. 1990), and diets containing as little as 30 mg B/kg fresh weight fed to mallard adults adversely affected the growth rate of their ducklings (Smith and Anders 1989). Resource managers must now consider boron, as well as selenium, and their possible interactions, as a toxic hazard to wildlife populations throughout areas of the western United States (Smith and Anders 1989).

Table 29.8 Lethal and Sublethal Effects of Boron on Birds

Species, Dose, and Other Variables	Effect (Reference[a])
MALLARD, *Anas platyrhynchos*	
Adults fed diets containing various concentrations of B, as boric acid, for 3 weeks, then mated. Resultant ducklings continued on same diets for 21 days. Data collected on reproduction, survival, residues, and histopathology when ducklings were age 21 days.	
8 mg B/kg diet fresh weight (controls). Diets contained about 10% moisture	Boron residues in egg, liver, and brain of adults and ducklings were always <3 mg B/kg dry weight (8)
30 mg B/kg diet FW	Duckling weight gain reduced compared to controls. Residues in egg and duckling liver and brain about 3–4 mg B/kg DW; residues <3 in adult liver and brain (8)
300 mg B/kg diet FW	Duckling body weights at hatch significantly lower than controls; duckling weight gain reduced. Mean residues, in mg B/kg dry weight, were 13 in egg, 15 in adult liver (Max. 24), 17 in duckling liver (Max. 36), 14 in adult brain (Max. 24), and 19 in duckling brain (Max. 44) (8)
1000 mg B/kg diet FW	No observable effect on adults. No effect on egg fertility or shell thickness. Significantly reduced hatching success; duckling mortality through age 7 days significantly greater than controls, and body weight lower. Total number of 21-day-old ducklings produced per female and brain:body weight ratios were significantly higher than controls. Mean B residues, in mg/kg DW, were 49 in egg, 33 in adult liver (Max. 74), 51 in duckling liver (Max. 89), 41 in adult brain (Max. 89), and 66 in duckling brain (Max. 110). No histopathology evident in liver, brain, kidney, or heart (8)
Breeding adults fed diets containing 0, 450, or 900 mg B/kg ration as boric acid for as long as 24 weeks. Ducklings produced received the same treatment as their parents for 14 days, then killed	No histopathology and no adverse effects on survival. B residues in the 900 mg B/kg group, in mg/kg B/kg FW, were 11 in egg vs. 2 in controls, 8.5 in adult liver vs. 0.6 in controls, and 13 in duckling liver vs. 0.3 in controls; values for the 450 mg/kg group were intermediate. Adults in the high-dose group had weight loss and decreased hemoglobin; eggs produced were lower in weight and fertility than controls. Hatching success and duckling weight in the high-dose group were reduced; duckling growth was reduced; altered duckling liver biochemistry (14)
Adult males fed diet containing 1600 mg B/kg ration as boric acid for 32 days	Boron accumulated in blood, brain, and liver, reaching 30–67 mg B/kg DW in 2–15 days. Boron was eliminated rapidly, with few detectable residues after 1 day on a boron-free diet (12)
Ducklings, age 1 day, fed diets containing 0, 100, 400, or 1600 mg B/kg ration as boric acid for 9 weeks	Only the high-dose group had consistently altered activity schedules, including decreases in amount of time spent in alert behaviors and in the water. Overall feeding time was increased, but did not result in an increase in the amount of food consumed (13)
Ducklings, age 1 day, 2-week dietary exposure	
1000 mg B/kg FW diet, as boric acid	Adverse effects on growth (9)
5000 mg B/kg diet, as boric acid	Some deaths (9)
Ducklings, age 1 day, 4-week dietary exposure. Diets contained as much as 1000 mg B/kg ration as boric acid alone or in combination with 15 or 60 mg Se/kg ration as selenomethionine. Diets contained either 22% or 7% protein	Boron alone caused growth reduction, with effect exacerbated by selenium and low protein (11)

Table 29.8 (continued) Lethal and Sublethal Effects of Boron on Birds

Species, Dose, and Other Variables	Effect (Reference[a])
Ducklings age 1 day, 10-week dietary exposure to boric acid	
Controls, 13 mg B/kg FW diet. Diets contained 12% to 14% moisture	Brain B concentration of 2 mg/kg DW (9)
100 mg B/kg FW diet	Delayed growth of females, plasma triglyceride levels elevated, abnormal liver metabolism, brain residue of 4 mg/kg DW (9)
400 mg B/kg FW diet	Delayed growth of females, plasma triglyceride elevated, brain B residue of 5 mg/kg DW, decrease in brain ATP, altered duckling behavior in bathing and resting (9)
1600 mg B/kg FW diet	Some deaths (10%), delayed growth, decreased food consumption, plasma triglyceride elevated, brain B residue of 51 mg/kg DW (Max. 99), decrease in brain calcium and ATP, reduction in time spent bathing and standing, increase in time spent resting, increased serum calcium, lower hematocrit and hemoglobin; no histopathology of brain, liver, or kidney (9)
DOMESTIC CHICKEN, *Gallus domesticus*	
Embryo, yolk injection	
Boric acid	
0.01 mg B/kg body weight (BW)	LD1 (1)
1.0 mg B/kg BW	LD50 (1)
1.0 mg at age 28 h	Developmental abnormalities (2)
2.0 mg at 28 h of development	Malformations of nervous system, eyes, and spinal cord (3)
2.5 mg at 24 h of development	Rumplessness (7)
2.5 mg at 84 h of development	Feet defects (7)
2.5 mg at 96 h of development	Skeletal deformities, cleft palate, missing toes, eye deformities (4–6)
15.8 mg B/kg egg at 96 h of development	LD50 (96 h). Most (70 to 85%) of the survivors at age 18 days had edema, inhibited feather growth, pale body coloration, and reduced body weight (10)
Borax	
0.01 mg B/kg BW	LD1 (1)
0.5 mg B/kg BW	LD50 (1)
Embryo, chorioallantoic membrane injection on day 8 of incubation with 0.1, 0.5, or 1.0 mg B/egg as sodium tetraborate	Chicks had a dose-dependent decrease in bone organic matrix, hatchability, and bone growth (16)
Adult	
Basal diets (11 mg B/kg ration) supplemented with 100 mg B/kg ration for 2 weeks, then reduced to 60 mg B/kg ration for 3 additional weeks	Egg production reduced (15)
875 mg B/kg diet, as boric acid, for 6 days	Egg production ceased; production normal 14 days after B withdrawn (1)

[a] **1,** Birge and Black 1977; **2,** Schowing and Cuevas 1975; **3,** Schowing et al. 1976; **4,** Landauer 1953a; **5,** Landauer 1953b; **6,** Landauer 1953c; **7,** Landauer 1952; **8,** Smith and Anders 1989; **9,** Hoffman et al. 1990; **10,** Ridgway and Karnofsky 1952; **11,** Hoffman et al. 1991; **12,** Pendleton et al. 1995; **13,** Whitworth et al. 1991; **14,** Stanley et al. 1996; **15,** Qin and Klandorf 1991; **16,** King et al. 1991.

29.4.6 Mammals

No requirement for boron in mammals is proven, although evidence is accumulating suggesting that boron may be an essential nutrient. Boron is related to normal energy utilization, immune function, and metabolism of bone, minerals, and lipids (Penland 1998). Boron deficiency (<0.04

mg B/kg ration of dams) impairs early embryonic development in rodents. These effects were not observed at 2 mg B/kg ration (Lanoue et al. 1998). Boron deprivation in animals and humans results in decreased brain electrical activity similar to that observed in nonspecific malnutrition, and reduced cognitive and psychomotor function (Penland 1998). Learning performance (manual dexterity, eye–hand coordination, memory, attention, perception) in humans was significantly higher when the daily boron ingestion rate was 3 mg vs. 0.23 mg (Penland 1994). Boron dietary supplements to postmenopausal women age 48 to 82 years induced changes consistent with the prevention of calcium loss and bone demineralization (Nielson et al. 1987; Nielsen 1994). In rats, adequate dietary boron protected against premature senescence (Massie 1994) and alleviated the signs of Vitamin D_3 deficiency through improved absorption and retention of calcium and phosphorus, and retention of femur magnesium (Hunt 1994). In cattle, increases in boron ingestion were associated with elevated boron levels in plasma and urine, increased boron excretion, decreased plasma phosphate concentrations, and increased renal and urinary clearance of phosphates (Weeth et al. 1981). Boron accumulations in rat testes were associated with progressive germ cell depletion that persisted long after toxic exposure to boron had occurred (Lee et al. 1978).

Boron effectively counteracts symptoms of fluoride intoxication in humans (Zhou et al. 1987) and in rabbits poisoned experimentally (Elsair et al. 1980a, 1980b, 1981). Humans suffering from skeletal fluorosis experienced 50 to 80% improvement after drinking solutions containing 300 to 1100 mg of borax per liter daily, 3 weeks a month for 3 months (Zhou et al. 1987). Boron enhances sequestration of fluoride from bone and excretion through kidneys and possibly the intestinal tract (Elsair et al. 1980a, 1981).

Inorganic borates, including boric acid, and sodium, ammonium, potassium, and zinc borates display low acute toxicity to mammals via oral, dermal, and inhalation routes of exposure. The critical effects in several species of mammals during chronic exposure to boron compounds are male reproductive toxicity and developmental abnormalities (Hubbard 1998). For example, prenatal exposure to elevated levels of boric acid causes reduced incidences of supernumerary ribs and a shortening or absence of the 13th rib in several species of laboratory animals (Narotsky et al. 1998). The doses that cause these effects are far higher than any levels to which human populations could be exposed. Humans would need to consume 3.3 g of boric acid or 5.0 g of borax to ingest the same dose level at the lowest animal NOAEL (Hubbard 1998). Boron has no measurable effect on human fertility or reproduction among workers exposed to borates or to populations exposed to high environmental borate levels (Hubbard 1998; Sayli 1998; Tuccar et al. 1998). Adult Turkish females, for example, residing in boron-rich areas (29 mg B/L drinking water) or boron-poor areas (0.3 to 0.5 mg B/L drinking water) did not differ in rate of spontaneous abortions, stillbirths, or congenital malformations (Tuccar et al. 1998).

Long-term exposure of humans to airborne boron dust may cause irritation of the nose, throat, and eyes, and large amounts of boron ingested over short periods of time can adversely affect the gastrointestinal tract, liver, kidney, and brain, and may lead to death (USPHS 1991). However, borax mean air exposures of 18 mg/m^3 measured for high-exposure workers together with dietary boron, resulted in an estimated absorption of only 0.38 mg B/kg BW daily. At this level, there was no progressive accumulation across the work week (Culver et al. 1994). Epidemics and sporadic cases of oral intoxication in people are often due to inadvertent addition of boric acid to infant formulas (Siegel and Wason 1986). Five of 11 human infants died within 3 days of exposure after ingesting formula prepared with a 2.5% aqueous solution of boric acid, equivalent to 4.5 to 14.0 g of boron ingested. Prior to death, these infants were lethargic and vomiting; postmortem degenerative changes were observed in liver, kidney, and brain (USPHS 1991). Some products containing boron compounds, such as pacifiers, have been sold in Ireland despite a recommendation from the Pharmaceutical Society of Great Britain that they should not be sold because of hazards to infants (O'Sullivan and Taylor 1983). Fatal cases of boron poisoning have involved misuse of boron compounds in hospitals, either from accidental substitution of boric acid solution for water in infant formula or from accidental use of boric acid as a diapering powder (USEPA 1975). In an adult fatality, the victim died after inundation by borax solution (USEPA 1975). In one case, a 12-month-old girl developed violent vomiting, coughing, irritability, tremors, seizures, and a delirious reaction after accidentally swallowing a mixture containing 3 g boric acid and 300 mg cinchocaine chloride prescribed due to a painful dental protrusion (Egfjord et al. 1988). Her plasma boric acid level 6 h later was 26 mg/L; the half-time persistence (Tb 1/2) for boric acid in plasma is about 7 h (Egfjord et al. 1988). The lethal dose of boric acid varies according to the species. In mammals, it ranges from 210 to 603 mg B/kg BW, and death is due to CNS paralysis and gastrointestinal irritation (Table 29.9; NAS 1980; USPHS 1991). Human newborns are especially sensitive, and accidental deaths have been recorded at doses between 50 and 140 mg B/kg BW (Table 29.9).

In mammals, excessive boron consumption results in a reduced growth rate and sometimes loss in body weight. These may not be due entirely to reduced feed and water consumption (Table 29.9; Seal and Weeth 1980). Growth retardation has been reported in cattle given 150 mg B/L drinking water (about 15 mg B/kg BW daily), in dogs consuming diets containing 1750 mg B/kg, in rabbits eating rations equivalent to >140 mg B/kg BW daily, and in rats given 150 mg B/L drinking water or 1060 mg B/kg diet (Table 29.9). In some cases, animals will avoid B-contaminated drinking water if given a choice. Rats, for example, will reject drinking water containing as little as 1.0 mg B/L (Dixon et al. 1976), and cattle will avoid water containing >29 mg B/L (Green and Weeth 1977).

Male workers engaged in boric acid production showed weakened sexual activity, decreased seminal volume, low sperm count and motility, and increased seminal fructose (USEPA 1975). Animal studies demonstrated that the testes atrophy or degenerate if large amounts of boron are eaten or drunk; these effects have not been reported in humans (USPHS 1991). Adverse effects on reproduction of laboratory animals have been reported in sensitive species fed diets containing more than 1000 mg B/kg, or given drinking water containing 1.0 mg B/L (equivalent to about 0.3 mg B/kg BW daily), or given a single oral dose of 3000 mg B/kg BW on the first day of pregnancy (Table 29.9). Boric acid caused developmental toxicity — including fetal weight reduction, prenatal mortality and malformations, decreased survival — in rats, mice, and rabbits in the range of 16 to 80 mg B/kg BW daily given either throughout gestation or only during major organogenesis (Heindel et al. 1994).

Volatile boron compounds, especially boranes, are usually more toxic than boric acid or soluble borates (Table 29.9) (NAS 1980). However, there is little commercial production of synthetic boranes, except for sodium borohydride — one of the least toxic boranes (Sprague 1972). Boron trifluoride is a gas used as a catalyst in several industrial systems, but on exposure to moisture in air, it reacts to form a stable dihydride (Rusch et al. 1986). For boric oxide dusts, occupational exposures to 4.1 mg/m^3 (range 1.2 to 8.5) are associated with eye irritation; dryness of mouth, nose and throat; sore throat; and cough (Garabrant et al. 1984).

Table 29.9 Lethal and Sublethal Effects of Boron on Mammals

Organism, Route of Administration, Dose, and Other Variables	Effect (Reference[a])
CATTLE, *Bos* spp.	
Drinking water	
Supplemented with 15, 30, 60, or 120 mg B/L for 10 days	Boron levels in plasma rose from 2.7 mg/L in controls to 4.4 (15 mg/L group) to 5.3 (30 mg/L group) to 8.3 (60 mg/L group) to 13.4 mg/L in the 120 mg B/L drinking water supplement (5)
29 mg B/L, and higher	When given choice, cattle preferred tap water to drinking water supplemented with B compounds (1)
120 mg B/L, as borax, for 10 days	No effect on feed or water consumption; no overt signs of toxicosis (2)
150 mg B/L, as borax, for 30 days, equivalent to 15.3 mg B/kg BW daily	Decreased feed consumption, weight loss, edema, inflammation of legs, daily elevated plasma B levels of 1.2 mg/L vs. 0.5 in controls, abnormal blood chemistry (1–5)
Diet	
Consumed feed containing 157 mg B/kg, as borax, for 42 days	No adverse effects (3)

Table 29.9 (continued) Lethal and Sublethal Effects of Boron on Mammals

Organism, Route of Administration, Dose, and Other Variables	Effect (Reference[a])
Fed 2–2.5 g boron daily as borax, for 40 days	No observable adverse effects; all B excreted, mostly in urine (6)
Fed 20 g borax daily	Milk B residues increased from <1.0 mg/L to >3 mg/L (5)
Ingested total dose of 100–300 g boron, equivalent to 200–600 mg B/kg BW	Toxic dose (7)
Found dead after consuming 1 kg borax, or about 250 g B	Residues, in mg B/kg FW, were 1300 in ruminal fluids, 1900 in abomasal fluids, 24 in liver, 19 in rumen, and 21 in abomasum (7)
DOG, *Canis familiaris*	
Diet	
350 mg B/kg feed, 2 years	Tolerated (8)
1540 mg borax/kg or 3000 mg boric acid/kg, chronic study (174–524 mg B/kg diet)	No adverse effects (6, 9)
1170 mg B/kg, 38 weeks	Testicular degeneration, spermatogenesis cessation (5, 8)
Inhalation	
92 mg pentaborane/m^3 for 15 min	LC50 (9)
GUINEA PIG, *Cavia* spp.	
Inhalation	
0.018 mg decaborane/m^3, 6 h daily, 5–6 exposures	Eye inflammation, listlessness, emaciation, convulsions (3)
HUMAN, *Homo sapiens*	
Dermal	
7-month-old infant treated for dermatitis with 3% boric acid powder	Fatal. Boron concentrations elevated in bile, intestinal contents, and spleen (9)
Adult administered about 645 g of boric acid dermally	Toxicosis observed (10)
Inhalation	
Borax dust, 1.1–14.4 mg/m^3, occupational exposure for at least 5 years	At 14.4 mg/m^3, 33% of workers noted dryness of mouth, nose, or throat; 28% had eye irritation problems; 15% had nosebleeds and cough; 13% had sore throat or shortness of breath and chest tightness. At 4.0 and 1.1 mg/m^3, no symptoms except eye irritation were noted by more than 5% and 3% of exposed participants (11)
Boron dust, >4.1 mg/m^3 for 11 years	Irritation of nose, throat, and eyes (27)
Boranes, various	Pulmonary irritation, headache, nausea, fatigue, muscular weakness, liver and kidney pathology (3)
Oral	
3 mg B daily for 119 days to diet containing 0.25 mg B	Reduction in urinary excretion of calcium and magnesium by postmenopausal women (12)
20 mg B daily	Normal adult intake (6)
Solutions >88 mg B/L or >500 mg boric acid/L	Fatal to infants (13)
895 mg B/kg BW; single attempted suicidal dose of a boric-acid-containing insecticide by adult female	Vomiting; hospitalized for 96h; asymptomatic after release (27)
1–3 g boric acid, or 0.3–0.8 g/kg BW	Lethal to newborns. (14, 27)
2–4.5 g boric acid or 0.5–1.2 g/kg BW	Nonfatal to infants, but serum levels elevated from 20–150 mg borate/L (14)
>3.5 g boric acid daily	Probably harmful or lethal to infants and newborns (10)
4 g boric acid or borates daily	No toxicosis in adults (6, 9)
4.5–15 g boric acid, equivalent to 1.25–4.2 g/kg BW, in accidentally contaminated formula in newborn nursery	Death preceded by severe symptomology; serum levels of 400–1000 mg borate/L (13, 14)
5–6 g of borates, or 0.7 g/kg BW	Fatal to infants (14, 15)
15–20 g of boric acid, equivalent to 0.25–0.3 g/kg BW	Fatal to adults (9, 14, 15, 27)

Table 29.9 (continued) Lethal and Sublethal Effects of Boron on Mammals

Organism, Route of Administration, Dose, and Other Variables	Effect (Reference[a])
Infants, age 6–16 weeks, given pacifiers dipped in a proprietary borax (107 g/L) and honey compound. Dose during 1-month-exposure period estimated at 3–9 g borax	Some developed seizure disorders characterized by vomiting, loose stools, irritability, diarrhea; elevated blood B values of 2.6–8.5 mg B/L vs. <0.6 in controls. When preparation withheld, seizures stopped and children remained well for at least 5 years (13)
Injection, intravenous	
Adult males, age 22–28 years, given single infusion of 562–611 mg boric acid, equivalent to 8.0–8.7 mg B/kg BW	Tb 1/2 persistence of boric acid was 21 h. Most excreted in urine in 24 h, 94% in 96 h, and ~99% in 120 h; plasma boric acid concentrations after infusion was about 16 mg/L vs. 0.5 at start; no discomfort during or after infusion (16)
Adults given total dose of 20 g boric acid	No permanent adverse effects (9)
MONKEY, *Macaca* spp.	
Inhalation	
Pentaborane, 640 mg/m^3, 2 min	LC50 (9)
Intraperitoneal injection	
Decaborane, 1 mg/kg BW daily, multiple injections	Altered brain wave activity (9)
Decaborane, 6 mg/kg BW, single injection	LC50 (9)
MICE, *Mus* sp.	
Drinking water	
5 mg B/L, lifetime exposure	No effect on growth, longevity, or tumor incidence (2, 5)
Ingestion	
3 g B/kg BW, first day of pregnancy	94% of embryos did not develop past blastocyst stage vs. 9% in controls (9)
Diet	
Equivalent to 27 mg B/kg BW daily	No adverse reproductive effects (33)
48 mg B/kg BW daily for 103 weeks	40% dead (27)
96 mg/kg BW daily for 103 weeks	Testicular atrophy (27)
Equivalent to 111 mg B/kg BW daily	Reduced fertility (33)
144 mg/kg BW daily for 13 weeks	Decreased survival (27)
288 mg/kg BW daily for 13 weeks	Testicular degeneration (27)
1500 mg boric acid/kg (262 mg B/kg) daily	All dead within 10 days (17)
Injection, intravenous	
1.32 g sodium borate per kg BW, single dose	LD50 (18)
Injection, intraperitoneal	
25.2 mg decaborane/kg BW, single dose	LD50 (9)
44.7 mg decaborane/kg BW, single dose, prior treatment for 8 days at 250 mg/kg BW with pyridoxine hydrochloride	LD50 (9)
2817 mg sodium borate/kg BW, single dose	LD50 (18)
Inhalation	
Pentaborane	
0.011 mg/m^3 for 4 h	LC50 (3)
50 mg/m^3, 15 min	LC50 (9)
342 mg/m^3, 2 min	LC50 (9)
1034 mg/m^3, 30 seconds	LC50 (9)
RABBIT, *Oryctolagus* sp.	
Diet	
Equivalent to 800–1000 mg borates/kg BW daily for 4 days	Growth retardation (18)
Intragastric route	
Daily dose of 100 mg calcium borate for 4 months	Altered serum chemistry (9)
Intravenous injection	
Single dose of 800–900 mg boric acid/kg BW	LD50 (18)

Table 29.9 (continued) Lethal and Sublethal Effects of Boron on Mammals

Organism, Route of Administration, Dose, and Other Variables	Effect (Reference[a])
Intraperitoneal injection	
30 mg decaborane/kg BW	Death within 24 h (3)
Dermal	
25–200 mg boric acid/kg BW daily	Not irritating or toxic when applied to intact skin (18)
Sodium borate solutions of 50,000 or 100,000 mg borates/L applied to skin	Mildly or moderately irritating (18)
Boron oxide dust	Application to skin produced erythema that lasted 2–3 days; instillation in eyes produced immediate conjunctivitis as a result of exothermic hydration of boron oxide to boric acid (19)
Inhalation	
120–150 mg calcium borate/m^3, 2 h daily, 10 weeks exposure	Respiratory tract pathology, growth inhibition, enlarged liver (9)
RAT, *Rattus* sp.	
Drinking water	
Free access for 90 days to drinking water containing 0.3, 1.0, or 6.0 mg B/L	Rats refused to drink water at 1.0 or 6.0 mg/L (15)
0.3 mg boric acid/L for 6 months	No effect on gonadotoxicity (20)
1.0 mg boric acid/L for 6 months, equivalent to 0.05 mg B/kg BW daily	Decreased spermatozoid count, reduction in spermatozoid activity (20)
6 mg B/L, 90 days	No toxic effect on male reproductive system, blood chemistry, or growth (5, 15)
6 mg B/L for 6 months, equivalent to 0.3 mg B/kg BW daily	Gonadotoxicity in male rats; altered enzyme activity levels in blood and liver (20, 21)
75 mg B/L, as borax, for 45 days	No effect on growth or reproduction (3)
100 mg B/L for 21 days	Tissue B levels in kidney, liver, brain, and blood increased for first 9 days but returned to normal by day 21 except for blood, which continued to rise (21)
150 mg B/L for 70 days, or 170 mg B/L for 25 days	Slight reduction in growth rate (5)
>150 mg B/L for extended periods	Adverse effects probable (5)
300 mg B/L for 49–70 days	Growth rate reduced 21%, but no change in food consumption; coarse coat; atrophied scrotal sacs (4)
440 mg B/L for 25 days	Growth inhibition (4)
880 mg B/L for 70 days	Inhibited sperm production (27)
3 g sodium tetraborate/L for 10–14 weeks	Increase in activity of cerebral succinic dehydrogenase and brain acid proteinase, and in brain RNA concentration; decrease in liver cytochrome P-450 activity (22)
Diet	
Females fed diets with 0.04 (low) or 2 (adequate) mg B/kg ration for 6 weeks before breeding and through pregnancy; reproductive outcome monitored on gestation day 20	Low dietary B significantly lowered maternal blood, liver, and bone B concentrations; however, it had no clear effects on fetal growth or development (35)
Day 10 embryos from dams fed either the low (0.04 mg/kg ration) or adequate (2 mg/kg ration) boron diets for at least 12 weeks were cultured in serum collected from male rats exposed to the low or adequate dietary B treatments	Dams fed the low B diet had a significant reduction in number of implantation sites when compared to dams fed the B-adequate diet; however, embryonic growth *in vitro* was not affected by B treatment (35)
Pregnant rats fed diets with boric acid equivalent to <0.35 (controls), 3, 6, 10, 13, or 25 mg B/kg BW daily from gestational days 0–20. About half the dams in each group were killed on gestational day 20 and blood and prenatal outcome evaluated. Remaining dams received a control diet beginning on gestational day 20 and their litters monitored throughout lactation	Maternal blood boron concentrations were elevated in all boron groups in a dose-dependent manner. On gestational day 20, blood B concentrations of 1.3 mg/kg FW were associated with the no-observed-adverse-effect level (NOAEL) and 1.53 mg/kg FW with the lowest-observed-adverse effect level (LOAEL), equivalent to dietary intakes of 10 and 13 mg B/kg BW daily, respectively, for developmental toxicity. Developmental toxicity persisted postnatally only at 25 mg B/kg BW daily, a dose associated with more than a 10-fold increase in maternal blood B (2.8 mg B/kg FW) vs. 0.23 mg/kg FW in controls (34)

Table 29.9 (continued) Lethal and Sublethal Effects of Boron on Mammals

Organism, Route of Administration, Dose, and Other Variables	Effect (Reference[a])
Equivalent to 17.5 mg B/kg BW daily	No adverse reproductive effects (33)
Equivalent to 26 mg B/kg BW daily	Mild reversible inhibition of spermiation (33)
Equivalent to 58.5 mg B/kg BW daily	Testicular atrophy; reduced fertility (33)
Fed diets containing 0, 200, 1000, 3000, or 9000 mg B/kg ration as boric acid for up to 12 weeks, equivalent to <0.2 (control), 1.7, 8.5, 26, or 68 mg B/kg BW daily; resistance to destructive testing was measured on femurs, tibias, and lumbar vertebrae	Vertebral resistance to a crushing force was increased by about 10% at all boron dose levels (200–9000 mg B/kg ration); no effect on femurs and tibias. Dietary loadings of 3000 and 9000 mg/kg were reproductively toxic to males and the developing fetus (31)
0.09–1.71 mg boric acid/kg BW daily for 6 months (0.015–0.3 mg B/kg BW daily)	Adverse changes in testes (18)
0.16 or 2.7 mg B/kg ration for 12 weeks to Vitamin D-deprived rats	Abnormal mineral balance in low-dose diet; normal calcium, magnesium, and phosphorus balance in 2.7 mg/kg supplement (28)
350 or 525 mg B/kg diet, as borax or boric acid, for 2 years	No observable adverse effects on fertility, lactation, litter size, weight, or appearance (6)
500, 1000, or 2000 mg B/kg diet, as borax, for 30–60 days, equivalent to 12, 25, or 50 mg B ingested daily	No adverse effects at 500 mg B/kg diet for 60 days. At 1000 and 2000 mg B/kg, adverse effects measured on male reproductive capacity, including germinal aplasia and infertility; effects persisted for at least 8 months following B exposure at highest dose (23)
525 mg B/kg diet for 90 days	Tolerated (8); testes damage (27)
1000 mg boric acid or borax/kg BW daily	Weight loss after 1 week on borax diet, or 2 weeks on boric acid diet; toxic signs after 3 weeks on both diets (24)
1050 mg B/kg diet, as borax or boric acid, for 2 years	Testicular degeneration (6)
1060 mg B/kg diet, as sodium borate, chronic exposure	Growth retardation and testicular atrophy (18)
1170 mg B/kg diet for 2 months	Coarse coat, scaly tails, hunched position, bloody discharge from eyes, depressed hemoglobin and hematocrit (5)
1170 mg B/kg diet, as borax or boric acid, 2 years	Sterility in males and females (6, 8)
1575 mg B/kg ration for 28 days as boric acid, adult males	Testicular lesions after 7 days, atrophy after 28 days (29); no return of spermatogenesis after resumption of normal diet for as long as 32 weeks (30)
1750 mg B/kg diet, 25 days	Reduction of 50% in growth rate (4)
1750 mg B/kg diet, as sodium borate, chronic	Severe testicular atrophy (18)
10,250 mg B/kg ration	Dead after 1 day (27)
Oral, single dose, except where noted	
450 mg B/kg BW	No effect on male fertility (15)
500 mg B/kg BW by gavage on gestation days 5–9, 6–9, 6–10, or on single days between gestation days 6 and 11	After multiday exposures, there was an increased frequency of malformations of the axial skeleton involving the head, sternum, ribs, and vertebrae. About 90% of the fetuses exposed on gestation day 9 had only 6 cervical vertebrae; 60% of the fetuses exposed on gestation day 10 had reduced survival and a reduction in the number of thoracic and lumbar vertebrae (32)
510–690 mg B/kg BW, as borax	LD50 (8, 9, 24)
550–710 mg B/kg BW, as boric acid	LD50 (8, 9, 24)
600 mg B/kg BW	LD50 (2)
3.45–5.14 g sodium borate/kg BW	LD50 (18)
5.1 g boric acid/kg BW	LD50 (6)
6.1 g borax/kg BW	LD50 (6)
Injection, subcutaneous	
1.4 g boric acid/kg BW	LD50 (18)
Injection, intravenous	
5–75 mg boric acid/kg BW	Slight reduction in arterial blood pressure (21)

Table 29.9 (continued) Lethal and Sublethal Effects of Boron on Mammals

Organism, Route of Administration, Dose, and Other Variables	Effect (Reference[a])
Injection, intraperitoneal	
42 mg sodium borate/kg BW, single injection	Tissue residues after 30 min, in mg B/kg FW, were 25 in blood, 30 in liver, and 50 in kidney vs. <5 in all control tissues. After 3 months, residues were 20 mg B/kg FW in brain, 45 in heart, 60 in liver, and 75 in kidney (21)
Inhalation, boron trifluoride	
2, 6, or 17 mg BF_3/m^3, 6 h daily, 5 days weekly, 13 weeks	At 17 mg/m^3, altered proximal tubular epithelium of kidney and abnormal serum chemistry. At 6 mg/m^3, elevated fluoride levels in urine, serum, and bone, but no toxic response. No difference from controls at 2 mg/m^3 (26)
24 or 66 mg/m^3, 6 h daily, 9 days	Clinical signs of respiratory irritation, nasal discharge, weight loss, increased lung weight, depressed liver weight, kidney pathology at 66 but not 24 mg/m^3 (26)
55 mg/m^3, 4–7 h daily, 5 days weekly, 6 weeks	Some deaths in rats and other rodent species tested, but no deaths in nonrodent species (26)
180 mg/m^3, 6 hours daily, consecutive days	All dead prior to sixth exposure (26)
259 mg/m^3, 4–7 h daily, 2 days	All dead. Mortality was lower for guinea pigs, dogs, rabbits, mice, and cats. Lung and kidney damage in all species (26)
1210 mg/m^3, 4 h	50% dead (26)
Inhalation, boron oxide	
470 mg/m^3, 10 weeks	Reddish exudates from nose, but no deaths or signs of lung damage (19)
470 mg/m^3, 24 weeks	No signs of toxicosis (9)
Inhalation, decaborane	
20 mg/m^3, 6 h daily, 5 days weekly	Tremors, convulsions, nervousness, restlessness, weight loss, belligerence (3)
36 mg/m^3, 4 h	LC50 (3)
Inhalation, pentaborane	
3 mg/m^3, 6 h daily, 5 days weekly	Extreme belligerence, tremors, weight loss (3)
18 mg/m^3, 4 h	LC50 (3)

[a] **1,** Green and Weeth 1977; **2,** Weeth et al. 1981; **3,** NAS 1980; **4,** Seal and Weeth 1980; **5,** Nielsen 1986; **6,** Sprague 1972; **7,** Brockman et al. 1985; **8,** Weir and Fisher 1972; **9,** USEPA 1975; **10,** Gupta and Parrish 1984; **11,** Garabrant et al. 1985; **12,** Nielsen et al. 1987; **13,** O'Sullivan and Taylor 1983; **14,** Siegel and Wason 1986; **15,** Dixon et al. 1976; **16,** Jansen et al. 1984; **17,** Lizzio 1986; **18,** Anonymous 1983; **19,** Garabrant et al. 1984; **20,** Krasovskii et al. 1976; **21,** Magour et al. 1982; **22,** Settimi et al. 1982; **23,** Lee et al. 1978; **24,** Dani et al. 1971; **25,** Benson et al. 1984; **26,** Rusch et al. 1986; **27,** USPHS 1991; **28,** Dupre et al. 1994; **29,** Ku and Chapin 1994; **30,** Chapin and Ku 1994; **31,** Chapin et al. 1998; **32,** Narotsky et al. 1998; **33,** Hubbard 1998; **34,** Price et al. 1998; **35,** Lenoue et al. 1998.

29.5 RECOMMENDATIONS

Many boron criteria have been proposed for the protection of crops, aquatic life, waterfowl, livestock, and human health (Table 29.10). The risk to aquatic ecosystems from boron is low (Howe 1998). Boron concentrations in contaminated industrial effluents seldom exceed 1.0 mg B/L, a level considered nonhazardous to aquatic life (Table 29.10) (Thompson et al. 1976). In a few boron-rich areas, natural levels will be higher, although organisms may adapt to local conditions (Howe 1998). However, future accumulations of boron in groundwater through wider uses of boron-containing cleansing agents may adversely affect aquatic organisms and other species of plants and animals, as now occurs in areas where natural boron deposits exist (USEPA 1975). Long-term monitoring of groundwaters and surface waters for boron levels seems warranted.

Results of chronic feeding studies using mallards demonstrate that diets containing 13 mg B/kg FW produce no adverse effects, but those containing 30 or 100 mg B/kg FW are associated with elevated tissue boron residues and growth reduction, and diets containing 1000 mg B/kg are fatal

(Table 29.10). More research is needed on the fate and effects of boron on waterfowl and raptors, especially in those areas where high dietary boron loadings are encountered as a result of agricultural drainwater disposal practices.

Minimum concentrations of dietary boron needed to maintain animal health are not known with certainty. However, diets containing <0.4 mg B/kg fresh weight may adversely affect metabolism of rats and chicks. Accordingly, animal diets should contain >0.3 mg B/kg fresh weight until necessary feeding data become available (Nielsen 1986). Also, the defensible boron maximum for livestock drinking water may be considerably higher than 5 mg/L (Table 29.10) because several "safe" water sources in Nevada exceeded this upper maximum and approached 80 mg B/L (Green and Weeth 1977). Data are unavailable on boron effects on terrestrial wildlife. Until these data become available, it seems reasonable to apply the same criteria proposed for livestock protection (Table 29.10) to mammalian wildlife, that is, diets should contain more than 0.4 mg B/kg DW but less than 100 mg/kg, and drinking water <5 mg/L.

Medicinal use of boric acid and borax for babies has resulted in anorexia, nausea, vomiting, diarrhea, marked cardiac weakness, a red eruption over the entire body, and (rarely) death (NAS 1980). The medical community has since abandoned the use of boric acid solutions as irrigants and antiseptics (Siegel and Wason 1986), abandoned all medical uses in Denmark (Egfjord et al. 1988), and severely limited availability (prescription only) in Ireland (O'Sullivan and Taylor 1983). Increased use of boric acid as a household pesticide should be viewed with concern, especially in households where children have access to non-safety-capped boric acid containers (Siegel and Wason 1986). However, the amine-carboxyborane derivatives show promise as therapeutic agents for a number of disease states. More research is needed on medical aspects of amine-carboxyborane compounds and their ability to reduce serum cholesterol and to relieve, through their anti-inflammatory properties, the effects of chronic arthritis (Hall et al. 1994; Newnham 1994). This group of compounds were effective antineoplastic agents with selective activity against single cell and solid tumors derived from human and rodent leukemias, lymphomas, sarcomas, and carcinomas (Hall et al. 1994). Health benefits of borates and boron compounds and their role in fertility and pregnancy merit additional investigation (Mastromatteo and Sullivan 1994). Boron, for example, may be essential to normal bone growth and composition and protect against bone loss associated with aging (McCoy et al. 1994).

The fact that boron is essential to plants is firmly established (NAS 1980; USPHS 1991). However, when boron concentrations in irrigation waters exceed 2 mg/L, extensive plant toxicity should be expected (Pagenkopf and Connolly 1982). High concentrations of boron in some potential irrigation waters in parts of the western United States at levels capable of causing crop damage have prompted implementation of boron criteria for irrigation waters (Table 29.10), although no legally enforceable boron standards have been promulgated (USEPA 1975). More information is needed on crop plants in the following subjects: interaction of boron with other elements in the soil and its effects on boron availability to plants, the role of boron on pollination as it affects seed yield and sugar content of crops, and distinguishing signs of boron deficiency in plants from similar signs of molybdenum deficiency (Gupta and Macleod 1982; Mastromatteo and Sullivan 1994).

More research is needed on the accurate measurement of boron in biological materials when the concentrations are <1.0 mg B/kg (Sullivan and Culver 1998). Standard biological reference materials with low boron levels need to be produced for use in interlaboratory comparisons. This becomes especially important in studies on boron-deficiency states and the ability of the organism to conserve boron at very low intakes (Sullivan and Culver 1998). More research is needed on homeostatic regulation of boron and functional markers of boron metabolism (Sutherland et al. 1998). Sullivan and Culver (1998) recommend additional studies to establish:

- The availability of boron from the diet and its distribution to the tissues
- Boron essentiality in higher organisms
- The beneficial effects of boron on health
- The role of borates in behavioral disorders and cognitive performance

New advances in boron nutrition research should include better characterization of the mechanisms through which boron modulates immune function and insulin release (Hunt 1998). Epidemiological studies should be initiated to identify health conditions associated with inadequate dietary boron (Sutherland et al. 1998). Finally, Dourson et al. (1998) recommend more research on uncertainty factors used in establishing tolerable daily intake values for the protection of human health, with emphasis on variations in interspecies and intraspecies differences in resistance to boron.

Table 29.10 Proposed Boron Criteria for the Protection of Natural Resources and Human Health

Resource and Other Variables	Criterion	Reference[a]
CROPS		
Irrigation waters		
Sensitive crops	0.3–<0.75 mg B/L	1–3, 19
Semitolerant crops	0.67–2.5 mg B/L	1–3
Tolerant crops	1–4 mg B/L	1–3
Maximum safe concentration	4 mg B/L	2
Residues in crops		
Boron deficiency	<15 mg B/kg dry weight (DW) plant	4, 5
Toxicosis	>200 mg B/kg DW plant	4, 5
Soil concentrations		
Optimal growth of several species	>0.1–<0.5 mg B/kg DW soil	19
Deficiency, Bangladesh	<0.2 mg B/kg DW surface layer	26
FOREST TREES		
Conifers, sensitive species		
Deficient	<4 mg B/kg DW foliage	23
Low	>4–<8 mg B/kg DW foliage	23
Intermediate	13–20 mg B/kg DW foliage	23
Toxic	>75 mg B/kg DW foliage	23
Angiosperms, sensitive species		
Deficient	8–16 mg B/kg DW foliage	23
Toxic	>180 mg B/kg DW foliage	23
AQUATIC ORGANISMS		
Nonhazardous levels in water		
Fish, oysters	<1–5 mg B/L	2, 6, 25
Aquatic communities	1–2 mg/L	25
Aquatic plants	4 mg B/L	2
Aquatic invertebrates	6–10 mg B/L	25
"Safe" levels in water		
Largemouth bass, *Micropterus salmoides*	<30 mg B/L	1, 7
Bluegill, *Lepomis macrochirus*	<33 mg B/L	1, 7
Rainbow trout, *Oncorhynchus mykiss*, embryos and larvae	0.75–1.0 mg B/L	20
Adverse effects, sensitive species	10–12 mg B/L	6, 20
WATERFOWL		
Diet		
No observed adverse effect	<13 mg B/kg fresh weight (FW)	17
Adverse effects	30–100 mg B/kg FW	17, 18
Fatal	1000 mg B/kg FW	18
LABORATORY ANIMALS		
No observed adverse effect		
Rat	<15.6 mg/kg body weight (BW) daily during gestation	22
Rabbit	<25 mg B/kg BW daily	22

Table 29.10 (continued) Proposed Boron Criteria for the Protection of Natural Resources and Human Health

Resource and Other Variables	Criterion	Reference[a]
Mouse	<50 mg B/kg BW daily	22
Adverse effect level		
Rat	>15.6 mg B/kg BW daily during gestation	22
Rabbit	>50 mg B/kg BW daily	22
Mouse	90 mg B/kg BW daily during gestation	22
LIVESTOCK		
Diet		
Boron deficiency	<0.4 mg B/kg DW	5
Toxic signs probable	>100 mg B/kg DW	5
Maximum tolerable level, as borax	150 mg B/kg DW	4, 5
Total dose, toxic	100–300 g of B (equivalent to 200–600 mg B/kg BW)	9
Drinking water		
Maximum allowable	5 mg B/L	4, 8, 10, 11
Maximum tolerated	40 mg B/L	10
"Safe"	40–150 mg B/L	11
Adverse effects	>150 mg B/L	5
PESTICIDE APPLICATIONS		
Boric acid, 99% powder	Effective for control of household cockroaches, ants, and fleas	2
Boric acid, 8% solution	Fungicide for vegetables, fruits, and trees	12
HUMAN HEALTH		
Air		
Threshold Limit Value (8 h daily, 5 days weekly)		
Pentaborane	0.01 mg/L	4
Diborane	0.1 mg/L	4
Decaborane	0.5 mg/L	4
Sodium borate	1–5 mg/m^3	16, 17
Boron trifluoride	<3 mg/m^3	19
Calcium borate	4–6 mg/m^3	3
Boron tribromide	10 mg/m^3	19
Boron oxide		
Total dust	10 mg/m^3	19
Respirable fraction	<5 mg/m^3	19
Sodium tetraborate	10 mg/m^3	19
Decahydrate	<5 mg/m^3	19
Anhydrous and pentahydrate	<1 mg/m^3	19
Borate dusts		
Safe	<1 mg B/m^3 daily	21
Infrequent effects	1.1 mg B/m^3 daily	21
Adverse effects	4–14.6 mg B/m^3 daily	21
Daily intake		
Total tolerable	0.4 mg/kg BW[b]	24
Worldwide	Range 0.3–41 mg B, means usually 10–20 mg B[c]	4, 5, 13
Finland	1.7 mg B	5
England	2.8 mg B	5
U.S.	3 mg B	4
No effect level	4 g boric acid	14
Adverse effect level		
Chronic intoxication	4–5 g boric acid	14
Lethal to infants and small children	5–6 g boric acid	14

Table 29.10 (continued) Proposed Boron Criteria for the Protection of Natural Resources and Human Health

Resource and Other Variables	Criterion	Reference[a]
Lethal to adults	18–20 g boric acid, single dose	14
Dermal, ocular		
Sodium borate and boric acid	Safe as cosmetic ingredients at <5% concentrations; not recommended on infant skin or injured skin	12, 16
Diet		
Citrus fruits	<8 mg B/kg FW	19
Cottonseed	<30 mg B/kg FW	19
Hop extracts	<310 mg B/kg FW	19
Minimal risk level	<3.2 mg B/kg ration FW	19
Adverse effects, including death	>4161 mg B/kg ration FW	19
Drinking water		
Recommended	<0.3 mg B/L	15
Former Soviet Union	<0.5 mg B/L	10
U.S.	<1.0 mg B/L	4, 11
"Safe"	<20 mg B/L	2, 10
No toxic effects	20–30 mg B/L	2
Tissue residues		
Blood, children and infants		
Normal	<1.25 mg B/L FW	19
Adverse systemic effects	20–150 mg B/L FW	19
Fatal	200–1600 mg B/L FW	19
Serum, adults		
No significant toxicity	<2320 mg B/L FW	19
Urine, adults		
Normal	0.7–1.5 mg B/L FW	19

[a] **1,** Sprague 1972; **2,** Papachristou et al. 1987; **3,** USEPA 1975; **4,** NAS 1980; **5,** Nielsen 1986; **6,** Thompson et al. 1976; **7,** Birge and Black 1977; **8,** Weeth et al. 1981; **9,** Brockman et al. 1985; **10,** Seal and Weeth 1980; **11,** Green and Weeth 1977; **12,** Siegel and Wason 1986; **13,** Benson et al. 1984; **14,** Schillinger et al. 1982; **15,** Krasovskii et al. 1976; **16,** Anonymous 1983; **17,** Hoffman et al. 1989; **18,** Smith and Anders 1989; **19,** USPHS 1991; **20,** Black et al. 1993; **21,** Wegman et al. 1994; **22,** Heindel et al. 1994; **23,** Stone 1990; **24,** Becking and Chen 1998; **25,** Howe 1998; **26,** Miah 1999.

[b] Based on NOAEL of 9.6 mg B/kg BW daily for reproductive effects in rats and an uncertainty factor of 25.

[c] Becking and Chen (1998) estimate global mean daily intake of B by humans as 1.9 mg, mostly from food (65%) and drinking water (30%). For a 70-kg adult, this is equivalent to 0.027 mg B/kg BW daily.

29.6 SUMMARY

The United States is the major global producer of boron compounds and supplies about 70% of the annual demand. Although boron is ubiquitous in the environment, human activities such as mining, coal burning, drainwater disposal, and use of borax laundry detergents have resulted in elevated boron loadings in the atmosphere and in irrigation waters. The chemistry of boron is complex and rivals that of carbon in its diversity. However, most boron compounds enter or degrade in the environment to B–O compounds (borates) — such as borax and boric acid — and these are considered to be the most significant ecologically.

Boron is an essential trace element for the growth of terrestrial crop plants and for some species of fungi, bacteria, and algae, but excess boron is phytotoxic. Representative species of aquatic organisms, including plants, invertebrates, fishes, and amphibians, usually tolerated up to 10 mg B/L medium for extended periods without harm. In waterfowl, growth was adversely affected at dietary levels of 30 to 100 mg B/kg fresh weight, tissue boron concentrations were elevated at 100 to 300 mg B/kg diet, and survival was reduced at dietary levels of 1000 mg B/kg. All of these dietary levels currently exist near agricultural drainwater disposal sites in the western United States. Boron is not now considered essential in mammalian nutrition, although low dietary levels protect against fluorosis and bone demineralization. Excessive consumption (i.e., >1000 mg B/kg diet, >15 mg B/kg body weight daily, >1.0 mg B/L drinking water, or >210 mg B/kg body weight in a single dose) adversely affects growth, survival, or reproduction in sensitive mammals. Boron and its compounds are potent teratogens when applied directly to the mammalian embryo, but there is no evidence of mutagenicity or carcinogenicity. Boron's unique affinity for cancerous tissues has been exploited in neutron capture radiation therapy of malignant human brain tumors.

Boron criteria recommended for the protection of sensitive species include:

- <0.3 mg B/L in water for irrigation of crops
- <1.0 mg B/L for aquatic life
- <5.0 mg B/L in livestock drinking waters
- <30 mg B/kg in diets of waterfowl
- <100 mg B/kg in diets of livestock.

29.7 LITERATURE CITED

Ahl, T. and E. Jonsson. 1972. Boron in Swedish and Norwegian fresh waters. *Ambio* 1:66-70.

Aitken, R.L. and L.C. Bell. 1985. Plant uptake and phytotoxicity of boron in Australian fly ashes. *Plant Soil* 84:245-257.

Anonymous. 1983. Final report on the safety assessment of sodium borate and boric acid. *Jour. Amer. Coll. Toxicol.* 2:87-125.

Antia, N.J. and J.Y. Cheng. 1975. Culture studies on the effects from borate pollution on the growth of marine phytoplankters. *Jour. Fish. Res. Board Canada* 32:2487-2494.

Argust, P. 1998. Distribution of boron in the environment. *Biol. Trace Elem. Res.* 66:131-143.

Becking, G.C. and B.H. Chen. 1998. International programme on chemical safety (IPCS) environmental health criteria on human health risk assessment. *Biol. Trace Elem. Res.* 66:439-452.

Belver, A. and J.P. Donaire. 1987. Phospholipid metabolism in roots and microsomes of sunflower seedlings: inhibition of choline phosphotransferase activity by boron. *Phytochem.* 26:2923-2927.

Benson, W.H., W.J. Birge, and H.W. Dorough. 1984. Absence of mutagenic activity of sodium borate (borax) and boric acid in the Salmonella preincubation test. *Environ. Toxicol. Chem.* 3:209-214.

Birge, W.J. and J.A. Black. 1977. Sensitivity of Vertebrate Embryos to Boron Compounds. U.S. Environ. Protection Agen. Rep. 560/1-76-008. 66 pp.

Black, J.A., J.B. Barnum, and W.J. Birge. 1993. An integrated assessment of the biological effects of boron to the rainbow trout. *Chemosphere* 26:1383-1413.

Blevins, D.G. and K.M. Lukaszewski. 1994. Proposed physiologic functions of boron in plants pertinent to animal and human metabolism. *Environ. Health Perspec.* 102(Suppl. 7):31-33.

Boyd, C.E. and W.W. Walley. 1972. Studies of the biogeochemistry of boron. I. Concentrations in surface waters, rainfall and aquatic plants. *Amer. Midl. Nat.* 88:1-14.

Brockman, R.P., R.J. Audette, and M. Gray. 1985. Borax toxicity. *Canad. Vet. Jour.* 26:147.

Brotherton, R.J. and H. Steinberg (eds.). 1970a. *Progress in Boron Chemistry. Vol. 2.* Pergamon Press, New York. 299 pp.

Brotherton, R.J. and H. Steinberg (eds.). 1970b. *Progress in Boron Chemistry. Vol. 3.* Pergamon Press, New York. 392 pp.

Butterwick, L., N. De Oude, and K. Raymond. 1989. Safety assessment of boron in aquatic and terrestrial environments. *Ecotoxicol. Environ. Safety* 17:339-371.

Cayton, M.T.C. 1985. Boron toxicity in rice. *Inter. Rice Res. Inst., Res. Pap. Ser.* 113:3-10.

Chapin, R.E. and W.W. Ku. 1994. The reproductive toxicity of boric acid. *Environ. Health Perspec.* 102(Suppl. 7):87-91.

Chapin, R.E., W.W. Ku, M.A. Kenney, and H. McCoy. 1998. The effects of dietary boric acid on bone strength in rats. *Biol. Trace Elem. Res.* 66:395-399.

Chatterjee, C., P. Sinha, N. Nautiyal, S.C. Agarwala, and C.P. Sharma. 1987. Metabolic changes associated with boron-calcium interactions in maize. *Soil Sci. Plant Nutr.* 33:607-617.

Combrink, N.J.J. and E.A. Davies. 1987. Effect of Ca, K and B supply on growth of cotton (*Gossypium hirsutum* L.) and tobacco (*Nicotiana tabacum* L.). *S. Afr. Jour. Plant Soil* 4:143-144.

Coughlin, J.R. 1998. Sources of human exposure. Overview of water supplies as sources of boron. *Biol. Trace Elem. Res.* 66:87-100.

Cox, J.A., G.C. Lundquist, A. Przyjazmy, and C.D. Schmulback. 1978. Leaching of boron from coal ash. *Environ. Sci. Technol.* 12:722-723.

Crandall, P.C., J.D. Chamberlain, and J.K.L. Garth. 1981. Toxicity symptoms and tissue levels associated with excess boron in trees. Comm. *Soil Sci. Plant. Anal.* 12:1047-1057.

Culver, B.D., P.T. Shen, T.H. Taylor, A. Lee-Feldstein, H. Anton-Culver, and P.L. Strong. 1994. The relationship of blood- and urine-boron to boron exposure in borax-workers and the usefulness of urine-boron as an exposure marker. *Environ. Health Perspec.* 102(Suppl. 7):133-137.

Culver, B.D., F.M. Sullivan, F.J. Murray, J.R. Coughlin, and P.L. Strong (eds.). 1998. Proceedings of the second international symposium on the health effects of boron and its compounds. *Biol. Trace Elem. Res.* 66:1-473.

Dani, H.M., H.S. Saini, I.S. Allag, B. Singh, and K. Sareen. 1971. Effect of boron toxicity on protein and nucleic acid contents of rat tissues. *Res. Bull. (N.S.) Panjab Univ.* 22:229-235.

Dear, B. S. and J. Lipsett. 1987. The effect of boron supply on the growth and seed production of subterranean clover (*Trifolium subterraneum* L.). *Aust. Jour. Agric. Res.* 38:537-546.

Dieter, M.P. 1994. Toxicity and carcinogenicity studies of boric acid in male and female $B6C3F_1$ mice. *Environ. Health Perspec.* 102(Suppl. 7):93-97.

Dixon, R.L., I.P. Lee, and R.J. Sherins. 1976. Methods to assess reproductive effects of environmental chemicals: studies of cadmium and boron administered orally. *Environ. Health Perspect.* 13:59-67.

Dourson, M., A. Maier, B. Meek, A. Renwick, E. Ohanian, and K. Poirier. 1998. Boron tolerable intake. Re-evaluation of toxicokinetics for data-derived uncertainty factors. *Biol. Trace Elem. Res.* 66:453-463.

Downing, R.G., P.L. Strong, B.M. Hovanec, and J. Northington. 1998. Considerations in the determination of boron at low concentrations. *Biol. Trace Elem. Res.* 66:3-21.

Dupre, J.N., M.J. Keenan, M. Hegsted, and A.M. Brudevold. 1994. Effect of dietary boron in rats fed a Vitamin D-deficient diet. *Environ. Health Perspec.* 102(Suppl. 7):55-58.

Dye, M.H., L. Buchanan, F.D. Dorofaeff, and F.G. Beecroft. 1983. Die-back of apricot trees following soil application of boron. *N.Z. Jour. Exp. Agric.* 11:331-342.

Eaton, F.M. 1944. Deficiency, toxicity, and accumulation of boron in plants. *Jour. Agric. Res.* 69:237-277.

Egfjord, M., J.A. Jansen, H. Flachs, and J.S. Schou. 1988. Combined boric acid and cinchocaine chloride poisoning in a 12-month-old infant: evaluation of haemodialysis. *Human Toxicol.* 7:175-178.

Eisler, R. 1981. *Trace Metal Concentrations in Marine Organisms.* Pergamon Press, New York. 687 pp.

Eisler, R. 1990. Boron Hazards to Fish, Wildlife, and Invertebrates: A Synoptic Review. U.S. Fish Wildl. Serv. Biol. Rep. 85 (1.20). 32 pp.

Elsair, J., R. Merad, R. Denine, M. Azzouz, K. Khelfat, M. Hamrour, B. Alamir, S. Benali, and M. Reggabi. 1981. Boron as antidote to fluoride: effect on bones and claws in subacute intoxication of rabbits. *Fluoride* 14:21-29.

Elsair, J., R. Merad, R. Denine, M. Reggabi, B. Alamir, S. Benali, M. Azzouz, and K. Khelfat. 1980b. Boron as a preventive antidote in acute and subacute fluoride intoxication in rabbits: its action on fluoride and calcium-phosphorus metabolism. *Fluoride* 13:129-138.

Elsair, J., R. Merad, R. Denine, M. Reggabi, S. Benali, M. Azzouz, K. Khelfat, and M.T. Aoul. 1980a. Boron as an antidote in acute fluoride intoxication in rabbits: its action on the fluoride and calcium-phosphorus metabolism. *Fluoride* 13:30-38.

Emin, D., T. Aselage, C.L. Beckel, I.A. Howard, and C. Wood (eds.). 1986. *Boron-Rich Solids.* American Inst. Physics, 21 Congress St., Salem, MA. 389 pp.

Evans, C.M. and D.L. Sparks. 1983. On the chemistry and mineralogy of boron in pure and in mixed systems: a review. *Comm. Soil Sci. Plant Anal.* 14:827-846.

Fairbrother, A., M. Fix, T. O'Hara, and C.A. Ribic. 1994. Impairment of growth and immune function of avocet chicks from sites with elevated selenium, arsenic, and boron. *Jour. Wildl. Dis.* 30:222-233.

Fannin, T.E. 1991. Contaminant residues in sandhill cranes killed upon striking powerlines in central Nebraska. *Proc. North Amer. Crane Workshop* 6:166-170.

Fernandez, E., E. Sanchez, I. Bonilla, P. Mateo, and P. Ortega. 1984. Effect of boron on the growth and cell composition of *Chlorella pyrenoidosa*. *Phyton* 44:125-131.

Fort, D.J., T.L. Propst, E.L. Stover, P.L. Strong, and F.J. Murray. 1998. Adverse reproductive and developmental effects in *Xenopus* from insufficient boron. *Biol. Trace Elem. Res.* 66:237-259.

Francois, L.E. 1986. Effect of excess boron on broccoli, cauliflower, and radish. *Jour. Amer. Hort. Soc.* 111:494-498.

Frick, H. 1985. Boron tolerance and accumulation in the duckweed, *Lemna minor. Jour. Plant Nutr.* 8:1123-1129.

Garabrant, D.H., L. Bernstein, J.M. Peters, and T.J. Smith. 1984. Respiratory and eye irritation from boron oxide and boric acid dusts. *Jour. Occup. Med.* 26:584-586.

Garabrant, D.H., L. Bernstein, J.M. Peters, T.J. Smith, and W.E. Wright. 1985. Respiratory effects of borax dust. *Brit. Jour. Ind. Med.* 42:831-837.

Gersich, F.M. 1984. Evaluation of a static renewal chronic toxicity test method for *Daphnia magna* Straus using boric acid. *Environ. Toxicol. Chem.* 3:89-94.

Gestring, W.D. and P.N. Soltanpour. 1987. Comparison of soil tests for assessing boron toxicity to alfalfa. *Soil Sci. Soc. Amer. Jour.* 51:1214-1219.

Gladney, E.S., L.E. Wangen, D.B. Curtis, and E.T. Jurney. 1978. Observations on boron release from coal-fired power plants. *Environ. Sci. Technol.* 12:1084-1085.

Glandon, R.P. and C.D. McNabb. 1978. The uptake of boron by *Lemna minor. Aquat. Botany* 4:53-64.

Glaubig, B.A. and F.T. Bingham. 1985. Boron toxicity characteristics of four northern California endemic tree species. *Jour. Environ. Qual.* 14:72-77.

Goldbach, H. and A. Amberger. 1986. Influence of boron nutrition on cell wall polysaccharides in cell cultures of *Daucus carota* L. Jour. *Plant Physiol.* 123:263-269.

Goldberg, S. and R.A. Glaubig. 1986. Boron adsorption on California soils. *Soil Sci. Soc. Amer. Jour.* 50:1173-1176.

Graham, R.D., R.M. Welch, D.L. Grunes, E.E. Cary, and W.A. Norvell. 1987. Effect of zinc deficiency on the accumulation of boron and other mineral nutrients in barley. *Soil Sci. Soc. Amer. Jour.* 51:652-657.

Green, G.H. and H.J. Weeth. 1977. Responses of heifers ingesting boron in water. *Jour. Anim. Sci.* 46:812-818.

Greenwood, N.N. and B.S. Thomas. 1973. *The Chemistry of Boron.* Pergamon Press, Elmsford, New York. 326 pp.

Grimes, R.N. (ed.). 1982. *Metal Interactions with Boron Clusters.* Plenum Press, New York. 327 pp.

Gupta, A.P. and M.D. Parrish. 1984. Effectiveness of a new boric acid bait (roach killer cream) on German cockroach (*Blattella germanica*) populations in urban dwellings. *Uttar Pradesh Jour. Zool.* 4:51-56.

Gupta, U.C. 1979. Boron nutrition of crops. *Adv. Agron.* 31:273-307.

Gupta, U.C. 1983. Boron deficiency and toxicity symptoms for several crops as related to tissue boron levels. *Jour. Plant Nutr.* 6:387-395.

Gupta, U.C. and J.A. Cutcliffe. 1985. Boron nutrition of carrots and table beets grown in a boron deficient soil. *Comm. Soil Sci. Plant Anal.* 16:509-516.

Gupta, U.C., Y.W. Jame, C.A. Campbell, A.J. Leyshon, and W. Nicholaichuk. 1985. Boron toxicity and deficiency: a review. *Canad. Jour. Soil Sci.* 65:381-409.

Gupta, U.C. and J.A. Macleod. 1982. Role of boron in crop production. *Ind. Jour. Agric. Chem.* 15:117-132.

Hall, I.H., S.Y. Chen, K.G. Rajendran, A. Sood, B.F. Spielvogel, and J. Shih. 1994. Hypolipidemic, anti-obesity, anti-inflammatory, anti-osteoporotic, and anti-neoplastic properties of amine carboxyboranes. *Environ. Health Perspec.* 102(Suppl. 7):21-30.

Hamada, T., K. Aoki, T. Kobayashi, and K. Kanda. 1983. The in vivo measurement of the time-dependent ^{10}B movement in tumor of hamsters. *Annu. Rep. Res. Reactor Inst. Kyoto Univ.* 16:112-116.

Hamilton, S.J. 1995. Hazard assessment of inorganics to three endangered fish in the Green River, Utah. *Ecotoxicol. Environ. Safety* 30:134-142.

Hamilton, S.J. and K.J. Buhl. 1990. Acute toxicity of boron, molybdenum, and selenium to fry of chinook salmon and coho salmon. *Arch. Environ. Contam. Toxicol.* 19:366-373.

Hamilton, S.J. and R.H. Wiedmeyer. 1990. Concentrations of boron, molybdenum, and selenium in chinook salmon. *Trans. Amer. Fish. Soc.* 119:500-510.

Hatanaka, H. (ed.). 1986. *Boron-Neutron Capture Therapy for Tumors.* MTP Press, 101 Philip Dr., Norwell, MA. 463 pp.

Heindel, J.J., C.J. Price, and B.A. Schwetz. 1994. The developmental toxicity of boric acid in mice, rats, and rabbits. *Environ. Health Perspec.* 102(Suppl. 7):107-112.

Heller, G. 1986. A survey of structural types of borates and polyborates. Pages 39-100 *in* F.L. Boschke (ed.). *Structural Chemistry of Boron and Silicon.* Springer-Verlag, Berlin.

Hermanek, S. (ed.). 1987. *Sixth International Meeting on Boron Chemistry.* World Scientific Publ. Co., Teaneck, NJ. 509 pp.

Hingston, F.J. 1986. Biogeochemical cycling of boron in native eucalypt forests of south-western Australia. *Aust. For. Res.* 16:73-83.

Hoffman, D.L., M.B. Camardese, L.J. LeCaptain, and G.W. Pendleton. 1990. Effects of boron on growth and physiology in mallard ducklings. *Environ. Toxicol. Chem.* 9:335-346.

Hoffman, D.J., H.M. Ohlendorf, and T.W. Aldrich. 1988. Selenium teratogenesis in natural populations of aquatic birds in central California. *Arch. Environ. Contam. Toxicol.* 17:519-525.

Hoffman, D.J., C.J. Sanderson, L.J. LeCaptain, E. Cromartie, and G.W. Pendleton. 1991. Interactive effects of boron, selenium, and dietary protein on survival, growth, and physiology in mallard ducklings. *Arch. Environ. Contam. Toxicol.* 20:288-294.

Hopmans, P. and D.W. Flinn. 1984. Boron deficiency in *Pinus radiata* D. Don and the effect of applied boron on height growth and nutrient uptake. *Plant Soil* 79:295-298.

Hothem, R.L. and H.M. Ohlendorf. 1989. Contaminants in foods of aquatic birds at Kesterson Reservoir, California, 1985. *Arch. Environ. Contam. Toxicol.* 18:773-786.

Howe, P.D. 1998. A review of boron effects in the environment. *Biol. Trace Elem. Res.* 66:153-166.

Hubbard, S.A. 1998. Comparative toxicology of borates. *Biol. Trace Elem. Res.* 66:343-357.

Hui, C.A. and W.N. Beyer. 1998. Sediment ingestion of two sympatric shorebird species. *Sci. Total Environ.* 224:227-233.

Hunt, C.D. 1994. The biochemical effects of physiologic amounts of dietary boron in animal nutrition models. *Environ. Health Perspec.* 102(Suppl. 7):35-43.

Hunt, C.D. 1998. Regulation of enzymatic activity. One possible role of dietary boron in higher animals and humans. *Biol. Trace Elem. Res.* 66:205-225.

Jansen, J.A., J. Andersen, and J.S. Schou. 1984. Boric acid single dose pharmacokinetics after intravenous administration to man. *Arch. Toxicol.* 55:64-67.

Jenkins, D.W. 1980. *Biological Monitoring of Toxic Trace Metals. Vol. 2. Toxic Trace Metals in Plants and Animals of the World.* Part I. U.S. Environ. Protection Agen. Rep. 600/3-80-090:150-171.

King, N., T.W. Odom, H.W. Sampson, and S.L. Pardue. 1991. *In ovo* administration of boron alters bone mineralization of the chicken embryo. *Biol. Trace Elem. Res.* 30:47-58.

Klaassen, C.D., M.O. Amdur, and J. Doull (eds.). 1986. *Casarett and Doull's Toxicology. Third Edition.* Macmillan Publ., New York. 974 pp.

Klasing, S.A. and S.M. Pilch. 1988. Agricultural Drainage Water Contamination in the San Joaquin Valley: A Public Health Perspective for Selenium, Boron, and Molybdenum. Avail. from San Joaquin Valley Drainage Program, 2800 Cottage Way, Sacramento, CA 98525. 135 pp.

Kluge, R. and W. Podlesak. 1985. Plant critical levels for the evaluation of boron toxicity in spring barley (*Hordeum vulgare* L.). *Plant Soil* 83:381-388.

Kobayashi, N. 1971. Fertilized sea urchin eggs as an indicatory material for marine pollution bioassay, preliminary experiments. *Publ. Seto Mar. Biol. Lab.* 18:379-406.

Korty, P. and W.N. Scott. 1970. Effects of boranes upon tissues of the rat. II. Tissue amino acid content in rats on a normal diet. *Proc. Soc. Exp. Biol. Med.* 135:629-632.

Krasovskii, G.N., S.P. Varshavskaya, and A.I. Borisov. 1976. Toxic and gonadotropic effects of cadmium and boron relative to standards for these substances in drinking water. *Environ. Health Perspect.* 13:69-75.

Ku, W.W. and R.E. Chapin. 1994. Mechanism of the testicular toxicity of boric acid in rats: *in vivo* and *in vitro* studies. *Environ. Health Perspec.* 102(Suppl. 7):99-105.

Landauer, W. 1952. Malformations of chicken embryos produced by boric acid and the probable role of riboflavin in their origin. *Jour. Exp. Zool.* 120:469-508.

Landauer, W. 1953a. Genetic and environmental factors in the teratogenic effects of boric acid on chicken embryos. *Genetics* 38:216-228.

Landauer, W. 1953b. Complex formation and chemical specificity of boric acid in production of chicken embryo malformations. *Proc. Soc. Exp. Biol. Med.* 82:633-636.

Landauer, W. 1953c. Abnormality of down pigmentation associated with skeletal defects of chicks. *Proc. Nat. Acad. Sci. USA* 39:54-58.

Landauer, W. and E.M. Clark. 1964. On the role of riboflavin in the teratogenic activity of boric acid. *Jour. Exp. Zool.* 156:307-312.

Landolph, J.R. 1985. Cytotoxicity and negligible genotoxicity of borax ores to cultured mammalian cells. *Amer. Jour. Ind. Med.* 7:31-43.

Lanoue, L., M.W. Taubeneck, J. Muniz, L.A. Hanna, P.L. Strong, F.J. Murray, F.H. Nielsen, C.D. Hunt, and C.L. Keen. 1998. Assessing the effects of low boron diets on embryonic and fetal development in rodents using in vitro and *in vivo* model systems. *Biol. Trace Elem. Res.* 66:271-298.

Laposata, M.M. and W.A. Dunson. 1998. Effects of boron and nitrate on hatching success of amphibian eggs. *Arch. Environ. Contam. Toxicol.* 35:615-619.

Lee, I.P., R.J. Sherins, and R.L. Dixon. 1978. Evidence for induction of germinal aplasia in male rats by environmental exposure to boron. *Toxicol. Appl. Pharmacol.* 45:577-590.

Lewis, J.C., R.C. Drewien, E. Kuyt, and C. Sanchez, Jr. 1991. Contaminants in habitat, tissues, and eggs of whooping cranes. *Proc. North Amer. Crane Workshop* 6:159-165.

Lewis, M.A. and L.C. Valentine. 1981. Acute and chronic toxicities of boric acid to *Daphnia magna* Straus. *Bull. Environ. Contam. Toxicol.* 27:309-315.

Liddicoat, M.I., D.R. Turner, and M. Whitfield. 1983. Conservative behaviour of boron in the Tamar estuary. *Estuar. Coastal Shelf Sci.* 17:467-472.

Lipsett, J., A. Pinkerton, and D.J. David. 1979. Boron deficiency as a factor in the reclamation by liming of a soil contaminated by mine waste. *Environ. Pollut.* 20:231-240.

Lizzio, E.F. 1986. A boric acid-rodenticide mixture used in the control of coexisting rodent-cockroach infestations. *Lab. Anim. Sci.* 36:74-76.

Maeso, E.S., E.F. Valiente, I. Bonilla, and P. Mateo. 1985. Accumulation of proteins in giant cells, induced by high boron concentrations in *Chlorella pyrenoidosa. Jour. Plant Physiol.* 121:301-311.

Magour, S., P. Schramel, J. Ovcar, and H. Maser. 1982. Uptake and distribution of boron in rats: interaction with ethanol and hexobarbital in the brain. *Arch. Environ. Contam. Toxicol.* 11:521-525.

Maier, K.J. and A.W. Knight. 1991. The toxicity of waterborne boron to *Daphnia magna* and *Chironomus decorus* and the effects of water hardness and sulfate on boron toxicity. *Arch. Environ. Contam. Toxicol.* 20:282-287.

Martinez, F., P. Matio, I. Bonilla, and E. Fernandez-Valiente. 1986a. Cellular changes due to boron toxicity in the blue-green alga *Anacystis nidulans. Phyton* 46:145-152.

Martinez, F., P. Mateo, I. Bonilla, E. Fernandez-Valiente, and A. Garate. 1986b. Growth of *Anacystis nidulans* in relation to boron supply. *Israel Jour. Bot.* 35:17-21.

Massie, H.R. 1994. Effect of dietary boron on the aging process. *Environ. Health Perspec.* 102(Suppl. 7):45-48.

Mastromatteo, E. and F. Sullivan. 1994. Summary: international symposium on the health effects of boron and its compounds. *Environ. Health Perspec.* 102(Suppl. 7):139-141.

Mateo, P., F. Martinez, I. Bonilla, E.F. Valiente, and E.S. Maeso. 1987. Effects of high boron concentrations on nitrate utilization and photosynthesis in blue-green algae *Anabaena* PCC 7119 and *Anacystis nidulans. Jour. Plant Physiol.* 128:161-168.

McCoy, H., M.A. Kenney, C. Montgomery, A. Irwin, L. Williams, and R. Orrell. 1994. Relation of boron to the composition and mechanical properties of bone. *Environ. Health Perspec.* 102(Suppl. 7):49-53.

Meacham, S.L. and C.D. Hunt. 1998. Dietary boron intakes of selected populations in the United States. *Biol. Trace Elem. Res.* 66:65-78.

Miah, M.M.U. 1999. Evaluation of critical limits of trace elements Zn, B, and Mo in Bangladesh agriculture. Pages 432-433 in W.W. Wenzel, D.C. Adriano, B. Alloway, H.E. Doner, C. Keller, N.W. Lepp, M. Mench, R. Naidu, and G. M. Pierzynski (eds.). Proceedings of the Fifth International Conference on the Biogeochemistry of Trace Elements. July 11-15, 1999, Vienna, Austria.

Mitchell, R.J., H.E. Garrett, G.S. Cox, A. Atalay, and R.K. Dixon. 1987. Boron fertilization, ecotomycorrhizal colonization, and growth of *Pinus echinata* seedlings. *Canad. Jour. For. Res.* 17:1153-1156.

Mora, M.A. and D.W. Anderson. 1995. Selenium, boron, and heavy metals in birds from the Mexicali Valley, Baja California, Mexico. *Bull. Environ. Contam. Toxicol.* 54:198-206.

Moseman, R.F. 1994. Chemical disposition of boron in animals and humans. *Environ. Health Perspec.* 102(Suppl. 7):113-117.

Mozafar, A. 1987. Effect of boron on ear formation and yield components of two maize (*Zea mays* L.) hybrids. *Jour. Plant Nutr.* 10:319-332.

Murray, F.J. 1998. A comparative review of the pharmacokinetics of boric acid in rodents and humans. *Biol. Trace Elem. Res.* 66:331-341.

Naeger, L.L. and K.C. Leibman. 1972. Mechanisms of decaborane toxicity. *Toxicol. Appl. Pharmacol.* 22:517-527.

Narotsky, M.G., J.E. Schmid, J.E. Andrews, and R.J. Kavlock. 1998. Effects of boric acid on axial skeletal development in rats. *Biol. Trace Elem. Res.* 66:373-394.

Narvekar, P.V., M.D. Zingde, and V.N.K. Dalal. 1983. Behaviour of boron, calcium and magnesium in a polluted estuary. *Estuar. Coastal Shelf Sci.* 16:9-16.

National Academy of Sciences (NAS). 1980. Boron. Pages 71-83 in *Mineral Tolerance of Domestic Animals.* Natl. Acad. Sci., Natl. Res. Coun., Comm. Anim. Nutr. Washington, D.C.

Newnham, R.E. 1994. Essentiality of boron for healthy bones and joints. *Environ. Health Perspec.* 102(Suppl. 7):83-85.

Nicholaichuk, W., A.J. Leyshon, Y.W. Jame, and C.A. Campbell. 1988. Boron and salinity survey of irrigated projects and the boron adsorption characteristics of some Saskatchewan soils. *Canad. Jour. Soil Sci.* 68:77-90.

Niedenzu, K. and S. Trofimenko. 1986. Pyrazole derivatives of boron. Pages 1-37 in F.L. Boschke (ed.). *Structural Chemistry of Boron and Silicon.* Springer-Verlag, Berlin.

Nielsen, F.H. 1986. Other elements: Sb, Ba, B, Br, Cs, Ge, Rb, Ag, Sr, Sn, Ti, Zr, Be, Bi, Ga, Au, In, Nb, Sc, Te, Tl, W. Pages 415-463 in W. Mertz (ed.). *Trace Elements in Human and Animal Nutrition. Vol. 2.* Academic Press, New York.

Nielsen, F.H. 1994. Biochemical and physiologic consequences of boron deprivation in humans. *Environ. Health Perspec.* 102(Suppl. 7):59-63.

Nielsen, F.H., C.D. Hunt, L.M. Mullen, and J.R. Hunt. 1987. Effect of dietary boron on mineral, estrogen, and testosterone metabolism in postmenopausal women. *FASEB Jour.* 1:394-397.

Nuttall, W.F., H. Ukrainetz, J.W.B. Stewart, and D.T. Spurr. 1987. The effect of nitrogen, sulphur and boron on yield and quality of rapeseed (*Brassica napus* L. and *B. campestris* L.). *Canad. Jour. Soil Sci.* 67:545-559.

Ohlendorf, H.M., D.J. Hoffman, M.K. Saiki, and T.W. Aldrich. 1986. Embryonic mortality and abnormalities of aquatic birds: apparent impacts of selenium from irrigation drainwater. *Sci. Total Environ.* 52:49-63.

Okay, O., H. Guclu, E. Soner, and T. Balkas. 1985. Boron pollution in the Simav River, Turkey and various methods of boron removal. *Water Res.* 19:857-862.

O'Sullivan, K. and M. Taylor. 1983. Chronic boric acid poisoning in infants. *Arch. Dis. Child.* 58:737-739.

Pagenkopf, G.K. and J.M. Connolly. 1982. Retention of boron by coal ash. *Environ. Sci. Technol.* 16:609-613.

Papachristou, E., R. Tsitouridou, and B. Kabasakalis. 1987. Boron levels in some groundwaters of Halkidiki (a land at northern Aegean Sea). *Chemosphere* 16:419-427.

Parry, R.W. and G. Kodama (eds.). 1980. *Boron Chemistry — 4.* Pergamon Press, New York. 161 pp.

Paveglio, F.L., C.M. Bunck, and G.H. Heinz. 1992. Selenium and boron in aquatic birds from central California. *Jour. Wildl. Manage.* 56:31-42.

Pendleton, G.W., M.R. Whitworth, and G.H. Olsen. 1995. Accumulation and loss of arsenic and boron, alone and in combination, in mallard ducks. *Environ. Toxicol. Chem.* 14:1357-1364.

Penland, J.G. 1994. Dietary boron, brain function, and cognitive performance. *Environ. Health Perspec.* 102(Suppl. 7):65-72.

Penland, J.G. 1998. The importance of boron nutrition for brain and psychological function. *Biol. Trace Elem. Res.* 66:299-317.

Petracek, P.D. and C.E. Sams. 1987. The influence of boron on the development of broccoli plants. *Jour. Plant Nutr.* 10:2095-2107.

Pilbeam, D.J. and E.A. Kirkby. 1983. The physiological role of boron in plants. *Jour. Plant Nutr.* 6:563-582.

Price, C.J., P.L. Strong, F.J. Murray, and M.M. Goldberg. 1998. Developmental effects of boric acid in rats related to maternal blood boron concentrations. *Biol. Trace Elem. Res.* 66:359-372.

Qin, X. and H. Klandorf. 1991. Effect of dietary boron supplementation on egg production, shell quality, and calcium metabolism in aged breeder hens. *Poultry Sci.* 70:2131-2138.

Rainey, C. and L. Nyquist. 1998. Multicountry estimation of dietary boron intake. *Biol. Trace Elem. Res.* 66:79-86.

Ransome, C.S. and R.H. Dowdy. 1987. Soybean growth and boron distribution in a sandy soil amended with scrubber sludge. *Jour. Environ. Qual.* 16:171-175.

Rao, D.V.S. 1981. Effect of boron on primary production of nanoplankton. *Canad. Jour. Fish. Aquat. Sci.* 38:52-58.

Rerkasem, B., R. Netsangtip, R.W. Bell, J.F. Loneragen, and N. Hiranburana. 1988. Comparative species responses to boron on a typic tropaqualf in northern Thailand. *Plant Soil* 106:15-21.

Ridgway, L.P. and D.A. Karnofsky. 1952. The effects of metals on the chick embryo: toxicity and production of abnormalities in development. *Ann. N.Y. Acad. Sci.* 55:203-215.

Riley, M.M. 1987. Boron toxicity in barley. *Jour. Plant Nutr.* 10:2109-2115.

Rowe, R.I., C. Bouzan, S. Nabili, and C.D. Eckhert. 1998. The response of trout and zebrafish embryos to low and high boron concentrations is U-shaped. *Biol. Trace Elem. Res.* 66:261-270.

Rusch, G.M., G.M. Hoffman, R.F. McConnell, and W.E. Rinehart. 1986. Inhalation toxicity studies with boron trifluoride. *Toxicol. Appl. Pharmacol.* 83:69-78.

Sage, R.F., S.L. Ustin, and S.J. Manning. 1989. Boron toxicity in the rare serpentine plant, *Streptanthus morrisonii*. *Environ. Pollut.* 61:77-93.

Saiki, M.K., M.R. Jennings, and W.G. Brumbaugh. 1993. Boron, molybdenum, and selenium in aquatic food chains from the lower San Joaquin River and its tributaries, California. *Arch. Environ. Contam. Toxicol.* 24:307-319.

Saiki, M.K., M.R. Jennings, and R.H. Wiedmeyer. 1992. Toxicity of agricultural subsurface drainwater from the San Joaquin Valley, California, to juvenile chinook salmon and striped bass. *Trans. Amer. Fish. Soc.* 121:78-93.

Samman, S., M.R. Naghii, P.M.L. Wall, and A.P. Verus. 1998. The nutritional and metabolic effects of boron in humans and animals. *Biol. Trace Elem. Res.* 66:227-235.

Sayli, B.S. 1998. An assessment of fertility in boron-exposed Turkish subpopulations. 2. Evidence that boron has no effect on human reproduction. *Biol. Trace Elem. Res.* 66:409-422.

Schillinger, B.M., M. Berstein, L.A. Goldberg, and A.R. Shalita. 1982. Boric acid poisoning. *Jour. Amer. Acad. Dermatol.* 7:667-673.

Schowing, J. and P. Cuevas. 1975. Teratogenic effects of boric acid upon the chick. Macroscopic results. *Teratology* 12:334.

Schowing, J., P. Cuevas, and J. Ventosa. 1976. Influence de l'acide borique sur le developpment de l'embryon de poulet traite a un stade precoce resultats preliminaires. *Arch. Biol. (Brussels)* 87:385-392.

Schuler, C.A. 1987. Impacts of Agricultural Drainwater and Contaminants on Wetlands at Kesterson Reservoir, California. M.S. Thesis. Oregon State Univ., Corvallis, OR. 136 pp.

Seal, B.S. and H.J. Weeth. 1980. Effect of boron in drinking water on the male laboratory rat. *Bull. Environ. Contam. Toxicol.* 25:782-789.

Settimi, L., E. Elovaara, and H. Savolaninen. 1982. Effects of extended peroral borate ingestion on rat liver and brain. *Toxicol. Lett.* 10:219-223.

Shapiro, M. and R.A. Bell. 1982. Enhanced effectiveness of *Lymantria dispar* (Lepidoptera: Lymantriidae) nucleopolyhedrosis virus formulated with boric acid. *Ann. Entomol. Soc. Am.* 75:346-349.

Shorrocks, V.M. and D.D. Nicholson. 1980. The influence of boron deficiency on fruit quality. *Acta Hortic.* 92:103-108.

Siegel, E. and S. Wason. 1986. Boric acid toxicity. *Pediatr. Clin. North Amer.* 33:363-367.

Slatkin, D., P. Micca, A. Forman, D. Gabel, L. Wielopolski, and R. Fairchild. 1986. Boron uptake in melanoma, cerebrum and blood from $Na_2B_{12}H_{11}SH$ and $Na_4B_{24}H_{22}S_2$ administered to mice. *Biochem. Pharmacol.* 35:1771-1776.

Smith, G.J. and V.P. Anders. 1989. Toxic effects of boron on mallard reproduction. *Environ. Toxicol. Chem.* 8:943-950.

Smith, K. (ed.). 1985. *Organometallic Compounds of Boron.* Chapman and Hall, New York. 304 pp.

Smyth, D.A. and W.M. Dugger. 1980. Effects of boron deficiency on ^{86}rubidium uptake and photosynthesis in the diatom *Cylindrotheca fusiformis*. *Plant Physiol.* 66:692-695.

Smyth, D.A. and W.M. Dugger. 1981. Cellular changes during boron-deficient culture of the diatom *Cylindrotheca fusiformis*. *Physiol. Plant.* 51:111-117.

Sprague, R.W. 1972. The Ecological Significance of Boron. United States Borax and Chemical Corp., Los Angeles, CA. 58 pp.

Stanley, T.R., Jr. G.J. Smith, D.J. Hoffman, G.H. Heinz, and R. Rosscoe. 1996. Effects of boron and selenium on mallard reproduction and duckling growth and survival. *Environ. Toxicol. Chem.* 15:1124-1132.

Steinberg, H. and A.L. McCloskey (eds.). 1964. *Progress in Boron Chemistry. Vol. 1.* Pergamon Press, New York. 487 pp.

Stone, E.L. 1990. Boron deficiency and excess in forest trees: a review. *Forest Ecol. Manage.* 37:49-75.

Sullivan, F.M. and B.D. Culver. 1998. Summary of research needs. *Biol. Trace. Elem. Res.* 66:465-468.

Sutherland, B., P. Strong, and J.C. King. 1998. Determining human dietary requirements for boron. *Biol. Trace Elem. Res.* 66:193-204.

Taylor, D., B.G. Maddock, and G. Mance. 1985. The acute toxicity of nine grey list metals (arsenic, boron, chromium, copper, lead, nickel, tin, vanadium and zinc) to two marine fish species: dab (*Limanda limanda*) and grey mullet (*Chelon labrosus*). *Aquat. Toxicol.* 7:135-144.

Thompson, J.A.J., J.C. Davis, and R.E. Drew. 1976. Toxicity, uptake and survey studies of boron in the marine environment. *Water Res.* 10:869-875.

Tuccar, E, A.H. Elhan, Y. Yavuz, and B.S. Sayli. 1998. Comparison of infertility rates in communities from boron-rich and boron-poor territories. *Biol. Trace Elem. Res.* 66:401-407.

U.S. Environmental Protection Agency (USEPA). 1975. *Preliminary Investigation of Effects on the Environment of Boron, Indium, Nickel, Selenium, Tin, Vanadium and Their Compounds. Vol. 1.* Boron. U.S. Environ. Protection Agen. Rep. 56/2-75-005A. 111 pp.

U.S. Public Health Service (USPHS). 1991. Toxicological Profile for Boron. Draft for Public Comment. U.S. Dept. Health Human Serv., Agen. Toxic Subst. Dis. Regis. 98 pp.

Vanderpool, R.A., D. Hoff, and P.E. Johnson. 1994. Use of inductively coupled plasma-mass spectrometry in boron-10 stable isotope experiments with plants, rats, and animals. *Environ. Health Perspec.* 102(Suppl. 7):13-20.

Vengosh, A. 1998. The isotopic composition of anthropogenic boron and its potential impact on the environment. *Biol. Trace Elem. Res.* 66:145-151.

Walker, W.M., D.W. Graffis, and C.D. Faulkner. 1987. Effect of potassium and boron upon yield and nutrient concentration of alfalfa. *Jour. Plant Nutr.* 10:2169-2180.

Weeth, H.J., C.F. Speth, and D.R. Hanks. 1981. Boron content of plasma and urine as indicators of boron intake in cattle. *Amer. Jour. Vet. Res.* 42:474-477.

Wegman, D.H., E.A. Eisen, X. Hu, S.R. Woskie, R.G. Smith, and D.H. Garabrant. 1994. Acute and chronic respiratory effects of sodium borate particulate exposures. *Environ. Health Perspec.* 102(Suppl. 7):119-128.

Weir, R.J., Jr. and R.S. Fisher. 1972. Toxicologic studies on borax and boric acid. *Toxicol. Appl. Pharmacol.* 23:351-364.

Wester, R.C., T. Hartway, H.I. Maibach, M.J. Schell, D.J. Northington, B.D. Culver, and P.L. Strong. 1998a. *In vitro* percutaneous absorption of boron as boric acid, borax, and disodium octaborate tetrahydrate in human skin. A summary. *Biol. Trace Elem. Res.* 66:111-120.

Wester, R.C., X. Hui, H.I. Maibach, K. Bell, M.J. Schell, D.J. Northington, P. Strong, and B.D. Culver. 1998b. *In vivo* percutaneous absorption of boron and boric acid, borax, and disodium octaborate tetrahydrate in humans. A summary. *Biol. Trace Elem. Res.* 66:101-109.

Whitworth, M.R., G.W. Pendleton, D.J. Hoffman, and M.B. Camardese. 1991. Effects of dietary boron and arsenic on the behavior of mallard ducklings. *Environ. Toxicol. Chem.* 10:911-916.

Willett, I.R., P. Jakobsen, and B.A. Zarcinas. 1985. Nitrogen-induced boron deficiency in lucerne. *Plant Soil* 86:443-446.

Woods, W.G. 1994. An introduction to boron: history, sources, uses, and chemistry. *Environ. Health Perspect.* 102(Suppl. 7):5-11.

Wright, C.G. and H.E. Dupree, Jr. 1982. Efficacy of experimental formulations of acephate, boric acid, encapsulated diazinon, permethrin, pirimphos-methyl, and propetamphos in control of German cockroaches. *Jour. Georgia Entomol. Soc.* 17:26-32.

Zhou, L.Y., Z.D. Wei, and S.Z. Ldu. 1987. Effect of borax in treatment of skeletal fluorosis. *Fluoride* 20:24-27.

CHAPTER 30

Molybdenum

30.1 INTRODUCTION

Molybdenum (Mo) is present in all plant, human, and animal tissues, and is considered an essential micronutrient for most life forms (Schroeder et al. 1970; Underwood 1971; Chappell and Peterson 1976; Chappell et al. 1979; Goyer 1986). The first indication of an essential role for molybdenum in animal nutrition came in 1953 when it was discovered that a flavoprotein enzyme, xanthine oxidase, was dependent on molybdenum for its activity (Underwood 1971). It was later determined that molybdenum is essential in the diet of lambs, chicks, and turkey poults (Underwood 1971). Molybdenum compounds are now routinely added to soils, plants, and waters to achieve various enrichment or balance effects (Friberg et al. 1975; Friberg and Lener 1986).

There are certain locations where plants will not grow optimally because of a deficiency in molybdenum, and other places where the levels of molybdenum in plants are toxic to livestock grazing on the plants (Chappell and Peterson 1976). Molybdenum poisoning in cattle was first diagnosed in England in 1938; molybdenosis was shown to be associated with consumption of herbage containing large amounts of this element, and to be controllable by treatment with copper sulfate (Underwood 1971). Molybdenum poisoning of ruminants, especially cattle, has been reported in at least 15 states, and in Canada, England, Australia, New Zealand, Ireland, the Netherlands, Japan, and Hungary. Molybdenosis was most pronounced in areas where soils were alkaline, high in molybdenum and low in copper, or near industrial point sources such as coal, aluminum, uranium, or molybdenum mines; steel alloy mills; or oil refineries (Dollahite et al. 1972; Alloway 1973; Kubota 1975; Buck 1978; Ward 1978; Chappell et al. 1979; Kincaid 1980; King et al. 1984; Kume et al. 1984; Sas 1987). All cattle are susceptible to molybdenosis, milking cows and young stock being the most sensitive (Underwood 1971). Industrial molybdenosis in domestic cattle and sheep, which usually involved a single farm or pasture, has been widely documented (Ward 1978):

- In Colorado in 1958 from contaminated river waters used in irrigation
- In Alabama in 1960 from mine spoil erosion
- In North Dakota in 1968 from flyash from a lignite burning plant
- In Missouri in 1970-1972 from clay pit erosion
- In Pennsylvania in 1971 from aerial contamination by a molybdenum smelter
- In South Dakota in 1975 from molybdenum-contaminated magnesium oxide
- In Texas in 1965-1972 from uranium mine waste leachate

In humans, a gout-like disease in two villages in Armenia was attributed to the ingestion of local foods high in molybdenum and grown in soils high in molybdenum (Friberg and Lener 1986).

Esophageal cancer was prevalent in various parts of southern Africa where food was grown in low molybdenum soils; it was reported in China in a low frequency rate that was significantly correlated with increasing molybdenum concentrations in cereals and drinking water (Luo et al. 1983). Additional and more extensive data on ecological and toxicological aspects of molybdenum in the environment were reviewed by Schroeder et al. (1970), Underwood (1971), Friberg et al. (1975), Chappell and Peterson (1976, 1977), Ward (1978), Chappell et al. (1979), Gupta and Lipsett (1981), Friberg and Lener (1986), and Eisler (1989).

30.2 ENVIRONMENTAL CHEMISTRY

30.2.1 General

Molybdenum is a comparatively rare element that is used primarily in the manufacture of steel alloys for the aircraft and weapons industries. Most of the global production of about 100,000 tons annually comes from the United States — primarily Colorado. Anthropogenic activities that have contributed to environmental molybdenum contamination include combustion of fossil fuels, and smelting, mining, and milling operations for steel, copper, and uranium, as well as for molybdenum. In general, the chemistry of molybdenum is complex and inadequately known. Its toxicological properties are governed to a remarkable extent by interactions with copper and sulfur, although other metals and compounds may confound this interrelation.

30.2.2 Sources and Uses

Molybdenum is used in the manufacture of high-strength, low-alloy steels and other steel alloys in the aircraft and weapons industries, and in the production of spark plugs, X-ray tubes and electrodes, catalysts, pigments, and chemical reagents (Friberg et al. 1975; Kummer 1980; Goyer 1986). The most important industrial compound is the trioxide, MoO_3, which is resistant to most acids and is oxidized in air at >500°C (Shamberger 1979).

Molybdenum, discovered about 200 years ago, entered the commercial market in the 1920s as a result of extensive metallurgical research into its alloying properties and to the finding at Climax, Colorado, of the largest proven reserves of molybdenum worldwide (King et al. 1973). Molybdenum does not occur free in nature and is found only in combination with sulfur, oxygen, tungsten, lead, uranium, iron, magnesium, cobalt, vanadium, bismuth, or calcium. The most economically important ores are molybdenite (MoS_2), jordisite (amorphous MoS_2), and ferrimolybdate ($FeMoO_3 \cdot H_2O$); less important are wulfenite ($PbMoO_4$), powellite ($CaMoO_4$), and ilsemannite (Mo_3O_8) (Friberg et al. 1975; Chappell et al. 1979; Friberg and Lener 1986; Goyer 1986).

World molybdenum production has increased from about 90 metric tons in 1900 — half from Australia and Norway, half from the United States — to 136 tons in 1906, 1364 in 1932 (an order of magnitude increase in 26 years), 10,909 in 1946, and 91,000 tons in 1973. Through the years, molybdenum has been produced in about 30 countries. In 1973, about 60% of the worldwide production was from the United States, 15% from Canada, 15% from the U.S.S.R. and China combined, and 10% from other nations — Chile, Japan, Korea, Norway, and Mexico (King et al. 1973). By 1979, the United States produced about 62% of the world production of 103,000 metric tons, and exported about half, chiefly to western Europe and Japan; other major producers in 1979 were Canada, Chile, and the U.S.S.R. (Kummer 1980). In the United States, only three mines in Colorado account for almost 70% of domestic production. Other active molybdenum mining sites in North America are in Arizona, Nevada, New Mexico, Utah, and California; molybdenum reserves have also been proven in Idaho, Alaska, Pennsylvania, and British Columbia (Kummer 1980). About 65% of domestic molybdenum is recovered from ores rich in molybdenum; the rest is a by-product from ores of copper, tungsten, and uranium (Chappell et al. 1979).

As a result of various human activities, molybdenum enters the environment from many sources (King et al. 1973; Friberg et al. 1975; Chappell et al. 1979). Coal combustion is the largest atmospheric source of molybdenum, contributing about 550 metric tons annually, or 61% of all atmospheric molybdenum worldwide that comes from anthropogenic sources. In Sweden alone, about 2.5 tons molybdenum are emitted into the atmosphere yearly from oil combustion (Friberg et al. 1975). Molybdenum mining and milling are the source of about 100 metric tons annually to aquatic systems. At the world's largest molybdenum mine in Climax, Colorado, where about 36,000 tons of tailings are generated daily, the operation releases up to 100 tons of molybdenum annually as aqueous effluent. Other sources are molybdenum smelting, uranium mining and milling, steel and copper milling, oil refining, shale oil production, and claypit mining.

30.2.3 Chemical Properties

Molybdenum, which can function both as a metal and metalloid, is an essential component in a large number of biochemical systems — including xanthine oxidase. At least four metalloenzymes are known to be molybdenum dependent, and all are molybdoflavoproteins (Schroeder et al. 1970). Molybdenum is characterized by the following physical and chemical properties: atomic number 42; atomic weight 95.94; density 10.2; melting point 2617°C; boiling point 4612°C; oxidation states 0, +2, +3, +4, +5, and +6; crystalline forms as gray-black powder, or silver-white metal; mass numbers (percent contribution of naturally occurring molybdenum) of 92 (15.86%), 94 (9.12%), 95 (15.7%), 96 (16.5%), 97 (9.45%), 98 (23.75%), and 100 (9.62%.); and radioactive isotopes of mass number 90, 91, 93, 99 (Tb 1/2 of 67 h, frequently used as a tracer), 101, 102, and 105 (Busev 1969; Schroeder et al. 1970; Shamberger 1979; Friberg and Lener 1986). In water at pH >7, molybdenum exists primarily as the molybdate ion, MoO_4^{2-}; at pH <7, various polymeric compounds are formed, including the paramolybdate ion, $Mo_7O_{24}^{6-}$ (Busev 1969). In soils, molybdate was sorbed most readily to alkaline, high calcium, high chloride soils; retention was least in low pH, low sulfate soils (Smith et al. 1987). There is general agreement that molybdenum chemistry is complex and inadequately known. Additional and more extensive information on its properties was summarized in major reviews by Busev (1969), Boschke (1978), Brewer (1980), Coughlan (1980), Newton and Otsuka (1980), Parker (1983), and Mitchell and Sykes (1986).

30.2.4 Mode of Action

Interactions among some trace metals are so pervading and so biologically influential that the results of nutritional and toxicological studies conducted with a single element can be misleading unless the dietary and body tissue levels of interacting elements are clearly defined (Underwood 1979). For molybdenum, interactions are so dominant — especially in ruminant species — that a particular level of intake in the diet can lead to molybdenum deficiency or to molybdenum toxicity in the animal, depending on the relative intakes of copper and inorganic sulfate (Schroeder et al. 1970; Underwood 1971, 1979; Clawson et al. 1972; Suttle 1973, 1983a; Friberg et al. 1975; Buck 1978; Ward 1978; Chappell et al. 1979; Shamberger 1979; Van Ryssen and Stielau 1980; Gupta and Lipsett 1981; Ivan and Veira 1985; Friberg and Lener 1986; Goyer 1986; Kincaid et al. 1986; Osman and Sykes 1989).

The first indications of interaction between copper and molybdenum came more than 40 years ago from studies of grazing cattle in certain areas of England. Afflicted animals lost weight, developed severe diarrhea, and (in extreme cases) died. The disease is sometimes called teart (rhymes with heart) or molybdenosis, and is caused by eating herbage rich in molybdenum (i.e., 20 to 100 mg/kg dry weight diet compared to <5 mg/kg in nearby healthy pastures) and low or deficient in copper and inorganic sulfate (Underwood 1979). Molybdenosis is a copper deficiency disease that occurs particularly in cattle and sheep and is usually caused by the depressing effect of molybdenum on the physiological availability of copper (Clawson et al. 1972; Dollahite et al. 1972; Alloway 1973; Erdman et al. 1978; Mills and Breamer 1980; Van Ryssen and Stielau 1980; Nederbragt 1982; Suttle 1983a; Goyer 1986; Osman and Sykes 1989). The disease was treated successfully with copper sulfate at 1 to 2 g daily in the diet, or 200 to 300 mg daily by intravenous injection (Buck 1978; Underwood 1979; Ivan and Veira 1985). When ruminant diets contained copper at 8 to 11 mg/kg weight — a normal range — cattle were poisoned at molybdenum levels of 5 to 6 mg/kg and sheep at 10 to 12 mg/kg. When dietary copper was low (i.e., <8 mg/kg) or sulfate ion level was high, molybdenum at 1 to 2 mg/kg ration was sometimes toxic to cattle. Increasing the copper in diets to 13 to 16 mg/kg protected cattle against concentrations up to 150 mg/kg of dietary molybdenum (Buck 1978). Studies of molybdenum metabolism are of limited value unless one knows the status in the diet of inorganic sulfate, which alleviates molybdenum toxicity in all known species by increasing urinary molybdenum excretion (Underwood 1971, 1979).

Copper prevents the accumulation of molybdenum in the liver and may antagonize the absorption of molybdenum from food. The antagonism of copper to molybdenum depends on sulfate, which may displace molybdate (Goyer 1986). In certain sheep pastures, for example, the herbage may contain up to 15 mg copper/kg dry weight and <0.2 mg Mo/kg dry weight — conditions favoring the development of a high copper status that may lead to copper poisoning. Treatment consists of providing molybdate salt licks, which are highly effective in reducing copper levels in grazing sheep (Buck 1978; Underwood 1979). A low copper:molybdenum ratio (i.e., <2), rather than the absolute dietary concentration of molybdenum, is the primary determinant of susceptibility to molybdenum poisoning; molybdenosis is not expected when this ratio is near 5 (Buck 1978; Ward 1978; Mills and Breamer 1980). Ratios of copper to molybdenum in sweet clover (*Melilotus* spp., a known molybdenum accumulator plant) growing in coal mine spoils in the Dakotas, Montana, and Wyoming ranged from 0.4 to 5, suggesting that molybdenosis can be expected to occur in cattle and sheep grazing in low Cu:Mo areas (Erdman et al. 1978). A similar situation existed in British Columbia, where 19% of all fodders and grains had a Cu:Mo ratio <2 (Underwood 1979).

There are several explanations for the high sensitivity of ruminants to increased dietary molybdenum and sulfur, the most plausible being the role of thiomolybdates (Penumarthy and Oehme 1978; Lamand et al. 1980; Nederbragt 1980, 1982; Suttle 1980, 1983b; Mills et al. 1981; Suttle and Field 1983; Weber et al. 1983; Hynes et al. 1985; Friberg and Lener 1986; Allen and Gawthorne 1987; Sas 1987; Strickland et al. 1987). Thiomolybdates are compounds formed by the progressive substitution for sulfur and oxygen in the molybdate (MoO_4^{2-}) anion when hydrogen sulfide and MoO_4^{2-} interact *in vitro* at neutral pH. Di-, tri-, and tetrathiomolybdates are formed, but only the last of these effectively impairs copper absorption. When sufficient tetrathiomolybdate (MoS_4) is formed in the rumen, it and copper in the gut combine and the resultant complex is bound strongly to proteins of high molecular weight. The molybdoproteins so formed are strong chelators of copper, and may be the agents responsible for copper deficiency through formation of biologically unavailable copper complexes in gut, blood, and tissues of animals that consume diets containing high concentrations of molybdenum. To confound matters, the complex molybdenum–copper–sulfur interrelationship can be modified, or disrupted entirely, by many compounds or mixtures. These include the salts of tungsten (Schroeder et al. 1970; Underwood 1971; Mills and Breamer 1980; Luo et al. 1983; Goyer 1986), zinc (Penumarthy and Oehme 1978; Parada 1981; Alary et al. 1983), lead and manganese (Underwood 1971), iron (Phillippo et al. 1987b), vanadium (Vaishampayan 1983), chromium (Vaishampayan 1983; Chung et al. 1985), phosphorus (Underwood 1971; Baldwin et al. 1981), cystine and methionine (Underwood 1971, 1979), fluoride (Goyer 1986), and proteins (Underwood 1971, 1979; Friberg and Lener 1986; Kincaid et al. 1986).

30.3 CONCENTRATIONS IN FIELD COLLECTIONS

30.3.1 General

Molybdenum levels tend to be elevated in nonbiological materials and in terrestrial flora in the vicinity of molybdenum mining and reclamation activities, fossil-fuel power plants, and disposal areas for molybdenum-contaminated sewage sludge, flyash, and irrigation waters. Concentrations of molybdenum in fish, wildlife, and invertebrates were low when compared to those in terrestrial plants, although certain aquatic invertebrates were capable of high bioconcentration. Concentrations of molybdenum alone, however, were not sufficient to diagnose molybdenum deficiency or toxicosis.

30.3.2 Nonbiological Samples

Elevated levels of molybdenum in nonbiological materials have been reported near certain mines, power plants, and oil shale deposits, as well as in various sewage sludges, fertilizers, and agricultural drainwaters (Table 30.1). Molybdenum is concentrated in coal and petroleum, and the burning of these fuels contributes heavily to atmospheric molybdenum (King et al. 1973). Combustion of fossil fuels contributes about 5000 metric tons of molybdenum annually to the atmosphere; atmospheric particulates contain about 0.001 μg Mo/m^3 air (Goyer 1986).

Natural molybdenum concentrations in ground and surface waters rarely exceed 20 μg/L; significantly higher concentrations are probably due to industrial contamination. Existing wastewater and water treatment facilities remove less than 20% of the molybdenum; accordingly, drinking water concentrations are near those of the untreated source (Chappell et al. 1979). Molybdenum concentrations in saline waters appear to be directly related to salinity (Prange and Kremling 1985; Sloot et al. 1985). In the Wadden Sea, for example, molybdenum concentrations were 0.08, 0.4, and 1.0 μg/L at salinities of 0.07, 1.2, and 3.3%, respectively (Sloot et al. 1985).

The molybdenum content of soil may vary by more than an order of magnitude, causing both deficient and excessive concentrations for plants and ruminants in some parts of the world (Friberg et al. 1975). Native soils may contain enough molybdenum to cause molybdenosis in range livestock in some areas of the United States, particularly in Oregon, Nevada, and California (Kubota et al. 1967; Erdman et al. 1978). Elevated soil molybdenum levels can result from both natural and industrial sources. Usually when soil molybdenum levels exceed 5 mg/kg dry weight, a geological anomaly or industrial contamination is the likely explanation (Chappell et al. 1979). Molybdenum is more available biologically to herbage plants in alkaline soils than in neutral or acidic soils (Underwood 1971; Friberg et al. 1975; Shacklette et al. 1978; Wright and Hossner 1984). Liming of acidic soils or treatment with molybdenum-containing fertilizers can effectively raise the molybdenum content of herbage (Underwood 1971; Pierzynski and Jacobs 1986).

The disposal of sewage sludge, flyash from coal combustion, and molybdenum-contaminated irrigation waters to agricultural fields may result in the production of molybdenum-rich herbage. Sewage sludges rich in molybdenum and applied to agricultural soils resulted in elevated molybdenum content in corn and soybeans in a dose-dependent pattern (Pierzynski and Jacobs 1986). Similarly, flyash from coal combustion applied to pasture and croplands at rates sufficient to provide molybdenum at concentrations of 40 g/kg and higher resulted in potentially hazardous levels in vegetation to ruminant grazers. Molybdenum in flyash applied to soils remained biologically available for extended periods, especially in calcareous soils (Elseewi and Page 1984). Irrigation has also been proposed as a possible disposal method for large quantities of water having molybdenum concentrations of 5 to 100 mg/L that result from mining and reclamation activities. This method of disposal is not recommended unless all animals are kept off irrigated sites and the vegetation can be harvested and destroyed until molybdenum levels in the plants remain below 10 mg/kg dry weight (Smith et al. 1987).

Table 30.1 Molybdenum Concentrations in Selected Nonbiological Materials

Material, Unit, and Location	Concentration[a]	Reference[b]
SEAWATER (μg/L)		
Worldwide	<1–10	1–3
Worldwide	4–12	4
Pacific Ocean, all depths	10.3	5
DRINKING WATER (μg/L)		
U.S.S.R.		
Winter	0.03–0.06	1
Summer	0.11–0.15	1
U.S.	0.1–6.2	2
U.S.	Usually <5, Max. 500	1, 4
Switzerland	Usually <1, Max. 29	1
SURFACE WATER (μg/L)		
North American rivers	0.4	4
California lakes	0.4 (<3–100)	1
U.S. rivers	1.2–4.1	1
Mineral waters	2–3	2
Near Mo mine and mill, Colorado	(100–10,000)	4
Ash pond effluent from coal-fired power plant, New Mexico	170	4
Power station effluent, Victoria, Australia	330	6
Near Mo tailings pile, New Mexico	600	4
Evaporation ponds, California, 1985–86	1100 (630–2600)	7
Leachate from oil shale retort, Colorado	4100 (2500–8300)	4
Irrigation water from Mo mining and reclamation	5000–100,000	8
GROUNDWATER (μg/L)		
U.S.	Usually <1	4
U.S.S.R.	3	4
California, agricultural drainwater, 1985–86	1200–5500	7
Colorado		
Mining areas	Max. 25,000	1
Near uranium mill	50,000	4
SEDIMENTS (mg/kg, dry weight)		
U.S. rivers	5–57	1
Evaporation ponds, California	18 (<2–22)	7
Near Mo tailings pile, Colorado	21	4
Baltic Sea	80	3
Near Mo mine and mill, Colorado	530, Max. 1800	1, 4
SOILS (mg/kg, dry weight)		
Natural soils		
Worldwide	0.1–10, usually 0.2–0.7	1, 2
Worldwide	1–2 (0.6–3.5)	4
U.S.	1.2 (0.1–40)	4
Molybdenosis areas	2–>6	9
Elevated Mo	12–76 (2–190)	4
Economic Mo deposits	>200	4
Impacted soils		
In upper 5 cm at 0.3 or 3 km from Mo ore processing plant in 1982 and 1983		
0.3 km		
1982, Total	28	10
1982, Extractable	5	10

Table 30.1 (continued) Molybdenum Concentrations in Selected Nonbiological Materials

Material, Unit, and Location	Concentration[a]	Reference[b]
1983, Total	73	10
1983, Extractable	3	10
3 km		
1982, Total	3	10
1982, Extractable	0.4	10
1983, Total	8	10
1983, Extractable	0.8	10
Near Mo mine and mill, Colorado, irrigated with Mo-contaminated effluent from uranium mill	61 (49–72)	4
Ireland, highly mineralized	170 (11–4000)	17
SEWAGE SLUDGE (mg/kg, dry weight)		
Iowa	<1–75	11
U.S.	2–30	2
Most states, U.S.	5–39	11
North America	<10 (2–100)	1, 12
Michigan	32 (6–3700)	11
AIR (mg/m^3)		
Rural, U.S.	0.0001–0.003	1, 2
Urban, U.S.	0.01–0.03	1, 2
Worldwide	<0.0005	13
FERTILIZERS (mg/kg, dry weight)		
Domestic	3–6	1, 14
OIL, OIL SHALE, COAL, AND WASTE PRODUCTS (liquids, mg/L; solids, mg/kg dry weight)		
Coal conversion process waters	0.001–0.5	15
Oil shale retort water	0.06–0.3	15
Light oil	<0.1	1
Heavy oil	Max. 0.5	1
Spent oil shale	0.6	15
Coal	1–73	15
Coal	3 (0.3–15)	1, 16
Oil shale	5–87	15
Coal ash	7–160	16
Flyash from power stations	Usually 10–40, Max. 180	1

[a] Concentrations are shown as means, range (in parentheses), and maximum (Max.).
[b] **1,** Friberg et al. 1975; **2,** Friberg and Lener 1986; **3,** Prange and Kremling 1985; **4,** Chappell et al. 1979; **5,** Collier 1985; **6,** Ahsanullah 1982; **7,** Fujii 1988; **8,** Smith et al. 1987; **9,** Kubota et al. 1967; **10,** Schalscha et al. 1987; **11,** Pierzynski and Jacobs 1986; **12,** Lahann 1976; **13,** Schroeder et al. 1970; **14,** Goyer 1986; **15,** Birge et al. 1980; **16,** Elseewi and Page 1984; **17,** Talbot and Ryan 1988.

30.3.3 Biological Samples

All plants contain molybdenum, and it is essential for the growth of all terrestrial flora (Schroeder et al. 1970). Molybdenum concentrations were elevated in terrestrial plants, especially in those collected from soils amended with flyash, liquid sludge, or molybdenum-contaminated irrigation waters, in naturally occurring teart pastures, and in the vicinity of molybdenum mining and ore processing activities, steelworks, and other metal processors. Molybdenum concentrations greater than 20 mg/kg dry weight were frequently documented in plants from contaminated areas (Table 30.2). Legumes, especially trefoil clovers (*Lotus* sp.) selectively accumulated molybdenum; concentrations of 5 to 30 mg/kg dry weight were common in molybdenum-contaminated areas (Friberg et al. 1975; Shacklette et al. 1978). The molybdenum levels were sometimes high and potentially toxic in legumes from poorly drained acidic soils (Kubota et al. 1967; Underwood 1971). Some terrestrial grasses displayed copper:molybdenum ratios between 0.5 and 3.7. Since ratios greater than 2 were within the range where molybdenosis is likely, and since most of the molybdenum concentrations were greater than the maximum tolerable level of 6 mg/kg dry weight, hypocuprosis (molybdenosis) in cattle was expected (Schalscha et al. 1987). Major sources of molybdenum overload in fodder were in plants grown on high-molybdenum alkaline soils and from industrial contamination by coal and uranium mines and alloy mills (Sas 1987). Variations in molybdenum content of pasture species ranged from 0.1 to 200 mg/kg dry weight, and most variations were due to soil and species differences (Underwood 1971). Pasture plants collected from mountainous areas of southern Norway were usually deficient in copper, and low to partly deficient in molybdenum. As a result, the copper:molybdenum ratios were generally high and may explain the occurrence of chronic copper poisoning in grazing sheep in that region (Garmo et al. 1986).

Except in terrestrial plants, molybdenum concentrations were low in all groups examined; maximum concentrations reported from all sampling locales were about 6 mg/kg dry weight in aquatic plants, about 4 mg/kg fresh weight in aquatic invertebrates, 2 mg/kg fresh weight in fishes (except for rainbow trout liver and kidney — 26 to 43 mg/kg fresh weight — from fish collected near a molybdenum tailings outfall), 4 mg/kg dry weight in birds, 30 mg/kg dry weight in domestic ruminant liver, 85 mg/kg dry weight in the horse, and <4 mg/kg dry weight in mammalian wildlife and humans (Table 30.2).

No food chain biomagnification of molybdenum was found in 1987 in aquatic organisms from the San Joaquin River, California (Saiki et al. 1993). Maximum concentrations of molybdenum recorded in the San Joaquin River in 1987 were 10 μg/L in water, and — in mg/kg DW — 3.1 in detritus, 1.4 in algae, 0.54 in chironomid larvae, 0.64 in whole crustaceans, and 0.51 in bluegills (Saiki et al. 1993). Molybdenum did not biomagnify in the fish/crustacean/hump-backed dolphin (*Sousa chinensis*) food chain (Parsons 1998). Molybdenum concentrations from cetacean liver tissues were 4.1 times lower than those of whole prey organisms, i.e., 0.8 vs. 3.3 mg Mo/kg DW. Hump-backed dolphins from Hong Kong consume about 4.3 mg of molybdenum daily, equivalent to about 0.02 mg Mo/kg BW (Parsons 1998). Large interspecies differences were evident among aquatic organisms in their ability to accumulate molybdenum from the medium. Marine bivalve molluscs usually contained 30 to 90 times more molybdenum than the ambient seawater; however, some species from Greek waters had bioconcentration factors up to 1300 (Eisler 1981). Marine plankton accumulated molybdenum from seawater by factors up to 25 (Goyer 1986). But growth in aquatic phytoplankton populations was inhibited under conditions of low or missing molybdenum, nitrogen, and organic matter concentrations. The role of molybdenum in this process requires clarification (Paerl et al. 1987). In rainbow trout (*Oncorhynchus mykiss*), residues of molybdenum in tissues were affected only slightly by the concentrations in water; tissue residues ranged from 5 to 118 μg/kg fresh weight in water containing trace (<6 μg/L) concentrations; 10 to 146 μg/kg in water containing low (6 μg/L) concentrations; and from 13 to 322 μg/kg in water containing high (300 μg/L) concentrations (Ward 1973). A similar pattern was reported for kokanee salmon, *Oncorhynchus nerka* (Ward 1973). Rainbow trout held for 2 weeks in live traps 1.6 km downstream from a molybdenum mine tailings outfall survived, but liver and kidney had significantly elevated levels of molybdenum, calcium, manganese, iron, zinc, strontium, and zirconium, and 10% less potassium. The observed mineral changes may have been due to outfalls from nonmolybdenum mines discharged into the river system (Kienholz 1977).

In moles (*Talpa europaea*), adults had significantly higher concentrations of molybdenum in liver than did juveniles (Pankakoski et al. 1993). The significance of this observation is imperfectly understood. Molybdenum mining operations are not detrimental to mammalian wildlife, as judged by normal appearance and low molybdenum levels in liver and kidney of nine species — including deer, squirrel, chipmunk, badger, beaver, marmot, and pika — collected from areas with high environmental molybdenum levels (Kienholz 1977). It is emphasized that molybdenum concentrations in animal tissues give little indication of the dietary molybdenum status, and are of little diagnostic value for this purpose unless the sulfate, protein, and copper status of the diet are also known. This point is discussed in greater detail later.

Table 30.2 Molybdenum Concentrations in Field Collections of Selected Species of Animals and Plants (Values shown are in mg Mo/kg [ppm Mo] fresh weight [FW], dry weight [DW], or ash weight [AW].)

Taxonomic Group, Organism, Tissue, and Other Variables	Concentration[a] (mg/kg)	Reference[b]
TERRESTRIAL PLANTS		
Bermuda grass, *Cynodon dactylon*		
Soil amended with molybdenum-contaminated irrigation water		
Control	5 DW	1
6 mg Mo/kg soil	225 DW	1
13 mg Mo/kg soil	309 DW	1
26 mg Mo/kg soil	447 DW	1
Herbage (forage)		
Normal	1–3 DW	2
Teart pastures	20–100 DW	2
Barley, *Hordeum vulgare*		
Soil amended with flyash		
40 g/kg soil	6 DW	3
80 g/kg soil	11 DW	3
Moss, *Hypnum cupressiforme*, Sweden		
Normal	1 DW	4
Near waste disposal plant	8 DW	4
Near metal processor	400 DW	4
Near steelworks	560 DW	4
Legumes		
From molybdenosis areas	17–125 DW	5
From nonmolybdenosis areas	6–28 DW	5
Black medic, *Medicago lupulina,* Carson Valley, Nevada	Max. 372 DW	6
Alfalfa, *Medicago sativa*		
Soil amended with flyash		
40 g/kg soil	10 DW	3
80 g/kg soil	12 DW	3
Pasture plants, southern Norway	0.3 (0.01–4) DW	7
Peas, *Pisum sativum*		
Canada	0.2 FW	4
U.S.	0.3–5 FW	4
India	0.7–2 FW	4
Romania, Germany	1 FW	4
Russia	6 FW	4
Ballica grass, *Lolium perenne*		
Distance from Mo ore processing plant		
1982		
0.3 km	29–40 DW	8
1.0 km	8–10 DW	8
1983		
0.3 km	6–10 DW	8
1.0 km	7–10 DW	8
9.0 km	4–5 DW	8
In soil amended with liquid sludge to contain 410 mg Mo/ha	20 DW	9
White clover, *Trifoleum repens*		
Soil amended with flyash		
40 g/kg soil	27 DW	3
80 g/kg soil	36 DW	3
Soil amended with liquid sludge		
17 mg Mo/ha	31 DW	9
410 mg Mo/ha	90 DW	9
Wheat, *Triticum aestivum*		
Germany, Romania, Russia	0.2–0.8 FW	4

Table 30.2 (continued) Molybdenum Concentrations in Field Collections of Selected Species of Animals and Plants (Values shown are in mg Mo/kg [ppm Mo] fresh weight [FW], dry weight [DW], or ash weight [AW].)

Taxonomic Group, Organism, Tissue, and Other Variables	Concentration[a] (mg/kg)	Reference[b]
India	0.5 FW	4
U.S.	0.6–6 FW	4
Vegetables		
Mo symptoms in humans	11–82 DW	4
Control site	3–5 DW	4
Vegetation		
Near Mo mine	Max. 5400 AW	6
Normal	<2–500 AW	6
AQUATIC PLANTS		
Algae, whole		
Marine	0.03–0.2 FW; 0.1–1.3 DW	4
Canada, 11 species	0.2–1.4 DW	10
Marine plants	0.5 FW	11
Marsh plants, whole		
Texas, 14 species	0.4–2.5 DW	10
Seaweeds, whole		
U.K., 5 species	0.2–1.3 DW; 0.04–0.2 FW	10
Norway, 11 species	0.3–6 DW	4
AQUATIC INVERTEBRATES		
Aquatic insects, 4 species		
Near low Mo waters (<1.0 μg Mo/L)	0.3–1.4 DW	12
Upstream	Max. 0.2 DW	12
Downstream	Max. 0.3 DW	12
Corals, marine, 34 species	<2 DW	13
Crustaceans, marine		
Tissues sold for human consumption, 16 species	0.1–0.4 FW	14
Molluscs, marine		
Soft parts		
15 species	<0.1–0.6 FW	14
3 species	0.7–4 FW	14
Common mussel, *Mytilus edulis*; southeast Alaska; 1980–82; soft parts	<1.9 DW	27
Mussel, *Mytilus edulis aoteanus*		
Soft parts	0.6 DW	15
Gill	0.6 DW	15
Visceral mass	2 DW	15
Shell	11 DW	15
Other tissues	<0.1 DW	15
Scallop, *Pecten novae-zelandiae*		
Soft parts	0.9 DW	15
Mantle	2 DW	15
Gill	3 DW	15
Intestine	4 DW	15
Kidney	3 DW	15
Foot	0.4 DW	15
Plankton, Baltic Sea	2 DW	16
FISH		
Fishes, marine		
Liver		
43 species	0.1–0.3 FW	14
29 species	0.4–2.0 FW	14
2 species	0.4–1.0 DW	17

Table 30.2 (continued) Molybdenum Concentrations in Field Collections of Selected Species of Animals and Plants (Values shown are in mg Mo/kg [ppm Mo] fresh weight [FW], dry weight [DW], or ash weight [AW].)

Taxonomic Group, Organism, Tissue, and Other Variables	Concentration[a] (mg/kg)	Reference[b]
Muscle		
130 species	0.1–0.3 FW	14
29 species	0.4–0.6 FW	14
Various	Max. 0.04 FW	11
Whole		
17 species	0.1–0.6 FW	14
8 species	0.012–0.15 FW	18
Rainbow trout, *Oncorhynchus mykiss*		
From waters with <6 μg Mo/L		
Liver	0.04–0.1 FW	19
Spleen	0.05–0.9 FW	19
Kidney	0.1 FW	19
Skin	0.07 FW	19
Bone	0.1–0.15 FW	19
Muscle	0.01 FW	19
Intestine	0.01–0.07 FW	19
Stomach	0.04 FW	19
Brain	0.02 FW	19
From waters with 300 μg Mo/L		
Liver	0.2 FW	19
Spleen	0.2 FW	19
Kidney	0.15 FW	19
Skin	0.1 FW	19
Bone	0.2 FW	19
Muscle	0.01 FW	19
Intestine	0.1 FW	19
Stomach	0.3 FW	19
Brain	0.09 FW	19
Held 2 weeks in live traps 1.6 km downstream from molybdenum tailings outfall		
Liver	43 DW	20
Kidney	26 DW	20
Control location		
Liver	1 DW	20
Kidney	<2 DW	20
BIRDS		
Alaska, near Ketchikan; 1980–82		
Barrows goldeneye, *Bucephala islandica*		
Kidney	5.2–7.8 DW	27
Liver	4.8–6.2 DW	27
Common merganser, *Mergus merganser*		
Kidney	2.4–8.0 DW	27
Liver	<1.9–5.7 DW	27
Chicken, *Gallus* sp.		
Liver	3.6 DW	21
Kidney	4.4 DW	21
Muscle	0.1 DW	21
Robin, *Turdus migratorius*, from molybdenum mine site		
Liver	1.6 DW	20
Kidney	1.9 DW	20
MAMMALS		
Alaskan moose, *Alces alces gigas*		
Hair	0.1–0.6 DW	22

Table 30.2 (continued) Molybdenum Concentrations in Field Collections of Selected Species of Animals and Plants (Values shown are in mg Mo/kg [ppm Mo] fresh weight [FW], dry weight [DW], or ash weight [AW].)

Taxonomic Group, Organism, Tissue, and Other Variables	Concentration[a] (mg/kg)	Reference[b]
Cattle, cows, *Bos* spp.		
Normal		
Blood	0.06 FW	2
Milk	0.07 (0.02–0.2) FW	2, 4
Liver	0.7–2 FW; 2.9–5.4 DW	2, 11, 23
Kidney	0.3 FW; 1.3–2.7 DW	2, 23
Muscle	0.1 FW; 0.5 DW	2, 23
Feces	1.1–2.1 DW	23
Elevated or poisoned		
Blood	0.6–0.8 FW	2
Kidney	21 FW	11
Rumen contents	21–28 DW	24
Horse, *Equus caballus*		
Liver	3–85 DW	21
Human, *Homo sapiens*		
Liver	0.5–1.0 FW; 3.2 DW	21, 25
Liver cortex	0.9 FW	21, 25
Kidney	0.2–0.3 FW; 1.6 DW	21, 25
Kidney cortex	0.2 FW	21, 25
Adrenal	0.7 FW	21, 25
Amnion	3.5 FW	21, 25
Chorion	0.6 FW	21, 25
Spleen	0.2 DW	21, 25
Lung	0.15 DW	21, 25
Brain	0.14 DW	21, 25
Muscle	0.14 DW	21, 25
Hair	0.06 (0.02–0.13) DW	21, 25
Blood	<0.005–0.1 FW	21, 25
Mule deer, *Odocoileus hemionus;* liver		
Molybdenum mining area	1.0 FW	26
Control site	0.6 FW	26
Healthy	1.3 FW	26
Sheep, *Ovis aries*		
Wool	0.2 (0.03–0.6) DW	21
Liver		
Normal diet		
Adults	2–4 DW	21
Newborn lambs	2–4 DW	21
High molybdenum diet		
Adults	25–30 DW	21
Newborns	12–20 DW	21
Milk		
Grazing on low molybdenum (<1 mg Mo/kg) pasture	<0.01 FW	21
Grazing on high molybdenum (13 mg Mo/kg) pasture	>1 FW	21
Grazing on high molybdenum (25 mg Mo/kg) pasture, and given high sulfate (23 g/daily) for 3 days	0.1 FW	21
As above, without sulfate administration	1 FW	21
Rat, *Rattus* sp.		
Liver	2 DW	21
Kidney	1 DW	21
Spleen	0.5 DW	21
Lung	0.4 DW	21
Brain	0.2 DW	21
Muscle	0.06 DW	21

Table 30.2 (continued) Molybdenum Concentrations in Field Collections of Selected Species of Animals and Plants (Values shown are in mg Mo/kg [ppm Mo] fresh weight [FW], dry weight [DW], or ash weight [AW].)

Taxonomic Group, Organism, Tissue, and Other Variables	Concentration[a] (mg/kg)	Reference[b]
Hump-backed dolphin, *Sousa chinensis*; Hong Kong		
Stomach contents		
Whole decapod crustaceans	<0.9 –17.2 DW	29
Whole fishes	<0.9–16.4 DW	29
Diet, all sources	3.3 DW	29
Liver	0.8 DW; Max. 1.6 DW	29
Kidney	0.09 DW; Max. 0.8 DW	29
Blubber	0.3 DW; Max. 1.2 DW	29
Mole, *Talpa europaea*; Finland; 1986–87; liver		
Adults	1.7 DW	28
Juveniles	1.5 DW	28
Wildlife, 9 species		
From areas of high environmental molybdenum levels		
Liver	0.1–4 DW	20
Kidney	0.3–3 DW	20

[a] Concentrations are listed as means, minimum–maximum (in parentheses), and maximum (Max.).
[b] **1,** Smith et al. 1987; **2,** Penumarthy and Oehme 1978; **3,** Elseewi and Page 1984; **4,** Friberg et al. 1975; **5,** Kubota et al. 1967; **6,** Shacklette et al. 1978; **7,** Garmo et al. 1986; **8,** Schalscha et al. 1987; **9,** Pierzynski and Jacobs 1986; **10,** Eisler 1981; **11,** Schroeder et al. 1970; **12,** Colborn 1982; **13,** Livingston and Thompson 1971; **14,** Hall et al. 1978; **15,** Brooks and Rumsby 1965; **16,** Prange and Kremling 1985; **17,** Papadopoulou et al. 1981; **18,** Rao 1984; **19,** Ward 1973; **20,** Kienholz 1977; **21,** Underwood 1971; **22,** Flynn et al. 1976; **23,** Kume et al. 1984; **24,** Sas 1987; **25,** Friberg and Lener 1986; **26,** King et al. 1984; **27,** Franson et al. 1995; **28,** Pankakoski et al. 1993; **29,** Parsons 1998.

30.4 EFFECTS

30.4.1 General

Trace quantities of molybdenum are beneficial and perhaps essential for normal growth and development of plants and animals. In mammals, molybdenum can protect against poisoning by copper, mercury, and probably other metals, and may have anticarcinogenic properties. For all organisms, the interpretation of molybdenum residues depends on knowledge of molybdenum, copper, and inorganic sulfate concentrations in diet and in tissues. Some molybdenum compounds have insecticidal properties at low concentrations and have been proposed as selective termite control agents.

Aquatic flora and fauna seem to be comparatively resistant to molybdenum salts. Adverse effects on growth and survival were usually noted only at water concentrations of 50 mg Mo/L and higher. However, one study with newly fertilized eggs of rainbow trout produced an LC50 (28 day) value of 0.79 mg Mo/L compared to an LC50 (96 h) value of 500 mg/L for adults. Also, bioconcentration of molybdenum by selected species of algae and invertebrates (up to 20 g/kg dry weight) poses questions on risk to higher trophic level organisms.

In birds, adverse effects of molybdenum have been reported on growth at dietary concentrations of 200 to 300 mg/kg, on reproduction at 500 mg/kg, and on survival at 6000 mg/kg. In mammals, cattle are especially sensitive to molybdenum poisoning, followed by sheep, under conditions of copper and inorganic sulfate deficiency. Cattle were adversely affected when grazing pastures with a copper:molybdenum ratio <3, when fed low copper diets containing 2 to 20 mg Mo/kg diet, or when total daily intake approaches 141 mg molybdenum. Cattle usually die at doses of 10 mg Mo/kg body weight. Other mammals, including horses, pigs, rodents, and ruminant and nonruminant wildlife, are comparatively tolerant to molybdenum. Deer, for example, are at least 10 times more resistant than domestic ruminants to molybdenum. No adverse effects in deer were noted at dietary levels of 1000 mg/kg after 8 days, slight effects at 2500 mg/kg after 25 days, and reduction in food intake and diarrhea at 5000 mg/kg diet after 15 days.

30.4.2 Terrestrial Plants

In a major literature review, Gupta and Lipsett (1981) concluded that molybdenum was essential for plant growth due to its role in the fixation of nitrogen by bacteria using the enzymes nitrogenase and nitrate reductase, and that plants readily accumulated MoO_4^{2-} except under conditions of low pH, high sulfate, and low phosphate, and in some highly organic soils. Molybdenum deficiency has been recorded in a variety of crops worldwide, but there is an extremely narrow range between adequacy and deficiency. In lettuce (*Lactuca sativa*), for example, adverse effects were noted at 0.06 mg/kg (dry weight) in plants, but sufficiency was attained at 0.08 to 0.14 mg/kg. A similar case is made for *Brassica* spp. (i.e., Brussels sprouts, cabbage, and cauliflower) (Gupta and Lipsett 1981). In certain species, such as beets (*Beta vulgaris*) and corn (*Zea mays*), the ratio between deficiency and sufficiency may differ by more than 10 times (Gupta and Lipsett 1981).

Okra (*Abelmoschus esculentus*), grown in soils supplemented with molybdenum at 1, 2, or 3 mg/kg, as sodium molybdate, showed increasing growth and yields when compared to nonsupplemented soils. Fruiting occurred earlier and persisted longer with increasing molybdenum concentration (Singh and Mourya 1983). The cashew (*Anacardium occidentale*) — one of the most valuable plantation crops in India — developed yellow-leaf spots accompanied by low molybdenum levels and excess manganese in low pH soils; in extreme cases the tree was defoliated (Subbaiah et al. 1986). The disorder was corrected by foliar spraying of molybdenum salts or by liming the soil. A similar case was reported for Florida citrus in the 1950s, which was shown to be due to molybdenum deficiency (Subbaiah et al. 1986).

Soils amended with sewage sludge containing 12 to 39 mg Mo/kg dry weight (soil contained 2 mg Mo/kg dry weight at start and 4.8 to 6 mg/kg after treatment) were planted with corn and bromegrass (*Bromus inermis*). A lime-treated sludge increased molybdenum concentrations in plant tissues after several years of sludge application; maximum values recorded were 1.9 mg Mo/kg dry weight in bromegrass and 3.7 in corn (Soon and Bates 1985). No toxicity of molybdenum has yet been observed in field-grown crops, although forages containing 10 to 20 mg/kg dry weight are considered toxic to cattle and sheep (Soon and Bates 1985).

30.4.3 Terrestrial Invertebrates

Sodium molybdate and other molybdenum compounds in toxic baits have potential for termite control (Brill et al. 1987). Baits containing 1000 mg Mo/kg were fatal to 99% of the termite *Reticulitermes flavipes* in 48 days. After 8 to 10 days, termites became steel-gray in color, but appeared otherwise normal. Mortality began only after day 16. Termites did not avoid the poisoned bait, even at concentrations of 5000 mg Mo/kg. Yoshimura et al. (1987) reported similar results with another species of termite; sodium molybdate killed 100% of the workers in a colony of *Copotermes formosanus* within 24 h after eating filter paper treated with a 5% solution. Some other species of insects — including fire ants (*Solenopsis* sp.) and various species of beetles and cockroaches — were not affected when exposed to baits containing 5000 mg Mo/kg for 48 days (Brill et al. 1987).

30.4.4 Aquatic Organisms

Aquatic plants are comparatively resistant to molybdenum. In sensitive species, adverse effects were evident on growth at 50 mg/L and on development at 108 mg/L (Table 30.3). Bioconcentration of molybdenum from the medium by certain freshwater algae can result in residues up to 20 g/kg dry weight without apparent damage (Table 30.3). The implications of this phenomenon for waterfowl and other species that consume molybdenum-laden algae need to be explored.

Molybdenum is considered essential for aquatic plant growth, but the concentrations required are not known with certainty and are considered lower than those for any other essential element (Schroeder et al. 1970; Henry and Tundusi 1982). Molybdenum starvation restricts nitrogen fixation in algae, thereby limiting photosynthetic production during depleted conditions. Blue-green alga (*Anabaena oscillaroides*) cultured in molybdenum-deficient media containing 0.004 to 0.005 μg Mo/L rapidly depleted molybdenum in the medium; this ability was lost at higher concentrations of added molybdenum, when *Anabaena* began to accumulate the element (Steeg et al. 1986). The addition of tungstate to molybdenum-deficient media enhances dinitrogenase inactivation, resulting in inhibited algal growth. This process is reversed at molybdenum levels of 0.005 to 0.04 μg/L (Steeg et al. 1986). On the other hand, algal growth was significantly enhanced when vanadium (V) was present at 12.5 μg/L, although higher concentrations of V were growth inhibitory in 7 days (Vaishampayan 1983). Algal uptake of molybdenum is rapid during the first 2 h, and slower thereafter. The sequential biological reduction of hexavalent to pentavalent to trivalent molybdenum occurs intracellularly in green algae (Sakaguchi et al. 1981). Uptake is greater in freshwater than in seawater, greater at increased doses, and greater at reduced algal densities (Sakaguchi et al. 1981); it is also greater at elevated temperatures (Penot and Videau 1975).

Molybdenum occurs naturally in seawater as molybdate ion, MoO_4^{2-}, at about 10 μg/L (Abbott 1977). Despite the high concentrations of dissolved molybdenum in offshore seawater, phytoplankton from offshore locales contain extremely low molybdenum residues, almost typical of molybdenum-deficient terrestrial plants (Howarth and Cole 1985). This phenomenon is attributed to the high concentrations of sulfate in seawater; sulfate inhibits molybdate assimilation by phytoplankton, making it less available in seawater than in freshwater. As one result, nitrogen fixation and nitrate assimilation — processes that require molybdenum — may require greater energy expenditure in marine than in freshwaters and may explain, in part, why marine ecosystems are usually nitrogen-limited and lakes are not (Howarth and Cole 1985). Experimentally increasing the ratio of sulfate to molybdate inhibits molybdate uptake by marine algae, slows nitrogen fixation rates, and slows the growth of organisms that use nitrate as a nitrogen source (Howarth and Cole 1985).

Limited data suggested that aquatic invertebrates were very resistant to molybdenum. Adverse effects were observed on survival at >60 mg Mo/L and on growth at >1000 mg Mo/L (Table 30.3). Bioconcentration factors were low, but depending on initial dose, measured residues (mg/kg fresh weight) were as high as 16 in amphipods, and were 3 in clams, 18 in crayfish muscle, and 32 in crayfish carapace (Short et al. 1971). The host organisms seemed unaffected under these molybdenum burdens, but effects on upper trophic level consumers were not clear. Tailings from a pilot molybdenum mine on the North American Pacific coast were acutely lethal at concentrations of >61,000 mg tailings solids/L seawater to larvae of the mussel *Mytilus edulis*, and to adults of the amphipod *Rhepoxynius abronius* and the euphausiid *Euphausia pacifica*. Acute sublethal effects were observed at >277,000 mg/L (Mitchell et al. 1986). All species of invertebrates tested in this preliminary study were more sensitive than juvenile coho salmon, *Oncorhynchus kisutch* (Mitchell et al. 1986). In another study, zooplankton exposed to molybdenum mine tailings <8 μm in diameter at high sublethal concentrations ingested and excreted these particles (Anderson and Mackas 1986). The lowest tailing concentration tested at which a deleterious effect was observed was 100 mg/L for depression of respiration in the copepod *Calanus marshallae*, and 560 mg/L for increased mortality in copepods and the euphausiid *Euphausia pacifica*; concentrations of molybdenum mine tailings were always <15 mg/L at 0.5 km downstream from a molybdenum tailings outfall (Anderson and Mackas 1986).

Freshwater and marine fishes were — with one exception — extremely resistant to molybdenum. LC50 (96 h) values ranged between 70 mg/L and <3000 mg/L (Table 30.3). The exception was newly fertilized eggs of rainbow trout exposed for 28 days through day 4 posthatch; the LC50 (28 day) value was only 0.79 mg/L (Birge et al. 1980), suggesting that additional research is needed on the sensitivity of early life stages to molybdenum. In general, molybdenum was more toxic to teleosts in freshwater than in seawater, and more toxic to younger fish than to older fish. In rainbow trout, it bioconcentrated up to 16 mg/kg fresh weight in liver, 18 in spleen, 7 in muscle, 6 in gill, and 2 in gastrointestinal tract (Table 30.3) (Short et al. 1971). Environmental levels of molybdenum as molybdate measured in the Mo mining areas of Colorado were not considered harmful to rainbow trout (McConnell 1977). Molybdenum enrichment of Castle Lake, California (a high mountain lake in which molybdenum was determined to be the limiting micronutrient), coupled with favorable environmental conditions, led to record high yields of trout. The addition of 16 kg sodium molybdate, or 6.4 kg molybdenum, to Castle Lake in July 1963 was followed by larger standing crops of zooplankton and bottom fauna, which probably promoted survival of the 1965 year class and resulted in record yields to the angler of rainbow trout and brook trout (*Salvelinus fontinalis*) in 1967 (Cordone and Nicola 1970). Enrichment of molybdenum-deficient waters to improve angler success merits additional research.

Table 30.3 Molybdenum Effects on Selected Species of Aquatic Organisms

Organism, Mo Concentration, and Other Variables	Effect	Reference[a]
AQUATIC PLANTS		
Blue-green alga, *Anabaena oscillaroides*		
0.005 µg/L	Bioconcentration factor (BCF) of 3300 in 60 min	1
0.073 µg/L	BCF of 550 in 60 min	1
25 µg/L	BCF of 7–24 in 60 min	1
Green alga, *Chlorella vulgaris*		
10 mg/L	Residues of about 20,000 mg/kg dry weight in 20 h. Dead (heat-killed) *Chlorella* contained 4902 mg/kg dry weight in 1 h vs. 3264 mg/kg in live cells	2
20 mg/L	Normal growth in 96 h	2
50 mg/L	Reduced growth in 96 h	2
Euglena, *Euglena gracilis*		
5.4 mg/L	Normal growth and reproduction	3
96 mg/L	No abnormal cells in 48 h	3
108 mg/L	Abnormal development, cells forming clusters. Culture is photosensitive with blue color	3
>960 mg/L	No growth	3
Freshwater alga, *Nitella flexilis*		
0.014 µg/L	BCF of 628 in 25 days	4
3.3 mg/L	BCF of 39 in 24 days; elevated residues of 130 mg/kg fresh weight	4
Blue-green alga, *Nostoc muscorum*		
17.7 µg/L	Required for growth	5
INVERTEBRATES		
Amphipod, *Allorchestes compressa*		
60 mg/L	No deaths (= LC0) in 96 h	6
247 mg/L	50% dead (= LC50) in 96 h	6
450 mg/L	97% dead (= LC97) in 96 h	6
Starfish, *Asterias rubens*		
127 mg/L	LC50 (24 h) at pH 5.8	7
254 mg/L	LC50 (24 h) at pH 8.2	7
Copepod, *Calanus marshallae*		
20 mg/L	Minor increase in oxygen consumption in 24 h	8
100 mg/L	Decreased oxygen consumption in 24 h	8
560 mg/L	LC50 (19 days)	8
Green crab, *Carcinus maenus*		
1018 mg/L	LC50 (48 h)	7

Table 30.3 (continued) Molybdenum Effects on Selected Species of Aquatic Organisms

Organism, Mo Concentration, and Other Variables	Effect	Reference[a]
American oyster, *Crassostrea virginica*		
1375 mg/L	Reduction of 50% in shell growth in 96 h	9
Hermit crab, *Eupagurus bernhardus*		
100 mg/L	LC0 (50 days)	7
222 mg/L	LC50 (48 h)	7
Euphausiid, *Euphausia pacifica*		
560 mg/L	LC50 (112 h)	8
Amphipod, *Gammarus* sp.		
3.3 mg/L	BCF of 4.8 in whole animal in 24 days	4
Lake periphyton		
0.014 µg/L	BCF of 3570 in 24 days	4
Clam, *Margaretifera margaretifera*		
3.3 mg/L	Maximum BCF values in 15–24 days were 1.8 in shell, 0.9 in soft parts, and 0.3 in muscle	4
Mysid shrimp, *Mysidopsis bahia*		
1205 mg/L	LC50 (96 h)	9
Mussel, *Mytilus edulis*, larvae		
147 mg/L	Development reduced 50% in 48 h, based on survival and abnormalities	10
Crayfish, *Pacifiastacus leniusculus*		
3.3 mg/L	BCF in 24 days of 5.7 for muscle and 9.8 for carapace	4
Pink shrimp, *Penaeus duorarum*		
1909 mg/L	LC50 (96 h)	9
Pullet-shell (clam), *Venerupis pallustra*		
381 mg/L	LC50 (24 h)	7
FISH		
Flannelmouth sucker, *Catostomus latipinnis*; larvae; age 12–13 days; molybdenum as sodium molybdate		
>2800 mg Mo/L	LC50 (48 h)	15
1940 (95% CI = 1680–2370) mg Mo/L	LC50 (96 h)	15
Sheepshead minnow, *Cyprinodon variegatus*		
3057 mg/L	LC50 (96 h)	9
Bluegill, *Lepomis macrochirus*		
1320 mg/L	LC50 (96 h)	11
Rainbow trout, *Oncorhynchus mykiss*		
Embryos and larvae exposed for 28 days starting at fertilization through 4 days posthatch		
28 µg/L	LC1 (28 days)	12
125 µg/L	LC10 (28 days)	12
790 (610–990) µg/L	LC50 (28 days)	12
17.0–18.5 mg/L, exposed continuously for 1 year from eyed eggs to juvenile stage	No significant effect on survival, growth, or blood hematocrit	10, 11
500 mg/L	LC25 (96 h), mean length 20 mm	11
800 mg/L	LC50 (96 h), mean length 20 mm	11
1320 mg/L	LC50 (96 h), mean length 55 mm	11
Steelhead trout, *Oncorhynchus mykiss*		
0.014 µg/L	Max. BCF of 1143 in liver and gastrointestinal tract after chronic exposure	4
3.3 mg/L	Max. BCF in 24 days of 5.4 in spleen, 4.5 in liver, 2.3 in muscle, 1.8 in gill, and 0.6 in gastrointestinal tract	4
Coho salmon, *Oncorhynchus kisutch*		
>1000 mg/L	LC50 (96 h); fry	13

Table 30.3 (continued) Molybdenum Effects on Selected Species of Aquatic Organisms

Organism, Mo Concentration, and Other Variables	Effect	Reference[a]
Chinook salmon, *Oncorhynchus tshawytscha*		
Juveniles exposed to 193 μg Mo/L, as sodium molybdate, for 90 days	No accumulation; whole-body content of <0.7 mg Mo/kg DW	14
More than 1000 mg/L	LC50 (96 h); fry	13
Fathead minnow, *Pimephales promelas*		
70 mg/L	LC50 (96 h), soft water	11
360 mg/L	LC50 (96 h), hard water	11

[a] **1,** Steeg et al. 1986; **2,** Sakaguchi et al. 1981; **3,** Colmano 1973; **4,** Short et al. 1971; **5,** Vaishampayan 1983; **6,** Ahsanullah 1982; **7,** Abbott 1977; **8,** Anderson and Mackas 1986; **9,** Knothe and Van Riper 1988; **10,** Morgan et al. 1986; **11,** McConnell 1977; **12,** Birge et al. 1980; **13,** Hamilton and Buhl 1990; **14,** Hamilton and Wiedmeyer 1990; **15,** Hamilton and Buhl 1997.

30.4.5 Birds

Data are missing on the effects of molybdenum on avian wildlife under controlled conditions. All studies conducted with birds have been restricted to domestic poultry. Signs of molybdenum deficiency in domestic chickens included loss of feathers, lowered tissue molybdenum concentrations, reduced xanthine dehydrogenase activity in various organs, decreased uric acid excretion, disorders in ossification of long bones, and changes in joint cartilage that led to complete immobility. Signs were eliminated when diets were supplemented with molybdenum at concentrations of 0.2 to 2.5 mg/kg (Reid et al. 1956; Friberg and Lener 1986). Efforts to produce a molybdenum deficiency syndrome in birds and mammals by feeding diets low in molybdenum have been unsuccessful (Friberg et al. 1975). Thus, it has been necessary to introduce a compound with a known property of inhibiting molybdenum, namely wolframate (Na_2WO_4), a tungsten compound. Wolframate increases molybdenum excretion, leading to molybdenum deficiency in rats and chickens. With this technique it has been possible to produce an assumed molybdenum deficiency in chicks consisting of reduced weight gain and sometimes death (Friberg et al. 1975). Dietary requirements to maintain normal growth in rats and chicks were probably less than 1 mg Mo/kg food, and thus substantially less than that of any other trace element recognized as essential (Mills and Breamer 1980). In fact, birds may require molybdenum at concentrations up to 6 mg/kg in their diets for optimal growth (Kienholz 1977). Dietary molybdenum counteracts adverse effects in chicks on growth and survival induced by hexavalent chromium. Chicks fed 900 mg chromium/kg ration for 4 weeks showed significantly depressed growth, 25% mortality, and elevated liver chromium. However, diet supplementation to 150 mg Mo/kg resulted in normal growth and liver chromium values, and no deaths (Chung et al. 1985).

Early studies with chicks and turkey poults showed that the addition of only 13 to 25 μg Mo/kg — as molybdate or molybdic acid — to basal diets containing 1.0 to 1.5 mg Mo/kg resulted in a growth advantage of 14 to 19% in 4 weeks over that in unsupplemented groups (Reid et al. 1956, 1957). Roosters given daily dietary supplements of 100 or 400 μg molybdenum per bird for 4 weeks to basal diets containing 0.51 mg Mo/kg had reduced serum uric acid values when compared to those of controls. The significance of this finding is not clear (Karring et al. 1981). Birds are relatively resistant to molybdenum. For example, day-old chicks fed diets containing 20% molybdenum mine tailings for 23 days were unaffected, and those fed diets containing 40% molybdenum mine tailings showed only a slight reduction in body weight during the same period (Kienholz 1977). Dietary levels of 200 mg Mo/kg ration results in minor growth inhibition of chicks; and at 300 mg/kg feed, the growth of turkey poults was reduced (Underwood 1971). Dietary supplements of 500 mg Mo/kg ration produced a slight decrease in growth rate of chicks after 4 weeks; hens, however, laid 15% fewer eggs than controls, and all eggs contained embryolethal concentrations of 16 to 20 mg Mo/kg (Friberg et al. 1975). At dietary supplements of 1000 mg

Mo/kg, egg production was reduced 50% in domestic chickens (Friberg et al. 1975). Dietary loadings of 2000 mg/kg induced severe growth depression and a 100-fold increase in molybdenum content in tibia (Underwood 1971) and an 80% reduction in egg production (Friberg et al. 1975). At 4000 mg/kg diet, severe anemia was reported in chickens (Underwood 1971). Mortality of chicks fed 6000 mg Mo/kg diet for 4 weeks was 33%; at 8000 mg Mo/kg diet for 4 weeks, 61% of the chicks died and survivors weighed only 16% as much as the controls (Friberg et al. 1975). Chicks, unlike mammals, did not experience molybdenum reduction in tissues after sulfate administration — although sulfate markedly reduced the signs of Mo toxicity (Underwood 1971).

30.4.6 Mammals

Almost all studies conducted to date on molybdenum effects under controlled conditions have been on livestock, especially cattle and sheep. Molybdenum is beneficial and perhaps essential to adequate mammalian nutrition. Moreover, it can protect against poisoning by copper or mercury, and may be useful in controlling cancer. Evidence of functional roles for molybdenum in the enzymes xanthine oxidase, aldehyde oxidase, and sulfite oxidase suggests that molybdenum is an essential trace nutrient for animals (Underwood 1971; Earl and Vish 1979; Mills and Breamer 1980). Signs of molybdenum deficiency include decreased intestinal and liver xanthine oxidase activity (Mills and Breamer 1980). Molybdenum prevents damage to the liver in sheep receiving excess copper; accumulations of copper and molybdenum in kidney were present in a biologically unavailable form and of negligible physiological significance (Van Ryssen et al. 1982). Dietary supplements of 70 mg molybdenum per day for a restricted period is recommended for reduction of liver Cu in sheep, provided dietary Cu levels are simultaneously reduced (Van Ryssen et al. 1986). Molybdenum, as sodium molybdate, protects against acute inorganic mercury toxicity in rats by altering the metabolism of cysteine-containing proteins in the cytoplasm of liver and kidney, resulting in lowered mercury content in these organs (Yamane and Koizumi 1982; Koizumi and Yamane 1984). Anticarcinogenic properties of molybdenum in rats have been reported, although the mechanisms of action are unknown. In one study, 2 or 20 mg Mo/L in drinking water significantly inhibited cancer of the esophagus and forestomach experimentally induced by *N*-nitrososarcosine ethyl ester (Luo et al. 1983). In another study with virgin female rats, 10 mg Mo/L in drinking water reduced by half the number of mammary carcinomas experimentally induced by *N*-nitroso-*N*-methylurea (Wei et al. 1985). Additional research seems warranted on the role of molybdenum in cancer inhibition.

Molybdenosis has been produced experimentally in many species of mammals, including cattle, sheep, rabbits, and guinea pigs (Friberg et al. 1975). Signs of molybdenum poisoning vary greatly among species, but generally include the following: copper deficiency, especially in serum; reduced food intake and growth rate; liver and kidney pathology; diarrhea and dark-colored feces; anemia; dull, wiry, and depigmented hair; reproductive impairment, including delayed puberty, female infertility, testicular degeneration, and abnormal or delayed estrus cycle; decreased milk production; joint and connective tissue lesions; bone abnormalities; and loosening and loss of teeth (Underwood 1971; Dollahite et al. 1972; Friberg et al. 1975; Erdman et al. 1978; Penumarthy and Oehme 1978; Ward 1978; Chappell et al. 1979; Mills and Breamer 1980; Alary et al. 1981; Baldwin et al. 1981; Friberg and Lener 1986; Van Ryssen et al. 1986; Phillippo et al. 1987a). These authorities also agree on three additional points:

- Early signs of molybdenosis are often irreversible, especially in young animals
- The severity of the signs depends on the level of molybdenum intake relative to that of copper and inorganic sulfate
- If afflicted animals are not removed promptly from molybdenum-contaminated diets and given copper sulfate therapy, death may result.

Molybdenum poisoning in ruminants, or teart disease, has been known since the mid-1800s and affects only ruminants of special pastures. Degree of teartness varies from field to field and season to season, and is usually proportional to the molybdenum content in herbage. Molybdenum levels in typical teart pastures range from 10 to 100 mg/kg dry weight compared to normal levels of 3 to 5 mg/kg (Friberg and Lener 1986). If herbage contains more than 12 mg Mo/kg dry weight, problems should be expected in cattle, and to a lesser extent in sheep (Friberg et al. 1975). In situations where cattle are accidentally exposed to high molybdenum levels, the administration of copper sulfate should result in molybdenum excretion, up to 50% in 10 days (Penumarthy and Oehme 1978). Aside from cattle and sheep, all evidence indicates that other mammals are comparatively tolerant of high dietary intakes of molybdenum, including horses, pigs, small laboratory animals, and mammalian wildlife (Underwood 1971; Buck 1978; Chappell et al. 1979; Friberg and Lener 1986; Osman and Sykes 1989; Table 30.4). Cattle excrete molybdenum primarily through feces, but other (more tolerant) species such as pigs, rats, and humans, rapidly excrete molybdenum through urine, and this may account, in part, for the comparative sensitivity of cattle to molybdenum (Underwood 1971). Cattle normally excrete about 67% of all administered MoO_3 in feces and urine in 7 days; guinea pigs excreted 100% in urine in 8 days; and swine excreted 75% in urine in 5 days (Penumarthy and Oehme 1978). Cattle are adversely affected when:

- They graze copper-deficient pastures containing 2 to 20 mg/kg molybdenum, and the copper to molybdenum ratio is less than 3
- They are fed low copper diets containing 5 mg (or more) Mo/kg dry weight
- The total daily intake approaches 141 mg molybdenum
- The body weight residues exceed (a fatal) 10 mg Mo/kg (Table 30.4).

It is clear that both the form of molybdenum administered and the route of exposure affect molybdenum metabolism and survival (Table 30.4). By comparison, adverse effects (some deaths) were noted at 250 mg Mo/kg body weight (BW) (in guinea pigs), at 50 mg/kg BW in domestic cats (central nervous system impairment), at 10 mg/L drinking water in mice (survival), at 10 to 15 mg total daily intake in humans (high incidence of gout-like disease), and at to 3 mg/m^3 air in humans for 5 years (respiratory difficulties), or 6 to 19 mg/m^3 in humans for 4 years (Table 30.4).

In newborn lambs from ewes that consumed high-molybdenum diets during pregnancy, demyelinization of the central nervous system was severe, accompanied by low copper contents in the liver (Earl and Vish 1979). Sheep are more tolerant than cattle to molybdenum poisoning due, in part, to a lower turnover of ceruloplasmin, a copper-transporting enzyme that is inhibited by molybdenum. However, this characteristic makes sheep more sensitive than cattle to copper poisoning (Ward 1978). For example, chronic copper poisoning in sheep in several districts in Norway is probably due to molybdenum-deficient forages rather than to excess copper intake (Froslie et al. 1983). Swayback is a spastic paralysis in lambs born of ewes that were copper deficient during pregnancy (Todd 1976). In northern Ireland, where cases have been reported, pastures were not copper deficient, and swayback was due to an imbalance of copper, molybdenum, and sulfur. Very severely affected lambs were paralyzed in all limbs and died shortly after birth because they were unable to stand and suckle. Lambs less severely affected developed signs in about 2 weeks, but usually only the hind limbs were affected. Brain and spinal cord lesions were present, resulting in demyelination of spinal cord and cavitation of brain tissues; lesions were irreversible, but death might have been avoided with adequate copper therapy (Todd 1976).

Horses are generally considered to be tolerant of dietary copper deficiencies and of copper and molybdenum excesses that affected ruminants. Yet molybdenum accumulated in equine liver and has been implicated as a possible contributory factor in bone disorders in foals and yearlings grazing pastures containing 5 to 25 mg Mo/kg (Cymbaluk et al. 1981; Strickland et al. 1987). Cattle and horses are highly susceptible to pyrrolizidine alkaloids, an ingredient in certain poisonous plants such as tansy ragwort (*Senecio jacobaea*). Signs of poisoning included elevated copper levels in liver followed by fatal hemolytic crisis. Sheep are more resistant to alkaloids than equines or bovines, and sheep grazing has been recommended as a means of controlling tansy ragwort. However, dietary supplements of 10 mg Mo/kg increased the susceptibility of sheep to tansy ragwort intoxication, despite the observed increase in copper excretion (White et al. 1984).

In rodents, molybdenum is neither teratogenic nor embryocidal to golden hamsters at doses up to 100 mg/kg body weight, and has no measurable effect on fertility or gestation of female rats given similar high doses (Earl and Vish 1979). Voluntary rejection of high-molybdenum diets by rats results in anorexia. This phenomenon implies sensory, probably olfactory, recognition of molybdate in combination with other dietary constituents to form compounds with a characteristic odor detectable by rats (Underwood 1971). The ability to reject high-molybdenum diets requires a learning or conditioning period because it is lacking or weak with freshly prepared diets and extends to a discrimination between a toxic (high molybdenum) and nontoxic (high molybdenum plus sulfate) diet. Rats may associate a gastrointestinal disturbance with a sensory attribute of diets containing toxic levels of molybdenum (Underwood 1971).

Data on molybdenum effects on mammalian wildlife are scarce, although those available strongly suggest that domestic livestock are at far greater risk (Osman and Sykes 1989) (Table 30.4). Studies with mule deer (*Odocoileus hemionus*) showed that this species was at least an order of magnitude more tolerant to high levels of dietary molybdenum than were domestic ruminants, and at least as resistant as swine, horses, and rabbits (Nagy et al. 1975; Ward and Nagy 1976; Ward 1978; Chappell et al. 1979). Female mule deer showed no visible effects after 33 days on diets containing up to 200 mg Mo/kg feed, or after 8 days at 1000 mg/kg. Only slight effects — some reduction in food intake and some animals with diarrhea — were observed at diets of 2500 mg/kg for 25 days. At feeding levels of 5000 and 7000 mg/kg for periods of 3 to 15 days, signs were more pronounced; however, recovery began almost immediately after transfer to uncontaminated feed. Signs of copper deficiency and of molybdenosis are very similar, and careful diagnosis is necessary to ensure use of the correct remedial action. For example, some populations of Alaskan moose (*Alces alces gigas*) showed faulty hoof keratinization and decreased reproductive rates, but this was attributed to copper-deficient browse growing on low copper soils, and not to increased molybdenum levels in herbage (Flynn et al. 1977). In another case, a high proportion of white-tailed deer (*Odocoileus virginianus*) feeding near uranium-mine spoil deposits in several Texas counties — areas in which extreme molybdenosis has been documented in grazing cattle — had antlers that were stunted, twisted, and broadened or knobby at the tips (King et al. 1984). However, the copper levels in liver of these deer were similar to those of deer in a control area — 16.7 mg/kg fresh weight vs. 18.0 — and only 1 of 19 deer examined from the mining district had a detectable molybdenum concentration in liver (0.7 mg/kg fresh weight) vs. none in any control sample. On the basis of low contents of copper in soils and vegetation, it was concluded that white-tailed deer examined were experiencing copper deficiency (hypocuprosis), with signs similar to molybdenosis (King et al. 1984).

In humans, molybdenum is low at birth, increases until age 20 years, and declines thereafter (Goyer 1986). Although conclusive evidence that molybdenum is required by humans is lacking, there is general agreement that it should be considered as one of the essential trace elements. The absence of any documented deficiencies in man indicates that the required level is much less than the average daily intake of 180 μg molybdenum in the United States (Chappell et al. 1979). Human discomfort has been reported in workers from copper–molybdenum mines, and in those eating food products containing 10 to 15 mg Mo/kg and <10 mg copper/kg and grown on soils containing elevated molybdenum of 77 mg/kg and 39 mg copper/kg. Symptoms included general weakness, fatigue, headache, irritability, lack of appetite, epigastric pain, pain in joints and muscles, weight loss, red and moist skin, tremors of the hands, sweating, dizziness (Friberg et al. 1975), renal xanthine calculi, uric acid disturbances (Schroeder et al. 1970), and increased serum ceruloplasmin (Friberg and Lener 1986). The typical human adult contains only 9 mg molybdenum, primarily in liver, kidney, adrenal, and omentum (Goyer 1986). Most of the ingested molybdenum is easily absorbed from the GI tract and excreted within hours or days in urine, mostly as molybdate; excesses may be excreted also by the bile, particularly as hexavalent molybdenum (Friberg et al. 1975; Goyer 1986; Friberg and Lener 1986). At high dietary levels, molybdenum reportedly prevents dental caries (Schroeder et al. 1970), but this requires verification.

Table 30.4 Molybdenum Effects on Selected Species of Mammals

Species, Dose, and Other Variables	Effects (Reference)[a]
CATTLE, COWS, *Bos* spp.	
Near steelworks, 20 kg Mo as MoO_3 emitted daily in gaseous form; fallout deposits ~2 mg/m^2 monthly, corresponding to a pasture Mo content from 2–20 mg Mo/kg dry weight (DW). Pasture had slight Cu deficiency of natural origin, with copper:Mo ratio in pastures <3	About 40% of 5000 grazing cows with signs of molybdenosis. No signs of poisoning before steelworks began operations. Signs evident almost immediately in first grazing season; most pronounced in younger animals closest to source. Remedial actions included copper glycine and installation of additional emission filters at the steel-works (1)
Low dietary Mo (<5 mg Mo/kg), adequate copper	Growth and fertilization rate normal; liver copper >70 mg/kg DW; 63% of embryos developed normally (2)
Fed diets containing 5 mg Mo/kg DW and 4 mg copper/kg DW for 84 weeks	Reduced food intake and efficiency of food use, altered iron metabolism, clinical signs of copper deficiency. Onset of puberty delayed 10 weeks, decreased conception rate (fertility 12–33% vs. 57–80% in controls), disrupted estrus cycle (67% were anestrus vs. 7% in controls), and other signs consistent with decreased releases of luteinizing hormones associated with altered ovarian secretion (3, 4)
High dietary Mo (15–20 mg Mo/kg), copper deficient	Growth and fertilization normal; liver copper 10 mg/kg DW; only 16% of embryos developed normally (2)
Fed diets of normal copper and high Mo (30 mg/kg feed)	Blood Mo level of 0.6–0.8 mg/L (5)
Diets containing 40 mg Mo/kg and 6 mg copper/kg fed to lactating cows for 9 weeks	Reduction of 30% in milk yield; rapid decline in plasma copper; milk Mo levels of 1.6 mg/L; growth reduction in nursing calves (6)
Fed diets of 60 mg Mo/kg DW	Low liver copper, intestinal disturbances, brittle bones prone to fracture (7)
Dairy herd fed pelleted feed containing 140 mg Mo/kg FW and up to 10 mg copper/kg FW	Molybdenosis. Contaminated magnesium oxide (12,200 mg Mo/kg) added to ration at 1% was the source of the excess Mo (8)
Drinking water with Mo as ammonium molybdate. Basal diet with 13 mg copper/kg and 2900 mg sulfur/kg	
1 or 10 mg Mo/L in drinking water for 21 days	In 5-week-old calves, there was no effect on liver or plasma copper levels (9)
50 mg Mo/L in drinking water for 21 days	Copper liver burden reduced to 201 mg/kg DW vs. 346 in controls; copper in plasma elevated to 1100 µg/L vs. 690 in controls. No effect on growth, or food and water consumption (9)
Total daily intake of 100 mg Mo	Normal milk Mo level of 0.06 mg/L (5)
Total daily intake of 141 mg Mo	Anorexia, diarrhea, and weight loss in Swiss beef cattle (10)
Total daily intake of 500 mg Mo	Milk Mo level of 0.37 mg/L (5)
Total daily intake of 1360 mg Mo daily as soluble molybdate	Signs of molybdenum poisoning (7)
10 mg Mo/kg BW	Lethal dose (11)
GUINEA PIG, *Cavia* sp.	
Chronic exposure, daily dose in mg Mo/animal	
25, as MoO_3	75% mortality (12)
200, as calcium molybdate	25% mortality (12)
Air concentrations of 28–285 mg Mo/m^3	Hexavalent Mo compounds absorbed appreciably, but not disulfide compounds (10, 13)
Dose, in mg Mo/kg BW, various administration routes	
80	LD0 (12)
250	Some deaths (11)
400	LD75 (4 days) (12)
800	LD100 (4 months) (12)

Table 30.4 (continued) Molybdenum Effects on Selected Species of Mammals

Species, Dose, and Other Variables	Effects (Reference)[a]
DOMESTIC RUMINANTS	
Pastures containing 10–20 mg Mo/kg DW	"Risk" zone for molybdenosis (12)
Pastures containing 20–100 mg Mo/kg DW	"Teart" disease characterized by anemia, poor growth, diarrhea; prolonged exposure resulted in joint deformities and death (13)
HORSES, PONIES, *Equus* sp.	
Feeding on teart pastures with elevated Mo content	No effect (5)
Given single oral dose of radio Mo-99, as molybdate, or about 20–28 mg	Mo-99 appeared rapidly in plasma as molybdate, but quickly cleared with Tb 1/2 of 7–10 h (14)
Fed diets containing 20 mg Mo/kg DW for 4.5 months; diet supplemented with sulfur for 1 month at 1.2 g/kg feed	Animals remained healthy. No decline in total plasma copper or increase in plasma insoluble copper (14)
Fed diets containing up to 107 mg Mo/kg for 14 days	Increasing dietary Mo resulted in decreasing copper retention due to increasing excretion of copper in feces; up to 1.45 g Mo/kg BW absorbed and retained with no obvious adverse effects (15)
DOMESTIC CAT, *Felis domesticus*	
Intravenous injection, in mg Mo/kg BW	
25	Increased arterial blood pressure (12)
50	Central nervous system impairment (12)
HUMAN, *Homo sapiens*	
Drinking water, in µg/L	
50	No effect (11)
200	Increased urinary excretion, normal serum Mo levels, no change in copper metabolism (11)
Total intake, in mg Mo daily	
0.18	Average intake in U.S. (11)
0.5–1	Increased urinary copper excretion (11)
10	Increase in blood and urine Mo levels, increases in serum ceruloplasmin, increased xanthine oxidase activity (11)
10–15	Increased uric acid, decreased copper excretion, high incidence of gout-like disease (11)
Atmospheric concentrations, in mg Mo/m^3 air	
1–3; 5-year exposure	Respiratory difficulties (12)
6–19; 4-year exposure	Respiratory difficulties (12)
MOUSE, *Mus* spp.	
10 mg Mo/L in drinking water of breeding mice	Decrease in survival of F_2 and F_3 generations (16)
SHEEP, *Ovis* spp.	
Molybdenum-deficient diet of 0.03 mg/kg	High incidence of renal xanthine calculi (5)
Adequate diet of 0.4 mg Mo/kg, due to resowing of pasture and lime treatment	Zero incidence of renal calculi (5)
Content of pasture 0.4–1.5 mg Mo/kg DW	Mo concentrations, in mg/kg FW, were 0.0–0.03 in plasma, 2.0–2.4 in liver, and 0.4–0.5 in kidney. No lameness or connective tissue lesions (17)
2.4 mg Mo/kg diet in lambs	Significantly enhanced growth when compared to sheep fed 0.36 mg Mo/kg diet; growth associated with increased cellulose digestibility by rumen biota (5)
Grazing pastures treated 3 times with 420 g Mo/ha: at start, and weeks 45 and 72. Mo content of pasture usually 5.5–12.5 mg/kg DW	Mo concentrations, in mg/kg FW, were 1.7–2.4 in plasma, 6.0–6.4 in liver, and 6.9–8.1 in kidney. Lameness and connective tissue lesions in most sheep (17)

Table 30.4 (continued) Molybdenum Effects on Selected Species of Mammals

Species, Dose, and Other Variables	Effects (Reference)[a]
Given diets of high copper (82 mg/kg) and sulfur (3.8 g/kg), and Mo at 20, 40, or 60 mg/kg for 193 days	Liver damage due to copper at low Mo (20 mg/kg) diets; at 40 and 60 mg Mo/kg, both metals accumulated in kidney cortex but no evidence of liver histopathology or kidney damage (18)
Breeding ewes fed diets of normal Cu, high Mo (30 mg/kg feed)	Blood Mo level of 2.4–3.4 mg/L (5)
Diets of 50 mg Mo/kg	Avoidance by lambs; may be learned olfactory recognition (19)
Lambs grazing on soils where copper:Mo ratio is <0.4	Swayback observed in 15–39% (10)
Ram lambs fed diets of adequate sulfate and copper (7.7 mg/kg DW). Copper to Mo ratios of 5.5, 5.3, 1.1, or 0.7 for 105 days	No significant measurable effects at ratios of 5.5 and 5.3. Secondary Cu deficiency (molybdenosis) at 1.1 ratio evident in blood and plasma, and in liver at 0.7 (20)
Lambs fed daily intake of 8 mg Mo, 36.3 mg Cu, and 3.7 g S for 125 days	No effect on growth of food intake; significant increases in levels of kidney cortex copper, liver Mo, and plasma copper; major differences in responses among breeds tested (21)
Total intake raised from 0.4 mg daily to 96 mg daily	Blood Mo level of 4.95 mg/L (5)
Fed 75 mg Cu daily for 50 days, followed by 140 mg Mo and 4 g S daily for 13 days with no added Cu	Molybdenosis within 8 days (22)
As above, but 70 mg Mo daily at day 13 for 34 days	40% reduction in liver copper (22)
WHITE RABBIT, *Oryctolagus* sp.	
Dietary Mo concentrations, in mg Mo/kg ration	
100; lifetime exposure	Reduced growth, hair loss, dermatosis, anemia, skeletal and joint deformities, decreased thyroxin (11)
500; 12 weeks	No obvious effects (12)
1000; 12 weeks	Some growth retardation (12)
2000–4000	Many deaths of weanlings in about 37 days, and of adults in 53 days. Survivors were anorexic, diarrheic, anemic, and had front-leg abnormalities; successful recovery after copper therapy (12)
5000	Thyroid dysfunction (11)
RAT, *Rattus* spp.	
Drinking water, in mg Mo/L	
10; exposure for 3 years	Disrupted calcium metabolism (29). Increased sensitivity to cold stress, elevated tissue residues of 50–60 mg Mo/kg DW (23, 24)
20; 30 weeks exposure	No effect on growth or organ histology (25)
50; lifetime exposure	Some growth retardation (11)
1000; lifetime exposure	No severe signs observed in breeding adults. Resultant pups, however, maintained on this regimen were stunted, rough haired, sterile (males), and hyperactive (11)
Dose, in mg Mo per animal daily for up to 232 days	
10	LD25 to LD50 for hexavalent Mo compounds (12)
100	LD50 for calcium molybdate (12)
125	LD50 for MoO_3 (12)
333	LD50 for ammonium molybdate (12)
Atmospheric concentrations, in mg Mo/m^3	
64; 2 h	Outwardly normal, some microscopic damage due to MoO_3 exposure (12)
Up to 5000 ammonium paramolybdate, 12,000 Mo dioxide, 15,000 Mo trioxide, or 30,000 metallic molybdenum; exposure for 1 h	At 4 weeks postexposure, there were no adverse effects except for irritation of upper respiratory passage (12)

Table 30.4 (continued) Molybdenum Effects on Selected Species of Mammals

Species, Dose, and Other Variables	Effects (Reference)[a]
Feeding levels, in mg Mo/kg diet	
50	Diet avoidance (19). In low sulfate diets and 5 weeks exposure, rats had reduced growth and mandibular exostoses (10)
80; copper deficient	Inhibited growth and reduced survival (5, 26)
80; 35 mg $CuSO_4$/kg	No measurable effects (5, 26)
100; lifetime exposure	Appetite loss, weight loss, reduced growth, anemia, mandibular exostoses, bone deformities, liver and kidney histopathology, increased liver copper residues, male sterility (11)
400	After 5 weeks, growth depression, anemia, mandibular exostoses; some deaths at lifetime exposures (12)
500 or 800	No deaths in 6 weeks; growth retardation and anemia (12)
500 or 1000; 77 mg Cu/kg	Poor growth (5, 26)
5000	Lethal in 2 weeks (11, 12)
Dose, in mg Mo/kg BW	
0.00002–0.001	50% excretion (Tb 1/2) in 60–113 h for kidney, liver, spleen, small intestine, and skin (10)
0.003	Tb 1/2 in 47 h (10)
>0.003	Tb 1/2 in 3 h when administered subcutaneously, 6 h for intragastric application (10)
4.5, intravenous injection	Biliary excretion of Mo^{+6} compounds was more rapid than Mo^{+5} compounds (27)
100	When inhaled as MoO_3, irritating to eyes and mucous membranes and eventually lethal. Repeated oral administration leads to histopathology of liver and kidney (13)
100–150	Lethal (11)
114	All recovered after intraperitoneal injection of sodium molybdate (12)
117	All dead within a few hours after intraperitoneal injection of sodium molybdate (12)
500; daily	Tolerated when given as disulfide (13)
500; 28 days	Reduced growth, disrupted blood and enzyme chemistry, histopathology of liver and kidney; partly reversed by 20% protein diet (28)
DOMESTIC PIG, *Sus* sp.	
Fed diets containing 1000 mg Mo/kg for 3 months	No effect (5)

[a] **1,** Alary et al. 1981; **2,** O'Gorman et al. 1987; **3,** Phillippo et al. 1987a; **4,** Phillippo et al. 1987b; **5,** Underwood 1971; **6,** Wittenberg and Devlin 1987; **7,** Penumarthy and Oehme 1978; **8,** Lloyd et al. 1976; **9,** Kincaid 1980; **10,** Friberg and Lener 1986; **11,** Chappell et al. 1979; **12,** Friberg et al. 1975; **13,** Goyer 1986; **14,** Strickland et al. 1987; **15,** Cymbaluk et al. 1981; **16,** Earl and Vish 1979; **17,** Pitt et al. 1980; **18,** Van Ryssen et al. 1982; **19,** White et al. 1984; **20,** Robinson et al. 1987; **21,** Harrison et al. 1987; **22,** Van Ryssen et al. 1986; **23,** Winston et al. 1973; **24,** Winston et al. 1976; **25,** Luo et al. 1983; **26,** Underwood 1979; **27,** Lener and Bibr 1979; **28,** Bandyopadhyay et al. 1981; **29,** Solomons et al. 1973.

30.5 RECOMMENDATIONS

Although molybdenum is generally recognized as an essential trace metal for plants and animals, and may reduce the incidence and severity of carcinomas in rats (Luo et al. 1983; Wei et al. 1985) and dental caries in humans (Shamberger 1979), there is no direct evidence of molybdenum deficiency being detrimental to animal health. The minimum daily molybdenum requirements in diets are not yet established due to problems in preparing molybdenum-free rations (Chappell et al. 1979). As a consequence, no regulatory agency recognizes molybdenum as safe and necessary, and molybdenum cannot be legally incorporated into animal feeds (Penumarthy and Oehme 1978).

The richest natural sources of molybdenum (i.e., 1.1 to 4.7 mg Mo/kg fresh weight) are plants unusually high in purines such as legumes and whole grains (Schroeder et al. 1970), followed by leafy vegetables, liver, and kidney (Shamberger 1979). The poorest sources are fruits, sugars, oils, and fat (Schroeder et al. 1970).

The greatest economic importance of molybdenosis is associated with subclinical manifestations of copper deficiency resulting from forages containing a low copper:molybdenum ratio. Unfortunately, these conditions are often difficult to diagnose accurately, and animal response to copper may be difficult to demonstrate (Ward 1978). One recommended treatment for afflicted cattle is 2 g daily of copper sulfate to cows and 1 g daily to young stock, or intravenous injection of 200 to 300 mg copper sulfate daily for several days (Underwood 1971).

The animals most sensitive to molybdenum insult are domestic ruminants, especially cattle. Diets containing more than 15 mg Mo/kg dry weight and with a low copper:molybdenum ratio, or drinking water levels more than 10 mg Mo/L were frequently associated with molybdenosis in cattle (Table 30.5). By contrast, adverse effects were documented in birds at dietary levels more than 200 mg Mo/kg ration, in ruminant wildlife at dietary levels greater than 2500 mg Mo/kg, and in aquatic organisms — with one exception — at more than 50 mg Mo/L (Table 30.5). The exception was newly fertilized eggs of rainbow trout, which were about 21 times more sensitive to molybdenum than were zygotes approximately one third through embryonic development, and about 90 times more sensitive than adult fish (Table 30.5).

Proposed criteria for human health protection include drinking water concentrations less than 50 μg Mo/L, and daily dietary intakes less than 7 μg Mo/kg food — based on a 70-kg adult (Table 30.5). Molybdenum concentrations in blood of "healthy" people averaged 14.7 μg Mo/L, distributed between the plasma and erythrocytes. Anemic people had significantly lower blood molybdenum levels. In leukemia patients, molybdenum levels increased significantly in whole blood and erythrocytes but not in plasma (Shamberger 1979). Additional work is recommended on the use of blood in fish and wildlife as an indicator of molybdenum stress and metabolism (Eisler 1989).

Increasing problems associated with marginal mineral deficiencies and unfavorable mineral interaction — as has been the case in the older agricultural areas of northern Europe — can be anticipated as pasture and forage production becomes more intensive (Ward 1978). Research has been recommended in areas having a high molybdenum content in soils and vegetation, and also in noncontaminated areas where consumption habits favor a high molybdenum intake and an imbalance in relation to other dietary constituents of importance, such as copper (Friberg et al. 1975). In some parts of the world where molybdenum has been substituted for lime, the soils have become more acidic, thus making them difficult to farm. Liming under these conditions may elevate soil molybdenum from levels previously considered safe, to levels potentially hazardous to grazing animals through high-molybdenum herbage (Gupta and Lipsett 1981). The addition of molybdenum fertilizers to sheep pastures resulted in small increments in molybdenum content with negligible risk of induced copper deficiency. But it would be unwise to apply molybdenum fertilizers to temperate grasslands grazed by animals of low initial copper status, as judged by growth retardation of lambs from pastures supplemented with molybdenum (Suttle 1983a).

Table 30.5 Proposed Molybdenum Criteria for the Protection of Living Resources and Human Health

Resource, Criterion, and Other Variables	Concentration	Reference[a]
TERRESTRIAL PLANTS		
Okra, *Abelmoschus esculentus*		
Increased growth	3 mg/kg soil	1
Lettuce, *Lactuca sativa*		
Molybdenum deficiency	~0.06 mg/kg dry weight (DW) plant	2
Molybdenum sufficiency	More than 0.08 mg/kg DW	2

Table 30.5 (continued) Proposed Molybdenum Criteria for the Protection of Living Resources and Human Health

Resource, Criterion, and Other Variables	Concentration	Reference[a]
Corn, *Zea mays*		
No adverse effect	3.7 mg/kg DW plant	3
Agricultural soils, Bangladesh		
Deficiency	<0.1 mg/kg DW surface soil	38
TERRESTRIAL INVERTEBRATES		
Toxic baits		
Termites	~1000 mg/kg	4
Other insect species	>5000 mg/kg	4
AQUATIC LIFE		
Algae		
Deficiency levels	<0.005–17.7 µg/L	5, 6
High bioconcentration	>0.014 µg/L	7
Growth reduction	>50 mg/L	8
Invertebrates		
Reduced survival	>60 mg/L	9
Fish		
Adults		
High bioconcentration	>0.014 µg/L	7
Reduced survival	>70 mg/L	10
Eggs		
Newly fertilized		
Reduced survival	>0.79 mg/L	11
No adverse effects	<28 µg/L	11
Eyed		
Adverse effects	>17.0 mg/L	10, 12
BIRDS		
Molybdenum deficiency	13–200 µg/kg diet	13–15
Normal growth	~1.0 mg/kg diet	16
Optimal growth	6.0 mg/kg diet	17
Growth reduction	200–300 mg/kg diet	18
Reproductive impairment	500 mg/kg diet	19
Reduced survival	6000 mg/kg diet	19
MAMMALS		
Cattle, cows (*Bos* spp.)		
Forage		
Healthy pasture	3–5 mg/kg DW	18
Possibility of molybdenosis	10–20 mg/kg DW	19
Probability of molybdenosis	20–100 mg/kg DW	18, 19
Toxic	15–30 mg/kg DW	20
Maximum tolerable level	6 mg/kg DW	20, 21
Recommended	0.1–0.5 mg/kg DW	22
Ratio of copper to molybdenum in diet		
Molybdenosis probable	<0.4	23
Critical	<2.0	20
Critical	>20.0	22
Optimal for growth and reproduction	6.1–10.1	22, 23
Drinking Water		
Safe level	<10 mg/L	24
Minimum toxic concentration for calves	10–50 mg/L	24

Table 30.5 (continued) Proposed Molybdenum Criteria for the Protection of Living Resources and Human Health

Resource, Criterion, and Other Variables	Concentration	Reference[a]
Guinea pig, *Cavia* sp.		
No effect on survival	80 mg/kg BW	19
Cat, *Felis domesticus*		
Adverse nonlethal effects	25–50 mg/kg BW	19
Mule deer, *Odocoileus hemionus*		
No effect	200–1000 mg/kg diet	25–28
Reduction in food intake	2500 mg/kg diet	25–28
Nonlethal adverse effects	5000–7000 mg/kg diet	25–28
Sheep, *Ovis* sp.		
Forage, recommended	<0.5 mg/kg DW	22
Rat, *Rattus* sp.		
Minimum daily need	0.5 µg	29
Disrupted calcium metabolism, elevated tissue residues	10 mg/L drinking water	30–32
Cancer inhibition	2–20 mg/L drinking water	33, 34
Food avoidance	50 mg/kg diet	35
HUMAN HEALTH		
Total daily intake, 70-kg adult		
Minimal need	120 µg	29
Average range	100–500 µg	18, 19, 28, 29, 36, 37
Maximum	10–15 mg	28
In molybdenum mining areas	>1 mg	19
From food		
United States	170 µg	28
United States	335 (210–460) µg	19
Former Soviet Union		
Children	159 µg	19
Adults	353 µg	19
United Kingdom	128 (110–1000) µg	19
From drinking water	<5 µg	28
No effect level	<500 µg daily	28
Adverse effects		
Biochemical	0.5–10 mg daily	28
Clinical	10–15 mg daily	28
Drinking water		
Safe level	<50 µg/L	28
Irrigation water		
Safe level	<10 µg/L	28
Air		
Maximum permissible concentration		
Former Soviet Union	6 mg/m^3	3
United States, 8 h daily, 5 days weekly	9.5–10 mg/m^3	15, 36
Blood		
"Normal"	14.7 µg/L	37

[a] **1,** Singh and Mourya 1983; **2,** Gupta and Lipsett 1981; **3,** Soon and Bates 1985; **4,** Brill et al. 1987; **5,** Vaishampayan 1983; **6,** Steeg et al. 1986; **7,** Short et al. 1971; **8,** Sakaguchi et al. 1981; **9,** Ahsanullah 1982; **10,** McConnell 1977; **11,** Birge et al. 1980; **12,** Morgan et al. 1986; **13,** Reid et al. 1956; **14,** Reid et al. 1957; **15,** Friberg and Lener 1986; **16,** Mills and Bremner 1980; **17,** Kienholz 1977; **18,** Underwood 1971; **19,** Friberg et al. 1975; **20,** Schalscha et al. 1987; **21,** Kume et al. 1984; **22,** Garmo et al. 1986; **23,** Baldwin et al. 1981; **24,** Kincaid 1980; **25,** Nagy et al. 1975; **26,** Ward and Nagy 1976; **27,** Ward 1978; **28,** Chappell et al. 1979; **29,** Schroeder et al. 1970; **30,** Solomons et al. 1973; **31,** Winston et al. 1973; **32,** Winston et al. 1976; **33,** Luo et al. 1983; **34,** Wei et al. 1985; **35,** White et al. 1984; **36,** Goyer 1986; **37,** Shamberger 1979; **38,** Miah 1999.

30.6 SUMMARY

The element molybdenum (Mo) is found in all living organisms and is considered to be an essential or beneficial micronutrient. However, molybdenum poisoning of ruminants has been reported in at least 15 states and 8 foreign countries. Molybdenum is used primarily in the manufacture of steel alloys. Its residues tend to be elevated in plants and soils near molybdenum mining and reclamation sites, fossil-fuel power plants, and molybdenum disposal areas. Concentrations of molybdenum are usually lower in fish and wildlife than in terrestrial macrophytes.

Aquatic organisms are comparatively resistant to molybdenum salts: adverse effects on growth and survival usually appeared only at water concentrations >50 mg Mo/L. But in one study, 50% of newly fertilized eggs of rainbow trout (*Oncorhynchus mykiss*) died in 28 days at only 0.79 mg Mo/L. High bioconcentration of molybdenum by certain species of aquatic algae and invertebrates — up to 20 g Mo/kg dry weight — has been recorded without apparent harm to the accumulator. However, hazard potential to upper trophic organisms (such as waterfowl) that may feed on bioconcentrators is not clear. Data on molybdenum effects are missing for avian wildlife and are inadequate for mammalian wildlife. In domestic birds, adverse effects on growth have been reported at dietary molybdenum concentrations of 200 mg Mo/kg, on reproduction at 500 mg/kg, and on survival at 6000 mg/kg.

Molybdenum chemistry is complex and inadequately known. Its toxicological properties in mammals are governed to a remarkable extent through interaction with copper and sulfur; residues of molybdenum alone are not sufficient to diagnose molybdenum poisoning. Domestic ruminants, especially cattle, are especially sensitive to molybdenum poisoning when copper and inorganic sulfate are deficient. Cattle are adversely affected — and die if not removed — when grazing on pastures where the ratio of copper to molybdenum is <3, or if they are fed low copper diets containing molybdenum at 2 to 20 mg/kg diet; death usually occurs when tissue residues exceed 10 mg/kg body weight. The resistance of other species of mammals tested, including domestic livestock, small laboratory animals, and wildlife, was at least tenfold higher than that of cattle. Mule deer (*Odocoileus hemionus*), for example, showed no adverse effects at dietary levels of 1000 mg/kg.

Additional research is needed in several fields, including:

- The role of molybdenum on inhibition of carcinomas and dental caries
- The establishment of minimum, optimal, and upper daily requirements of molybdenum in aquatic and wildlife species of concern
- The improvement in diagnostic abilities to distinguish molybdenum poisoning from copper deficiency
- The determination of sensitivity of early developmental stages of fishes to molybdenum insult.

30.7 LITERATURE CITED

Abbott, O.J. 1977. The toxicity of ammonium molybdate to marine invertebrates. *Mar. Pollut. Bull.* 8:204-205.

Ahsanullah, M. 1982. Acute toxicity of chromium, mercury, molybdenum and nickel to the amphipod *Allorchestes compressa. Austral. Jour. Mar. Freshwater Res.* 33:465-474.

Alary, J., P. Bourbon, J. Esclassan, J.C. Lepert, J. Vandaele, and F. Klein. 1983. Zinc, lead, molybdenum contamination in the vicinity of an electric steelworks and environmental response to pollution abatement by bag filter. *Water Air Soil Pollut.* 20:137-145.

Alary, J., P. Bourbon, J. Esclassan, J.C. Lepert, J. Vandaele, J.M. Lecuire, and F. Klein. 1981. Environmental molybdenum levels in industrial molybdenosis of grazing cattle. *Sci. Total Environ.* 19:111-119.

Allen, J.D. and J.M. Gawthorne. 1987. Involvement of the solid phase of rumen digesta in the interaction between copper, molybdenum and sulphur in sheep. *Brit. Jour. Nutr.* 58:265-276.

Alloway, B.J. 1973. Copper and molybdenum in swayback pastures. *Jour. Agric. Sci. (Cambridge)* 80:521-524.

Anderson, E.P. and D.L. Mackas. 1986. Lethal and sublethal effects of a molybdenum mine tailing on marine zooplankton: mortality, respiration, feeding and swimming behavior in *Calanus marshallae*, *Metridia pacifica* and *Euphausia pacifica*. *Mar. Environ. Res.* 19:131-155.

Baldwin, W.K., D.W. Hamar, M.L. Gerlach, and L.D. Lewis. 1981. Copper-molybdenum imbalance in range cattle. *Bovine Practice* 2:9-16.

Bandyopadhyay, S.K., K. Chatterjee, R.K. Tiwari, A. Mitra, A. Banerjee, K.K. Ghosh, and G.C. Chatterjee. 1981. Biochemical studies on molybdenum toxicity in rats: effects of high protein feeding. *Int. Jour. Vit. Nutr. Res.* 51:401-409.

Birge, W.J., J.A. Black, A.G. Westerman, and J.E. Hudson. 1980. Aquatic toxicity tests on inorganic elements occurring in oil shale. Pages 519-534 in C. Gale (ed.). *Oil Shale Symposium: Sampling, Analysis and Quality Assurance.* U.S. Environ. Protection Agency Rep. 600/9-80-022.

Boschke, F.L. (ed.). 1978. Aspects of molybdenum and related chemistry. *Topics in Current Chemistry 76*. Springer-Verlag, New York. 159 pp.

Brewer, L. (ed.). 1980. *Molybdenum: Physico-Chemical Properties of Its Compounds and Alloys.* Atom. Ener. Rev. Spec. issue No. 7. 714 pp. Avail. from UNIPUB, 345 Park Ave. S., New York. 10010.

Brill, W.J., S.W. Ela, and J.A. Breznak. 1987. Termite killing by molybdenum and tungsten compounds. *Naturwissenschaften* 74:494-495.

Brooks, R.R. and M.G. Rumsby. 1965. The biogeochemistry of trace element uptake by some New Zealand bivalves. *Limnol. Oceanogr.* 10:521-527.

Buck, W.B. 1978. Copper/molybdenum toxicity in animals. Pages 491-515 in F.W. Oehme (ed.). *Toxicity of Heavy Metals in the Environment. Part I.* Marcel Dekker, New York.

Busev, A.I. 1969. *Analytical Chemistry of Molybdenum.* Ann Arbor-Humphey Sci. Publ., Ann Arbor, MI. 283 pp.

Chappell, W.R., R.R. Meglen, R. Moure-Eraso, C.C. Solomons, T.A. Tsongas, P.A. Walravens, and P.W. Winston. 1979. Human Health Effects of Molybdenum in Drinking Water. U.S. Environ. Protection Agency Rep. 600/1-79-006. 101 pp.

Chappell, W.R. and K.K. Petersen (eds.). 1976. *Molybdenum in the Environment. Vol. 1. The Biology of Molybdenum.* Marcel Dekker, New York. 1-315.

Chappell, W.R. and K.K. Peterson (eds.). 1977. *Molybdenum in the Environment. Vol. 2. The Geochemistry, Cycling, and Industrial Uses of Molybdenum.* Marcel Dekker, New York. 316-812.

Chung, K.H., Y.O. Suk, and M.H. Kang. 1985. The toxicity of chromium and its interaction with manganese and molybdenum in the chicks. *Korean Jour. Anim. Sci.* 27:391-395.

Clawson, W.J., A.L. Lesperance, V.R. Bohman, and D.C. Layhee. 1972. Interrelationship of dietary molybdenum and copper on growth and tissue composition of cattle. *Jour. Anim. Sci.* 34:516-520.

Colborn, T. 1982. Measurement of low levels of molybdenum in the environment by using aquatic insects. *Bull. Environ. Contam. Toxicol.* 29:422-428.

Collier, R.L. 1985. Molybdenum in the northeast Pacific Ocean. *Limnol. Oceanogr.* 30:1351-1354.

Colmano, G. 1973. Molybdenum toxicity: abnormal cellular division of teratogenic appearance in *Euglena gracilis*. *Bull. Environ. Contam. Toxicol.* 9:361-364.

Cordone, A.J. and S.J. Nicola. 1970. Influence of molybdenum on the trout and trout fishing of Castle Lake. *Calif. Fish Game* 56:96-108.

Coughlan, M.P. (ed.). 1980. *Molybdenum and Molybdenum-Containing Enzymes.* Pergamon Press, Elmsford, NY. 577 pp.

Cymbaluk, N.F., H.F. Schryver, H.F. Hintz, D.F. Smith, and J.E. Lowe. 1981. Influence of dietary molybdenum on copper metabolism in ponies. *Jour. Nutr.* 111:96-106.

Dollahite, J.W., L.D. Rowe, L.M. Cook, D. Hightower, E.M. Bailey, and J.R. Kyzar. 1972. Copper deficiency and molybdenosis intoxication associated with grazing near a uranium mine. *Southwest. Vet.* 26:47-50.

Earl, F.L. and T.J. Vish. 1979. Teratogenicity of heavy metals. Pages 617-639 in F.W. Oehme (ed.). *Toxicity of Heavy Metals in the Environment. Part 2.* Marcel Dekker, New York.

Eisler, R. 1981. *Trace Metal Concentrations in Marine Organisms.* Pergamon Press, New York. 687 pp.

Eisler, R. 1989. Molybdenum Hazards to Fish, Wildlife, and Invertebrates: A Synoptic Review. U.S. Fish Wildl. Serv. Biol Rep. 85 (1.19). 61 pp.

Elseewi, A.A. and A.L. Page. 1984. Molybdenum enrichment of plants grown on fly ash-treated soils. *Jour. Environ. Qual.* 13:394-398.

Erdman, J.A., R.J. Ebens, and A.A. Case. 1978. Molybdenosis: a potential problem in ruminants grazing on coal mine spoils. *Jour. Range Manage.* 31:34-36.

Flynn, A., A.W. Franzmann, and P.D. Arneson. 1976. Molybdenum-sulfur interactions in the utilization of marginal dietary copper in Alaskan moose (*Alces alces gigas*). Pages 115-124 in W.R. Chappell and K.K. Peterson (eds.). *Molybdenum in the Environment. Vol. 1. The Biology of Molybdenum.* Marcel Dekker, New York.

Flynn, A., A.W. Franzmann, P.O. Arneson, and J.L. Oldemeyer. 1977. Indications of copper deficiency in a subpopulation of Alaskan moose. *Jour. Nutr.* 107:1182-1189.

Franson, J.C., P.S. Koehl, D.V. Derksen, T.C. Rothe, C.M. Bunck, and J.F. Moore. 1995. Heavy metals in seaducks and mussels from Misty Fjords National Monument in southeast Alaska. *Environ. Monitor. Assess.* 36:149-167.

Friberg, L., P. Boston, G. Nordberg, M. Piscator, and K.H. Robert. 1975. Molybdenum — A Toxicological Appraisal. U.S. Environ. Protection Agency Rep. 600/1-75-004. 142 pp.

Friberg, L. and J. Lener. 1986. Molybdenum. Pages 446-461 in L. Friberg, G.F. Nordberg, and V.B. Vouk (eds.). *Handbook of the Toxicology of Metals. Vol. II. Specific Metals.* Elsevier, New York.

Froslie, A., G. Norheim, and N.E. Soli. 1983. Levels of copper, molybdenum, zinc, and sulphur in concentrates and mineral feeding stuffs in relation to chronic copper poisoning in sheep in Norway. *Acta Agric. Scand.* 33:261-267.

Fujii, R. 1988. Water-Quality and Sediment-Chemistry Data of Drain Water and Evaporation Ponds from Tulare Lake Drainage District, Kings County, California, March 1985 to March 1986. U.S. Geol. Surv. Open-File Rep. 87-700. 19 pp. Avail. from U.S. Geological Survey, Federal Center, Building 810, Box 25425, Denver, CO 80225.

Garmo, T.H., A. Froslie, and R. Hole. 1986. Levels of copper, molybdenum, sulphur, zinc, selenium, iron and manganese in native pasture plants from a mountain area in southern Norway. *Acta Agric. Scand.* 36:147-161.

Goyer, R.A. 1986. Toxic effects of metals. Pages 582-635 in C.D. Klassen, M.O. Amdur, and J. Doull (eds.). *Casarett and Doull's Toxicology. Third Edition.* Macmillan, New York.

Gupta, U.C. and J. Lipsett. 1981. Molybdenum in soils, plants, and animals. *Adv. Agron.* 34:73-115.

Hall, R.A., E.G. Zook, and G.M. Meaburn. 1978. National Marine Fisheries Service Survey of Trace Elements in the Fishery Resources. U.S. Dep. Commerce NOAA Tech. Rep. NMFS SSRF-721. 313 pp.

Hamilton, S.J. and K.J. Buhl. 1990. Acute toxicity of boron, molybdenum, and selenium to fry of chinook salmon and coho salmon. *Arch. Environ. Contam. Toxicol.* 19:366-373.

Hamilton, S.J. and K.J. Buhl. 1997. Hazard evaluation of inorganics, singly and in mixtures, to flannelmouth sucker *Catostomus latipinnis* in the San Juan River, New Mexico. *Ecotoxicol. Environ. Safety* 38:296-308.

Hamilton, S.J. and R.H. Wiedmeyer. 1990. Concentrations of boron, molybdenum, and selenium in chinook salmon. *Trans. Amer. Fish. Soc.* 119:500-510.

Harrison, T.J., J.B.J. van Ryssen, and P.R. Barrowman. 1987. The influence of breed and dietary molybdenum on the concentration of copper in tissues of sheep. *S. Afr. Jour. Anim. Sci.* 17:104-110.

Henry, R. and J.G. Tundisi. 1982. Evidence of limitation by molybdenum and nitrogen on the growth of the phytoplankton community of the Lobo Reservoir (Sao Paulo, Brazil). *Rev. Hydrobiol. Trop.* 15:201-208.

Howarth, R.W. and J.J. Cole. 1985. Molybdenum availability, nitrogen limitation, and phytoplankton growth in natural waters. *Science* 229:653-655.

Hynes, M., M. Woods, D. Poole, P. Rogers, and J. Mason. 1985. Some studies on the metabolism of labelled molybdenum compounds in cattle. *Jour. Inorg. Biochem.* 24:279-288.

Ivan, M. and D.M. Veira. 1985. Effects of copper sulfate supplement on growth, tissue concentration, and ruminal solubilities of molybdenum and copper in sheep fed low and high molybdenum diets. *Jour. Dairy Sci.* 68:891-896.

Karring, M., R. Pohjanvirta, T. Rahko, and H. Korpela. 1981. The influence of dietary molybdenum and copper supplementation on the contents of serum uric acid and some trace elements in cocks. *Acta Vet. Scand.* 22:289-295.

Kienholz, E.W. 1977. Effects of environmental molybdenum levels upon wildlife. Pages 731-737 in W.R. Chappell and K.K. Peterson (eds.). *Molybdenum in the Environment. Vol. 2. The Geochemistry, Cycling, and Industrial Uses of Molybdenum.* Marcel Dekker, New York.

Kincaid, R.L. 1980. Toxicity of ammonium molybdate added to drinking water of calves. *Jour. Dairy Sci.* 63:608-610.

Kincaid, R.L., R.M. Blauwiekel, and J.D. Cronrath. 1986. Supplementation of copper as copper sulfate or copper proteinate for growing calves fed forages containing molybdenum. *Jour. Dairy Sci.* 69:160-163.

King, K.A., J. Leleux, and B.M. Mulhern. 1984. Molybdenum and copper levels in white-tailed deer near uranium mines in Texas. *Jour. Wildl. Manage.* 48:267-270.

King, R.U., D.R. Shawe, and E.M. MacKevett, Jr. 1973. Molybdenum. Pages 425-435 in D.A. Brobst and W.P. Pratt (eds.). United States Mineral Resources. U.S. Geol. Surv. Prof. Paper 820. Avail. from U.S. Govt. Printing Office, Washington, D.C. 20402.

Knothe, D.W. and G.G. Van Riper. 1988. Acute toxicity of sodium molybdate dehydrate (Molhibit 100) to selected saltwater organisms. *Bull. Environ. Contam. Toxicol.* 40:785-790.

Koizumi, T. and Y. Yamane. 1984. Protective effect of molybdenum on the acute toxicity of mercuric chloride. III. *Chem. Pharm. Bull.* 32:2316-2324.

Kubota, J. 1975. Areas of molybdenum toxicity to grazing animals in the western United States. *Jour. Range Manage.* 28:252-256.

Kubota, J., V.A. Lazar, G.H. Simonson, and W.W. Hill. 1967. The relationship of soils to molybdenum toxicity in grazing animals in Oregon. *Soil Sci. Soc. Amer. Proc.* 31:667-671.

Kume, S., A. Mukai, and M. Shibata. 1984. Effects of dietary copper and molybdenum levels on liver and kidney minerals in Holstein cattle. *Japan. Jour. Zootech. Sci.* 55:670-676.

Kummer, J.T. 1980. Molybdenum. Pages 615-628 in U.S. Bur. Mines. *Minerals Yearbook 1978–79. Vol. I. Metals and Minerals.* Avail. from Sup. Doc., U.S. Govt. Printing Office, Washington, D.C. 20402.

Lahann, R.W. 1976. Molybdenum hazard in land disposal of sewage sludge. *Water Air Soil Pollut.* 6:3-8.

Lamand, M., C. Lab, J.C. Tressal, and J. Mason. 1980. Biochemical parameters useful for the diagnosis of mild molybdenosis in sheep. *Ann. Rech. Vet.* 11:141-145.

Lener, J. and B. Bibr. 1979. Biliary excretion and tissue distribution of penta- and hexavalent molybdenum in rats. *Toxicol. Appl. Pharmacol.* 51:259-263.

Livingston, H.D. and G.H. Thompson. 1971. Trace element concentrations in some modern corals. *Limnol. Oceanogr.* 16:786-795.

Lloyd, W.E., H.T. Hill, and G.L. Meerdink. 1976. Observations of a case of molybdenosis-copper deficiency in a South Dakota dairy herd. Pages 85-95 in W.R. Chappell and K.K. Peterson (eds.). *Molybdenum in the Environment. Vol. 1. The Biology of Molybdenum.* Marcel Dekker, New York.

Luo, X.M., H.J. Wei, and S.P. Yang. 1983. Inhibitory effects of molybdenum on esophageal and forestomach carcinogenesis in rats. *Jour. Natl. Cancer Inst.* 71:75-80.

McConnell, R.P. 1977. Toxicity of molybdenum to rainbow trout under laboratory conditions. Pages 725-730 in W.R. Chappell and K.K. Peterson (eds.). *Molybdenum in the Environment. Vol. 2. The Geochemistry, Cycling, and Industrial Uses of Molybdenum.* Marcel Dekker, New York.

Miah, M.M.U. 1999. Evaluation of critical limits of trace elements Zn, B, and Mo in Bangladesh agriculture. Pages 432-433 in W.W. Wenzel, D.C. Adriano, B. Alloway, H.E. Doner, C. Keller, N.W. Lepp, M. Mench, R. Naidu, and G. M. Pierzynski (eds.). Proceedings of the Fifth International Conference on the Biogeochemistry of Trace Elements. July 11-15, 1999, Vienna, Austria.

Mills, C.F. and I. Breamer. 1980. Nutritional aspects of molybdenum in animals. Pages 517-542 in M.P. Coughlan (ed.). *Molybdenum and Molybdenum-Containing Enzymes.* Pergamon Press, Elmsford, New York.

Mills, C.F., T.T. El-Gallad, I. Breamer, and G. Wenham. 1981. Copper and molybdenum absorption by rats given ammonium tetrathiomolybdate. *Jour. Inorg. Biochem.* 14:163-175.

Mitchell, D.G., J.D. Morgan, J.C. Cronin, D.A. Cobb, G.A. Vigers, and P.M. Chapman. 1986. Acute Lethal Marine Bioassay Studies for the U.S. Borax Quartz Hill Project. Canad. Tech. Rep. Fish. Aquat. Sci. No. 1480:48-49.

Mitchell, P.C.H. and A.G. Sykes (eds.). 1986. The chemistry and uses of molybdenum. Proceedings of the Climax Fifth International Conference. *Polyhedron* 5(1/2):1-606.

Morgan, J.D., D.G. Mitchell, and P.M. Chapman. 1986. Individual and combined toxicity of manganese and molybdenum to mussel, *Mytilus edulis*, larvae. *Bull. Environ. Contam. Toxicol.* 37:303-307.

Nagy, J.G., W. Chappell, and G.M. Ward. 1975. Effects of high molybdenum intake in mule deer. *Jour. Anim. Sci.* 41:412.

Nederbragt, H. 1980. The influence of molybdenum on the copper metabolism of the rat at different Cu levels of the diet. *Brit. Jour. Nutr.* 43:329-338.

Neberbragt, H. 1982. Changes in the distribution of copper and molybdenum after Mo administration and subsequent additional oral or intraperitoneal Cu administration to rats. *Brit. Jour. Nutr.* 48:353-364.

Newton, W.E. and S. Otsuka (eds.). 1980. *Molybdenum Chemistry of Biological Significance.* Plenum Press, New York. 425 pp.

O'Gorman, J., F.H. Smith, D.B.R. Poole, M.P. Boland, and J.F. Roche. 1987. The effect of molybdenum-induced copper deficiency on reproduction in beef heifers. *Theriogenology* 27:265.

Osman, N.H.I. and A.R. Sykes. 1989. Comparative effects of dietary molybdenum concentration on distribution of copper in plasma in sheep and red deer (*Cervus elaphus*). *Proc. N.Z. Soc. Anim. Product.* 49:15-19.

Paerl, H.W., K.M. Crocker, and L.E. Prufert. 1987. Limitation of N fixation in coastal marine waters: relative importance of molybdenum, iron, phosphorus, and organic matter availability. *Limnol. Oceanogr.* 32:525-536.

Pankakoski, E., H. Hyvarinen, M. Jalkanen, and I. Koivisto. 1993. Accumulation of heavy metals in the mole in Finland. *Environ. Pollut.* 80:9-16.

Papadopoulou, C., I. Hadjistelios, M. Ziaka, and D. Zafiropoulos. 1981. Stable molybdenum in plankton and pelagic fish from the Aegean Sea. *Rapp. P.-V. Reun. Comm. Int. Explor. Sci. Mer Mediterr.* 27:135-138.

Parada, R. 1981. Zinc deficiency in molybdenum poisoned cattle. *Vet. Human Toxicol.* 23:16-21.

Parker, G.A. 1983. *Analytical Chemistry of Molybdenum.* Springer-Verlag, New York. 175 pp.

Parsons, E.C.M. 1998. Trace metal pollution in Hong Kong: implications for the health of Hong Kong's Indo-Pacific hump-backed dolphins (*Sousa chinensis*). *Science Total Environ.* 214:175-184.

Penot, M. and C. Videau. 1975. Absorption du ^{86}Rb et du ^{99}Mo par deux algues marines: le *Laminaria digitata* et le *Fucus serratus*. *Z. Pflanzenphysiol.* Bd. 76(Suppl.):285-293.

Penumarthy, L. and F.W. Oehme. 1978. Molybdenum toxicosis in cattle. *Vet. Human Toxicol.* 20:11-12.

Phillippo, M., W.R. Humpheries, T. Atkinson, G.D. Henderson, and P.H. Garthwaite. 1987a. The effect of dietary molybdenum and iron on copper status, puberty, fertility and oestrous cycles in cattle. *Jour. Agric. Sci. (Cambridge)* 109:321-326.

Phillippo, M., W.R. Humphries, and P.H. Garthwaite. 1987b. The effect of dietary molybdenum and iron on copper status and growth in cattle. *Jour. Agric. Sci. (Cambridge)* 109:315-320.

Pierzynski, G.M. and L.W. Jacobs. 1986. Molybdenum accumulation by corn and soybeans from a molybdenum-rich sewage sludge. *Jour. Environ. Qual.* 15:394-398.

Pitt, M., J. Fraser, and D.C. Thurley. 1980. Molybdenum toxicity in sheep: epiphysiolysis, exostoses and biochemical changes. *Jour. Compar. Pathol.* 90:567-576.

Prange, A. and K. Kremling. 1985. Distribution of dissolved molybdenum, uranium and vanadium in Baltic Sea waters. *Mar. Chem.* 16:259-274.

Rao, T.A. 1984. Iron, copper and molybdenum in the different body parts of some clupeoids. *Indian Jour. Fish.* 31:357-360.

Reid, B.L., A.A. Kurnick, R.N. Burroughs, R.L. Svacha, and J.R. Couch. 1957. Molybdenum in poult nutrition. *Proc. Soc. Exp. Biol. Med.* 94:737-740.

Reid, B.L., A.A. Kurnick, R.L. Svacha, and J.R. Couch. 1956. The effect of molybdenum on chick and poult growth. *Proc. Soc. Exp. Biol. Med.* 93:245-248.

Robinson, J.A., T.J. Devlin, K.M. Wittenberg, and N.E. Stanger. 1987. The influence of molybdenum and sulfur on various copper parameters of afaunated ram lambs of different sire breeds. *Canad. Jour. Anim. Sci.* 67:65-74.

Saiki, M.K., M.R. Jennings, and W.G. Brumbaugh. 1993. Boron, molybdenum, and selenium in aquatic food chains from the lower San Joaquin River and its tributaries, California. *Arch. Environ. Contam. Toxicol.* 24:307-319.

Sakaguchi, T., A. Nakajima, and T. Horikoshi. 1981. Studies on the accumulation of heavy metal elements in biological systems. Accumulation of molybdenum by green microalgae. *European Jour. Appl. Microbiol. Biotechnol.* 12:84-89.

Sas, B. 1987. Accidental molybdenum contamination induced fatal secondary copper deficiency in cattle. *Acta Vet. Hung.* 35:281-289.

Schalscha, E.B., M. Morales, and P.F. Pratt. 1987. Lead and molybdenum in soils and forage near an atmospheric source. *Jour. Environ. Qual.* 16:313-315.

Schroeder, H.A., J.J. Balassa, and I.H. Tipton. 1970. Essential trace metals in man: molybdenum. *Jour. Chron. Dis.* 23:481-499.

Shacklette, H.T., J.A. Erdman, T.F. Harms, and C.S.E. Papp. 1978. Trace elements in plant food stuffs. Pages 25-68 in F.W. Oehme (ed.) *Toxicity of Heavy Metals in the Environment. Part 1.* Marcel Dekker, New York.

Shamberger, R.J. 1979. Beneficial effects of trace elements. Pages 689-796 in F.W. Oehme (ed.). *Toxicity of Heavy Metals in the Environment. Part 2.* Marcel Dekker, New York.

Short, Z.F., P.R. Olson, R.F. Palumbo, J.R. Donaldson, and F.G. Lowman. 1971. Uptake of molybdenum, marked with ^{99}Mo, by the biota of Fern Lake, Washington, in a laboratory and a field experiment. Pages 474-485 in D.J. Nelson (ed.). *Radionuclides in Ecosystems. Proceedings of the Third National Symposium on Radioecology.* Vol. 1. May 10–12, 1971, Oak Ridge, TN.

Singh, S.S. and A.N. Mourya. 1983. Effect of molybdenum on growth, yield, and quality of okra (*Abelmoschus esculentus* (L.) Moench). *Agric. Sci. Dig.* 3:105-107.

Sloot, H.A. van der, D. Hoede, J. Wijkstra, J.C. Duinker, and R.F. Nolting. 1985. Anionic species of V, As, Se, Mo, Sb, Te, and W in the Scheldt and Rhine estuaries and the Southern Bight (North Sea). *Estuar. Coastal Shelf Sci.* 21:633-651.

Smith, C., K.W. Brown, and L.E. Deuel, Jr. 1987. Plant availability and uptake of molybdenum as influenced by soil type and competing ions. *Jour. Environ. Qual.* 16:377-382.

Solomons, C.C., D.J. Ernisse, and E.M. Handrich. 1973. Skeletal biology of molybdenum. Pages 233-239 in D.D. Hemphill (ed.). *Trace Substances in Environmental Health. VII.* Univ. Missouri, Columbia, MO.

Soon, Y.K. and T.E. Bates. 1985. Molybdenum, cobalt and boron uptake from sewage-sludge-amended soils. *Canad. Jour. Soil Sci.* 65:507-517.

Steeg, P.F. ter, P.J. Hanson, and H.W. Paerl. 1986. Growth-limiting quantities and accumulation of molybdenum in *Anabaena oscillarioides* (Cyanobacteria). *Hydrobiologia* 140:143-147.

Strickland, K., F. Smith, M. Woods, and J. Mason. 1987. Dietary molybdenum as a putative copper antagonist in the horse. *Equine Vet. Jour.* 19:50-54

Subbaiah, C.C., P. Manikandan, and Y. Joshi. 1986. Yellow leaf spot of cashew: a case of molybdenum deficiency. *Plant Soil* 94:35-42.

Suttle, N.F. 1973. The nutritional significance of the Cu:Mo interrelationship to ruminants and non-ruminants. Pages 245-249 in D.D. Hemphill (ed.). *Trace Substances in Environmental Health. VII.* Univ. Missouri, Columbia, MO.

Suttle, N.F. 1980. The role of thiomolybdates in the nutritional interactions of copper, molybdenum, and sulfur: fact or fantasy? *Ann. N.Y. Acad. Sci.* 355:195-207.

Suttle, N.F. 1983a. Effects of molybdenum concentration in fresh herbage, hay and semi-purified diets on the copper metabolism of sheep. *Jour. Agric. Sci. (Cambridge)* 100:651-656.

Suttle, N.F. 1983b. A role for thiomolybdates in the Cu × Mo × S interaction in ruminant nutrition. Pages 599-610 in P. Bratter and P. Schramel (eds.). *Trace Element-Analytical Chemistry in Medicine and Biology, Vol. 2.* Walter de Gruyter, New York.

Suttle, N.F. and A.C. Field. 1983. Effects of dietary supplements of thiomolybdates on copper and molybdenum metabolism in sheep. *Jour. Comp. Pathol.* 93:379-389.

Talbot, V. and P. Ryan. 1988. High molybdenum land values in Ireland: possible implications. *Sci. Total Environ.* 76:217-228.

Todd, J.R. 1976. Problems of copper-molybdenum imbalance in the nutrition of ruminants in northern Ireland. Pages 33-49 *in* W.R. Chappell and K.K. Peterson (eds.). *Molybdenum in the Environment. Vol. 1. The Biology of Molybdenum.* Marcel Dekker, New York.

Underwood, E.J. 1971. *Trace Elements in Human and Animal Nutrition.* Chapter 4. Molybdenum. Academic Press, New York. 116-140.

Underwood, E.J. 1979. Interactions of trace elements. Pages 641-668 in F.W. Oehme (ed.). *Toxicity of Heavy Metals in the Environment. Part 2.* Marcel Dekker, New York.

Vaishampayan, A. 1983. Mo-V interactions during N_2- and NO_3-metabolism in a N_2-fixing blue-green alga *Nostoc muscorum. Experientia* 39:358-360.

Van Ryssen, J.B.J., W.S. Botha, and W.J. Stielau. 1982. Effect of a high copper intake and different levels of molybdenum on the health of sheep. *Jour. S. Afr. Vet. Assoc.* 53:167-170.

Van Ryssen, J.B.J., S. van Malsen, and P.R. Barrowman. 1986. Effect of dietary molybdenum and sulphur on the copper status of hypercuprotic sheep after withdrawal of dietary copper. *S. Afr. Jour. Anim. Sci.* 16:77-82.

Van Ryssen, J.B.J. and W.J. Stielau. 1980. The effect of various levels of dietary copper and molybdenum on copper and molybdenum metabolism in sheep. *S. Afr. Jour. Anim. Sci.* 10:37-47.

Ward, G.M. 1978. Molybdenum toxicity and hypocuprosis in ruminants: a review. *Jour. Anim. Sci.* 46:1078-1085.

Ward, G.M. and J.G. Nagy. 1976. Molybdenum and copper in Colorado forages, molybdenum toxicity in deer, and copper supplementation in cattle. Pages 97-113 in W.R. Chappell and K.K. Peterson (eds.). *Molybdenum in the Environment. Vol. 1. The Biology of Molybdenum.* Marcel Dekker, New York.

Ward, J.V. 1973. Molybdenum concentrations in tissues of rainbow trout (*Salmo gairdneri*) and kokanee salmon (*Oncorhynchus nerka*) from waters differing widely in molybdenum content. *Jour. Fish. Res. Board Canada* 30:841-842.

Weber, K.M., R.C. Boston, and D.D. Leaver. 1983. The effect of molybdenum and sulfur on the kinetics of copper metabolism in sheep. *Austral. Jour. Agric. Res.* 34:295-306.

Wei, H.-J., X.-M. Luo, and S.P. Yang. 1985. Effects of molybdenum and tungsten on mammary carcinogenesis in SD rats. *Jour. Natl. Cancer Inst.* 74:469-473.

White, R.D., R.A. Swick, and P.R. Cheeke. 1984. Effects of dietary copper and molybdenum on tansy ragwort (*Senecio jacobaea*) toxicity in sheep. *Amer. Jour. Vet. Res.* 45:159-161.

Winston, P.W., M.S. Heppe, L. Hoffman, L.J. Kosarek, and R.S. Spangler. 1976. Physiological adaptations to excess molybdenum ingested over several generations. Pages 185-200 in W.R. Chappell and K.K. Peterson (eds). *Molybdenum in the Environment. Vol. 1. The Biology of Molybdenum.* Marcel Dekker, New York.

Winston, P.W., L. Hoffman, and W. Smith. 1973. Increased weight loss in molybdenum-treated rats in the cold. Pages 241-244 *in* D.D. Hemphill (ed.). *Trace Substances in Environmental Health. VII.* Univ. Missouri, Columbia, MO.

Wittenberg, K.M. and T.J. Devlin. 1987. Effects of dietary molybdenum on productivity and metabolic parameters of lactating beef cows and their offspring. *Canad. Jour. Anim. Sci.* 67:1055-1066.

Wright, R.J. and L.R. Hossner. 1984. Molybdenum release from three Texas soils. *Soil Sci.* 138:374-377.

Yamane, Y. and T. Koizumi. 1982. Protective effect of molybdenum on the acute toxicity of mercuric chloride. *Toxicol. Appl. Pharmacol.* 65:214-221.

Yoshimura, T., K. Tsunoda, and K. Nishimoto. 1987. Effect of molybdenum and tungsten compounds on the survival of *Coptotermes formosanus* Shiraki (Isoptera: Rhinotermitidae) in laboratory experiments. *Mater. Org. (Berl.)* 22:47-56.

CHAPTER 31

Selenium

31.1 INTRODUCTION

Selenium poisoning is an ancient and well-documented disease (Rosenfeld and Beath 1964). Signs of it were reported among domestic livestock by Marco Polo in western China near the borders of Turkestan and Tibet in about the year 1295; among livestock, chickens, and children in Colombia, South America, by Father Pedro Simon in 1560; among human adults in Irapuato, Mexico, in about 1764; and among horses of the U.S. Cavalry in South Dakota in 1857 and again in 1893 (Rosenfeld and Beath 1964). In 1907/08, more than 15,000 sheep died in a region north of Medicine Bow, Wyoming, after grazing on seleniferous plants. The incidents have continued, and recent technical literature abounds with isolated examples of selenosis among domestic animals and wildlife.

Selenium (Se) was first identified as an element in 1817 by the Swedish chemist Berzelius. It is now firmly established that selenium is beneficial or essential in amounts from trace to μg/kg (ppb) concentrations for humans and some plants and animals, but toxic at some concentrations present in the environment (Rosenfeld and Beath 1964). Selenium deficiency was reported among cattle grazing in the Florida Everglades, which showed evidence of anemia, slow growth, and reduced fertility (Morris et al. 1984). Selenium deficiency has been demonstrated in Atlantic salmon, *Salmo salar* (Lorentzen et al. 1994), in various species of deer in Florida and Washington (Hein et al. 1994; McDowell et al. 1995), and in free-ranging ungulates in Washington state, including moose (*Alces alces*) and bighorn sheep (*Ovis canadensis*) (Hein et al. 1994). Conversely, calves of Indian buffaloes died of selenium poisoning after eating rice husks grown in naturally seleniferous soils (Prasad et al. 1982). Adverse effects of excess selenium are reported on reproduction of cattle, monkeys, sheep, swine, rats, and hamsters, including fetal and maternal death, and a dramatic increase in developmental abnormalities (Domingo 1994). Severe reproductive and developmental abnormalities were observed in aquatic birds nesting at selenium-contaminated irrigation drainwater ponds in the San Joaquin Valley, California (Ohlendorf et al. 1986, 1986a, 1987, 1989, 1990; Hoffman et al. 1988; Schuler et al. 1990; Besser et al. 1993; Lemly 1996b). Accumulation of more than 8 mg Se/kg dry weight in fish gonads is the probable cause of reduced reproduction and subsequent species disappearances in Belews Lake, North Carolina, and the endangered razorback sucker (*Xyrauchen texanus*) from the Green River, Utah, in 1991 (Cumbie and Van Horn 1978; Hamilton and Waddell 1994; Waddell and May 1995).

Selenium has been the subject of many reviews (Rosenfeld and Beath 1964; Frost 1972; Sandholm 1973; Zingaro and Cooper 1974; Frost and Ingvoldstad 1975; Anonymous 1975; National Academy of Sciences [NAS] 1976; Harr 1978; U.S. Environmental Protection Agency [USEPA] 1980, 1987; Lo and Sandi 1980; Shamberger 1981; Wilber 1980, 1983; Fishbein 1977, 1983; National Research Council [NRC] 1983; Reddy and Massaro 1983; Eisler 1985; Lemly and Smith

1987; Ohlendorf 1989; Hodson 1990; Goede 1993; Lemly 1993, 1996a, 1996b; Heinz 1996; U.S. Public Health Service [USPHS] 1996). These authorities agree that selenium is widely distributed in nature, being especially abundant with sulfide minerals of various metals, such as iron, lead, and copper. The major source of environmental selenium is the weathering of natural rock. The amount of selenium entering the atmosphere as a result of anthropogenic activities is estimated to be 3500 metric tons annually, of which most is attributed to combustion of coal and the irrigation of high-selenium soils for crop production. However, aside from highly localized contamination, the contribution of selenium by human activities is small in comparison with that attributable to natural sources. Collectively, all authorities agree that selenium may favorably or adversely affect growth, survival, and reproduction of algae and higher plants, bacteria and yeasts, crustaceans, molluscs, insects, fish, birds, and mammals (including humans). Most acknowledge that sensitivity to selenium and its compounds is extremely variable in all classes of organisms and, except for some instances of selenium deficiency or of selenosis, metabolic pathways and modes of action are imperfectly understood. For example, selenium indicator plants can accumulate selenium to concentrations of thousands of parts per million (mg/kg) without ill effects. In these plants, selenium promotes growth; whereas in crop plants, accumulations as low as 25 to 50 mg/kg may be toxic. Thus, plants and waters high in selenium are considered potentially hazardous to livestock and to aquatic life and other natural resources in seleniferous zones.

31.2 ENVIRONMENTAL CHEMISTRY

Selenium is characterized by an atomic weight of 78.96, an atomic number of 34, a melting point of 271°C, a boiling point of 685°C, and a density of 4.26 to 4.79. Chemical properties, uses, and environmental persistence of selenium were documented by a number of researchers whose works constitute the major source material for this section: Rosenfeld and Beath (1964); Bowen (1966); Lakin (1973); Stadtman (1974, 1977); Frost and Ingvoldstad (1975); Chau et al. (1976); Harr (1978); Wilber (1980, 1983); Zieve and Peterson (1981); Robberecht and Von Grieken (1982); Cappon and Smith (1982); Nriagu and Wong (1983); Eisler (1985); USPHS (1996).

There was general agreement on four points.

1. Selenium chemistry is complex, and additional research is warranted on chemical and biochemical transformations among valence states, allotropic forms, and isomers of selenium.
2. Selenium metabolism and degradation are significantly modified by interaction with heavy metals, agricultural chemicals, microorganisms, and a variety of physicochemical factors.
3. Anthropogenic activities (including fossil fuel combustion and metal smelting) and naturally seleniferous areas pose the greatest hazards to fish and wildlife.
4. Selenium deficiency is not as well documented as selenium poisoning, but may be equally significant.

Selenium chemistry is complex (Rosenfeld and Beath 1964; Harr 1978; Wilber 1983; Porcella et al. 1991; Wiedmeyer and May 1993; Besser et al. 1994; USPHS 1996). In nature, selenium exists: as six stable isotopes (Se-74, -76, -77, -78, -80, and -82), of which Se-80 and -78 are the most common, accounting for 50% and 23.5%, respectively; in three allotropic forms; and in five valence states. Changes in the valence state of selenium from –2 (hydrogen selenide) through 0 (elemental selenium), +2 (selenium dioxide), +4 (selenite), and +6 (selenate) are associated with its geologic distribution, redistribution, and use. Soluble selenates occur in alkaline soils, are slowly reduced to selenites, and are then readily taken up by plants and converted into organoselenium compounds, including selenomethionine, selenocysteine, dimethyl selenide, and dimethyl diselenide. In drinking water, selenates represent the dominant chemical species. Selenites are less soluble than the corresponding selenates and are easily reduced to elemental selenium. In seawater, selenites are the dominant chemical species under some conditions (Cappon and Smith 1981). Selenium dioxide is formed by combustion of elemental selenium present in fossil fuels or rubbish.

Selenium is the most strongly enriched element in coal, being present as an organoselenium compound, a chelated species, or as an adsorbed element. On combustion of fossil fuels, the sulfur dioxide formed reduces the selenium to elemental selenium. Elemental selenium is insoluble and largely unavailable to the biosphere, although it is still capable of satisfying metabolic nutritional requirements. Hydrogen selenide is highly toxic (at 1 to 4 μg/L in air), unstable, acidic, and irritative. Selenides of mercury, silver, copper, and cadmium are very insoluble, although their insolubility may be the basis for the reported detoxification of methylmercury by dietary selenite, and for the decreased heavy metal toxicity associated with selenite. Metallic selenides are thus biologically important in sequestering both Se and heavy metals in a largely unavailable form.

In areas of acid or neutral soils, the amount of biologically available selenium should steadily decline. The decline may be accelerated by active agricultural or industrial practices. In dry areas, with alkaline soils and oxidizing conditions, elemental selenium and selenides in rocks and volcanic soils may oxidize sufficiently to maintain the availability of biologically active selenium. Concentrations of selenium in water are a function of selenium levels in the drainage system and of water pH. In Colorado, for example, streams with pH 6.1 to 6.9 usually contain <1 μg Se/L, but those with pH 7.8 to 8.2 may contain 270 to 400 μg/L (Lakin 1973). Selenium volatilizes from soils at rates that are modified by temperature, moisture, time, season of year, concentration of water-soluble selenium, and microbiological activity. Conversion of inorganic and organic selenium compounds to volatile selenium compounds (such as dimethyl selenide, dimethyl diselenide, and an unidentified compound) by microorganisms has been observed in lake sediments of the Sudbury area of Ontario. This conversion may have been effected by pure cultures of *Aeromonas*, *Flavobacterium*, *Pseudomonas*, or an unidentified fungus, all of which are found in methylated lake sediments. Production of volatile selenium is temperature dependent. Compared with the amount of $(CH_3)_2Se$ produced at an incubation temperature of 20°C, 25% less was produced at 10°C and 90% less at 4°C. Details of selenium reduction and oxidation by microorganisms are not clear. One suggested mechanism for selenite reduction in certain microorganisms involves attachment to a carrier protein and transformation from selenite to elemental selenium, which in turn may be oxidized to selenite by the action of *Bacillus* spp., as one example. It is apparent that much additional research on this problem is warranted. It now appears that selenates and selenites are absorbed by plants, reduced, and then incorporated in amino acid synthesis. The biological availability of selenium is higher in plant foods than in foods of animal origin (Lo and Sandi 1980). The net effect of soil, plant, and animal metabolism is to convert selenium to inert and insoluble forms such as elemental selenium, metallic selenides, and complexes of selenite with ferric oxides.

Selenium was used in the early 1900s as a pesticide to control plant pests, and is still used sparingly to control pests of greenhouse chrysanthemums and carnations (Rosenfeld and Beath 1964). It has been used to control cotton pests (in Trinidad), mites and spiders that attack citrus, and mites that damage apples. Although no insect-resistant strains have developed, the use of selenium pesticides has been discontinued, owing to their stability in soils and resultant contamination of food crops, their high price, and their proven toxicity to mammals and birds (Rosenfeld and Beath 1964; Eisler 1985). In Canada and France, sodium selenite applied to the soil to discourage deer from browsing conifer seedlings when deer numbers were high was unsuccessful and should be avoided (Jobidon and Prevost 1994). Selenium shampoos, which contain about 1% selenium sulfide, are still used to control dandruff in humans and dermatitis and mange in dogs. Selenium is used extensively in the manufacture and production of glass, pigments, rubber, metal alloys, textiles, petroleum products, medical therapeutic agents, and photographic emulsions (Eisler 1985).

Domestic consumption of selenium in 1981 exceeded 453,000 kg. About 50% was used in electronic and copier components, 22% in glass manufacturing, 20% in chemicals and pigments, and 8% miscellaneous (Cleveland et al. 1993). In 1987, world production of selenium was about 1.4 million kg (USPHS 1996). In 1986, 46% of the global selenium produced was used in the semiconductor and photoelectric industries; 27% in the glass industry to counter coloration impurities from iron; 14% in pigments; and 13% in medicine, in antidandruff shampoos, as catalysts in pharmaceutical preparations, in nutritional feed additives for poultry and livestock, and in pesticide formulations (USPHS 1996).

Air and surface waters generally contain nonhazardous concentrations of selenium. Significant increases of selenium in specific areas are attributed exclusively to industrial sources, and to leaching of groundwater from seleniferous soils. In the United States, about 4.6 million kg selenium are released annually into the environment: 33% from combustion of fossil fuels, 59% from industrial losses, and 8% from municipal wastes. Of the total, about 25% is in the form of atmospheric emissions, and the rest in ash. Mining and smelting of copper–nickel ores at Sudbury, Ontario, Canada, alone releases about 2 metric tons selenium to the environment daily, and probably represents the greatest point source of selenium release in the world. In 1977, 680,000 kg selenium was produced at Sudbury, but only about 10% was recovered, suggesting that about 90% was lost to the environment. Of the amount lost, perhaps 50 metric tons were dispensed into the atmosphere, probably as selenium dioxide (airborne Se levels 1 to 3 km from Sudbury were as high as 6.0 $\mu g/m^3$). The rest was probably associated with mine tailings, wastewater, and scoria, and is a local source of selenium contamination, most notably in lakes. The present annual rates of selenium accumulation in lake sediments in the Sudbury area range from 0.3 to 12.0 mg/m^2. These deposition rates exceed those of pre-colonial times by factors of 3 to 18, and are among the highest recorded in North America (Nriagu and Wong 1983).

Selenium is a serious hazard to livestock and probably to people in a wide semiarid belt that extends from inside Canada southward across the United States into Mexico (NRC 1983). Selenium tends to be present in large amounts in areas where the soils have been derived from Cretaceous rocks. Total selenium in such soils averages about 5 mg/kg, but is sometimes as high as 80 mg/kg. Lack of rainfall has prevented the solution of the selenium minerals and the removal of their salts in drainage waters. In some areas, modern fertilization practices and the buildup of sulfates in the soil due to acid precipitation partly lessen the availability of selenium to plants and forage crops. In the United States, highly seleniferous natural areas (200 to 300 μg/kg in forage) are most abundant in the Rocky Mountain and High Central Plains areas. Areas with lower concentrations (20 to 30 μg/kg) in forage are typical in the Pacific Northwest and the Southeast. However, huge variations are not uncommon from one specific location to another. Among plants, primary and secondary selenium accumulators are almost always implicated in cases of acute or chronic selenium poisoning of livestock. Primary selenium accumulator plants, such as various species of *Astragalus*, *Oonopsis*, *Stanelya*, *Zylorhiza*, and *Machaeranthera*, may require 1 to 50 mg Se/kg in either soil or water for growth, and may contain 100 to 10,000 mg Se/kg as a glutamyl dipeptide or selenocystanthionine. Secondary accumulator plants (representative genera: *Aster*, *Gutierrezia*, *Atriplex*, *Grindelia*, *Castillaja*, and *Comandra*) grow in either seleniferous or nonseleniferous soils and may contain 25 to 100 mg Se/kg. Nonaccumulator plants growing on seleniferous soils contain 1 to 25 mg Se/kg fresh weight. Meat and eggs of domestic animals may contain 8 to 9 mg Se/kg in seleniferous areas, compared with 0.01 to 1.0 mg/kg in nonseleniferous areas. Tissues from animals maintained on high-Se feeds generally contain 3 to 5 mg Se/kg fresh weight vs. up to 20 mg/kg in animals dying of selenium poisoning (Harr 1978).

Selenium is nutritionally important as an essential trace element, but is harmful at slightly higher concentrations. Although normal selenium dietary levels required to ensure human health range from 0.04 to 0.1 mg/kg, toxicity may occur if food contains as little as 4.0 mg/kg. Minimum selenium concentrations required are usually higher in livestock than in humans. In areas with highly seleniferous soils, excess selenium is adsorbed onto a variety of plants and grains and can be fatal to grazing livestock. There is general agreement, however, that selenium inadequacy can be of greater concern to health than selenium toxicity. Selenium has a comparatively short effectual biological life in various species of organisms for which data are available. Studies with radioselenium-75 indicated that its biological half-life is 10 to 64 days: 10 in pheasants, 13 in guppies and voles, 15 in ants, 27 in eels, 28 in leeches, and 64 in earthworms (Wilber 1983). Many investigators concluded that the greatest current and direct use of selenium is in the transportation of grains grown in seleniferous areas to selenium-deficient areas as animal and human food.

31.3 CONCENTRATIONS IN FIELD COLLECTIONS

Selenium concentrations in nonbiological materials extend over several orders of magnitude (Table 31.1; Lemly 1996b). In terrestrial materials, concentrations in excess of 5 mg Se/kg are routinely recorded in meteorites, copper–nickel ores, coal and other fossil fuels, lake sediments in the vicinity of a nickel–copper smeltery, and in sediments of flyash settling ponds. Water concentrations exceeding 50 µg Se/L have been documented in groundwater, especially in areas with seleniferous soils, in sewage wastes, in irrigation drain water, and in water of flyash settling ponds. Selenium concentrations in air samples were >0.5 µg/m^3 in the vicinity of selenium production plants, and these were at least 500 times higher than in a control area (Table 31.1).

Table 31.1 Selenium Concentrations in Nonbiological Materials

Sample and Unit of Concentration	Concentration	Reference[a]
TERRESTRIAL (mg/kg)		
Earth's crust	0.05	1
Soils	0.2	2
Limestones	0.08	2
Sandstones	Up to 0.05	2
Shales	0.6	2
Chondrites	8.0	2
Ocean sediments	0.34–4.8	3
Coal	3.4 (0.5–10.7)	4, 5, 15
Fossil fuels	1–10	6
Petroleum	500–1650	15
Lake sediments		
NY, Lake George	0.22	7
Great Lakes	0.35–0.75	8
Freshwater lakes, Canada	0.2–14.5	9
AQUATIC (µg/L)		
Drinking water		
Worldwide	0.12–0.44	10
Groundwater		
Nebraska, U.S.	<1–480	10
Argentina	48–67	10
Australia	0.008–0.33	10
France	<5–75	10
Israel	0.9–27	10
Italy	<0.002–1.9	10
Sewage waters		
United States		
Raw sewage	280	4
Primary effluent	45	4
Secondary effluent	50	4
Worldwide		
Japan	480–700	2
United States	10–280	2
Former Soviet Union	1.8–2.7	2
Germany	1.5	2
River waters		
Japan	0.03–0.09	11
Germany	0.015	2
Amazon River	0.021	2
United States		
Ohio River	<0.01	10
Mississippi River	0.14	10

Table 31.1 (continued) Selenium Concentrations in Nonbiological Materials

Sample and Unit of Concentration	Concentration	Reference[a]
Michigan	0.8–10	10
Nebraska	<1–20	10
Colorado River	30	10
Lake waters		
Sweden	0.04–0.21	11
United States	0.04–1.4	4
Great Lakes	0.001–0.036	8
Seawater		
Worldwide	0.009–0.045	1, 2
Worldwide	0.09–0.45	12, 15
Worldwide	0.09–<6.0	4
Israel, Dead Sea	0.8	10
Japan	0.04–0.08	10
AIR (μg/m³)		
Near Se industrial plant	0.7	4
Control area	0.001	4
Near Sudbury (Canada) smelter Max.	6.0	11
United States	Usually <0.01	15
INTEGRATED STUDIES (μg/kg or μg/L)		
California		
Rainwater	0.05	10
Lake water	0.018	10
Seawater	0.058–0.08	10
Irrigation drain water		
Subsurface	300–1400	13
Surface	300	13
Vicinity of nickel–copper smelter, Sudbury, Ontario		
Lake waters	0.1–0.4	11
Lake sediments	2000–6000	11
Lake sediments 240 km south of Sudbury	1000–3000	11
Cu–Ni ores	20,000–80,000	11
Flyash ponds		
Sediment	14,000	14
Water	350	14

[a] **1,** Frost and Ingvoldstad 1975; **2,** Ebens and Shacklette 1982; **3,** de Goeij et al. 1974; **4,** NAS 1976; **5,** Kuhn et al. 1980; **6,** Harr 1978; **7,** Heit et al. 1980; **8,** Adams and Johnson 1977; **9,** Speyer 1980; **10,** Robberecht and Von Grieken 1982; **11,** Nriagu and Wong 1983; **12,** Whittle et al. 1977; **13,** Ohlendorf et al. 1986; **14,** Furr et al. 1979; **15,** USPHS 1996.

As a result of natural and anthropogenic processes, comparatively high concentrations of selenium in nonbiological materials may offer protection or pose significant risks to fish and wildlife. In Finland, for example, agricultural fertilizers were supplemented with 6 to 16 mg Se/kg beginning in 1985. In Finnish lakes, selenium concentrations have increased in sediments and fish muscle from this activity and from atmospheric fallout, but no adverse biological effects were observed (Wang et al. 1995). In Sweden, selenium treatment raised the lake water concentrations from 3 μg Se/L at the start to 5 μg/L. This treatment lowered mercury concentrations in mercury-contaminated northern pike (*Esox lucius*) and yellow perch (*Perca flavescens*) in treated Swedish lakes by 60 to 85% (Paulsson and Lundbergh 1991). Selenium is normally present in surface waters at about 0.1 to 0.3 μg/L. However, at 1 to 5 μg/L, it can biomagnify in aquatic food chains and pose a concentrated dietary source of selenium that is toxic to fish and wildlife (Lemly 1993c, 1996a).

Selenium concentrations in representative species of freshwater, marine, and terrestrial flora and fauna are listed in Table 31.2. Additional information on body and tissue burdens of selenium was given by Birkner (1978), Jenkins (1980), Lo and Sandi (1980), Eisler (1981, 1985), and Wilber (1983). It is emphasized that selenium concentrations in all organisms tended to be significantly higher when collected from locales having certain characteristics: highly seleniferous soils or sediments (de Goeij et al. 1974; Birkner 1978; Speyer 1980; Wilber 1983); high human population densities (Beal 1974); heavy accumulations of selenium-laden wastes, such as effluents from systems used to collect flyash scrubber sludge or bottom ash (Cumbie and Van Horn 1978; Sorensen et al. 1982, 1984); and selenium-contaminated subsurface irrigation drainwater (Ohlendorf et al. 1986, 1987; Presser and Ohlendorf 1987; Schuler et al. 1990; Lemly 1993c; Hothem and Welsh 1994). Accumulation, transfer, and release of selenium by aquatic biota may affect the speciation and toxicity of dissolved selenium in aquatic environments (Besser et al. 1994). Depletion of dissolved selenite and increased concentrations of organoselenium compounds occur during seasonal peaks in phytoplankton abundance in freshwater and marine systems. For example, green algae (*Chlamydomonas reinhardtii*) previously exposed to inorganic radioselenium-75 produced increased concentrations of organoselenium species during population blooms and crashes (Besser et al. 1994).

Among terrestrial plants, selenium accumulations in species of *Aster*, *Astragalus*, and several other genera are sometimes spectacularly high (Table 31.2). *Astragalus* is the most widely distributed. About 24 of its more than 200 species are selenium accumulators that require selenium to grow well. The highest reported concentration in plants was 15,000 mg Se/kg DW, in loco weed (*Astragalus racemosus*) (Wilber 1983). Consumption of these and other selenium-accumulating forage plants by livestock has induced illness and death from selenium poisoning. Even at much lower concentrations, selenium may harm animals that eat considerable amounts of the forage. Plants that accumulate selenium tend to be deeper rooted than the grasses and survive more severe aridity, thus remaining as the principal forage for grazing in time of drought (Wilber 1983). There is little danger to human health of selenium toxicity from consuming game that foraged in high-selenium environments (Medeiros et al. 1993).

Selenium levels in freshwater biota are relatively low compared with those in their marine counterparts. In freshwater organisms, about 36% of the total selenium was present as selenate, and the rest as selenite and selenide. In marine samples, only 24% of the total selenium was present as selenate (Cappon and Smith 1982). The implications of this difference are not now understood, but have relevance in the ability of selenium to complex and detoxify various potentially toxic heavy metals, such as mercury and cadmium. In a nationwide monitoring of selenium and other contaminants in freshwater fishes, selenium ranged from 0.05 to 2.9 mg/kg FW whole fish and averaged about 0.6 mg/kg. Stations where concentrations in fish exceeded 0.82 mg/kg (>85th percentile) were in three areas: Atlantic coastal streams, Mississippi River system, and California (Table 31.2) (May and McKinney 1981). Among fish from Atlantic coastal streams, those from the Delaware River near Camden, New Jersey, had elevated whole-body concentrations (i.e., >1.0 and <3.0 mg Se/kg FW), which were attributed to the industrialized character of the river. In the Big Horn and Yellowstone Rivers, high selenium concentrations in fish may result from geologic sources of the element, including coal, phosphate, and sedimentary rock. Fish from the South Platte River near Denver, Colorado, may receive selenium from industrial effluents, or from natural and anthropogenic activities associated with the removal of deposits of coal, barite, and sulfur (May and McKinney 1981). These same trends persisted in more recent nationwide monitoring of freshwater fishes, with selenium concentrations usually highest in whole fish from stations in Utah, Nevada, Texas, California, Hawaii, and in arid locations of the western United States (Table 31.2; Schmitt and Brumbaugh 1990). In California, where selenium was elevated in fish from the San Joaquin River, it was speculated that Selocide, a selenium-containing pesticide registered for use on citrus fruits in the 1960s, may have been a source, although contaminated irrigation drainwater was considered a more likely possibility (Ohlendorf et al. 1986). Of seven species of fishes analyzed from the San Joaquin Valley, California, in 1986/1987, mosquitofish (*Gambusia affinis*) had the highest concentrations (11.1 mg Se/kg DW whole body); these fish were collected from canals and sloughs in the Grasslands Water District that received large inflows of subsurface agricultural drainage water (Saiki et al. 1991, 1992). Selenium persisted in the biota of the Grasslands drainage regions for at least 1 year after the switch to uncontaminated drainage water (Hothem and Welsh 1994).

Selenium bioconcentrates and biomagnifies in aquatic food chains from invertebrates to birds (Rusk 1991; Saiki et al. 1993). Maximum selenium concentrations reported in Cibola Lake in the lower Colorado River Valley in 1989/90 were 5.0 µg/L in water, and — in mg Se/kg DW — 3.3 in sediments, 1.2 in aquatic plants (*Myriophyllum*, *Ceratophyllum*), 4.6 in crayfish (*Procambarus clarki*), and 9.2 in bluegills (*Lepomis macrochirus*) (Welsh 1992). Diet is the primary source of selenium to fish, as judged by radioselenium-75 uptake studies in Canadian oligotrophic lakes (Harrison et al. 1990). Hatchery-reared smolts and adults of silver salmon (*Oncorhynchus kisutch*) had less selenium in livers than did wild fish, and this could account for the higher survival and better health of wild fish (Felton et al. 1990).

Belews Lake in North Carolina was contaminated with selenium during the 1970s from coal-fired power plant wastewater, causing mortality and reproductive failure in the fish population (Lemly 1993a, 1993c). Selenium concentrations in fish tissues were as high as 125 mg Se/kg DW and were as much as 100 times higher than those from nearby reference sites. There was a positive relation between tissue selenium concentrations and frequency of developmental malformations for largemouth bass and bluegill over the range 1 to 80 mg Se/kg DW tissue and 0% to 70% deformities. In 1992, selenium residues had declined to less than 20 mg/kg DW, but were still 5 to 18 times higher than those in reference lakes, and deformity frequency was 7 times higher (Lemly 1993a). Alterations in zooplankton species densities and dominance — but not diversity — were observed in Belews Lake between 1970 (uncontaminated), 1976/77 (selenium contamination), and 1984 to 1986. Observed changes are attributed to the dominance of planktivorous fishes (Marcogliese et al. 1992).

All reported selenium levels in tissues of marine invertebrates and plants were less than 2 mg Se/kg on a fresh weight (FW) basis, or 12 mg/kg dry weight (DW). In marine algae, most of the selenium accumulated was associated with proteins and may represent a form of storage prior to detoxification (Boisson et al. 1995). Higher levels are routinely recorded in liver and kidney tissues of marine and coastal vertebrates, including teleosts, birds, and mammals. Livers from adult seals were comparatively rich in selenium (Table 31.2); however, high concentrations in liver of maternal California sea lions were not reflected in the livers of newborn pups (Martin et al. 1976). In marine mammals, selenium concentrations are positively correlated with increasing age (Teigen et al. 1993; Mackey et al. 1996) and with increasing mercury residues in piscivorous mammals (Reijnders 1980; Wren 1984; Leonzio et al. 1992; Teigen et al. 1993). The mercury:selenium ratio was close to 1.0 in tissues of marine mammals at mercury concentrations >15 mg Hg/kg FW (Skaare et al. 1994). Increasing mercury concentrations in tissues of marine teleosts are also positively correlated with selenium (Ganther et al. 1972; Leonzio et al. 1982), although the evidence is conflicting (Tamura et al. 1975; Speyer 1980; Maher 1983). Selenium varies seasonally in crustaceans (Zafiropoulos and Grimanis 1977). In general, concentrations of selenium in various tissues are usually higher in older than in younger organisms. Among marine vertebrates, selenium increases were especially pronounced among the older specimens of predatory, long-lived species (Eisler 1984).

Selenium concentrations in avian tissues are modified by the age, condition, and diet of the organism, the presence of other metals, and other variables. Fish-eating birds had the highest selenium concentrations in livers, and herbivorous species the lowest; omnivores were intermediate (Mora and Anderson 1995). Selenium concentrations were elevated in livers of molting birds compared to nonmolting conspecifics (Jenny et al. 1990), elevated in feathers of older terns and egrets when compared to younger stages (Burger et al. 1994), and elevated in tissues of marine birds that consume invertebrate prey animals with elevated selenium burdens (Goede et al. 1993).

Selenium concentrations in tissues of shorebirds were positively correlated with concentrations of copper, zinc (Wenzel and Gabrielsen 1995), and iron (Goede and Wolterbeek 1994). Feathers have been proposed as indicators of selenium exposure. However, variability in selenium concentrations in whole feathers is considerable (Goede 1991) (Table 31.2). In shorebirds, for example, the highest selenium concentrations are found in wing feathers, specifically in the outer primaries, notably primary 8. Moreover, within the vane of a single feather, the highest selenium concentrations are in the tip and the lowest at the basis. All of these differences need to be considered before feathers are routinely used as indicators of selenium exposure (Goede 1991).

Subsurface agricultural drainage waters from the western San Joaquin Valley, California, had elevated selenium concentrations, as selenate. In 1978, these drainage waters were diverted to Kesterson Reservoir, a pond system within the Kesterson National Wildlife Refuge (KNWR), with diversion complete by 1982. In 1983, aquatic birds at KNWR had unusual rates of death and developmental abnormalities attributed to selenium (Presser and Ohlendorf 1987). In 1984/85, selenium-induced recruitment failure was observed at KNWR in American avocets (*Recurvirostra americana*) and black-necked stilts (*Himantopus mexicanus*); unlike a nearby reference area, chicks at KNWR of either species did not survive to fledging (Williams et al. 1989). Selenium concentrations in livers of diving ducks from San Francisco Bay in 1982 were similar to those of dabbling ducks in the nearby San Joaquin Valley where reproduction was severely impaired (Ohlendorf et al. 1986b). Mean concentrations of selenium in kidneys of seven species of coastal birds collected from the highly industrialized Corpus Christi, Texas, area usually varied between 1.7 and 5.6 mg Se/kg FW, but were 10.2 mg/kg FW in one bird. According to White et al. (1980), selenium concentrations of this magnitude may be sufficient to impair reproduction in shorebirds. Barn swallows (*Hirundo rustica*) nesting at a selenium-contaminated lake in Texas had elevated concentrations of selenium in eggs and tissues when compared to conspecifics at a reference site. However, nest success of barn swallows was significantly higher at the contaminated site; development was normal at both sites (King et al. 1994).

Table 31.2 Selenium Concentrations in Field Populations of Selected Species of Flora and Fauna (Values shown are in mg total Se/kg [ppm] fresh weight [FW], dry weight [DW], or ash weight [AW]. Hyphenated numbers show range, and single numbers the mean; where both appear, the range is in parentheses.)

Ecosystem, Taxonomic Group, Organism, Tissue, Location, and Other Variables	Concentration (ppm)	Reference[a]
MARINE		
Algae and macrophytes		
Whole	0.04–0.24 DW	1–3
Edible seaweeds, whole	0.16–0.39 DW; 0.047 FW	4
Molluscs		
American oyster, *Crassostrea virginica*		
Redwood Creek, San Francisco, CA		
Mantle	4.8 DW	5
Digestive gland	8.8 DW	5
Kidney	4.7 DW	5
Tomales Bay, CA		
Mantle	2.2 DW	5
Digestive gland	6.4 DW	5
Kidney	2.5 DW	5
Transferred from Redwood Creek to Tomales Bay for 56 days		
Mantle	3.5 DW	5
Digestive gland	6.5 DW	5
Kidney	3.8 DW	5
Bivalve molluscs		
Shell	0.03–0.06 DW	6
Soft parts	0.1–0.9 FW; 1.3–9.9 DW	2, 6–10

Table 31.2 (continued) Selenium Concentrations in Field Populations of Selected Species of Flora and Fauna (Values shown are in mg total Se/kg [ppm] fresh weight [FW], dry weight [DW], or ash weight [AW]. Hyphenated numbers show range, and single numbers the mean; where both appear, the range is in parentheses.)

Ecosystem, Taxonomic Group, Organism, Tissue, Location, and Other Variables	Concentration (ppm)	Reference[a]
Muscle	1.1–2.3 DW	11
Viscera	1.6–2.5 DW	11
Edible flesh, 3 species		
Total Se	0.22 (0.16–0.31) FW	12
As selenate	0.05 FW	12
As selenite and selenide	0.17 FW	12
Common mussel, *Mytilus edulis*		
Gills	2.0–16.0 DW	13
Viscera	Up to 5.0 DW	13
Mussel, *Mytilus galloprovincialis*		
Gills	7.0 DW	14
Soft parts	6.0 DW	14
Mantle	5.2 DW	14
Viscera	3.2 DW	14
Muscle	1.9 DW	14
Shell	<0.05 DW	14
Echinoderms		
Whole, 7 species	0.8–4.4 DW	15
Crustaceans		
Digestive system, 3 species	3.0–3.5 DW	11
Edible tissues sold for human consumption, 17 species	0.2–2.0 FW	4, 8, 10
Edible tissues, 2 species		
Total Se	0.21 FW	12
As selenate	0.05 FW	12
As selenite and selenide	0.16 FW	12
Muscle, 5 species	2.4–4.4 DW	11
Soft tissues, 2 species	2.0–2.8 DW	11
Copepods, whole	1.8–3.4 DW	3, 16
Euphausid, *Meganyctiphanes norvegica*		
Viscera	11.7 DW	17
Eyes	7.8 DW	17
Muscle	1.8 DW	17
Exoskeleton	0.8 DW	17
Shrimp, *Lysmata seticaudata*		
Viscera	7.0 DW	14
Eyes	4.8 DW	14
Whole	2.6 DW	14
Muscle	1.9 DW	14
Exoskeleton	1.5 DW	14
Molts	0.3 DW	14
Sharks		
Muscle	0.2–0.8 FW	18
Fishes		
Digestive system, 4 species	1.0–2.4 DW	11
Liver		
2 species	2.6–6.6 DW	19
74 species	0.6–5.0 FW	8, 10, 20
13 species	5.0–30.0 FW	8
Meals, 3 species	1.0–4.0 DW	21
Muscle		
4 species	0.5–1.5 DW	11
182 species	0.1–2.0 FW	4, 8, 10, 19, 22, 23

Table 31.2 (continued) Selenium Concentrations in Field Populations of Selected Species of Flora and Fauna (Values shown are in mg total Se/kg [ppm] fresh weight [FW], dry weight [DW], or ash weight [AW]. Hyphenated numbers show range, and single numbers the mean; where both appear, the range is in parentheses.)

Ecosystem, Taxonomic Group, Organism, Tissue, Location, and Other Variables	Concentration (ppm)	Reference[a]
5 species, Total Se	0.4 (0.2–0.6) FW	12
As selenate	0.1 FW	12
As selenite and selenide	0.3 FW	12
Whole, 21 species	0.3–2.0 FW	8, 23, 24
Japanese tunas, 4 species		
Liver	10.0–15.0 FW	25
White muscle	0.5–1.3 FW	25
Red muscle	3.5–9.1 FW	25
Snapper, *Chrysophrys auratus*; Australia; 1976; muscle	Max. 0.85 FW	73
Black marlin, *Makaira indica*		
Muscle	0.4–4.3 FW	26
Liver	1.4–13.5 FW	26
Blue marlin, *Makaira nigricans*		
Kidney	23.0 FW	27
Blood	1.0 FW	27
Gill	1.0 FW	27
Muscle		
Total Se	3.3 (2.5–4.1) FW	12
As selenate	0.2 (0.09–0.3) FW	12
As selenite and selenide	3.1 (2.4–3.8) FW	12
Striped bass, *Morone saxatilis*		
Muscle	0.3 FW	28
Liver	0.6 FW	28
Tuna, canned	1.9–2.9 FW	29
Swordfish, *Xiphias gladius*		
Muscle	0.3–1.3 FW	30
Birds		
Kidney, 12 species	1.2–5.6 FW	31, 32
Liver; 5 species; Baja California; 1986	0.7–5.1 (0.2–7.3) FW	101
Western grebe, *Aechmophorus occidentalis*; Puget Sound, Washington; 1986; liver	7.6–9.3 (2.2–24.0) DW	109
Great Skua, *Catharacta skua*		
Kidney	32.8 (13.3–89.1) DW	37
Liver	19.7 (6.7–34.6) DW	37
Oystercatcher, *Haematopus ostralegus*		
Kidney	12.7 (2.3–17.5) DW	37
Liver	12.8 (5.0–20.5) DW	37
Herring gull, *Larus argentatus*		
Kidney	14.1 (8.6–19.4) DW	37
Liver	7.9 (6.9–9.3) DW	37
Eggs; Long Island, New York; 1989–94	1.0–2.1 DW	102
Franklin's gull, *Larus pipixcan*; Minnesota; 1994		
Feathers, males vs. females	0.5 DW vs. 0.6 DW	103
Eggs	3.0 DW	103
Diet (earthworms)	4.9 DW	103
Brown pelican, *Pelecanus occidentalis*		
Egg	0.19–0.38 FW	36
Liver	1.0–4.2 FW	36
White-faced ibis, *Plegadis chihi*		
Egg	0.3–1.1 FW	33
Wedge-tailed shearwater, *Puffinus pacificus*		
Egg	1.1–1.3 FW	35
Black skimmer, *Rynchops niger*; breast feathers; New York	1.2–1.3 DW	104

Table 31.2 (continued) Selenium Concentrations in Field Populations of Selected Species of Flora and Fauna (Values shown are in mg total Se/kg [ppm] fresh weight [FW], dry weight [DW], or ash weight [AW]. Hyphenated numbers show range, and single numbers the mean; where both appear, the range is in parentheses.)

Ecosystem, Taxonomic Group, Organism, Tissue, Location, and Other Variables	Concentration (ppm)	Reference[a]
Seabirds		
Liver, 11 species	Means 17–107 DW	128
Most tissues, 3 species	Max. 2–34 DW	128
Seabirds; northern Norway; summer, 1992–93		
Kittiwake, *Rissa tridactyla*; adults vs. fledglings		
Feather	1.8 FW vs. 1.3 FW	96
Liver	16.9 DW vs. 8.9 DW	96
Common guillemot, *Uria aalge*		
Feather	2.6 FW	96
Gonad	21.9 DW	96
Kidney	43.7 DW	96
Liver	17.6 DW	96
Brunnich's guillemot, *Uria lomvia*		
Feather	2.7 FW	96
Gonad	12.2 DW	96
Kidney	15.6 DW	96
Liver	17.6 DW	96
Shorebirds		
New Jersey, Cape May; 3 species; feathers; 1991–92	1.3–6.2 DW	106
Pacific coast; U.S.; 1984–85; livers		
Dunlin, *Calidris alpina*	12.9 (7.0–20.2) DW	105
Long-billed dowitcher, *Limnodromus scolopaceus*	11.1 (5.5–15.3) DW	105
Black-bellied plover, *Pluvialis squatorola*	9.4 (5.8–29.9) DW	105
Texas; 1984; mercury-contaminated bay vs. reference site; eggs less shell		
Forster's tern, *Sterna forsteri*	0.71 FW vs. 0.68 FW	108
Black skimmer, *Rynchops niger*	0.8 FW vs. 0.3 FW	108
Sooty tern, *Sterna fuscata*		
Egg	1.1–1.4 FW	35
Common tern, *Sterna hirundo*; feathers; Massachusetts; May–June		
Adults		
Age 2–3 years	1.3 DW	107
Age 9–10 years	2.1 DW	107
Age 16–21 years	2.4 DW	107
Fledglings, age 20–23 days	0.8 DW	107
Royal tern, *Sterna maxima*		
Egg	0.4–2.1 FW	34
Red-footed booby, *Sula sula*		
Egg	0.8–0.9 FW	35
Mammals		
Alaska; liver		
Bowhead whale, *Balaena mysticetus*	0.5–1.2 FW	85
Beluga whale, *Delphinapterus leucas*	4–75 FW; usually <20.0 FW	85
Bearded seal, *Erignathus barbatus*	0.5–5.3 FW	85
Ringed seal, *Phoca hispida*	1.2–5.7 FW	85
Minke whale, *Balaenoptera rostrata*; Antarctic Ocean; 1990–91; urine	1.5 FW	84
Pilot whale, *Globicephala macrorhynchus*		
Blubber	0.8–1.4 FW	38
Liver	22.8–61.6 FW	38
Kidney	3.0–10.0 FW	38

Table 31.2 (continued) Selenium Concentrations in Field Populations of Selected Species of Flora and Fauna (Values shown are in mg total Se/kg [ppm] fresh weight [FW], dry weight [DW], or ash weight [AW]. Hyphenated numbers show range, and single numbers the mean; where both appear, the range is in parentheses.)

Ecosystem, Taxonomic Group, Organism, Tissue, Location, and Other Variables	Concentration (ppm)	Reference[a]
Italy; 1987–89; found stranded		
Striped dolphin, *Stenella coeruleoalba*		
Brain	9 (5–36) DW	88
Kidney	25 (50–101) DW	88
Liver	106 (2–960) DW	88
Muscle	(10–55) DW	88
Bottle-nosed dolphin, *Tursiops truncatus*		
Brain	4.9 (3.3–5.2) DW	88
Kidney	53 (21–186) DW	88
Liver	139 (2–2400) DW	88
Muscle	Max. 48 DW	88
Saimaa ringed seal, *Phoca hispida saimensis*		
Muscle	0.2–2.8 FW	23
Liver	29.0–170.0 FW	23
Kidney	0.3–3.0 FW	23
Blubber	0.06–0.11 FW	23
Harbor seal, *Phoca vitulina*		
Juveniles		
Kidney	0.6 (0.0–1.3) FW	39
Liver	2.8 (2.6–6.5) FW	39
Brain	1.1 (0.0–7.4) FW	39
Adults		
Kidney	3.5 (1.9–7.3) FW	39
Liver	109.0 (Max. 409.0) FW	39
Brain	3.7 (1.5–8.2) FW	39
Norway, 1989–90		
Brain, 4 species	<0.01 FW	86
Grey seal, *Halichoerus grypus*		
Kidney	Max. 4.1 FW	86
Liver	Max. 21.8 FW	86
Ringed seal, *Phoca hispida*		
Kidney	Max. 5.7 FW	86
Liver	Max. 3.7 FW	86
Harp seal, *Phoca groenlandica*		
Kidney	Max. 7.2 FW	86
Liver	Max. 3.4 FW	86
Harbor seal, *Phoca vitulina*		
Kidney	Max. 7.7 FW	86
Liver	Max. 7.8 FW	86
Harbor porpoise, *Phocaena phocoena*; Norway; 1989–90; ages 1–5 years		
Kidney	0.6–8.6 FW	87
Liver	0.7–14.2 FW	87
Seals		
Liver, 4 species	6.1–170.0 FW	23, 40, 41
California sea lion, *Zalophus californianus*		
Mothers with normal pups		
Liver	260.0 DW	42
Kidney	22.0 DW	42
Normal pups		
Liver	4.1 DW	42
Kidney	6.1 DW	42

Table 31.2 (continued) Selenium Concentrations in Field Populations of Selected Species of Flora and Fauna (Values shown are in mg total Se/kg [ppm] fresh weight [FW], dry weight [DW], or ash weight [AW]. Hyphenated numbers show range, and single numbers the mean; where both appear, the range is in parentheses.)

Ecosystem, Taxonomic Group, Organism, Tissue, Location, and Other Variables	Concentration (ppm)	Reference[a]
Mothers with premature pups		
Liver	79.0 DW	42
Kidney	12.0 DW	42
Pups born prematurely		
Liver	2.9 DW	42
Kidney	3.7 DW	42
FRESHWATER		
Algae and higher plants		
Algae, whole	<2.0 DW	43
Higher plants	0.1 DW	44
Aquatic mosses	0.8 DW	44
Filamentous algae		
Se-contaminated area	35.2 (12–68) DW	57
Control area	<0.5 DW	57
Rooted plants		
Se-contaminated area	52.1 (18–79) DW	57
Control area	0.4 DW	57
Molluscs		
Mussels, 3 species, NY state		
Soft parts	2.0–4.0 DW	71
Asiatic clam, *Corbicula fluminea*		
Whole, Florida	0.7 FW	45
Arthropoda		
Zooplankton	0.8–3.9 DW	46
Plankton		
Se-contaminated area	85 (58–124) DW	57
Control site	2 (1.4–2.9) DW	57
Insects		
Se-contaminated area	20–218 DW	57
Control site	1.1–3.0 DW	57
Mayfly, *Hexagenia* sp.	0.3–0.5 FW	45
Fishes		
California		
San Joaquin River; whole		
1984–85; 5 species	Max. 23.0 DW	75
1986; 5 species	Max. 11.0 DW	75
1986–87; 7 species; whole		
Sacramento River	Max. 2.1 DW	76
San Francisco Bay	Max. 3.3 DW	76
San Joaquin River	Max. 11.1 DW	76
Reservoirs receiving ash pond effluent		
Bluegill, *Lepomis macrochirus*; ovary	Max. 12.0 FW	89
Largemouth bass, *Micropterus salmoides*; ovary	Max. 7.0 FW	89
Common carp, *Cyprinus carpio*		
Whole	1.0 (0.7–1.4) DW	47
Liver	3.6 (2.2–5.2) DW	47
Bluegill, *Lepomis macrochirus*		
From water containing <5 µg Se/L		
White muscle	0.04 FW	48
Liver	0.7 FW	48
Spleen	1.6 FW	48
Erythrocytes	0.04 FW	48
Heart	1.0 FW	48

Table 31.2 (continued) Selenium Concentrations in Field Populations of Selected Species of Flora and Fauna (Values shown are in mg total Se/kg [ppm] fresh weight [FW], dry weight [DW], or ash weight [AW]. Hyphenated numbers show range, and single numbers the mean; where both appear, the range is in parentheses.)

Ecosystem, Taxonomic Group, Organism, Tissue, Location, and Other Variables	Concentration (ppm)	Reference[a]
From water containing 22.6 µg Se/L		
White muscle	3.1 FW	48
Liver	11.2 FW	48
Spleen	17.7 FW	48
Erythrocytes	7.2 FW	48
Heart	12.8 FW	48
Upper Mississippi River		
Whole	1.2 (0.7–1.4) DW	47
California, San Joaquin Valley, 1988		
Carcass		
Males	2.0 (1.5–3.1) DW	80
Females	1.9 (1.2–3.0) DW	80
Gonads		
Males	3.8 (3.2–4.1) DW	80
Females	3.2 (2.3–4.3) DW	80
Largemouth bass, *Micropterus salmoides*		
From water containing <5 µg Se/L		
White muscle	0.05 FW	48
Liver	0.8 FW	48
Spleen	1.8 FW	48
Erythrocytes	0.07 FW	48
Heart	1.2 FW	48
From water containing 22.6 µg Se/L		
White muscle	1.7 FW	48
Liver	10.2 FW	48
Spleen	16.6 FW	48
Erythrocytes	8.0 FW	48
Heart	12.0 FW	48
Colorado River Valley; Cibola Lake, 1989–90		
Liver	8.4–18.0 DW	81
Muscle	4.5–5.9 DW	81
Ovary	5.4–7.8 DW	81
Apalachicola River, Florida		
Whole		
Females	0.3–0.4 FW	45
Males	0.3–0.5 FW	45
Juveniles	0.3–0.5 FW	45
Eggs	0.7–1.0 FW	45
Striped bass, *Morone saxatilis*; California; whole; juveniles		
San Joaquin River		
1984	6.5 DW	74
1985	4.1 DW	74
1986	3.5 DW	74
San Joaquin Valley vs. San Francisco estuary; 1986	Max. 7.9 DW vs. Max. 3.3 DW	74
Channel catfish, *Ictalurus punctatus*		
Apalachicola River, Florida		
Whole		
Females	1.1–0.3 FW	45
Males	0.2–0.3 FW	45
Juveniles	0.4–0.6 FW	45
Eggs	0.8–2.1 FW	45
Threadfin shad, *Dorosoma petenense*		
Whole	0.3–0.5 FW	45

Table 31.2 (continued) Selenium Concentrations in Field Populations of Selected Species of Flora and Fauna (Values shown are in mg total Se/kg [ppm] fresh weight [FW], dry weight [DW], or ash weight [AW]. Hyphenated numbers show range, and single numbers the mean; where both appear, the range is in parentheses.)

Ecosystem, Taxonomic Group, Organism, Tissue, Location, and Other Variables	Concentration (ppm)	Reference[a]
Green sunfish, *Lepomis cyanellus*		
From water containing 13 µg Se/L		
Liver	7.0–21.4 FW	49
Muscle	2.3–12.9 FW	49
From control site		
Liver	1.3 FW	49
Muscle	1.3 FW	49
Trout, 2 species, Wyoming		
From water containing 12.3–13.3 µg Se/L		
Liver	50.0–70.0 FW	50
Muscle	<2.0 FW	50
Skin	Max. 4.8 FW	50
Mosquitofish, *Gambusia affinis*, whole		
From Se-contaminated irrigation drainwater pond	170 (65–360) DW	57, 77
Control site	1.3 (1.2–1.4) DW	57
Coho salmon, *Oncorhynchus kisutch*		
Muscle	0.7–1.0 DW	51
Liver	3.8 DW	51
Adults, liver		
Ocean caught	8.8 DW; 1.7 FW	79
Estuary caught	8.1 DW; 1.6 FW	79
Hatchery return	7.6 DW; 1.6 FW	79
Farmed fish	4.8 DW; 1.0 FW	79
Smolts, liver		
Hatchery reared	2.0 DW; 0.4 FW	79
Naturally reared	3.6 DW; 0.7 FW	79
Chinook salmon, *Oncorhynchus tshawytscha*		
Muscle	1.6 DW	51
Razorback sucker, *Xyrauchen texanus*; Green River, Utah		
Eggs		
1988	4.9 DW	82
1991	28.0 DW	82
1992	3.7–10.2 DW	82
Milt, 1992	<1.1–6.7 DW	82
Fry, 1992	2.2 DW; 0.54 FW	82
Muscle, 1992		
Females	4.4–32.0 DW	82
Males	3.6–26.0 DW	82
Maximum concentrations	11.5–54.1 DW	83
Fish		
Muscle		
25 species	0.0–0.5 FW	52–56
18 species	0.5–1.0 FW	53–55
3 species	1.0–2.0 FW	55, 58
3 species		
Total Se	0.25 (0.15–0.34) FW	12
As selenate	0.09 FW	12
As selenate and selenide	0.16 FW	12
10 species, Western Lake Erie	0.4–1.5 FW; 1.8–8.1 DW	46
Liver		
17 species	0.0–0.5 FW	59
7 species	0.5–1.0 FW	59
4 species	0.6–5.0 FW	10

Table 31.2 (continued) Selenium Concentrations in Field Populations of Selected Species of Flora and Fauna (Values shown are in mg total Se/kg [ppm] fresh weight [FW], dry weight [DW], or ash weight [AW]. Hyphenated numbers show range, and single numbers the mean; where both appear, the range is in parentheses.)

Ecosystem, Taxonomic Group, Organism, Tissue, Location, and Other Variables	Concentration (ppm)	Reference[a]
Whole		
Nationwide, U.S.		
1972	0.60 (0.57–0.64) FW	60
1973	0.46 (0.42–0.49) FW	60
1976–77	0.58 (0.53–0.62) FW	60
1978–79	0.48 FW	78
1980–81	0.46 FW	78
1984–85	0.42 FW	78
1976–84	Max. 2.3–3.6 FW	78
5 species	0.5–1.9 FW	61
4 species	0.2–0.3 DW	56
6 species	0.0–0.5 FW	62
12 species	0.5–1.0 FW	62
6 species	1.0–2.0 FW	62
6 species	2.1–6.0 FW	62
Amphibians		
Southern toad, *Bufo terrestris*; adults; whole		
From coal ash settling basins vs. reference site (sediments 4.4 mg Se/kg DW vs. 0.1 mg/kg)	17.4 DW vs. 2.1 DW	130
Transferred from reference site to settling basin for 7–12 weeks	2.1 DW vs. 3.5–5.5 DW	130
Bullfrog, *Rana catesbeiana*; tadpoles; South Carolina; 1997		
With digestive tract		
Body	9.7 DW; 1.9 FW	131
Tail	11.5 DW; 1.8 FW	131
Whole	9.3 DW; 1.9 FW	131
Without digestive tract		
Body without gut	10.0 DW	131
Tail	12.6 DW	131
Whole	6.7 DW	131
Digestive tract	18.2 DW	131
Various species, liver	0.7–4.7 FW	43
Reptiles		
Water snake, *Natrix* sp.		
Whole, Florida	0.3–0.5 FW	45
Birds		
Little green heron, *Butorides virescens*; Apalachicola River, Florida; whole	0.1–0.5 FW	45
California		
Aquatic birds; 4 species; 1985–88; Grasslands area; livers; North Grasslands vs. South Grasslands (more Se-contaminated area)		
1985	Max. 14 DW vs. Max. 25 DW	97
1987	Max. 10 DW vs. Max. 24 DW	97
1988	Max. 12 DW vs. Max. 14 DW	97
Grasslands drainage area (Se-contaminated prior to 1985); 1986 vs. 1987; eggs		
Mallard, *Anas platyrhynchos*	6 DW vs. 20 DW	98
Cinnamon teal, *Anas cyanoptera*	0.2 DW vs. 12 DW	98
Gadwall, *Anas strepera*	5.3 DW vs. 8 DW	98
Near Salton Sea; 1985; eggs		
Black-crowned night heron, *Nycticorax nycticorax*	1.1 (0.9–1.4) FW	99
Great egret, *Casmerodius albus*	0.6 (0.5–0.8) FW	99

Table 31.2 (continued) Selenium Concentrations in Field Populations of Selected Species of Flora and Fauna (Values shown are in mg total Se/kg [ppm] fresh weight [FW], dry weight [DW], or ash weight [AW]. Hyphenated numbers show range, and single numbers the mean; where both appear, the range is in parentheses.)

Ecosystem, Taxonomic Group, Organism, Tissue, Location, and Other Variables	Concentration (ppm)	Reference[a]
San Francisco Bay, 1982		
Greater scaup, *Aythya marila*; liver	19 (7–31) DW	100
Surf scoter, *Melanitta perspicillata*; liver	34 (16–59) DW	100
From Kesterson National Wildlife Refuge (KNWR), California, nesting on Se-contaminated irrigation drainwater ponds — 1983		
American coot, *Fulica americana*		
Liver	37 (21–63) DW	57
Egg	54 (34–110) DW	57
Ducks, *Anas* spp.		
Liver	28.6 (19–43) DW	57
Egg	9.9 (2.2–46) DW	57
Black-necked stilt, *Himantopus mexicanus*		
Egg	32.7 (12–74) DW	57
American avocet, *Recurvirostra americana*		
Egg	9.1 DW	57
Eared grebe, *Podiceps nigricollis*		
Liver	130.0 DW	57
Egg	81.4 (72–110) DW	57
From Volta Wildlife Area, California, control site — 1983		
American coot		
Liver	5.0 (4.4–5.6) DW	57
Ducks, 2 species		
Liver	4.1 (3.9–4.4) DW	57
Black-necked stilt		
Liver	6.1 DW	57
KNWR vs. reference site; 1983–85		
Diets	>50 DW vs. <2 DW	125
Livers		
Adults, 8 species	26–101 DW vs. 3–11 DW	125
Juveniles, 5 species	21–95 DW vs. 2–4 DW	125
Canada; eastern section; from freezer archives; total Se		
Common loon, *Gavia immer*		
Kidney	15 DW	129
Liver	15 DW	129
Muscle	2.8 DW	129
Common merganser, *Mergus merganser*		
Kidney	8.5 DW	129
Liver	9.7 DW	129
Muscle	1.8 DW	129
Willet, *Catoptrophorus semipalmatus*; 1986; south Texas; liver	2.8–8.3 DW	110
China; feathers; 1992		
Herons; chicks; 2 species	1.0–2.0 DW	111
Egrets; chicks; 3 species	1.2–2.8 DW	111
Common loon, *Gavia immer*		
Egg	0.4 (0.3–0.7) FW	63
Red-breasted merganser, *Mergus serrator*		
Egg	0.47–1.0 FW	72
Mallard, *Anas platyrhynchos*		
Egg	0.28–0.81 FW	72
Flamingo, *Phoenicopterus ruber*; feather; France; 1988		
Adults	7.1 (1.6–33.0) FW	112
Juveniles	0.8 FW; Max. 2.3 FW	112

Table 31.2 (continued) Selenium Concentrations in Field Populations of Selected Species of Flora and Fauna (Values shown are in mg total Se/kg [ppm] fresh weight [FW], dry weight [DW], or ash weight [AW]. Hyphenated numbers show range, and single numbers the mean; where both appear, the range is in parentheses.)

Ecosystem, Taxonomic Group, Organism, Tissue, Location, and Other Variables	Concentration (ppm)	Reference[a]
White-faced ibis, *Plegadis chihi*; Carson Lake, Nevada; 1985–86		
Eggs	1.9–5.4 DW	113
Livers	9.6 (5–27) DW	113
Eared grebe, *Podiceps nigricollis*; Stewart Lake, North Dakota; 1991; eggs	4.5 (2.9–6.5) DW; no deformities	114
TERRESTRIAL		
Fungi	<2.0 DW	43
Macrophytes		
Western wheat grass, *Agropyron smithii*, South Dakota, plant top	0.0–8.4 DW	43
Little bluestem, *Andropogon scoparius*, plant top	0.0–6.0 DW	43
Asparagus, *Asparagus officinale,* western U.S.	2.7–11.0 DW	43
Aster, whole		
Aster caerulescens	560.0 DW	43
A. commutatus	Max. 590.0 DW	43
A. multiflora	Max. 320.0 DW	43
A. occidentalis	284.0 DW	43
Milk vetch, *Astragalus argillosus*		
Top	385.0 DW	43
Root	27.0 DW	43
A. beathii		
Top	1963.0 DW	43
Root	6.0 DW	43
A. bisulcatus		
Top	Max. 10,239.0 DW	43
Seed	305.1 DW	43
A. confertiflorus		
Top	1372.0 DW	43
A. crotulariae		
Top	2000.0 DW	43
Root	45.0 DW	43
Loco weed, *Astragalus* spp.	Max. 46,000.0 AW; Max. 6000.0 DW	43
Saltbush, *Atriplex* spp.	300.0–1734.0 DW	43
Oats, *Avena sativa*	2.0–15.0 DW	43
Buffalo grass, *Bouteloua dactyloides*	2.7 (0.0–12.0) DW	43
Indian paint brush, *Castillaja* spp.	0.0–1812.0 DW	43
Gumweed, *Grindelia squarrosa*	38 (0.0–2160) DW	43
Broomweeds		
Gutierrezia spp.	Max. 723.0 DW	43
Haplopappus spp.	Max. 4800.0 DW	43
Tobacco, *Nicotiana tabacum*		
Leaf	5.8 DW	43
Stem	44.2 DW	43
Rice, *Oryza sativa*		
Grain	0.09–0.11 FW	43
Pear, *Pyrus communis*		
Fruit	0.02 FW	43
Rye, *Secale cereale*	0.9–25.0 DW	43
Potato, *Solanum tuberosum*		
Tuber	0.2–0.9 DW	43

Table 31.2 (continued) Selenium Concentrations in Field Populations of Selected Species of Flora and Fauna (Values shown are in mg total Se/kg [ppm] fresh weight [FW], dry weight [DW], or ash weight [AW]. Hyphenated numbers show range, and single numbers the mean; where both appear, the range is in parentheses.)

Ecosystem, Taxonomic Group, Organism, Tissue, Location, and Other Variables	Concentration (ppm)	Reference[a]
Wheat, *Triticum aestivum*		
Grain	1.1–35.0 DW	43
Stem and leaf	17.0 DW	43
Root	36.0 DW	43
Corn, *Zea mays*		
Grain	1.0–20.0 DW	43
Grape, *Vitis* sp.		
Raisin	<0.001 FW	43
Annelids		
Earthworms, whole		
From normal soil	2.2 FW	64
From selenite-enriched soil	7.5 FW	64
From soil amended with sewage sludge		
Whole	15.0–22.4 DW	65
Casts	0.6–0.7 DW	65
From control field		
Whole	22.1 DW	65
Casts	0.6 DW	65
Arthropods		
Sow bug, *Porcellio* sp.	0.9 FW	43
Crane fly, larva, *Tipula* sp.	0.9 FW	43
"Fly larvae," whole, from *Astragalus* plant with 1800 mg Se/kg	20.0 FW	43
Birds		
Cattle egret, *Bubulcus ibis*; juveniles; 1989–91; feathers		
New York	1.3 DW	115
Delaware	1.6 DW	115
Puerto Rico	1.3 DW	115
Cairo, Egypt	0.3 DW	115
Aswan, Egypt	1.0 DW	115
Barn swallow, *Hirundo rustica*; 1986–87; Martin Lake, Texas (selenium-contaminated) vs. reference site		
Eggs	Max. 12 DW vs. Max. 4.5 DW	116
Kidneys	Max. 14 DW vs. Max. 5.8 DW	116
Wood stork, *Mycteria americana*; feathers		
Florida, juveniles, 1991	1.8 DW	117
Costa Rica		
Adults, 1992	3.4 DW	117
Juveniles		
1990	2.2 DW	117
1992	1.5 DW	117
House sparrow, *Passer domesticus*		
Whole	0.6 DW	66
Ring-necked pheasant, *Phasianus colchicus*		
Whole	0.6 DW	66
Common blackbird, *Turdus merula*		
Whole	2.1 DW	66
Mammals		
California; San Francisco Bay; 1989; livers		
California vole, *Microtus californicus*	0.5–1.6 DW	90
House mouse, *Mus musculus*	1.5–4.8 DW	90
Deer mouse, *Peromyscus maniculatus*	2.3–3.5 DW	90

Table 31.2 (continued) Selenium Concentrations in Field Populations of Selected Species of Flora and Fauna (Values shown are in mg total Se/kg [ppm] fresh weight [FW], dry weight [DW], or ash weight [AW]. Hyphenated numbers show range, and single numbers the mean; where both appear, the range is in parentheses.)

Ecosystem, Taxonomic Group, Organism, Tissue, Location, and Other Variables	Concentration (ppm)	Reference[a]
California; Kesterson National Wildlife Refuge vs. control site; May 1984; liver		
California vole, *Microtus californicus*	Max. 250.0 DW vs. Max. 1.4 DW	120
House mouse, *Mus musculus*	Max. 41.0 DW vs. Max. 3.7 DW	120
Desert cottontail, *Sylvilagus audubonii*	Max. 3.3 DW vs. Max. 0.1 DW	120
South Dakota; various species; muscle	0.2–1.1 FW	95
Washington State; 1992–93; blood		
Moose, *Alces alces*	0.015 (0.01–0.02) FW	94
Elk, *Cervus elaphus*	0.04–0.16 FW; Max. 0.49 FW	94
Mule deer, *Odocoileus hemionus*	0.08 (0.06–0.15) FW	94
California bighorn sheep, *Ovis canadensis californiana*	0.09 (0.04–0.13) FW	94
Livestock	>0.1 FW (adequate selenium)	94
Wyoming; muscle; animals had access to forage grown on medium up to 5.0 mg/L water-soluble selenium) to high (>10 mg Se/L)		
American bison, *Bison bison*	0.5 FW; 1.7 DW	95
Cattle, *Bos* sp.	0.1 FW; 0.5 DW	95
Elk, *Cervus elaphus*	0.4 FW; 1.6 DW	95
Mule deer	0.6 FW; 2.5 DW	95
Common beaver, *Castor canadensis*		
Liver	0.2 FW	67
Kidney	0.9 FW	67
Intestine	0.04 FW	67
Muscle	0.09 FW	67
Human, *Homo sapiens*		
Fetus, United States		
Most tissues	<0.4 FW	118
Thyroid, blood	<1.0 FW	118
Liver	2.8 FW	118
Adults, United States		
Urine	0.002–0.113 FW	118
Milk	0.007–0.053 FW	118
Semen	0.016–0.131 FW	118
Erythrocytes	0.02–0.52 FW	118
Whole blood	0.08–0.3 FW	118
Nails	0.08–3.8 FW	118
Hair	0.6 FW	118
Pancreas, kidney	0.6–0.9 FW	118
Liver	0.6–1.7 FW	118
Diet		
Fruits and vegetables	Usually <0.03 FW	118
Grains and cereals	0.03–1.4 FW	118
Brazil nuts	14.7 (0.2–253.0) FW	118
Dairy products	0.01–0.1 FW	118
Meat and poultry		
Muscle	0.1–0.5 FW	118
Organ meats	0.4–2.3 FW	118
Seafood	0.2–3.4 FW	118

Table 31.2 (continued) Selenium Concentrations in Field Populations of Selected Species of Flora and Fauna (Values shown are in mg total Se/kg [ppm] fresh weight [FW], dry weight [DW], or ash weight [AW]. Hyphenated numbers show range, and single numbers the mean; where both appear, the range is in parentheses.)

Ecosystem, Taxonomic Group, Organism, Tissue, Location, and Other Variables	Concentration (ppm)	Reference[a]
River otter, *Lutra canadensis*		
Liver	2.1 FW	67
Kidney	1.9 FW	67
Intestine	1.1 FW	67
Muscle	0.2 FW	67
Woodchuck, *Marmota monax*		
From flyash landfill vicinity		
Adults		
Liver	2.2–10.7 DW	68
Lung	1.4–4.4 DW	68
Juveniles		
Liver	3.9–6.4 DW	68
Lung	2.1–2.8 DW	68
From control area		
Adults		
Liver	0.4 DW	68
Lung	0.4 DW	68
Juveniles		
Liver	0.2–0.4 DW	68
Lung	0.2 DW	68
Field vole, *Microtus agrestis*		
Whole	0.5 DW	66
Mule deer, *Odocoileus hemionus*; California; whole blood; 1981–88		
Winter	0.07 FW	91
Spring	0.05 FW	91
Summer	0.09 FW	91
Fall	0.02 FW	91
Range	0.02–0.17 FW	91
Most deer	<0.1 FW (Se-deficient)	92
White-tailed deer, *Odocoileus virginianus*		
Muscle	0.16 (0.05–0.49) DW	69
Southern Florida; 1984–88		
Heart	0.37 (0.01–1.0) DW	93
Kidney	3.7 (1.2–11.3) DW	93
Liver	0.7 (0.1–4.3) DW	93
Serum	0.05 (0.01–0.5) FW	93
Raccoon, *Procyon lotor*		
Liver	1.8 FW	67
Kidney	1.9 FW	67
Muscle	0.2 FW	67
Norway rat, *Rattus norvegicus*		
Whole	0.4 DW	66
Rock squirrel, *Spermophilus variegatus*		
Kidney	8.9–53.0 DW; Max. 90.0 DW	70
Mole, *Talpa europaea*		
Whole	2.6 DW	66
INTEGRATED STUDIES		
California; Kesterson NWR vs. control site; 1984–85; liver		
Gopher snake, *Pituophis melanoleucas*	11.1 (8.2–32.0) DW vs. 2.1 (1.3–3.6) DW	121
Bullfrog, *Rana catesbeiana*	45.0 DW vs. 6.2 DW	121

Table 31.2 (continued) Selenium Concentrations in Field Populations of Selected Species of Flora and Fauna (Values shown are in mg total Se/kg [ppm] fresh weight [FW], dry weight [DW], or ash weight [AW]. Hyphenated numbers show range, and single numbers the mean; where both appear, the range is in parentheses.)

Ecosystem, Taxonomic Group, Organism, Tissue, Location, and Other Variables	Concentration (ppm)	Reference[a]
California; Kesterson NWR vs. control site; August 1983; maximum values recorded		
Water	0.3 FW vs. 0.0005 FW	124
Sediments	8.8 DW vs. not detectable (ND)	124
Detritus	80 DW vs. 2 DW	124
Algae	330 DW vs. <2 DW	124
Rooted plants	300 DW vs. <1 DW	124
Aquatic insects	290 DW vs. <3 DW	124
Mosquitofish	290 DW vs. ND	124
California; San Joaquin River; 1987; maximum concentrations recorded		
Water	0.025 FW	122
Sediments	3 DW	122
Invertebrates	14 DW	122
Fishes	17 DW	122
Greenland; 1975–91		
Bivalve molluscs; 3 species; soft parts	0.2–0.9 FW	127
Crustaceans; 6 species; whole	0.2–3.2 FW	127
Fish; 5 species		
Liver	0.3–0.9 FW	127
Muscle	<0.2–0.7 FW	127
Seabirds; 9 species		
Kidney	3.7–17.6 FW	127
Liver	1.9–14.1 FW	127
Muscle	0.5–4.4 FW	127
Seals; 2 species		
Kidney	2.6–4.2 FW	127
Liver	1.0–7.6 FW	127
Muscle	0.2–0.4 FW	127
Whales; 4 species		
Kidney	1.5–6.3 FW	127
Liver	1.7–5.0 FW	127
Muscle	Max. 0.2 FW	127
Skin	Max. 47.9 FW	127
Polar bear, *Ursus maritimus*		
Kidney	6.0–11.6 FW	127
Liver	3.1–9.1 FW	127
Muscle	<0.2–1.3 FW	127
Norway; 1990–91; near Russian nickel smelter vs. reference site; liver		
Willow ptarmigan, *Lagopus lagopus*	0.9 DW vs. 0.5 DW	119
Blue hare, *Lepus timidus*	Max. 1.8 DW vs. Max. 0.6 DW	119
Common shrew, *Sorex araneus*	Max. 4.6 DW vs. 4.5 DW	119
Grey-sided vole, *Clethrionomus rufocanus*	Max. 1.8 DW vs. no data	119
Texas; shoalgrass community; Lower Laguna Madre; 1986–87		
Sediments	2.8 (0.5–4.5) DW	123
Shoalgrass, *Halodule wrightii*; rhizomes	Not detected	123
Grass shrimp, *Palaemonetes* sp.; whole	1.2 (0.7–1.8) DW	123
Brown shrimp, *Penaeus aztecus*; whole	1.8 (0.6–3.2) DW	123
Blue crab, *Callinectes sapidus*; whole except legs, carapace, and abdomen	1.3 (0.4–3.9) DW	123
Pinfish, *Lagodon rhomboides*; whole	1.4 (0.8–2.2) DW	123

Table 31.2 (continued) Selenium Concentrations in Field Populations of Selected Species of Flora and Fauna (Values shown are in mg total Se/kg [ppm] fresh weight [FW], dry weight [DW], or ash weight [AW]. Hyphenated numbers show range, and single numbers the mean; where both appear, the range is in parentheses.)

Ecosystem, Taxonomic Group, Organism, Tissue, Location, and Other Variables	Concentration (ppm)	Reference[a]
Lower Colorado River Valley; 1985–91		
Sediments	1.2 (0.3–3.9) DW	126
Crayfish	3.5 (1.5–3.9) DW	126
Marsh birds, livers		
Rails, 3 species	13.1–26.0 DW	126
Others, 7 species	2.6–8.5 DW	126

[a] **1,** Chau and Riley 1970; **2,** Lunde 1970; **3,** Tijoe et al. 1977; **4,** Noda et al. 1979; **5,** Okazaki and Panietz 1981; **6,** Bertine and Goldberg 1972; **7,** Karbe et al. 1977; **8,** Hall et al. 1978; **9,** Fukai et al. 1978; **10,** Luten et al. 1980; **11,** Maher 1983; **12,** Cappon and Smith 1982; **13,** Stump et al. 1979; **14,** Fowler and Benayoun 1976c; **15,** Papadopoulu et al. 1976; **16,** Zafiropoulos and Grimanis 1977; **17,** Fowler and Benayoun 1976a; **18,** Glover 1979; **19,** Grimanis et al. 1978; **20,** de Goeij et al. 1974; **21,** Kifer and Payne 1968; **22,** Bebbington et al. 1977; **23,** Kari and Kauranen 1978; **24,** United Nations 1979; **25,** Tamura et al. 1975; **26,** MacKay et al. 1975; **27,** Schultz and Ito 1979; **28,** Heit 1979; **29,** Ganther et al. 1982; **30,** Freeman et al. 1978; **31,** Turner et al. 1978; **32,** White et al. 1980; **33,** King et al. 1980; **34,** King et al. 1983; **35,** Ohlendorf and Harrison 1986; **36,** Blus et al. 1977; **37,** Hutton 1981; **38,** Stoneburner 1978; **39,** Reijnders 1980; **40,** Smith and Armstrong 1978; **41,** van de Ven et al. 1979; **42,** Martin et al. 1976; **43,** Jenkins 1980; **44,** Rossi et al. 1976; **45,** Winger et al. 1984; **46,** Adams and Johnson 1977; **47,** Wiener et al., 1984; **48,** Lemly 1982b; **49,** Sorensen et al. 1984; **50,** Kaiser et al. 1979; **51,** Rancitelli et al. 1968; **52,** Uthe and Bligh 1971; **53,** Willford 1971; **54,** Tong et al. 1971; **55,** Pakkala et al. 1972; **56,** Rossi et al. 1976; **57,** Ohlendorf et al. 1986; **58,** Schroeder et al. 1970; **59,** Lucas et al. 1970; **60,** May and McKinney 1981; **61,** Pratt et al. 1972; **62,** Walsh et al. 1977; **63,** Haseltine et al. 1983; **64,** Birkner 1978; **65,** Helmke et al. 1979; **66,** Nielsen and Gissel-Nielsen 1975; **67,** Wren 1984; **68,** Fleming et al. 1979; **69,** Ullrey et al. 1981; **70,** Sharma and Shupe 1977; **71,** Heit et al. 1980; **72,** Haseltine et al., 1981; **73,** Chvojka et al. 1990; **74,** Saiki and Palawski 1990; **75,** Saiki et al. 1992; **76,** Saiki et al. 1991; **77,** Saiki 1987; **78,** Schmitt and Brumbaugh 1990; **79,** Felton et al. 1990; **80,** Nakamoto and Hassler 1992; **81,** Welsh 1992; **82,** Hamilton and Waddell 1994; **83,** Waddell and May 1995; **84,** Hasunuma et al. 1993; **85,** Mackey et al. 1996; **86,** Skaare et al. 1994; **87,** Teigen et al. 1993; **88,** Leonzio et al. 1992; **89,** Baumann and Gillespie 1986; **90,** Clark et al. 1992; **91,** Dierenfeld and Jessup 1990; **92,** Oliver et al. 1990b; **93,** McDowell et al. 1995; **94,** Hein et al. 1994; **95,** Medeiros et al. 1993; **96,** Wenzel and Gabrielsen 1995; **97,** Paveglio et al. 1992; **98,** Hothem and Welsh 1994; **99,** Ohlendorf and Marois 1990; **100,** Ohlendorf et al. 1986b; **101,** Mora and Anderson 1995; **102,** Burger and Gochfeld 1995; **103,** Burger and Gochfeld 1996; **104,** Burger and Gochfeld 1992; **105,** Custer and Meyers 1990; **106,** Burger et al. 1993b; **107,** Burger et al. 1994; **108,** King et al. 1991; **109,** Henny et al. 1990; **110,** Custer and Mitchell 1991; **111,** Burger and Gochfeld 1993; **112,** Amiard-Triquet et al. 1991; **113,** Henny and Herron 1989; **114,** Olson and Welsh 1993; **115,** Burger et al. 1992; **116,** King et al. 1994; **117,** Burger et al. 1993a; **118,** USPHS 1996; **119,** Kalas et al. 1995; **120,** Clark 1987; **121,** Ohlendorf et al. 1988; **122,** Saiki et al. 1993; **123,** Custer and Mitchell 1993; **124,** Saiki and Lowe 1987; **125,** Ohlendorf et al. 1990; **126,** Rusk 1991; **127,** Dietz et al. 1996; **128,** Kim et al. 1998; **129,** Scheuhammer et al. 1998; **130,** Hopkins et al. 1998; **131,** Burger and Snodgrass 1998.

31.4 DEFICIENCY AND PROTECTIVE EFFECTS

Selenium is an essential nutrient for most plants and animals. It constitutes an integral part of the enzyme glutathione peroxidase and may have a role in other biologically active compounds, especially Vitamin E and the enzyme formic dehydrogenase. Some animals require selenium-containing amino acids (*viz*. selenocysteine, selenocystine, selenomethionine, selenocystathionine, selenium-methylselenocysteine, and selenium-methylselenomethionine), but reportedly are incapable of producing them. Selenium also forms part of certain proteins, including cytochrome c, hemoglobin, myoglobin, myosin, and various ribonucleoproteins (Rosenfeld and Beath 1964; Eisler 1985; USPHS 1996).

The availability of selenium to plants may be lessened by modern agricultural practices, eventually contributing to selenium deficiency in animal consumers. For example, fertilizers containing nitrogen, sulfur, and phosphorus all influence selenium uptake by plants through different modes of action, the net effect being a reduction in selenium uptake (Frost and Ingvoldstad 1975). The buildup of sulfur (as sulfates) in the soil — due to acid rain, fertilizers, and other sources — interferes with selenium accumulation by crops (Frost and Ingvoldstad 1975). In addition, high dietary levels of various heavy metals (including copper, zinc, silver, and mercury) contribute to selenium deficiency in animals (Frost and Ingvoldstad 1975; Harr 1978), presumably as a result of selenium binding with the metal into biologically unavailable forms (Harr 1978; Kaiser et al. 1979).

Clinical selenium deficiency in ruminants is expressed as white muscle disease, lethargy, impaired reproduction, weight loss and reduced growth, shedding, decreased immune response, decreased erythrocyte glutathione peroxidase, and sudden death (Knox et al. 1987; Flueck and Flueck-Smith 1990). Selenium deficiency — as judged by blood concentrations <0.1 mg Se/L — has been documented in California among domestic cattle, mule deer (*Odocoileus hemionus*), pronghorn antelope (*Antilocapra americana*), elk (*Cervus elaphus*), and bighorn sheep (*Ovis canadensis*) (Oliver et al. 1990a). More than 95% of black-tailed deer (*Odocoileus hemionus columbianus*) in northern California had inadequate blood selenium levels (37 μg/kg whole blood FW), as judged by recommended levels for cattle (>40 μg/kg) and sheep (>50 μg/kg) (Flueck 1994). Selenium deficiency in red deer was reversed with subcutaneous injection of 50 mg Se/mL as barium selenate at 2 mL per 50 kg body weight (Knox et al. 1987). Selenium boluses calibrated to release 1.0 mg selenium daily, given orally to selenium-deficient adult female black-tailed deer, effectively raised whole blood levels to 121 μg Se/kg FW. These selenium-supplemented females produced fawns with increased survival (0.83 fawns/female) when compared to untreated does (0.32 fawns/female) (Flueck 1994). Supplementation of selenium-deficient mule deer does with intrarumenal selenium pellets can triple fawn survivability (Oliver et al. 1990a). Some feral animals selectively prefer plants with comparatively elevated selenium content. The black rhinoceros (*Diceros bicornis*) in Kenya, for example, prefers 10 of 103 plants ingested; preferred vegetation contained 3.0 to 6.3 μg Se/kg FW vs. 1.8 to 2.7 μg/kg FW in nonpreferred plants (Ghebremeshel et al. 1991).

There is a general consensus that selenium deficiency in livestock is increasing in many countries, resulting in a need for added selenium in the food. Selenium deficiency is considered by some researchers to constitute a greater threat to health than selenium poisoning. Studies with animals and humans have suggested that selenium deficiency, in part, underlies susceptibility to cancer, arthritis, hypertension, heart disease, and possibly periodontal disease and cataracts (Frost and Ingvoldstad 1975; Shamberger 1981; Robberecht and Von Grieken 1982). These linkages have not yet been demonstrated conclusively. For example, eye lens cataract was induced in 10-day-old male rats by selenate, selenite, selenomethionine, and selenocystine, presumably through interference with glutathione metabolism (Ostadalova and Babicky 1980). On the other hand, adverse effects of selenium inadequacy have been clearly documented for a wide variety of organisms, including bacteria, protozoans, Atlantic salmon, rainbow trout, Japanese quail, ducks, poultry, rats, dogs, horses, domestic sheep, bighorn sheep, swine, cattle, antelopes, gazelles, deer, monkeys, and humans (Jensen 1968; Frost and Ingvoldstad 1975; Jones and Stadtman 1975; Fishbein 1977; Harr 1978; Kaiser et al. 1979; Hilton et al. 1980; Shamberger 1981; Bovee and O'Brien 1982; Robberecht and Von Grieken 1982; NRC 1983; Levander 1983, 1984; Knox et al. 1987; Morris et al. 1984; Flueck and Smith-Flueck 1990; Oliver et al. 1990a; USPHS 1996). Selenium deficiency, whether induced experimentally by use of low-selenium feeds supplemented with alpha-tocopherol or by chronic ingestion of low-selenium diets, has caused a number of maladies:

- High embryonic mortality in cattle and sheep
- Anemia in cattle
- Poor growth and reproduction in sheep and rats
- Reduced viability of newly hatched quail

- Nutritional myopathy (white muscle disease) in sheep, swine, and cattle
- Hepatic necrosis and lameness in dogs, horses, and breeding bulls
- Hair loss and sterility in rat offspring
- Spermatozoan abnormalities in rats

Deficiencies were usually prevented or reversed by supplements with sodium selenate or selenite at 100 μg Se/kg ration, or 20 μg Se/kg body weight administered parenterally.

The protective action of selenium against the adverse or lethal effects induced by mercury, cadmium, arsenic, thallium, copper, zinc, silver, and various pesticides is well documented for a wide variety of plant and animal species (Hill 1976; Wilber 1985; Eisler 1985; USPHS 1996). Among marine organisms, for example, selenium protects against toxic levels of mercury in algae (Gotsis 1982), shrimp (Lucu and Skreblin 1982), crabs and oysters (Glickstein 1978), fish (Sheline and Schmidt-Nielsen 1977), and mammals (Koeman et al. 1975). Similar observations have been recorded for copper and marine algae (Gotsis 1982); cadmium and freshwater snails (Wilber 1983), marine crabs (Bjerragaard 1982), earthworms (Helmke et al. 1979; Beyer et al. 1982), and rats (Harr 1978); mercury or methylmercury and rats (Cappon and Smith 1982), eggs of lake trout (Klaverkamp et al. 1983b), freshwater teleosts (Kim et al. 1975, 1977) and (temporarily) Japanese quail (El-Bergearmi et al. 1977, 1982; Beijer and Jernelov 1978); and arsenic and freshwater and marine teleosts (Luten et al. 1980; Orvini et al. 1980). Not all tests were conclusive. Studies with some species of freshwater teleosts demonstrated negligible antagonism of selenium against mercury (Klaverkamp et al. 1983a) or cadmium (Duncan and Klaverkamp 1983). Selenium reportedly protects mammals and poikilotherms against poisoning by thallium, the herbicide paraquat, cadmium, mercury, lead, arsenic, and copper (Wilber 1983; USPHS 1996). Selenium also protects against fatal biological agents. Juvenile chinook salmon (*Oncorhynchus tshawytscha*) naturally infected with *Renibacterium salmoninarum*, the causative agent of bacterial kidney disease, were protected when diets were supplemented with 2.5 mg Se/kg DW ration and Vitamin E (Thorarinsson et al. 1994).

Reasons to account for the antagonism of selenium and heavy metals (here, mercury is used as an example) include the dietary source and chemical form of selenium, influence of sulfur, biological translocation of selenium or mercury to less-critical body parts, and chemical linkage of selenium to mercury on a linear basis. The exact mode of interaction is probably complex and has not yet been resolved. In regard to diet, selenium of animal origin and in the form of selenate is less effective than selenium from plant and inorganic sources in preventing methylmercury neurotoxicity in experimental animals (Cappon and Smith 1982). Disruption of sulfur metabolism by selenium, the sulfur being replaced by seleno-amino acids and other cell constituents containing selenium in living organisms, is one probable cause of selenosis. It is conceivable that Se–Hg compounds formed within the organism would be sufficiently nonreactive biologically to interfere with sulfur kinetics, presumably –SH groups (Koeman et al. 1975; Beijer and Jernelov 1978; Cappon and Smith 1982; Gotsis 1982). Differential redistribution of selenium or mercury to less-critical body parts may partly account for observed antagonisms. Pretreatment of marine minnows with selenium protects against mercury poisoning and causes a marked redistribution of mercury among organs, presumably to noncritical body parts, and this transfer may partly account for the observed Se–Hg antagonisms in that species (Sheline and Schmidt-Nielsen 1977). Some investigators have reported that selenium results in increased mercury accumulations. Increased retention of mercury and other metals may lead to a higher level of biomagnification in the food chain and higher body burden in the individual, which might counteract the positive effect of decreased intoxication (Beijer and Jernelov 1978). Extensive research is under way on the chemical linkage of selenium and mercury. In marine mammals and humans, selenium and mercury concentrations are closely related, almost linearly in a 1:1 molar ratio, but this relation blurs in teleosts in which selenium is in abundance, and fails in birds (Koeman et al. 1975; Beijer and Jernelov 1978; Orvini et al. 1980; Cappon and Smith 1982).

31.5 LETHAL EFFECTS

31.5.1 Aquatic Organisms

Among representative species of aquatic organisms, death was observed at water concentrations between 60 and 600 μg Se/L; early life history stages that were subjected to comparatively lengthy exposures accounted for most of these data (Hamilton and Wiedmeyer 1990; Cleveland et al. 1993; Table 31.3). Sensitive species of fishes had reduced survival after extended exposure to 10 to 47 μg Se/L. Adult bluegills (*Lepomis macrochirus*), for example, had reduced survival after exposure to 10 μg Se/L for 1 year; those exposed to 30 μg Se/L all died (Hermanutz et al. 1992). Adult bluegills exposed for 60 days to 10 μg Se/L as selenite, plus 33.3 mg Se/kg ration as seleno-L-methionine, were normal but produced fry with reduced survival (Coyle et al. 1993). Exposure for 1 year to 25 μg/L caused reduced survival and reproduction of perch and grass carp (Crane et al. 1992). Mortality of 35% occurred at 47 μg Se/L, as selenite, in chinook salmon exposed for 90 days at the yolk-sac stage; 70% died at 100 μg/L (Lemly et al. 1993). Survival was reduced in chinook salmon fingerlings when their diets contained >9.6 mg Se/kg ration (Hamilton et al. 1990; Lemly et al. 1993).

Latent mortality after exposure to comparatively high selenium concentrations has been documented, but not extensively. For example, all embryos of the zebrafish (*Brachydanio rerio*) survived exposure to 3000 μg Se/L during development, but more than 90% of the resultant larvae died soon after hatching. At 1000 μg/L, survival was similar to that in controls (Niimi and LaHam 1975). It has been suggested (USEPA 1980) that selenite is more toxic than selenate and is preferentially concentrated over selenate by mussels, *Mytilus galloprovincialis* (Measures and Burton 1980). Selenite is generally more toxic to early life history stages, and effects are most pronounced at elevated temperatures (Klaverkamp et al. 1983a). Also, selenium salts may be converted to methylated forms by microorganisms, and these are readily accumulated by aquatic vertebrates (Klaverkamp et al. 1983a). Among freshwater algae species, it has been demonstrated that selenite, selenate, selenomethionine, and selenopurine are all toxic, but that sulfur, as sulfate, has a significant protective role against selenium toxicity (Kumar and Prakash 1971). Numerous additional chemical compounds and mixtures probably protect against selenium toxicity, much as selenium protects against toxic effects of mercury salts and other chemicals, but data are sparse on selenium protective agents.

More than 60 years ago, Ellis et al. (1937) recorded a long list of signs of selenium poisoning in teleosts: loss of equilibrium, lethargy, contraction of dermal chromatophores, loss of coordination, muscle spasms, protruding eyes, swollen abdomen, liver degeneration, reduction in blood hemoglobin and erythrocyte number, and increase in white blood cells. Sorensen et al. (1984) observed most of these signs in selenium-poisoned green sunfish (*Lepomis cyanellus*), together with elevated liver selenium concentrations, reduced blood hematocrit, enlarged liver, histopathology of kidney and heart, swollen gill lamellae with extensive cellular vacuolization, and necrotic and degenerating ovarian follicles. Other signs of selenosis in freshwater fishes include loss of osmotic control and liver histopathology (Hodson 1990).

Table 31.3 Toxicity of Selenium Salts to Aquatic Biota (Values shown are in μg/L [ppb] in medium fatal to 50% of the organisms during exposure for various intervals.)

Medium, Taxonomic Group, and Species	Exposure Interval (hours [h], days [d], or life cycle [LC])	LC50 (μg/L)	Reference[a]
FRESHWATER			
Algae			
Anabaena variabilis	96 h	15,000–17,000	1
Anacystis nidulans	96 h	30,000–40,000	1
Oedogonium cardiacum	48 h	<100	2

Table 31.3 (continued) Toxicity of Selenium Salts to Aquatic Biota (Values shown are in μg/L [ppb] in medium fatal to 50% of the organisms during exposure for various intervals.)

Medium, Taxonomic Group, and Species	Exposure Interval (hours [h], days [d], or life cycle [LC])	LC50 (μg/L)	Reference[a]
Coelenterates			
Hydra sp.			
Selenite	96 h	1700	17
Selenate	96 h	7300	17
Molluscs			
Snail, *Physa* sp.	96 h	24,000	3
Snail, *Physa* sp.	48 h	>10,000	2
Insects			
Mosquito larvae, *Culex fatigans*	48 h	<3100	2
Midge, *Tanytarsus dissimilis*	96 h	42,400	4
Crustaceans			
Daphnid, *Ceriodaphnia affinis*	96 h	480–720	17
Cladoceran, *Daphnia magna*			
Selenite	48 h	700	18
Selenate	48 h	2560	18
Selenite	MATC[b]	110–237	17
Selenate	MATC[b]	1730–2310	17
D. magna	96 h	710	5
D. magna	14 d	430	5
D. magna	28 d	240	4
Scud, *Hyallela azteca*	96 h	760	6
H. azteca	96 h	340	7
H. azteca	14 d	70	5
Cladoceran, *Daphnia pulex*	96 h	3870	3
D. pulex	LC	600–800	4
Amphibians			
South African clawed frog, *Xenopus laevis*			
Embryo	27 h	20,000	8
Embryo	61 h	10,000	8
Embryo	96 h	4000	8
Embryo	113 h	2000	8
Tadpole	3 d	8000	8
Tadpole	5 d	2600	8
Tadpole	7 d	1500	8
Fishes			
Goldfish, *Carassius auratus*	96 h	26,100	9
C. auratus	14 d	6300	9
White sucker, *Catostomus commersoni*	48 h	48,600	10
C. commersoni	96 h	31,400	10
Common carp, *Cyprinus carpio*	24 h	72,000	11
C. carpio	96 h	35,000	12
Northern pike, *Esox lucius*	75 h	11,100	13
Mosquitofish, *Gambusia affinis*	48 h	>6000	2
G. affinis	96 h	12,600	3
Green River, Utah; 3 endangered species (Colorado squawfish, *Ptychocheilus lucius*; razorback sucker, *Xyrauchen texanus*; bonytail, *Gila elegans*); fry and juveniles			
Selenite	96 h	18,000 (14,000–22,000)	19
Selenate	96 h	97,000 (75,000–129,000)	19
Channel catfish, *Ictalurus punctatus*	96 h	13,600	9
Flagfish, *Jordanella floridae*	96 h	6500	9

Table 31.3 (continued) Toxicity of Selenium Salts to Aquatic Biota (Values shown are in µg/L [ppb] in medium fatal to 50% of the organisms during exposure for various intervals.)

Medium, Taxonomic Group, and Species	Exposure Interval (hours [h], days [d], or life cycle [LC])	LC50 (µg/L)	Reference[a]
Bluegill, *Lepomis macrochirus*			
Seleno[DL]methionine	96 h	13	18
Seleno[L]methionine	96 h	13	18
Selenite	96 h	7800–13,000	18
Selenate	96 h	98,000	18
Striped bass, *Morone saxatilis*, fingerlings			
Seleno[L]methionine	96 h	4	18
Selenite	96 h	1000	18
Selenate	96 h	39,000	18
Coho salmon, *Oncorhynchus kisutch*			
Fry	43 d	160	6
Yellow perch, *Perca flavescens*	10 d	4800	13
Fathead minnow, *Pimephales promelas*			
Fry	96 h	2100	9
Juvenile	96 h	5200	9
Adult	96 h	620–12,500	4–6
Adult	9 d	2100	9
Adult	48 d	1100	6
Selenite	MATC[b]	83–153	17
Selenate	MATC[b]	390–820	17
Rainbow trout, *Oncorhynchus mykiss*	96 h	4200–12,500	4, 6, 12, 14
O. mykiss	9 d	5400–7000	14
O. mykiss			
Selenite	MATC[b]	60–130	17
Selenate	MATC[b]	2200–3800	17
Chinook salmon, *Oncorhynchus tshawytscha*, fry			
Selenite	96 h	13,800	18
Selenate	96 h	115,000	18
Colorado squawfish, *Ptychocheilus lucius*			
Larva, selenite	96 h	12,800	20
Larva, selenate	96 h	24,600	20
Juveniles, selenite	96 h	27,900	20
Juveniles, selenate	96 h	77,500	20
Brook trout, *Salvelinus fontinalis*	96 h	10,200	9
MARINE			
Molluscs			
Pacific oyster, *Crassostrea gigas*			
Larvae	48 h	>10,000	15
Crustaceans			
Copepod, *Acartia clausi*	96 h	1740	4
Copepod, *A. tonsa*	96 h	800	4
Blue crab, *Callinectes sapidus*	96 h	4600	16
Dungeness crab, *Cancer magister*			
Larvae	96 h	1040	15
Mysid shrimp, *Mysidopsis bahia*			
Adult	96 h	1500	16
Juvenile	96 h	600	4
Egg	LC	27–143	4
Brown shrimp, *Penaeus aztecus*	96 h	1200	16

Table 31.3 (continued) Toxicity of Selenium Salts to Aquatic Biota (Values shown are in µg/L [ppb] in medium fatal to 50% of the organisms during exposure for various intervals.)

Medium, Taxonomic Group, and Species	Exposure Interval (hours [h], days [d], or life cycle [LC])	LC50 (µg/L)	Reference[a]
Fish			
Fourspine stickleback, *Apeltes quadracus*	96 h	17,350	4
Sheepshead minnow, *Cyprinodon variegatus*			
Adult	96 h	7400–67,100	4
Egg through juvenile	MATC[b]	470–970	17
Pinfish, *Lagodon rhomboides*	96 h	4400	16
Haddock, *Melanogrammus aeglefinus*			
Larvae	96 h	600	4
Atlantic silverside, *Menidia menidia*	96 h	9725	4
Striped bass, *Morone saxatilis*			
Selenite	96 h	1600	17
Selenate	96 h	9800	17
Summer flounder, *Paralichthys dentatus*			
Larvae	96 h	3500	4
Winter flounder, *Pleuronectes americanus*			
Larvae	96 h	4250–15,100	4

[a] **1,** Kumar and Prakash 1971; **2,** Nassos et al. 1981; **3,** Reading 1979; **4,** USEPA 1980; **5,** Halter et al. 1980; **6,** Adams 1976; **7,** Murphy 1971; **8,** Browne and Dumont 1979; **9,** Cardwell et al. 1976; **10,** Duncan and Klaverkamp 1983; **11,** Sato et al. 1980; **12,** Spehar et al. 1982; **13,** Klaverkamp et al. 1983a; **14,** Hodson et al. 1980; **15,** Glickstein 1978; **16,** Ward et al. 1981; **17,** USEPA 1987; **18,** Lemly et al. 1993; **19,** Hamilton 1995; **20,** Buhl and Hamilton 1995.

[b] MATC = maximum acceptable toxicant concentration. Lower value in each MATC pair indicates highest concentration tested producing no measurable effect on growth, survival, reproduction, and metabolism during chronic exposure. Higher value indicates lowest concentration tested producing a measurable effect.

31.5.2 Mammals and Birds

"The element selenium can be traced in an orderly sequence from its origin in the Earth's crust to specific geological formation, to distribution of specific genera and groups of plants which require the element for their growth, to the accumulation in vegetation, and to its subsequent toxicity to birds or mammals that consume the seleniferous foods" (Rosenfeld and Beath 1964). Selenosis in warm-blooded organisms is modified by numerous factors, including method of administration, chemical form of selenium, dietary composition, and age and needs of the organism. Concurrent ingestion of minerals and rough or high protein feeds reduces selenium toxicity, and exposure by diet is less toxic than exposure parenterally or by inhalation. Many compounds are known to prevent or reduce toxic effects of subacute and chronic selenosis in pigs, beef cattle, and other warm-blooded organisms. A partial list includes arsenic, strychnine sulfate, tungsten, germanium, antimony, beet pectin, high-fat diets, ACTH injections, sulfate, increased dietary proteins, lactalbumin, ovalbumin, wheat protein, dried brewer's yeast, desiccated liver, linseed oil meal, glucosamine, hemocysteine, creatine, methionine, and choline. Not all of these compounds afforded equal protection against various selenium formulations. The reasons for the difference are not clear, but it appears that the subject of selenoprotective agents warrants additional research effort. Selenium poisoning in livestock, discussed here, was largely extracted from reviews by Rosenfeld and Beath (1964), Frost (1972), Fishbein (1977), Harr (1978), Shamberger (1981), NRC (1983), Wilber (1983), and USPHS (1996).

In livestock, there are three basic types of selenium poisoning:

- Acute, resulting from consumption (usually in a single feeding) of a sufficient quantity of highly seleniferous weeds
- "Blind staggers," from consumption of moderately toxic amounts of seleniferous weeds over an extended period of time
- "Alkali disease," caused by the consumption of moderately seleniferous grains and forage grasses over a period of several weeks to months

Acute poisoning is associated with plant materials containing 400 to 800 mg Se/kg: sheep died when fed amounts of plant material ranging from 8 to 16 g/kg BW, or about 3.2 to 12.8 mg Se/kg BW. The minimum lethal dose of Se administered orally as selenite (mg Se per kg body weight) ranged from 3.3 for horses and mules, to 11 for cattle, and 15 for swine. Other modes of administration were more toxic; for example, 2 and 1.2 mg Se/kg BW given subcutaneously killed swine in 4 h and 5 days, respectively; and 1.5 to 6.0 mg Se/kg BW given intravenously or intraperitoneally to rats and rabbits were fatal. Accidental toxicosis of sheep and cattle from overtreatment with commercial mixtures of Se salts and Vitamin E are also documented for Australia and New Zealand. Acute Se poisoning in domestic livestock is characterized by abnormal movements, lowered head, drooped ears, diarrhea, elevated temperature, rapid pulse, labored breathing, bloating with abdominal pain, increased urination, and dilated pupils. Before death, which is due to respiratory failure, there is complete prostration and lethargy. Duration of illness extends from a few hours to several days, depending on the toxicity of plant material ingested. In these cases, selenium is distributed by the circulatory system to all body organs, the concentrations being highest in liver, blood, kidney, spleen, and brain, and lowest in muscle, skin, hair, and bone. Elimination is primarily in the urine; smaller quantities are excreted with the feces, breath, perspiration, and bile. Postmortem examinations indicate many pathological changes in the heart, lungs, rumen, liver, kidney, and other organs. No effective treatment is known for counteracting toxic effects of large amounts of ingested selenium.

Chronic selenosis in mammals may be induced by dietary exposure to natural selenite, selenate, or seleniferous feedstuffs at dietary concentrations between 1 mg/kg (rat) and 44 mg/kg (horse), or from water containing 0.5 to 2.0 mg Se/L. Cattle fed 0.5 mg Se/kg BW three times weekly lost their appetite; sheep fed up to 75 mg selenite daily developed myocardial degeneration and fibrosis, pulmonary congestion, and edema. The minimum toxic concentration of selenium in lifetime exposure of rats (a comparatively sensitive species) fed Se-deficient diets fortified with selenium was 0.35 mg Se/kg diet, as judged by changes in liver chemistry; and 0.75 mg Se/kg diet, as judged by longevity, and histological changes in heart, kidney, and spleen. These concentrations are 10 times the nutritional threshold for selenium, and about 25% of the minimum lifetime exposure to selenium in natural feedstuffs that produces similar effects under the same experimental conditions. Signs of chronic selenosis include skin lesions, lymph channel inflammation, loss of hair and nails, anemia, enlarged organs (spleen, pancreas, liver), fatigue, lassitude, and dizziness. "Blind staggers" is characterized by anorexia, emaciation, and sudden collapse, followed by death. Typically, the upper intestinal tract is ulcerated. In "alkali disease" in cattle, hogs, and horses that had eaten seleniferous grains, the signs were deformation and sloughing of the hooves, hair loss, lassitude, erosion of the articular cartilages, reduced conception, increased reabsorption of fetuses, and degeneration of heart, kidney, and liver. It is likely that selenium displaces sulfur in keratin, resulting in structural changes in hair, nails, and hooves (Fishbein 1977).

Elevated selenium concentrations were measured in tissues and diet of two captive California sea lions (*Zalophus californianus*) that died shortly after performing at a show in 1988. Selenium concentrations, in mg/kg FW, were 49 and 88 in liver, 42 and 47 in kidney, and 5.1 and 5.2 in blood. Selenium concentrations in their fish diet was 2.5 mg/kg FW, and in thawed fish fluids 45 mg/kg FW (Alexander et al. 1990).

Fatal chronic selenosis in aquatic birds is characterized by low body weight or emaciation, liver necrosis, enlarged kidneys (up to 40% heavier than normal), and more than 66 mg Se/kg DW liver

(Albers et al. 1996). Selenomethionine was the most toxic form of selenium tested against mallards (Lemly et al. 1994). Mallard ducklings fed 8 mg Se/kg ration as selenomethionine for 120 days had hepatotoxicity as adults; 10 mg/kg ration for 120 days inhibited reproduction; 15 mg/kg ration for 28 days inhibited growth; and 60 mg/kg ration for 60 days was fatal to all ducklings (Lemly et al. 1994) (Table 31.4). All fatal cases of selenomethionine-induced poisoning in mallards were characterized by histologic lesions of the liver, pancreas, spleen, and lymph nodes, and severe atrophy and degeneration of fat (Green and Albers 1997).

Table 31.4 Selenium Effects on Birds

Species, Dose, and Other Variables	Effect	Reference[a]
MALLARD, *Anas platyrhynchos*		
Adults given drinking water containing 0, 0.5, or 3.5 mg Se/L as sodium selenite, or 2.2 mg Se/L as selenomethionine for 12 weeks	Selenomethionine group had altered immune function, altered serum enzyme activities, and elevated concentrations of selenium in liver (4 times control values) and breast muscle (14 times). Sodium selenite-treated birds had normal immune function and selenium tissue burdens; however, serum enzyme activity was disrupted in the 3.5 mg/L group	1
Breeding adults fed diets for 100 days containing 0, 1, 2, 4, 8, or 16 mg Se/kg fresh weight (FW) ration as seleno-DL-methionine, or 16 mg Se/kg ration as seleno-DL-cysteine	Adults normal. Impaired reproduction (reduced survival of ducklings, increased developmental abnormalities) for selenomethionine occurs between 4 and 8 mg/kg ration; selenocysteine did not impair reproduction at 16 mg Se/kg ration	2
Breeding adults fed diets containing 0, 1, 5, 10, 25, or 100 mg Se/kg ration as sodium selenite for as long as 12 weeks, or 10 mg Se/kg ration as seleno-DL-methionine for as long as 12 weeks	Sodium selenite groups had normal growth, survival, and reproduction at 10 mg/kg ration and lower; growth and reproduction inhibited in the 25- and 100-mg/kg groups. Selenomethionine birds produced fewer ducklings with a high incidence of developmental abnormalities; surviving ducklings had impaired growth	3
Breeding adults fed diets containing 0, 3.5, or 7 mg Se/kg ration as seleno-DL-methionine for as long as 21.8 weeks. Ducklings produced received the same treatment as their parents for 14 days, then killed	No deaths or histopathology in any group. Dose-dependent decrease in adult growth, duckling weight, and hatching success; dose-dependent increase in selenium concentrations in adult liver, eggs, and duckling liver	4
Adults and resultant ducklings fed diets supplemented with up to 400 mg As (as sodium arsenate)/kg ration and zero or 10 mg Se (as seleno-DL-methionine)/kg ration	Arsenic or selenium at dose levels and forms given adversely affect mallard reproduction and duckling growth and survival. In mixtures, arsenic reduced selenium accumulation in liver and eggs and alleviated adverse affects of selenium on hatching success and embryo deformities	5
Day-old mallard ducklings fed diets for 2 weeks containing 0, 15, or 30 mg Se/kg ration in a 75% wheat diet (22% protein). Selenium given as seleno-DL-methionine, seleno-L-methionine, or selenized yeast	All forms of selenium caused significant increases in plasma and hepatic glutathione peroxidase activities. Seleno-L-methionine at 30 mg/kg ration was the most toxic form, resulting in high mortality (64%), impairing growth more than 50% in survivors, and the greatest increase in the ratio of oxidized to reduced hepatic glutathione. When the basal diet was a commercial duck feed (22% protein), survival was not adversely affected and oxidative effects were less pronounced	6
Adult males fed diets containing 0.2 (control), 1, 2, 4, 8, 16, or 32 mg Se/kg ration as selenomethionine for 14 weeks	Plasma glutathione peroxidase activity increased at 2 mg/kg ration and greater; altered liver enzyme activity in 8 mg/kg ration and higher; dose-dependent increase in liver selenium, reaching 29 mg Se/kg FW in the 32-mg/kg group; the high-dose group had decreased survival, altered blood chemistry, and hepatotoxicity	7
Adults fed diets 6 weeks before egg laying through day 7 of incubation containing either sodium selenite — at 1, 5, 10, or 25 mg Se/kg ration — or selenomethionine at 10 or 16 mg Se/kg ration	The 25 mg/kg group of sodium selenite had 42% fewer eggs than controls that hatched, more birth defects (4.2%), and decreased embryo weights. Developmental abnormalities in the selenomethionine groups were 13.1% in the 10-mg/kg group and 68.0% in the 16-mg/kg group	8

Table 31.4 (continued) Selenium Effects on Birds

Species, Dose, and Other Variables	Effect	Reference[a]
Ducklings fed diets containing 0, 10, 20, 40, or 80 mg Se/kg ration as selenomethionine or sodium selenite from hatching to age 6 weeks	At 80 mg/kg ration, all ducklings were dead in the selenomethionine group and 98% were dead in the selenite group. At 40 mg/kg ration, 25% were dead in the selenomethionine group and 13% in the selenite group. Survival was normal in other groups. Growth was decreased at 20 mg/kg ration and higher — regardless of chemical form — due to decreases in food consumption. The 10-mg/kg selenite group had significantly heavier livers than controls	9
Adult males fed diets containing 0, 10, 20, 40, or 80 mg Se/kg ration as seleno-DL-methionine for 16 weeks	All dead at 80 mg/kg ration. Survival reduced, growth impaired, and molt delayed at 40 mg/kg ration. Dose-dependent increase in tissue selenium concentrations. Dead birds had consistent histologic lesions in the liver, kidneys, and organs of the immune system	10, 13
Adults males given a choice between a control diet or diets with 5, 10, or 20 mg Se/kg ration as selenomethionine	Mallards avoided diets containing 10 or 20 mg Se/kg; avoidance may not be due to aversion to the taste of selenium	11
Adult males were fed diets supplemented with 0, 10, 20, 40, or 80 mg Se/kg ration as selenomethionine for 16-week exposure that began in November. Survivors were fed untreated diets for 4 more weeks	No deaths in controls or 10-mg/kg group, 25% dead in the 20-mg/kg group, 95% in the 40-mg/kg group, and 100% in the high-dose group. Body weights depressed in the 20-, 40-, and 80-mg/kg groups; but after 4 weeks on untreated diet, the 20-mg/kg group was the same as controls	12
Mallards fed diet containing 0 or 15 mg Se/kg ration as seleno-DL-methionine for 21 weeks, untreated food for 12 weeks, followed by 100 mg Se/kg ration for 5 weeks for both groups	No difference between groups in survival (85%), weight loss in survivors (40%), or liver burdens (35–40 mg Se/kg FW)	14
Adults fed diet containing 10 mg Se/kg ration as selenomethionine for 6 weeks followed by 6 weeks on untreated diet	Equilibrium reached in liver in 7.8 days and in muscle in 81 days with half-time persistence of 19 days in liver and 30 days in muscle	15
Fed diet with 15 mg Se/kg ration as selenomethionine for 21 weeks during winter, ending with onset of reproductive season	Selenium group had elevated concentrations in eggs, decreased survival, and higher incidence of deformed embryos. Reproduction and survival normal after 2 weeks on an uncontaminated diet	16
Ducklings fed diets for 4 weeks containing 15 or 60 mg Se/kg ration as selenomethionine, with and without 1000 mg boron/kg ration	Severe adverse effects of 60 mg Se/kg ration on survival, growth, and liver histology; effects exacerbated by the addition of boron	17
Females that had just initiated egg laying were fed diet containing 20 mg Se/kg ration as selenomethionine for 20 days, then an untreated diet for 20 days	Selenium concentrations in eggs reached a maximum of 20 mg Se/kg FW in about 2 weeks on the treated diet. Concentrations fell to less than 5 mg/kg FW after 10 days on the untreated diet	18
Ducklings fed diet containing 22% protein and 60 mg Se/kg ration as selenomethionine for 4 weeks	Selenium-induced reduction in growth and survival, and increase in liver histopathology. Effects exacerbated at 7% and 44% protein. Effects alleviated by addition of 200 mg As/kg ration as sodium arsenate, and partially alleviated by methionine dietary supplement	19, 20
AMERICAN KESTREL, *Falco sparverius*		
Adults fed diets for 11 weeks containing 5 or 9 mg Se/kg DW ration as seleno-L-methionine or naturally incorporated selenium (mammals from Kesterson National Wildlife Refuge, California)	All birds seemed normal during the study. Maximal Se concentrations in blood were measured at week 5 in the seleno-L-methionine groups: 4.3 mg Se/kg DW (low Se group) and 8.4 mg Se/kg DW (high Se group) vs. 1.6 mg/kg in controls. Excreta Se levels were 5.8 and 1.8 mg/kg DW in the low and high seleno-L-methionine groups, respectively, vs. 1.4 in controls. All treatment groups had reduction of Se concentration in excreta, but not in blood, to baseline values 4 weeks after treatment ended	22

Table 31.4 (continued) Selenium Effects on Birds

Species, Dose, and Other Variables	Effect	Reference[a]
EASTERN SCREECH-OWL, *Otus asio*		
Breeding adults fed diets containing 0, 4.4, or 13.2 mg Se/kg ration as seleno-DL-methionine for 3 months (equivalent to 0, 10, and 30 mg seleno-DL-methionine/kg ration)	Growth and reproduction inhibited at high dose. No malformed nestlings at low dose, but femur lengths were shorter than controls. Altered liver biochemistry of nestlings from parents fed low dose	21

[a] **1,** Fairbrother and Fowles 1990; **2,** Heinz et al. 1989; **3,** Heinz et al. 1987; **4,** Stanley et al. 1996; **5,** Stanley et al. 1994; **6,** Hoffman et al. 1996; **7,** Hoffman et al. 1991a; **8,** Hoffman and Heinz 1988; **9,** Heinz et al. 1988; **10,** Albers et al. 1996; **11,** Heinz and Sanderson 1990; **12,** Heinz and Fitzgerald 1993a; **13,** Green and Albers 1997; **14,** Heinz 1993b; **15,** Heinz et al. 1990; **16,** Heinz and Fitzgerald 1993b; **17,** Hoffman et al. 1991b; **18,** Heinz 1993a; **19,** Hoffman et al. 1992a; **20,** Hoffman et al. 1992b; **21,** Wiemeyer and Hoffman 1996; **22,** Yamamoto et al. 1998.

31.6 SUBLETHAL AND LATENT EFFECTS

Results of laboratory studies and field investigations with fish, mammals, and birds have led to general agreement that elevated concentrations of selenium in diet or water were associated with reproductive abnormalities, including congenital malformations, selective bioaccumulation by the organism, and growth retardation. Not as extensively documented, but nevertheless important, are reports of selenium-induced chromosomal aberrations, intestinal lesions, shifts in species composition of freshwater algal communities, swimming impairment of protozoans, and behavioral modifications.

31.6.1 Aquatic Organisms

Adverse effects on reproduction are among the most insidious effects of selenium in freshwater ecosystems. Adult bluegills exposed to 2.5 μg Se/L as sodium selenite for 319 days in an experimental stream ecosystem produced fry with a high incidence of edema, lordosis, and hemorrhaging (Coyle et al. 1993). Fathead minnows (*Pimephales promelas*) held in ecosystems containing 10 or 30 μg Se/L for 1 year produced a high incidence of malformed progeny with humped backs, missing scales, and malformations of the jaw, head, operculum, snout, and mandible. The frequency of malformations in the controls was 0.3%; in the 10- and 30-μg/L groups, these frequencies were 8% and 29.8%, respectively (Hermanutz 1992). Selenium concentrations in whole fathead minnows from the 10-μg Se/L group were 3.9 mg/kg FW vs. 0.3 in the controls (Schultz and Hermanutz 1990). Mosquitofish reproduction was inhibited when whole-body selenium concentrations were >100 mg Se/kg DW vs. normal reproduction in a reference area at 1.5 mg Se/kg DW whole body (Saiki and Ogle 1995).

In green sunfish from a lake in North Carolina receiving selenium (as flyash wastes from a coal-fired power station), reproduction failed and the population declined markedly. In these fish, selenium levels were elevated in liver (up to 21.4 mg/kg FW) and other tissues; kidney, heart, liver, and gill showed histopathology; and blood chemistry was altered. Ovaries of fish had numerous necrotic and ruptured egg follicles that may have contributed to the population extinction (Sorensen et al. 1984). It is probable that selenium uptake by plankton (containing 41 to 97 mg/kg DW) from lake water (9 to 12 μg/L) introduced selenium to the food chain, where it ultimately reached elevated levels in fish through biomagnification (Cumbie and Van Horn 1978). In laboratory tests, however, eggs of common carp hatched normally when incubated in media containing 5000 μg Se/L (Huckabee and Griffith 1974), as did eggs of lake trout (*Salvelinus namaycush*) at 10,000 μg Se/L (Klaverkamp et al. 1983b). In frogs (*Xenopus laevis*), cranial and vertebral deformities and lowered survival were documented during development in water with concentrations of 2000 μg Se/L or higher (Browne and Dumont 1979).

Reduced growth of freshwater fishes was associated with tissue concentrations of 3.6 to 6.7 mg Se/kg DW (Hamilton and Wiedmeyer 1990), and dietary concentrations between 5.3 and 25 mg Se/kg DW ration (Hamilton et al. 1990; Cleveland et al. 1993; Lemly et al. 1993). Toxic sublethal effects of selenium in aquatic systems were more pronounced for organoselenium compounds than inorganic compounds and more pronounced for inorganic selenite than inorganic selenate compounds (USEPA 1987; Hamilton and Buhl 1990; Chapman 1992; Boisson et al. 1995) (Table 31.3). In salmon, younger stages were more sensitive than older stages (Hamilton and Buhl 1990; Chapman 1992). Adverse effects of selenium stress to freshwater fishes were reduced with increasing water hardness (Hamilton and Buhl 1990; Hamilton et al. 1990), and fishes were more sensitive to selenium stress under conditions of reduced temperature and photoperiod (Lemly 1993c).

At water concentrations of 47 to 53 μg/L, selenium was associated with anemia and reduced hatch of rainbow trout (Hodson et al. 1980), growth retardation of freshwater green algae (Hutchinson and Stokes 1975; Klaverkamp et al. 1983a), and shifts in species composition of freshwater algal communities (Patrick 1978). At 250 μg Se/L, growth was reduced in rainbow trout fry after exposure for 21 days (Adams 1976), and goldfish demonstrated an avoidance response after 48 h (Weir and Hine 1970). At water concentrations of 7930 to 11,000 μg Se/L, growth was inhibited in freshwater and marine algae (Patrick 1978; USEPA 1980), and swimming rate was reduced in the protozoan *Tetrahymena pyriformis* (Bovee and O'Brien 1982). Eggs of channel catfish exposed to certain metals (including cadmium, mercury, and copper) produced an increased percentage of albino fry; however, eggs exposed to 250 μg Se/L produced fry with normal pigmentation (Westernman and Birge 1978).

A significant number of chromosomal aberrations were induced in the edible goby (*Boleophthalmus dussumieri*) by selenium after intramuscular and water exposures (Krishnaja and Rege 1982). Intramuscular injections as low as 0.1 mg Se/kg BW, or 3200 μg/L in the water column, were associated with a marked enhancement of polyploid cells 76 to 96 h postadministration. Some deaths were recorded at higher test concentrations. Selenite was more effective than selenate in inducing chromosomal aberrations. The authors concluded that a relatively narrow range of selenium concentrations leads to a mutagenic rather than lethal effect.

Accumulation of selenium by aquatic organisms is highly variable. In short-term (48-h) laboratory tests at water concentrations of 0.015 to 3.3 μg Se/L, Nassos et al. (1980) reported biological concentration factors (BCFs) of 460 for mosquitofish to 32,000 for a freshwater gastropod. Values were intermediate for daphnids (2100), plankton (2600), and *Fundulus kansae* (3300), the freshwater killifish. High BCFs (>680) were recorded for freshwater diatoms subjected to maximum concentrations of 40 μg Se/L (Patrick 1978). Livers from rainbow trout and brown trout may contain from 50 to 70 mg Se/kg FW during lifetime exposure in seleniferous (12.3 to 13.3 μg/L) water, and have BCF values of 3759 to 5691 (Kaiser et al. 1979). BCF values were 361 to 390 for skin, and about 180 for muscle (Kaiser et al. 1979). In short-term exposures, most of the selenium was probably adsorbed to the body surface (Fowler and Benayoun 1976c), and then rapidly lost on transfer to selenium-free media (Browne and Dumont 1979). In longer exposures, the BCF values in aquatic organisms were lower after immersion in high ambient selenium concentrations over extended periods. Thus, marine crabs exposed to a water concentration of 250 μg Se/L for 29 days accumulated selenium over water concentration level by a factor of 25 for carapace, and 3.8 for gill; accumulations in muscle and hepatopancreas were negligible. Cadmium in solution enhanced selenium uptake (Bjerragaard 1982). Exposure of common carp to 1000 μg Se/L for 85 days resulted in a whole-body BCF of 6; additional studies of 7 weeks exposure plus 7 weeks postexposure at concentrations between 500 and 5000 μg Se/L (Sato et al. 1980) yielded a BCF range of 0.6 (5000 μg/L) to 1.8 (500 μg/L). Highest BCF values in carp were 50 for kidney and 80 for liver after exposure of the fish to 100 μg Se/L for 7 weeks plus 7 weeks in selenium-free media. For carp, selenium tended to accumulate in kidney, liver, gill, gall bladder, heart, bone, and muscle, in that general order (Sato et al. 1980). Studies with freshwater organisms collected from a farm pond contaminated by flyash with high selenium levels (Furr et al. 1979), and with marine bivalves and nereid worms held for 4 months in seawater flowing through coal flyash containing 6200 μg Se/L (Ryther et al. 1979), showed that accumulation was slight. Contrasted to this are the observations of Cherry et al. (1976) and Ohlendorf et al. (1986). Cherry and co-workers collected mosquitofish from a drainage system that received high coal flyash concentrations at one end and thermal discharges at the other. Mosquitofish contained up to 9.0 mg Se/kg whole-body fresh weight. Of 40 elements examined, only selenium, zinc, and calcium were accumulated in excess of the levels measured in the water. Ohlendorf et al. found mean residues of 172 mg/kg DW (range 110 to 280 mg/kg) in whole mosquitofish from irrigation drainwater ponds contaminated by about 300 μg Se/L; based on a wet/dry factor of 4, the BCF for whole mosquitofish was >91.

Selenium accumulation is modified by water temperature, age of the organism, organ or tissue specificity, mode of administration, chemical species, and other factors. In freshwater fishes, selenium concentrations in tissues increased with increasing exposure (Hamilton and Wiedmeyer 1990) and increasing dose (Hamilton and Wiedmeyer 1990; Cleveland et al. 1993), and decreased with increasing water hardness (Hamilton and Wiedmeyer 1990). Food chain accumulation of selenium can severely affect reproductive success of bluegills (Coyle et al. 1993). Bluegills tend to accumulate inorganic selenium compounds through the diet, and organoselenium compounds through the medium as well as the diet (Besser et al. 1993). Dietary selenite accumulates in liver and gonads, and selenomethionine in muscle and whole fish (Gillespie et al. 1988; Lorentzen et al. 1994).

Diet is the major route of selenium intake by mussels (*Mytilus edulis*). Studies with radioselenium-75 demonstrated that selenium efficiency was 28 to 34% from the diet and 0.03% from the medium (Wang et al. 1996). Authors concluded that 96% of the selenium in mussels is obtained from ingested food under conditions typical of coastal waters (Wang et al. 1996). In the marine mussel *Mytilus galloprovincialis*, an increase in water temperature from 13 to 29°C doubled the bioconcentration factor (BCF) in 13 days (Fowler and Benayoun 1976b). Mussels preferentially accumulated selenite over selenate (Fowler and Benayoun 1976b); however, mussels did not reach a steady state in 63 days (Fowler and Benayoun 1976c), indicating that selenium kinetics in some species are difficult to elucidate in short-term studies. Accumulation rates were higher in small than in large mussels (Fowler and Benayoun 1976b), as they were in freshwater teleosts (Furr et al. 1979). However, the reverse was documented for marine mammals and teleosts (Eisler 1984). When selenium was available from both the diet and the medium, concentrations were highest in liver, kidney, and gills of teleosts (Sorensen et al. 1982; Furr et al. 1979; Kaiser et al. 1979), exoskeleton of crustaceans (Fowler and Benayoun 1976c; Bjerragaard 1982), and visceral mass and gills of molluscs (Fowler and Benayoun 1976c). When selenium was administered in food to marine shrimps, concentrations were highest in viscera and exoskeleton, suggesting that ingested selenium is readily translocated from internal to external tissues (Fowler and Benayoun 1976c). Concentrations of selenium in crustaceans usually were higher in fecal pellets than in the diet; fecal pellets may represent a possible biological mechanism for downward vertical transport of selenium in the sea (Fowler and Benayoun 1976a), as well as in freshwater environments.

The time for 50% excretion of accumulated selenium was found to range from 13 to 181 days in various species of marine and freshwater fauna. Biological half-life of selenium accumulated from the medium was estimated at 28 days for carp (Sato et al. 1980), 37 days for the marine euphausid crustacean *Meganctiphanes norvegica* (Fowler and Benayoun 1976a), 63 to 81 days for the marine mussel *Mytilus galloprovincialis* (Fowler and Benayoun 1976b), 58 to 60 days for the marine shrimp *Lysmata caudata* (Fowler and Benayoun 1976b), and, as reviewed by Stadtman (1974, 1977), 13 days for guppies, 27 days for eels, and 28 days for leeches. Studies by Lemly (1982a) with bluegills and largemouth bass showed elevated tissue levels after exposure to 10 μg Se/L for 120 days. Time for 50% excretion in 30-day elimination trials was about 15 days from gill and erythrocytes; however, there was essentially no elimination from spleen, liver, kidney, or muscle. It appears that research is needed on preferential tissue retention of selenium and its implications for biochemical and metabolic transport mechanisms. Urine is a major excretory route for selenium in marine mammals. Urine of minke whales (*Balaenoptera acutorostrata*) contains 1.5 mg Se/L, or about 30 times more selenium than human urine (Hasunuma et al. 1993). There are at least five selenium components in urine of minke whales, including trimethylselonium ion; the significance of this observation is imperfectly understood (Hasunuma et al. 1993).

31.6.2 Terrestrial Invertebrates

Concentrations of selenium decreased in whole earthworms from 22.4 to 15.0 mg/kg (dry weight) as the rate of sludge application increased from 15 to 60 metric tons/ha. Concentrations of selenium in soil and sludge were 0.3 and 0.5 mg/kg dry weight, respectively (Helmke et al. 1979). Other studies indicated that some metals, notably cadmium, decreased in worms living in soils amended with sewage sludge but that selenium concentrations were not affected (Beyer et al. 1982). The biological half-life of selenium in earthworms is estimated to be 64 days, a period consistent with values of 10 to 81 days documented for ants, birds, mammals, and aquatic biota (Stadtman 1974, 1977).

31.6.3 Birds

Embryos of the domestic chicken (*Gallus domesticus)* are extremely sensitive to selenium. The hatchability of eggs is reduced by concentrations of selenium in feeds (6 to 9 mg/kg) that were too low to produce poisoning in other avian species. Dietary selenium excess was associated with decreased egg weight, decreased egg production and hatchability, anemia, elevated kidney selenium residues in chicks, and a high incidence of grossly deformed embryos with missing or distorted eyes, beaks, wings and feet (Ort and Latshaw 1978; Harr 1979). Similar results were observed by El-Bergearmi et al. (1977) in Japanese quail at 6 and 12 mg/kg dietary selenite. Ohlendorf et al. (1986) reported severe reproductive effects in ducks (*Anas* spp.), American coot (*Fulica americana*), and other species of aquatic birds nesting at irrigation drainwater ponds in the San Joaquin Valley, California. Water in these ponds contained abnormally high concentrations of about 300 μg Se/L, but low or nondetectable levels of silver, chromium, arsenic, cadmium, mercury, lead, and zinc. Of 347 nests examined from this site, about 40% had at least one dead embryo, and about 20% had at least one embryo or chick with obvious external anomalies, including missing or abnormal beaks, eyes, wings, legs, or feet. In addition, brain, heart, liver, and skeletal anomalies were recorded. Concentrations of selenium (mg/kg, dry weight) were 2 to 110 in eggs and 19 to 130 in livers of birds, 12 to 79 in plants, 23 to 200 in invertebrates, and 110 to 280 in fish from the ponds, or 7 to 130 times those found at a nearby control area. It was concluded that selenium was the probable cause of poor reproduction and developmental abnormalities in the aquatic nesting birds, due to interference with their reproductive processes. The concentrations of selenium in breast muscle of coots were sufficiently high (up to 11.0 mg Se/kg FW) to induce state agencies to post the area, advising against the consumption of more than one meal per week of this species, or of any coots by children or pregnant women (Ohlendorf et al. 1986).

Selenomethionine is the predominant form of selenium in commercial grains (Heinz et al. 1989), usually as seleno-L-methionine (Heinz et al. 1996). Selenomethionine is more teratogenic and embryotoxic to avian waterfowl than inorganic forms of selenium tested (Heinz and Fitzgerald 1993b; Heinz et al. 1996; Hoffman et al. 1996) (Table 31.4), possibly due to higher uptake of organoselenium (Hoffman and Heinz 1988). Blood selenium concentrations in American kestrels (*Falco sparverius*) seemed to reflect dietary concentrations of seleno-L-methionine (Yamamoto et al. 1998). Embryotoxic and teratogenic effects of selenomethionine were observed in mallards at dietary concentrations exceeding 4 mg Se/kg FW ration in the laboratory, causing effects similar to those found in field studies (Heinz et al. 1987, 1989; Hoffman and Heinz 1988). Excess dietary selenium, as seleno-DL-methionine has a more pronounced effect on hepatic glutathione metabolism and lipid peroxidation than selenite (Hoffman et al. 1989), and may enhance selenium accumulation (Moksnes 1983; Hoffman et al. 1989). In aquatic birds, developmental malformations were associated with lipid peroxidation in livers. Selenomethionine causes lipid peroxidation in livers of aquatic birds, and this is consistent with the observation that selenomethionine is the primary causative agent of selenium-induced embryonic mortality and overt teratogenesis in waterfowl at Kesterson Reservoir (Hoffman et al. 1988). Dietary selenomethionine effects were modified by salts of boron (Hoffman et al. 1991b; Stanley et al. 1996), arsenic (Hoffman et al. 1992a; Stanley et al. 1994), and protein composition (Hoffman et al. 1992b, 1996), with significant interactions between mixtures.

31.6.4 Mammals

Pregnant long-tailed macaques (*Macaca fascicularis*) given L-selenomethionine for 30 days at doses of 25, 150, or 300 μg Se/kg body weight daily showed dose-dependent increases in erythrocyte and plasma selenium, glutathione peroxidase activities, hair and fecal selenium, and urinary selenium excretion. Adverse effects, including body weight loss, were associated with daily doses of 150 and 300 μg Se/kg BW and concentrations of erythrocyte selenium >2.3 mg/L, plasma selenium >2.8 mg/L, and hair selenium >27 mg/kg FW (Hawkes et al. 1992). Young adult female mice (*Mus* sp.) given intraperitoneal injections of sodium selenate or selenomethionine (in each case, three injections of 2 mg Se/kg BW at 2-day intervals) had altered blood composition 24 days after the last injection. Both forms of selenium induced a transient, marked decrease in the number of circulating leukocytes (a condition known as leukopenia) following serial injections. Leukopenia was more extensive and of greater duration for selenomethionine-treated mice (Hogan 1998).

Harr (1978) and NRC (1983) summarized nonlethal effects of selenium on mammals, including reproductive anomalies. Selenosis caused congenital malformations in rats, mice, swine, and cattle. In general, young born to females with selenosis were emaciated and unable to nurse. Mice given selenium in drinking water reproduced normally for three generations, but litters were fewer and smaller when compared to controls, pups were runts with high mortality before weaning, and most survivors were infertile.

In rats, selenium did not induce cirrhosis or neoplasia; however, intestinal lesions were observed among those fed diets containing 0.8 to 1.0 mg Se/kg ration during lifetime exposure. The threshold requirement for optimal rat nutrition under similar conditions is about 0.08 mg Se/kg ration, again demonstrating the relatively narrow range separating selenium deficiency from selenium poisoning. Absorption of oral radioselenite by rats was as high as 95 to 100%. A single dose of radioselenite concentrated, in descending order of accumulation, in pancreas, intestine, erythrocytes, liver, kidney, and testes; tissue distributions from chronic exposure were similar. As expected, levels of selenium in poisoned rats were highest in liver and kidney. Rats, and probably other mammals, can regulate dietary selenium accumulations. Dietary concentrations in excess of 54 to 84 μg/kg ration were usually excreted in urine (Harr 1978; NRC 1983). Urine is the major excretory route for selenium. In urine of selenium-challenged rats, the trimethylselonium ion — a metabolic product of selenite or selenoamino acid, such as selenomethionine — is dominant (Hasunuma et al. 1993). However, when selenium intake exceeded 1000 μg/kg ration, pulmonary excretion was active (Harr 1978; NRC 1983). Excretion of selenium in feces, bile, saliva, and hair appears to be relatively constant, regardless of the amount of exposure. Yonemoto et al. (1983) demonstrated that some selenotoxic effects in mice, including abortion and maternal death, were prevented by prior treatment with Vitamin E, but exacerbated by reduced glutathione. The mechanisms for these interactions are unknown, and merit additional research.

31.7 RECOMMENDATIONS

All investigators appear to agree on four points:

1. Insufficient selenium in the diet may have harmful and sometimes fatal consequences.
2. Exposure to grossly elevated levels of selenium in the diet or water is inevitably fatal over time to terrestrial and aquatic organisms.
3. There is a comparatively narrow concentration range separating effects of selenium deficiency from those of selenosis.
4. Additional fundamental and basic research is required on selenium metabolism, physiology, recycling, interactions with other compounds or formulations, and chemical speciation in order to elucidate its nutritive role as well as its toxic effects.

Accordingly, the proposed selenium criteria shown in Table 31.5 for prevention of selenium deficiency and for protection of aquatic life, livestock, crops, and human health, should be viewed as guidelines, pending acquisition of additional, more definitive, data.

Table 31.5 Proposed Criteria for Prevention of Selenium Deficiency and for Protection Against Selenosis (Values are in mg of selenium/kg fresh weight [FW] or dry weight [DW].)

Criterion	Selenium Concentration	Reference[a]
PREVENTION OF SELENIUM DEFICIENCY		
Rainbow trout (water levels of 0.4 μg Se/L)		
Diet	>70 μg/kg FW	1
Diet	150–380 μg/kg DW	26
Poultry		
Diet	>30–50 μg/kg FW	2
Dietary supplement allowed		
Chickens, ducks	100 μg/kg DW ration as sodium selenate or selenite	37
Turkeys	200 μg/kg DW ration as sodium selenate or selenite	37
Rats		
Diet	>54–84 μg/kg FW	3
Livestock		
Diet	>20 μg/kg FW	2
Dietary supplement allowed		
Cattle, sheep, adult swine	100 μg/kg DW ration as sodium selenate or selenite	37
Weanling swine	300 μg/kg DW ration as sodium selenate or selenite	37
Forage, grazing sheep and cattle	>100 μg/kg DW	4
Blood	>40–80 μg/L FW	17
White-tailed deer		
Heart	>150 μg/kg DW	17
Kidney	>3000 μg/kg DW	17
Liver	>250 μg/kg DW	17
Serum	>30 μg/L FW	17
Humans		
Diet	40–200 μg/kg FW	3, 5
Daily intake		
Recommended		
Females	55 μg (= 1.0 μg/kg BW daily)	18, 40
Males	70 μg (= 1.0 μg/kg BW daily)	18, 40
Children	0.004 μg/kg BW	18
Maximum	500 μg	40

Table 31.5 (continued) Proposed Criteria for Prevention of Selenium Deficiency and for Protection Against Selenosis (Values are in mg of selenium/kg fresh weight [FW] or dry weight [DW].)

Criterion	Selenium Concentration	Reference[a]
Drinking water		
Bottled water	<10 µg/L FW	18
Assuming water is sole selenium source, 2 L daily	20 µg/L FW	6
PROTECTION AGAINST SELENOSIS		
Crop protection		
Irrigation water	<50 µg/L	5, 12
Aquatic life protection		
Freshwater		
Total dissolved	Average 4-day concentration not to exceed 5 µg acid-soluble Se/L more than once every 3 years; average 1-h concentration not to exceed 20 µg/L more than once every 3 years	7, 18–20
Total recoverable[b]	After filtration through 0.45-µ filter, <2 µg total Se/L, sometimes <1 µg/L for organoselenium compounds	27
Waterborne selenium	<2 µg/L	42
Adverse effects on fish reproduction possible	>2 µg inorganic Se/L; >1 µg organic Se/L; sometimes, <1 µg organic Se/L	42
Fish		
Diet		
Freshwater fishes		
Acceptable	<3000 µg/kg DW ration	42
Lethal	>6500–54,000 µg/kg DW feed	42
Growth reduction	>5000–20,000 µg/kg DW feed	42
Kidney damage	>11,000 µg/kg DW feed	42
Juvenile chinook salmon		
Adverse effects	>3000–5000 µg total Se/kg DW	21
Safe	<3000 µg total Se/kg DW	21
Rainbow trout, safe	<3000 µg/kg DW	26
Tissue residues, acceptable		
Whole body	<12,000 µg/kg DW	22
Gonads	<8000–<10,000 µg/kg DW	23, 25, 27
Carcass or ovaries, bluegills	<6000 µg/kg FW	24
Freshwater and anadromous fishes		
Whole body	<4000 µg/kg DW	27, 42
Liver	<12,000 µg/kg DW	27, 42
Muscle	<8000 µg/kg DW	27, 42
Ovaries and eggs	<10,000 µg/kg DW	42
Great Lakes		
Water	<10 µg/L	8, 18
Saltwater		
Total dissolved	Average 4-day concentration not to exceed 71 µg acid-soluble Se/L more than once every 3 years and 1-h concentration does not exceed 300 µg/L more than once every 3 years	7, 18
Birds		
Diet		
Mallard		
Maximum tolerated	<10,000 µg/kg DW	31
Reproduction inhibited	7000–11,000 µg/kg DW, as selenomethionine	33, 34, 41, 42
Malformed embryos	16,000 µg/kg DW, as selenomethionine	34

Table 31.5 (continued) Proposed Criteria for Prevention of Selenium Deficiency and for Protection Against Selenosis (Values are in mg of selenium/kg fresh weight [FW] or dry weight [DW].)

Criterion	Selenium Concentration	Reference[a]
Fatal	10,000–20,000 μg/kg, as selenomethionine	33
Water		
Sensitive fish-eating species	<0.8–<1.9 μg dissolved Se/L	28
Mallard	<2.1 μg dissolved Se/L	28
Aquatic birds, most species		
Minimal hazard	<2.3 μg dissolved Se/L	29
Hazardous	3–20 μg dissolved Se/L	29
Maximum allowable	<10.0 μg dissolved Se/L	29
Tissues		
Eggs		
Reproductive impairment		
Unlikely	<2000 μg/kg DW; <3000–<3300 μg/kg FW	27, 31, 35, 41
Possible	>1000 μg/kg FW	30
Probable	>5000 μg/kg FW; >15,000 μg/kg DW	30, 35
Liver		
Acceptable	<3000 μg/kg FW	41
Acceptable	<5200–<10,000 μg/kg DW	27, 31
Adverse effects	>10,000 μg/kg FW	41
Poisoned	>66,000 μg/kg DW	32
Lethal	>20,000 μg/kg FW	41
Mammals (non-humans)		
Livestock protection		
Drinking water	<50 μg/L FW	5, 9
Diet (total)	<2000 μg/kg DW	10
Diet (natural)	<4000 μg/kg DW	5, 39
Feeds (natural)	<2000 μg/kg DW	11
Forage (natural	<5000 μg/kg DW	11
Tissues		
Blood		
Adequate	80 μg/L FW	17
Toxic	>3000 μg/L FW	17, 36
Kidney		
Toxic	>3000–6000 μg/kg FW	36
Liver		
Toxic	>12,000–15,000 μg/kg FW	36
Monkeys		
Adverse effects		
Erythrocytes	>2300 μg/kg FW	38
Hair	>27,000 μg/kg FW	38
Plasma	>2800 μg/L FW	38
Drinking water		
River otter	<0.7 μg dissolved Se/L	28
Bats and shrews	<0.9 μg dissolved Se/L	28
Mink	<1.1 μg dissolved Se/L	28
Human health protection		
Seafood	Not to exceed 2000 μg/kg FW	13
Drinking water		
Most states	<10 μg/L FW	18
Minnesota	<20 μg/L FW	18
International	<10 μg/L FW	18
Maximum permissible[c]	<50 μg/L FW	18
Health advisory, chronic		
Child	<31 μg/L FW	18
Adult	<107 μg/L FW	18

Table 31.5 (continued) Proposed Criteria for Prevention of Selenium Deficiency and for Protection Against Selenosis (Values are in mg of selenium/kg fresh weight [FW] or dry weight [DW].)

Criterion	Selenium Concentration	Reference[a]
Food (natural)	<5000 μg/kg FW; <850 μg daily	5, 18
Milk or water	<500 μg/L FW	5
Daily intake (all sources)		
Adults		
Safe	<200 μg	4
Safe, chronic[d]	<5 μg/kg BW (= <350 μg for a 70-kg adult)	18
Normal	60–250 μg	14
Maximum tolerable level	<500 μg	15
Infants	4–<35 μg	14
Children		
Age 1–3	20–<80 μg	15
Age 4–6	30–<120 μg	15
Age 7–11+	50–<200 μg	15
Air		
Japan	<100 μg/m³	16
Russia	<100 μg/m³	16
United States	Usually <200 μg/m³; some states 2–<5 μg/m³	16, 18

[a] **1,** Hilton et al. 1980; **2,** Fishbein 1977; **3,** Harr 1978; **4,** Shamberger 1981; **5,** Wilber 1983; **6,** Robberecht and Von Grieken 1982; **7,** USEPA 1987; **8,** Wong et al. 1980; **9,** NAS 1973; **10,** NRC 1983; **11,** Frost 1972; **12,** Birkner 1978; **13,** Bebbington et al. 1977; **14,** Lo and Sandi 1980; **15,** Levander 1984; **16,** NAS 1976; **17,** Oliver et al. 1990; **18,** USPHS 1996; **19,** Allen and Wilson 1990; **20,** Coyle et al. 1993; **21,** Hamilton et al. 1990; **22,** Saiki et al. 1992; **23,** Waddell and May 1995; **24,** Gillespie 1986; **25,** Lemly 1993c; **26,** Lorentzen et al. 1994; **27,** Lemly 1993b; **28,** Peterson and Nebeker 1991; **29,** Skorupa and Ohlendorf 1991; **30,** Heinz et al. 1989; **31,** Wiemeyer and Hoffman 1996; **32,** Albers et al. 1996; **33,** Heinz and Fitzgerald 1993a; **34,** Heinz and Fitzgerald 1993b; **35,** Ohlendorf et al. 1986a; **36,** Alexander et al. 1990; **37,** U.S. Food and Drug Administration 1993; **38,** Hawkes et al. 1992; **39,** Hoffman et al. 1989; **40,** Medeiros et al. 1993; **41,** Heinz 1996; **42,** Lemly 1996b.

[b] High potential for biomagnification in aquatic food chains, dietary toxicity, and reproductive toxicity.

[c] Based on a NOAEL (no observable adverse effect level) of 400 μg Se daily, equivalent to 5.7 μg/kg BW daily for a 70-kg person.

[d] Based on a NOAEL of 15 μg/kg BW daily for dermal effects and an uncertainty factor of 3 to account for sensitive individuals (USPHS 1996).

Regarding selenium deficiency, it appears that diets containing 50 to 100 μg Se/kg ration provide adequate protection to humans and to various species of fish, small laboratory mammals, and livestock (Table 31.5). Factors contributing to selenium deficiency in crops include increasing use of agricultural fertilizers and increasing atmospheric fallout of sulfur. Furthermore, foliar applications of selenate, although efficient in raising selenium levels in plants, have only short-term value (Frost and Ingvoldstad 1975). There is a general consensus that selenium deficiency in livestock in many countries is increasing, resulting in a need for added selenium in the food chain.

Recommendations for protection of freshwater aquatic life include acid-soluble total selenium concentrations in the water of less than 5 μg/L on a daily average, or 20 μg/L at any time (Table 31.5). These values are higher for saltwater life: 71 μg/L daily average, 300 μg/L at any time. The concentration range of 5 to 20 μg/L recommended for protection of freshwater aquatic life is below the range of 60 to 600 μg/L that is fatal to various sensitive species of marine and freshwater fauna, and in this respect affords an adequate measure of protection. It is also below the range (47 to 53 μg/L) that has been associated with growth inhibition of freshwater algae, anemia and reduced hatch in rainbow trout, and shifts in species composition of freshwater algal communities. But studies by Cumbie and Van Horn (1978) and Sorensen et al. (1984) showed that water concentrations of 9 to 12 μg Se/L were associated with reduced reproduction of freshwater fishes, and their results strongly indicated that some downward modification of the selenium freshwater aquatic life protection criterion may be appropriate. Furthermore, high bioconcentration and accumulation of selenium from the water column by numerous species of algae, fish, and invertebrates is well documented at levels between 0.015 and 3.3 µg Se/L, which is substantially below the recommended range of 5 to 20 µg/L in freshwater. The significance of selenium residues in aquatic biota in terms of bioavailability and selenium receptor sites is imperfectly understood, and it appears that much additional research is warranted on formulating suitable models of selenium biogeochemistry and pharmacokinetics in aquatic environments (Hodson 1990; Bowie and Grieb 1991; Hermanutz et al. 1992).

Resource managers and aquatic biologists need data on selenium concentrations in water, food-chain organisms, and fish and wildlife tissues in order to adequately assess the overall selenium status and health of aquatic ecosystems (Lemly 1993b). Because selenium is depurated rapidly in aquatic birds, resource managers should be concerned primarily about current exposure in nature and not previous exposures (Heinz 1993b). The potential for food-chain biomagnification and reproductive impairment in fish and birds is the most sensitive biological response for estimating ecosystem-level impacts of selenium contamination (Baumann and Gillespie 1986; Lemly 1996a), and these are best reflected in selenium concentrations in gravid ovaries and eggs of adult fish and aquatic bird populations (Ohlendorf et al. 1986a; Lemly 1993b). More research is recommended on biomarkers of selenium exposure and effect (USPHS 1996).

Field studies demonstrated that migratory waterfowl were heavily and adversely affected while nesting at selenium-contaminated irrigation drainwater ponds in California, where food chain organisms contained between 12,000 and 280,000 µg Se/kg. The source and fate of selenium in irrigation drainwater ponds are largely unknown; they must be determined so that alternate technologies for selenium control can be implemented to protect waterfowl in that geographical region.

Livestock appear to be protected against selenosis provided that their diets contain less than 4000 µg Se/kg of natural (i.e., nonsupplemented) selenium (Table 31.5). This concentration is somewhat higher than levels reported for rats (Wilber 1983). Minimum toxic concentrations of selenium in lifetime exposure of rats given diets containing natural selenium were 1400 µg/kg ration as judged by evidence of liver changes, and 3000 µg/kg ration as estimated from longevity and histological changes in heart, kidney, and spleen. These values were only 350 and 750 µg/kg ration, respectively, when rat diets contained purified, rather than natural, selenium. This relationship emphasizes that accidental poisoning of livestock, and presumably fish and wildlife, may occur when soils are deliberately supplemented with purified selenium, or when soils or aquifers are contaminated as a result of faulty waste disposal practices. Although the concentration of <50 µg Se/L for livestock drinking water and irrigation water for crop protection is inconsistent with that of <35 µg Se/L for aquatic life protection, neither livestock nor crops appear threatened at the higher level. Since many waterways that abut agricultural lands or areas of high anthropogenic loadings of selenium contain valuable and desirable aquatic species, it would appear that some downward modification of the current livestock drinking water concentration of selenium is necessary. Selenium is a proven teratogen in fishes and birds, but this has not been established in mammals (USPHS 1996). More research is needed on possible teratogenic effects of organoselenium compounds in mammals.

Acute lethal doses for livestock species ranged from 3300 µg Se/kg body weight for horses and mules to 15,000 µg/kg for swine; appetite loss in cattle was noted at 500 µg/kg. For humans, the maximum tolerance level is usually set at 500 µg Se daily (Lo and Sandi 1980), and the "safe" level at 200 µg/day (Shamberger 1981). Selenium dietary levels for humans should not exceed 5000 µg/kg ration; however, recommended maximum dietary levels in other mammals ranged from 1000 µg/kg for rats to 4000 µg/kg for horses. For all species, including humans, there is a tendency to list selenium dosage levels in terms of "natural" and "supplemented" levels, with the tacit understanding that "natural" levels are about one fourth as toxic as "supplemented" values. Given the complexities of selenium metabolism and speciation, it appears that greater precision and clarity are necessary in formulating selenium criteria if these criteria are to become administratively enforced standards through passage of appropriate legislation.

Aerosol concentrations in excess of 4.0 μg Se/m^3 are potentially harmful to human health (Harr 1978). Concentrations in excess of this value (6.0 μg Se/m^3) were regularly encountered in the vicinity of the smeltery at Sudbury, Ontario, Canada (Nriagu and Wong 1983). It is not now known whether respiration rates of wildlife, particularly birds, are comparable to those of humans, whether selenium absorption energetics are similar, or whether wildlife species that frequent point sources of air contaminated by high selenium levels for protracted periods are at greater risk than humans. Until additional and more conclusive data become available, aerosol concentrations of less than 4.0 μg Se/m^3 are recommended for the protection of sensitive wildlife species.

31.8 SUMMARY

Most authorities agree on five points.

1. Selenium deficiency is not as well documented as selenosis, but may be equally significant.
2. Selenium released as a result of anthropogenic activities (including fossil fuel combustion and metal smelting), as well as that in naturally seleniferous areas, poses the greatest threat of poisoning to fish and wildlife.
3. Additional research is required on chemical and biological transformations among valence states, allotropic forms, and isomers of selenium.
4. Metabolism and degradation of selenium are both significantly modified by interaction with various heavy metals, agricultural chemicals, microorganisms, and numerous physicochemical factors; and until these interactions are resolved, it will be difficult to meaningfully interpret selenium residues in various tissues.
5. Documented biological responses to selenium deficiency or to selenosis vary widely, even among closely related taxonomic groups.

It is generally agreed that selenium deficiency can be prevented in fish, small laboratory mammals, and livestock by feeding diets containing 50 to 100 μg Se/kg. The concentration range of total acid-soluble selenium currently recommended for aquatic life protection — 5 μg/L in freshwater to 71 μg/L in marine waters — is below the range of 60 to 600 μg/L that is fatal to sensitive aquatic species. In freshwater, it is also below the range of 47 to 53 μg/L associated with growth inhibition of freshwater algae, anemia and reduced hatching in trout, and shifts in species composition of freshwater algae communities. Accordingly, current recommendations for selenium with respect to aquatic life appear to afford an adequate measure of protection. However, some studies have shown that water concentrations of 9 to 12 μg Se/L are associated with inhibited reproduction of certain freshwater teleosts, suggesting that selenium criteria for protection of freshwater life should be revised downward. Also, high bioconcentration and accumulation of selenium from water by numerous species of algae, fish, and invertebrates is well documented at levels of 0.015 to 3.3 μg Se/L, which are substantially below the recommended range of 5 to 71 μg Se/L. The significance of selenium residues in aquatic biota is still unclear, and more research appears to be needed on selenium pharmacokinetics in aquatic environments.

Aerosol concentrations exceeding 4.0 μg Se/m^3 are considered potentially harmful to human health. However, no comparable database for birds and other wildlife species is available at this time. Selenium poisoning in livestock is prevented if diets do not exceed 5.0 mg Se/kg natural forage, or 2.0 mg Se/kg in feeds supplemented with purified selenium. Minimum toxic concentrations of selenium in the rat (a sensitive species) fed diets containing natural selenium were 1400 μg Se/kg as judged by evidence of liver changes, and 3000 μg Se/kg as estimated from longevity and histopathology. These values were only 350 and 750 μg/kg, respectively, when diets low in natural selenium were fortified with purified selenium. The evidence is incomplete for migratory waterfowl and other birds, but diets containing more than 3.0 mg Se/kg are demonstrably harmful, as are total selenium concentrations in excess of 5 mg/kg FW in eggs and 10 mg/kg DW in livers.

31.9 LITERATURE CITED

Adams, W.J. 1976. The Toxicity and Residue Dynamics of Selenium in Fish and Aquatic Invertebrates. Ph.D. thesis, Michigan State Univ., East Lansing. 109 pp.

Adams, W.J. and H.E. Johnson. 1977. Survey of the selenium content in the aquatic biota of western Lake Erie. *Jour. Great Lakes Res.* 3:10-14.

Albers, P.H., D.E. Green, and C.J. Sanderson. 1996. Diagnostic criteria for selenium toxicosis in aquatic birds: dietary exposure, tissue concentrations, and macroscopic effects. *Jour. Wildl. Dis.* 32:468-485; *Correction. Jour. Wildl. Dis.* 32:725-726.

Alexander, J.W., M.A. Solangi, W.C. Edwards, and D. Whitenack. 1990. Selenium toxicosis in two California sea lions (*Zalophus californianus*). *Int. Assoc. Aquatic Animal Med. (IAAAM) Proc.* 21:25-28.

Allen, G.T. and R.M. Wilson. 1990. Selenium in the aquatic environment of Quivra National Wildlife Refuge. *Prairie Natural.* 22:129-135.

Amiard-Triquet, C., D. Pain, and H.T. Delves. 1991. Exposure to trace elements of flamingos living in a biosphere reserve, The Camargue (France). *Environ. Pollut.* 69:193-201.

Anonymous. 1975. Preliminary Investigation of Effects on the Environment of Boron, Indium, Nickel, Selenium, Tin, Vanadium and Their Compounds. Vol. IV. Selenium. U.S. Environ. Protection Agency Rep. 560/2-75-005d. Available as PB — 245 987 from Nat. Tech. Inform. Serv., U.S. Dept. Commerce, Springfield, VA. 92 pp.

Baumann, P.C. and R.B. Gillespie. 1986. Selenium bioaccumulation in gonads of largemouth bass and bluegill from three power plant cooling reservoirs. *Environ. Toxicol. Chem.* 5:695-701.

Beal, A.R. 1974. A Study of Selenium Levels in Freshwater Fishes of Canada's Central Region. Canada Dep. Environ. Fish. Mar. Serv., Tech. Rep. Ser. CEN/T-74-6:14 pp.

Bebbington, G.N., N.J. Mackay, R. Chvojka, R.J. Williams, A. Dunn, and E.H. Auty. 1977. Heavy metals, selenium and arsenic in nine species of Australian commercial fish. *Austral. Jour. Mar. Freshwater Res.* 28:277-286.

Beijer, K. and A. Jernelov. 1978. Ecological aspects of mercury-selenium interactions in the marine environment. *Environ. Health Perspec.* 25:43-45.

Bertine, K.K. and E.D. Goldberg. 1972. Trace elements in clams, mussels, and shrimp. *Limnol. Ocean.* 17:877-884.

Besser, J.M., T.J. Canfield, and T.W. La Point. 1993. Bioaccumulation of organic and inorganic selenium in a laboratory food chain. *Environ. Toxicol. Chem.* 12:57-72.

Besser, J.M., J.N. Huckins, and R.C. Clark. 1994. Separation of selenium species released from Se-exposed algae. *Chemosphere* 29:771-780.

Beyer, W.N., R.L. Chaney, and B.M. Mulhern. 1982. Heavy metal concentrations in earthworms from soil amended with sewage sludge. *Jour. Environ. Qual.* 11:381-385.

Birkner, J.H. 1978. Selenium in Aquatic Organisms from Seleniferous Habitats. Ph.D. thesis, Colorado State Univ., Fort Collins. 121 pp.

Bjerragaard, P. 1982. Accumulation of cadmium and selenium and their mutual interaction in the shore crab *Carcinus maenus* (L.). *Aquat. Toxicol.* 2:113-125.

Blus, L.J., B.S. Neely, Jr., T.G. Lamont, and B. Mulhern. 1977. Residues of organochlorines and heavy metals in tissues and eggs of brown pelicans, 1969-73. *Pestic. Monitor. Jour.* 11:40-53.

Boisson, F., M. Gnassia-Barelli, and M. Romeo. 1995. Toxicity and accumulation of selenite and selenate in the unicellular marine algae *Cricosphaera elongata. Arch. Environ. Contamin. Toxicol.* 28:497-493.

Bovee, E.C. and T.L. O'Brien. 1982. Some effects of selenium, vanadium and zirconium on the swimming rate of *Tetrahymena pyriformis*: a bioassay study. *Univ. Kans. Sci. Bull.* 52 (4):39-44.

Bowen, H.J.M. 1966. *Trace Elements in Biochemistry.* Academic Press, New York. 241 pp.

Bowie, G.L. and T.M. Grieb. 1991. A model framework for assessing the effects of selenium on aquatic ecosystems. *Water Air Soil Pollut.* 57-58:13-22.

Browne, C.L. and J.N. Dumont. 1979. Toxicity of selenium to developing *Xenopus laevis* embryos. *Jour. Toxicol. Environ. Health* 5:699-709.

Buhl, K.J. and S.J. Hamilton. 1996. Toxicity of inorganic contaminants, individually and in environmental mixtures, to three endangered fishes (Colorado squawfish, bonytail, and razorback sucker). *Arch. Environ. Contamin. Toxicol.* 30:84-92.

Burger, J. and M. Gochfeld. 1992. Heavy metal and selenium concentrations in black skimmers (*Rynchops niger*): gender differences. *Arch. Environ. Contamin. Toxicol.* 23:431-434.

Burger, J. and M. Gochfeld. 1993. Heavy metal and selenium levels in feathers of young egrets and herons from Hong Kong and Szechuan, China. *Arch. Environ. Contamin. Toxicol.* 25:322-327.

Burger, J. and M. Gochfeld. 1995. 1995. Heavy metal and selenium concentrations in eggs of herring gulls (*Larus argentatus*): temporal differences from 1989 to 1994. *Arch. Environ. Contamin. Toxicol.* 29:192-197.

Burger, J. and M. Gochfeld. 1996. Heavy metal and selenium levels in Franklin's gull (*Larus pipixcan*) parents and their eggs. *Arch. Environ. Contamin. Toxicol.* 30:487-491.

Burger, J., I.C.T. Nisbet, and M. Gochfeld. 1994. Heavy metal and selenium levels in feathers of known-aged common terns (*Sterna hirundo*). *Arch. Environ. Contamin. Toxicol.* 26:351-355.

Burger, J., K. Parsons, T. Benson, T. Shukla, D. Rothstein, and M. Gochfeld. 1992. Heavy metal and selenium levels in young cattle egrets from nesting colonies in the northeastern United States, Puerto Rico and Egypt. *Arch. Environ. Contamin. Toxicol.* 23:435-439.

Burger, J., J.A. Rodgers, Jr., and M. Gochfeld. 1993a. Heavy metal and selenium levels in endangered wood storks *Mycteria americana* from nesting colonies in Florida and Costa Rica. *Arch. Environ. Contamin. Toxicol.* 24:417-420.

Burger, J., S. Seyboldt, N. Morganstein, and K. Clark. 1993b. Heavy metals and selenium in feathers of three shorebird species from Delaware Bay. *Environ. Monitor. Assess.* 28:189-198.

Burger, J. and J. Snodgrass. 1998. Heavy metals in bullfrog (*Rana catesbeiana*) tadpoles: effects of depuration before analysis. *Environ. Toxicol. Chem.* 17:2203-2209.

Cappon, C.J. and J.C. Smith. 1981. Mercury and selenium content and chemical form in fish muscle. *Arch. Environ. Contam. Toxicol.* 10:305-319.

Cappon, C.J. and J.C. Smith. 1982. Chemical form and distribution of selenium in edible seafood. *Jour. Anal. Toxicol.* 6:10-21.

Cardwell, R.D., D.G. Foreman, T.R. Payne, and D.J. Wilbur. 1976. Acute toxicity of selenium dioxide to freshwater fishes. *Arch. Environ. Contam. Toxicol.* 4:129-144.

Chapman, D.C. 1992. Failure of gas bladder inflation in striped bass: effect on selenium toxicity. *Arch. Environ. Contamin. Toxicol.* 22:296-299.

Chau, Y.K. and J.P. Riley. 1965. The determination of selenium in sea water, silicates and marine organisms. *Anal. Chim. Acta* 33:36-49.

Chau, Y.K., P.T.S. Wong, B.A. Silverberg, P.L. Luxon, and G.A. Bengert. 1976. Methylation of selenium in the aquatic environment. *Science* 192:1130-1131.

Cherry, D.S., R.K. Guthrie, J.H. Rodgers, Jr., J. Cairns, Jr., and K.L. Dickson. 1976. Responses of mosquitofish (*Gambusia affinis*) to ash effluent and thermal stress. *Trans. Amer. Fish. Soc.* 105:686-694.

Chvojka, R., R.J. Williams, and S. Fredrickson. 1990. Methylmercury, total mercury, and selenium in snapper from two areas of the New South Wales Coast, Australia. *Mar. Pollut. Bull.* 21:570-573.

Clark, D.R. Jr. 1987. Selenium accumulation in mammals exposed to contaminated California irrigation drainwater. *Sci. Total Environ.* 66:147-168.

Clark, D.R., Jr. K.S. Foerster, C.M. Marn, and R.L. Hothem. 1992. Uptake of environmental contaminants by small mammals in pickleweed habitats at San Francisco Bay, California. *Arch. Environ. Contamin. Toxicol.* 22:389-396.

Cleveland, L., E.E. Little, D.R. Buckler, and R.H. Wiedmeyer. 1993. Toxicity and bioaccumulation of waterborne and dietary selenium in juvenile bluegill (*Lepomis macrochirus*). *Aquat. Toxicol.* 27:265-280.

Coyle, J.J., D.R. Buckler, C.G. Ingersoll, J.F. Fairchild, and T.W. May. 1993. Effect of dietary selenium on the reproductive success of bluegills (*Lepomis macrochirus*). *Environ. Toxicol. Chem.* 12:551-565.

Crane, M., T. flower, D. Holmes, and S. Watson. 1992. The toxicity of selenium in experimental freshwater ponds. *Arch. Environ. Contamin. Toxicol.* 23:440-452.

Cumbie, P.M. and S.L. Van Horn. 1978. Selenium accumulation associated with fish mortality and reproductive failure. *Proc. Annu. Conf. Southeast Assoc. Fish Wildl. Agen.* 32:612-624.

Custer, T.W. and J.P. Meyers. 1990. Organochlorines, mercury, and selenium in wintering shorebirds from Washington and California. *Calif. Fish Game* 76:118-125.

Custer, T.W. and C.A. Mitchell. 1991. Contaminant exposure of willets feeding in agricultural drainages of the lower Rio Grande Valley of South Texas. *Environ. Monitor. Assess.* 16:189-200.

Custer, T.W. and C.A. Mitchell. 1993. Trace elements and organochlorines in the shoalgrass community of the Lower Laguna Madre. *Environ. Monitor. Assess.* 25:235-246.

de Goeij, J.J.M., V.P. Guinn, D.R. Young, and A.J. Mearns. 1974. Neutron activation analysis trace-element, studies of dover sole liver and marine sediments. Pages 189-200 in *Comparative Studies of Food and Environmental Contamination.* Int. Atom. Ener. Agen., Vienna.

Dierenfeld, E.S. and D.A. Jessup. 1990. Variation in serum of alpha-tocopherol, retinol, and selenium of free-ranging mule deer (*Odocoileus hemionus*). *Jour. Zoo Wildl. Medic.* 21:425-432.
Dietz, R., F. Riget, and P. Johansen. 1996. Lead, cadmium, mercury and selenium in Greenland marine animals. *Sci. Total Environ.* 186:67-93.
Domingo, J.L. 1994. Metal-induced developmental toxicity in mammals: a review. *Jour. Toxicol. Environ. Health* 42:123-141.
Duncan, D.A. and J.F. Klaverkamp. 1983. Tolerance and resistance to cadmium in white suckers *Catostomus commersoni* previously exposed to cadmium, mercury, zinc, or selenium. *Canad. Jour. Fish. Aquat. Sci.* 40:128-138.
Ebens, R.J. and H.T. Shacklette. 1982. Geochemistry of Some Rocks, Mine Spoils, Stream Sediments, Soils, Plants, and Waters in the Western Energy Region of the Conterminous United States. U.S. Geol. Surv. Prof. Paper 1237. U.S. Govt. Printing Off., Washington, D.C. 173 pp.
Eisler, R. 1981. *Trace Metal Concentrations in Marine Organisms.* Pergamon Press, New York. 687 pp.
Eisler, R. 1984. Trace metal changes associated with age of marine vertebrates. *Biol. Trace Elem. Res.* 6:165-180.
Eisler, R. 1985. Selenium Hazards to Fish, Wildlife, and Invertebrates: A Synoptic Review. U.S. Fish Wildl. Serv. Biol. Rep. 85 (1.5). 57 pp.
El-Begearmi, M.M., M.L. Sunde, and H. E. Ganther. 1977. A mutual protective effect of mercury and selenium in Japanese quail. *Poult. Sci.* 56:313-322.
El-Begearmi, M.M., H.E. Ganther, and M.L. Sunde. 1982. Dietary interaction between methylmercury, selenium, arsenic, and sulfur amino acids in Japanese quail. *Poult. Sci.* 61:272-279.
Ellis, M.M., H.L. Motley, M.D. Ellis, and R.O. Jones. 1937. Selenium poisoning in fishes. *Proc. Soc. Exp. Biol. Med.* 36:519-522.
Fairbrother, A. and J. Fowles. 1990. Subchronic effects of sodium selenite and selenomethionine on several immune-functions in mallards. *Arch. Environ. Contamin. Toxicol.* 19:836-844.
Felton, S.P., W. Ji, and S.B. Mathews. 1990. Selenium concentrations in coho salmon outmigrant smolts and returning adults: a comparison of wild versus hatchery-reared fish. *Dis. Aquat. Organ.* 9:157-161.
Fishbein, L. 1977. Toxicology of selenium and tellurium. Pages 191-240 in R.A. Goyer and M.A. Mehlman (eds.). *Advances in Modern Toxicology. Vol. 2, Toxicology of Trace Elements.* Hemisphere Publ. Corp., Washington, D.C.
Fishbein, L. 1983. Environmental selenium and its significance. *Fundam. Appl. Toxicol.* 3:411-419.
Fleming, W.J., W.H. Gutenmann, and D.J. Lisk. 1979. Selenium in tissues of woodchucks inhabiting fly ash landfills. *Bull. Environ. Contam. Toxicol.* 21:1-3.
Flueck, W.T. 1994. Effect of trace elements on population dynamics: selenium deficiency in free-ranging black-tailed deer. *Ecology* 75:807-812.
Flueck, W.T. and J.M. Smith-Flueck. 1990. Selenium deficiency in deer: the effect of a declining selenium cycle? *Trans. 19th Inter. Union Game Biol.,* Trondheim, Norway 1989:292-309.
Fowler, S.W. and G. Benayoun. 1976a. Selenium kinetics in marine zooplankton. *Mar. Sci. Comm.* 2:43-67.
Fowler, S.W. and G. Benayoun. 1976b. Influence of environmental factors on selenium flux in two marine invertebrates. *Mar. Biol.* 37:59-68.
Fowler, S.W. and G. Benayoun. 1976c. Accumulation and distribution of selenium in mussel and shrimp tissues. *Bull. Environ. Contam. Toxicol.* 16:339-346.
Freeman, H.C., G. Shum, and J.F. Lithe. 1978. The selenium content in swordfish (*Xiphias gladius*) in relation to total mercury content. *Jour. Environ. Sci. Health* A13(3):235-240.
Frost, D.V. 1972. The two faces of selenium. Can selenophobia be cured? *CRC Crit. Rev. Toxicol.* 1:467-514.
Frost, D.V. and D. Ingvoldstad. 1975. Ecological aspects of selenium and tellurium in human and animal health. *Chem. Scr.* 8A:96-107.
Fukai, R., B. Oregioni, and D. Vas. 1978. Interlaboratory comparability of measurements of trace elements in marine organisms: results of intercalibration exercise on oyster homogenate. *Oceanol. Acta* 1:391-396.
Furr, A.K., T.F. Parkinson, W.D. Youngs, C.O. Berg, W.H. Gutenmann, I.S. Pakkala, and D.J. Lisk. 1979. Elemental content of aquatic organisms inhabiting a pond contaminated with coal fly ash. *N.Y. Fish Game Jour.* 26:154-161.
Ganther, H.E., C. Goudie, M.L. Sunde, M.J. Kopecky, P. Wagner, S.H. Oh, and W.G. Hoekstra.1972. Selenium: relation to decreased toxicity of methylmercury added to diets containing tuna. *Science* 175:1122-1124.
Ghebremeskel, K., G. Williams, R.A. Brett, R. Burek, and L.S. Harbige. 1991. Nutrient composition of plants most favoured by black rhinoceros (*Diceros bicornis*) in the wild. *Comp. Biochem. Physiol.* 98A:529-534.

Gillespie, R.B. 1986. Effects of high tissue concentrations of selenium on reproduction by bluegills. *Trans. Amer. Fish. Soc.* 115:208-213.

Gillespie, R.B., P.C. Baumann, and C.T. Singley. 1988. Dietary exposure of bluegills (*Lepomis macrochirus*) to (75)Se: uptake and distribution in organs and tissues. *Bull. Environ. Contamin. Toxicol.* 40:771-778.

Glickstein, N. 1978. Acute toxicity of mercury and selenium to *Crassostrea gigas* embryos and *Cancer magister* larvae. *Mar. Biol.* 49:113-117.

Glover, J.W. 1979. Concentrations of arsenic, selenium and ten heavy metals in school shark, *Galeorhinus australis* (Macleay), and gummy shark, *Mustelus antarcticus* Gunt, in south-eastern Australian waters. *Austral. Jour. Mar. Freshwater Res.* 30:505-510.

Goede, A.A. 1991. The variability and significance of selenium concentrations in shorebird feathers. *Environ. Monitor. Assess.* 18:203-210.

Goede, A.A. 1993. Selenium in marine waders. Ph.D. thesis. Delft Univ. Technology, The Netherlands. 159 pp.

Goede, A.A. and H.T. Wolterbeek. 1994. The possible role of selenium in antioxidation in marine waders; a preliminary study. *Sci. Total Environ.* 144:241-246.

Goede, A.A., H.T. Wolterbeek, and M.J. Koese. 1993. Selenium concentrations in the marine invertebrates *Macoma balthica*, *Mytilus edulis*, and *Nereis diversicolor*. *Arch. Environ. Contamin. Toxicol.* 25:85-89.

Gotsis, O. 1982. Combined effects of selenium/mercury and selenium/copper on the cell population of the alga *Dunaliella minuta*. *Mar. Biol.* 71:217-222.

Green, D.E. and P.H. Albers. 1997. Diagnostic criteria for selenium toxicosis in aquatic birds: histologic lesions. *Jour. Wildl. Dis.* 33:385-404.

Grimanis, A.P., D. Zafiropoulos, and M. Vassilaki-Grimani. 1978. Trace elements in the flesh and liver of two fish species from polluted and unpolluted areas of the Aegean Sea. *Environ. Sci. Technol.* 12:723-726.

Hall, R.A., E.G. Zook, and G.M. Meaburn. 1978. National Marine Fisheries Service Survey of Trace Elements in the Fishery Resource. U.S. Dept. Commer. NOAA Tech. Rep. NMFS SSRF-721. 313 pp.

Halter, M.T., W.J. Adams, and H.E. Johnson. 1980. Selenium toxicity to *Daphnia magna*, *Hyallela azteca*, and the fathead minnow in hard water. *Bull. Environ. Contam. Toxicol.* 24:102-107.

Hamilton, S.J. 1995. Hazard assessment of inorganics to three endangered fish in the Green River, Utah. *Ecotoxicol. Environ. Safety* 30:134-142.

Hamilton, S.J. and K.J. Buhl. 1990. Acute toxicity of boron, molybdenum, and selenium to fry of chinook salmon and coho salmon. *Arch. Environ. Contamin. Toxicol.* 19:366-373.

Hamilton, S.J., K.J. Buhl, N.L. Faerber, R.H. Wiedmeyer, and F.A. Bullard. 1990. Toxicity of organic selenium in the diet to chinook salmon. *Environ. Toxicol. Chem.* 9:347-358.

Hamilton, S.J. and B. Waddell. 1994. Selenium in eggs and milt of razorback sucker (*Xyrauchen texanus*) in the middle Green River, Utah. *Arch. Environ. Contamin. Toxicol.* 27:195-201.

Hamilton, S.J. and R.H. Wiedmeyer. 1990. Concentrations of boron, molybdenum, and selenium in chinook salmon. *Trans. Amer. Fish. Soc.* 119:500-510.

Harr, J.R. 1978. Biological effects of selenium. Pages 393-426 in F.W. Oehme (ed.). *Toxicity of Heavy Metals in the Environment. Part l.* Marcel Dekker, New York.

Harrison, S.E., J.F. Klaverkamp, and R.J. Hesslein. 1990. Fates of metal radiotracers added to a whole lake: accumulation in fathead minnow (*Pimephales promelas*) and lake trout (*Salvelinus namaycush*). *Water Air Soil Pollut.* 52:277-293.

Haseltine, S.D., G.H. Heinz, W.L. Reichel, and J.F. Moore. 1981. Organochlorine and metal residues in eggs of waterfowl nesting on islands in Lake Michigan off Door County, Wisconsin, 1977–78. *Pestic. Monitor. Jour.* 15:90-97.

Haseltine, S.D., J.S. Fair, S.A. Sutcliffe, and D.M. Swineford. 1983. Trends in organochlorine and mercury residues in common loon (*Gavia immer*) eggs from New Hampshire. Pages 131-141 in R.H. Yahner (ed.), *Trans. Northeast Sec., Wildl. Soc., 40th Northeast Fish Wildl. Confer.*, May 15–18, 1983, West Dover, VT.

Hasunuma, R., T. Ogawa, Y. Fujise, and Y. Kawanashi. 1993. Analysis of selenium metabolites in urine samples of minke whales (*Balaenoptera acutorostrata*) using ion exchange chromatography. *Comp. Biochem. Physiol.* 104C:87-89.

Hawkes, W.C., C.C. Willhite, K.A. Craig, S.T. Omaye, D.N. Cox, W.N. Choy, and A.G. Hendrickx. 1992. Effects of excess selenomethionine on selenium status indicators in pregnant long-tailed macaques (*Macaca fascicularis*). *Biol. Trace Elem. Res.* 35:281-297.

Hein, R.G., P.A. Talcott, J.L. Smith, and W.L. Myers. 1994. Blood selenium values of selected wildlife populations in Washington. *Northwest Sci.* 68:185-188.

Heinz, G.H. 1993a. Selenium accumulation and loss in mallard eggs. *Environ. Toxicol. Chem.* 12:775-778.

Heinz, G.H. 1993b. Re-exposure of mallards to selenium after chronic exposure. *Environ. Toxicol. Chem.* 12:1691-1694.

Heinz, G.H. 1996. Selenium in birds. Pages 447-458 in W.N. Beyer, G.H. Heinz, and A.W. Redmon-Norwood (eds.). *Environmental Contaminants in Wildlife: Interpreting Tissue Concentrations.* CRC Press, Boca Raton, FL.

Heinz, G.H. and M.A. Fitzgerald. 1993a. Overwinter survival of mallards fed selenium. *Arch. Environ. Contamin. Toxicol.* 25:90-94.

Heinz, G.H. and M.A. Fitzgerald. 1993b. Reproduction of mallards following overwinter exposure to selenium. *Environ. Pollut.* 81:117-122.

Heinz, G.H., D.J. Hoffman, and L.G. Gold. 1988. Toxicity of organic and inorganic selenium to mallard ducklings. *Arch. Environ. Contamin. Toxicol.* 17:561-568.

Heinz, G.H., D.J. Hoffman, and L.G. Gold. 1989. Impaired reproduction of mallards fed an organic form of selenium. *Jour. Wildl. Manage.* 53:418-428.

Heinz, G.H., D.J. Hoffman, and L.J. LeCaptain. 1996. Toxicity of seleno-L-methionine, seleno-DL-methionine, high selenium wheat, and selenized yeast to mallard ducklings. *Arch. Environ. Contamin. Toxicol.* 30:93-99.

Heinz, G.H., D.J. Hoffman, A.J. Krynitsky, and D.M.G. Weller. 1987. Reproduction in mallards fed selenium. *Environ. Toxicol. Chem.* 6:423-433.

Heinz, G.H., G.W. Pendleton, A.J. Krynitsky, and L.G. Gold. 1990. Selenium accumulation and elimination in mallards. *Arch. Environ. Contamin. Toxicol.* 19:374-379.

Heinz, G.H. and C.J. Sanderson. 1990. Avoidance of selenium-treated food by mallards. *Environ. Toxicol. Chem.* 9:1155-1158.

Heit, M. 1979. Variability of the concentrations of seventeen trace elements in the muscle and liver of a single striped bass, *Morone saxatilis. Bull. Environ. Contam. Toxicol.* 23:1-5.

Heit, M., C.S. Klusek, and K.M. Miller. 1980. Trace element, radionuclide, and polynuclear aromatic hydrocarbon concentrations in Unionidae mussels from northern Lake George. *Environ. Sci. Technol.* 14:465-468.

Helmke, P.A., W.P. Robarge, R.L. Korotev, and P.J. Schomberg. 1979. Effects of soil-applied sewage sludge on concentrations of elements in earthworms. *Jour. Environ. Qual.* 8:322-327.

Henny, C.J., L.J. Blus, and R.A. Grove. 1990. Western grebe, *Aechmophorus occidentalis*, wintering biology and contaminant accumulation in Commencement Bay, Puget Sound, Washington. *Canad. Field-Natural.* 104:460-472.

Henny, C.J. and G.B. Herron. 1989. DDE, selenium, mercury, and white-faced ibis reproduction at Carson Lake, Nevada. *Jour. Wildl. Manage.* 53:1032-1045.

Hermanutz, R.O. 1992. Malformation of the fathead minnow (*Pimephales promelas*) in an ecosystem with elevated selenium concentrations. *Bull. Environ. Contamin. Toxicol.* 49:290-294.

Hermanutz, R.O., K.N. Allen, T.H. Roush, and S.F. Hedtke. 1992. Effects of elevated selenium concentrations on bluegills (*Lepomis macrochirus*) in outdoor experimental streams. *Environ. Toxicol. Chem.* 11:217-224.

Hill, C.H. 1975. Interrelationships of selenium with other trace elements. *Fed. Proc.* 34:2096-2100.

Hilton, J.W., P.V. Hodson, and S.J. Slinger. 1980. The requirement and toxicity of selenium in rainbow trout (*Salmo gairdneri*). *Jour. Nutr.* 110:2527-2535.

Hodson, P. 1990. Indicators of ecosystem health at the species level and the example of selenium effects on fish. *Environ. Monitor. Assess.* 15:241-254.

Hodson, P.V., D.J. Spry, and B.R. Blunt. 1980. Effects on rainbow trout (*Salmo gairdneri*) of a chronic exposure to waterborne selenium. *Canad. Jour. Fish. Aquat. Sci.* 37:233-240.

Hoffman, D.J. and G.H. Heinz. 1988. Embryotoxic and teratogenic effects of selenium in the diet of mallards. *Jour. Toxicol. Environ. Health* 24:477-490.

Hoffman, D.J., G.H. Heinz, and A.J. Krynitsky. 1989. Hepatic glutathione metabolism and lipid peroxidation in response to excess dietary selenomethionine and selenite in mallard ducklings. *Jour. Toxicol. Environ. Health* 27:263-271.

Hoffman, D.J., G.H. Heinz, L.J. LeCaptain, and C.M. Bunck. 1991a. Subchronic hepatotoxicity of selenomethionine ingestion in mallard ducks. *Jour. Toxicol. Environ. Health* 32:449-464.

Hoffman, D.J., G.H. Heinz, L.J. LeCaptain, J.D. Eisemann, and G.W. Pendleton. 1996. Toxicity and oxidative stress of organic selenium and dietary protein in mallard ducklings. *Arch. Environ. Contamin. Toxicol.* 31:120-127.

Hoffman, D.J., C.J. Sanderson, L.J. LeCaptain, E. Cromartie, and G.W. Pendleton. 1991b. Interactive effects of boron, selenium, and dietary protein on survival, growth, and physiology in mallard ducklings. *Arch. Environ. Contamin. Toxicol.* 20:288-294.

Hoffman, D.J., C.J. Sanderson, L.J. LeCaptain, E. Cromartie, and G.W. Pendleton. 1992a. Interactive effects of arsenate, selenium, and dietary protein on survival, growth, and physiology in mallard ducklings. *Arch. Environ. Contamin. Toxicol.* 22:55-62.

Hoffman, D.J., C.J. Sanderson, L.J. LeCaptain, E. Cromartie, and G.W. Pendleton. 1992b. Interactive effects of selenium, methionine, and dietary protein on survival, growth, and physiology in mallard ducklings. *Arch. Environ. Contamin. Toxicol.* 23:163-171.

Hogan, G.R. 1998. Selenate- and selenomethionine-induced leukopenia in ICR female mice. *Jour. Toxicol. Environ. Health* 53A:113-119.

Hopkins, W.A., M.T. Mendonca, C.L. Rowe, and J.D. Congdon. 1998. Elevated trace metal concentrations in southern toads, *Bufo terrestris*, exposed to coal combustion waste. *Arch. Environ. Contam. Toxicol.* 35:325-329.

Hothem, R.L. and D. Welsh. 1994. Contaminants in eggs of aquatic birds from the grasslands of central California. *Arch. Environ. Contamin. Toxicol.* 27:180-185.

Huckabee, J.W. and N.A. Griffith. 1974. Toxicity of mercury and selenium to the eggs of carp (*Cyprinus carpio*). *Trans. Amer. Fish. Soc.* 103:822-825.

Hutchinson, T.C. and P.M. Stokes. 1975. Heavy metal toxicity and algal bioassays. Pages 320-343 in S. Barabas (ed.). *Water Quality Parameters.* ASTM Spec. Tech. Publ. 573. Amer. Soc. Testing Mater., 1916 Race Street, Philadelphia, PA.

Hutton, M. 1981. Accumulation of heavy metals and selenium in three seabird species from the United Kingdom. *Environ. Pollut.* 26A:129-145.

Jenkins, D.W. 1980. *Biological Monitoring of Toxic Trace Metals.* Vol. 2. Toxic Trace Metals in Plants and Animals of the World. Part III. U.S. Environ. Protect. Agen. Rep. 600/3-80-0921:1090-1129.

Jensen, L.S. 1968. Selenium deficiency and impaired reproduction in Japanese quail. *Proc. Soc. Exp. Biol. Med.* 128:970-972.

Jobidon, R. and M. Prevost. 1994. Potential use of quadrivalent selenium as a systemic deer-browsing repellent: a cautionary note. *Northern Jour. Appl. Forestry* 11:63-64.

Jones, J.B. and T.C. Stadtman. 1977. *Methanococcus vanniellii*: culture and effects of selenium and tungsten on growth. *Jour. Bacteriol.* 130:1404-1406.

Kaiser, I.I., P.A. Young, and J.D. Johnson. 1979. Chronic exposure of trout to waters with naturally high selenium levels: effects on transfer RNA methylation. *Jour. Fish. Res. Board Canada* 36:689-694.

Kalas, J.A., T.H. Ringsby, and S. Lierhagen. 1995. Metals and selenium in wild animals from Norwegian areas close to Russian nickel smelters. *Environ. Monitor. Assess.* 36:251-270.

Karbe, L., C. Schnier, and H.O. Slewers. 1977. Trace elements in mussels (*Mytilus edulis*) from coastal areas of the North Sea and the Baltic. Multielement analyses using instrumental neutron activation analysis. *Jour. Radioanal. Chem.* 37:927-943.

Kari, T. and P. Kauranen. 1978. Mercury and selenium contents of seals from fresh and brackish water in Finland. *Bull. Environ. Contam. Toxicol.* 19:273-280.

Kifer, R.R. and W.L. Payne. 1968. Selenium content of fish meal. *Feedstuffs* 31(Aug 68):32.

Kim, E.Y., R. Goto, S. Tanabe, H. Tanaka, and R. Tatsukawa. 1998. Distribution of 14 elements in tissues and organs of oceanic seabirds. *Arch. Environ. Contam. Toxicol.* 35:638-645.

Kim, J.H., E. Birks, and J.F. Heisinger. 1977. Protective action of selenium against mercury in northern creek chubs. *Bull. Environ. Contam. Toxicol.* 17:132-136.

King, K.A., T.W. Custer, and J.S. Quinn. 1991. Effects of mercury, selenium, and organochlorine contaminants on reproduction of Forster's terns and black skimmers nesting in a contaminated Texas Bay. *Arch. Environ. Contamin. Toxicol.* 20:32-40.

King, K.A., T.W. Custer, and D.A. Weaver. 1994. Reproductive success of barn swallows nesting near a selenium-contaminated lake in east Texas, USA. *Environ. Pollut.* 84:53-58.

King, K.A., C.A. Lefever, and B.M. Mulhern. 1983. Organochlorine and metal residues in royal terns nesting on the central Texas coast. *Jour. Field Ornithol.* 54:295-303.

King, K.A., D.L. Meeker, and D.M. Swineford. 1980. White-faced ibis populations and pollutants in Texas, 1969–1976. *Southwest. Nat.* 25:225-240.

Klaverkamp, J.F., D.A. Hodgkins, and A. Lutz. 1983a. Selenite toxicity and mercury-selenium interactions in juvenile fish. *Arch. Environ. Contam. Toxicol.* 12:405-413.

Klaverkamp, J.F., W.A. Macdonald, W.R. Lillie, and A. Lutz. 1983b. Joint toxicity of mercury and selenium in salmonid eggs. *Arch. Environ. Contam. Toxicol.* 12:415-419.

Knox, D.P., H.W. Reid, and J.G. Peters. 1987. An outbreak of selenium responsive unthriftiness in farmed red deer (*Cervus elephus*). *Veterin. Rec.* 120(4):91-92.

Koeman, J.H., W.S.M. van de Ven, J.J.M. de Goeij, and P.S. Tijoe. 1975. Mercury and selenium in marine mammals and birds. *Sci. Total Environ.* 3:279-287.

Krishnaja, A.P. and M.S. Rege. 1982. Induction of chromosomal aberrations in fish *Boleophthalmus dussumieri* after exposure *in vivo* to mitomycin C and heavy metals mercury, selenium and chromium. *Mutat. Res.* 102:71-82.

Kuhn, J.K., F.L. Fiene, R.A. Cahill, H.J. Gluskoter, and N.F. Shimp. 1980. Abundance of trace and minor elements in organic and mineral fractions of coal. *Environ. Geol. Notes 88.* Illinois State Geol. Surv. Div., Urbana, IL. 67 pp.

Kumar, H.D. and G. Prakash. 1971. Toxicity of selenium to the blue-green algae, *Anacystis nidulans* and *Anabaena variabilis. Ann. Bot.* 35:697-705.

Lakin, H.W. 1973. Selenium in our environment. Pages 96-111 in E.L. Kothny (ed.). *Trace Elements in the Environment.* Adv. Chemistry Ser., Amer. Chem. Soc., Washington, D.C.

Lemly, A.D. 1982a. Response of juvenile centrarchids to sublethal concentrations of waterborne selenium. I. Uptake, tissue distribution, and retention. *Aquat. Toxicol.* 2:235-282.

Lemly, A.D. 1982b. Determination of selenium, in fish tissues with differential pulse polarography. *Environ. Technol. Lett.* 3:497-502.

Lemly, A.D. 1993a. Teratogenic effects of selenium in natural populations of freshwater fish. *Ecotoxicol. Environ. Safety* 26:181-204.

Lemly, A.D. 1993b. Guidelines for evaluating selenium data from aquatic monitoring and assessment studies. *Environ. Monitor. Assess.* 28:83-100.

Lemly, A.D. 1993c. Metabolic stress during winter increases the toxicity of selenium to fish. *Aquat. Toxicol.* 27:133-158.

Lemly, A.D. 1996a. Assessing the toxic threat of selenium to fish and aquatic birds. *Environ. Monitor. Assess.* 43:19-35.

Lemly, A.D. 1996b. Selenium in aquatic organisms. Pages 427-445 in W.N. Beyer, G.H. Heinz, and A. Redmon-Norwood (eds.). *Environmental Contaminants in Wildlife: Interpreting Tissue Concentrations.* CRC Press, Boca Raton, FL.

Lemly, A.D., S.E. Finger, and M.K. Nelson. 1993. Sources and impacts of irrigation drainwater contaminants in arid wetlands. *Environ. Toxicol. Chem.* 12:2265-2279.

Lemly, A.D. and G.J. Smith. 1987. Aquatic Cycling of Selenium: Implications for Fish and Wildlife. U.S. Fish Wildl. Serv., Fish Wildl. Leafl. 12. 9 pp.

Leonzio, C., S. Focardi, and E. Bacci. 1982. Complementary accumulation of selenium and mercury in fish muscle. *Sci. Total Environ.* 24:249-254.

Leonzio, C., S. Focardi, and C. Fossi. 1992. Heavy metals and selenium in stranded dolphins of the northern Tyrrhenian (NW Mediterranean). *Sci. Total Environ.* 119:77-84.

Levander, O.A. 1983. Recent developments in selenium nutrition. Pages 147-162 in J. Weininger and C. Briggs (eds.). *Nutrition Update, Vol. 1.* John Wiley, New York.

Levander, O.A. 1984. The importance of selenium in total parenteral nutrition. *Bull. N.Y. Acad. Med.* 60:144-155.

Lo, M.T. and E. Sandi. 1980. Selenium: occurrence in foods and its toxicological significance. A review. *Jour. Environ. Pathol. Toxicol.* 4:193-218.

Lorentzen, M., A. Maage, and K. Julshamn. 1994. Effects of dietary selenite or selenomethionine on tissue selenium levels of Atlantic salmon (*Salmo salar*). *Aquaculture* 121:359-367.

Lucas, H.F., D.N. Edgington, and P.J. Colby. 1970. Concentrations of trace elements in Great Lakes fishes. *Jour. Fish. Res. Board Canada* 27:677-684.

Lucu, C. and M. Skreblin. 1981. Evidence of the interaction of mercury and selenium in the shrimp *Palaemon elegans. Mar. Environ. Res.* 5:265-274.

Lunde, G. 1970. Analysis of arsenic and selenium in marine raw materials. *Jour. Sci. Food Agric.* 21:242-247.

Luten, J.B., A. Ruiter, T.M. Ritskes, A.B. Rauchbaar, and G. Riekwel-Booy. 1980. Mercury and selenium in marine and freshwater fish. *Jour. Food Sci.* 45:416-419.

Mackay, N.J., M.N. Kazacos, R.J. Williams, and M.I. Leedow. 1975. Selenium and heavy metals in black marlin. *Mar. Pollut. Bull.* 6:57-60.

Mackey, E.A., P.R. Becker, R. Demiralp, R.R. Greenberg, B.J. Koster, and S.A. Wise. 1996. Bioaccumulation of vanadium and other trace metals in livers of Alaskan cetaceans and pinnipeds. *Arch. Environ. Contamin. Toxicol.* 30:503-512.

Maher, W.A. 1983. Selenium in marine organisms from St. Vincent's Gulf, South Australia. *Mar. Pollut. Bull.* 14:35-36.

Marcogliese, D.J., G.W. Esch, and R.V. Dimock, Jr. 1992. Alterations in zooplankton community structure after selenium-induced replacement of a fish community: a natural whole-lake experiment. *Hydrobiologia* 242:19-32.

Martin, J.H., P.D. Elliot, V.C. Anderlini, D. Girvin, S.A. Jacobs, R.W. Risebrough, R.L. Delong, and W.G. Gilmartin. 1976. Mercury-selenium bromine imbalance in premature parturient California sea lions. *Mar. Biol.* 35:91-104.

May, T.W. and G.L. McKinney. 1981. Cadmium, lead, mercury, arsenic, and selenium concentrations in freshwater fish, 1976–77 — National Pesticide Monitoring Program. *Pestic. Monitor. Jour.* 15:14-38.

McDowell, L.R., D.J. Forrester, S.B. Linda, S.D. Wright, and N.S. Wilkinson. 1995. Selenium status of white-tailed deer in Florida. *Jour. Wildl. Dis.* 31:205-211.

Measures, C.I. and J.D. Burton. 1980. The vertical distribution and oxidation states of dissolved selenium in the northeast Atlantic ocean and their relationship to biological processes. *Earth Plan. Sci. Lett.* 46:385-396.

Medeiros, L.C., R.P. Belden, and E.S. Williams. 1993. Selenium content of bison, elk and mule deer. *Jour. Food Sci.* 58:731-733.

Moksnes, K. 1983. Selenium deposition in tissues and eggs of laying hens given surplus of selenium as selenomethionine. *Acta Vet. Scand.* 24:34-44.

Mora, M.A. and D.W. Anderson. 1995. Selenium, boron, and heavy metals in birds from the Mexicali Valley, Baja California, Mexico. *Bull. Environ. Contamin. Toxicol.* 54:198-206.

Morris, J.G., W.S. Cripe, H.L. Chapman, Jr., D.F. Walker, J.B. Armstrong, J.D. Alexander, Jr., R. Miranda, A. Sanchez, Jr., B. Sanchez, J.R. Blair-West, and D.A. Denton. 1984. Selenium deficiency in cattle associated with Heinz bodies and anemia. *Science* 223:491-493.

Murphy, C.P. 1981. Bioaccumulation and toxicity of heavy metals and related trace elements. *Jour. Water Pollut. Control Fed.* 53:993-999.

Nakamoto, R.J. and T.J. Hassler. 1992. Selenium and other trace elements in bluegills from agricultural return flows in the San Joaquin Valley, California. *Arch. Environ. Contamin. Toxicol.* 22:88-98.

Nassos, P.A., J.R. Coats, R.L. Metcalf, D.D. Brown, and L.G. Hansen. 1980. Model ecosystem, toxicity, and uptake evaluation of ^{75}Se-selenite. *Bull. Environ. Contam. Toxicol.* 24:752-758.

National Academy of Sciences (NAS). 1973. Water Quality Criteria 1972. U.S. Environ. Protect. Agen. Rep. R3-73-033. 594 pp.

National Academy of Sciences (NAS). 1976. Selenium. Comm. Med. Biol. Eff. *Environ. Poll.*, Div. Med. Sci., Natl. Res. Counc., Nat. Acad. Sci., Washington, D.C. 203 pp.

National Research Council (NRC). 1983. *Selenium in Nutrition.* Subcomm. on selenium, Comm. animal nutrition, Board on Agricul., Natl. Res. Counc., National Academy Press, Washington, D.C. 174 pp.

Nielsen, M.G. and G. Gissel-Nielsen. 1975. Selenium in soil-animal relationships. *Pedobiologia* 15:65-67.

Niimi, A.J. and Q.N. LaHam. 1975. Selenium toxicity on the early life stages of zebrafish (*Brachydanio rerio*). *Jour. Fish. Res. Board Canada* 32:803-806.

Noda, K., S. Hirai, K. Sunayashiki, and H. Danbara. 1979. Neutron activation analyses of selenium and mercury in marine products from along the coast of Shikoku Island. *Agric. Biol. Chem.* 47:1381-1386.

Nriagu, J.O. and H.K. Wong. 1983. Selenium pollution of lakes near the smelters at Sudbury, Ontario. *Nature* 301:55-57.

Ohlendorf, H.M. 1989. Bioaccumulation and effects of selenium in wildlife. Pages 133-177 in L.W. Jacobs, (ed.). *Selenium in Agriculture and the Environment.* SSSA Special Publ. No. 23. *Soil Sci.* Soc. Amer. and Amer. Soc. Agron., Madison, WI.

Ohlendorf, H.M. and C.S. Harrison. 1986. Mercury, selenium, cadmium, and organochlorines in eggs of three Hawaiian seabird species. *Environ. Pollut.* 11B:169-191.

Ohlendorf, H.M., D.J. Hoffman, M.K. Saiki, and T.W. Aldrich. 1986. Embryonic mortality and abnormalities of aquatic birds: apparent impacts of selenium from irrigation drainwater. *Sci. Total Environ.* 52:49-63.

Ohlendorf, H.M.R.L. Hothem, and T.W. Aldrich. 1988. Bioaccumulation of selenium by snakes and frogs in the San Joaquin Valley, California. *Copeia* 1988:704-710.

Ohlendorf, H.M., R.L. Hothem, T.W. Aldrich, and A.J. Krynitsky. 1987. Selenium contamination of the grasslands, a major California waterfowl area. *Sci. Total Environ.* 66:169-183.

Ohlendorf, H.M., R.L. Hothem, C.M. Bunck, T.W. Aldrich, and J.F. Moore. 1986a. Relationships between selenium concentrations and avian reproduction. *Trans. North Amer. Wildl. Nat. Resour. Conf.* 51:330-342.

Ohlendorf, H.M., R.L. Hothem, C.M. Bunck, and K.C. Marois. 1990. Bioaccumulation of selenium in birds at Kesterson Reservoir, California. *Arch. Environ. Contamin. Toxicol.* 19:495-507.

Ohlendorf, H.M., R.L. Hothem, and D. Welsh. 1989. Nest success, cause-specific nest failure, and hatchability of aquatic birds at selenium-contaminated Kesterson Reservoir and a reference site. *Condor* 91:787-796.

Ohlendorf, H.M., R.W. Lowe, P.R. Kelly, and T.E. Harvey. 1986b. Selenium and heavy metals in San Francisco Bay diving ducks. *Jour. Wildl. Manage.* 50:64-71.

Ohlendorf, H.M. and K.C. Marois. 1990. Organochlorines and selenium in California night-heron and egret eggs. *Environ. Monitor. Assess.* 15:91-104.

Okazaki, R.K. and M.H. Panietz. 1981. Depuration of twelve trace metals in tissues of the oysters *Crassostrea gigas* and *C. virginica. Mar. Biol.* 63:113-120.

Oliver, M.N., D.A. Jessup, and B.B. Norman. 1990a. Selenium supplementation of mule deer in California. *Trans. Western Sec. Wildl. Soc.* 26:87-90.

Oliver, M.N., G. Ros-Mcgauran, D.A. Jessup, B.B. Norman, and C.E. Franti. 1990b. Selenium concentrations in blood of free-ranging mule deer in California. *Trans. Western Sec. Wildl. Soc.* 26:80-86.

Olson, M.M. and D. Welsh. 1993. Selenium in eared grebe embryos from Stewart Lake National Wildlife Refuge, North Dakota. *Prairie Natural.* 25:119-126.

Ort, J.F. and J.D. Latshaw. 1978. The toxic level of sodium selenite in the diet of laying chickens. *Jour. Nutr.* 108:1114-1120.

Orvini, E., V. Caramella-Crespi, and N. Genova. 1980. Activation analysis of As, Hg and Se in some marine organism. Pages 441-448 in J. Albaiges (ed.). *Analytical Techniques in Environmental Chemistry.* Pergamon Press, New York.

Ostadalova, I. and A. Babicky. 1980. Toxic effect of various selenium compounds on the rat in the early postnatal period. *Arch. Toxicol.* 45:207-211.

Pakkala, I.S., W.H. Gutenmann, D.J. Lisk, G.E. Burdick, and E.J. Harris. 1972. A survey of the selenium content of fish from 49 New York State waters. *Pestic. Monitor. Jour.* 6:107-114.

Papadopoulu, C., G.D. Kanias, and E.M. Kassimati. 1976. Stable elements of radioecological importance in certain echinoderm species. *Mar. Pollut. Bull.* 7:143-144.

Patrick, R. 1978. Effects of trace metals in the aquatic ecosystem. *Amer. Sci.* 66:185-191.

Paulsson, K. and K. Lundbergh. 1991. Treatment of mercury contaminated fish by selenium addition. *Water Air Soil Pollut.* 56:833-841.

Paveglio, F.L., C.M. Bunck, and G.H. Heinz. 1992. Selenium and boron in aquatic birds from central California. *Jour. Wildl. Manage.* 56:31-42.

Peterson, J.A. and A.V. Nebeker. 1992. Estimation of waterborne selenium concentrations that are toxicity thresholds for wildlife. *Arch. Environ. Contamin. Toxicol.* 23:154-162.

Porcella, D.B., G.L. Bowie, J.G. Sanders, and G.A. Cutter. 1991. Assessing Se cycling and toxicity in aquatic ecosystems. *Water Air Soil Pollut.* 57-58:3-11.

Prasad, T., S.P. Arora, and R.C. Chopra. 1982. Selenium toxicity as induced by feeding rice husk to buffalo calves: a clinical case report. *Indian Vet. Jour.* 59:235-237.

Pratt, D.R., J.S. Bradshaw, and B. West. 1972. Arsenic and selenium analyses in fish. *Utah Acad. Arts Sci. Proc., Part 1.* 49:23-26.

Presser, T.S. and H.M. Ohlendorf. 1987. Biogeochemical cycling of selenium in the San Joaquin Valley, California, USA. *Environ. Manage.* 11:805-821.

Rancitelli, L.A., W.A. Haller, and J.A. Cooper. 1968. Trace element variations in silver salmon and king salmon muscle tissue. *Rapp. Americain BNWL* — 715, Part 2:42-47.

Reading, J.T. 1979. Acute and Chronic Effects of Selenium on *Daphnia pulex*. Thesis. Virginia Polytech. Inst., Blacksburg, VA. 91 pp.

Reddy, C.C. and E.J. Massaro. 1983. Biochemistry of selenium: a brief overview. *Fundam. Appl. Toxicol.* 3:431-436.

Reijnders, P.J.H. 1980. organochlorine and heavy metal residues in harbour seals from the Wadden Sea and their possible effects on reproduction. *Neth. Jour. Sea Res.* 14:30-65.

Robberecht, H. and R. Von Grieken. 1982. Selenium in environmental waters: determination, speciation and concentration levels. *Talanta* 29:823-844.

Rosenfeld, I. and O.A. Beath. 1964. Selenium. *Geobotany, Biochemistry, Toxicity, and Nutrition.* Academic Press, New York. 411 pp.

Rossi, L.C., G.F. Clemente, and G. Santaroni. 1976. Mercury and selenium distribution in a defined area and in its population. *Arch. Environ. Health* 31:160-165.

Rusk, M.K. 1991. Selenium Risk to Yuma Clapper Rails and Other Marsh Birds of the Lower Colorado River. M.S. thesis, Univ. Arizona. 75 pp.

Ryther, J., T.M. Losordo, A.K. Furr, T.F. Parkinson, W.H. Gutenmann, I.S. Pakkala, and D.J. Lisk. 1979. Concentration of elements in marine organisms cultured in seawater flowing through coal-fly ash. *Bull. Environ. Contam. Toxicol.* 23:207-210.

Saiki, M.K. 1987. Relation of length and sex to selenium concentrations in mosquitofish. *Environ. Pollut.* 47:171-186.

Saiki, M.K., M.R. Jennings, and W.G. Brumbaugh. 1993. Boron, molybdenum, and selenium in aquatic food chains from the lower San Joaquin River and its tributaries, California. *Arch. Environ. Contamin. Toxicol.* 24:307-319.

Saiki, M.K., M.R. Jennings, and S.J. Hamilton. 1991. Preliminary assessment of the effects of selenium in agricultural drainage on fish in the San Joaquin Valley. Pages 369-385 in A. Dinar and D. Zilberman (eds.). *The Economics and Management of Water and Drainage in Agriculture.* Kluwer Acad. Publ., Boston.

Saiki, M.K., M.R. Jennings, and T.W. May. 1992. Selenium and other elements in freshwater fishes from the irrigated San Joaquin valley, California. *Sci. Total Environ.* 126:109-137.

Saiki, M.K. and T.P. Lowe. 1987. Selenium in aquatic organisms from subsurface agricultural drainage water, San Joaquin Valley, California. *Arch. Environ. Contamin. Toxicol.* 16:657-670.

Saiki, M.K. and R.S. Ogle. 1995. Evidence of impaired reproduction by western mosquitofish inhabiting seleniferous agricultural drainwater. *Trans. Amer. Fish. Soc.* 124:578-587.

Saiki, M.K. and D.U. Palawski. 1990. Selenium and other elements in juvenile striped bass from the San Joaquin Valley and San Francisco estuary, California. *Arch. Environ. Contamin. Toxicol.* 19:717-730.

Sandholm, M. 1973. Biological Aspects of Selenium: Uptake of Selenium by SH-Groups and Different Organic Materials in the Ecosystem. Thesis, Dept. Medicine, Coll. Veterin. Medic., Helsinki, Finland. 35 pp.

Sato, T., Y. Ose, and T. Sakai. 1980. Toxicological effect of selenium on fish. *Environ. Pollut.* 21A:217-224.

Scheuhammer, A.M., A.H.K. Wong, and D. Bond. 1998. Mercury and selenium accumulation in common loons (*Gavia immer*) and common mergansers (*Mergus merganser*) from eastern Canada. *Environ. Toxicol. Chem.* 17:197-201.

Schmitt, C.J. and W.G. Brumbaugh. 1990. National Contaminant Biomonitoring Program: concentrations of arsenic, cadmium, copper, lead, mercury, selenium, and zinc in U.S. freshwater fish, 1976-1984. *Arch. Environ. Contamin. Toxicol.* 19:731-747.

Schroeder, H.A., D.V. Frost, and J.J. Balassa. 1970. Essential trace metals in man: selenium. *Jour. Chron. Dis.* 23:227-243.

Schuler, C.A., R.G. Anthony, and H.M. Ohlendorf. 1990. Selenium in wetlands and waterfowl foods at Kesterson Reservoir, California, 1984. *Arch. Environ. Contamin. Toxicol.* 19:845-853.

Schultz, C.D. and B.M. Ito. 1979. Mercury and selenium in blue marlin, *Makaira nigricans*, from the Hawaiian Islands. U.S. Natl. Mar. Fish. Serv., *Fish Bull.* 76:872-879.

Schultz, R. and R. Hermanutz. 1990. Transfer of toxic concentrations of selenium from parent to progeny in the fathead minnow (*Pimephales promelas*). *Bull. Environ. Contamin. Toxicol.* 45:568-573.

Shamberger, R.J. 1981. Selenium in the environment. *Sci. Total Environ.* 17:59-74.

Sharma, R.P. and J.L. Shupe. 1977. Trace metals in ecosystems: relationships of residues of copper, molybdenum, selenium, and zinc in animal tissues to those in vegetation and soil in the surrounding environment. Pages 595-608 in H. Drucker and R.E. Wildung (eds.). *Biological Implications of Metals in the Environment.* Avail. as CONF 750929 from Nat. Tech. Infor. Serv., U.S. Dep. Commer., Springfield, VA.

Sheline, J. and B. Schmidt-Nielsen. 1977. Methylmercury-selenium interaction in the killifish, *Fundulus heteroclitus*. Pages 119-130 in F.J. Vernberg, A. Calabrese, F.P. Thurberg and W.B. Vernberg (eds.). *Physiological Responses of Marine Biota to Pollutants.* Academic Press, New York.

Skaare, J.U., E. Degre, P.E. Aspholm, and K.I. Ugland. 1994. Mercury and selenium in Arctic and coastal seals off the coast of Norway. *Environ. Pollut.* 85:153-160.

Skorupa, J.P. and H.M. Ohlendorf. 1991. Contaminants in drainage water and avian risk thresholds. Pages 345-368 in A. Dinar and D. Zilberman (eds.). *The Economics and Management of Water and Drainage in Agriculture.* Kluwer Acad. Publ., Boston.

Smith, T.G. and F.A.J. Armstrong. 1978. Mercury and selenium in ringed and bearded seal tissues from Arctic Canada. *Arctic* 31:75-84.

Sorensen, E.B., T.L. Bauer, J.S. Bell, and C.W. Harlan. 1982. Selenium accumulation and cytotoxicity in teleosts following chronic, environmental exposure. *Bull. Environ. Toxicol. Contam.* 29:688-696.

Sorensen, E.M.B., P.M. Cumbie, T.L. Bauer, J.S. Bell, and C.W. Harlan. 1984. Histopathological, hematological, condition-factor, and organ weight changes associated with selenium accumulation in fish from Belews Lake, North Carolina. *Arch. Environ. Contam. Toxicol.* 13:153-162.

Spehar, R.L., G.M. Christensen, C. Curtis, A.E. Lemke, T.J. Norberg, and Q.H. Pickering. 1982. Effects of pollution on freshwater fish. *Jour. Water Pollut. Control Fed.* 54:877-922.

Speyer, M.R. 1980. Mercury and selenium concentrations in fish, sediments, and water of two northwestern Quebec lakes. *Bull. Environ. Contam. Toxicol.* 24:427-432.

Stadtman, T.C. 1974. Selenium biochemistry. *Science* 183:915-921.

Stadtman, T.C. 1977. Biological function of selenium. *Nutr. Rev.* 35:161-166.

Stanley, T.R., Jr. G.J. Smith, D.J. Hoffman, G.H. Heinz, and R. Rosscoe. 1996. Effects of boron and selenium on mallard reproduction and duckling growth and survival. *Environ. Toxicol. Chem.* 15:1124-1132.

Stanley, T.R., Jr. J.W. Spann, G.J. Smith, and R. Rosscoe. 1994. Main and interactive effects of arsenic and selenium on mallard reproduction and duckling growth and survival. *Arch. Environ. Contamin. Toxicol.* 26:444-451.

Stoneburner, D.L. 1978. Heavy metals in tissues of stranded short-finned pilot whales. *Sci. Total Environ.* 9:293-297.

Stump, I.G., J. Kearney, J.M. D'Auria, and J.D. Popham. 1979. Monitoring trace elements in the mussel, *Mytilus edulis* using X-ray energy spectroscopy. *Mar. Pollut. Bull.* 10:270-274.

Tamura, Y., T. Maki, H. Yamada, Y. Shimamura, S. Ochiai, S. Nishigaki, and Y. Kimura.1975. Studies on the behavior of accumulation of trace elements in fishes. III. Accumulation of selenium and mercury in various tissues of tuna. *Tokyo Toritsu Eisei Kenkyusho Nenpo* 26:200-204. (In Japanese. Translation available from multilingual Serv., Dept. Secr. State Canada, Fish. Mar. Serv. No. 3994, 1977:11 pp.)

Teigen S.W., J.U. Skaare, A. Bjorge, E. Degre, and G. Sand. 1993. Mercury and selenium in harbor porpoise (*Phocoena phocoena*) in Norwegian waters. *Environ. Toxicol. Chem.* 12:1251-1259.

Thorarinsson, R., M.L. Landholt, D.G. Elliott, R.J. Pascho, and R.W. Hardy. 1994. Effect of dietary Vitamin E and selenium on growth, survival and the prevalence of *Renibacterium salmoninarium* in chinook salmon (*Oncorhynchus tshawytscha*). *Aquaculture* 121:343-358.

Tijoe, P.S., J.J.M. de Goeij, and M. de Bruin. 1977. Determination of Trace Elements in Dried Sea-Plant Homogenate (SP-M-1) and in Dried Copepod Homogenate (MA-A-1) by Means of Neutron Activation Analysis. Interuniv. Reactor Inst. Rept. 133-77-05, Delft, Nederlands:14 pp.

Tong, S.S.C., W.H. Gutenmann, D.J. Lisk, G.E. Burdick, and E.J. Harris. 1972. Trace metals in New York State fish. *N.Y. Fish Game Jour.* 19:123-131.

Turner, J.S., S.R.B. Solly, J.C.M. Mol-Krijnen, and V. Shanks. 1978. Organochlorine, fluorine, and heavy-metal levels in some birds from N.Z. estuaries. *N.Z. Jour. Sci.* 21:99-102.

Ullrey, D.E., W.G. Youatt, and P.A. Whetter. 1981. Muscle selenium concentrations in Michigan deer. *Jour. Wildl. Manage.* 45:534-536.

United Nations. 1979. Review of Harmful Substances. Reports and Studies. No. 2. Jnt. Group Experts Sci. Aspect. Mar. Pollut. GESAMP-76-98245. N.Y.:79 pp.

U.S. Environmental Protection Agency (USEPA). 1980. Ambient Water Quality Criteria for Selenium. U.S. Environ. Protect. Agen. Rep. 440/5-80-070. 123 pp.

U.S. Environmental Protection Agency (USEPA). 1987. Ambient Water Quality Criteria for Selenium — 1987. U.S. Environ. Protect. Agen. Rep. 440/5-87-006. 121 pp.

U.S. Food and Drug Administration. 1993. Food additives permitted in feed and drinking water of animals: selenium; stay of the 1987 amendments; final rule. *Feder. Regis.* 58 (175):47962-47973.

U.S. Public Health Service (USPHS). 1996. Toxicological Profile for Selenium (Update). U.S. Dept. Health Human Serv., PHS, Agen. Toxic Subst. Dis. Regis., Atlanta, GA. 324 pp.

Uthe, J.F. and E.G. Bligh. 1971. Preliminary survey of heavy metal contamination of Canadian freshwater fish. *Jour. Fish. Res. Board Canada* 28:786-788.

van de Ven, W.S.M., J.H. Koeman, and A. Svenson. 1979. Mercury and selenium in wild and experimental seals. *Chemosphere* 8:539-555.

Waddell, B. and T. May. 1995. Selenium concentrations in the razorback sucker (*Xyrauchen texanus*): substitution of non-lethal muscle plugs for muscle tissue in contaminant assessment. *Arch. Environ. Contamin. Toxicol.* 28:321-326.

Walsh, D.F., B.L. Berger, and J.R. Bean. 1977. Mercury, arsenic, lead, cadmium and selenium residues in fish, 1971-73. National Pesticide Monitoring Program. *Pestic. Monitor. Jour.* 11:5-34.

Wang, D., G. Alfthan, A. Aro, A. Makela, S. Knuuttila, and T. Hammar. 1995. The impact of selenium supplemented fertilization on selenium in lake ecosystems in Finland. *Agricul. Ecosys. Environ.* 54:137-148.

Wang, W.X., N.S. Fisher, and S.N. Luoma. 1996. Kinetic determinations of trace element bioaccumulation in the mussel *Mytilus edulis*. *Mar. Ecol. Prog. Ser.* 140:91-113.

Ward, G.S., T.A. Hollister, P.T. Heitmuller, and P.R. Parrish. 1981. Acute and chronic toxicity of selenium to estuarine organisms. *Northeast Gulf Sci.* 4:73-78.

Welsh, D. 1992. Selenium in Aquatic Habitats at Cibola National Wildlife Refuge. Ph.D. thesis, Univ. Arizona. 132 pp.

Wenzel, C. and G.W. Gabrielsen. 1995. Trace element accumulation in three seabird species from Hornoya, Norway. *Arch. Environ. Contamin. Toxicol.* 29:198-206.

Westernman, A.G. and W.J. Birge. 1978. Accelerated rate of albinism in channel catfish exposed to metals. *Prog. Fish-Cult.* 40:143-146.

White, D.H., K.A. King, and R.M. Prouty. 1980. Significance of organochlorine and heavy metal residues in wintering shorebirds at Corpus Christi, Texas, 1976–77. *Pestic. Monitor. Jour.* 14:58-63.

Whittle, K.J., R. Hardy, A.V. Holden, R. Johnston, and R.J. Pentreath. 1977. Occurrence and fate of organic and inorganic contaminants in marine animals. *Ann. N.Y. Acad. Sci.* 298:47-79.

Wiedmeyer, R.H. and T.W. May. 1993. Storage characteristics of three selenium species in water. *Arch. Environ. Contamin. Toxicol.* 25:67-71.

Wiemeyer, S.N. and D.J. Hoffman. 1996. Reproduction in eastern screech-owls fed selenium. *Jour. Wildl. Manage.* 60:332-341.

Wiener, J. G, G.A. Jackson, T.W. May, and B.P. Cole. 1984. Longitudinal distribution of trace elements (As, Cd, Cr, Hg, Pb, and Se) in fishes and sediments in the upper Mississippi River. Pages 139-170 in J.G. Weiner, R.V. Anderson, and D.R. McConville (eds.). *Contaminants in the Upper Mississippi River.* Butterworth Publ., Stoneham, MA.

Wilber, C.G. 1980. Toxicology of selenium: A review. *Clin. Toxicol.* 17:171-230.

Wilber, C.G. 1983. Selenium. *A Potential Environmental Poison and A Necessary Food Constituent.* Charles C Thomas, Springfield, IL. 126 pp.

Willford, W.A. 1971. Heavy metals research in the Great Lakes, 1970-1971. Pages 53-65 in *Prevalence and Effects of Toxic Metals in the Aquatic Environment.* Proc. Univ. North Carolina, Rep. 57.

Williams, M.L., R.L. Hothem, and H.M. Ohlendorf. 1989. Recruitment failure in American avocets and black-necked stilts nesting at Kesterson Reservoir, California, 1984–1985. *Condor* 91:797-802.

Winger, P.V., C. Sieckman, T.W. May, and W.W. Johnson. 1984. Residues of organochlorine insecticides, polychlorinated biphenyls, and heavy metals in biota from Apalachicola River, Florida, 1978. *Jour. Assoc. Off. Anal. Chem.* 67:325-333.

Wong, P.T.S., Y.K. Chau, and P.L. Luxon. 1978. Toxicity of a mixture of metals on freshwater algae. *Jour. Fish. Res. Board Canada* 35:479-481.

Wren, C.D. 1984 Distribution of metals in tissues of beaver, raccoon and otter from Ontario, *Canad. Sci. Total Environ.* 34:177-184.

Wrench, J.J. and N.C. Campbell. 1981. Protein bound selenium in some marine organisms. *Chemosphere* 10:1155-1161.

Yamamoto, J.T., G.M. Santolo, and B.W. Wilson. 1998. Selenium accumulation in captive American kestrels (*Falco sparverius*) fed selenomethionine and naturally incorporated selenium. *Environ. Toxicol. Chem.* 17:2494-2497.

Yonemoto, J., H. Satoh, S. Himeno, and T. Suzuki. 1983. Toxic effects of sodium selenite on pregnant mice and modification of the effects by Vitamin E or reduced glutathione. *Teratology* 28:333-340.

Zafiropoulos, D. and A.P. Grimanis. 1977. Trace elements in *Acartia clausi* from Elefsis Bay of the upper Saronikos Gulf, Greece. *Mar. Pollut. Bull.* 8:79-81.

Zieve, R. and P.J. Peterson. 1981. Factors influencing the volatilization of selenium from soil. *Sci. Total Environ.* 19:277-284.

Zingaro, R.A. and W.C. Cooper (eds.). 1974. *Selenium*. Van Nostrand Reinhold Co., New York. 835 pp.

CHAPTER 32

Radiation

32.1 INTRODUCTION

Life on Earth has evolved under the ubiquitous presence of environmental solar, X-ray, gamma, and charged-particle radiation. On a global basis, radiation from natural sources is a far more important contributor to radiation dose to living organisms than radiation from anthropogenic sources (Aarkrog 1990). However, ionizing radiation can harm biological systems (Aarkrog 1990; Nozaki 1991; Severa and Bar 1991), and this harm can be expressed (1) in a range of syndromes from prompt lethality to reduced vigor, shortened life span, and diminished reproductive rate by the irradiated organism and (2) by the genetic transmission of radiation-altered genes that are most commonly recessive and almost always disadvantageous to their carriers (Bowen et al. 1971). Direct effects of radiation were documented in lampreys in 1896 — soon after H. Becquerel discovered radioactivity — and in brine shrimp (*Artemia* sp.) in 1923 (Whicker and Schultz 1982a). Genetic effects of ionizing radiation, and thus X-rays, as a mutagenic agent were first documented in 1927 in fruit flies, *Drosophila melanogaster* (Evans 1990). The discovery of radioactivity of nuclear particles and the discovery of uranium fission resulted in a great upsurge of nuclear research. During and shortly after World War II, nuclear reactors, testing of nuclear weapons, and use of radionuclides as tracers in almost all scientific and technical fields were developed rapidly (Severa and Bar 1991). Environmental radiation from anthropogenic sources caused serious concerns beginning in the early 1940s when fission of uranium and transuranic nuclei became possible in reactors and in explosions of nuclear weapons (Aarkrog 1990). The first nuclear explosion resulted from a 19-kiloton (TNT-equivalent) source in New Mexico in July 1945 (Whicker and Schultz 1982a). On August 6, 1945, about 75,000 people were killed when the United States Army Air Corps dropped a uranium nuclear bomb on Hiroshima, Japan; on August 9, 1945, about 78,000 Japanese were killed and more than 100,000 injured when a plutonium nuclear bomb was detonated at Nagasaki (Kudo et al. 1991). The former Soviet Union detonated its first nuclear device in August 1949; and in 1952, the United Kingdom exploded a device in Australia (Whicker and Schultz 1982a). Since 1960, nuclear devices have also been detonated by France, India, Pakistan, The People's Republic of China, and possibly others. Nuclear devices have been developed that can release energy in the megaton range. The first such device was detonated by the United States in 1954 at Bikini Atoll and accidentally contaminated Japanese fishermen and Marshall Island natives. Between 1945 and 1973, an estimated 963 nuclear tests were conducted by The People's Republic of China, France, the former Soviet Union, the United Kingdom, and the United States; 47% of them were atmospheric and 53% subsurface (Whicker and Schultz 1982a).

Today, the most important environmentally damaging anthropogenic radiation comes from atmospheric testing of nuclear weapons conducted 20 to 30 years ago, authorized discharges to the sea from nuclear reprocessing plants, and from the Chernobyl accident in 1986 (Aarkrog 1990).

By the year 2000, the United States will have an estimated 40,000 tons of spent nuclear fuel stored at some 70 sites and awaiting disposal. By 2035, after all existing nuclear plants have completed 40 years of operation, about 85,000 metric tons will be awaiting disposal (Slovic et al. 1991).

Ecological and toxicological information on radiation is especially voluminous, and the reader is strongly advised to consult several of the reviews listed below.*

32.2 PHYSICAL PROPERTIES OF RADIATION

32.2.1 General

Radiation is usually defined as the emission and propagation of energy through space in the form of waves and subatomic particles (Weast 1985; Kiefer 1990). For regulatory purposes in the United States, radiation is narrowly defined as α, β, γ, or X-rays; neutrons; and high-energy electrons, protons, or other atomic particles; but not radio-waves nor visible, infrared, or ultraviolet light (U.S. Code of Federal Regulations [USCFR] 1990). Readers may wish to consult the glossary at the end of this chapter.

In current atomic theory, all elementary forms of matter consist of small units called atoms. All atoms of the same element have the same size and weight. Atoms of different elements differ in size and weight. Atoms of the same or different elements may unite to form compound substances called molecules. Each atom consists of a central nucleus and several negatively charged electrons in a cloud around the nucleus. The nucleus is composed of positively charged particles called protons, and particles without charge called neutrons. Electrons are arranged in successive energy levels around the nucleus, and the extranuclear electronic structure of the atom is characteristic of the element. Electrons in the inner shells are tightly bound to the nucleus but can be altered by high-energy waves and particles (Weast 1985). Atoms are classified chemically into 92 naturally occurring elements and another dozen or so artificial elements based on the number of protons in their nucleus (= the atomic number) (Rose et al. 1990; Severa and Bar 1991). Atoms of the same element may occur as isotopes that differ in the number of neutrons accompanying the protons in the nucleus. The sum of the number of protons and neutrons in the nucleus is called the mass number (see Glossary), and is indicated by a superscript that precedes the chemical symbol of the element. For example, three isotopes of hydrogen (one proton) are denoted as ^{1}H (no neutrons), ^{2}H (1 neutron, also known as deuterium), and ^{3}H (2 neutrons, also known as tritium). A nuclide is an elemental form distinguished from others by its atomic and mass numbers. Some nuclides, such as ^{238}U and ^{137}Cs, are radioactive and spontaneously decay to a different nuclide with the emission of characteristic energy particles or electromagnetic waves; isomers of a given nuclide that differ in energy content are metastable (i.e., ^{115m}Cd) and characterized, in part, by the half-life of the isomer (Rose et al. 1990; Severa and Bar 1991).

Chemical forms with at least one radioactive atomic nucleus are radioactive substances. The capability of atomic nuclei to undergo spontaneous nuclear transformation is called radioactivity. Nuclear transformations are accompanied by emission of nuclear radiation (Severa and Bar 1991). The average number of nuclei that disintegrate per unit time (= activity) is directly proportional to the total number of radioactive nuclei. The time for 50% of the original nuclei to disintegrate (= half-life or Tb 1/2) is equal to ln 2/decay constant for that element (Kiefer 1990). Radiations

* National Academy of Sciences [NAS] 1957, 1971; Glasstone 1958; Schultz and Klement 1963; Nelson and Evans 1969; Nelson 1971; Polikarpov 1973; Cushing 1976; Nelson 1976; International Atomic Energy Agency [IAEA] 1976, 1992; International Commission on Radiological Protection [ICRP] 1977, 1991a, 1991b; Luckey 1980; Whicker and Schultz 1982a, 1982b; League of Women Voters [LWV] 1985; Hobbs and McClellan 1986; United Nations Scientific Committee on the Effects of Atomic Radiation [UNSCEAR] 1988; Becker 1990; Kiefer 1990; Majumdar et al. 1990; Brisbin 1991; Kershaw and Woodhead 1991; Sankaranarayanan 1991a, 1991b, 1991c; National Council on Radiation Protection and Measurements [NCRP] 1991; Severa and Bar 1991; Eisler 1994; Talmage and Meyers-Schone 1995.

that have sufficient energy to interact with matter to produce charged particles are called ionizing radiations (Hobbs and McClellan 1986; UNSCEAR 1988). Radiation injury is related to the production of ions inside the cell. Ionizing radiations include electromagnetic radiation such as gamma (γ) and X-rays and particulate or corpuscular radiation such as alpha (α) particles, beta (β) particles, electrons, positrons, and neutrons. Ionizing radiation may be produced from manufactured devices such as X-ray tubes or from the disintegration of radioactive nuclides. Some nuclides occur naturally, but others may be produced artificially, for example, in nuclear reactors. The basic reaction of ionizing radiation with molecules is either ionization or excitation. In ionization, an orbital electron is ejected from the molecule and forms an ion pair. Directly ionizing particles are charged and possess the energy to produce ionizations along their path from impulses imparted to orbital electrons via electrical forces between the charged particles and electrons. In excitation, an electron is raised to a higher energy level. Indirectly ionizing radiations are not charged and penetrate a medium until they collide with elements of the atom and liberate energetically charged ionizing particles.

32.2.2 Electromagnetic Spectrum

The electromagnetic spectrum is defined as the ordered array of known electromagnetic radiations, including cosmic rays; gamma rays; X-rays; ultraviolet, visible, and infrared radiations; and radio-waves (Weast 1985). The energy transfer by electromagnetic waves can be described by discrete processes with elementary units called photons (Kiefer 1990). Their energy, E, is given by $E = hv$, where h is Planck's constant and v the frequency. Because velocity C, wavelength λ, and frequency v are related ($C = \lambda v$), $E = hc/\lambda$ (Kiefer 1990). The relationships between E, v, and λ for parts of the total spectrum of the electromagnetic waves are shown in Figure 32.1. The high-energy radiation that enters the Earth's atmosphere from outer space is known as primary cosmic rays. On interaction with the nuclei of atoms in the air, secondary cosmic rays and a variety of reaction products (cosmogenic nuclides) such as ^{3}H, ^{7}Be, ^{10}Be, ^{14}C, ^{22}Na, and ^{24}Na are produced (UNSCEAR 1988).

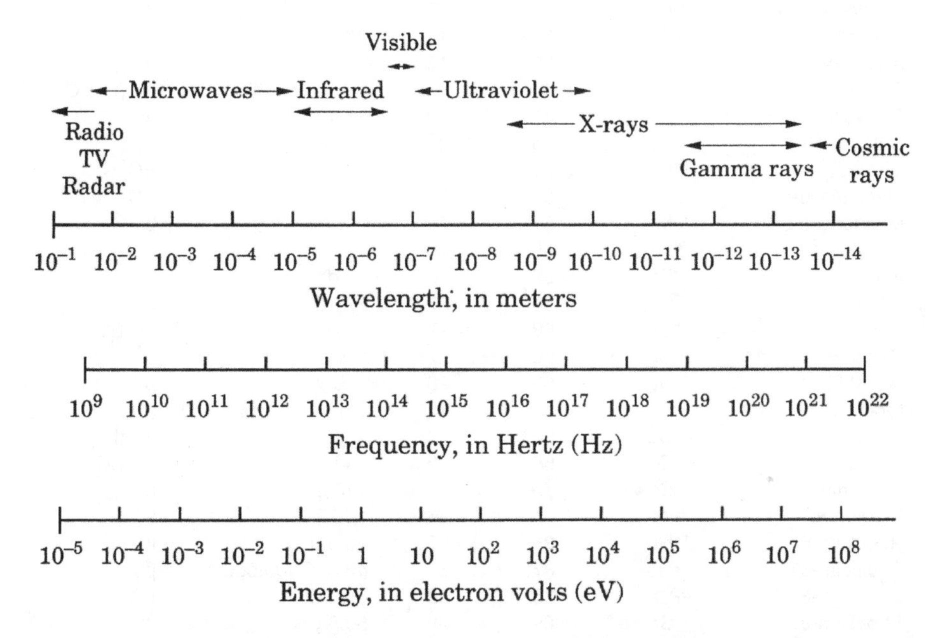

Figure 32.1 The spectrum of electromagnetic waves, showing the relationship between wavelength, frequency, and energy. (Modified from Kiefer, J. 1990. *Biological Radiation Effects.* Springer-Verlag, Berlin. 444 pp.)

32.2.3 Radionuclides

Radioactive nuclides contain atoms that disintegrate by emission of subatomic particles and gamma or X-ray photons (Weast 1985; Hobbs and McClellan 1986; Kiefer 1990; Rose et al. 1990). In alpha decay, a helium nucleus of two protons and two neutrons is emitted and reduces the mass number by 4 and the atomic number by 2. In beta decay, an electron — produced by the disintegration of a neutron into a proton, an electron, and an antineutrino — is emitted from the nucleus and increases the atomic number by 1 without changing the mass number. Sometimes, a positron together with a neutrino is emitted. And sometimes an electron may be captured from the K (outermost) shell of the atom; the resultant electron hole in the K shell is filled by electrons from outer orbits and causes the emission of X-rays. Alpha and beta decay generally leave the resultant daughter nuclei in an excited state that is deactivated by emission of γ photons. Although γ emission accompanies most decays, it is not always detected, especially not with light β emitters such as ^{3}H, ^{14}C, ^{32}P, and ^{35}S. The half-life of individual radionuclides can be measured (i.e., the time during which half the atoms of the radionuclide spontaneously decay to a daughter nuclide). Another form of nuclear breakdown is fission, in which the nucleus breaks into two nuclides of approximately half the parent's size (Rose et al. 1990). The symbol, mass number, atomic number, half-life, and decay mode of all radionuclides mentioned herein are listed in Table 32.1.

Table 32.1 Selected Radionuclides: Symbol, Mass Number, Atomic Number, Half-Life, and Decay Mode

Nuclide	Symbol	Mass Number	Atomic Number	Half-life[a]	Major Decay Mode[b]
Hydrogen-3	^{3}H	3	1	12.26 y	β^-
Beryllium-7	^{7}Be	7	4	53.3 d	EC
Beryllium-10	^{10}Be	10	4	1,600,000 y	β^-
Carbon-14	^{14}C	14	6	5730 y	β^-
Sodium-22	^{22}Na	22	11	2.6 y	β^+, EC
Sodium-24	^{24}Na	24	11	15 h	β^-
Phosphorus-32	^{32}P	32	15	14.3 d	β^-
Sulfur-35	^{35}S	35	16	87 d	β^-
Argon-39	^{39}Ar	39	18	261 y	β^-
Potassium-40	^{40}K	40	19	1,250,000,000 y	β^-, β^+, EC
Potassium-42	^{42}K	42	19	12.4 h	β^-
Calcium-45	^{45}Ca	45	20	164 d	β^-
Chromium-51	^{51}Cr	51	24	28 d	EC
Manganese-54	^{54}Mn	54	25	312 d	EC
Manganese-56	^{56}Mn	56	25	2.6 h	β^-
Iron-55	^{55}Fe	55	26	2.7 y	EC
Iron-59	^{59}Fe	59	26	45 d	β^-
Cobalt-57	^{57}Co	57	27	271 d	EC
Cobalt-58	^{58}Co	58	27	71 d	β^+, EC
Cobalt-60	^{60}Co	60	27	5.3 y	β^-
Nickel-63	^{63}Ni	63	28	100 y	β^-
Nickel-65	^{65}Ni	65	28	2.5 h	β^-
Copper-64	^{64}Cu	64	29	12.7 h	β^-, β^+, EC
Zinc-65	^{65}Zn	65	30	244 d	β^+, EC
Selenium-75	^{75}Se	75	34	118 d	EC
Krypton-85	^{85}Kr	85	36	10.72 y	β^-
Rubidium-86	^{86}Rb	86	37	18.6 d	β^-
Rubidium-87	^{87}Rb	87	37	49,000,000,000 y	β^-
Strontium-85	^{85}Sr	85	38	64.8 d	EC
Strontium-89	^{89}Sr	89	38	50.5 d	β^-
Strontium-90	^{90}Sr	90	38	29 y	β^-

Table 32.1 (continued) Selected Radionuclides: Symbol, Mass Number, Atomic Number, Half-Life, and Decay Mode

Nuclide	Symbol	Mass Number	Atomic Number	Half-life[a]	Major Decay Mode[b]
Yttrium-90	^{90}Y	90	39	64 h	β^-
Yttrium-91	^{91}Y	91	39	59 d	β^-
Zirconium-95	^{95}Zr	95	40	65 d	β^-
Niobium-95	^{95}Nb	95	41	35 d	β^-
Molybdenum-99	^{99}Mo	99	42	66 h	β^-
Technetium-99	^{99}Tc	99	43	213,000 y	β^-
Technetium-99m	^{99m}Tc	99	43	6 h	IT
Ruthenium-103	^{103}Ru	103	44	40 d	β^-
Ruthenium-106	^{106}Ru	106	44	373 d	β^-
Rhodium-106	^{106}Rh	106	45	29.8 s	β^-
Palladium-109	^{109}Pd	109	46	14 h	β^-
Silver-108m	^{108m}Ag	108	47	130 y	EC, IT
Silver-110m	^{110m}Ag	110	47	250 d	β^-, IT
Silver-110	^{110}Ag	110	47	24.6 s	β^-
Silver-111	^{111}Ag	111	47	7.5 d	β^-
Silver-113	^{113}Ag	113	47	5.3 h	β^-
Cadmium-109	^{109}Cd	109	48	462 d	EC
Cadmium-113m	^{113m}Cd	113	48	13.7 y	β^-
Cadmium-115m	^{115m}Cd	115	48	44.6 d	β^-
Cadmium-115	^{115}Cd	115	48	54 h	β^-
Tin-123	^{123}Sn	123	50	129 d	β^-
Tin-126	^{126}Sn	126	50	100,000 y	β^-
Antimony-124	^{124}Sb	124	51	60 d	β^-
Antimony-125	^{125}Sb	125	51	2.7 y	β^-
Antimony-127	^{127}Sb	127	51	3.8 d	β^-
Tellurium-127m	^{127m}Te	127	52	109 d	IT, β^-
Tellurium-129m	^{129m}Te	129	52	33 d	IT, β^-
Tellurium-129	^{129}Te	129	52	69.5 m	β^-
Tellurium-132	^{132}Te	132	52	78.2 h	β^-
Iodine-125	^{125}I	125	53	60 d	β^-, EC
Iodine-129	^{129}I	129	53	16,000,000 y	β^-
Iodine-130	^{130}I	130	53	12.4 h	β^-
Iodine-131	^{131}I	131	53	8 d	β^-
Xenon-131	^{131}Xe	131	54	11.9 d	IT
Xenon-133	^{133}Xe	133	54	5.3 d	β^-
Xenon-135	^{135}Xe	135	54	9.1 h	β^-
Cesium-134	^{134}Cs	134	55	2.06 y	β^-
Cesium-135	^{135}Cs	135	55	3,000,000 y	β^-
Cesium-137	^{137}Cs	137	55	30.2 y	β^-
Barium-140	^{140}Ba	140	56	12.8 d	β^-
Lanthanum-140	^{140}La	140	57	40 h	β^-
Cerium-141	^{141}Ce	141	58	33 d	β^-
Cerium-143	^{143}Ce	143	58	33 h	β^-
Cerium-144	^{144}Ce	144	58	284 d	β^-
Praseodymium-143	^{143}Pr	143	59	13.6 d	β^-
Praseodymium-144	^{144}Pr	144	59	7.2 m	IT
Praseodymium-147	^{147}Pr	147	59	13.4 m	β^-
Neodymium-147	^{147}Nd	147	60	11 d	β^-
Promethium-147	^{147}Pm	147	61	2.6 y	β^-
Samarium-143	^{143}Sm	143	62	8.8 m	β^+, EC
Samarium-151	^{151}Sm	151	62	90 y	β^-
Europium-152	^{152}Eu	152	63	13.4 y	EC, β^-
Europium-155	^{155}Eu	155	63	15.2 d	β^-
Tungsten-181	^{181}W	181	74	121 d	EC

Table 32.1 (continued) Selected Radionuclides: Symbol, Mass Number, Atomic Number, Half-Life, and Decay Mode

Nuclide	Symbol	Mass Number	Atomic Number	Half-life[a]	Major Decay Mode[b]
Tungsten-185	^{185}W	185	74	75 d	β^-
Tungsten-187	^{187}W	187	74	24 h	β^-
Gold-198	^{198}Au	198	79	2.7 d	β^-
Mercury-203	^{203}Hg	203	80	47 d	β^-
Mercury-206	^{206}Hg	206	80	8.1 m	β^-
Thallium-206	^{206}Tl	206	81	4.3 m	β^-
Thallium-207	^{207}Tl	207	81	4.8 m	β^-
Thallium-208	^{208}Tl	208	81	3 m	β^-
Thallium-210	^{210}Tl	210	81	1.3 m	β^-
Lead-210	^{210}Pb	210	82	22.3 y	β^-
Lead-211	^{211}Pb	211	82	36.1 m	β^-
Lead-212	^{212}Pb	212	82	10.6 h	β^-
Lead-214	^{214}Pb	214	82	26.8 m	β^-
Bismuth-210	^{210}Bi	210	83	5.0 d	β^-
Bismuth-211	^{211}Bi	211	83	2.2 m	α
Bismuth-212	^{212}Bi	212	83	1.0 h	β^-, α
Bismuth-214	^{214}Bi	214	83	19.9 m	β^-
Bismuth-215	^{215}Bi	215	83	7.4 m	β^-
Polonium-210	^{210}Po	210	84	138.4 d	α
Polonium-211	^{211}Po	211	84	0.52 s	α
Polonium-212	^{212}Po	212	84	0.0000003 s	α
Polonium-214	^{214}Po	214	84	0.000163 s	α
Polonium-215	^{215}Po	215	84	0.00178 s	α
Polonium-216	^{216}Po	216	84	0.15 s	α
Polonium-218	^{218}Po	218	84	3.1 m	α
Astatine-215	^{215}At	215	85	0.0001 s	α
Astatine-218	^{218}At	218	85	1.6 s	α
Astatine-219	^{219}At	219	85	0.9 m	α
Radon-218	^{218}Rn	218	86	0.0356 s	α
Radon-219	^{219}Rn	219	86	3.96 s	α
Radon-220	^{220}Rn	220	86	56 s	α
Radon-222	^{222}Rn	222	86	3.8 d	α
Francium-223	^{223}Fr	223	87	21.8 m	β^-
Radium-223	^{223}Ra	223	88	11.4 d	α
Radium-224	^{224}Ra	224	88	3.7 d	α
Radium-226	^{226}Ra	226	88	1620 y	α
Radium-228	^{228}Ra	228	88	5.75 y	β^-
Actinium-227	^{227}Ac	227	89	21.8 y	β^-
Actinium-228	^{228}Ac	228	89	6.13 h	β^-
Thorium-227	^{227}Th	227	90	18.8 d	α
Thorium-228	^{228}Th	228	90	1.91 y	α
Thorium-230	^{230}Th	230	90	75,400 y	α
Thorium-231	^{231}Th	231	90	25.6 h	β^-
Thorium-232	^{232}Th	232	90	14,000,000,000 y	α
Thorium-234	^{234}Th	234	90	24 d	β^-
Protactinium-231	^{231}Pa	231	91	32,700 y	α
Protactinium-234	^{234}Pa	234	91	6.7 h	β^-
Protactinium-234m	^{234m}Pa	234	91	1.17 m	β^-, IT
Uranium-233	^{233}U	233	92	160,000 y	α
Uranium-234	^{234}U	234	92	245,000 y	α
Uranium-235	^{235}U	235	92	710,000,000 y	α
Uranium-236	^{236}U	236	92	23,400,000 y	α
Uranium-238	^{238}U	238	92	4,470,000,000 y	α
Neptunium-235	^{235}Np	235	93	1.08 y	EC

Table 32.1 (continued) Selected Radionuclides: Symbol, Mass Number, Atomic Number, Half-Life, and Decay Mode

Nuclide	Symbol	Mass Number	Atomic Number	Half-life[a]	Major Decay Mode[b]
Neptunium-237	^{237}Np	237	93	2,140,000 y	α
Neptunium-239	^{239}Np	239	93	2.35 d	β^-
Neptunium-241	^{241}Np	241	93	13.9 m	β^-
Plutonium-238	^{238}Pu	238	94	87.7 y	α
Plutonium-239	^{239}Pu	239	94	24,110 y	α
Plutonium-240	^{240}Pu	240	94	6537 y	α
Plutonium-241	^{241}Pu	241	94	14.4 y	β^-
Plutonium-242	^{242}Pu	242	94	376,000 y	α
Plutonium-244	^{244}Pu	244	94	82,000,000 y	α
Americium-241	^{241}Am	241	95	458 y	α
Americium-243	^{243}Am	243	95	7370 y	α
Curium-241	^{241}Cm	241	96	33 d	EC
Curium-242	^{242}Cm	242	96	463 d	α
Curium-243	^{243}Cm	243	96	28.5 y	α
Curium-244	^{244}Cm	244	96	18.1 y	α
Curium-247	^{247}Cm	247	96	15,600,000 y	α
Curium-248	^{248}Cm	248	96	340,000 y	α
Curium-250	^{250}Cm	250	96	7400 y	SF
Californium-252	^{252}Cf	252	98	2.6 y	α, SF

[a] s = seconds; m = minutes; h = hours; d = days; y = years

[b] Observed modes of decay for all radioactive species: α = particle emission; β^- = beta emission, β^+ = positron emission; EC = electron capture resulting in X-ray emission; IT = isomeric transition from higher to lower energy state; SF = spontaneous fission.

Modified from Whicker and Schultz 1982a, 1982b; Weast 1985; Kiefer 1990; Severa and Bar 1991.

Four general groups of radionuclides are distinguished:

1. A long half-life group (i.e., Tb 1/2 > 10^9 years) of elements, including ^{238}U, ^{235}U, ^{232}Th, ^{40}K, ^{87}Rb, and ^{143}Sm, formed about 4.5 billion years ago
2. Shorter-lived daughters of U and Th such as Ra and Rn that form as a result of the decay of their long-lived parents
3. Nuclides (i.e., ^{14}C and ^{3}H) formed by continuing natural nuclear transformations driven by cosmic rays, natural sources of neutrons, or energetic particles that are formed in the upper atmosphere by cosmic rays
4. Nuclides formed as a result of nuclear weapons tests, nuclear reactor operations, and other human activities. Important members of this group include ^{90}Sr, ^{137}Cs, ^{14}C, and ^{3}H; note that many members of the third group — such as ^{14}C and ^{3}H — are also formed in this fourth fashion (Rose et al. 1990).

Radioactive decay usually does not immediately lead to a stable end product, but to other unstable nuclei that form a decay series (Kiefer 1990). The most important examples of unstable nuclei are started by very heavy, naturally occurring nuclei. Because the mass number changes only with α decay, all members of a series can be classified according to their mass numbers (see the uranium-238 decay series in Figure 32.2). A total of three natural decay series — formed at the birth of our planet — are named after their parent isotope: ^{232}Th, ^{235}U, and ^{238}U (Figure 32.3). Several shorter decay series also exist. For example, ^{90}Sr decays with a Tb 1/2 of 28 years by β emission to ^{90}Y, which in turn disintegrates (β emission) with a Tb 1/2 of 64 h to the stable ^{90}Zr (Kiefer 1990). Other examples of known radionuclides since the Earth's origin include ^{40}K and ^{87}Rb. In hazard assessments, all members of a decay series must be considered.

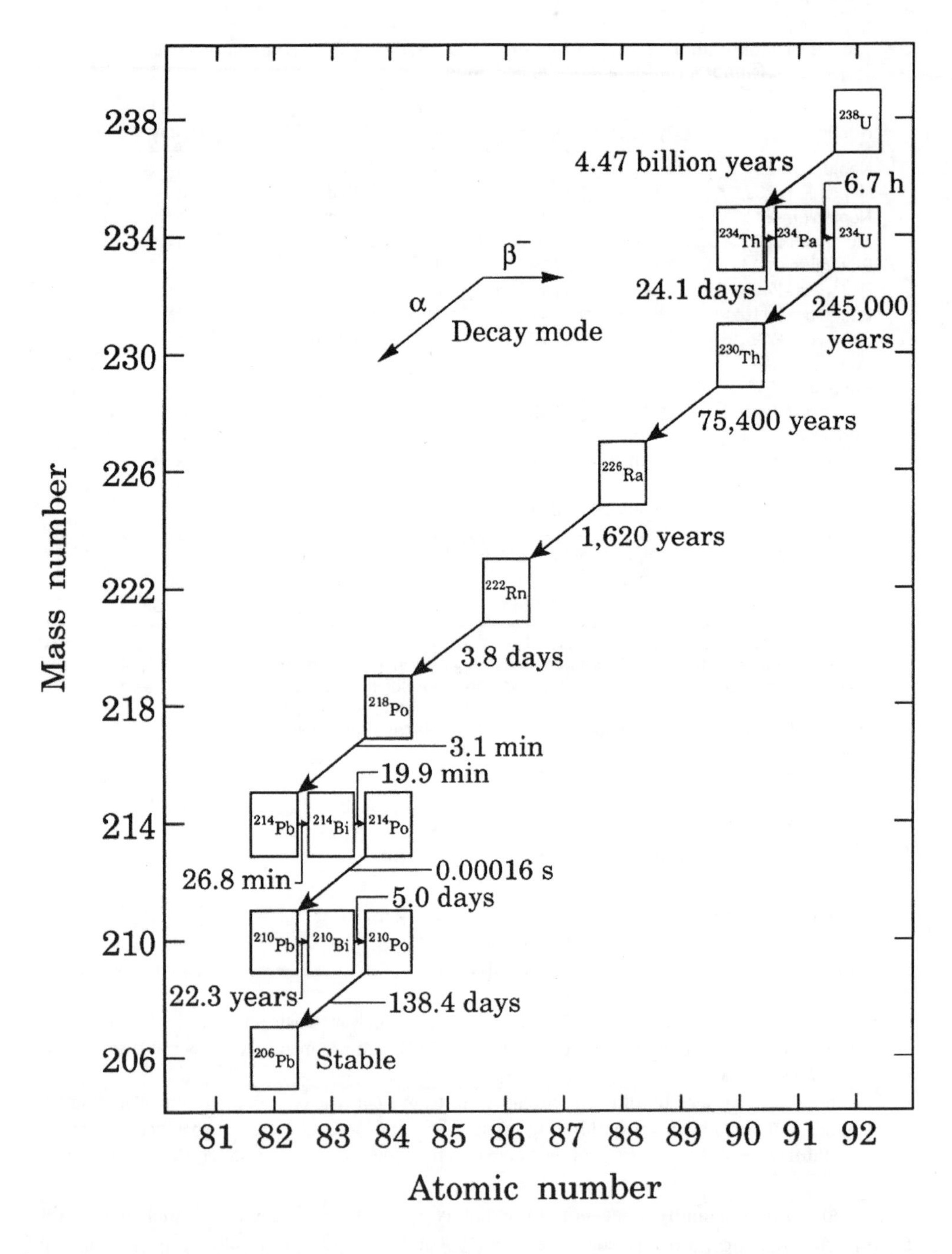

Figure 32.2 The principal uranium-238 decay series, indicating major decay mode and physical half-time of persistence. (Modified from Cecil, L.D. and T.F. Gesell. 1992. Sampling and analysis for radon-222 dissolved in ground water and surface water. *Environ. Monitor. Assess.* 20:55-66.)

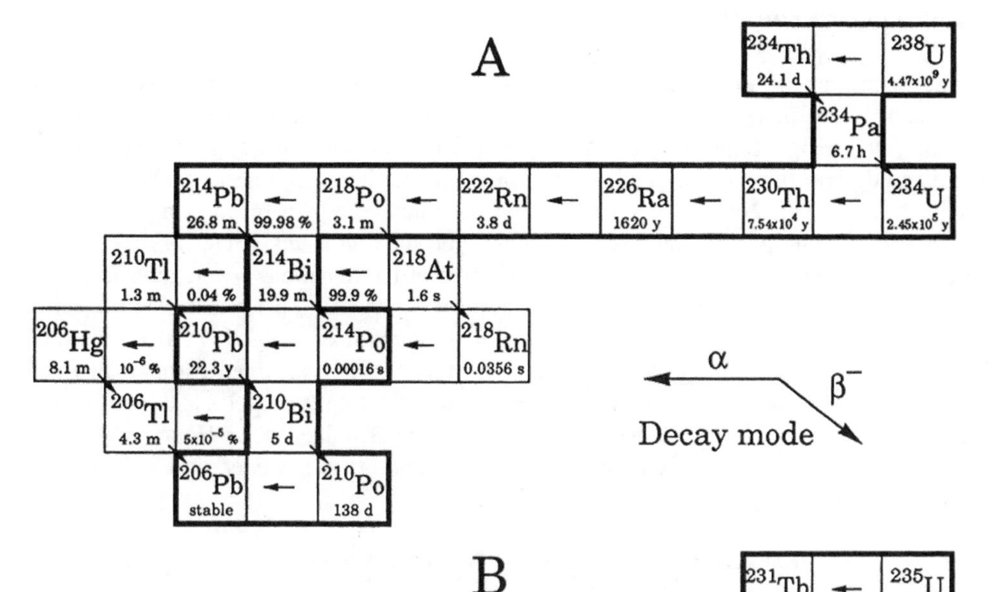

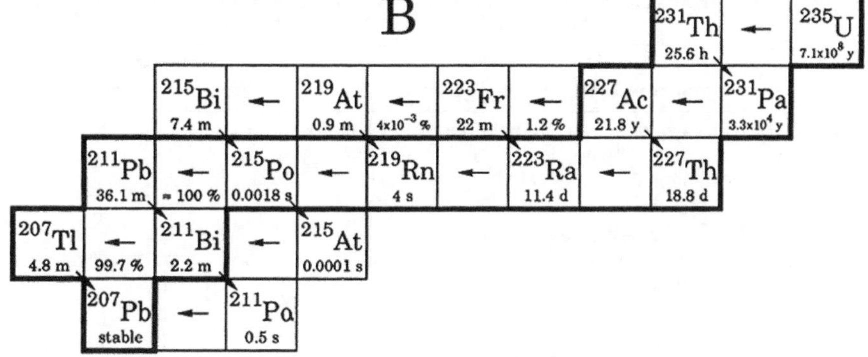

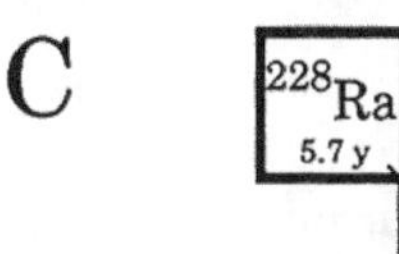

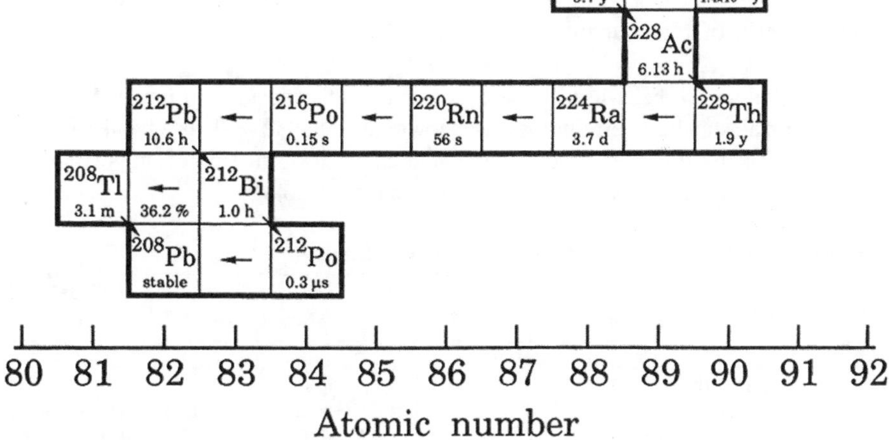

Figure 32.3 The three still-existing natural decay series. **A.** Uranium-238; **B.** Uranium-235; and **C.** Thorium-232. (Modified from Holtzman 1969; LWV 1985; UNSCEAR 1988; Kiefer 1990; Rose et al. 1990). Principal decay products occur within the heavy borders outlined.

32.2.4 Linear Energy Transfer

The deposition of energy in an exposed body is mediated almost exclusively by charged particles. These particles cause ionizations but lose energy with each ionization until they reach the end of their range. Depending on the type of particle, the ionizations are more or less closely spaced and described by the energy loss of a traversing particle. The linear energy transfer (LET) is defined as the amount of locally absorbed energy per unit length; that is, only the energy fraction that leads to ionizations or excitations in the considered site is counted (Kiefer 1990; ICRP 1991a). Because radiation effects are dependent on the nature of the radiation, a weighting factor is used to modify the absorbed dose and to define the dose equivalent. This factor — now called the Radiation Weighting Factor — is a function of LET. Approximate weighting values range from 1 (X-rays, electrons, gamma rays) to 10 (neutrons, protons, singly charged particles of rest mass greater than one atomic mass of unknown energy), and to 20 (alpha particles and multiply charged particles of unknown energy) (ICRP 1977; Whicker and Schultz 1982a; Hobbs and McClellan 1986; Severa and Bar 1991). The relation between radiation type and energy to weighting factors is shown in Table 32.2.

Table 32.2 Radiation Weighting Factors for Various Types of Ionizing Radiations

Radiation Type and Energy Range	Radiation Weighting Factor
X-rays, gamma rays, beta particles, electrons, muons; all energies	1
Neutrons	
10 keV	5
10 keV–100 keV	10
>100 keV–2 MeV	20
>2 MeV–20 MeV	10
>20 MeV	5
Protons	5
Alpha particles, fission fragments, heavy nuclei	20

Data from International Commission on Radiological Protection 1991a.

32.2.5 New Units of Measurement

A variety of units have been used for the assessment of exposures to ionizing radiation. The current international standard terminology is shown in Table 32.3. This chapter uses the new terminology exclusively; this frequently necessitated data transformation of units from early published accounts into the currently accepted international terminology.

Table 32.3 New Units for Use with Radiation and Radioactivity

Variable	Old Unit	New Unit	Old Unit in Terms of New Unit
Activity	Curie (Ci) = 3.7×10^{10} disintegrations per s (dps)	Becquerel (Bq) = 1 dps	1 Ci = 3.7×10^{10} Bq
Exposure	Roentgen (R) = 2.58×10^{-4} Coulombs/kg	Coulomb/kg (C/kg)	1 R = 2.58×10^{-4} C/kg
Absorbed dose	Rad = 100 erg/g	Gray (Gy) = 1 J/kg	1 Rad = 0.01 Gy
Dose equivalent	Rem = damage effects of 1 R	Sievert (Sv) = 1 J/kg	1 Rem = 0.01 Sv

Note: See Glossary (Section 32.11).

Data from International Commission on Radiological Protection [ICRP] 1977, 1991a; Hobbs and McClellan 1986; United Nations Scientific Committee on the Effects of Atomic Radiation [UNSCEAR] 1988.

32.3 SOURCES AND USES

32.3.1 General

Most external exposure of living organisms to radiation is from naturally occurring electromagnetic waves, and most internal exposure from naturally occurring radionuclides, such as potassium-40. Natural radiation doses vary significantly with altitude, radionuclide concentrations in the biogeophysical environment, and uptake kinetics. The major source of global anthropogenic radioactivity is fallout from military atmospheric weapons testing; locally, radiation levels tend to be elevated near nuclear power production facilities, nuclear fuel reprocessing plants, and nuclear waste disposal sites. Dispersion of radioactive materials is governed by a variety of physical, chemical, and biological vectors, including winds, water currents, plankton, and avian and terrestrial wildlife.

32.3.2 Natural Radioactivity

Exposure to natural sources of radiation is unavoidable. Externally, individuals receive cosmic rays, terrestrial X-rays, and gamma radiation. Internally, naturally occurring radionuclides of Pb, Po, Bi, Ra, Rn, K, C, H, U, and Th contribute to the natural radiation dose from inhalation and ingestion. Potassium-40 is the most abundant radionuclide in foods and in all tissues. The mean effective human dose equivalent from natural radiations is 2.4 milliSieverts (mSv). This value includes the lung dose from radon daughter products and is about 20% higher than a 1982 estimate that did not take lung dose into account (Table 32.4).

Table 32.4 Annual Effective Dose Equivalent to Humans from Natural Sources of Ionizing Radiation

Source of Radiation	Dose Equivalents (mSv)
Cosmic rays	
Ionizing component	0.30
Neutron component	0.06
Cosmogenic radionuclides (mainly ^{3}H and ^{14}C)	0.02
Primordial radionuclides	
Potassium-40	0.33
Rubidium-87	0.01
Uranium-238 series	1.34
Thorium-232 series	0.34
Total	2.4

Data from Whicker and Schultz 1982a; Hobbs and McClellan 1986; United Nations Scientific Committee on the Effects of Atomic Radiation [UNSCEAR] 1988; Aarkrog 1990.

The dose of natural radiation that an organism receives depends on height above sea level, amount and type of radionuclides in the soil of its neighborhood, and the amount taken up from air, water, and food (ICRP 1977; Whicker and Schultz 1982a; Hobbs and McClellan 1986; UNSCEAR 1988; Aarkrog 1990; Kiefer 1990; Nozaki 1991). Natural radiations in various ecosystems result in radiation dose equivalents that usually range between <0.005 and 2.07 mSv annually (Figure 32.4). Radiation doses are substantially higher at atypically elevated local sites (Table 32.5), such as Denver, and sometimes exceed 17 mSv annually in mountainous regions of Brazil and the former Soviet Union (Whicker and Schultz 1982a).

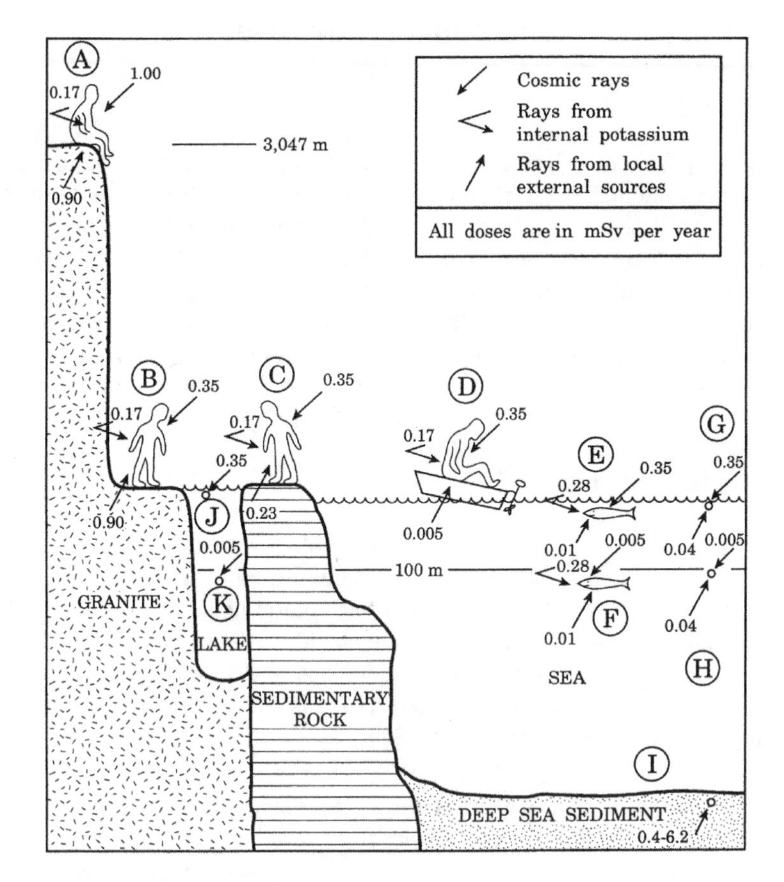

Figure 32.4 Natural radiations in selected radiological domains. (Modified from Folsom, T.R. and J.H. Harley. 1957. Comparisons of some natural radiations received by selected organisms. Pages 28-33 in National Academy of Sciences. The Effects of Atomic Radiation on Oceanography and Fisheries. Publ. No. 551, Natl. Acad. Sci.-Natl. Res. Coun., Washington, D.C.) **A.** Human over granite at 3047 m (10,000 feet) elevation above sea level; total annual dose equivalent of 2.07 mSv (cosmic rays 1.00, granite 0.90, internal emitters 0.17). **B.** Human over granite at sea surface; total annual dose of 1.42 mSv. **C.** Human over sedimentary rock at sea level; total annual dose of 0.75 mSv. **D.** Human over sea; total annual dose of 0.525 mSv. **E.** Large fish in sea near surface; total annual dose of 0.64 mSv. **F.** Large fish in sea at depth of 100 m; total annual dose of 0.295 mSv. **G.** Microorganism in water near sea surface; total annual dose of 0.39 mSv. **H.** Microorganism in water >100 m deep in sea; total annual dose of 0.045 mSv. **I.** Microorganism buried in deep sea sediments; total annual dose between 0.4 and 6.2 mSv. **J.** Microorganism near freshwater surface; total annual dose of 0.35 mSv. **K.** Microorganism 100 m deep in a freshwater lake; total annual dose of 0.005 mSv.

32.3.3 Anthropogenic Radioactivity

Nuclear explosions and nuclear power production are the major sources of anthropogenic activity in the environment. But radionuclide use in medicine, industry, agriculture, education, and production and transport, use, and disposal from these activities present opportunities for wastes to enter the environment (Whicker and Schultz 1982a; Table 32.6). Radiation was used as early as 1902 in the treatment of diseases, including enlarged thymus, tinea capitis, acne, and cancers of childhood and adolescence (Bowden et al. 1990). The use of X-rays by physicians and dentists represents the largest source of annual dose equivalent of the U.S. population to artificial radiation: 0.78 to 1.01 mSv to bone marrow and 0.016 mSv to the upper GI tract; radiopharmaceuticals contribute an additional 0.14 mSv or a yearly total mean dose of 0.94 to 1.17 mSv to bone marrow (Hobbs and McClellan 1986).

Table 32.5 Annual Whole-Body Radiation Doses to Humans from Various Sources

Source of Radiation	Dose (mSv)
Natural external background	
Denver, Colorado	1.65
Washington state, mean	0.88
United States, average	0.84
Hanford, Washington	0.59
Average medical dose per capita, United States	0.36
Average internal dose from natural radioactivity, United States	0.25
Global weapons fallout	0.05
Consumer product radiation (TV, smoke detector, and other sources)	0.02
Total	1.27–2.33

Data from Gray, R.H., R.E. Jaquish, P.J. Mitchell, and W.H. Rickard. 1989. Environmental monitoring at Hanford, Washington, USA: a brief site history and summary of recent results. *Environ. Manage.* 13:563-572.

Table 32.6 Sources and Applications of Atomic Energy

Source and Output	Application
Nuclear reactor	
Steam, electricity	Electric power (stationary or portable plants), desalination, propulsion of submarines and surface ships
Heat, electricity, neutrons	Spacecraft and satellite power, spacecraft propulsion, research and special materials production
Nuclear explosives, kinetic energy	Military and civilian applications: large-scale earth moving, subsurface excavation, mineral extraction from underground
Encapsulated radioisotopes	
Electricity	Marine navigation aids, unmanned weather stations, spacecraft project power, artificial human organs
Beta and gamma radiation	Food preservation, polymerization, sterilization of medical supplies, thickness gauges
Radionuclides, beta and gamma radiation	Medical uses, tracers in scientific research, measures of manufacturing processes

Data from Joseph, A.B., P.F. Gustafson, I.R. Russell, E.A. Schuert, H.L. Volchok, and A. Tamplin. 1971. Sources of radioactivity and their characteristics. Pages 6-41 in *National Academy of Sciences. Radioactivity in the Marine Environment.* Natl. Acad. Sci., Panel on Radioactivity in the Marine Environment, Washington, D.C.

Atmospheric testing of nuclear weapons is an important human source of environmental radiation (Hobbs and McClellan 1986; UNSCEAR 1988; Aarkrog 1990) (Table 32.7). The first test explosion of a nuclear weapon took place in 1945. Atmospheric tests by the United States, the former Soviet Union, and the United Kingdom continued until they were banned in 1963. France and The People's Republic of China continued to conduct limited atmospheric tests, although no atmospheric nuclear explosions have taken place since 1980. Large nuclear explosions in the atmosphere carry most of the radioactive material into the stratosphere where it remains for 1 to 5 years, depending on the altitude and latitude. Fallout can occur years after an explosion injected material into the atmosphere. Smaller explosions carry the radioactive material only into the troposphere, and fallout occurs within days or weeks. Fallout was highest in the temperate regions and in the northern hemisphere where most of the testing was done. The most abundant radionuclides from atmospheric tests to date are $^{14}C > ^{137}Cs > ^{95}Zr > ^{90}Sr > ^{106}Ru > ^{144}Ce > ^{3}H$. Of the many radionuclides produced in nuclear and thermonuclear explosions, the primary contributors to human radiation exposure include ^{14}C, $^{89+90}Sr$, ^{95}Zr, ^{106}Ru, ^{131}I, ^{137}Cs, ^{141}Ce, and ^{144}Ce. Isotopes of plutonium and americium — although present in quantity — are not significant contributors

Table 32.7 Annual Effective Dose Equivalent from Nuclear Weapons Testing to Humans in the North Temperate Zone

Nuclide	Dose (mSv[a])
^{3}H	0.05
^{14}C	2.6
^{90}Sr	0.18
^{95}Zr	0.29
^{106}Ru	0.14
^{131}I	0.05
^{137}Cs	0.88
^{144}Ce	0.09
Pu and Am nuclides	0.09
Other nuclides	0.08
Total	4.45[b]

[a] External, 24%; inhalation, 5%; ingestion, 71%.
[b] Equivalent to 1.85 times the natural background dose.

Data from Aarkrog, A. 1990. Environmental radiation and radiation releases. *Inter. Jour. Radiation Biol.* 57:619-631.

because of their low solubility. The primary dose from fallout radiation is through external gamma radiation, assimilation through the food chain, or beta radiation of the skin.

Radioisotope thermoelectric generators (RTGs) are sometimes used as power sources for space systems. In April 1964, a United States RTG navigational satellite, SNAP 9A, reentered the atmosphere and burned up at high altitude over the Mozambique Channel, releasing 629 trillion becquerels (TBq), equivalent to 17,000 Ci, of ^{238}Pu and 0.48 TBq of ^{239}Pu (Whicker and Schultz 1982a; Richmond 1989). In January 1978, a Soviet RTG satellite, Kosmos 954, reentered the atmosphere over Canada and spread radiouranium across parts of that country (Richmond 1989). The amount of radioactive materials in space applications is expected to increase (Richmond 1989).

Significant amounts of radioactivity are present in the Great Lakes basin, which has numerous nuclear reactors and uranium-mine waste areas (Joshi 1991). The prevailing low levels of artificially produced radionuclides, arising largely from previous fallout (Table 32.8), provide small doses of radiation to residents who consume lake water. Radionuclides enter the Great Lakes ecosystem from natural and anthropogenic processes. The main natural processes that introduce radioactivity are the weathering of rocks, which contain uranium- and thorium-series radionuclides, and fallout of cosmic ray-produced radionuclides such as ^{3}H, ^{7}Be, and ^{14}C. Anthropogenic radioactivity is created, for example, by uranium mining, milling, and fuel fabrication; releases of artificially

Table 32.8 Estimated Fallout of ^{90}Sr and ^{137}Cs over the Great Lakes, 1954–1983, in Cumulative Millions of Bq/km^2

Great Lake	Cesium-137	Strontium-90	Total
Superior	2429	1491	3920
Michigan	2738	1680	4418
Huron	2670	1638	4308
Erie	2859	1754	4613
Ontario	2773	1701	4474
Total	13,469	8264	21,733

Data from Joshi, S.R. 1991. Radioactivity in the Great Lakes. *Sci. Total Environ.* 100:61-104.

produced radionuclides through nuclear power reactors and nuclear fuel processing plants; medical uses of radioisotopes; and coal-fired electrical generating plants (Joshi 1991).

Production of power from nuclear reactors involves uranium mining, fuel fabrication, the reactor operations, and storage of wastes. All of these processes may expose humans and the environment to radiation (Hobbs and McClellan 1986). Uranium production in the United States was 12,300 tons U_3O_8 in 1977, primarily from western states, Texas, and Florida (Whicker and Schultz 1982a). Mining from deep shafts or open pits is the preferred method of uranium extraction, although in Florida it is produced as a by-product of phosphate mining. Mines disperse radionuclides of uranium, thorium, and radium, which are associated with dust particles, and radon, which emanates from ore as a gas and decays to create a series of radioactive daughters. Groundwater also contains radionuclides of the uranium series. As many as 18 uranium mills were in operation, located close to major mining centers in the western states. Collectively, these mills process or processed about 30,000 tons of ore daily and use acid or alkali leach methods to extract 90 to 95% of the uranium from ore. Uranium is barreled at the mill for shipment as uranium oxide or as salt concentrates (yellowcake) that contain 70 to 90% U_3O_8 by weight. Residues of the uranium extraction process are usually pumped as a slurry to liquid-retention impoundments; about 0.55 TBq of ^{230}Th and ^{226}Ra enter tailings each day from milling operations. Radium-226 produces gaseous ^{222}Rn; daughters of ^{222}Rn, such as ^{210}Pb, expose the surrounding biota to measurable radiation. Purification of yellowcake to UF_6 (uranium hexafluoride) and its enrichment to ^{235}U causes a loss of about 0.55 TBq annually. Nuclear reactor fuel contains about 3% ^{235}U. A nuclear explosion in a nuclear reactor is highly unlikely because the nuclear fuel suitable for weapons must contain >90% ^{235}U. Following enrichment, UF_6 is hydrolyzed to uranyl fluoride, converted to ammonium diuranate, and calcined to the dioxide UO_2. Uranium dioxide pellets at one time were prepared by as many as ten commercial fuel fabrication plants and subsequently transported to nuclear reactors (Whicker and Schultz 1982a). In the current light-water-cooled reactors, the most abundant radionuclides in the reactor effluents under normal conditions are ^{3}H, ^{58}Co, ^{60}Co, ^{85}Kr, ^{85}Sr, ^{90}Sr, ^{130}I, ^{131}I, ^{131}Xe, ^{133}Xe, ^{134}Cs, ^{137}Cs, and ^{140}Ba (Hobbs and McClellan 1986). Gaseous and volatile radionuclides, such as ^{85}Kr, ^{131}Xe, and ^{133}Xe, contribute to the external gamma dose, whereas the others contribute to the dose externally by surface deposition and internally by way of the food chain. The mean dose from environmental releases of all radionuclides from nuclear reactors in the United States is <0.01 mSv/year (Hobbs and McClellan 1986). Nuclear fission follows the capture of a neutron by an atom of fissionable material, such as ^{235}U or ^{239}Pu. The fission releases 1 to 3 neutrons and, if additional fissionable material is present in sufficient quantity and in the right configuration, a chain reaction occurs (Hobbs and McClellan 1986). Radionuclides formed per megaton of fission include fission products (^{89}Sr, ^{90}Sr, ^{95}Zr, ^{103}Ru, ^{106}Ru, ^{131}I, ^{137}Cs, ^{144}Ce) and activation products in air (^{3}H, ^{14}C, ^{39}Ar) and soil (^{24}Na, ^{32}P, ^{42}K, ^{45}Ca, ^{55}Fe, ^{59}Fe) (Whicker and Schultz 1982a). Fission-product radionuclides of potential biological importance include ^{90}Sr, ^{137}Cs, ^{131}I, ^{129}I, ^{144}Ce, ^{103}Ru, ^{106}Ru, ^{95}Zr, ^{140}Ba, ^{91}Y, ^{143}Ce, ^{147}Nd (Kahn 1971; Whicker and Schultz 1982a), and others (Table 32.9).

Most of the world's supply of uranium consists of about 0.7% ^{235}U and 99% ^{238}U. In theory, about 2.27 kg of ^{235}U can release energy equivalent to 20,000 tons of TNT (Hobbs and McClellan 1986). ^{238}U and ^{232}Th can be converted into fissionable material following neutron capture. Radionuclides of biological significance produced by neutron activation in nuclear reactors include ^{3}H, ^{14}C, ^{24}Na, ^{32}P, ^{35}S, ^{45}Ca, ^{54}Mn, ^{55}Fe, $^{57+58+60}$Co, ^{65}Zn, ^{239}Pu, ^{239}Np, ^{241}Am, and ^{242}Cm (Whicker and Schultz 1982a). Nuclear energy can also be released by fusion of smaller nuclei into larger nuclei that is accompanied by a decrease in mass (Hobbs and McClellan 1986). Fusion reactors — which do not yet exist — require very high temperatures of several million degrees; no fission products are produced in the fusion process (Whicker and Schultz 1982a).

Radioactive wastes are usually stored in underground tanks or in temporary storage at reactor sites for recycling or disposal (Whicker and Schultz 1982a). For low-level wastes, containment and isolation are the preferred disposal options, including burial, hydraulic injection into deep geological strata, and ocean disposal (Table 32.10). Options for the disposal of high-level wastes include

Table 32.9 Fission Products per kg ^{235}U Reactor Charge at 100 Days Cooling

Product	Grams	Trillions of becquerels per kg ^{235}U	
		Beta	Gamma
Short-lived[a]	15.93	7217	6002
Lonγ–lived[b]	16.61	698	755
Inactive fission products	230.00	—	—
Total	262.54	8045	6757

[a] ^{90}Y, ^{106}Rh, ^{144}Ce, ^{95}Zr, ^{95}Nb, ^{91}Y, ^{89}Sr, ^{103}Ru, ^{141}Ce, ^{137}Ba, ^{106}Ru, ^{143}Pr, ^{140}Ba, ^{140}La, ^{131}I.
[b] ^{137}Cs, ^{90}Sr, ^{144}Pr, ^{129}Te.

Modified from Renn, C.E. 1957. Physical and chemical properties of wastes produced by atomic power industry. Pages 26-27 in National Academy of Sciences. The Effects of Atomic Radiation on Oceanography and Fisheries. Publ. No. 551, NAS-Natl. Res. Coun., Washington, D.C.

Table 32.10 Radioactive Waste Disposal at Sea

Disposer and Other Variables	Quantity (trillions of becquerels [TBq])	Reference[a]
UNITED STATES		
Atlantic Ocean, 1951–60 vs. 1961–67	2939 vs. 2	1
Pacific Ocean, 1951–60 vs. 1961–67	527 vs. 16	1
UNITED KINGDOM		
1951–67, alpha vs. beta	123 vs. 1631	1
Sellafield, alpha (primarily Pu and Am)		
1968–70	50–61	2
1971 vs. 1972	99 vs. 143	2
1973 vs. 1974	181 vs. 17	2
Sellafield reprocessing plant[b]		
1980 vs. 1981	5145 vs. 4451	3
1982 vs. 1983	4005 vs. 3112	3
1984 vs. 1985	1835 vs. 646	3
EUROPE		
Germany, Netherlands, Belgium, France; 1961; alpha vs. beta plus gamma	6 vs. 220	1
France, Cap de la Hague reprocessing plant[c]		
1980 vs. 1981	503 vs. 455	3
1982 vs. 1983	694 vs. 683	3
1984 vs. 1985	670 vs. 674	3

[a] **1,** Joseph et al. 1971; **2,** Hetherington et al. 1976; **3,** UNSCEAR 1988.
[b] Effluent composition primarily ^{137}Cs and ^{241}Pu.
[c] Effluent composition primarily ^{106}Ru and ^{125}Sb.

retrievable surface storage and entombment in deep geological strata; many risks are associated with these options, and more suitable alternative disposals are needed. Spent nuclear fuel elements are usually stored for about 3 months to allow the decay of shorter-lived radionuclides before reprocessing or disposal. Reprocessing involves extractions to separate uranium and plutonium from the fission products into UF_6 and plutonium dioxide. Longer-lived fission products, such as ^{90}Sr and ^{137}Cs, are sometimes chemically separated and encapsulated for storage or disposal. Fuel reprocessing tends to release measurable quantities of various radionuclides that are detected in fish, wildlife, and food for humans (Whicker and Schultz 1982a). Liquid discharges from the Sellafield reprocessing plant (Table 32.10) have been reduced by a factor of more than 100 since the mid-1970s (Aarkrog 1990). Human populations that consume higher than average quantities of marine fish and shellfish from the Sellafield area theoretically receive about 3.5 mSv annually from radioactivity associated with nuclear power production. Human populations in the vicinity of nuclear power production discharging directly into the marine environment — except for Sellafield — generally receive <0.05 mSv annually from this source (Aarkrog 1990).

Radioactive transuranic elements with atomic numbers that are greater than 92 have been introduced into the environment since the 1940s from atmospheric testing of nuclear weapons, discharges of nuclear wastes, and nuclear fuel reprocessing (Noshkin et al. 1971; Hetherington et al. 1976; Sibley and Stohr 1990; Morse and Choppin 1991). Transuranic isotopes with half-lives of more than 10,000 years (i.e., ^{247}Cm, ^{248}Cm, ^{239}Pu, ^{242}Pu, ^{244}Pu, ^{237}Np) will persist over geologically significant time periods. Transuranics at detectable but considered nonhazardous levels to biota are now widely dispersed throughout the environment in most waters, soils, sediments, and living organisms including humans. Of current primary concern are ^{244}Cm, ^{241}Am, $^{238+239+240+241}$Pu, and ^{237}Np — especially ^{241}Am, which is increasing globally as a result of ^{241}Pu decay (Sibley and Stohr 1990; Morse and Choppin 1991). However, the estimated peak dose received from Pu and Am radioisotopes seems to be decreasing in the vicinity of the Sellafield nuclear fuel reprocessor (Table 32.11). Miscellaneous exposures include radiations from television sets, luminous dial watches, smoke detectors, electron microscopes, building materials, and air travel (Hobbs and McClellan 1986). Most of the exposure in building materials is due to naturally occurring radionuclides. Similarly, air travel increases radiation exposure of travelers from increased exposure to cosmic radiations. Cigarette smokers may receive dose-equivalent rates up to 3 times higher than nonsmokers because of inhalation of ^{210}Po and ^{210}Pb from the cigarette. Some of the lung dose is also received from radionuclides released during combustion of fossil fuels, which contain small quantities of naturally occurring radionuclides (Hobbs and McClellan 1986).

Table 32.11 Theoretical Peak Dose, in microsieverts per year, Received from Plutonium and Americium by Three Human Populations

Population	Year		
	1973	1987	2000
Average person near Sellafield nuclear fuel reprocessor	24	4	2
Critical group, mainly agricultural workers	35	20	16
Heavy consumers of Irish Sea fish and shellfish in local fishing communities	—	250	55–90

Data from McKay, W.A. and N.J. Pattenden. 1990. The transfer of radionuclides from sea to land via the air: a review. *Jour. Environ. Radioactiv.* 12:49-77.

32.3.4 Dispersion

Radioactive materials are cycled throughout the environment by a variety of physical, chemical, and biological vectors. Dispersion through the atmosphere is governed by the magnitude, frequency, and direction of the wind. In the hydrosphere, transport is modified by water depth, motion, temperature, winds, tides, and groundwater (Whicker and Schultz 1982b). Deposition from the atmosphere is a function of particle size, precipitation, and dry deposition. Small radioactive particles may be elevated into the airstream from the ground surface; resuspension is a function of disturbances by wind at the soil surface, atmospheric variables (i.e., velocity, turbulence, density, viscosity), and soil–ground variables such as texture, cohesiveness, moisture content, density, vegetation cover, ground surface roughness, and topography (Whicker and Schultz 1982b). Only 1 kg of the original 15 kg Pu was fissioned from the dropping of the plutonium nuclear bomb on Nagasaki, Japan, on August 9, 1945 (Kudo et al. 1991). The remaining 14 kg Pu escaped into the environment. Local fallout accounted for about 37 g or 0.26% of the total global fallout; the highest $^{239+240}$Pu concentration measured was 64 Bq/kg soil about 2.8 km from ground zero (Kudo et al. 1991).

Biological agents can also transport radioactive wastes. Birds, especially waterfowl, disperse accumulated radiocesium and other radionuclides along their migratory flyways (Brisbin 1991). Native mammalian herbivores and their predators that have come in contact with radioactivity in food or soils disperse the material in their feces, urine, or regurgitated pellets (O'Farrell and Gilbert 1975). For example, the black-tailed jackrabbit (*Lepus californicus*) in the vicinity of radioactive waste-disposal trenches dispersed radioactive fecal pellets over an area of 15 km^2. Elevated radioactivity readings were recorded in jackrabbits and in their predators, including feces of coyotes (*Canis latrans*) and bones of hawks (O'Farrell and Gilbert 1975).

Biological transport of trace elements and radionuclides in the sea is provided mainly through phytoplankton and zooplankton because of their (1) ability to accumulate these elements to high levels, (2) diurnal vertical migration, and (3) production of detritus in the form of fecal pellets, molts, and carcasses (Lowman et al. 1971). Considerations related to biomass, feeding rates, conversion efficiencies, migratory habits of zooplankton, and the chemical properties of trace elements suggest that the major downward transport of these elements and radionuclides is through gravitational action on fecal pellets, molts, and carcasses; direct biological transport accounts for <10% of the total downward movement. In estuarine and near-shore regions, the bottom sediments and their associated epiphyton often significantly influence the distribution of added radionuclides. Large populations of sessile filter feeders may drastically increase the rate of sedimentation of added trace elements and radionuclides (Table 32.12).

Table 32.12 Time Required to Transport Selected Radionuclides Added into Marine Waters at Surface out of the Upper Mixed Layer by Biological Transport (Processes include diurnal vertical migration, fecal pellets, and sinking of dead matter.)

	Time required to transport radionuclides (years)		
Radionuclide	Eastern North Pacific	Coastal Areas	Upwelling Areas
^{54}Mn	74	7	3
$^{55+59}Fe$	7.2	0.7	0.3
$^{57+58+60}Co$	220	20	8.8
^{65}Zn	12	1.1	0.5
^{95}Zr	5.4	0.5	0.2
^{210}Pb	7.3	0.7	0.3

Data from Lowman, F.G., T.R. Rice, and F.A. Richards. 1971. Accumulation and redistribution of radionuclides by marine organisms. Pages 161-199 in *National Academy of Sciences. Radioactivity in the Marine Environment.* Natl. Acad. Sci., panel on radioactivity in the marine environment. Washington, D.C.

In some coastal areas, some of the radionuclides discharged into coastal waters from industrial establishments are recycled via the air/sea interface back onto land (McKay and Pattenden 1990). At the sea surface, aerosol is generated by bubble bursting and wave shearing. The aerosol is advected to land by onshore winds and deposited in coastal regions. Sea-to-land transfer has been documented from the vicinity of nuclear fuel reprocessing facilities in England, Scotland, and France; however, the sea-to-land transfer pathway was only about 8% of that from the seafood pathway (McKay and Pattenden 1990). The solubility of different radionuclides at the sediment/seawater interface is variable. Plutonium solubility, for example, depends on pH, Eh, ionic strength, complexing ions, organic chelators, living accumulator organisms, and oxidation state (Mo and Lowman 1976). The oceanic distributions of many nuclides are strongly controlled by interactions with particulate matter (Nozaki 1991). Thorium is an extreme case; the high reactivity of this element accounts for its residence of only a few decades in the ocean from which it is removed largely by vertical transport in association with settling particulate matter. ^{210}Pb and ^{231}Pa are also particle-reactive but to a lesser extent than Th. Their oceanic mean residence time is about 100 years.

The mean oceanic residence time of ^{227}Ac and Ra isotopes is about 1000 years because of particulate scavenging; these nuclides are supplied by insoluble parents in underlying sediments and are released to overlying waters by porewater diffusion. ^{228}Ra can serve as a novel tracer in ocean circulation for about 30 years; ^{227}Ac can be used for about 100 years. The distribution of ^{226}Ra is largely governed by biogeochemical cycling, much like dissolved silica (Nozaki 1991).

32.4 RADIONUCLIDE CONCENTRATIONS IN FIELD COLLECTIONS

32.4.1 General

The wide dispersion of anthropogenic radiocontaminants has significantly altered natural background levels of radioactivity in many parts of the globe. Radionuclide concentrations in selected abiotic materials and living organisms were usually elevated in samples from the vicinity of human nuclear activities, especially atmospheric military tests. Radionuclide concentrations in organisms were significantly modified by the organism's age, sex, diet, metabolism, trophic level, proximity to point source, and many other biological, chemical, and physical variables, as discussed later. Additional and more detailed data on environmental radionuclide concentrations and isotopic composition and levels of radioactive wastes discharged into the biosphere from nuclear plants and other anthropogenic activities are given in Schultz and Klement (1963), Nelson and Evans (1969), Nelson (1971), IAEA (1976), Whicker and Schultz (1982a, 1982b), and UNSCEAR (1988).

32.4.2 Abiotic Materials

Radionuclide concentrations in selected nonliving materials (Table 32.13) show that concentrations are elevated in samples from the site of repeated nuclear detonations, near nuclear fuel reprocessing and waste facilities, and from locations receiving radioactive fallout from atmospheric military tests. Rocks, especially granite, had high levels of naturally occurring radionuclides such as ^{40}K. Concentrations were usually low or negligible in drinking water and cow's milk for human consumption. Nuclear-weapons testing has resulted in large environmental releases of radionuclides. Between 1961 and 1966, for example, the Republic of Korea received fallout from nuclear tests by the former Soviet Union in 1961 and by the United States in 1962 and from three explosions by The People's Republic of China (Bai 1969). The highest levels of total combined β and γ activity in various Korean samples during 1962 to 1964, in Bq/L or Bq/kg, were: 0.0002 in air, 133 in water, 1572 in milk, 2023 in rain, 16,428 in plants, and 99,345 in soils (Bai 1969).

Water in the Great Lakes in 1981 contained measurable concentrations of ^{137}Cs, ^{3}H, and ^{90}Sr, and detectable — but extremely low — concentrations of ^{241}Am, ^{113m}Cd, ^{144}Ce, ^{210}Pb, $^{239+240}$Pu, ^{226}Ra, ^{125}Sb, and ^{228}Th (Joshi 1991). Radiocesium-137 in water from the Hudson River estuary, New York, decreased tenfold between 1964 and 1970, but the ^{137}Cs content in fish and in sediments remained relatively constant (Wrenn et al. 1971). The effluent from the United Kingdom's Atomic Energy Agency Sellafield facility on the Cumberland Coast of the Irish Sea contained ^{90}Sr and ^{137}Cs, which are soluble in seawater and tend to remain in solution, and ^{106}Ru, ^{144}Ce, and ^{95}Zr/^{95}Nb, which are relatively insoluble in seawater and coprecipitate or adsorb on free inorganic and organic surfaces (Pentreath et al. 1971).

Soils in the vicinity of an English nuclear fuel reprocessing facility in the period 1979 to 1985 contained as much as 42 times more ^{241}Am, 12 times more ^{137}Cs, 13 times more ^{90}Sr, and 87 times more $^{239+240}$Pu than soils from a reference site (Curtis et al. 1991). In the United States, radiological trends in abiotic materials were difficult to interpret. For example, one nationwide monitoring program for radionuclide concentrations in air, drinking water, milk, groundwater, and precipitation (Table 32.13) was not consistent in the selection of measured radionuclides, frequency of sampling, and types of samples analyzed.

Table 32.13 Radionuclide Concentrations in Field Collections of Selected Abiotic Materials (Concentrations are in becquerels per kilogram fresh weight [FW], or dry weight [DW].)

Material, Radionuclide, and Other Variables	Concentration (Bq/kg or Bq/L)	Reference[a]
COMMON ROCK TYPES		
Shale, limestone, sandstone, basalt		
^{40}K	63–518 DW	2
^{232}Th	4–48 DW	2
^{238}U	6–44 DW	2
Granite vs. beach sands		
^{40}K	1184 DW vs. 100 DW	2
^{232}Th	74 DW vs. 25 DW	2
^{238}U	62 DW vs. 37 DW	2
DRINKING WATER		
Mol, Belgium, 1983, near former nuclear fuel reprocessing plant closed in 1974, ^{129}I	Max. 0.000082 FW	3
United States, nationwide		
1977 vs. 1981		
^{238}Pu	Max. 0.00004 FW vs. Max. 0.0004 FW	1, 4
^{239}Pu	Max. 0.0004 FW vs. Max. 0.0003 FW	1, 4
^{234}U	Max. 0.093 vs. Max. 2.19 FW	1, 4
^{235}U	Max. 0.0026 FW vs. Max. 0.027 FW	1, 4
^{238}U	Max. 0.067 FW vs. Max. 0.562 FW	1, 4
1988		
^{131}I	Max. 0.011 FW	5
^{238}Pu	Max. 0.002 FW	6
$^{239+240}Pu$	Max. 0.0003 FW	6
^{226}Ra	Usually <0.007 FW; Max. 0.24 FW	6
^{90}Sr	Max. 0.018 FW	6
^{234}U	Max. 0.090 FW	6
^{235}U	Max. 0.007 FW	6
^{238}U	Max. 0.183 FW	6
1989, ^{131}I	Max. 0.022 FW	7
1990, ^{131}I	Max. 0.022 FW	8
FRESHWATER		
Vicinity of nuclear weapons tests and operation of nuclear reactors, maximum values		
^{141}Ce	0.08 FW	2
^{144}Ce	0.41 FW	2
^{137}Cs	0.18 FW	2
^{131}I	5.2 FW	2
^{54}Mn	0.05 FW	2
^{103}Ru	0.25 FW	2
^{106}Ru	1.1 FW	2
^{89}Sr	1.9 FW	2
^{90}Sr	0.66 FW	2
$^{95}Zr/^{95}Nb$	2.4 FW	2
Typical maximum concentrations		
^{3}H	0.6 FW	2
^{40}K	0.2 FW	2
^{210}Pb	0.01 FW	2
^{210}Po	0.008 FW	2
^{226}Ra	0.11 FW	2
^{87}Rb	0.00007 FW	2
^{222}Rn	6.7 FW	2

Table 32.13 (continued) Radionuclide Concentrations in Field Collections of Selected Abiotic Materials (Concentrations are in becquerels per kilogram fresh weight [FW], or dry weight [DW].)

Material, Radionuclide, and Other Variables	Concentration (Bq/kg or Bq/L)	Reference[a]
^{232}Th	0.0002 FW	2
^{234}U	0.12 FW	2
^{235}U	0.002 FW	2
^{238}U	0.06 FW	2
GROUNDWATER		
United States, nationwide, ^{222}Rn, 1981 vs. 1982	Usually <10 FW; Max. 388 FW vs. Max. 90 FW	1, 9
LAKEWATER		
Canada 1984–87, ^{226}Ra		
Near uranium tailings area, dissolved vs. total	0.12 FW vs. 0.56 FW	10
Control site, dissolved vs. total	0.012 FW vs. 0.009 FW	10
Great Lakes, 1973 vs. 1981		
^{137}Cs	0.003 FW vs. 0.0006–0.002 FW	11
^{3}H	12.6 FW vs. 6.7–13.5 FW	11
^{90}Sr	0.019–0.047 FW vs. 0.016–0.024 FW	11
MILK, (COW) PASTEURIZED		
Mol, Belgium, 1983, near former nuclear fuel reprocessing plant, ^{129}I	Max. 0.0005 FW	3
United States, nationwide		
1975 vs. 1977		
^{14}C	17.7–18.8 FW vs. —[b]	12
^{137}Cs	Max. 1.07 FW vs. Max. 1.04 FW	4, 12
^{129}I	—[b] vs. ND[c]	4
^{131}I	ND vs. Max. 0.59 FW	4, 12
^{89}Sr	ND vs. Max. 0.22 FW	4, 12
^{90}Sr	Max. 0.17 FW vs. Max. 0.27 FW	4, 12
1978 vs. 1981		
^{137}Cs	Max. 0.92 FW vs. Max. 0.66 FW	9, 13
^{131}I	Max. 0.29 FW vs. Max. 0.48 FW	9, 13
^{89}Sr	Max. 0.15 FW vs. Max. 0.07 FW	9, 13
^{90}Sr	Max. 0.32 FW vs. Max. 0.14 FW	9, 13
1982 vs. 1988		
^{137}Cs	Max. 0.67 FW vs. Max. 0.70 FW	1, 15, 16
^{131}I	Max. 0.25 FW vs. Max. 0.48 FW	1, 15, 16
^{89}Sr	Max. 0.07 FW vs. 0.007–0.09 FW	1, 16
^{90}Sr	Max. 0.13 FW vs. Max. 0.07 FW	1, 16
1983, ^{14}C	16.1–17.5 FW	14
1989 vs. 1990		
^{137}Cs	Max. 0.78 FW vs. Max. 0.67 FW	5, 7, 8, 14, 17–19
^{131}I	Max. 0.66 FW vs. Max. 0.48 FW	5, 7, 8, 14, 17–19
^{89}Sr	Max. 0.11 FW vs. —[b]	5, 14, 17, 18
^{90}Sr	Max. 0.18 FW vs. —[b]	5, 14, 17, 18
PRECIPITATION		
United States, nationwide		
1978		
^{238}Pu	Max. 0.0004 FW	13
^{239}Pu	Max. 0.0006 FW	13
^{234}U	Max. 0.004 FW	13
^{235}U	Max. 0.0001 FW	13
^{238}U	Max. 0.003 FW	13

Table 32.13 (continued) Radionuclide Concentrations in Field Collections of Selected Abiotic Materials (Concentrations are in becquerels per kilogram fresh weight [FW], or dry weight [DW].)

Material, Radionuclide, and Other Variables	Concentration (Bq/kg or Bq/L)	Reference[a]
1987 vs. 1988		
^{238}Pu	Max. 0.0007 vs. Max. 0.001 FW	5, 15
$^{239+240}Pu$	Max. 0.0003 FW vs. Max. 0.0005 FW	5, 15
^{234}U	Max. 0.013 FW vs. Max. 0.002 FW	5, 15
^{235}U	Max. 0.0004 FW vs. Max. 0.0003 FW	5, 15
^{238}U	Max. 0.0026 FW vs. Max. 0.002 FW	5, 15
SEAWATER		
Major fallout radionuclides in surface seawater, typical concentrations		
^{14}C	0.0004–0.001 FW	2
^{137}Cs	0.005–0.04 FW	2
^{3}H	0.3–1.8 FW	2
^{90}Sr	0.003–0.026 FW	2
^{239}Pu	0.000004–0.00005 FW	2
Natural radionuclides in surface seawater, typical concentrations		
^{3}H	0.022–0.111 FW	2
^{14}C	0.007 FW	2
^{40}K	11.8 FW	2
^{210}Pb	<0.0003 FW	2
^{210}Po	0.0002–0.001 FW	2
^{226}Ra	0.0016 FW	2
^{228}Ra	0.00004–0.004 FW	2
^{87}Rb	0.107 FW	2
^{228}Th	0.00007–0.0001 FW	2
^{230}Th	<0.00005 FW	2
^{232}Th	<0.00003 FW	2
^{234}U	0.048 FW	2
^{235}U	<0.002 FW	2
^{238}U	0.044 FW	2
SEDIMENTS		
Deep Ocean		
^{232}Th	1–74 DW	2
^{238}U	5–37 DW	2
Hanford, Washington, 1973, plutonium processing waste pond		
^{241}Am	2627 DW	20
^{238}Pu	4144 DW	20
$^{239+240}Pu$	4477 DW	20
Hudson River estuary, 1970, ^{137}Cs, bottom sediments vs. suspended sediments	75 DW vs. 152 DW	21
SOILS		
Belgium, Mol, near former nuclear fuel reprocessing plant, 1983, ^{129}I	Max. 0.2 DW	3
Tennessee, 1974, ^{137}Cs; 12–22 cm depth; accidentally contaminated in 1944 vs. control site	Usually near 185,000 DW, Max. 740,000 DW vs. <222 DW	22
WATER, VARIOUS LOCATIONS		
Hanford, Washington; plutonium processing waste ponds		
^{241}Am	0.04 FW	20
^{238}Pu	0.0003 FW	20
$^{239+240}Pu$	0.00007 FW	20

Table 32.13 (continued) Radionuclide Concentrations in Field Collections of Selected Abiotic Materials (Concentrations are in becquerels per kilogram fresh weight [FW], or dry weight [DW].)

Material, Radionuclide, and Other Variables	Concentration (Bq/kg or Bq/L)	Reference[a]
Hudson River estuary, 1970, ^{137}Cs, dissolved vs. suspended	0.01 FW vs. 0.005 FW	21
Italy, 1971, nuclear power station		
^{60}Co	Max. 0.06 FW	23
^{137}Cs	Max. 0.33 FW	23

[a] **1,** U.S. Environmental Protection Agency (USEPA) 1982b; **2,** International Atomic Energy Agency (IAEA) 1976; **3,** Handl et al. 1990; **4,** USEPA 1977; **5,** USEPA 1989c; **6,** USEPA 1990a; **7,** USEPA 1990c; **8,** USEPA 1991; **9,** USEPA 1982a; **10,** Clulow et al. 1991; **11,** Joshi 1991; **12,** USEPA 1975; **13,** USEPA 1979; **14,** USEPA 1990a; **15,** USEPA 1989a; **16,** USEPA 1989b; **17,** USEPA 1989d; **18,** USEPA 1990b; **19,** USEPA 1990d; **20,** Emery et al. 1976; **21,** Wrenn et al. 1971; **22,** Dahlman and Voris 1976; **23,** Smedile and Queirazza 1976.

[b] — = no data.

[c] ND = not detectable.

32.4.3 Aquatic Ecosystems

Field studies indicate that effects of radiation on marine ecosystems cannot be demonstrated at prevailing dose rates (Templeton et al. 1971). Two major periods of worldwide fallout occurred in Arctic ecosystems. The first and most sustained occurred from 1953 to 1959, and the second from 1961 to 1964, reflecting the atmospheric nuclear-weapons test regimes of Great Britain, the former Soviet Union, and the United States (Hanson 1976). Military accidents created localized radiocontamination of the Arctic environment. In one case, a B-52 aircraft from the U.S. Air Force crashed on the ice in northwestern Greenland in January 1968. Plutonium from the nuclear weapons onboard contaminated the benthos (Figure 32.5). The $^{239+240}Pu$ concentrations in various environmental samples declined at a much faster rate than the physical half-life of ^{239}Pu (24,000 years), suggesting that Pu becomes increasingly unavailable to the benthos over time as a result of dispersion from the epicenter and a dilution effect (Aarkrog 1990).

In marine environments, the major portion of the background dose rate in plankton and fish arises from the incorporated activity of natural alpha emitters, such as ^{210}Po, and from ^{40}K; in molluscs, crustaceans, and benthos, the gamma radiation from the seabed provides the major background dose (IAEA 1976). The situation is similar in freshwater environments, although water containing appreciable levels of ^{222}Rn and its daughter radionuclides may exert an additional burden, especially to phytoplankton. Artificial radionuclides that contribute significantly to background concentrations of marine organisms include ^{239}Pu and ^{90}Sr; of freshwater organisms, ^{137}Cs and ^{90}Sr (IAEA 1976). The total natural radiation received by a marine flounder (*Pleuronectes platessa*) in the Irish Sea consisted of 63% from radiations from seabed sediments, 16% from ^{40}K in seawater, 15% from internal ^{40}K, and 6% from cosmic radiation (Templeton et al. 1971). The estimated dose rates in aquatic environments from natural background are as high as 3.5 mGy annually and of the same order as those in most terrestrial environments. By 1976, the estimated dose rates from global fallout had declined to the same range as natural dose rates, although environments receiving radioactive wastes had variable responses (IAEA 1976).

Muscle of largemouth bass (*Micropterus salmoides*) collected from lakes in South Carolina in May and June of 1993 contained 109 to 4607 Bq ^{137}Cs/kg DW. Increasing concentrations of ^{137}Cs were correlated with increasing DNA damage (Sugg et al. 1995). Increasing levels of ^{137}Cs in the muscle of fish in Minnesota between 1954 and 1966 reflect fallout from atmospheric nuclear testing. The effective half-time for ^{137}Cs in these lakes, as judged from small fish, is about 30 months (Gustafson 1969). In game fish from Colorado, ^{137}Cs in muscle was up to 7 times higher in 1968 than in 1965; higher in fish in mountain lakes than in fish from reservoirs in the plains, foothills, lakes, and rivers; and highest in trout from alpine lakes and reservoirs (Nelson and Whicker 1969).

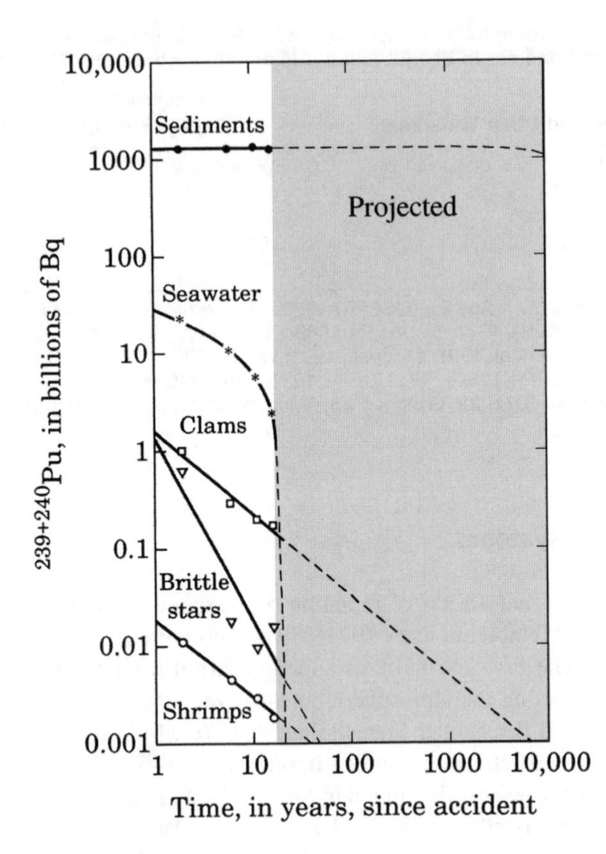

Figure 32.5 Plutonium-239+240 in environmental samples at Thule, Greenland, between 1970 and 1984, after a military accident in 1968. (Modified from Aarkrog, A. 1990. Environmental radiation and radiation releases. *Inter. Jour. Radiation Biol.* 57:619-631.) Within the contaminated area of 3.2×10^9 m^2, the fresh weight biomass of shrimps was 0.11×10^9 kg; of brittle star echinoderms 0.06×10^9 kg; and of clam (*Macoma balthica*) soft parts 0.32×10^9 kg. The seawater mass was 3×10^{14} kg, and the dry weight of the upper 15-cm sediment layer was 3×10^{11} kg.

In 1966, ^{137}Cs levels in trout from Colorado alpine lakes were 8 to 18 times higher than mean levels in muscle of deer from Colorado during the same period, and 20 to 300 times higher than domestic meat products (Nelson and Whicker 1969). Radionuclides in livers of tunas from southern California during the period 1964 to 1970 originated mainly from weapons tests in 1961/62, although ^{65}Zn may have reached southern California waters from nuclear reactors in Hanford (Washington) and from French or Chinese nuclear tests (Folsom et al. 1971).

Many variables are known to modify radionuclide concentrations in biota. In general, lower trophic levels of aquatic organisms are likely to have greater concentrations of radionuclides than higher trophic levels (Bowen et al. 1971). However, radionuclide concentrations in biota are modified significantly by the organism's age, size, sex, tissue, season of collection, and other variables — and these have to be acknowledged when integrating radiological analyses. For example, older *Fucus vesiculosus* had higher radioactivity concentrations than younger algae; concentrations of ^{60}Co and ^{54}Mn were highest in older parts of plants during spring and summer; and ^{137}Cs and ^{40}K were highest in receptacles and new vegetative fronds (Carlson and Erlandsson 1991). Changes in concentrations of ^{60}Co and ^{137}Cs in freshwater plankton from the discharge canal of an Italian nuclear power station seem to reflect changes in water concentrations of these isotopes; changes were lowest in winter and highest in summer (Smedile and Queirazza 1976). Marine bivalve molluscs and algae from Connecticut in 1960 and 1961 had the highest levels of gross beta radioactivity in spring and summer and the lowest in winter (Table 32.14) (Hatfield et al. 1963); natural ^{40}K probably accounted for most of the beta radioactivity. Similar seasonal variations in gross beta radioactivity in other species of marine algae and molluscs were documented, suggesting a correspondence with periods of dormancy and activity (Hatfield et al. 1963). Although fat in the liver of crabs accounted for 47% of the fresh weight (74% on a dry weight basis), the gross beta activity of the fat fraction amounted to <0.5% of the total radioactivity, suggesting that radiological liver analyses be compared on the basis of nonfat solids (Chakravarti and Eisler 1961). In mosquitofish (*Gambusia holbrooki*) from some locations in a ^{137}Cs-contaminated reservoir, males contained higher ^{137}Cs concentrations than females, and smaller females contained more ^{137}Cs than larger females (Newman and Brisbin 1990). Strontium-90 concentrations in the carapace bone of turtles from five southwestern states in 1970 were used as indicators of ^{90}Sr fallout. However, older turtles tended to have lower concentrations of ^{90}Sr and concentrations differed geographically. Concentrations were highest in Georgia and increasingly lower in Tennessee, Mississippi, Arkansas, and Florida (Table 32.14) (Holcomb et al. 1971).

Consumption of shellfish represents a negligible radiological risk to humans (Crowley et al. 1990), although bivalve molluscs seem to be effective accumulators of radioisotopes. After the Chinese nuclear tests in May and December 1966, concentrations of ^{144}Ce, ^{103}Ru, ^{95}Zr, ^{95}Nb, ^{140}Ba, and ^{140}La in three species of bivalves in the Neuse River, North Carolina, increased suddenly (Wolfe and Schelske 1969). In 1973, Pacific oysters (*Crassostrea gigas*) from the discharge canal of a nuclear power plant in Humboldt Bay, California, rapidly accumulated ^{54}Mn, ^{60}Co, ^{65}Zn, and ^{137}Cs within 30 min of release. Isotope uptake correlated positively with particulates in the water, including living microorganisms, organic detritus, inorganic materials, and especially resuspended bottom sediments (Harrison et al. 1976). Although concentrations of cesium and plutonium in mussels (*Mytilus edulis*) from most Irish estuaries are essentially the same as global fallout levels, concentrations were elevated in mussels from the northeast coast (Crowley et al. 1990).

32.4.4 Birds

Television and newspaper reporters attributed radionuclides to a decline in bird numbers at the Ravenglass estuary, England, particularly of the black-headed gull (*Larus ridibundus*), although the concentrations of radionuclides in the avian diet, body tissues, and general environment were at least 1000 times too low to have had any effect (Table 32.14) (Lowe 1991). Although oystercatchers (*Haematopus ostralegus*) and shelducks (*Tadorna tadorna*) had the highest concentrations of ^{137}Cs in their tissues, the breeding success and population size of these birds were not affected. Black-headed gulls had less radiocontamination than other birds at Ravenglass, but their population continued to decline. The most likely cause was a combination of an uncontrolled fox population, a severe outbreak of myxomatosis in rabbits (normal fox prey), and a drought — all in the same year (Lowe 1991). Nesting success of birds was unaffected in the vicinity of nuclear power plants. For example, nesting barn swallows (*Hirundo rustica*) near radioactive leaching ponds had normal nesting success despite their consumption of arthropods from the pond and use of contaminated mud for nest construction (Millard and Whicker 1990; Table 32.14). Adult swallows received a total internal dose rate of 219 μGy/day, mostly (72%) from ^{24}Na; daily dose rates for eggs and nestlings during the nesting season were 840 μGy and 2200 μGy. The total dose to eggs and nestlings (54 mGy) and adults (450 mGy) had no measurable effect on survival and was below accumulated doses reported to cause death of passerines (Millard and Whicker 1990).

Strontium-90 behaves much like calcium in the biological environment. In birds, ^{90}Sr is expected to occur in bone and in the calcium-rich eggshell. In one case, a positive relation was demonstrated between reactor releases of ^{90}Sr to the Columbia River and ^{90}Sr concentrations in reed canary grass (*Phalaris arundinacea*) and eggshells of the Canada goose (*Branta canadensis moffitti*) (Rickard and Price 1990).

No human health problem is anticipated from consumption of ruffed grouse (*Bonasa umbellus*) contaminated with ^{226}Ra in Canada or American coot (*Fulica americana*) contaminated with ^{137}Cs in Washington state. Tissues of ruffed grouse collected near discharged uranium tailings in Canada in 1987/88 did not contain grossly elevated levels of ^{226}Ra over controls; consumption of grouse by humans did not present a radiological health problem (Clulow et al. 1992). Based on ^{137}Cs alone, humans who consume a single contaminated American coot captured at Hanford, Washington, would receive about 1.1% of the annual radiation protection dose of 1.70 mSv for individuals and populations in uncontrolled areas (Cadwell et al. 1979).

32.4.5 Mammals

Diets in Denmark contained elevated loadings of ^{137}Cs in 1964 because of the intensive atmospheric nuclear test series by the United States and the former Soviet Union in 1961 and 1962. Total ^{137}Cs intake declined in the Danish population from 72 Bq/kg BW in 1964 to <2 in 1985, but rose to about 13 in 1986 from the effects of debris from Chernobyl on dietary ^{137}Cs during the first year after the accident (Aarkrog 1990). The estimated dose equivalent from ^{137}Cs to human consumers of fish from the Great Lakes is about 0.01 μSv/kg fresh weight (FW) muscle from fish in Lakes Erie and Ontario, and 0.06 to 0.07 μSv/kg from fish in Lakes Superior and Huron (Joshi 1991). The guide for the protection of the general public from radiation is <5 mSv annually, and consumption of fish containing a dose equivalent greater than 0.02 μSv/kg fish flesh is not recommended (Joshi 1991). Some Scandinavians now receive a dose equivalent of about 5 mSv/year from intake of radiocesium in the diet (Johanson 1990). In Finland, uptake of radionuclides by humans in Finnish Lapland and in other areas with an arctic climate is attributed to ecological factors and to a high amount of local fallout. For example, reindeer-herding Finnish Lapps contained about 50 times more ^{137}Cs and 10 times more ^{55}Fe than other Finns during 1961 to 1967. For ^{137}Cs, this disparity is attributed mainly to the reliance by Finns on reindeer meat — which contains high levels of ^{137}Cs as a result of reindeer feeding on lichens — and secondarily, on freshwater fish and cow's milk (Miettinen 1969).

In the United States, the estimated annual whole-body human radiation dose equivalent is 1.61 mSv, mostly from natural sources (0.85 mSv) and medical sources (0.70 mSv), but also from fallout (0.03 mSv), miscellaneous sources (0.02 mSv), occupational hazards (0.008 mSv), and nuclear power (0.0001 mSv) (League of Women Voters [LWV] 1985). Radiation doses to people living near the Hanford nuclear industrial and research site in the state of Washington are well below existing regulatory standards. Only trace amounts of radionuclides from Hanford have been detected in the offsite environment (Gray et al. 1989). In December 1984, radon levels up to 130 times greater than considered safe under the current guideline for underground uranium miners were discovered in human residences in eastern Pennsylvania, New Jersey, and New York. About 25% of all residences in ten states exceeded the action level for radon of 0.185 Bq/L air (Cross 1990; Oge and Dickson 1990). The significance of this observation to avian and terrestrial wildlife merits investigation.

As a result of nuclear weapons testing, mandibles of Columbian black-tailed deer (*Odocoileus hemionus columbianus*) from California increased from <9 Bq ^{90}Sr/kg ash weight (AW) to >204 Bq/kg AW between 1952 and 1960 (Table 32.14) (Schultz and Longhurst 1963). Age and season affected strontium kinetics in male mule deer (*Odocoileus hemionus hemionus*) during the period of antler growth; these variables did not affect strontium kinetics in females (Schreckhise and Whicker 1976). The concentrations of ^{90}Sr in forage of mule deer were higher in summer than in winter, and the differences were of sufficient magnitude to account for the ^{90}Sr variations in mule deer antlers (Farris et al. 1969); ^{137}Cs concentrations were similar in the forage and flesh of the white-tailed deer (*Odocoileus virginianus*) (Cummings et al. 1971). Levels of iodine-129 in thyroids of mule deer and pronghorns (*Antilocapra americana*) increased with proximity to nuclear fuel reprocessing plants in Colorado, Idaho, New Mexico, and Wyoming during 1972 to 1976, although levels were considered of no consequence to the health of the animals (Markham et al. 1983).

Radium-226, a bone-seeking α emitter with a half-life of 1600 years, may cause tissue damage and possibly subsequent osteosarcoma. Elevated ^{226}Ra concentrations have been reported in tissues of the common beaver (*Castor canadensis*) from the Serpent River watershed, Canada, the recipient of uranium tailings during 1984 to 1987 (Table 32.14). Measurable levels of ^{226}Ra were also found in feces of snowshoe hares (*Lepus americanus*) from this area and in black cutworms (*Agrotis ipsilon*) eaten by herring gulls (*Larus argentatus*) on the tailings (Clulow et al. 1991). Maximum levels in tissues of beavers from this watershed were <5 Bq ^{232}Th/kg dry weight (DW) in all tissues, 15 Bq ^{228}Th/kg DW bone, <5 Bq ^{228}Th/kg DW muscle and liver, 70 to 160 Bq ^{210}Po/kg DW bone, 11 to 75 Bq ^{210}Po/kg DW muscle, and 35 to 65 Bq ^{210}Po/kg DW liver. Consumption of these beavers would not be hazardous to human health. In the worst case, humans who consume substantial (71 kg) amounts of the flesh of beavers from the Serpent River drainage system would receive <10% of the annual limits set by Canadian regulatory authorities (Clulow et al. 1991).

Cesium-137 levels in grey seals (*Halichoerus grypus*) in 1987 seem to reflect ^{137}Cs levels in their fish diet, but there is no biomagnification of ^{137}Cs and other radionuclides. An estimated 29% of the ^{137}Cs in the diets of grey seals is from the Chernobyl accident and 71% from the nuclear facility at Sellafield, United Kingdom. The dose to grey seals from their diet is about 36 mSv annually and higher than the permissible dose limit of 5 mSv/year allowed the general public, but below the current limit for radiation workers of 50 mSv/year (S.S. Anderson et al. 1990).

The weekly dose rates from internal radionuclides were markedly different in muskrats (*Ondatra zibethicus*) and cotton rats (*Sigmodon hispidus*) collected at Oak Ridge, Tennessee, in August 1960 (20 to 1112 mSv for muskrats vs. 3 mSv for cotton rats). The difference is probably due to differences in diets and habitats (Kaye and Dunaway 1963). Foxes and wildcats contain 2 to 16 times more ^{137}Cs than their prey organisms, such as rats and rabbits (Jenkins et al. 1969), suggesting food-chain magnification. The biological half-time of ^{137}Cs is about 30 days in foxes, dogs, and pigs but about 60 days in humans (Jenkins et al. 1969). Jackrabbits (*Lepus californicus*) in 1958, one year after contamination at the Nevada test site, averaged 1908 Bq ^{90}Sr/kg AW bone within a 160-km radius from ground zero. In 1961, the average for the same population was only 984 Bq ^{90}Sr/kg AW bone, and the few higher values were restricted to older animals (Neel and Larson 1963). The authors concluded that ^{90}Sr from fallout in jackrabbits is at its maximum at an early time after contamination and that biological availability is later reduced by natural (unspecified) mechanisms. Jackrabbits at the Nevada test site also contained certain neutron activation products, including isotopes of Co, Mn, and W (Romney et al. 1971).

Radionuclide concentrations in sheep and cattle grazing near a nuclear fuel reprocessing facility amounted to a small fraction of the recommended limits. ^{241}Am, ^{137}Cs, and $^{239+240}Pu$ in the bone, liver, lung, and muscle of beef cattle were quite low from the vicinity of a nuclear fuel reprocessing facility in England between September and December 1986 and practically indistinguishable from control samples. Maximum concentrations, in Bq/kg FW, were 0.0015 $^{239+240}Pu$ in lung, 0.019 ^{241}Am in liver, and 3.1 ^{137}Cs in muscle (Curtis et al. 1991). Levels of ^{129}I were elevated in thyroids of cows near Mol, Belgium, in 1978 in the vicinity of a nuclear reprocessing plant closed in 1974 (Table 32.14) (Handl et al. 1990).

Table 32.14 Radionuclide Concentrations in Field Collections of Selected Living Organisms (Concentrations are in becquerels per kilogram fresh weight [FW], dry weight [DW], or ash weight [AW].)

Ecosystem, Organism, Radionuclide, and Other Variables	Concentration (Bq/kg[a])	Reference[b]
TERRESTRIAL PLANTS		
Sweet potato, *Ipomoea batatas*, Nagasaki, Japan, 1945, postatomic detonation		
^{137}Cs	0.09 DW	1
$^{239+240}Pu$	0.01 DW	1
Lichens, various species, Alaska and Greenland		
^{238}Pu		
1971 vs. 1972	0.4 DW vs. 0.9 DW	2
1973	0.4 DW	2
$^{239+240}Pu$		
1971 vs. 1972	7.4 DW vs. 10.3 DW	2
1973 vs. 1974	5.4 DW vs. 9.6 DW	2
Reed canarygrass, *Phalaris arundinacea*, Columbia River Washington, 1985–87, near reactor, ^{90}Sr	Max. 1480–1850 DW	3
Largetooth aspen, *Populus grandidentata*, ^{226}Ra		
Near uranium tailing plant vs. control site		
Leaves	53 DW vs. 4 DW	4
Stems	99 DW vs. 5 DW	4
Elliot Lake, Canada, 1984–87 vs. control site		
Leaves	252 DW vs. 46 DW	5
Stems	223 DW vs. 4 DW	5
Trembling aspen, *Populus tremuloides*, ^{226}Ra		
Near uranium tailings plant vs. control site		
Leaves	42 DW vs. 11–15 DW	4
Stems	69 DW vs. 3–11 DW	4
Vegetation		
Belgium, near former nuclear fuel reprocessing plant, 1983, ^{129}I	Max. 0.09 FW	6
California, deer forage plants, three spp., 1968–69, ^{137}Cs	414–514 DW	7
Colorado, mule deer diet, all plants, ^{90}Sr		
1962–63 vs. 1963–64	2242 AW vs. 4499 AW	8
1964–65 vs. 1965–66	3492 AW vs. 2257 AW	8
Colorado, mule deer diet, 8 species of forage plants, ^{90}Sr, 1963–64 vs. 1964–65	1258–17,412 AW vs. 828–16,620 AW	8
Finland, reindeer forage plants, 1961, Lapland		
Lichen, *Cladonia alpestris*		
^{137}Cs	466,200 AW	11
^{90}Sr	53,428 AW	11
Lichen mixture		
^{137}Cs	133,200 AW	11
^{90}Sr	19,980 AW	11
Other forage plants		
^{137}Cs	962–8800 AW	11
^{90}Sr	266–1924 AW	11
Florida, April 1969, ^{137}Cs	Max. 0.65 DW	12
Georgia, deer browse, 29 species, 1965–66		
^{144}Ce	Max. 373 DW	13
^{60}Co	Max. 15 DW	13
^{137}Cs	Max. 104 DW	13
^{54}Mn	Max. 118 DW	13
^{106}Ru	Max. 226 DW	13
^{125}Sb	Max. 56 DW	13

Table 32.14 (continued) Radionuclide Concentrations in Field Collections of Selected Living Organisms (Concentrations are in becquerels per kilogram fresh weight [FW], dry weight [DW], or ash weight [AW].)

Ecosystem, Organism, Radionuclide, and Other Variables	Concentration (Bq/kg[a])	Reference[b]
^{90}Sr	Max. 377 DW; Max. 2005 AW	13
^{95}Zr	Max. 63 DW	13
^{65}Zn	Max. 22 DW	13
Tennessee, 1974, ^{137}Cs, from soil accidentally contaminated in 1944		
Roots	Usually ~3700 DW; Max. 111,000 DW	14
Trees	74–5920 DW	14
Ground vegetation	592–3996 DW	14
Ginger, *Zingiker officinale*, root, Nagasaki, Japan, 1945, postatomic detonation		
^{137}Cs	0.07 DW	1
$^{239+240}Pu$	0.04 DW	1
AQUATIC PLANTS		
Algae, decomposing; Hanford, Washington, 1973; plutonium processing pond		
^{241}Am	9472 DW	15
^{238}Pu	36,482 DW	15
$^{239+240}Pu$	22,755 DW	15
Algae and macrophytes, ^{137}Cs, Hudson River, 1970	1.5–5.6 FW	16
Algae, South Carolina, 1971–72, reactor discharge, ^{137}Cs	12,284 DW	17
Brown algae, *Fucus vesiculosus*		
Ireland, 1985–86, $^{239+240}Pu$, northeast coast vs. western seaboard	3.2 DW vs. 0.09 DW	18
Sweden, 1984, vicinity of nuclear plant		
^{58}Co	20–23 DW	19
^{60}Co	1700–2003 DW	19
^{137}Cs	7–16 DW	19
^{40}K	735–966 DW	19
^{54}Mn	36–60 DW	19
^{65}Zn	90–144 DW	19
Seaweed, *Porphyra* sp., 1974, Cumbrian coast, U.K., <2 km from beach		
^{241}Am	458 FW	20
^{242}Cm	18 FW	20
^{238}Pu	37 FW	20
$^{239+240}Pu$	162 FW	20
Sea lettuce, *Ulva lactuca*, whole, Connecticut, 1960, gross beta activity		
May	5402–6253 AW	21
August	5291–8066 AW	21
December	2183–3700 AW	21
AQUATIC INVERTEBRATES		
Clams, 15 species, freshwater, 1960, Tennessee River, near Oak Ridge, ^{90}Sr, shell	15–921 AW	23
Connecticut, 1960, gross beta activity		
American oyster, *Crassostrea virginica*, soft parts		
May	2553–3589 AW	21
August	2775–4551 AW	21
December	851–1850 AW	21

Table 32.14 (continued) Radionuclide Concentrations in Field Collections of Selected Living Organisms (Concentrations are in becquerels per kilogram fresh weight [FW], dry weight [DW], or ash weight [AW].)

Ecosystem, Organism, Radionuclide, and Other Variables	Concentration (Bq/kg[a])	Reference[b]
Mussel, *Mytilus edulis*, soft parts, August vs. December	3256–4551 AW vs. 3034–3108 AW	21
Crabs, Hudson River, 1970, ^{137}Cs	0.6 FW	16
Crustaceans, marine; fallout radionuclides, typical values		
^{110m}Ag	37.0 FW	24
^{60}Co	24.1 FW	24
^{54}Mn	2.2 FW	24
^{65}Zn	2.6 FW	24
Crustaceans, marine; natural radionuclides, typical values		
^{14}C	22.2 FW	24
^{3}H	0.1 FW	24
^{40}K	92.5 FW	24
^{210}Pb	2.2 FW	24
^{210}Po	37.0 FW	24
^{87}Rb	1.5 FW	24
Molluscs, bivalves, Hudson River, 1970, ^{137}Cs, soft parts	3 FW	16
Molluscs, freshwater; fallout radionuclides, typical values		
^{14}C	4–11 FW	24
^{3}H	0.1–159 FW	24
^{54}Mn	4–518 FW	24
Molluscs, marine; fallout radionuclides, typical values		
$^{141+144}Ce$	5–1813 FW	24
^{57}Co	2–16 FW	24
^{60}Co	1–26 FW	24
^{137}Cs	5–25 FW	24
^{55}Fe	14–5180 FW	24
^{54}Mn	2–222 FW	24
^{63}Ni	1–555 FW	24
^{239}Pu	Max. 0.02 FW	24
$^{103+106}Ru$	1–518 FW	24
^{110m}Ag	0.1–155 FW	24
^{65}Zn	0.7–425 FW	24
$^{95}Zr/^{95}Nb$	3–925 FW	24
Molluscs, marine; natural radionuclides		
^{14}C	18 FW	24
^{3}H	0.1 FW	24
^{40}K	107 FW	24
^{210}Pb	0.3 FW	24
^{210}Po	25 FW	24
^{87}Rb	2 FW	24
Mussel, *Mytilus edulis*, soft parts		
Irish coastal waters, August 1988		
^{134}Cs	<0.7 DW	18
^{137}Cs	Usually <3 DW; Max. 9 DW	18
^{40}K	182–355 DW	18
^{238}Pu	Usually <0.003 DW; Max. 0.21 DW	18
$^{239+240}Pu$	Usually <0.035 DW; Max. 1 DW	18
England, 1986–87, near nuclear plant		
^{110m}Ag	13 FW	25
^{241}Am	9–15 FW	25
^{144}Ce	6 FW	25

Table 32.14 (continued) Radionuclide Concentrations in Field Collections of Selected Living Organisms (Concentrations are in becquerels per kilogram fresh weight [FW], dry weight [DW], or ash weight [AW].)

Ecosystem, Organism, Radionuclide, and Other Variables	Concentration (Bq/kg[a])	Reference[b]
^{60}Co	3–7 FW	25
^{134}Cs	11 FW	25
^{137}Cs	5–31 FW	25
^{40}K	25–188 FW	25
^{95}Nb	3–106 FW	25
^{103}Ru	4–169 FW	25
^{106}Ru	64–151 FW	25
^{95}Zr	3–36 FW	25
Plankton, ^{137}Cs, Hudson River, 1970	2 FW	16
Plankton, Italy, 1971, near nuclear power station		
^{60}Co	Max. 203 FW	26
^{137}Cs	Max. 1113 FW	26
Plankton, marine; fallout radionuclides, typical values		
$^{141+144}Ce$	14–17,760 FW	24
^{57}Co	85 FW	24
^{60}Co	11–592 FW	24
^{137}Cs	18–1332 FW	24
^{155}Eu	14 FW	24
^{54}Mn	196 FW	24
^{63}Ni	4–14 FW	24
^{147}Pm	122 FW	24
$^{103+106}Ru$	11–1110 FW	24
^{125}Sb	33 FW	24
^{90}Sr	0.7–12 FW	24
$^{95}Zr/^{95}Nb$	74–29,600 FW	24
Plankton, marine; natural radionuclides, typical values		
^{14}C	11 FW	24
^{3}H	0.1 FW	24
^{40}K	92 FW	24
^{210}Pb	9–25 FW	24
^{210}Po	22–62 FW	24
^{226}Ra	0.7 FW	24
^{228}Th	0.4–2 FW	24
^{234}U	0.7–2 FW	24
^{235}U	0.02–0.07 FW	24
^{238}U	0.7–2 FW	24
Polychaete annelid worms, marine; England, 1984–86; near nuclear plant vs. control location		
Arenicola marina		
^{137}Cs	132–321 FW vs. 3 FW	25
^{40}K	162–307 FW vs. 90 FW	25
^{238}Pu	14–16 FW vs. <0.05 FW	25
$^{239+240}Pu$	60–72 FW vs. 0.01 FW	25
Nereis diversicolor		
^{137}Cs	41–358 FW vs. 6 FW	25
^{40}K	23–148 FW vs. 134 FW	25
^{238}Pu	6–11 FW vs. <0.02 FW	25
$^{239+240}Pu$	25–48 FW vs. 0.03 FW	25
Clam, *Rangia cuneata*, Neuse River, North Carolina, 1965–67, soft parts; before Chinese nuclear tests in May and December 1966 vs. posttest		
^{144}Ce	5.3 FW vs. 7.2 FW	27, 28
^{137}Cs	1.0 FW vs. 1.6 FW	27, 28

Table 32.14 (continued) Radionuclide Concentrations in Field Collections of Selected Living Organisms (Concentrations are in becquerels per kilogram fresh weight [FW], dry weight [DW], or ash weight [AW].)

Ecosystem, Organism, Radionuclide, and Other Variables	Concentration (Bq/kg^a)	Reference[b]
^{55}Fe	0.12 FW vs. 0.75 FW	27, 28
^{54}Mn	2.5 FW vs. 2.7 FW	27, 28
^{106}Ru	2.1 FW vs. 2.7 FW	27, 28
^{65}Zn	0.4 FW vs. 0.8 FW	27, 28
Sea urchin, *Strongylocentrotus purpuratus*, 1966		
^{210}Pb	Max. 2 AW	29
^{210}Po	Max. 7 AW	29
FISH		
Goldfish, *Carassius auratus* from plutonium processing waste pond, Hanford, Washington, 1973		
^{241}Am, whole vs. muscle	399 DW vs. 14 DW	15
$^{238+239+240}Pu$, whole vs. muscle	351 DW vs. 10 DW	15
Colorado, 1965–66, ^{137}Cs, muscle, maximum values		
Cutthroat trout, *Oncorhynchus clarki*	59 FW	30
Rainbow trout, *Oncorhynchus mykiss*	117 FW	30
Sockeye (kokanee) salmon, *Oncorhynchus nerka*	8 FW	30
Brook trout, *Salvelinus fontinalis*	215 FW	30
Lake trout, *Salvelinus namaycush*	25 FW	30
Brown trout, *Salmo trutta*	121 FW	30
Columbia River, Washington; near nuclear facility, 1961, ^{239}Np, muscle		
Chiselmouth, *Acrocheilus alutaceus*	Max. 14,900 FW	31
Bridgelip sucker, *Catostomus columbianus*	Max. 5600 FW	31
Largescale sucker, *Catostomus macrocheilus*	Max. 3600 FW	31
Mountain whitefish, *Prosopium williamsoni*	Max. 18,800 FW	31
Freshwater fish, whole body; fallout radionuclides, typical values		
^{14}C	4–7 FW	24
^{137}Cs	1–973 FW	24
^{55}Fe	1–3 FW	24
^{3}H	0.1–159 FW	24
^{54}Mn	11 FW	24
^{85}Sr	0.04–0.4 FW	24
^{89}Sr	0.2–40 FW	24
^{90}Sr	0.04–177 FW	24
$^{95}Zr/^{95}Nb$	2.2–2.6 FW	24
Freshwater fish, 1963–64, ^{210}Pb, bone vs. soft tissues	2.5 AW vs. 0.2 AW	29
Freshwater fish, whole body, ^{137}Cs; Red Lakes, Minnesota		
1954–57	0.7–2.4 FW	32
1959–62	3–12 FW	32
1963–66	8–22 FW	32
Freshwater fish, typical maximum concentrations, whole body		
^{3}H	0.5 FW	24
^{40}K	130 FW	24
^{87}Rb	8 FW	24
^{238}U	0.1 FW	24

Table 32.14 (continued) Radionuclide Concentrations in Field Collections of Selected Living Organisms (Concentrations are in becquerels per kilogram fresh weight [FW], dry weight [DW], or ash weight [AW].)

Ecosystem, Organism, Radionuclide, and Other Variables	Concentration (Bq/kg[a])	Reference[b]
^{234}U	0.2 FW	24
^{226}Ra	129 FW	24
^{210}Pb (bone)	3 FW	24
^{210}Po (liver)	18 FW	24
^{232}Th	0.05 FW	24
^{235}U	0.004 FW	24
Mosquitofish, *Gambusia holbrooki*, ^{137}Cs, April 1987; from South Carolina reservoir contaminated with ^{137}Cs between 1961 and 1964, whole body	Max. 2230 FW	33
Hudson River, 1970, ^{137}Cs		
Atlantic sturgeon, *Acipenser oxyrhynchus*, muscle	0.6 FW	16
American eel, *Anguilla rostrata*, muscle	1.3 FW	16
Mummichog, *Fundulus heteroclitus*, whole	2.0 FW	16
Catfish, *Ictalurus* sp., muscle	1.9 FW	16
White perch, *Morone americana*, muscle vs. whole body	0.8 FW vs. 0.8 FW	16
Striped bass, *Morone saxatilis*, muscle	0.9 FW	16
Yellow perch, *Perca flavescens*, muscle	1.5 FW	16
Italy, 1971, near nuclear power station, whole fish, various species		
^{60}Co	Max. 9 DW	26
^{137}Cs	Max. 104 DW	26
Lake Ontario, ^{137}Cs, 1981		
Common carp, *Cyprinus carpio*, bone vs. other tissues	5 FW vs. <5 FW	34
Northern pike, *Esox lucius*		
Bone, liver	5 FW	34
Roe	15 FW	34
Other tissues	<5 FW	34
Coho salmon, *Oncorhynchus kisutch*		
GI tract	5 FW	34
Liver	13 FW	34
Other tissues	<5 FW	34
Largemouth bass, *Micropterus salmoides*, South Carolina, reactor discharge, 1971–72, ^{137}Cs, whole	3677 DW	17
Micropterus salmoides, South Carolina, muscle, 5 lakes; May–June 1993, ^{137}Cs	109–4607 DW	60
Marine fishes, whole body, fallout radionuclides, typical values		
^{110m}Ag	2–3 FW	24
$^{141+144}Ce$	2–1036 FW	24
^{60}Co	1–13 FW	24
^{137}Cs	2–3 FW	24
^{55}Fe		
Gonad	8140–10,360 FW	24
Liver	59,940–68,820 FW	24
Muscle	37–3922 FW	24
^{54}Mn	0.07–2 FW	24
^{239}Pu	Max. 0.005 FW	24
$^{103+106}Ru$	2–244 FW	24
$^{95}Zr/^{95}Nb$	1–277 FW	24
^{65}Zn	2–7 FW	24

Table 32.14 (continued) Radionuclide Concentrations in Field Collections of Selected Living Organisms (Concentrations are in becquerels per kilogram fresh weight [FW], dry weight [DW], or ash weight [AW].)

Ecosystem, Organism, Radionuclide, and Other Variables	Concentration (Bq/kg[a])	Reference[b]
Marine fishes, whole body, natural radionuclides, typical values		
^{14}C	15 FW	24
^{3}H	0.1 FW	24
^{40}K	92 FW	24
^{210}Pb	5 FW	24
^{210}Po	33 FW	24
^{226}Ra	0.2 FW	24
^{87}Rb	1 FW	24
^{234}U	1 FW	24
^{235}U	0.05 FW	24
^{238}U	1 FW	24
Golden shiner, *Notemigonus crysoleucas*, whole, ^{137}Cs, Hudson River estuary		
1966 vs. 1968	0.9 FW vs. 0.8 FW	16
1969 vs. 1970	0.7 FW vs. 0.5 FW	16
Oceanic fishes, 1962–64, bone vs. soft parts		
^{210}Pb	10 AW vs. 0.06 AW	29
^{210}Po	12 AW vs. 0.1 AW	29
^{226}Ra	2 AW vs. 0.06 AW	29
Plaice, *Pleuronectes platessa*, near nuclear facility, England		
1968 vs. 1969, ^{137}Cs		
Gut contents	44–181 FW vs. 126–266 FW	35
Muscle	26–70 FW vs. 89–152 FW	35
1968, gut contents		
^{144}Ce	880–1150 FW	35
^{106}Ru	1343–5143 FW	35
$^{95}Zr/^{95}Nb$	3122–5794 FW	35
Albacore, *Thunnus alalunga*, southern California near San Diego, liver		
Summer 1964 vs. summer 1965		
^{110m}Ag	3 FW vs. 4 FW	36
^{60}Co	7 FW vs. 7 FW	36
^{40}K	71 FW vs. 72 FW	36
^{54}Mn	39 FW vs. 22 FW	36
^{65}Zn	46 FW vs.14 FW	36
Summer 1968 vs. summer 1970		
^{60}Co	2 FW vs. 2 FW	36
^{40}K	81 FW vs. 78 FW	36
^{54}Mn	2 FW vs. 0.6 FW	36
^{65}Zn	25 FW vs. 9 FW	36
Yellowfin tuna, *Thunnus albacares*, 1968, near San Diego, liver		
^{60}Co	1 FW	36
^{40}K	93 FW	36
^{54}Mn	1 FW	36
^{65}Zn	3 FW	36
Tunas, 1970–71, Hawaii, liver		
^{108m}Ag	0.03–2 FW	36
^{110m}Ag	0.01–7 FW	36
^{60}Co	0.9–3 FW	36
^{40}K	68–83 FW	36
^{65}Zn	5–27 FW	36

Table 32.14 (continued) Radionuclide Concentrations in Field Collections of Selected Living Organisms (Concentrations are in becquerels per kilogram fresh weight [FW], dry weight [DW], or ash weight [AW].)

Ecosystem, Organism, Radionuclide, and Other Variables	Concentration (Bq/kg[a])	Reference[b]
REPTILES		
Snakes, two species (*Elaphe obsoleta*, *Nerodia taxispilota*), Aiken, South Carolina; whole animal, ^{137}Cs		
Site contaminated with ^{137}Cs between 1961 and 1970, *Elaphe* vs. *Nerodia*		
1972	6037 FW vs. 7629 FW	58
1976	592 FW vs. 1333 FW	58
1980	296 FW vs. 1037 FW	58
Uncontaminated site, both species, 1972–80	<37 FW	58
Snakes, 19 species, whole, vicinity of Aiken, South Carolina, March 1971–November 1972, $^{134+137}Cs$		
Near reactor effluent stream	4870 FW, Max. 38,200 FW	9
Near reactor cooling reservoir	1025 FW, Max. 5159 FW	9
Uncontaminated habitats	92 FW	9
Slider turtle, *Trachemys scripta*, from radioactive reservoirs, Aiken, South Carolina, whole body		
High-level waste pond vs. low-level waste pond		
^{137}Cs	3020 FW vs. 1002 FW	37
^{90}Sr	94,030 FW vs. 2236 FW	37
Control sites		
^{137}Cs	0.001 FW	37
^{90}Sr	0.2 FW	37
Turtles, southeastern U.S., 1970, ^{90}Sr, exoskeleton		
Snapping turtle, *Chelydra serpentina*	784 (284–1283) AW	38
Gopher tortoise, *Gopherus polyphemus*	4765 AW	38
Common mud turtle, *Kinosternon sabrubrum*	1309 (569–2904) AW	38
Missouri slider, *Trachymys floridana hoyi*	1761 AW	38
Peninsula cooter, *Pseudemys floridana pennisularis*	33 (ND–48) AW	38
Pond slider, *Pseudemys scripta*	777 (188–2190) AW	38
Loggerhead musk turtle, *Sternotherus minor*	24 (ND–48) AW	38
Common musk turtle, *Sternotherus odoratus*	525 (52–999) AW	38
Common box turtle, *Terrapene carolina*	1087 (48–2856) AW	38
BIRDS		
Wood duck, *Aix sponsa*; 1991–92; from abandoned (in 1964) reactor cooling reservoir; eggs, ^{137}Cs		
Whole egg	113 FW	59
Albumin	1096 DW	59
Shell vs. yolk	132 DW vs. 98 DW	59
Ruffed grouse, *Bonasa umbellus*; near uranium tailings discharge, Canada, Elliot Lake, 1987–88, ^{226}Ra		
Bone vs. gut contents	10–28 DW vs. 7–22 DW	4
Liver vs. muscle	5–12 DW vs. 1.5–1.9 DW	4
Canada goose, *Branta canadensis moffitti*; Columbia River, Washington, 1985–87, near reactor; eggshell, ^{90}Sr	18–60 DW	39

Table 32.14 (continued) Radionuclide Concentrations in Field Collections of Selected Living Organisms (Concentrations are in becquerels per kilogram fresh weight [FW], dry weight [DW], or ash weight [AW].)

Ecosystem, Organism, Radionuclide, and Other Variables	Concentration (Bq/kg[a])	Reference[b]
American coot, *Fulica americana*; Hanford, Washington, June 1974–January 1977		
^{137}Cs (Hanford vs. control ponds)		
Bone	7400 vs. 37 DW	40
Gut contents	125,800 vs. 29 DW	40
Liver	16,280 vs. 26 DW	40
Muscle	21,090 vs. 0.7 DW	40
^{90}Sr (Hanford only)		
Bone	96 DW	40
Gut contents	159 DW	40
Liver	18 DW	40
Muscle	10 DW	40
Barn swallow, *Hirundo rustica*; Idaho, 1976–77, nesting near radioactive leaching ponds		
Whole adults		
^{140}Ba	800 FW	41
^{134}Cs	1300 FW	41
^{137}Cs	6400 FW	41
^{51}Cr	16,100 FW	41
^{60}Co	1480 FW	41
^{131}I, whole vs. thyroid	5500 FW vs. 3,330,000 FW	41
^{24}Na	8600 FW	41
^{75}Se	5000 FW	41
^{65}Zn	5900 FW	41
Nests		
^{140}Ba	1200 DW	41
^{134}Cs	13,800 DW	41
^{137}Cs	92,000 DW	41
^{141}Ce	1200 DW	41
^{144}Ce	4000 DW	41
^{51}Cr	230,000 DW	41
^{131}I	800 DW	41
^{65}Zn	1800 DW	41
Massachusetts, 1973–75, 15 passerine species, trapped near nuclear power station, whole body		
Common bobwhite, *Colinus virginianus*		
^{137}Cs	Max. 73 FW	42
^{131}I	Max. 6 FW	42
^{40}K	Max. 131 FW	42
^{95}Zr–^{95}Nb	Max. 4 FW	42
Bluejay, *Cyanocitta cristata*		
^{137}Cs	28 FW; Max. 65 FW	42
^{131}I	1 FW; Max. 9 FW	42
^{40}K	96 FW; Max. 181 FW	42
^{95}Zr–^{90}Nb	2 FW; Max. 6 FW	42
13 species		
^{137}Cs	Max. 82 FW	42
^{131}I	Max. 18 FW	42
^{40}K	Max. 268 FW	42
^{95}Zr–^{95}Nb	Max. 40 FW	42
United Kingdom, Ravenglass estuary, 1980–84, near nuclear plant		
Mallard, *Anas platyrhynchos*		
^{134}Cs, muscle	87 FW	25
^{137}Cs, muscle vs. liver	167 FW vs. 126 FW	25

Table 32.14 (continued) Radionuclide Concentrations in Field Collections of Selected Living Organisms (Concentrations are in becquerels per kilogram fresh weight [FW], dry weight [DW], or ash weight [AW].)

Ecosystem, Organism, Radionuclide, and Other Variables	Concentration (Bq/kg[a])	Reference[b]
$^{239+240}Pu$, liver	3.4 FW	25
^{238}Pu, liver	1.1 FW	25
Greylag goose, *Anser anser*		
^{137}Cs, muscle vs. liver	58 FW vs. 28 FW	25
^{238}Pu, muscle vs. liver	0.03 FW vs. 3 FW	25
$^{239+240}Pu$, muscle vs. liver	0.1 FW vs. 13 FW	25
Carrion crow, *Corvus corone*		
^{137}Cs, Ravenglass vs. control location		
Muscle	162 FW vs. 17 FW	25
Liver	131 FW vs. 8 FW	25
Lesser black-backed gull, *Larus marinus*		
^{137}Cs, muscle vs. liver	158 FW vs. 163 FW	25
$^{239+240}Pu$, muscle vs. liver	0.1 FW vs. 5 FW	25
Black-headed gull, *Larus ridibundus*, whole chick		
^{134}Cs	0.8 FW	25
^{137}Cs	25 FW	25
^{238}Pu	0.1 FW	25
$^{239+240}Pu$	0.5 FW	25
Oystercatcher, *Haematopus ostralegus*, Ravenglass vs. control location		
^{137}Cs		
Muscle	613 FW vs. 22 FW	25
Liver	463 FW vs. 20 FW	25
^{238}Pu		
Muscle	0.2 FW vs. <0.01 FW	25
Liver	1.8 FW vs. 0.04 FW	25
$^{239+240}Pu$		
Muscle	0.5 FW vs. 0.04 FW	25
Liver	4.1 FW vs. 0.09 FW	25
Bar-tailed godwit, *Limosa lapponica lapponica*		
^{137}Cs, muscle vs. liver	478 FW vs. 510 FW	25
^{238}Pu, muscle vs. liver	<0.02 FW vs. 0.2 FW	25
$^{239+240}Pu$, muscle vs. liver	0.03 FW vs. 0.9 FW	25
Merganser, *Mergus serrator*		
^{134}Cs, muscle vs. liver	8 FW vs. 13 FW	25
^{137}Cs, muscle vs. liver	144 FW vs. 251 FW	25
^{238}Pu, muscle vs. liver	<0.01 FW vs. <0.04 FW	25
$^{239+240}Pu$, muscle vs. liver	0.02 FW vs. <0.04 FW	25
Curlew, *Numenius arquata*		
^{137}Cs, Ravenglass vs. control location		
Muscle	140 FW vs. 49 FW	25
Liver	104 FW vs. 99 FW	25
^{238}Pu, Ravenglass vs. control location		
Muscle	0.09 FW vs. <0.02 FW	25
Liver	0.14 FW vs. <0.05 FW	25
$^{239+240}Pu$, Ravenglass vs. control location		
Muscle	0.09 FW vs. <0.02 FW	25
Liver	0.14 FW vs. <0.05 FW	25
MARINE MAMMALS		
Bearded seal, *Erignathus barbatus*; Alaska, 1963		
Bone		
^{210}Pb	Max. 2.7 AW	29
^{226}Ra	2.4 AW	29

Table 32.14 (continued) Radionuclide Concentrations in Field Collections of Selected Living Organisms (Concentrations are in becquerels per kilogram fresh weight [FW], dry weight [DW], or ash weight [AW].)

Ecosystem, Organism, Radionuclide, and Other Variables	Concentration (Bq/kg[a])	Reference[b]
Soft tissues, ^{210}Pb	Max. 0.2 AW	29
Grey seal, *Halichoerus grypus*, North Sea and northeast Atlantic Ocean, 1987		
Females, milk vs. muscle		
^{241}Am	<0.0002 FW vs. <0.0005 FW	43
^{134}Cs	0.6 (0.4–0.7) FW vs. <0.002 FW	43
^{137}Cs	2.9 (1.1–4.8) FW vs. 14.3 FW	43
^{40}K	107 (67–215) FW vs. 0.2 FW	43
^{238}Pu	<0.0002 FW vs. <0.0005 FW	43
$^{239+240}Pu$	<0.0002 FW vs. <0.0005 FW	43
Pup, muscle vs. liver		
^{241}Am	<0.0003 FW vs. <0.0003 FW	43
^{134}Cs	Max. 0.003 FW vs. Max. 0.001 FW	43
^{137}Cs	Max. 0.03 FW vs. Max. 0.02 FW	43
^{40}K	Max. 0.2 FW vs. Max. 0.2 FW	43
^{238}Pu	Max. 0.0005 FW vs. Max. 0.001 FW	43
$^{239+240}Pu$	Max. 0.002 FW vs. Max. 0.004 FW	43
Spotted seal, *Phoca largha*; Alaska, 1963, bone vs. soft tissues		
^{210}Pb	2 AW vs. 0.1 AW	29
^{226}Ra	3 AW vs. No Data	29
Sperm whale, *Physeter macrocephalus*; Alaska, 1965, bone vs. soft tissue		
^{210}Pb	135 AW vs. 0.37 AW	29
^{210}Po	114 AW vs. 23 AW	29
TERRESTRIAL MAMMALS		
Cattle, *Bos* sp.		
Nevada, 1973, grazing for 3 years in area contaminated in 1957 with transuranic radionuclides		
^{241}Am		
Bone vs. liver	Max. 1 FW vs. Max. 0.6 FW	44
Lymph nodes vs. lungs	Max. 24 FW vs. Max. 2 FW	44
Other tissues	<0.6 FW	44
^{238}Pu		
Lungs, lymph nodes	Max. 3 FW	44
Testes	Max. 0.8 FW	44
Other tissues	<0.6 FW	44
$^{239+240}Pu$		
Bone vs. liver	Max. 3 FW vs. Max. 34 FW	44
Lungs vs. lymph nodes	Max. 34 FW vs. Max. 85 FW	44
Muscle vs. other tissues	Max. 7 FW vs. <1.2 FW	44
Europe, ^{129}I, thyroids		
1978		
Belgium vs. Germany	0.017–3.7 FW vs. Max. 0.03 FW	6
Italy vs. Netherlands	Max. 0.05 FW vs. Max. 0.03 FW	6
1979, Netherlands	Max. 0.07 FW	6
1980, Netherlands	0.07–0.6 FW	6
1981, Germany	Max. 0.02 FW	6
Common beaver, *Castor canadensis*; Canada, 1984–87, adults, ^{226}Ra; from watershed containing uranium tailings vs. control site		
Bone	115 DW vs. 20 DW	5
Gut contents	62 DW vs. 9 DW	5

Table 32.14 (continued) Radionuclide Concentrations in Field Collections of Selected Living Organisms (Concentrations are in becquerels per kilogram fresh weight [FW], dry weight [DW], or ash weight [AW].)

Ecosystem, Organism, Radionuclide, and Other Variables	Concentration (Bq/kg[a])	Reference[b]
Kidney	9 DW vs. 2 DW	5
Liver	2.7 DW vs. 1.4 DW	5
Muscle	2.9 DW vs. 1.0 DW	5
Georgia and South Carolina, 1964–66, ^{137}Cs, whole organism		
Domestic dog, *Canis familiaris*	23 FW	45
Coyote, *Canis latrans*	26 FW	45
Bobcat, *Lynx rufus*	117–561 FW	45
Cotton rat, *Sigmodon hispidus*	16–29 FW	45
Eastern cottontail, *Sylvilagus floridanus*	19–35 FW	45
Gray fox, *Urocyon cinereoargentatus*	34–169 FW	45
Red fox, *Vulpes fulva*	23–60 FW	45
Humans, *Homo sapiens*; Denmark, ^{137}Cs, annual dietary loading		
1964	71.9 FW	46
1985	1.4 FW	46
1986–87	12.6 FW	46
Black-tailed jack rabbit, *Lepus californicus*; Nevada test site, bone, ^{90}Sr		
1952–66	74–476 AW	47
1958 (1-year postdetonation), ground zero vs. 32–700 km distant	373 AW vs. 88–198 AW	48
1959, ground zero	329 AW	48
1959, 32 km vs. 120–700 km	466 AW vs. 95–222 AW	48
1961, within 160 km of ground zero	143 AW	48
Mule deer, *Odocoileus hemionus*; 1961–65, Colorado, femur, ^{90}Sr		
1961–62 vs. 1962–63	Max. 215 AW vs. Max. 528 AW	8
1963–64 vs. 1964–65	Max. 777 AW vs. Max. 637 AW	8
Black-tailed deer, *Odocoileus hemionus columbianus*; California		
Muscle vs. rumen contents, 1968–69, ^{137}Cs		
Summer	37 DW vs. 48 DW	7
Fall	33 DW vs. 37 DW	7
Winter	48 DW vs. 67 DW	7
Mendocino County, California, mandible, yearlings, ^{90}Sr		
1952–53 vs. 1954	3–11 AW vs. 26–34 AW	49
1955 vs. 1956	29–124 AW vs. 112 AW	49
1957 vs. 1958	87–239 AW vs. 134–228 AW	49
1959 vs. 1960	243–533 AW vs. 204–332 AW	49
White-tailed deer, *Odocoileus virginianus*		
Georgia, 1965–66		
^{137}Cs		
Heart vs. kidney	127 FW vs. 149 FW	13
Liver vs. lung	70 FW vs. 73 FW	13
Muscle vs. spleen	126 FW vs. 126 FW	13
Tongue	172 FW	13
^{90}Sr, mandible		
Age 1.5 years	940 AW	13
Age 2.5 years	828 AW	13
Age 3.5 years	799 AW	13
Southeastern United States		
^{137}Cs, muscle, 1967–71		
Alluvial region (LA, MS, FL, SC, NC)	85 (9–650) FW	50
Lower Coastal Plain (SC, GA, FL, VA, NC)	1036 (9–5658) FW	50

Table 32.14 (continued) Radionuclide Concentrations in Field Collections of Selected Living Organisms (Concentrations are in becquerels per kilogram fresh weight [FW], dry weight [DW], or ash weight [AW].)

Ecosystem, Organism, Radionuclide, and Other Variables	Concentration (Bq/kg[a])	Reference[b]
Mountain region (WV, KY, MD, NC, TN, GA)	78 (9–401) FW	50
Piedmont region (GA, SC, AL)	105 (9–383) FW	50
Upper Coastal Plain region (MD, NC, GA, VA, MS, LA, AK)	154 (9–1752) FW	50
^{90}Sr, bone, 1969		
Lower Coastal Plain	1172 (376–1766) FW	50
Mountain region	499 (148–888) FW	50
Piedmont region	471 (263–683) FW	50
Muskrat, *Ondatra zibethicus*; August 1960, Oak Ridge, Tennessee, from settling basin for radioactive wastes, single most radioactive animal		
Brain vs. eyes		
^{60}Co	10,545 DW vs. 39,960 DW	51
^{137}Cs	392,200 DW vs. 640,100 DW	51
^{65}Zn	21,016 DW vs. 36,593 DW	51
Femur		
^{60}Co	5920 DW	51
^{137}Cs	121,360 DW	51
^{90}Sr	7,030,000 DW	51
^{65}Zn	28,601 DW	51
Kidney vs. spleen		
^{60}Co	279,720 DW vs. 47,730 DW	51
^{137}Cs	954,600 DW vs. 799,200 DW	51
Liver		
^{60}Co	156,880 DW	51
^{137}Cs	629,000 DW	51
^{65}Zn	78,440 DW	51
Muscle		
^{60}Co	8103 DW	51
^{134}Cs	13,949 DW	51
^{137}Cs	1,265,400 DW	51
^{65}Zn	19,610 DW	51
Teeth		
^{137}Cs	64,010 DW	51
^{90}Sr	9,916,000 DW	51
^{65}Zn	25,789 DW	51
Pelt		
^{60}Co	15,022 DW	51
^{137}Cs	204,980 DW	51
^{90}Sr	37,000 DW	51
^{65}Zn	26,196 DW	51
Domestic sheep, *Ovis aires*		
Near nuclear fuel reprocessing facility vs. control site, England, 1983		
Bone		
^{241}Am	1 FW vs. 0.003 FW	52
$^{239+240}$Pu	0.6 FW vs. 0.002 FW	52
Liver		
^{241}Am	1 FW vs. 0.002 FW	52
^{137}Cs	8 FW vs. 0.2 FW	52
$^{239+240}$Pu	2 FW vs. 0.008 FW	52
Lung		
^{241}Am	0.3 FW vs. 0.003 FW	52
$^{239+240}$Pu	0.4 FW vs. 0.002 FW	52

Table 32.14 (continued) Radionuclide Concentrations in Field Collections of Selected Living Organisms (Concentrations are in becquerels per kilogram fresh weight [FW], dry weight [DW], or ash weight [AW].)

Ecosystem, Organism, Radionuclide, and Other Variables	Concentration (Bq/kg[a])	Reference[b]
Muscle		
^{241}Am	0.03 FW vs. 0.0005 FW	52
^{137}Cs	49 FW vs. 0.2 FW	52
$^{239+240}Pu$	0.007 FW vs. 0.0008 FW	52
Near nuclear fuel reprocessing plant, England, winter 1986–87		
^{241}Am		
Bone vs. liver	0.03–0.7 FW vs. 0.03–0.8 FW	53
Lung vs. muscle	0.009–0.1 FW vs. 0.002–0.03 FW	53
Whole sheep	0.27–4 FW	53
^{137}Cs		
Bone vs. liver	1.3–14 FW vs. 1.8–30 FW	53
Lung vs. muscle	1.5–16 FW vs. 4.6–42 FW	53
Whole sheep	159–748 FW	53
$^{239+240}Pu$		
Bone vs. liver	0.024–0.2 FW vs. 0.07–0.9 FW	53
Lung vs. muscle	0.005–0.02 FW vs. 0.0005–0.005 FW	53
Whole sheep	0.02–2 FW	53
Serbia, 1988, wildlife		
Roe deer, *Capreolus* sp.; bone vs. muscle		
^{137}Cs	ND vs. 0.2 AW	54
^{40}K	23 AW vs. 39 AW	54
^{90}Sr	6 AW vs. 0.6 AW	54
Fallow deer, *Dama* sp.; bone vs. muscle		
^{137}Cs	ND vs. 0.1 AW	54
^{40}K	8 AW vs. 45 AW	54
^{90}Sr	10 AW vs. 0.3 AW	54
Wild hare, *Lepus* sp.; bone vs. muscle		
^{137}Cs	ND vs. 0.1 AW	54
^{40}K	26 AW vs. 52 AW	54
^{90}Sr	18 AW vs. ND	54
Wild boar, *Sus scrofa*; bone vs. muscle		
^{137}Cs	ND vs. 0.4 AW	54
^{40}K	21 AW vs. 56 AW	54
^{90}Sr	34 AW vs. 2 AW	54
Common shrew, *Sorex araneus*; 1988, England, muscle; shrews from mineral soils vs. peaty soils		
^{134}Cs	7 FW vs. 16 FW	55
^{137}Cs	58 FW vs. 161 FW	55
INTEGRATED STUDIES		
Brazil, site of radiological accident in September 1987 at Goiania wherein ^{137}Cs was deposited on soil for 3 weeks before remedial action. Rainwater runoff contaminated the waterways		
3 weeks postaccident, up to 12 km from accident area, ^{137}Cs		
Fish muscle	Max. 200 FW	56
Sediments	Max. 1300 DW	56
Surface waters and suspended particulates	<1 FW	56
10 months postaccident, up to 80 km downstream, ^{137}Cs		
Fish muscle		
Pike, *Hoplias* sp.	14 FW	56

Table 32.14 (continued) Radionuclide Concentrations in Field Collections of Selected Living Organisms (Concentrations are in becquerels per kilogram fresh weight [FW], dry weight [DW], or ash weight [AW].)

Ecosystem, Organism, Radionuclide, and Other Variables	Concentration (Bq/kg[a])	Reference[b]
Piranha, *Seerasalmus* sp.	10 FW	56
Sediments	100 DW	56
Water hyacinth, *Eichhornia* sp.	Max. 0.4 FW	56
Great Lakes, ^{137}Cs, 1981		
Aquatic plants vs. clams	1.4 FW vs. 0.3 FW	34
Fish vs. plankton	1.5 FW vs. 0.1 FW	34
Sediments vs. water	24 FW vs. 0.0007 FW	34
Irish Sea and North Sea, 1983, invertebrates vs. fish		
^{241}Am	Max. 75 FW vs. Max. 0.05 FW	57
^{242}Cm	Max. 2 FW vs. Max. 0.0003 FW	57
$^{243+244}Cm$	Max. 0.5 FW vs. 0.0003 FW	57
^{238}Pu	14 FW vs. 0.01 FW	57
$^{239+240}Pu$	54 FW vs. 0.04 FW	57
^{241}Pu, invertebrates only	Max. 1000 FW	57
Japan, Nagasaki, 1945 post atomic detonation		
Fish vs. snail		
^{137}Cs	0.01 DW vs. 0.02 DW	1
$^{239+240}Pu$	0.03 DW vs. 0.03 DW	1
South Carolina, watershed of a former reactor effluent stream, ^{137}Cs, 1971 vs. 1981		
Plants	14,000–19,000 DW vs. 2600–9600 DW	10
Arthropods	9600–16,000 DW vs. 700–3300 DW	10
South Carolina; reactor cooling impoundment accidentally contaminated in 1961–64 with ^{137}Cs, ^{90}Sr, and various transuranics; samples collected September 1983–February 1984		
^{137}Cs		
Water vs. sediments	0.76 FW vs. Max. near 40,000 DW	22
Aquatic macrophytes vs. benthic invertebrates	Max. near 30,000 DW vs. 930–14,000 DW	22
Fish muscle	2100–8000 FW; 21,000 DW	22
Turtle muscle	2100 FW	22
Waterfowl muscle	3100 FW; 15,000 DW	22
^{90}Sr		
Water vs. sediments	0.14 FW vs. Max. near 400 DW	22
Aquatic macrophytes vs. benthic invertebrates	Max. 2600 DW vs. 42–7900 DW	22
Fish bone ash vs. fish muscle	12,000–23,000 DW vs. 86–470 DW	22
Turtle shell and bone ash	12,000 DW	22
Waterfowl muscle vs. waterfowl bone ash	14 DW vs. 420 DW	22
^{238}Pu		
Water vs. sediments	0.0000034 FW vs. Max. 10 DW	22
Aquatic macrophytes vs. fish muscle	Max. 0.5 DW vs. 0.004 DW	22
Turtle shell ash vs. waterfowl bone ash	0.1 DW vs. 100 DW	22
Waterfowl muscle	0.013 DW	22
$^{239+240}Pu$		
Water	0.0000088 FW	22
Sediments	Max. near 85 DW	22
Aquatic macrophytes	Max. near 1.2 DW	22
Turtle shell ash	ND	22
Waterfowl muscle	0.008 DW	22
^{241}Am		
Water vs. sediments	0.000023 FW vs. Max. 40 DW	22
Turtle shell ash vs. waterfowl muscle	ND vs. 0.015 DW	22

Table 32.14 (continued) Radionuclide Concentrations in Field Collections of Selected Living Organisms (Concentrations are in becquerels per kilogram fresh weight [FW], dry weight [DW], or ash weight [AW].)

Ecosystem, Organism, Radionuclide, and Other Variables	Concentration (Bq/kg[a])	Reference[b]
^{244}Cm		
Water vs. sediments	0.00064 FW vs. Max. 18 DW	22
Fish liver vs. turtle shell ash	11 DW vs. 0.2 DW	22
Waterfowl muscle	0.071 DW	22

[a] Values originally expressed in strontium units (1 nCi ^{90}Sr/g calcium AW) were transformed to Bq/kg AW by a multiplication factor of 98.4.

[b] **1,** Kudo et al. 1991; **2,** Hanson 1976; **3,** Rickard and Price 1990; **4,** Clulow et al. 1992; **5,** Clulow et al. 1991; **6,** Handl et al. 1990; **7,** Book 1969; **8,** Farris et al. 1969; **9,** Brisbin et al. 1974; **10,** Brisbin et al. 1989; **11,** Miettinen 1969; **12,** Cummings et al. 1971; **13,** Plummer et al. 1969; **14,** Dahlman and Voris 1976; **15,** Emery et al. 1976; **16,** Wrenn et al. 1971; **17,** Shure and Gottschalk 1976; **18,** Crowley et al. 1990; **19,** Carlson and Erlandsson 1991; **20,** Hetherington et al. 1976; **21,** Hatfield et al. 1963; **22,** Whicker et al. 1990; **23,** Nelson 1963; **24,** IAEA 1976; **25,** Lowe 1991; **26,** Smedile and Queirazza 1976; **27,** Wolfe and Schelske 1969; **28,** Wolfe and Jennings 1971; **29,** Holtzman 1969; **30,** Nelson and Whicker 1969; **31,** Poston et al. 1990; **32,** Gustafson 1969; **33,** Newman and Brisbin 1990; **34,** Joshi 1991; **35,** Pentreath et al. 1971; **36,** Folsom et al. 1971; **37,** Lamb et al. 1991; **38,** Holcomb et al. 1971; **39,** Rickard and Price 1990; **40,** Cadwell et al. 1979; **41,** Millard and Whicker 1990; **42,** Levy et al. 1976; **43,** S. S. Anderson et al. 1990; **44,** Gilbert et al. 1989; **45,** Jenkins et al. 1969; **46,** Aarkrog 1990; **47,** Romney et al. 1971; **48,** Neel and Larson 1963; **49,** Schultz and Longhurst 1963; **50,** Jenkins and Fendley 1971; **51,** Kaye and Dunaway 1963; **52,** Curtis et al. 1991; **53,** Ham et al. 1989; **54,** Veskovic and Djuric 1990; **55,** Lowe and Horrill 1991; **56,** Godoy et al. 1991; **57,** United Nations Scientific Committee on the Effects of Atomic Radiation (UNSCEAR) 1988; **58,** Bagshaw and Brisbin 1984; **59,** Colwell et al. 1996; **60,** Sugg et al. 1995.

[c] ND = not detectable.

32.5 CASE HISTORIES

Military weapons tests conducted at the Pacific Proving Grounds in the 1940s and 1950s resulted in greatly elevated local concentrations of radionuclides, and an accident at the Chernobyl nuclear power plant in the former Soviet Union in 1986 resulted in comparatively low concentrations of radionuclides dispersed over a wide geographical area. Both cases are briefly reviewed.

32.5.1 Pacific Proving Grounds

The first artificial, large-scale introduction of radionuclides into a marine environment was at Bikini Atoll in 1946. In succeeding years through 1958, Bikini and Eniwetok became the Pacific Proving Grounds where 59 nuclear and thermonuclear devices were detonated between 1946 and 1958 (Welander 1969; Templeton et al. 1971; Bair et al. 1979) (Table 32.15). Gross radiation injury to marine organisms has not been documented, possibly because seriously injured individuals do not survive and the more subtle injuries are difficult to detect. On land, the roof rat (*Rattus rattus*) survived heavy initial radiation by remaining in deep burrows. Terrestrial vegetation was heavily damaged by heat and blast, although regrowth occurred in 6 months. The land-dwelling hermit crab (*Coenobita* sp.) and coconut crab (*Birgus latro*) were subjected to higher levels of chronic radiation from internally deposited radionuclides than any other atoll organism studied. Levels remained constant in *Coenobita* at 166,000 Bq ^{90}Sr/kg skeleton and 16,835 Bq ^{137}Cs/kg muscle over 2 years; *Birgus* contained 25,900 Bq ^{90}Sr/kg skeleton and 3700 Bq ^{137}Cs/kg muscle over 10 years (Templeton et al. 1971). A survey in August 1964 at Eniwetok and Bikini Atolls (Welander 1969) (Table 32.15) showed that general levels of radioactivity were comparatively elevated and highest in soils and increasingly lower in aquatic invertebrates, groundwater, shorebirds, plants, rats, zooplankton, algae, fish, sediments, seawater, and seabirds. Cobalt-60 was found in all samples of animals, plants, water, sediments, and soils and was the major radionuclide in the marine environment; on land, cesium-137 and ^{90}Sr predominated. All samples contained traces of ^{54}Mn; ^{106}Ru and ^{125}Sb were detected in groundwater and soil and trace concentrations in animals and plants. Trace amounts of ^{207}Bi and ^{144}Ce were usually detected in algae, soils, and land plants. Iron-55 was comparatively high in vertebrates, and ^{239}Pu was found in the soil and in the skin of rats and birds (Welander 1969).

Table 32.15 Radionuclide Concentrations in Selected Samples from the Pacific Proving Grounds (Concentrations are in becquerels/kg fresh weight [FW] or dry weight [DW].)

Location, Sample, Radionuclide, and Other Variables	Concentration (Bq/kg or Bq/L)	Reference[a]
BIKINI ATOLL		
Samples with highest concentrations, August 1964		
^{207}Bi, sediments	Max. 6660 DW	1
^{144}Ce, marine algae	Max. 1739 DW	1
^{137}Cs, land invertebrates	Max. 14,060 DW	1
^{57}Co, sediments	Max. 3400 DW	1
^{60}Co, marine invertebrates	Max. 35,150 DW	1
^{54}Mn, sediments	Max. 962 DW	1
^{106}Ru, sediments	Max. 10,360 DW	1
^{125}Sb, groundwater	Max. 12,950 DW	1
Seawater, 1972, ^{55}Fe	Max. 0.025 FW	2
Sediments		
1958 vs. 1972, ^{55}Fe	Max. 777,000 DW vs. 11,100 DW	2
August 1964, ground zero		
^{207}Bi	6660 DW	1
^{57}Co	3404 DW	1
^{60}Co	9620 DW	1
^{54}Mn	962 DW	1
^{106}Ru	10,360 DW	1
^{125}Sb	3663 DW	1
ENIWETOK ATOLL, AUGUST 1964		
Whole marine algae vs. whole marine fishes		
^{207}Bi	181 DW vs. 74 DW	1
^{144}Ce	814 DW vs. nondetectable (ND)	1
^{137}Cs	52 DW vs. 21 DW	1
^{60}Co	355 DW vs. 888 DW	1
^{54}Mn	48 DW vs. 70 DW	1
^{106}Ru	96 DW vs. ND	1
^{125}Sb	34 DW vs. ND	1
Terrestrial invertebrates vs. terrestrial vegetation		
^{207}Bi	6 DW vs. 10 DW	1
^{144}Ce	5 DW vs. 888 DW	1
^{137}Cs	No data vs. 12,580 DW	1
^{60}Co	888 DW vs. 141 DW	1
^{54}Mn	281 DW vs. 296 DW	1
^{106}Ru	15 DW vs. 19 DW	1
^{125}Sb	ND vs. 8 DW	1
Seabirds (whole) vs. shorebirds (whole)		
^{207}Bi	ND vs. ND	1
^{57}Co	12 DW vs. ND	1
^{60}Co	340 DW vs. 4810 DW	1
^{137}Cs	ND vs. 4440 DW	1
^{54}Mn	81 DW vs. ND	1
^{106}Ru, ^{125}Sb	ND vs. ND	1

Table 32.15 (continued) Radionuclide Concentrations in Selected Samples from the Pacific Proving Grounds (Concentrations are in becquerels/kg fresh weight [FW] or dry weight [DW].)

Location, Sample, Radionuclide, and Other Variables	Concentration (Bq/kg or Bq/L)	Reference[a]
Roof rat, *Rattus rattus*; whole		
^{207}Bi	5 DW	1
^{144}Ce	362 DW	1
^{60}Co	888 DW	1
^{137}Cs	19,980 DW	1
^{54}Mn	1 DW	1
^{106}Ru, ^{125}Sb	ND	1
Samples with highest concentrations		
^{207}Bi, marine plankton	Max. 333 DW	1
^{144}Ce, soils	Max. 2109 DW	1
^{137}Cs, rats	Max. 19,980 DW	1
^{57}Co, sediments	Max. 740 DW	1
^{60}Co, marine invertebrates	Max. 6290 DW	1
^{54}Mn, land plants	Max. 296 DW	1
^{106}Ru, soils	Max. 4440 DW	1
^{125}Sb, soils	Max. 703 DW	1
Soils vs. sediments		
^{207}Bi	20 DW vs. 218 DW	1
^{144}Ce	2109 DW vs. No Data	1
^{137}Cs	2072 DW vs. 814 DW	1
^{57}Co	No Data vs. 740 DW	1
^{60}Co	2849 DW vs. 1073 DW	1
^{54}Mn	44 DW vs. 148 DW	1
^{106}Ru	4440 DW vs. 3700 DW	1
^{125}Sb	703 DW vs. 407 DW	1
ENIWETOK ATOLL, RUNIT ISLAND (8 nuclear detonations between 1948 and 1958)		
Roof rat, whole		
Immediate vicinity of detonations; 1967 vs. 1973		
^{137}Cs		
Bone	21,978 DW vs. 81,363 DW	3
Intestine	137,344 DW vs. No Data	3
Kidney	189,958 DW vs. 126,799 DW	3
Liver	83,657 DW vs. 83,583 DW	3
Muscle	137,122 DW vs. 156,880 DW	3
Skin	13,209 DW vs. 77,256 DW	3
^{60}Co		
200 m vs. 2460 m; 1967		
Bone	185 DW vs. ND	3
Intestine	8251 DW vs. No Data	3
Kidney	110,223 DW vs. 333 DW	3
Muscle	499 DW vs. 266 DW	3
Skin	259 DW vs. ND	3
Soils		
^{137}Cs, 1967		
Ground zero vs. 200 m	1258 DW vs. 399 DW	3
1030 m vs. 2460 m	88 DW vs. 18 DW	3
^{137}Cs, 1971		
Ground zero vs. 200 m	4736 DW vs. 403 DW	3
1030 m	44 DW	3
^{60}Co, 1967		
Ground zero vs. 200 m	1221 DW vs. 66 DW	3
1030 m	25 DW	3

Table 32.15 (continued) Radionuclide Concentrations in Selected Samples from the Pacific Proving Grounds (Concentrations are in becquerels/kg fresh weight [FW] or dry weight [DW].)

Location, Sample, Radionuclide, and Other Variables	Concentration (Bq/kg or Bq/L)	Reference[a]
^{60}Co, 1971		
Ground zero vs. 200 m	1110 DW vs. 133 DW	3
1030 m vs. 2460 m	40 DW vs. 4 DW	3
1973, 2460 m		
^{137}Cs	11 DW	3
^{60}Co	52 DW	3
Terrestrial vegetation		
Ground zero, 1967 vs. 1971		
^{137}Cs	16,199–93,380 DW vs. 34,780–94,239 DW	3
^{60}Co	Max. 1221 DW vs. Max. 2775 DW	3
1030 m, 1967 vs. 1971		
^{137}Cs	296–2035 DW vs. 333–1961 DW	3
^{60}Co	Max. 14 DW vs. Max. 48 DW	3

[a] **1,** Welander 1969; **2,** Schell 1976; **3,** Bastian and Jackson 1976.

32.5.2 Chernobyl

32.5.2.1 General

Several accidents in nuclear facilities have been extensively analyzed and reported. The three most widely publicized accidents were at Windscale (now known as Sellafield), United Kingdom, in 1957; Three Mile Island, Pennsylvania, in 1979; and Chernobyl, Ukraine, in 1986 (UNSCEAR 1988; Severa and Bar 1991; Eisler 1995). From the accident at Windscale about 750 trillion (T)Bq ^{131}I, 22 TBq ^{137}Cs, 3 TBq ^{89}Sr, and 0.33 TBq ^{90}Sr were released and twice the amount of noble gases that were released at Chernobyl, but 2000 times less ^{131}I and ^{137}Cs. From the Three Mile Island accident, about 2% as much noble gases and 50,000 times less ^{131}I than from the Chernobyl accident were released. The most abundant released radionuclides at Three Mile Island were ^{133}Xe, ^{135}Xe, and ^{131}I, but the collective dose equivalent to the population during the first post-accident days was <1% of the dose accumulated from natural background radiation in a year.

The most serious accident of a nuclear reactor occurred on April 26, 1986, at one of the four units at Chernobyl when at least 3,000,000 TBq were released from the fuel during the accident (Table 32.16). The accident happened while a test was conducted during a normal scheduled shutdown and is attributed mainly to human error: "…the operators deliberately and in violation of rules, withdrew most control rods from the core and switched off some important safety systems…" (UNSCEAR 1988). The first power peak reached 100 times the nominal power within 4 seconds. Energy released in the fuel by the power excursion suddenly ruptured part of the fuel into minute pieces. Small, hot fuel particles caused a steam explosion. After 2 or 3 seconds, another explosion occurred, and hot pieces of the reactor were ejected. The damage to the reactor allowed air to enter, causing combustion of the graphite (UNSCEAR 1988). About 25% of the released radioactive materials escaped during the first day of the accident; the rest, during the next 9 days (UNSCEAR 1988). The initial explosions and heat from the fire carried some of the radioactive materials to an altitude of 1500 m where they were transported by prevailing winds (Figure 32.6) and caused widespread radioactive contamination of Europe and the former Soviet Union, initially with ^{131}I, ^{134}Cs, and ^{137}Cs (Smith and Clark 1986; Anspaugh et al. 1988; Clark and Smith 1988; UNSCEAR 1988; Aarkrog 1990; Johanson 1990; Brittain et al. 1991; Palo et al. 1991). Long-range

Table 32.16 Selected Fission Products in the Chernobyl Reactor Core, and Their Estimated Escape into the Environment

Radionuclide	Trillions of becquerels (TBq) In Core	Escaped[a]
^{85}Kr	33,000	33,000
^{133}Xe	1,700,000	1,700,000
^{131}I	1,300,000	260,000
^{132}Te	320,000	48,000
^{134}Cs	190,000	19,000
^{137}Cs	290,000	37,700
^{99}Mo	4,800,000	110,400
^{95}Zr	4,400,000	140,800
^{103}Ru	4,100,000	118,900
^{106}Ru	2,000,000	58,000
^{140}Ba	2,900,000	162,400
^{141}Ce	4,400,000	101,200
^{144}Ce	3,200,000	89,600
^{89}Sr	2,000,000	80,000
^{90}Sr	200,000	8000
^{239}Np	140,000	4200
^{238}Pu	1000	30
^{239}Pu	850	25
^{240}Pu	1200	36
^{241}Pu	170,000	5100
^{242}Cm	26,000	780

[a] Aarkrog (1990) estimates escapement of 100,000 TBq of ^{137}Cs; 50,000 TBq of ^{134}Cs; and 35,000 TBq of ^{106}Ru. Aarkrog (1990) also includes the following radionuclides in the Chernobyl escapement: 1500 TBq of ^{110}Ag, 3000 TBq of ^{125}Sb, 6 TBq of ^{241}Am, and 6 TBq of $^{243+244}Cm$.

Data from Severa, J. and J. Bar. 1991. *Handbook of Radioactive Contamination and Decontamination.* Studies in Environmental Science 47. Elsevier, New York. 363 pp.

atmospheric transport spread the radioactive materials through the northern hemisphere where it was first detected in Japan on May 2, in China on May 4, in India on May 5, and in Canada and the United States on May 5–6 1986 (UNSCEAR 1988). Airborne activity was also detected in Turkey, Kuwait, Monaco, and Israel in early May. No airborne activity from Chernobyl has been reported south of the equator (UNSCEAR 1988). Among the reactors now operating in the former Soviet Union are 13 identical to the one in Chernobyl, Ukraine, including units in Chernobyl, Leningrad, Kursk, and Smolensk (Mufson 1992).

Effective dose equivalents from the Chernobyl accident in various regions of the world were highest in southeastern Europe (1.2 mSv), northern Europe (0.97 mSv), and Central Europe (0.93 mSv) (Table 32.17). In the first year after the accident, whole-body effective dose equivalents were highest in Bulgaria, Austria, Greece, and Romania (0.5 to 0.8 mSv); Finland, Yugoslavia, Czechoslovakia, Italy (0.3 to 0.5 mSv); Switzerland, Poland, U.S.S.R., Hungary, Norway, Germany, and Turkey (0.2 to 0.3 mSv); and elsewhere (<0.2 mSv) (UNSCEAR 1988). Thyroid dose equivalents were significantly higher than whole-body effective dose equivalents because of significant amounts of ^{131}I in the released materials. Thyroid dose equivalents were as high as 25 mSv to infants in Bulgaria, 20 mSv in Greece, and 20 mSv in Romania; the adult thyroid dose equivalents were usually 80% lower than the infant dose equivalents (UNSCEAR 1988).

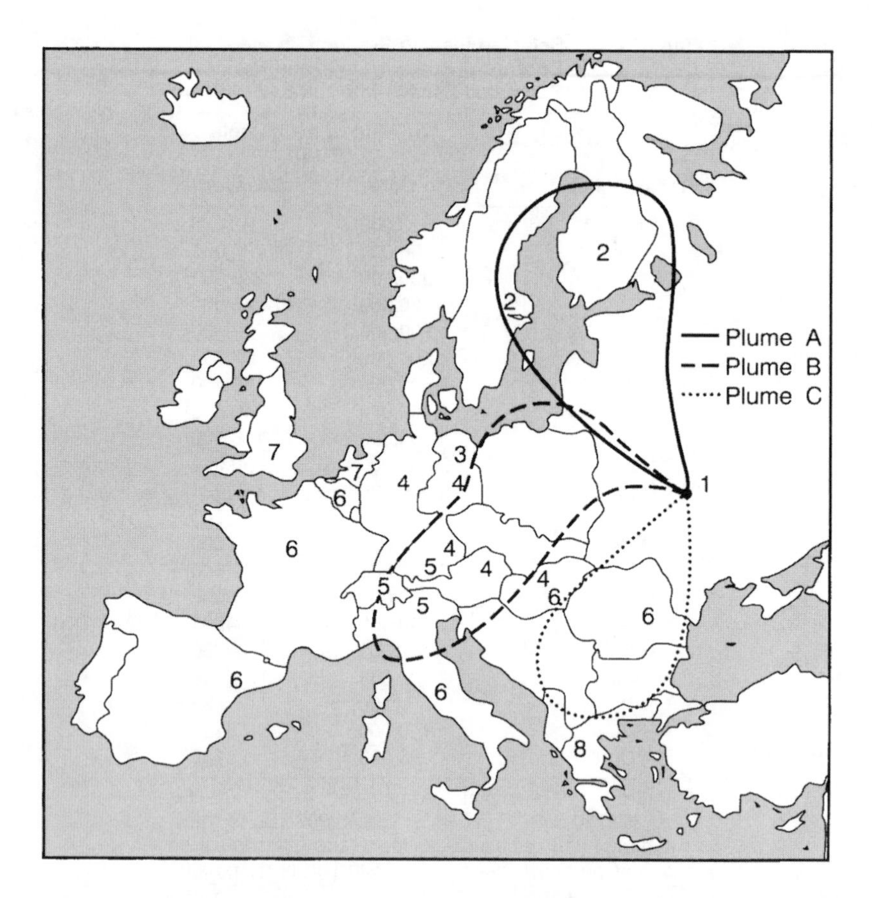

Figure 32.6 Chernobyl air plume behavior and reported initial arrival times of detectable radioactivity. Plume **A** originated from Chernobyl on April 26, 1986; Plume **B** on April 27–28; and Plume **C** on April 29–30. The numbers indicate initial arrival times: **1,** April 26; **2,** April 27; **3,** April 28; **4,** April 29; **5,** April 30; **6,** May 1; **7,** May 2; and **8,** May 3. (From United Nations Scientific Committee on the Effects of Atomic Radiation (UNSCEAR). 1988. *Sources, Effects and Risks of Ionizing Radiation.* United Nations, New York. 647 pp.)

Table 32.17 Regional Total Effective Human Dose Equivalent Commitment from the Chernobyl Accident

Region	Effective Dose Equivalent (mSv)
Southeastern Europe	1.2
Northern Europe	0.97
Central Europe	0.93
Former Soviet Union	0.81
Southwest Asia, West Europe	>0.1–<0.2
North Africa, Greenland, East Africa, Central Africa, South Asia, West Africa	>0.01–<0.1
East Asia, Southwest Europe, Southeast Asia, North America, Caribbean, South America, Central America	<0.01

Data from United Nations Scientific Committee on the Effects of Atomic Radiation (UNSCEAR). 1988. *Sources, Effects and Risks of Ionizing Radiation.* United Nations, New York. 647 pp.; Aarkrog, A. 1990. Environmental radiation and radiation releases. *Inter. Jour. Radiation Biol.* 57:619-631.

32.5.2.2 Local Effects

At Chernobyl, at least 115 humans received acute bone-marrow doses of greater than 1 Gy, as judged by lymphocyte aberrations (UNSCEAR 1988). The death toll within 3 months of the accident was at least 30 individuals, usually from groups receiving more than 4 Gy and including the reactor's operating staff and the fire-fighting crew. Local residents were evacuated from a 30-km exclusion zone around the reactor because of increasing radiation levels. More than 115,000 people, including 27,000 children, were evacuated from the Kiev region, Byelorussia, and the Ukraine. Tens of thousands of cattle were also removed from the contaminated area, and consumption of locally produced milk and other foods was banned. Agricultural activities were halted and a large-scale decontamination effort was undertaken (UNSCEAR 1988). The radiological effect of the accident to individual risk was insignificant outside a limited local region, either because contamination levels were generally low or because remedial actions to ban the consumption of highly contaminated foodstuffs prevented high exposures (UNSCEAR 1988).

Acute biological effects of the Chernobyl accident on local natural resources were documented by Sokolov et al. (1990). They concluded that the most sensitive ecosystems affected at Chernobyl were the soil fauna and pine forest communities and that the bulk of the terrestrial vertebrate community was not adversely affected by released ionizing radiation. Pine forests seemed to be the most sensitive ecosystem. One stand of 400 ha of *Pinus silvestris* died and probably received a dose of 80 to 100 Gy; other stands experienced heavy mortality of 10- to 12-year-old trees and up to 95% necrotization of young shoots. These pines received an estimated dose of 8 to 10 Gy. Abnormal top shoots developed in some *Pinus*, and these probably received 3 to 4 Gy. In contrast, leafed trees such as birch, oak, and aspen in the Chernobyl Atomic Power Station zone survived undamaged, probably because they are about 10 times more radioresistant than pines. There was no increase in the mutation rate of the spiderwort, (*Arabidopsis thaliana*) a radiosensitive plant, suggesting that the dose rate was less than 0.05 Gy/h in the Chernobyl locale.

Populations of soil mites were reduced in the Chernobyl area, but no population showed a catastrophic drop in numbers. By 1987, soil microfauna — even in the most heavily contaminated plots — were comparable to controls. Flies (*Drosophila* spp.) from various distances from the accident site and bred in the laboratory had higher incidences of dominant lethal mutations (14.7%, estimated dose of 0.8 mGy/h) at sites nearest the accident than controls (4.3%). Fish populations seemed unaffected in July/August 1987, and no grossly deformed individuals were found. However, $^{134+137}$Cs levels were elevated in young fishes. The most heavily contaminated teleost in May 1987 was the carp (*Carassius carassius*). But carp showed no evidence of mutagenesis, as judged by incidence of chromosomal aberrations in cells from the corneal epithelium of carp as far as 60 km from Chernobyl (Sokolov et al. 1990).

The most contaminated water body in the Chernobyl emergency zone was the Chernobyl cooling pond ecosystem (Kryshev 1995). On May 30, 1986, the total amount of radioactivity in the water of this system was estimated at 806 TBq, and in sediments 5657 TBq. In water, ^{131}I contributed about 31% of the total radioactivity, ^{140}Ba-^{140}La 25%, ^{95}Zr-Nb^{95} 15%, ^{134}Cs and ^{137}Cs 11%, ^{141}Ce and ^{144}Ce 10%, ^{103}Ru and ^{106}Ru 7%, and ^{90}Sr <1%. The distribution pattern in sediments was significantly different: about 41% of the total radioactivity was contributed by ^{95}Zr-^{95}Nb, 27% by ^{141}Ce and ^{144}Ce, 16% by ^{103}Ru and ^{106}Ru, 12% by ^{140}Ba-^{140}La, 3% by ^{134}Cs and ^{137}Cs, 1% by ^{90}Sr, and 0.5% by ^{131}I. Concentrations of radioactivity in water, sediments, and biota declined between 1986 and 1990, as judged by ^{137}Cs concentrations (Kryshev 1995). Cesium-137 concentrations in the Chernobyl cooling pond ecosystem declined in water from 210 Bq/L in 1986 to 14 Bq/L in 1990; for bottom sediments, these values were 170,000 Bq/kg FW in 1986 and 140,000 Bq/kg FW in 1990; for algae, 90,000 Bq/kg FW in 1986 to 19,000 Bq/kg FW in 1990; and for muscle of five species of fishes, ^{137}Cs concentrations ranged from 30,000 to 180,00 Bq/kg FW in 1986, and from

8000 to 80,000 Bq/kg FW in 1990 (Kryshev 1995). Silver carp (*Hypophthalmichthys molitrix*) born in 1989 from parents reared in Chernobyl cooling pond waters had a marked increase of 17 to 26% above controls in reproductive system anomalies in 1989 to 1992 (Makeyeva et al. 1995). Anomalies included degenerative changes in oocytes, spermatogonia, and spermatocytes, and the appearance of bisexual and sterile fish. The gonadal abnormalities are attributed to the high radiation dose of 7 to 10 Gy received by the parent fish during gonad formation and the continuing exposure of 0.2 Gy annually to this generation (Makeyeva et al. 1995). Silver carp from this ecosystem also had a dose-dependent decrease in hormonal control over Na^+, K^+-pump in erythrocytes, with increased passive permeability of the erythrocyte membrane to radioactive analogues of sodium and potassium (Kotelevtsev et al. 1996).

Several rodent species compose the most widely distributed and numerous mammals in the Chernobyl vicinity. It was estimated that about 90% of rodents died in an area that received 60 Gy and 50% in areas that received 6 to 60 Gy. Rodent populations seemed normal in spring 1987, and this was attributed to migration from adjacent nonpolluted areas. The most sensitive small mammal was the bank vole (*Clethrionomys glareolus*), which experienced embryonic mortality of 34%. The house mouse (*Mus musculus*) was one of the more radioresistant species. *Mus* from plots receiving 0.6 to 1 mG/h did not show signs of radiation sickness, were fertile with normal sperm, bred actively, and produced normal young. Some chromosomal aberrations were evident, namely, an increased frequency of reciprocal translocations (Sokolov et al. 1990). New data on the house mouse suggests that fertility was dramatically reduced in the 30-km zone around the Chernobyl nuclear power plant station in 1986/87 and that survivors had high frequencies of abnormal spermatozoa heads and dominant lethal mutations (Pomerantseva et al. 1996). Dose rates from soil to the house mouse between 1986 and 1993 ranged from 0.0002 to 2 mGy/h, and these were positively correlated with the frequency of reciprocal translocations in mouse spermatocytes. The frequency of mice heterozygous for recessive lethal mutations decreased over time after the accident (Pomerantseva et al. 1996).

During the early period after the accident, there was no evidence of increasing mortality, decline in fecundity, or migration of vertebrates as a result of the direct action of ionizing radiation. The numbers and distribution of wildlife species were somewhat affected by the death of the pine stand, the evacuation of people, the termination of cultivation of soils (the crop of 1986 remained standing), and the evacuation of domestic livestock. There were no recorded changes in survival or species composition of game animals and birds. In fact, because humans had evacuated and hunting pressure was negligible, many game species, including foxes, hares, deer, moose, wolves, and waterfowl, moved into the zone in fall 1986 to winter 1987 from the adjacent areas in a 50- to 60-km radius (Sokolov et al. 1990).

In 1991, 5 years after the accident, a female root vole (*Microtus oeconomus*) with an abnormal karyotype (reciprocal translocation) was found within the 30-km radius of the Chernobyl nuclear power plant. These chromosomal aberrations were probably inherited and did not affect the viability of vole populations (Nadzhafova et al. 1994). In 1994/95, the diversity and abundance of the small mammal population (12 species of rodents) at the most radioactive sites at Chernobyl were the same as reference sites (Baker et al. 1996). Rodents from the most radioactive areas did not show gross morphological features other than enlargement of the spleen. There were no gross chromosomal arrangements, as judged by examination of the karyotypes. Also observed within the most heavily contaminated site were red fox (*Vulpes vulpes*), gray wolf (*Canis lupus*), moose (*Alces alces*), river otter (*Lutra lutra*), roe deer (*Capreolus capreolus*), Russian wild boar (*Sus scrofa*), brown hare (*Lepus europaeus*), and feral dogs (Baker et al. 1996).

32.5.2.3 Nonlocal Effects

The partial meltdown of the 1000-megawatt reactor at Chernobyl on April 26, 1986, released large amounts of radionuclides into the environment — especially ^{131}I, ^{137}Cs, and ^{134}Cs — and resulted in widely dispersed and deposited radioactive material in Europe and throughout the northern hemisphere (UNSCEAR 1988; Palo et al. 1991) (Table 32.18). Transuranics and to some extent ^{90}Sr were deposited closer to the accident site than more volatile radionuclides such as radiocesium; accordingly, radiological problems changed quantitatively and qualitatively with increasing distance from the accident site (Aarkrog 1990).

Soil and Vegetation. The radiocesium fallout in Sweden was among the highest in western Europe — exceeding 60,000 Bq/m^2 on Sweden's Baltic coast — and involved mainly upland pastures and forests (Johanson 1990; Brittain et al. 1991; Palo et al. 1991). In Norway, radiocesium deposition from the Chernobyl accident ranged from $<$5000 to $>$200,000 Bq/m^2 and greatly exceeded the deposition from prior nuclear weapons tests (Hove et al. 1990a). In Italy, heavy rainfall coincident with the passage of the Chernobyl radioactive cloud caused high local deposition of radionuclides in the soil, grass, and plants (Battiston et al. 1991). The Chernobyl plume reached Greece on May 1, 1986. A total of 14 gamma emitters were identified in the soil and vegetation in May 1986, and three (^{134}Cs, ^{137}Cs, ^{131}I) were also detected in the milk of free-grazing animals in the area (Assimakopoulos et al. 1989). Radiocesium-134 and ^{137}Cs intake by humans in Germany during the period 1986/87 was mainly from rye, wheat, milk, and beef (Clooth and Aumann 1990). In the United Kingdom, elevated concentrations of radionuclides of iodine, cesium, ruthenium, and others were measured in the air and rainwater from May 2–5, 1986 (Smith and Clark 1986). The background activity concentrations were about 3 times normal levels in early May, and those of ^{131}I approached the derived emergency reference level (DERL) of drinking water of 5 mSv ^{131}I (equivalent to a thyroid dose of 50 mSv); however, ^{131}I levels were not elevated in foodstuffs or cow's milk (Smith and Clark 1986). Syria — 1800 km from Chernobyl — had measurable atmospheric concentrations of ^{137}Cs and ^{131}I and near-detection limit concentrations of ^{144}Ce, ^{134}Cs, ^{140}La, and ^{106}Ru (Othman 1990). The maximum ^{131}I thyroid dose equivalent received by Syrians was 116 μSv in adults and 210 μSv in children. One year later, these values were 25 μSv in adults and 70 μSv in a 10-year-old.

The amount of fallout radioactivity deposited on plant surfaces depends on the exposed surface area, the developmental season of the plants, and the external morphology. Mosses, which have a relatively large surface area, showed the highest concentrations of radiocesium (Table 32.18). In northern Sweden, most of the radiocesium fallout was deposited on plant surfaces in the forest ecosystem and was readily incorporated into living systems because of browsing by herbivores and cesium's chemical similarity to potassium (Palo et al. 1991). Forest plants seemed to show less decrease than agricultural crops in ^{137}Cs activity over time (Bothmer et al. 1990). For example, the effective retention half-time of ^{137}Cs from Chernobyl was 10 to 20 days in herbaceous plants and 180 days in chestnut trees, *Castanea* spp. (Tonelli et al. 1990). The radioactive fallout from the Chernobyl accident also resulted in high ^{137}Cs levels in Swedish pasture grass and other forage, although levels in grain were relatively low (Andersson et al. 1990). Radiocesium isotopes were still easily measurable in grass silage harvested in June 1986 and used as fodder for dairy cows in 1988 (Voors and Van Weers 1991). The rejection of the first harvests of radiocesium-contaminated perennial pasture and in particular of rye grass (*Lolium perenne*) does not constitute a safe practice because later harvests — even 1 year after the contamination of the field — may contain very high values, as in Greece (Douka and Xenoulis 1991).

Aquatic Life. After Chernobyl, the consumption of freshwater fishes by Europeans declined, fish license sales dropped by 25%, and the sale of fish from radiocesium-contaminated lakes was prohibited (Brittain et al. 1991). Many remedial measures have been attempted to reduce radiocesium loadings in fishes, but none have been effective to date (Hakanson and Andersson 1992).

Radiocesium concentrations in muscle of fishes from the southern Baltic Sea increased 3 to 4 times after Chernobyl (Grzybowska 1989), and $^{134+137}$Cs and ^{106}Ru in fishes from the Danube River increased by a factor of 5. However, these levels posed negligible risk to human consumers

(Conkic et al. 1990). Chernobyl radioactivity, in particular ^{141}Ce and ^{144}Ce, entering the Mediterranean as a single pulse, was rapidly removed from surface waters and transported to 200 m in a few days, primarily in fecal pellets of grazing zooplankton (Fowler et al. 1987). Bioconcentration factors (BCF) of ^{137}Cs in fishes from Lake Paijanne, Finland — a comparatively contaminated area — ranged between 1250 and 3800; the highest BCF values were measured in the predatory northern pike (*Esox lucius*) a full 3 years after the Chernobyl accident; consumption of these fishes was prohibited (Korhonen 1990).

After the Chernobyl accident, radiocesium isotopes were also elevated in trees and lichens bordering an alpine lake in Scandinavia and in lake sediments, invertebrates, and fishes (Table 32.18). Radiocesium levels in muscle of resident brown trout (*Salmo trutta*) remained elevated for at least 2 years (Brittain et al. 1991). People consuming food near this alpine lake derived about 90% of their effective dose equivalent from the consumption of freshwater fish, reindeer meat, and milk. The average effective dose equivalent of this group during the next 50 years is estimated at 6 to 9 mSv with a changed diet and 8 to 12 mSv without any dietary changes (Brittain et al. 1991).

Wildlife. Reindeer (*Rangifer tarandus*) — also known as caribou in North America — are recognized as a key species in the transfer of radioactivity from the environment to humans because (1) the transfer factor of radioactivity from reindeer feed to reindeer muscle is high, (2) lichens — which constitute a substantial portion of the reindeer diet — are efficient accumulators of strontium, cesium, and actinide radioisotopes, and (3) reindeer feed is not significantly supplemented with grain or other feeds low in contamination (Jones et al. 1989; Rissanen and Rahola 1989; 1990; Eikelmann et al. 1990; Skogland and Espelien 1990). During 1986/87, about 75% of all reindeer meat produced in Sweden was unfit for human consumption because ^{137}Cs exceeded 300 Bq/kg FW. In May 1987, the maximum permissible level of ^{137}Cs in Swedish reindeer, game, and freshwater fish was raised to 1500 Bq/kg FW; however, about 25% of slaughtered reindeer in 1987 to 1989 still exceeded this limit (Ahman et al. 1990b). Concentrations in excess of 100,000 Bq $^{134+137}Cs$/kg FW lichens have been recorded in the most contaminated areas and in the 1986/87 season was reflected in reindeer muscle concentrations >50,000 Bq/kg FW from the most contaminated areas of central Norway (Roed et al. 1991). Norwegian reindeer containing 60,000 to 70,000 Bq ^{137}Cs/kg FW in muscle receive an estimated yearly dose of 500 mSv (Jones 1990). The maximum radiation dose to reindeer in Sweden after the Chernobyl accident was about 200 mSv/year, with a daily dose rate of about 1 mSv during the winter period of maximum tissue concentrations (Jones et al. 1989). In general, reindeer calves had higher ^{137}Cs levels in muscle than adult females (4700 vs. 2700 Bq/kg FW) during September 1988, suggesting translocation to the fetus (Eikelmann et al. 1990). Two reindeer herds in Norway that were heavily contaminated with radiocesium had a 25% decline in survival of calves; survival was normal in a herd with low exposure (Skogland and Espelien 1990). Several compounds inhibit uptake and reduce retention of ^{137}Cs in reindeer muscle from contaminated diets, but the mechanisms of the action are largely unknown. These compounds include:

- Zeolite — a group of tectosilicate minerals — when fed at 25 to 50 g daily (Ahman et al. 1990a)
- Ammonium hexacyanoferrate — also known as Prussian Blue or Giese salt — at 0.3 to 1.5 g daily (Hove et al. 1990b; Mathiesen et al. 1990; Staaland et al. 1990)
- Bentonite — a montmorillonite clay — when fed at 2% of diet (Ahman et al. 1990a)
- High intakes of potassium (Ahman et al. 1990a)

Much additional work seems needed on chemical and other processes that hasten excretion and prevent uptake and accumulation of radionuclides in livestock and wildlife. Reindeer herding is the most important occupation in Finnish Lapland and portions of Sweden (Rissanen and Rahola

1989). Swedish Lapland reindeer herders have experienced a variety of sociocultural problems as a result of the Chernobyl accident. The variability of contamination has been compounded by the variability of expert statements about risk, the change in national limits of Bq concentrations set for meat marketability, and the variability of the compensation policy for slaughtered reindeer. These concerns may result in fewer Lapps becoming herders and a general decline in reindeer husbandry (Beach 1990).

Caribou in northern Quebec contained up to 1129 Bq ^{137}Cs/kg muscle FW in 1986/87, but only 10 to 15% of this amount originated from Chernobyl; the remainder is attributed to fallout from earlier atmospheric nuclear tests (Crete et al. 1990). The maximum concentration of ^{137}Cs in meat of caribou (*Rangifer tarandus granti*) from the Alaskan Porcupine herd after the Chernobyl accident did not exceed 232 Bq/kg FW, and this is substantially below the recommended level of 2260 Bq ^{137}Cs/kg FW (Allaye-Chan et al. 1990). Radiocesium transfer in an Alaskan lichen–reindeer–wolf (*Canis lupus*) food chain has been estimated. If reindeer forage contained 100 Bq/kg DW in lichens and 5 Bq/kg DW in vascular plants, the maximum winter concentrations — at an effective half-life of 8.2 years in lichens and 2.0 years in vascular plants — were estimated at 20 Bq/kg FW in reindeer–caribou skeletal muscle and 24 Bq/kg FW in wolf muscle (Holleman et al. 1990).

The radioactive body burden of exposed reindeer and the character of chromosomal aberrations — which was different in exposed and nonexposed reindeer — indicated a genetic effect of radiation from the Chernobyl accident (Roed et al. 1991). Chromosomal aberrations in Norwegian female reindeer positively correlated with increasing radiocesium concentrations in flesh (Skogland and Espelien 1990). The frequency of chromosomal aberrations in reindeer calves from central Norway were greatest in those born in 1987 when tissue loadings were equivalent to fetal doses of 70 to 80 mSv and lower in 1988 (50 to 60 mSv) and 1989 (40 to 50 mSv), strongly suggesting a dose-dependent induction (Roed et al. 1991). Mutagenicity tests have also been used successfully with feral rodents to evaluate the biological effects of the radiation exposure from the Chernobyl accident. Increased mutagenicity in mice (*Mus musculus domesticus*) was evident, as judged by tests of the bone-marrow micronucleus at 6 months and 1 year after the accident. Rodents with increased chromosomal aberrations also had ^{137}Cs burdens that were 70% higher 6 months after the accident and 55% higher after 1 year, but elevated radiocesium body burdens alone were not sufficient to account for the increase in mutagenicity (Cristaldi et al. 1990). In bank voles, however, mutagenicity (micronucleated polychromatic erythrocytes) correlated well with the ^{137}Cs content in muscle and in the soil of the collection locale (Cristaldi et al. 1991). The estimated daily absorbed doses (4.2 to 39.4 μGy) were far lower than those required to produce the same effect in the laboratory (Cristaldi et al. 1991).

For many households in Sweden, moose (*Alces alces*) are an important source of meat (Palo et al. 1991). Radiocesium concentrations in the foreleg muscle of moose in Sweden during 1987/88 were highest in autumn when the daily dietary intake of the animals was about 25,000 Bq ^{137}Cs and lowest during the rest of the year when the mean daily intake was about 800 Bq (Bothmer et al. 1990). Cesium-137 levels in moose flesh did not decrease significantly for about 2 years after the Chernobyl accident (Johanson 1990). The selection of food by moose is paramount to the uptake of environmental contaminants and the changes in tissue levels over time. Increased foraging on highly contaminated plant species, such as bilberry (*Vaccinium myrtillus*), aquatic plants, and mushrooms, might account for the increased ^{137}Cs radioactivity in moose (Palo et al. 1991). Habitat is a useful indicator of ^{137}Cs radioactivity in moose muscle; radioactivity was highest in moose captured in swamp and marsh habitats and lowest in farmlands (Nelin 1995). For reasons that are not yet clear, transfer coefficients of ^{137}Cs from diet to muscle were about the same in moose (0.03) and beef cattle (0.02), but were significantly higher in sheep (0.24) (Bothmer et al. 1990).

The songthrush (*Turdus philomelas*) collected in Spain in November 1986 had elevated concentrations of ^{134}Cs, ^{137}Cs, and ^{90}Sr. The contamination probably occurred in central and northern Europe before the birds' migration to Spain (Baeza et al. 1991). Spaniards who ate songthrushes contaminated with radiocesium isotopes usually received about 58 μSv/year, which is well below current international guidelines (Baeza et al. 1991). Consumption of game or wildlife in Great Britain after the Chernobyl accident probably also does not exceed the annual limits of intake (ALI) based on $^{134+137}$Cs concentrations in game and the numbers of animals that can be eaten in 1 year before ALI is exceeded (Lowe and Horrill 1991). For example, a person who eats hares containing 3114 Bq $^{134+137}$Cs/kg FW in muscle would have to consume 99 hares before exceeding the ALI. For the consumption of red grouse (3022 Bq/kg), this number is 441 grouse; and for the consumption of woodcock (55 Bq/kg), it is 45,455 woodcocks (Lowe and Horrill 1991). Rabbits (*Oryctolagus* sp.) from northeastern Italy that were fed Chernobyl-contaminated alfalfa meal (1215 Bq $^{134+137}$Cs/kg diet) had a maximum of 156 Bq/kg muscle FW of $^{134+137}$Cs, a value much lower than the current Italian guideline of 370 Bq/kg FW for milk and children's food and 600 Bq/kg FW for other food (Battiston et al. 1991). More than 85% of the ingested radiocesium was excreted by rabbits in their feces and urine; about 3% was retained (Battiston et al. 1991).

Cesium radioactivity in tissues and organs of the wolverine (*Gulo gulo*), lynx (*Felis lynx*), and Arctic fox (*Alopex logopus*) in central Norway after the Chernobyl accident was highly variable. In general, cesium-137 levels were substantially lower in these carnivores than in lower trophic levels (Ekker et al. 1990), suggesting little or no food-chain biomagnification, and at variance with results of studies of the omnivore and herbivore food chain.

Domestic Animals. Radiocesium isotopes from the Chernobyl accident transferred easily to grazing farm animals (Hove et al. 1990a). Both ^{134}Cs and ^{137}Cs were rapidly distributed throughout the soft tissues after dietary ingestion and were most highly concentrated in muscle (Book 1969; Van Den Hoek 1989). Radiocesium activity in milk and flesh of Norwegian sheep and goats increased three- to fivefold 2 years after the accident and coincided with an abundant growth and availability of fungal fruit bodies with $^{134+137}$Cs levels as much as 100 times greater than green vegetation (Hove et al. 1990a). In cattle, coefficients of radiocesium transfer from diet to muscle were about 2.5% in adults and 16% in calves. The higher value in calves was probably due to a high availability of cesium from the gastrointestinal tract and to daily uptake of potassium in growing animal muscle (Daburon et al. 1989). There was no correlation between the retention of ^{137}Cs and the pregnancy stage in cattle (Calamosca et al. 1990). Radiocesium concentrations in pork in Czechoslovakia did not decline between 1986 and 1987 because the feed of pigs during this period contained milk by-products contaminated with $^{134+137}$Cs (Kliment 1991).

Sheep farming is the main form of husbandry in the uplands of west Cumbria and north Wales, a region that received high levels of radiocesium fallout during the Chernobyl accident. Afterwards, typical vegetation activity concentrations were ~6000 Bq/kg (down to ~1000 Bq/kg in January 1989). But sheep muscle concentrations exceeded 1000 Bq ^{137}Cs/kg FW, which is the United Kingdom's dietary limit for human health protection (Crout et al. 1991). Contaminated lambs — which usually had higher concentrations of ^{137}Cs than ewes — that were removed to lowland pastures (<50 Bq/kg vegetation) rapidly excreted radiocesium in feces and urine, and cesium body burdens had an effective half-life of 11 days. This practice should not significantly increase radiocesium levels in the soil and vegetation of lowland pastures (Crout et al. 1991). The absorption and retention of radiocesium by suckling lambs is highly efficient, about 66%. Fecal excretion was an important pathway after the termination of ^{137}Cs ingestion. In weaned animals, the absorption of added ionic cesium was about twice that of cesium fallout after the accident at Chernobyl (Moss et al. 1989). Silver-110m was also detected in the brains and livers of ewes and lambs in the United Kingdom. The transfer of ^{110m}Ag was associated with perennial rye grass harvested soon after deposition in 1986. Silver-110m was taken up to a greater extent than ^{137}Cs in liver; but unlike ^{137}Cs, the ^{110m}Ag was not readily translocated to other tissues. Other than cesium isotopes and ^{131}I, ^{110m}Ag was the only detected nuclide in sheep tissues (Beresford 1989).

Atmospheric deposition of ^{137}Cs from Chernobyl to vegetation and eventually to the milk of sheep, cows, and goats on contaminated silage was reported in Italy, the Netherlands, Japan, and the United Kingdom (Book 1969; Belli et al. 1989; Pearce et al. 1989; Voors and Van Weers 1989; Aii et al. 1990; Monte 1990). The effective half-life of ^{137}Cs was 6.7 days in pasture grass and 13.6 days in milk (Spezzano and Giacomelli 1991). The average transfer coefficient of $^{134+137}Cs$ from Chernobyl from a 70% grass silage diet to milk of Dutch dairy cows was about 0.25%/liter/day (Voors and Van Weers 1991). In goats (*Capra* sp.), about 12% of orally administered ^{137}Cs was collected in milk within 7 days after dosing (Book 1969).

Iodine-131 was one of the most hazardous radionuclides released in the Chernobyl accident because it is easily transferred through the pasture–animal–milk pathway and rapidly concentrated in the thyroid gland to an extent unparalleled by any other organ. Because of its high specific activity, ^{131}I can transmit a high dose of radiation to the thyroid (Ionannides and Pakou 1991). Iodine-131 levels of 618,000 Bq/kg FW sheep thyroids from northwestern Greece on July 3, 1986, are similar to maximal ^{131}I concentrations in sheep thyroids in Tennessee in 1957 after global atmospheric fallout from military weapons tests and in London after the Windscale accident (Ionannides and Pakou 1991). Iodine-131 has an effective whole-body half-life of about 24 h and is rapidly excreted from sheep and cows (Assimakopoulos et al. 1989). The effective half-life of ^{131}I in pasture grass was 3.9 days and 5 days in cow's milk (Spezzano and Giacomelli 1991). The transfer coefficients of ^{131}I from vegetation to cow's milk was 0.007% day/L milk. This value was 57 times higher (0.4) in sheep (Monte 1990), but the mechanism to account for this large interspecies difference is not clear.

Table 32.18 Radionuclide Concentrations in Biotic and Abiotic Materials from Various Geographic Locales Before or After the Chernobyl Nuclear Accident on April 26, 1986 (All concentrations are in Bq/kg fresh weight [FW],or dry weight [DW], unless noted otherwise.)

Locale, Radionuclide, Sample, and Other Variables	Concentration	Reference[a]
ALASKA AND YUKON TERRITORIES		
Barren-ground caribou (*Rangifer tarandus granti*); porcupine herd; March–November 1987; ^{137}Cs		
Feces	Max. 802 DW	1
Muscle	133 (26–232) FW	1
Rumen contents	Max. 538 DW	1
ALBANIA		
^{137}Cs; May 2–19, 1986		
Air	Max. 1.8 Bq/m^3	2
Milk vs. wheat flour	Max. 380 FW vs. Max. 236 FW	2
^{131}I; cow's milk; May 2–19, 1986	Max. 3500 FW	2
CANADA		
Caribou, *Rangifer tarandus*; northern Quebec; 1986 (post-Chernobyl)–1987; muscle; ^{137}Cs	166–1129 FW	3
CZECHOSLOVAKIA		
$^{134+137}Cs$; 1986 (post-Chernobyl)		
Barley, *Hordeum vulgare*	7 DW	4
Cow, *Bos* sp., milk		
May	42 FW	4
July	10 FW	4
December	7 DW	4
Wheat, *Triticum* sp.	16 DW	4
$^{134+137}Cs$; domestic pig, *Sus* sp.; muscle; July 1986 vs. July 1987	15–22 FW vs. 22 FW	4

Table 32.18 (continued) Radionuclide Concentrations in Biotic and Abiotic Materials from Various Geographic Locales Before or After the Chernobyl Nuclear Accident on April 26, 1986 (All concentrations are in Bq/kg fresh weight [FW],or dry weight [DW], unless noted otherwise.)

Locale, Radionuclide, Sample, and Other Variables	Concentration	Reference[a]
DANUBE RIVER, HUNGARY–YUGOSLAVIA		
Water; 1986; post-Chernobyl		
^{134}Cs	0.015 FW	5
^{137}Cs	0.096 FW	5
^{103}Ru	0.070 FW	5
Fish, various species; 1986 (post-Chernobyl) vs. 1987		
^{134}Cs	8 FW vs. 4 FW	5
^{137}Cs	13 FW vs. 12 FW	5
^{103}Ru	1 FW vs. <1 FW	5
^{106}Ru	4 FW vs. 3 FW	5
Sediments; 1986 (post-Chernobyl) vs. 1988		
^{134}Cs	500 DW vs. 80 DW	5
^{137}Cs	750 DW vs. 200 DW	5
Algae; 1986 (post-Chernobyl) vs. 1988		
^{134}Cs	275 FW vs. 25 FW	5
^{137}Cs	625 FW vs. 100 FW	5
FINLAND		
Finnish Lapland; ^{137}Cs; 1979–84 vs. 1986 (post-Chernobyl)		
Arboreal lichens	120 DW vs. 590 DW	7
Ground lichens	230 DW vs. 900 DW	7
Birch, *Betula* sp.	68 DW vs. 51 DW	7
Horsetails, *Equisetum* sp.	203 DW vs. 280 DW	7
Bilberry, *Vaccinium* sp.	120 DW vs. 590 DW	7
Lichens; ^{137}Cs		
From reindeer herding areas; 1986 (post-Chernobyl) vs. 1987	900 DW vs. 800 DW	8
Isolated areas; 1986 (post-Chernobyl)–1987	3000–10,000 DW	8
Gulf of Finland; ^{137}Cs		
Seawater		
1985 (pre-Chernobyl) vs. 1986 (post-Chernobyl)	0.01 FW vs. 1.05 FW	53
1987 vs. 1990	0.23 FW vs. 0.05 FW	53
Bottom sediments		
1985 vs. 1986	1.2 FW vs. 40.0 FW	53
1987 vs. 1990	19.0 FW vs. 5.0 FW	53
Algae, whole		
1985 vs. 1986	3.9 FW vs. 175.0 FW	53
1987 vs. 1990	30.0 FW vs. 14.0 FW	53
Fish, 2 species, whole		
1985 (pre-Chernobyl)	1.4–3.5 FW	53
1986 (post-Chernobyl)	22.0–54.0 FW	53
1987	60.0–120.0 FW	53
1990	36.0–116.0 FW	53
Lake Paijanne (estimated ^{137}Cs Chernobyl loading of 20,000 Bq/m^2); ^{137}Cs; whole fish; three species (northern pike, *Esox lucius*; yellow perch, *Perca flavescens*; roach, *Rutilus rutilus*)		
1986; pre-Chernobyl vs. post-Chernobyl	580 FW vs. 1250 FW	6
1987	1000–2000 FW	6
1988	160–2000 FW	6
Reindeer, *Rangifer tarandus*; muscle; ^{137}Cs		
1964–65 (following nuclear tests) vs. 1985–86 (pre-Chernobyl)	Max. 2500–2600 FW vs. 300 FW	7, 8
1986–87 vs. 1987–88	720 FW, Max. 16,000 FW vs. 640 FW, Max. 9000 FW	8

Table 32.18 (continued) Radionuclide Concentrations in Biotic and Abiotic Materials from Various Geographic Locales Before or After the Chernobyl Nuclear Accident on April 26, 1986 (All concentrations are in Bq/kg fresh weight [FW],or dry weight [DW], unless noted otherwise.)

Locale, Radionuclide, Sample, and Other Variables	Concentration	Reference[a]
FRANCE		
Cows, fed hay (harvested post-Chernobyl) diet containing 5500 $^{134+137}$Cs/kg for mean daily intake of 15,900 Bq	A plateau was observed in milk after 15 days and in meat after 50–60 days; radiocesium transfer coefficients from diet were 1.1% for milk and 2.0–2.7% for meat	9
Calves fed $^{134+137}$Cs-contaminated milk from birth to age 80 days	Transfer coefficient from milk to meat was 16%	9
GERMANY		
Soils; June 24, 1986		
^{134}Cs	Max. 602 Bq/m^2 DW	10
^{137}Cs	Max. 1545 Bq/m^2 DW	10
^{103}Ru	Max. 808 Bq/m^2 DW	10
Pasture vegetation; May 1986		
^{134}Cs	20 FW	10
^{137}Cs	40 FW	10
^{131}I	75 FW	10
Cow; milk; May 1986		
^{134}Cs	140 FW	10
^{137}Cs	250 FW	10
^{131}I	250 FW	10
^{103}Ru	250 FW	10
Human, *Homo sapiens*		
Intake per person		
^{134}Cs; 1986 vs. 1987	354 Bq vs. 8 Bq	10
^{137}Cs; 1986 vs. 1987	728 Bq vs. 37 Bq	10
Whole-body dose (Bonn and vicinity); 1986 vs. 1987	0.0147 mSv (0.008 from ^{137}Cs, 0.0067 from ^{134}Cs) vs. 0.00056 mSv (0.0004 from ^{137}Cs, 0.00016 from ^{134}Cs)	10
Thyroid, ^{129}I	Negligible	11
GREECE		
Alfalfa, *Medicago sativa*; June 1986		
^{134}Cs	2303 DW	12
^{137}Cs	4551 DW	12
^{103}Ru	358 DW	12
^{106}Ru	1075 DW	12
Lichen, *Ramalina fraxinea* vs. moss, *Homalothecium sericium*; 1986 (post-Chernobyl); after decay of short-lived radionuclides		
^{134}Cs	426 FW vs. 1121 FW	13
^{137}Cs	951 FW vs. 2612 FW	13
^{40}K	222 FW vs. 278 FW	13
^{103}Ru	63 FW vs. 115 FW	13
^{106}Ru	436 FW vs. 1365 FW	13
Rye grass, *Lolium perenne*; June 1986		
^{134}Cs	3518 DW	12
^{137}Cs	7090 DW	12
^{103}Ru	708 DW	12
^{106}Ru	1747 DW	12
Plants, various; measured about 4 months post-Chernobyl; ^{137}Cs; values represent about 9% of initial Chernobyl radioactivity		
Aromatic plants; 11 species	22–11,344 FW; 26–22,000 DW	13
Cereals; 4 species	11–2257 FW; 11–2775 DW	13

Table 32.18 (continued) Radionuclide Concentrations in Biotic and Abiotic Materials from Various Geographic Locales Before or After the Chernobyl Nuclear Accident on April 26, 1986 (All concentrations are in Bq/kg fresh weight [FW],or dry weight [DW], unless noted otherwise.)

Locale, Radionuclide, Sample, and Other Variables	Concentration	Reference[a]
Fruit-bearing trees; 7 species	85–1572 FW; 122–2116 DW	13
Fungi; 4 species	103–5553 FW; 214–11,418 DW	13
Marine algae; 4 species	85–139 FW; 529–917 DW	13
Mosses and lichens; 6 species	1184–9413 FW; 1110–18,847 DW	13
Vegetables; 18 species	18–244 FW; 18–299 DW	13
Northern Greece; May 1986; ^{131}I		
Grasses	Max. 1500 FW	14
Milk; cow vs. domestic sheep, *Ovis aries*	Max. 300 FW vs. Max. 800 FW	14
Domestic sheep; thyroid; ^{131}I; maximum values; 1986		
June 27 vs. July 2	4000 FW vs. 15,600 FW	15
July 3 vs. July 5	618,000 FW vs. 9000 FW	15
July 29 vs. August 20	8500 FW vs. 600 FW	15
ITALY		
Honey bee, *Apis* spp.; honey; May 10, 1986		
^{134}Cs	Max. 171 FW	16
^{137}Cs	Max. 363 FW	16
^{131}I	Max. 1051 FW	16
^{103}Ru	Max. 575 FW	16
Cow		
Fed diets contaminated with Chernobyl ^{137}Cs for 8 months before slaughter		
Female vs. fetus		
Amniotic fluid	Max. 82 FW vs. —[b]	17
Blood	Max. 13 FW vs. Max. 44 FW	17
Muscle	Max. 179 FW vs. Max. 126 FW	17
Kidney	Max. 232 FW vs. Max. 139 FW	17
Liver	Max. 163 FW vs. Max. 115 FW	17
Placenta	Max. 93 FW vs. —[b]	17
Rodent, *Mus musculus domesticus*; carcass less internal organs; ^{137}Cs		
October–November 1981 vs. May 1986	5 DW vs. 43 DW	18
October–November 1986 vs. May 1987	20 DW vs. 18 DW	18
Northwest Saluggia, May 1986		
^{137}Cs, pasture grass vs. cow's milk	8000 DW vs. 180 FW	19
^{131}I, pasture grass vs. cow's milk	12,000 DW vs. 870 FW	19
Rabbit, *Oryctolagus* sp.; fed Chernobyl-contaminated alfalfa meal diet containing, in Bq/kg FW, 856 ^{137}Cs, 369 ^{134}Cs, and 540 ^{40}K; or normal diet (112 ^{137}Cs, 41 ^{134}Cs, 503 ^{40}K) for various intervals		
Control diet		
Whole animal	16 ^{137}Cs FW, 7 ^{134}Cs FW, 87 ^{40}K FW	20
Muscle	22 ^{137}Cs FW, 8 ^{134}Cs FW, 117 ^{40}K FW	20
21 days on contaminated diet followed by 21 days on control diet		
Whole animal	20 ^{137}Cs FW, 9 ^{134}Cs FW, 79 ^{40}K FW	20
Muscle	31 ^{137}Cs FW, 128 ^{134}Cs FW, 117 ^{40}K FW	20
42 days on contaminated diet		
Whole animal	81 ^{137}Cs FW, 32 ^{134}Cs FW, 85 ^{40}K FW	20
Muscle	112 ^{137}Cs FW, 44 ^{134}Cs FW, 124 ^{40}K FW	20
JAPAN		
^{137}Cs		
Milk; cow; May 1986	Max. 0.6 FW	21
Soil; estimated deposition from Chernobyl	180 Bq/m^2 DW	21

Table 32.18 (continued) Radionuclide Concentrations in Biotic and Abiotic Materials from Various Geographic Locales Before or After the Chernobyl Nuclear Accident on April 26, 1986 (All concentrations are in Bq/kg fresh weight [FW],or dry weight [DW], unless noted otherwise.)

Locale, Radionuclide, Sample, and Other Variables	Concentration	Reference[a]
$^{134+137}Cs$; humans, children; estimated internal dose through milk consumption		
1986	0.0006 mSv	21
1987	0.0003 mSv	21
1988	0.0001 mSv	21
^{131}I, grass vs. cow's milk		
May 10–11, 1986	65 FW vs. 4.3 FW	22
May 30, 1986	14 FW vs. ND[c]	22
MONACO		
Air, Bq/m^3, April 26, 1986; Monaco vs. Chernobyl (former Soviet Union)		
^{134}Cs	8.2 vs. 53	50
^{137}Cs	1.6 vs. 120	50
^{103}Ru	3.5 vs. 280	50
^{131}I	4.6 vs. 750	50
^{106}Ru	3.0 vs. 110	50
^{140}Ba	9.8 vs. 420	50
^{99}Mo	3.8 vs. 490	50
^{141}Ce	3.7 vs. 190	50
^{144}Ce	2.5 vs. 110	50
^{95}Zr	1.2 vs. 590	50
Marine copepods, 3 species; May 6, 1986; whole organism vs. fecal pellets		
^{103}Ru	280 DW vs. 16,000 DW	49
^{106}Ru	70 DW vs. 5800 DW	49
^{134}Cs	22 DW vs. 3400 DW	49
^{137}Cs	34 DW vs. 6300 DW	49
^{141}Ce	20 DW vs. 900 DW	49
^{144}Ce	100 DW vs. 2500 DW	49
Mussel, *Mytilus galloprovincialis*; soft parts; May 6 vs. August 14, 1986		
^{103}Ru	480 FW vs. 9.6 FW	51
^{106}Ru	121 FW vs. 11.2 FW	51
^{131}I	84 FW vs. <2 FW	51
^{134}Cs	6 FW vs. 0.1 FW	51
^{137}Cs	5.2 FW vs. 0.3 FW	51
NETHERLANDS		
^{134}Cs; grass silage; 1986 (post-Chernobyl) vs. 1987	Max. 50 DW vs. 2 DW	23
^{137}Cs; grass silage; 1986 (post-Chernobyl) vs. 1987	Max. 172 DW vs. 9 DW	23
^{137}Cs-contaminated roughage fed to lactating cows		
10.3 Bq ^{137}Cs/kg FW; grass	1.0–1.6 FW milk	24
173–180 Bq ^{137}Cs/kg FW; grass silage	12–28 FW milk	24
260–271 Bq ^{137}Cs/kg DW; grass	5.4–6.2 FW milk	24
^{40}K; grass silage; 1986 vs. 1987	910 DW vs. 1028 DW	23
NORWAY		
Alpine lake and vicinity; $^{134+137}Cs$		
Dwarf birch, *Betula nana*; leaves; August 1986	4000 FW	25
Lichens; August 1986	60,000 FW	25
Willow, *Salix* spp.; leaves; September 1980 vs. August 1986	<50 FW vs. 600 FW	25
Lake sediment; upper 10 cm; July–August 1986	1050 FW	25

Table 32.18 (continued) Radionuclide Concentrations in Biotic and Abiotic Materials from Various Geographic Locales Before or After the Chernobyl Nuclear Accident on April 26, 1986 (All concentrations are in Bq/kg fresh weight [FW],or dry weight [DW], unless noted otherwise.)

Locale, Radionuclide, Sample, and Other Variables	Concentration	Reference[a]
Aquatic organisms; July–August 1986		
Cladoceran, *Bosmina longispina*, whole	5300 FW	25
Amphipod, *Gammarus lacustris*, whole	6700 FW	25
Mayfly, *Siphlonurus lacustris*, whole	2800 FW	25
Stonefly, 2 spp., whole	1300–4120 FW	25
Minnow, *Phoxinus phoxinus*, whole	8800 FW	25
Brown trout, *Salmo trutta*		
Muscle		
1985 (pre-Chernobyl) vs. June 1986	<100 FW vs. 300 FW	25
August 1986 vs. June 1988	7000 FW vs. 4000 FW	25
Eggs vs. milt; July–August 1986	1740–3600 FW vs. 1300 FW	25
Dovrefjell, May 1986 vs. August 1990		
Earthworms (*Lumbricus rubellus*, *Allobophora caliginosa*), whole	121 FW vs. 74 FW	52
Eurasian woodcock, *Scolopax rusticola*, breast muscle	737 FW vs. 53 FW	52
Litter	14,400 DW vs. 2900 DW	52
Mushroom, *Lactarius* spp.; post-Chernobyl; $^{134+137}Cs$	Max. 445,000 FW	26
Reindeer; muscle; $^{134+137}Cs$		
1986; post-Chernobyl	10,000–50,000 FW	27
January 1987 vs. September 1988	Max. 56,000 FW vs. Max. 13,900 FW	28
Reindeer; 2 groups of adult females were fed lichen diets containing 45,000 Bq $^{134+137}Cs$/kg ration for 35 days; one group received daily oral administration of 250 mg ammonium hexacyanoferrate (Giese salt)	Both groups accumulated 400 Bq/kg FW daily in muscle. Retention time of Cs isotopes was 25 days without Giese salt and only 7–10 days when treated with Giese salt	29
POLAND		
Freshwater fish; 4 species; muscle; January 1987; $^{134+137}Cs$	4.5–6.1 FW	30
Southern Baltic Sea, $^{134+137}Cs$; pre-Chernobyl (1982–February 1986) vs. post-Chernobyl (June 1986–July 1987)		
Water	(13.8–19.8) Bq/m^3 vs. (59–100) Bq/m^3	30
Atlantic cod, *Gadus morhua*; muscle	(1.4–2.3) FW vs. (5.0–7.4) FW	30
Flounder, *Pleuronectes flesus*; muscle	(1.1–4.5) FW vs. (3.4–6.7) FW	30
SPAIN		
Songthrush, *Turdus philomelos*; edible tissues; November 1986 vs. November 1987		
^{134}Cs	Max. 90 DW for adults and young vs. Max. 7 DW for adults and 5 DW for young	31
^{137}Cs	Max. 208 DW vs. Max. 27 DW for adults and 22 for young	31
^{90}Sr	Max. 23 DW vs. Max. 7 DW	31
SWEDEN		
Moose, *Alces alces*; central Sweden; muscle; ^{137}Cs		
September 1986; adults vs. calves	300 FW vs. 500 FW	32
1986; all age groups	20–3000 FW	33
September 1987; adults vs. calves	201 FW vs. 401 FW	32
1987, all age groups	Max. 1600 FW	34
September 1988, adults vs. calves	640 FW vs. 1300 FW	32
1988, all age groups	Max. 2500 FW	34
1991, all age groups; May–September; various habitats	Mean 478 FW; Max. 1060 FW; highest in swamps and marshes and lowest in farmlands	54

Table 32.18 (continued) Radionuclide Concentrations in Biotic and Abiotic Materials from Various Geographic Locales Before or After the Chernobyl Nuclear Accident on April 26, 1986 (All concentrations are in Bq/kg fresh weight [FW],or dry weight [DW], unless noted otherwise.)

Locale, Radionuclide, Sample, and Other Variables	Concentration	Reference[a]
Moose dietary plants; 1986 (post-Chernobyl)–1988; ^{137}Cs		
Birch, *Betula* spp.; leaves	1200 DW	34
Heather, *Calluna vulgaris*; whole	13,000–32,000 DW	34, 35
Sedge, *Carex* spp.; whole	12,000 DW	35
Hair grass, *Deschampia flexuosa*; whole	1900 DW	34
Fireweed, *Epilobium angustifolium*; whole	400 DW	34
Grasses, various species; blades	2500 DW	34
Buckbean, *Menyanthes trifoliata*; whole	3800 DW	34
Pine, *Pinus sylvestris*; shoots	2500 DW	34
Aspen, *Populus tremula*; leaves	700 DW	34
Willow, *Solix* spp.; leaves	300 DW	34
Mountain ash, *Sorbus aucuparia*; leaves	1300 DW	34
Bilberry, *Vaccinium myrtillus*; leaves		
July 1986	2000 FW; 4000 DW	32, 34
July 1987 vs. July 1988	1138 FW vs. 600 FW	32
Bog whortleberry, *Vaccinium ulgmosum*; foliage	5900 DW	34
Cowberry, *Vaccinium vitis-idaea*; foliage	7500 DW	34
Cow's milk; ^{137}Cs; July 1986 vs. 1987	Usually <250 FW, Max. 375 FW vs. usually <70 FW, Max. 120 FW	36
Lichen, *Bryoria fuscescens*; ^{137}Cs; June 4, 1986	34,000–120,000 DW	35
Roe deer, *Capreolus* sp.; muscle; ^{137}Cs; 1986 (post-Chernobyl)	20–12,000 FW	33
Lichen, *Cladina* spp.; ^{137}Cs; 1986 (post-Chernobyl)	Max. 40,000 DW	35
Bank vole, *Clethrionomys glareolus*; collected from soil containing various concentrations of $^{134+137}Cs$; voles analyzed less skull and digestive organs		
1800 Bq/m^2 soil (control)	Voles had 9 Bq ^{134}Cs/kg FW and 39 of ^{137}Cs; mutation frequency of 1.3; total irradiation of 0.0042 mGy daily	37
22,000 Bq/m^2 soil	In Bq/kg FW, voles had 279 ^{134}Cs and 1031 ^{137}Cs; mutation frequency was 1.5; daily dose rate of 0.0088 mGy	37
90,000 Bq/m^2 soil	Voles had 1356 Bq ^{134}Cs/kg FW and 5119 of ^{137}Cs; mutation frequency 1.9; daily dose of 0.0268 mGy	37
145,000 Bq/m^2 soil	Voles had 2151 Bq ^{134}Cs/kg FW and 7784 ^{137}Cs; mutation frequency 2.6; daily dose of 0.0394 mGy	37
Buckbean, *Menyanthes trifoliata*; foliage; ^{137}Cs; 1985 vs. 1987	1800 DW vs. 3880 DW	35
Reindeer dietary lichens; ^{137}Cs; April 1986	Usually 40,000–60,000 DW; Max. 120,000 DW	38
Reindeer		
Moved in November 1986 from a highly contaminated area (>20,000 Bq ^{137}Cs/m^2) to a less-contaminated area (<3000 Bq/m^2) of natural pasture	^{137}Cs content in muscle declined from 12,000 FW in November to about 3000 FW in April	39
Reindeer; muscle; ^{137}Cs; 1986 (post-Chernobyl)	100–40,000 FW	33
Rodents and insectivores; July–August 1986; ^{137}Cs		
Control site, soil	1800 Bq/m^2	40
Bank vole; whole less skull, stomach, viscera	39 FW	40
Common shrew, *Sorex araneus*; whole less skull, stomach, viscera	48 FW	40
Site 2, soil	22,000 Bq/m^2	40
Bank vole vs. common shrew	676 FW vs. 751 FW	40
Site 3, soil	90,000 Bq/m^2	40
Bank vole vs. common shrew	5119 FW vs. 3233 FW	40
Site 4, soil	145,000 Bq/m^2	40
Bank vole vs. common shrew	7993 FW vs. 6289 FW	40

Table 32.18 (continued) Radionuclide Concentrations in Biotic and Abiotic Materials from Various Geographic Locales Before or After the Chernobyl Nuclear Accident on April 26, 1986 (All concentrations are in Bq/kg fresh weight [FW],or dry weight [DW], unless noted otherwise.)

Locale, Radionuclide, Sample, and Other Variables	Concentration	Reference[a]
SYRIA		
^{137}Cs; air; May 7–10, 1986	0.12 Bq/m^3	41
^{131}I; May 7–10, 1986; air vs. goat's milk	4 Bq/m^3 vs. 55 FW	41
UNITED KINGDOM		
Upland pastures		
Moss, *Sphagnum* sp.; September 1986		
^{110m}Ag	202 DW	42
^{144}Ce	202 DW	42
^{134}Cs	8226 DW	42
^{137}Cs	17,315 DW	42
^{106}Ru	1893 DW	42
^{125}Sb	294 DW	42
Vegetation; $^{134+137}Cs$; June 1986 vs. January 1989	About 6000 DW vs. 1000 DW	43
Marine molluscs; 7 species; near nuclear plant; 1984 (pre-Chernobyl) vs. 1986 (post-Chernobyl)		
^{110m}Ag	<77 FW vs. 13–77 FW	44
^{60}Co	<29 FW vs. 16–32 FW	44
^{134}Cs	<14 FW vs. 37–388 FW	44
^{137}Cs	<139 FW vs. 31–836 FW	44
^{40}K	<59 FW vs. 57–61 FW	44
^{238}Pu	<27 FW vs. 11–22 FW	44
$^{239+240}Pu$	<107 FW vs. 19–89 FW	44
^{106}Ru	<632 FW vs. 124–1648 FW	44
^{125}Sb	ND vs. 29 FW	44
European oystercatcher, *Haematopus ostralegas*; near nuclear reactor; June 1986; egg contents vs. egg shells		
^{134}Cs	4 FW vs. —[b]	44
^{137}Cs	18 FW vs. 6 FW	44
^{238}Pu	0.2 FW vs. 1.1 FW	44
$^{239+240}Pu$	0.05 FW vs. 4.6 FW	44
Red grouse, *Lagopus lagopus*; muscle; November 1986–February 1987		
^{134}Cs; cock vs. hen	325 FW vs. 602 FW	45
^{137}Cs; cock vs. hen	962 FW vs. 1684 FW	45
Black-headed gull, *Larus ridibundus ridibundus*; near nuclear reactor; 1980 vs. June 1986		
Egg contents		
^{134}Cs	ND vs. 22 FW	44
^{137}Cs	10 FW vs. 43 FW	44
^{238}Pu	0.02 FW vs. 0.01 FW	44
$^{239+240}Pu$	0.05 FW vs. 0.04 FW	44
Egg shells		
^{134}Cs	—[b] vs. 7 FW	44
^{137}Cs	—[b] vs. 16 FW	44
^{238}Pu	<0.17 FW vs. 0.4	44
$^{239+240}Pu$	0.6 FW vs. 1.6 FW	44
Woodcock, *Scolopax rusticola*; muscle; November 1986–February 1987		
^{134}Cs	13 FW	45
^{137}Cs	42 FW	45
Black grouse, *Tetrao tetrix*; ^{137}Cs, November 1986–February 1987; diet vs. muscle	167 FW vs. 270 FW	45
Cow's milk; May 5–8, 1986		
^{137}Cs	Max. 150 FW	46
^{131}I	Max. 127 FW	46

Table 32.18 (continued) Radionuclide Concentrations in Biotic and Abiotic Materials from Various Geographic Locales Before or After the Chernobyl Nuclear Accident on April 26, 1986 (All concentrations are in Bq/kg fresh weight [FW],or dry weight [DW], unless noted otherwise.)

Locale, Radionuclide, Sample, and Other Variables	Concentration	Reference[a]
Roe deer, *Capreolus capreolus*; ^{137}Cs; muscle; November 1986–February 1987		
Calves	711 FW	45
Hinds	375–586 FW	45
Stags	1564 FW	45
Red deer, *Cervus elephus*; muscle; November 1986–February 1987		
^{134}Cs; calf vs. hind	186 FW vs. 112 FW	45
^{137}Cs; calf vs. hind	535 FW vs. 311 FW	45
Brown hare, *Lepus capensis*; ^{137}Cs; female; November 1986–February 1987; diet vs. muscle	198 FW vs. 656 FW	45
Blue hare, *Lepus timidus*; ^{137}Cs; November 1986–February 1987		
Males; diet vs. muscle	808 FW vs. 1677 FW	45
Females; diet vs. muscle	577 FW vs. 1440 FW	45
Rabbit, *Oryctolagus* sp.; muscle; male; November 1986–February 1987		
^{134}Cs	6 FW	45
^{137}Cs	15 FW	45
Domestic sheep		
Muscle; ^{137}Cs; September 1986 vs. July 1987	1500 FW vs. 1170 FW	42
Liver; ^{110m}Ag; ewes vs. lambs		
September 1986	34 FW vs. 17 FW	47
July 1987	55 FW vs. <8 FW	47
Diet (rye grass and vegetation); ^{110m}Ag; 1986 vs. 1987	32 DW vs. 10–30 DW	47
Lambs fed a milk replacement diet containing 950 Bq ^{137}Cs/kg ration for 21 days. After weaning, lambs were fed silage contaminated with fallout radiocesium plus ionic $^{134}CsCl$ for 3 weeks	Absorption during the first 21 days was about 90%, equivalent to 975 Bq ^{137}Cs/kg BW. During the silage feeding period, uptake of ionic ^{134}Cs was about twice that of fallout ^{134}Cs present in silage	48
Red fox, *Vulpes vulpes*; muscle; November 1986–February 1987; vixen		
^{134}Cs	176 FW	45
^{137}Cs	461–643 FW	45

[a] **1,** Allaye-Chan et al. 1990; **2,** Kedhi 1990; **3,** Crete et al. 1990; **4,** Kliment 1991; **5,** Conkic et al. 1990; **6,** Korhonen 1990; **7,** Rissanen and Rahola 1989; **8,** Rissanen and Rahola 1990; **9,** Daburon et al. 1989; **10,** Clooth and Aumann 1990; **11,** Handl et al. 1990; **12,** Douka and Xenoulis 1991; **13,** Sawidis 1988; **14,** Assimakopoulos et al. 1989; **15,** Ionannides and Pakou 1991; **16,** Tonelli et al. 1990; **17,** Calamosca et al. 1990; **18,** Cristaldi et al. 1990; **19,** Spezzano and Giacomelli 1991; **20,** Battiston et al. 1991; **21,** Imanaka and Koide 1990; **22,** Aii et al. 1990; **23,** Voors and Van Weers 1991; **24,** Vreman et al. 1989; **25,** Brittain et al. 1991; **26,** Hove et al. 1990a; **27,** Skogland and Espelien 1990; **28,** Eikelmann et al. 1990; **29,** Mathiesen et al. 1990; **30,** Grzybowska 1989; **31,** Baeza et al. 1991; **32,** Palo et al. 1991; **33,** Johanson 1990; **34,** Bothmer et al. 1990; **35,** Eriksson 1990; **36,** Johanson et al. 1989; **37,** Cristaldi et al. 1991; **38,** Jones 1990; **39,** Jones et al. 1989; **40,** Mascanzoni et al. 1990; **41,** Othman 1990; **42,** Coughtrey et al. 1989; **43,** Crout et al. 1991; **44,** Lowe 1991; **45,** Lowe and Horrill 1991; **46,** Clark and Smith 1988; **47,** Beresford 1989; **48,** Moss et al. 1989; **49,** Fowler et al. 1987; **50,** Whitehead et al. 1988b; **51,** Whitehead et al. 1988a; **52,** Kalas et al. 1994; **53,** Kryshev 1995; **54,** Nelin 1995.

[b] — = no data.

[c] ND = not detectable.

32.6 EFFECTS: NONIONIZING RADIATIONS

Living organisms are constantly exposed to nonionizing electromagnetic radiations, including ultraviolet (UV), visible, infrared, radio, and other low-energy radiations that form an integral part of the biosphere. Emissions from anthropogenic sources such as radios, microwave ovens, television communications, and radar have significantly altered the character of our natural electromagnetic field (Garaj-Vrhovac et al. 1990). Although the primary focus of this review is on ionizing radiations, it would be misleading to assume that low-energy electromagnetic waves cannot elicit significant biological responses. For example, behavioral and biochemical changes are reported in rats, monkeys, rabbits, and other laboratory animals after exposure to nonionizing electromagnetic radiations. The severity of the effect is associated with the type and duration of the radiation and various physicochemical variables (Ghandi 1990). Selected examples follow.

Ultraviolet radiation should be considered a plausible factor contributing to amphibian malformations in field settings (Ankley et al. 1998). Ultraviolet radiation is linked to teratogenesis, growth inhibition, and DNA photodamage in larvae of the South African clawed frog, *Xenopus laevis* (Bruggeman et al. 1998). About 50% of newly fertilized eggs of the northern leopard frog (*Rana pipiens*) exposed to UV radiation for 24 h developed hindlimb malformations, such as missing limb segments, missing or reduced digits, and missing or malformed femurs (Ankley et al. 1998). The higher-energy portion of the ultraviolet spectrum (UV-B) was lethal to embryos of Canadian frogs, with death occurring in as little as 30 min; the lower-energy portion (UV-A) was not harmful to eggs or larvae at exposures twice that of ambient levels (Grant and Licht 1995). Exposure to solar radiation — even at low elevations — is lethal to amphibian eggs of species with relatively poor capacity to repair UV-damaged DNA (Blaustein et al. 1994, 1996). Increased UV-B radiation in the environment due to decreasing ozone levels has been suggested as a factor in the worldwide decline of sensitive species of amphibians (Blaustein et al. 1994; Ankley et al. 1998). However, UV-B radiation was not implicated in the decline of the endangered green and golden bell frog (*Litoria aurea*) in Australia (van de Mortel and Buttemer 1996).

UV radiation in mammals causes the aging of skin, making it wrinkled and leathery (Kligman and Kligman 1990). Dermatologists of the late 19th century described the devastating effects of sunlight on the skin of farmers and sailors when compared with indoor workers. Photoaged skin has a variety of neoplasms, deep furrows, extensive sagging, and profound structural alterations that are quite different from those in protected, intrinsically aged skin (Kligman and Kligman 1990). Similar results were documented for the skin of guinea pigs (Davidson et al. 1991) and rodents (Ananthaswamy and Pierceall 1990; Ronai et al. 1990) after exposure to UV radiation. UV radiation causes eye cancer in cattle (Anderson and Badzioch 1991), interferes with wound healing in guinea pig skin (Davidson et al. 1991), is a potent damaging agent of DNA and a known inducer of skin cancer in experimental animals (Ronai et al. 1990), and interferes with an immune defense mechanism that normally protects against skin cancer (Ananthaswamy and Pierceall 1990). Aquatic organisms exposed to UV radiation show disrupted orientation, decreased motility, and reduced pigmentation in *Peridinium gatunense*, a freshwater alga (Hader et al. 1990). Effects were similar in several species of marine algae (Lesser and Shick 1990; Hader and Hader 1991; Shick et al. 1991). Increased lipid peroxidation rates and a shortening of the life-span after UV exposure were reported in the rotifer *Asplanchna brightwelli* (Sawada et al. 1990). Cells of the goldfish (*Carassius auratus*) were damaged, presumably by DNA impairment, from UV exposure (Yasuhira et al. 1991).

Visible radiation adversely affected survival and growth of embryos of the chinook salmon (*Oncorhynchus tshawytscha*) (Eisler 1961), the chloroplast structure in the symbiotic marine dinoflagellate *Symiodinium* sp. (Lesser and Shick 1990), and *in vitro* growth of cultured mammalian cells (Karu 1990). Infrared radiation contributes significantly to skin photoaging, producing severe elastosis; the epidermis and the dermis were capable of self-restoration when the exogenous injury ceased (Kligman and Kligman 1990).

Investigations of the cellular effects of radiofrequency radiation provide evidence of damage to various types of avian and mammalian cells. These effects involve radiofrequency interactions with cell membranes, especially the plasma membrane. Effects include alterations in membrane cation transport, Na^+/K^+-ATPase activity, protein kinase activity, neutrophil precursor membrane receptors, firing rates and resting potentials of neurons, brain cell metabolism, DNA and RNA synthesis in glioma cells, and mitogenic effects on human lymphocytes (Cleary 1990).

Microwaves inhibit thymidine incorporation by DNA blockage in cultured cells of the Chinese hamster; irradiated cells had a higher frequency of chromosome lesions (Garaj-Vrhovac et al. 1990). Microwaves induce teratogenic effects in mice when the intensity of exposure places a thermal burden on the dams and fetuses, resulting in a reduction in fetal body mass and an increased number of resorptions (O'Connor 1990).

Extremely low frequency (ELF) electromagnetic fields — similar to fields that emanate from electrical appliances and the electrical power distribution network, usually <300 Hz — are used therapeutically in the healing of human nonunion bone fractures, in the promotion of nerve regeneration, and in the acceleration of wound healing (Anderson 1990). ELF electric and magnetic fields produce biological effects, usually subtle, and are of low hazard in short-term exposure. These effects include altered neuronal excitability, neurochemical changes, altered hormone levels, and changes in behavioral responses. For example, electric field perception has been reported in humans, mice, pigs, monkeys, pigeons, chickens, and insects; altered cardiovascular responses in dogs and chickens; and altered growth rate of chicks. No deleterious effects of ELF fields on mammalian reproduction and development or on carcinogenesis and mutagenesis have been documented (Anderson 1990). In the vicinity of a powerful radar station, some birds avoided nesting (flycatchers, *Myiarchus* spp.), and the percent of nesting boxes occupied by other species (tits, *Parus* spp.) increased significantly with increasing distance from the radar station (Liepa and Balodis 1994). ELF fields had no effect on the growth of bone in chicks (Coulton and Barker 1991). However, adult newts (*Notophthalmus viridescens*), regenerating amputated forelimbs, had grossly abnormal forelimbs 12% of the time when exposed for 30 days to ELF fields of the type reported to facilitate healing of human bone fractures (Landesman and Douglas 1990). Additional studies are recommended on the biological effects of nonionizing radiations on fishery and wildlife resources, especially on ELF radiations.

32.7 EFFECTS: IONIZING RADIATIONS

32.7.1 General

High acute doses of ionizing radiation produce adverse biological effects at every organizational level: molecular, cellular, tissue–organ, whole animal, population, community, and ecosystem (ICRP 1977; Whicker and Schultz 1982b; LWV 1985; Hobbs and McClellan 1986; UNSCEAR 1988; Kiefer 1990; Severa and Bar 1991). Typical adverse effects of ionizing radiation include:

- Cell death (McLean 1973; LWV 1985; Kiefer 1990)
- Decreased life expectancy (Lorenz et al. 1954; Brown 1966; Hobbs and McClellan 1986; Kiefer 1990; Rose 1992)
- Increased frequency of malignant tumors (Lorenz et al. 1954; ICRP 1977; Hobbs and McClellan 1986; UNSCEAR 1988; Hopewell 1990; Kim et al. 1990; Little 1990; Nagasawa et al. 1990; Raabe et al. 1990; Fry 1991)
- Inhibited reproduction (ICRP 1977; Barendson 1990; Kiefer 1990; Rose 1992)
- Increased frequency of gene mutations (ICRP 1977; Whicker and Schultz 1982b; Hobbs and McClellan 1986; Abrahamson 1990; Evans 1990; Kiefer 1990; Thacker 1990; Sankaranarayanan 1991a; 1991b; Rose 1992; Macdonald and Laverock 1998)
- Leukemia (ICRP 1977; Kiefer 1990)
- Altered blood–brain barrier function (Trnovec et al. 1990)
- Reduced growth and altered behavior (Rose 1992)

Species within kingdoms have a wide variation in sensitivity, and sometimes at low radiation exposures the response is considered beneficial (Luckey 1980; Rose 1992). Overall, the lowest

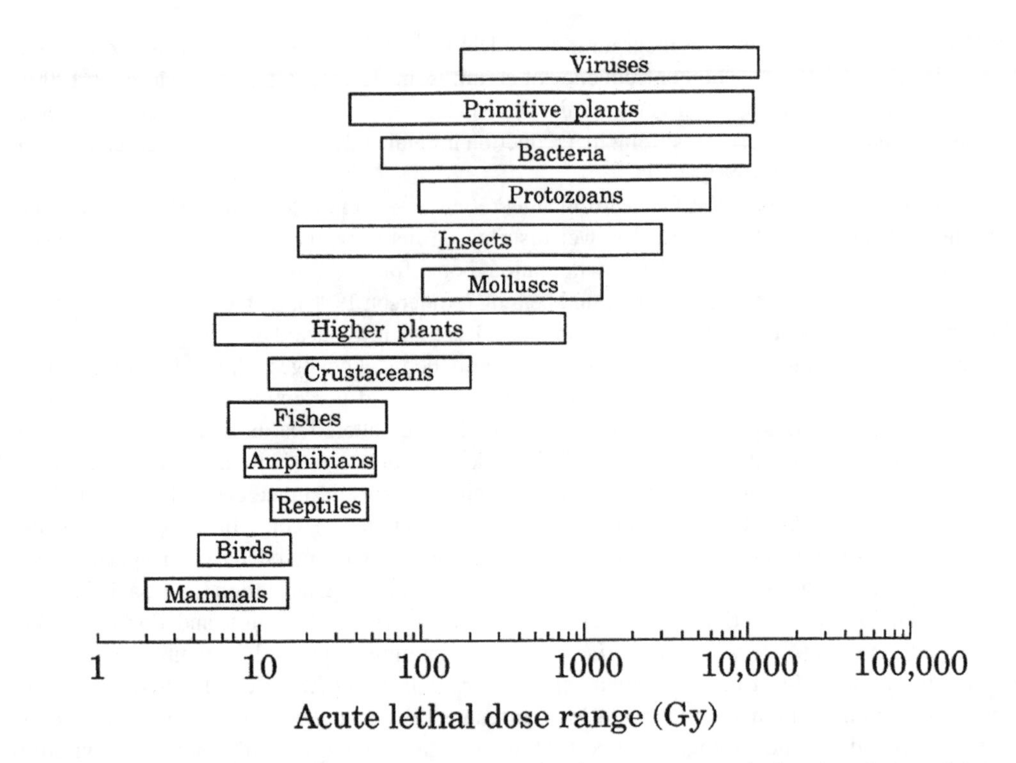

Figure 32.7 Acute radiation dose range fatal to 50% (30 days postexposure) of various taxonomic groups. (Modified from Whicker, F.W. and V. Schultz. 1982b. *Radioecology: Nuclear Energy and the Environment.* Vol. II. CRC Press LLC, Boca Raton, FL. 228 pp.; Hinton, T.G. and D.E. Scott. 1990. Radioecological techniques for herpetology, with an emphasis on freshwater turtles. Pages 267-287 in J.W. Gibbons (ed.). *Life History and Ecology of the Slider Turtle.* Smithsonian Instit. Press, Washington, D.C.)

dose rate at which harmful effects of chronic irradiation have been reliably observed in sensitive species is about 1 Gy/year; this value for acute radiation exposures is about 0.01 Gy (Rose 1992).

In general, the primitive organisms are the most radioresistant taxonomic groups, and the more advanced complex organisms — such as mammals — are the most radiosensitive (Figure 32.7). The early effects of exposure to ionizing radiation result primarily from cell death; cells that frequently undergo mitosis are the most radiosensitive, and cells that do not divide are the most radioresistant. Thus, embryos and fetuses are particularly susceptible to ionizing radiation, and very young animals are consistently more radiosensitive than adults (McLean 1973; Hobbs and McClellan 1986). In addition to the evolutionary position and cell mitotic index, many extrinsic and intrinsic factors modify the response of a living organism to a given dose of radiation. Abiotic variables include the type and energy of radiation, exposure rate, length of exposure, total exposure and absorbed dose, dose rate, spatial distribution of dose, season, temperature, day length, and environmental chemicals. Biotic variables include the species, type of cell or tissue, metabolism, sex, nutritional status, sensitizing or protective substances, competition, parasitism, and predation (Whicker and Schultz 1982b; Hobbs and McClellan 1986; UNSCEAR 1988; Kiefer 1990).

Radiosensitivity of cells is related directly to their reproductive capacity and indirectly to their degree of differentiation (Hobbs and McClellan 1986). Early adverse effects of exposure to ionizing radiation are due mainly to the killing of cells. Cell death may result from the loss of reproductive integrity, that is, when after irradiation a cell fails to pass through more than one or two mitoses. Reproductive death is important in rapidly dividing tissues such as bone marrow, skin, gut lining, and germinal epithelium. When the whole animal is exposed to a large dose of ionizing radiation, some tissues are more prone to damage than others. Death rates of mammalian reproductive cells from ionizing radiations are modified by variations in the linear energy transfer of the radiation, the stage in the cell cycle, cell culture conditions, and sensitizing and protecting compounds (Barendsen 1990). The chemical form of the main stage of the acute radiation syndrome depends on the size and distribution of the absorbed dose. It is determined mainly by damage to blood platelets and other blood-forming organs at 4 to 5 Gy, to epithelial cells lining the small intestine at 5 to 30 Gy, and to brain damage at >30 Gy; death usually occurs within 48 h at >30 Gy (McLean 1973).

Cellular DNA is extremely sensitive to ionizing radiation, although other cell constituents may approach DNA in sensitivity (IAEA 1976; Billen 1990; Kiefer 1990; Lett 1990; Lucke-Huhle et al. 1990; Woloschak et al. 1990a; Shadley et al. 1991). Radiation-induced mutations are explainable on the basis of chromatin and DNA organization in cells and the biophysical properties of ionizing radiation (Sankaranarayanan 1991b). Based on studies of spontaneous and radiation-induced mutations in the mouse (Sankaranarayanan 1991a), more than 67% of the ionizing radiation-induced mutations are lethal, and almost all mutations, including enzyme activity variants, dominant visibles, and dominant skeletal mutations, are lethal. These findings are consistent with the view that most radiation-induced mutations in germ cells of mice are due to DNA deletions (Sankaranarayanan 1991a).

Experimental animal data clearly demonstrate that ionizing radiation at relatively high doses and delivered at high dose rates is mutagenic (Hobbs and McClellan 1986). However, radiation-induced genetic damage in the offspring of exposed parents has not been credibly established in any study with humans (Abrahamson 1990). In one human population — the ethnically isolated Swedish reindeer-breeding Lapps — elevated concentrations of fallout products have been ingested via the lichen–reindeer–human food chain since the 1950s. However, from 1961 to 1984, no increased incidence of genetic damage was evident in Lapps (Wiklund et al. 1990).

Radiation is carcinogenic. The frequency of death from cancer of the thyroid, breast, lung, esophagus, stomach, and bladder was higher in Japanese survivors of the atomic bomb than in nonexposed individuals, and carcinogenesis seems to be the primary latent effect of ionizing radiation. The minimal latent period of most cancers was <15 years and depended on an individual's age at exposure and site of cancer. The relation of radiation-induced cancers to low doses and the shape of the dose-response curve (linear or nonlinear), the existence of a threshold, and the influence of dose rate and exposure period have to be determined (Hobbs and McClellan 1986).

Radioactive materials that gain entry to the body, typically through ingestion or inhalation, exert effects that are governed by their physical and chemical characteristics which, in turn, influence their distribution and retention inside the body. The effective half-life includes both physical and biologic half-times. In addition, the type of radiation (i.e., α, β, γ) and its retention and distribution kinetics govern the radiation dose pattern. In general, the radiation dose from internal emitters is a function of the effective half-time, energy released in the tissue, initial amount of introduced radioactivity, and mass of the organ (Hobbs and McClellan 1986). Retention of radionuclides by living organisms is quite variable and modified by numerous biologic and abiotic variables. For example, ^{137}Cs retention in selected animals varies significantly with the body weight, diet, and metabolism of an organism (Figure 32.8). The time for 50% persistence of ^{137}Cs ranges between 30 and 430 days in ectotherms, and it was longer at lower temperatures and shortest in summer and under conditions of inadequate nutrition (Hinton and Scott 1990). In mammals, the ^{137}Cs biological half-life was between 6 and 43 days in rodents, dogs, mule deer, reindeer, and monkeys. In humans, this value ranged from 60 to 160 days. The biological half-life of ^{90}Sr ranges from 122 to 6000 days in ectotherms and is longer at colder temperatures and under laboratory conditions. In mammals and under conditions of chronic intake, the ^{90}Sr biological half-life was 533 days in rat, 750 days in humans, and at least 848 days in beagles (Hinton and Scott 1990).

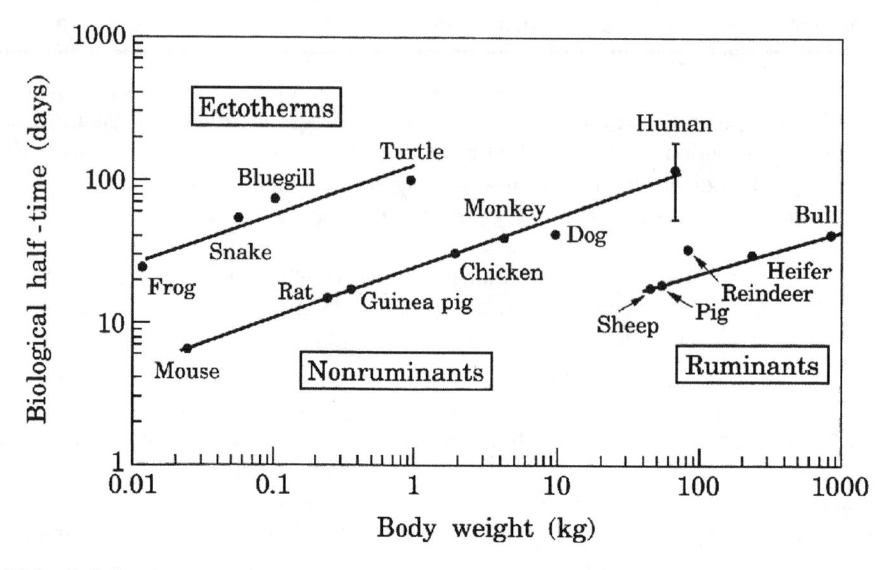

Figure 32.8 Relation between diet, metabolism, and body weight with half-time retention of longest-lived component of cesium-137. Data are shown for selected ruminant and nonruminant mammals (Richmond 1989) and ectotherms (Hinton and Scott 1990.)

32.7.2 Terrestrial Plants and Invertebrates

Radiosensitive terrestrial plants exposed to single doses of ionizing radiation had reduced growth at 0.5 to 1.0 Gy and reduced survival at 3.0 to 4.1 Gy (Table 32.19). Chronic exposures of 0.2 to 0.65 Gy/day adversely affected sensitive forest ecosystems (Table 32.19). Chronic gamma irradiation of 131 Gy/year and higher of mixed forest ecosystems caused the disappearance of trees and shrubs and subsequent erosion of the soil (Poinsot-Balaguer et al. 1991). The radiation sensitivity of five plant communities suggested that pine forests were the most sensitive and that deciduous evergreen forests, tropical rainforests, herbaceous rock outcrop communities, and abandoned cropland were increasingly less sensitive (McCormick 1969). Neutrons were 3 to 4 times more effective than gamma rays in root growth inhibition (Witherspoon 1969). Altitude affects the response of vegetation to ionizing radiation. Peas (*Pisum sativum*) in gardens 2225 to 3750 m above mean sea level and exposed to 0, 5, 10, or 50 Gy had reduced growth from all treatments at increasing altitudes; however, a dose-response growth curve was evident only at <3049 m altitude (Osburn 1963). Seeds of tobacco (*Nicotiana tabacum*) exposed to cosmic rays aboard a spacecraft had a higher mutation rate than controls; effects occurred at total doses as low as 0.1 to 0.2 Gy (Gaubin et al. 1990), but this needs verification.

Sometimes, irradiation prevents the usual colonizing vegetation from becoming established (Poinsot-Balaguer et al. 1991). Germination and survival of shrub seedlings have been much slower at nuclear test sites than at non-disturbed sites (Romney et al. 1971). The return to its original state of the perennial shrub vegetation takes decades on a radiation-disturbed site, although native annual species and grasses have grown abundantly within 12 months. Transplanting of shrubs into radiation-disturbed areas has been largely unsuccessful because of intense browsing by rabbits and other small mammals (Romney et al. 1971). A nuclear detonation damages terrestrial vegetation by heat, blast, or radiation. Plant injury from thermal or ionizing radiation at an above-ground detonation site varied with stem rigidity and stability of the substratum, although radiation effects are ordinarily masked by damage from blasts. A typical nuclear detonation at the Nevada test site — an airburst of a 20- to 40-kiloton yield — denuded a zone of desert within a 0.8-km radius of shrub vegetation. Recovery at the Nevada site seemed complete within 4 years, suggesting little relation between fatal injury, morphological aberration in vegetation, and ionizing radiation from nuclear detonations (Shields and Wells 1963). A northern Wisconsin forest experimentally subjected to a ^{137}Cs radiation source for 5 months showed several trends (Zavitokovski and Rudolph 1971):

1. Herbaceous and shrub species with a spreading form of growth are more radioresistant than upright forms.
2. Larger pine and oak trees are more radioresistant than smaller trees.
3. Perennial plants with shielded buds and vigorous asexual reproduction are relatively radioresistant.
4. Plants adapted to extreme habitats, such as old fields and granite rock outcrops, and plants typical of early successional stages are relatively radioresistant.
5. All plants are more radiosensitive during the growing season than during the dormant season.
6. Reproductive stages are always more radiosensitive than vegetative stages.

The recovery of vegetation in a tropical rainforest in Puerto Rico — after plants were deliberately subjected to lethal doses of gamma radiation — closely resembled secondary succession after other types of disturbances, such as mechanical stripping and treatment with the Picloram herbicide (Jordan 1969).

Table 32.19 Radiation Effects on Selected Terrestrial Plants

Species, Dose, and Other Variables	Effect	Reference[a]
Tropical rainforest tree, *Dacryodes excelsa*, 4–280 Gy per year	Growth stimulation	1
Deciduous evergreen forest		
40 Gy yearly	Minor effects	2
100 Gy yearly	Severe sublethal effects	2
350 Gy yearly	Lethal	2
Deciduous plants, 13 species		
4–15 Gy, single fast neutron doses	Shoot growth inhibited by >85%	3
60–85 Gy, single gamma radiation dose	Shoot growth inhibited by >85%	3
Forest ecosystem, northern Wisconsin, experimentally exposed to a ^{137}Cs point source for 5 months during a growing season. Distance from source (meters) and daily exposure (Gy)		
5 m, 15 Gy	No vegetation	4
5–10 m, 5–15 Gy	Lower plants present	4
10–15 m, 1.5 Gy	Resistant trees and shrubs present	4
10–15 m, 2.5–5.0 Gy	Some growth	4
20–30 m, 0.65–1.5 Gy	Resistant angiosperm trees	4
30–50 m, 0.2–0.65 Gy	Angiosperm trees present	4
50 m, 0.2 Gy	Original northern forest	4
Herbaceous rock outcrop community		
90 Gy yearly	Minor effects	2
400 Gy yearly	Severe sublethal effects	2
1000 Gy yearly	Lethal	2
Queensland mango, *Mangifera indica*, fruit irradiated postharvest, single dose, 250 or 750 Gy	At 250 Gy, skin and pulp color inhibited 50% due to irradiation-induced suppression of chlorophyll breakdown and reduction in carotenoid production. At 750 Gy, fruit respiration increased for 3–5 days, but no effect on fruit firmness	5
Mixed oak forest, southern France, experimentally irradiated for 18 years by a ^{137}Cs source at dose rates between 0.3 and 116 mGy/h, equivalent to a yearly rate between 2.6 and 1016 Gy	At 60–100 mGy/h (525–876 Gy yearly), all trees, shrubs, and litter were absent; low overall insect density; soil deficient in carbon, nitrogen, and water. At 15 mGy/h (131 Gy yearly), woody plants were present, but visibly abnormal	6

Table 32.19 (continued) Radiation Effects on Selected Terrestrial Plants

Species, Dose, and Other Variables	Effect	Reference[a]
Tobacco, *Nicotiana tabacum*, 55 Gy per year	Growth stimulation	1
Pine forest		
1–10 Gy yearly	Minor effects	2
20 Gy yearly	Severe sublethal effects	2
30 Gy yearly	Lethal	2
Slash pine, *Pinus ellottii*, acute single exposure of 3 Gy	50% dead 1–4 months after exposure; no other deaths in 2 years	2
Sugar pine, *Pinus lambertiana*, acute single exposure of 4.1 Gy	LD50 (30 days postexposure)	1
Longleaf pine, *Pinus palustris*		
0.5 Gy, single dose	Growth inhibition	1
8 Gy, acute single exposure	50% of trees <5 years old died in 1–4 months; others survived for at least 2 years	2
>28 Gy, acute single exposure	Fatal to 50% of trees >5 years old in 1–4 months; no other deaths in 2 years	2
Winter wheat, *Triticum aestivum*, acute single exposure of 1.0 Gy	Growth inhibition	1
Tropical rainforest		
70 Gy yearly	Minor effects	2
350 Gy yearly	Severe sublethal effects	2
400 Gy yearly	Lethal	2
Vegetation, abandoned crop land		
50 Gy yearly	Minor effects	2
450 Gy yearly	Severe sublethal effects	2
1500 Gy yearly	Lethal	2
Bean, *Vicia faba*, 58–100 Gy per year	Growth stimulation	1

[a] **1,** Rose 1992; **2,** McCormick 1969; **3,** Witherspoon 1969; **4,** Zavitokovski and Rudolph 1971; **5,** Boag et al. 1990; **6,** Poinsat-Balaguer et al. 1991.

Hormesis — the beneficial physiological stimulation by low doses of a potentially harmful agent — is documented for ionizing radiation and many species of terrestrial plants and invertebrates (Luckey 1980). Radiation hormesis in plants includes increased germination, growth, survival, and yield. Some species of terrestrial invertebrates had increased fecundity, growth, survival, disease resistance, and longevity after exposure to low sublethal doses of ionizing radiation (Luckey 1980). The growth and development of some terrestrial invertebrates are stimulated at comparatively high sublethal acute doses (i.e., 2 Gy in silkworm, *Bombyx mori*), but survival is reduced at 10 Gy. In all cases, younger stages were the most sensitive (Table 32.20). Cockroaches (*Blaberus giganteus*) adapted to the dark reportedly can visually detect radiation sources as low as 0.001 mGy (Rose 1992). However, the mechanisms are not understood.

Following the successful application of radiation to sterilize male screw-worm flies (*Cochliomyia hominivorax*), various insect pests became the target of similar techniques throughout the world (Al-Izzi et al. 1990). The technique has suppressed populations of the Mediterranean fruitfly (*Ceratitis capitata*), a major pest of fruits, although results have not been as spectacular as with the screw-worm fly (McInnes and Wong 1990). The pestiferous Caribbean fruitfly (*Anastrepha suspensa*), heavily parasitized by a beetle, became sterile after acute exposures to ionizing radiation, although beetles remained fecund. Mass rearing and inundative release of the radioresistant beetle parasite is now considered an option for control of the Caribbean fruitfly (Table 32.20).

Table 32.20 Radiation Effects on Selected Terrestrial Invertebrates

Species, Dose, and Other Variables	Effect	Reference[a]
Caribbean fruitfly, *Anastrepha suspensa*; larvae, heavily parasitized by the hymenopteron *Diachasmimorpha longicaudata*, exposed to single acute exposures of 10–70 Gy	50% of control flies developed into adults vs. 25% at 10 Gy, and <1% at 30 Gy. At 40 Gy and higher, no adults were recovered but parasite development was the same at all doses	1
Silkworm, *Bombyx mori*; eggs, acute single exposure of 2, 5, or 10 Gy	At 2 Gy, an average increase of 23% in larval mass, cocoon shell weight, and silk production; no stimulatory effect at 5 Gy; at 10 Gy, larval development inhibited	2
Mediterranean fruitfly, *Ceratitis capitata*; females, acute single exposure of 150–155 Gy	Inhibited oviposition	3
Moth, *Ectomyelois ceratoniae*; male pupae, age 3 or 5 days, acute single exposure of 50–500 Gy	Younger pupae were more sensitive than older pupae. Only 3% of pupae developed into adults at 500 Gy. At >250 Gy, progeny development reduced 50%. Normal fecundity at 100–250 Gy when mated with control females	4
Leafmining fly, *Liriomyza trifolii*; immature stages, on artificially infested bean seedlings; acute single exposure of 25–2000 Gy	All dead at >750 Gy; 80% dead at 250 Gy; eggs and prepupae were the most sensitive stages; no phytotoxic effects	5

[a] **1,** Sivinski and Smittle 1990; **2,** Yusifov et al. 1990; **3,** McInnes and Wong 1990; **4,** Al-Izzi et al. 1990; **5,** Yathom et al. 1990.

32.7.3 Aquatic Organisms

Among aquatic organisms, it is generally acknowledged that primitive forms are more radioresistant than complex vertebrates and that older organisms are more resistant than the young (Donaldson and Foster 1957; Bonham and Welander 1963; Templeton et al. 1971) (Table 32.21). Developing eggs and young of some freshwater fish species are among the most sensitive tested aquatic organisms. Death was observed at acute doses of 0.3 to 0.6 Gy, and minor effects on physiology or metabolism were observed at chronic daily dose rates of 0.01 Gy (Bonham and Welander 1963; Templeton et al. 1971; IAEA 1976) (Table 32.21). Radiosensitivity correlated positively with the metabolic rate of the dividing cell, which accounts for the radioresistance of dormant eggs of aquatic invertebrates and the general sensitivity of early embryonic stages of all aquatic species (Donaldson and Foster 1957) (Table 32.21).

Adverse effects on the fecundity of sensitive aquatic vertebrates were detected at dose rates as low as 0.4 mGy/h; adverse effects on fecundity of resistant species were measured only at dose rates greater than 1.0 mGy/h. Thus, deleterious effects in populations of aquatic vertebrates are probably not detected until the 0.4 to 1.0-mGy/h dose rate is exceeded (NCRP 1991). Organisms, such as estuarine organisms, that are exposed to variable physicochemical conditions are more radioresistant than those in buffered environments, and this may be due to a higher degree of genetic polymorphism in species of fluctuating environment (IAEA 1976). Dose-effect estimates for the induction of chromosomal aberrations in polychaete annelid worms were dependent on cell stage at time of irradiation (S.L. Anderson et al. 1990). For reproduction, dose-effect estimates were dependent on potential for regeneration of gonadal tissue (S.L. Anderson et al. 1990).

Radiation causes dominant lethal mutations in the medaka (*Oryzias latipes*) (Shima and Shimada 1991). Mosquitofish (*Gambusia* spp.) from radionuclide-contaminated ponds in South Carolina differed from conspecifics in reference ponds, as judged by the frequency of DNA markers, and this is consistent with the hypothesis that these DNA markers may originate from genetic elements that provide a selective advantage in contaminated habitats (Theodorakis et al. 1998). Ionizing radiation at low-level chronic exposure reportedly has no deleterious genetic effects on aquatic populations because exposure is compensated by density-dependent responses in fecundity (IAEA 1976). However, this needs verification.

Table 32.21. Radiation Effects on Selected Aquatic Organisms

Taxonomic Group, Organism, Dose, and Other Variables	Effect	Reference[a]
ALGAE		
Diatom, *Nitzchia closterium*; acute single exposure of 100 Gy	Lethal	1
Euglena, *Euglena gracilis*; acute single exposure of 550 Gy	Tolerated	1
Freshwater algae, 7 species, held in water containing 1110 Bq ^{226}Ra/L for as long as 14 days	After 24 h, 4 species (*Ankistrodesmus falcatus*, *Chlorella vulgaris*, *Coelastrium cambricum*, *Scenedesmus obliquus*) had decreased oxygen production by 22–37%; after 14 days, no effect on growth or protein content	2
Various species, single acute exposure		
80–1000 Gy	LD50, 45 days, postexposure	3
250–6000 Gy	LD100, 45 days after single exposure	3
PROTOZOANS		
Various species, acute single exposure		
100–300 Gy	LD50, up to 40 days postexposure	3
180–12,500 Gy	LD100, up to 40 days postexposure	3
COELENTERATES		
Sea anemone, *Anthopleura xanthogrammica*; 0.2 Gy, acute single exposure	Tentacles withdrawn	1
Jellyfish, *Aurelia aurita*; acute single exposure		
50–150 Gy	No deaths in 60 days	4
50–400 Gy	Dose-dependent increase in developmental abnormalities and abnormal budding rates and patterns	4
100 Gy	Metamorphosis and budding inhibited; reduction in pulsation rate	4
150 Gy	Inhibited reproduction	4
200 Gy	60% died in 60 days	4
400 Gy	90% died in 30 days	4
MOLLUSCS		
Water snail, *Physa heterostropha*; exposure of 2.4–5.5 Gy daily for 1 year	Increased growth rate	1
Various species, acute single exposure		
50–200 Gy	LD50, up to 2 years postexposure	3
100–500 Gy	LD100, up to 2 years postexposure	3
CRUSTACEANS		
Brine shrimp, *Artemia salina*; acute single exposure		
0.004 Gy	No adverse effects on development of cysts	5
0.1–9 Gy	Decreased development when exposed as cysts	5
4.5–9 Gy	LD50, nauplii, 20–25 days postexposure	6
130 Gy	LD50, adults, 25 days after exposure	6
486–2084 Gy	Dose-dependent delay in development of eggs	7
3000 Gy	LD50, cysts	5
Blue crab, *Callinectes sapidus*; continuous exposure to 0.76 Gy daily for 1 year	Increased growth rate	1

Table 32.21. (continued) Radiation Effects on Selected Aquatic Organisms

Taxonomic Group, Organism, Dose, and Other Variables	Effect	Reference[a]
Shore crab, *Carcinus maenus*		
Americium-241, dose unknown	After 8 days, bioconcentration factors (BCF) were 145 in whole crab, 960 in gills, and 240 in exoskeleton; 50% elimination in 45 days	8
Plutonium-237, dose unknown	After 8 days, BCF values were 75 in whole crab, 340 in gills, and 70 in exoskeleton; 50% elimination in 55 days	8
Daphnid, *Daphnia pulex*; daily exposure to 8.2–17.8 Gy for 1 year	Increased growth rate	1
Various species, single acute exposure		
5–900 Gy	LD50, up to 80 days postexposure	3
50–800 Gy	LD100, up to 80 days postexposure	3
ANNELIDS		
Polychaete, *Neanthes arenaceodentata*; acute single exposure		
1–4 Gy	Adverse effects on reproduction	9
2–100 Gy	Significant increase in frequency of chromosomal aberrations	9
>100 Gy	Decreased life span	9
>500 Gy	Death	9
FISH		
Common carp, *Cyprinus carpio*		
Adults, 3 Gy, acute single exposure	No effect on growth	1
Fertilized eggs, exposed through hatch		
144 million Bq ^{238}Pu/L	Increased abnormalities	10
277 million Bq ^{238}Pu/L	Decreased hatch	10
44 million Bq ^{232}U/L	Increased abnormalities	10
815 million Bq ^{232}U/L	Decreased hatch	10
Anchovy, *Engraulis* sp.; fertilized eggs, ^{90}Sr–^{90}Y, continuous exposure		
7.4 Bq/L	Increased developmental abnormalities	10
740 Bq/L	Decreased hatch, retarded growth rate	10
Fish, various species, acute single exposure		
6–30 Gy	LD50, up to 460 days postexposure	3
3.7–200 Gy	LD100, up to 460 days postexposure	3
Pinfish, *Lagodon rhomboides*; exposure to 0.197 Gy daily for 1 year	Increased growth rate	1
Bluegill, *Lepomis macrochirus*; acute single exposure of 10, 20, or 30 Gy	At 20 and 30 Gy, serum proteins were reduced more than 50% within 24 h; damage to the GI capillary system and injury to the gastroepithelium accounted for the excessive protein loss	12
Marine teleosts, 6 species, 10–55 Gy, acute single exposure	LD50	11
Silver salmon, *Oncorhynchus kisutch*; acute single exposure		
Early embryonic stages, 0.3–0.6 Gy	LD50 at hatch	13
Later embryonic stages, 9.2–18.7 Gy	LD50 at hatch	13
Rainbow trout, *Oncorhynchus mykiss*		
Embryos, acute single exposure		
0.6 Gy, 1-cell stage	LD50 by end of yolk resorption	10
0.8 Gy, 1-cell stage	LD50 at hatch	3
3.1 Gy, 32-cell stage	LD50 by end of yolk resorption	10
4.1 Gy, early eyed stage	LD50 by end of yolk resorption	3

Table 32.21. (continued) Radiation Effects on Selected Aquatic Organisms

Taxonomic Group, Organism, Dose, and Other Variables	Effect	Reference[a]
4.6 Gy, 32-cell stage	LD50 by hatch	3
9.0 Gy, late eyed stage	LD50 by end of yolk resorption	3
Embryos held in water containing 370 million Bq/L of ^{3}H from immediately after fertilization through hatching	No effect on hatching abnormalities	10
Embryos held in water containing 37 million Bq/L of ^{3}H from 6 h after fertilization through hatch	Suppressed immune response of fry	10
Immatures, single acute exposure of 0.2 Gy	Growth stimulation	1
Juveniles exposed for 27 days to radioneptunium isotopes and analyzed 2–15 days postexposure	Maximum BCF values were 8.7 for whole fish, 1.1 for skin, and 0.34 for muscle	14
Yearlings, force-fed 185,000, 1.85 million, or 18.5 million Bq ^{90}Sr–^{90}Y/kg BW daily for 21 weeks	At highest dose, adverse effects on growth (week 12) and survival (week 15); survivors had leucopenia and gut histopathology, and concentrations of 9.2 billion Bq/kg FW in bone and 9.99 million Bq/kg FW in muscle. Residues in the 1.85-million group were 1.04 billion Bq/kg in bone and 2.96 million Bq/kg in muscle. For the 185,000 group, these values were 77.7 million Bq/kg in bone and 74,000 Bq/kg in muscle	11, 15
Yearlings force-fed 370,000, 3.7 million, or 37 million Bq ^{65}Zn/kg BW daily for 17 weeks, or 370 million Bq ^{65}Zn/kg BW daily for 10 weeks	Adverse effects on growth, survival, or gut histology at any dose; leucopenia evident at week 10 at the highest dose. Residues, in Bq/kg FW, in the 37 million group at 17 weeks were 148 million in bone and 12.9 million in muscle	11, 15
Yearlings force-fed 222,000, 2.2 million, or 22.2 million Bq ^{32}P/kg BW daily for as long as 25 weeks	At highest dose tested, adverse effects on growth, survival, and gut histology between day 17 and 77. In the intermediate 2.2-million group, adverse effects on growth at 17 weeks; residues (Bq/kg FW) were 66.6 million in bone and 8.5 million in muscle. The 220,000 group had no adverse effects in 25 weeks on growth, survival, or tissue alterations	11
Gametes of adults, single acute exposure of 0.5–1.0 Gy	50% reduction in fecundity	3
Adults, single acute exposure of 15 Gy	LD50	3
Chinook salmon, *Oncorhynchus tshawytscha*		
0.0004 Gy/h, eggs, 81-day exposure, total dose of 0.78 Gy	Significant adverse effects on survival and development	21
0.005 Gy daily, continuous exposure from egg fertilization through yolk sac absorption; total dose of 0.35 Gy	No adverse effects on growth and survival or on numbers of returning adults after seaward migration	22
0.028 Gy daily, continuous exposure from egg fertilization through yolk sac absorption; total dose of 1.99 Gy	No adverse effects observed prior to seaward migration	22
0.2 Gy, single acute exposure	Growth increase	1
10 Gy, eyed eggs, single acute exposure	LD50	3
12.5–25 Gy, fingerlings, single acute exposure	LD50	3
Medaka, *Oryzias latipes*; adult males receiving single acute exposure of 0.64, 4.75, or 9.5 Gy	Dose-dependent increase in total mutations in sperm, spermatids, and spermatogonia	16
Sea lamprey, *Petromyzon marinus*; males captured during spawning run, single acute exposure		
20 Gy	LD50, 45 days postexposure; survivors sterile	17
30 Gy	All died before spawning	17

Table 32.21. (continued) Radiation Effects on Selected Aquatic Organisms

Taxonomic Group, Organism, Dose, and Other Variables	Effect	Reference[a]
Fathead minnow, *Pimephales promelas*; developing eggs, continuous exposure		
4440 Bq ^{144}Ce–^{144}Pr/L	No effect on embryonic development or hatch	10
9.6 million Bq ^{238}Pu/L	Increased abnormalities	10
48.1 million Bq ^{238}Pu/L	Decreased hatch	10
7.4 million Bq ^{232}U/L	Increased abnormalities	10
18.5 million Bq ^{232}U/L	Decreased hatch	10
Atlantic salmon, *Salmo salar*; fertilized eggs, continuous immersion in 92.5 Bq ^{137}Cs/L or 185 Bq ^{90}Sr/L	Increased mortality of embryos and fry	10
Brown trout, *Salmo trutta*		
Fertilized eggs continuously immersed in water containing 3.7 million Bq/L of ^{90}Sr–^{90}Y through hatch	No effect on hatch or developmental abnormalities	10
Juveniles held in water containing 30,000 Bq ^{110m}Ag/L for 57 days, then transferred to uncontaminated media for 28 days	At day 57, whole trout contained 105,000 Bq ^{110m}Ag/kg FW; about 70% was in liver. No detectable radioactivity after depuration for 28 days	18
Juveniles fed diet containing 3,343,000 Bq ^{110m}Ag/kg for 1 week (5 times weekly), then 269,000–296,000 Bq ^{110m}Ag/kg diet between weeks 2 and 5. Depuration for 28 days	At end of exposure, whole trout contained 27,400 Bq ^{110m}Ag/kg, equivalent to 11.7% of ingested radioactivity; liver accounted for 63%. No detectable radioactivity after depuration for 28 days	19
INTEGRATED STUDY		
Artificial stream simulating outfall from Czechoslovakian nuclear power plant, 28-day exposure, ^{90}Sr		
Water	894 Bq/L	20
Sediments	1589–2288 Bq/kg FW	20
Alga, *Cladophora glomerate*	Max. 22,106 Bq/kg FW	20
Snail, *Planorbis corneus*, shell vs. soft parts	760,588 Bq/kg FW vs. 27,468 Bq/kg FW	20
Common carp, *Cyprinus carpio*		
Bone	29,144 Bq/kg FW	20
Muscle	580 Bq/kg FW	20
Scales	13,101 Bq/kg FW	20
Uncontaminated site		
Water	0.002–0.005 Bq/L	20
Common carp, internal organs vs. scales	0.1–0.5 Bq/kg FW vs. 1.5–9.3 Bq/kg FW	20

[a] **1,** Rose 1992; **2,** Havlik and Robertson 1971; **3,** Donaldson and Foster 1957; **4,** Prokopchak et al. 1990; **5,** Gaubin et al. 1990; **6,** Engel and Davis 1976; **7,** Su et al. 1990; **8,** Guary and Fowler 1990; **9,** S. L. Anderson et al. 1990; **10,** Whicker and Schultz 1982b; **11,** Templeton et al. 1971; **12,** Ulrickson 1971; **13,** Bonham and Welander 1963; **14,** Poston et al. 1990; **15,** IAEA 1976; **16,** Shima and Shimada 1991; **17,** Hanson 1990; **18,** Garnier et al. 1990; **19,** Garnier and Baudin 1990; **20,** Stanek et al. 1990; **21,** National Council on Radiation Protection and Measurements (NCRP) 1991; **22,** Donaldson and Bonham 1970.

Accumulation of radionuclides from water by aquatic organisms varies substantially with ecosystem, radionuclide, and trophic level (Tables 32.22, 32.23, 32.24, respectively); with numerous biological, chemical, and physical variables; and with proximity to sources of radiation (Bowen et al. 1971; Lowman et al. 1971; Templeton et al. 1971; Mo and Lowman 1976; Shure and Gottschalk 1976; Whicker and Schultz 1982b; Becker 1990; Poston et al. 1990; Joshi 1991). Accumulated radionuclides within embryos of the scorpionfish (*Scorpaena porcus*) and turbot (*Scophthalmus maeoticus*) increased the frequency of nuclear disruptions in these species; ^{90}Sr-^{90}Y and ^{91}Y had greater cytogenetic effects than other radionuclides tested (Polikarpov 1973). In the absence of site-specific data, the U.S. Nuclear Regulatory Commission recommends the use of

Table 32.22 Concentration Factors for Cesium-137 and Strontium-90 in Aquatic Organisms

Radionuclide and Ecosystem	Molluscs, Whole	Crustaceans, Whole	Fish, Muscle
CESIUM-137			
Freshwater	600	4000	3000
Marine	8	23	15
STRONTIUM-90			
Freshwater	600	200	200
Marine	1	3	0.1

Note: Concentration factors given in Bq per gram fresh weight sample/Bq per mL medium

Data from Whicker, F.W. and V. Schultz. 1982a. *Radioecology: Nuclear Energy and the Environment.* Vol. I. CRC Press LLC, Boca Raton, FL. 212 pp.

Table 32.23 Approximate Maximum Concentration Factors for Selected Transuranics in Marine Sediments, Macroalgae, and Fish

	Concentration Factor		
Transuranic Nuclide	Sediments	Macroalgae	Fish
Neptunium	1000	5000	10
Plutonium	100,000	2000	40
Americium, curium, berkelium, californium	2,000,000	8000	50

Note: Concentration factors given in Bq per gram fresh weight sample/Bq per mL water.

Data from Morse, J.W. and G.R. Choppin. 1991. The chemistry of transuranic elements in natural waters. *Rev. Aquat. Sci.* 4:1-22.

listed concentration ratios — the concentration of the element in the organism (in mg/kg FW) divided by the concentration in the medium (in mg/L) — for various elements in marine and freshwater fishes and invertebrates (Whicker and Schultz 1982b). However, the Commission clearly indicates that these values are only approximations.

After more than 400 atmospheric nuclear test explosions and the fallout from Chernobyl, ^{137}Cs became the most frequently released nuclear fission product throughout central Europe (Jandl et al. 1991). Cesium behaves like potassium: it has a ubiquitous distribution inside the body, especially in soft tissues. In the gastropod *Helix pomatia*, the biological half-time after a single 24-h dietary dose was 2.5 days for the short-lived component and 28.5 days for the long-lived component (Jandl et al. 1991). Concentration factors (CF) of ^{137}Cs in muscle (ratio of Bq/kg FW muscle:Bq/L filtered seawater) of marine fishes from the North Sea between 1978 and 1985 ranged from a low of 39 in the plaice (*Pleuronectes platessa*) to a high of 150 in the whiting (*Merlangius merlangius*); CF values were intermediate in the haddock (*Melanogrammus aeglefinus*; CF of 58) and Atlantic cod (*Gadus morhua*; CF of 92). These data seem to support the use of a CF of 100 for ^{137}Cs in muscle of marine fishes in generalized assessments, although some adjustment is necessary when particular species, such as whiting, form the bulk of a consumer's diet (Steele 1990). In the Great Lakes, the maximum CF values of ^{137}Cs range from 1000 to 10,000 in algae, amphipods, and fishes, and from 100 to 1000 in zooplankton (Joshi 1991). Maximum concentration factors of ^{137}Cs in a contaminated creek in South Carolina were 4243 in suspended particulates, 938 in detritus, 4496 in algae and macrophytes, 997 in omnivores, 1292 in primary carnivores, and 1334 to 2595 in top carnivores, such as redbreast sunfish (*Lepomis auritis*), largemouth bass (*Micropterus salmoides*), and water

Table 32.24 Maximum Concentration Factors Reported for Selected Elements in Marine Organisms at Various Trophic Levels

Element	Algae	Grazers	Predators
Ag	1000	20,000	3000
Cd	6000	2,000,000	10,000
Ce	4500	300	12
Co	1000	10,000	50,000
Cr	600	300,000	3900
Cs	50	15	10
Fe	70,000	300,000	30,000
I	7000	70	10
Mo	200	175	200
Mn	20,000	60,000	100,000
Ni	1000	10,000	80
Pb	3,000,000	2,000,000	200,000
Ru	1000	16	10
Sr	90	85	5
Ti	30,000	20,000	3000
Zn	3000	100,000	20,000
Zr	20,000	30,000	40,000

Note: Concentration factors given in Bq per gram fresh weight tissue/Bq per mL seawater.

Data from Bowen, V.T., J.S. Olsen, C.L. Osterberg, and J. Ravera. 1971. Ecological interactions of marine radioactivity. Pages 200-222 in National Academy of Sciences. *Radioactivity in the Marine Environment.* Natl. Acad. Sci., Panel on Radioactivity in the Marine Environment. Washington, D.C.

snakes (*Natrix* spp.) (Shure and Gottschalk 1976). Cesium uptake by oligochaete worms (*Limnodrilus hoffmeisteri*) is inhibited by low temperatures, potassium concentrations >1 mg/L, and the presence of bacteria (*Escherichia coli*) that compete with the worms for ^{137}Cs (Steger and Goodnight 1976).

Atmospheric fallout from nuclear testing is the main pathway by which transuranic nuclides, such as Np, Pu, Cm, and Am, enter the aquatic environment (Guary and Fowler 1990). In general, transuranics are strongly partitioned onto particulates. Living organisms are less enriched than particulate matter by as much as 1000 times, and concentration factors by marine biota are similar for transuranics beyond neptunium (Morse and Choppin 1991). The uptake of ^{241}Am and ^{244}Cm from contaminated sediments by a freshwater amphipod (*Hyalella* sp.) and oligochaete (*Tubifex* sp.) is reported, presumably by way of adsorption, and this is considered the principal uptake pathway by benthic organisms in freshwater and marine ecosystems (Sibley and Stohr 1990). Transuranics ingested with food by various crabs were initially excreted with feces; the remaining transuranics entered a soluble radionuclide pool within the animal that was slowly excreted. Decapod crustaceans assimilate and retain 10 to 40% of the transuranic nuclides in their diets. Initially, absorbed radionuclides accumulate in the hepatopancreas but are then translocated to other tissues, particularly to tissues of the exoskeleton. Accordingly, molting strongly influences elimination in crustaceans (Guary and Fowler 1990). Neptunium isotopes have a higher potential for environmental transport in aquatic systems and groundwater than other actinides tested. Laboratory studies with ^{235}Np and ^{237}Np, for example, show concentration factors between 275 and 973 in a green alga (*Selenastrum capricornutum*); between 32 and 72 in a daphnid (*Daphnia magna*); 2 in an amphipod (*Gammarus* sp.); and in juvenile rainbow trout, 8.7 in carcass, 1.1 in skin, and 0.3 in muscle over a 96-h period (Poston et al. 1990). When the much higher biological effectiveness of alpha vs. beta or gamma radiation is considered, plutonium isotopes may contribute more artificial radiation dose equivalent to marine invertebrates than either ^{90}Sr or ^{137}Cs. Concentration factors of

^{239}Pu and marine organisms ranged from 300 to 100,000 in seaweeds, 250 to 690 in molluscs, 760 to 1020 in echinoderms, 2100 in sponges, and as much as 4100 in worms (Noshkin et al. 1971). Concentration factors of $^{239+240}$Pu in Lake Michigan ranged between 1 and 10 in predatory salmonids, between 10 and 300 in nonpredatory fish, between 900 and 1200 in amphipods and shrimp, about 200 in zooplankton, and about 6000 in algae (Joshi 1991).

Iodine-131 (half-life of 8 days) may cause deleterious effects in marine teleosts — although ^{131}I concentrations in tissues were not detectable. In one case, coral reef fishes from Eniwetok Atoll collected as long as 8 months after a nuclear explosion had thyroid necroalteration, suggesting a thyrotoxic level of ^{131}I in the environment. Laboratory studies with teleosts injected with ^{131}I showed similar signs of histopathology. Herbivorous fishes and species that habitually consumed bivalve molluscs were the most severely affected (Gorbman and James 1963).

Strontium-90 is an anthropogenic radionuclide in liquid effluents from some European nuclear power plants. Algae and sediments are the most important accumulators of ^{90}Sr, although levels in gastropods and fish bone and scales are also elevated, suggesting piscine uptake through gills and skin (Stanek et al. 1990). Fish tend to accumulate calcium more than strontium, even when Ca levels in food and water were low. Gill tissue was the most and gut the least discriminatory against Sr. Strontium assimilation was linked to the Sr:Ca ratio in food and water, amounts of Ca derived from each source, and biological discrimination against Sr relative to Ca (Ophel and Judd 1976). The ability of organisms to discriminate between strontium radioisotopes is also documented. In one case, ^{85}Sr was taken up rapidly in bluegill (*Lepomis macrochirus*) muscle and blood and quickly exchanged with stable strontium. However, ^{90}Sr was retained longer than 35 days in these tissues (Reed and Nelson 1969).

Ruthenium-106 appeared in clams from North Carolina within 2 weeks after the third and fifth Chinese nuclear tests in 1965 to 1967. Its retention was resolved into two rate functions with apparent effective half-lives of 40 days and 7 days (Wolfe and Jennings 1971). Iron-55 is a neutron-activation product produced in large quantities from ferrous materials in the immediate vicinity of a nuclear detonation. Concentration factors of ^{55}Fe and plankton in Bikini Atoll ranged from 15,000 to 25,000 (Schell 1976). Silver-110m has been detected in marine organisms after atmospheric weapon tests in the Pacific, in fishes from the Rhone River after the Chernobyl accident, and in fishes near reactor-waste outfalls (Garnier et al. 1990). Silver-110m is depurated rapidly by brown trout (*Salmo trutta*) after high intake exposures via the water or diet (Garnier et al. 1990; Garnier and Baudin 1990). Radiotungsten is produced in quantity by certain types of nuclear devices. In one case, tungsten was the most abundant radionuclide in the environment, accounting for about 90% of the total fallout activity 167 days after the detonation (Reed and Martinedes 1971). Tungsten-181 tended to concentrate in the hepatopancreas and gut of the crayfish (*Cambarus longulus longirostris*). Whole-body elimination consisted of two components: a rapid 1-day component and a second slower component with a biological half-time of 12.2 days (Reed and Martinedes 1971). Benthic organisms take up limited amounts of heavy metals and radionuclides associated with bottom sediments and recycle them to benthic and pelagic food webs. For example, polychaete worms (*Nereis diversicolor*) in contact with ^{65}Zn-contaminated sediments for 5 days lost 50% of accumulated ^{65}Zn in about 19 days on transfer to uncontaminated sediments (Renfro and Benayoun 1976).

32.7.4 Amphibians and Reptiles

Radiation adversely affects limb regeneration of amphibians, alters DNA metabolism, and increases the frequency of chromosomal aberrations and liver lesions (Table 32.25). In some species of amphibians and reptiles, as in many mammals, mortality rates after acute exposure to radiation do not stabilize within 30 days — effectively invalidating the conventional LD50 (30-day postexposure) value. In the rough-skinned newt (*Taricha granulosa*), for example, the minimal LD50 dose at 200 days after irradiation was 2.5 Gy, compared with 350 Gy at 30 days (Willis and

Table 32.25 Radiation Effects on Selected Amphibians and Reptiles

Species, Dose, and Other Variables	Effect	Reference[a]
Leopard lizard, *Crotaphytus wislizenii*; chronic field exposure of 0.04–0.06 Gy daily (4–5 Sv yearly) for 3–6 years	No female reproduction in years 3 and 4. In year 5, males were sterile and females had complete regression of ovaries, undeveloped oviductal walls, and hypertrophied fat bodies. In year 6, 75% of females lacked ovaries and 25% had normal ovaries with signs of recent egg deposition; males appeared normal	1
Mud puppy, *Necturus maculosus*; 1.1 Gy, single acute exposure	LD50, 30 days postexposure	2
Salamander, *Necturus* sp.; 0.8 Gy, single acute exposure	LD50, 200 days postexposure	3
Newt, *Notophthalmus viridescens*; adults, single acute exposure of 20 Gy, one limb shielded; or 22 Gy, whole body, no limbs shielded	Forelimb regeneration completely suppressed when limbs to be amputated were irradiated directly. Irradiated limbs had severe and protracted inflammation, with total resorption of the affected limbs in 85% of the cases. Shielded limbs subsequently amputated had delays — but not suppression — in rate of forelimb regeneration and skin graft rejection	4
Frog, *Rana* sp., single acute exposure		
7.0–7.2 Gy	LD50, 730 days after exposure	3
7.8 Gy	LD50, 150 days after exposure	3
Snakes, 2 species, 3–4 Gy, single acute exposure	LD50, 90 days after exposure	3
Rough-skinned newt, *Taricha granulosa*; single acute exposure		
2.5 Gy	LD50, 200 days after exposure; skin lesions and depigmentation	5
>6.5 Gy	Progressive anemia over 6-week postirradiation period; reduction in erythrocyte numbers and weight of spleen	5
80 Gy	LD50, 100 days after exposure	5
350 Gy	LD50, 30 days after exposure	5
Turtles, 4 species, <8–15 Sv, single acute exposure	LD50, 120 days postexposure	3
Slider turtle, *Trachemys scripta*; inhabiting a radioactive reservoir, in Aiken, South Carolina. Radionuclide concentrations, in Bq/kg, whole-body fresh weight, (FW) were 1002 for 137Cs and 550 for 90Sr. For controls, these values were 2 for 137Cs and 260 for 90Sr	Contaminated turtles, when compared with controls, had greater variation in DNA content of red blood cell nuclei, suggesting genetic damage. The biological half-life of 137Cs in soft tissues was 64 days; for 90Sr in shell and bone, it was 364 days	6
Spiny tailed lizard, *Uromastix hardwickii*; single acute exposure, held for up to 14 days after irradiation		
2.25 Gy	No lesions in liver	7
4.5 Gy	No liver lesions, but swollen hepatocytes, increases in bile pigmentation, and altered cytoplasmic degranulation; normal after 14 days	7
9.0 Gy	Some liver lesions, but all livers normal 14 days postexposure	7
Lizard, *Uta* sp.; 10–22 Gy, single acute exposure	LD50, 30 days postexposure	3

[a] **1,** Turner et al. 1971; **2,** Rose 1992; **3,** Hinton and Scott 1990; **4,** Sicard and Lombard 1990; **5,** Willis and Lappenbusch 1976; **6,** Lamb et al. 1991; **7,** Gupta and Umadevi 1990.

Lappenbusch 1976). Low temperatures seem to prolong the survival of amphibians exposed to ionizing radiation. The survival was greater of leopard frogs (*Rana pipiens*) held at low temperatures

(5 to 6°C) after total-body exposure to lethal doses of X-rays than of frogs held at higher temperatures. Prolonged survival at low temperatures was due to a prolongation of the latent period rather than to appreciable recovery (Patt and Swift 1948).

The South African clawed frog (*Xenopus laevis*) has been suggested as a bioindicator of radioactive contamination because of the greater radiosensitivity of amphibians than fishes, the ease of maintaining *Xenopus* in the laboratory, and the sensitivity of the *Xenopus* liver to radioactive contamination — including ^{45}Ca, which does not accumulate in the liver (Giannetti et al. 1990). *Xenopus* oocytes exposed to X-rays showed single- and double-strand breaks in DNA and oxidative-type base lesions at a frequency between 85 and 95%. *Xenopus* oocytes repaired X-ray induced damage in plasmid DNA; however, some X-ray lesions can stimulate homologous recombination in these cells (Sweigert and Carroll 1990). Slider turtles (*Trachemys scripta*) in a radioactive reservoir show evidence of genetic damage, and this was attributed to long-term exposure to low concentrations of long-lived radionuclides, including ^{137}Cs and ^{90}Sr (Lamb et al. 1991). Natural populations of toads (*Bufo valliceps*) reportedly can survive genetically damaging doses of ionizing radiation without impairment of population integrity (Whicker and Schultz 1982b). Toads and many other species share a high attrition on the large numbers of young produced each generation, and this provides an agency for intensive selection. Also under this regime, recessive mutants are eliminated as they are exposed through inbreeding in future generations (Whicker and Schultz 1982b). Sterility in field collections of the leopard lizard (*Crotaphytus wislizenii*) and the whip-tail lizard (*Cnemidophorus tigris*) was reported after long-term exposure of 3 to 5 years to various doses of gamma radiation (i.e., 4 to 5 Sv annually in *Crotaphytus* and 2.0 to 2.5 Sv annually in *Cnemidophorus*) (Turner et al. 1971). However, a third species of lizard in the study area (side-blotched uta, *Uta stansbunana*) reproduced normally (Turner et al. 1971).

The retention of selected isotopes by amphibians and reptiles is quite variable. For example, whole-body retention of ^{131}I after intraperitoneal injection in the rough-skinned newt showed two distinct loss components with biological half-lives of 2 and 210 days. The slower component accounted for 26% of the administered activity; thyroid contained 78% of the total ^{131}I and clearly accounted for the long-term component (Willis and Valett 1971). However, similar studies with ^{131}I and the leopard frog showed three distinct loss components (0.1 day; 1.4 to 2.9 days; 44.3 to 69.4 days); loss of each component was greater at 25°C than at 10°C. Also, the fast component probably represented plasma clearance through urinary excretion (Willis and Valett 1971).

32.7.5 Birds

Among birds, as in most other tested species, there is a direct relation between dose and mortality at single high doses of ionizing radiations (Whicker and Schultz 1982b; Table 32.26). For any given total dose, the survival of a bird is higher if the dose is delivered at a lower rate or over a longer period of time and suggests that biological repair processes compensate for radiation-induced cellular and tissue damage over a prolonged period or at a comparatively low dose rate (Brisbin 1991). Nestling bluebirds (*Sialia sialis*) were more resistant to gamma radiation than young domestic chickens (*Gallus* sp.), and nestling great crested flycatchers (*Myiarchus crinitus*) were more sensitive than bluebirds (Willard 1963). Passerine nestlings are more resistant to radiation stress than adults of larger-bodied precocial species (Brisbin 1991). But the comparatively resistant passerine nestlings frequently show a disproportional disturbance in radiation-induced growth, resulting in a reduction of overall survival. For example, if feather growth is stunted, death results from the inability to escape predators because of impaired flight (Brisbin 1991).

Free-living, resident bird populations in the vicinity of sites contaminated with low levels of ionizing radiations generally have negligible genotoxic effects (George et al. 1991). However, 14% of mallards (*Anas platyrhynchos*) from an abandoned South Carolina reactor cooling reservoir heavily contaminated with ^{137}Cs (mallards contained an average of 2520 Bq ^{137}Cs/kg whole-body FW) had abnormal chromosome numbers and unusual variability in the concentration of erythrocyte

DNA (George et al. 1991). Contaminated waterfowl rapidly eliminate accumulated radionuclides, suggesting inconsequential long-term damage to the birds and little hazard to human consumers of waterfowl flesh (Halford et al. 1983). This conclusion was from a study wherein mallards were held for 68 to 145 days on liquid radioactive waste ponds in southeastern Idaho and then transferred to an uncontaminated environment for 51 days. The biological half-life in mallards under these conditions was 10 days for ^{131}I and ^{134}Cs, 11 days for ^{137}Cs, 22 days for ^{140}Ba, 26 days for ^{75}Se, 32 days for ^{58}Co, 67 days for ^{60}Co and ^{65}Zn, and 86 days for ^{51}Cr. At the time of removal from the waste ponds, radionuclide concentrations were highest in gut, then feather, liver, and muscle, in that order. After 51 days in a radionuclide-free environment, decreasing order of radionuclide concentrations was feather, liver, muscle, and gut (Halford et al. 1983).

Zinc-65 in trace amounts is accumulated by migratory waterfowl in the Pacific flyway of North America from ^{65}Zn discharged into the Columbia River from water-cooled reactors at Hanford, Washington (Curnow 1971). The retention of ^{65}Zn in mallards was affected by sex and season, but not by the age of the duck. Biological retention of ^{65}Zn was greater in males (Tb 1/2 of 34.7 days) than in females (29.8 days), and greater in October (38 days) than in the spring (32 days). Egg production accounted for the elimination of 25% of the ^{65}Zn and feather molt of 2% to 8% (Curnow 1971). Retention of ^{60}Co and ^{137}Cs — but not ^{109}Cd — in the common bobwhite (*Colinus virginianus*) after either acute or chronic exposure to contaminated food is similar. The biological half-life in bobwhites during exposure for 21 days was 8 days of ^{109}Cd, 11 days of ^{137}Cs, and 13 days of ^{60}Co. When radioisotopes were administered during a single 4-h feeding, Tb 1/2 values were 3 days of ^{109}Cd, 10 days of ^{137}Cs, and 15 days of ^{60}Co (Anderson et al. 1976). The biological half-life of ^{137}Cs in avian tissues is about 6.0 days in domestic chickens (Andersson et al. 1990); 6.7 days in the bluejay (*Cyanocitta cristata*) (Levy et al. 1976); 5.6 days in the American wood duck (*Aix sponsa*); and 11.7 days in mallards (Cadwell et al. 1979). Domestic poultry, when compared with mammals, seem to accumulate a higher fraction of the daily ingested ^{137}Cs/kg muscle, but levels were effectively reduced by feeding an uncontaminated ration for at least 10 days prior to slaughter (Andersson et al. 1990).

Table 32.26 Radiation Effects on Selected Birds

Species, Dose, and Other Variables	Effect	Reference[a]
Green-winged teal, *Anas carolinensis*; 4.8 Gy, single acute exposure	LD50, 30 days postexposure	1
Northern shoveler, *Anas clypeata*; 8.9 Gy, single acute exposure	LD50, 30 days postexposure	1
Blue-winged teal, *Anas discors*; 7.2 Gy, single acute exposure	LD50, 30 days postexposure	1
Birds		
Eggs, passerine species, single acute exposure, 5–10 Gy	LD100	2
Nestlings, various species, 1 Gy daily	Growth retardation	3
Common quail, *Coturnix coturnix;* fertilized eggs, exposed first 9 days of incubation, single acute exposure		
5 Gy	Negligible effect on survival	4
7 Gy	Mortality >50%	4
9 Gy	All dead before hatch	4
Domestic chicken, *Gallus* sp.		
Single acute exposure		
Eggs of broilers exposed to 0.05–2.1 Gy before incubation	No adverse effects on embryonic development at 1.6 Gy and lower; at 2.1 Gy, adverse effects on development, survival, and body weight of hatched chicks	5
Chicks, age 15 days		
2.1 Gy	Reversible changes in blood chemistry within 60 days; no deaths	6
6.6 Gy	Irreversible and permanent damage to red blood cells, hemoglobin, and hematocrit; all dead within 7 days	6

Table 32.26 (continued) Radiation Effects on Selected Birds

Species, Dose, and Other Variables	Effect	Reference[a]
Dietary exposure		
Laying hens fed diet containing 400 Bq ^{137}Cs/kg ration for 4 weeks	Of total ^{137}Cs ingested, 3% was distributed in egg contents (29–33 Bq/kg egg; 2 Bq egg); 9% in muscle (171 Bq/kg FW); and 81% in excreta	7
Broiler chickens fed diets containing 400 Bq ^{137}Cs/kg ration for 40 days; some diets contained up to 5% bentonite	Feeding with bentonite reduced ^{137}Cs concentration in muscle by 32% from 155 to 105 Bq/kg FW	7
Black-headed gull, *Larus ridibundus ridibundis*; eggs, 9.6 Gy over 20 days	LD50	3
Great crested flycatcher, *Myiarchus crinitus*; nestlings, single acute exposure >8 Gy	All dead by fledging	8
Eastern bluebird, *Sialia sialis*, single acute exposure		
Nestlings, age 2 days		
3 Gy	Reduced growth after 16 days	8
3–5 Gy	Reduced growth and shorter primary feathers at fledging	8
4–12 Gy	Developed normally and fledged successfully	2
5–6 Gy	LD50, nestling to fledgling	8
25 Gy	LD50, 16 days postexposure	8
30 Gy	All dead 4 days postexposure	2, 8
Fertilized eggs, 6 Gy	All dead before hatch	2
European starling, *Sturnus vulgaris*; >2 Gy, single exposure	Fatal	9
Tree swallow, *Tachycineta bicolor*		
0.006 mGy/h during breeding season, equivalent to annual dose of about 50 mSv	No adverse effects on breeding performance of adults or growth performance of nestlings	10
0.9–4.5 Gy, single acute exposure, nestlings	Adverse effects on growth, survival, or both	2
1.0 Gy daily, chronic	Reduced hatch, depressed growth	2
House wren, *Troglodytes aedon*; fledglings, 0.9 Gy, single acute exposure	Growth reduction	9

[a] **1,** Hinton and Scott 1990; **2,** Millard and Whicker 1990; **3,** Lowe 1991; **4,** Wetherbee 1966; **5,** Zakaria 1991; **6,** Malhotra et al. 1990; **7,** Andersson et al. 1990; **8,** Willard 1963; **9,** Rose 1992; **10,** Zach et al. 1993.

32.7.6 Mammals

The mammalian sensitivity to acute and chronic exposures of ionizing radiation, ability to retain selected radionuclides, and effect of biological and abiotic variables on these parameters are briefly summarized in Table 32.27. These data clearly indicate a dose-dependent effect of radiation on growth, survival, organ development, mutagenicity, fatal neoplasms, kidney failure, skeletal development, behavior, and all other investigated parameters. In general, fetuses and embryos were most sensitive to ionizing radiation, and acute or chronic exposures between 0.011 and 0.022 Gy were demonstrably harmful to mice, rats, and guinea pigs.

Table 32.27 Radiation Effects on Selected Mammals

Species, Dose, and Other Variables	Effect	Reference[a]
Short-tailed shrew, *Blarina brevicauda*; 7.8 Gy, single acute exposure	LD50, 30 days postexposure	1
Cow, *Bos* sp.		
Oral intake of 0.89 Bq ^{129}I, whole animal	Thyroid contained 0.97 Bq ^{129}I/kg fresh weight (FW) vs. <0.0012 Bq/kg FW for all other tissues	2
Fed 6.4 Bq ^{129}I daily for 8 days	After 8 days, 22% of total dose of 51.2 Bq was in thyroid; after 63 days, thyroid contained 1 Bq/kg FW and other tissues <0.01 Bq/kg FW	3

Table 32.27 (continued) Radiation Effects on Selected Mammals

Species, Dose, and Other Variables	Effect	Reference[a]
Dog, *Canis familiaris*		
Beagle embryos age 55 days, or pups 2 days old, given single acute exposure of 0.16, 0.83, or 1.25 Gy	Dose-dependent increase in immature dysplastic glomeruli and other signs of progressive renal failure	4
Beagles, prenatal and early neonatal stages, given single acute dose of 0.2–1.0 Gy, then observed over 11-year life span	Irradiation at all ages was associated with increased risk of: decreased fertility; inhibited growth and development; lower brain weight; and increase in fatal neoplasms	5
Beagle embryos or pups. As above, 2.24–3.57 Gy	Reduction in total number of nephrons and progressive renal failure	4
Beagles, 17–20 months old, single intravenous injection of 200–440,000 Bq ^{226}Ra/kg body weight (BW)	Dose-dependent increase in skeletal malignancies in 36% of dogs during lifetime	6
Beagles, age 5 years, given single injection of ^{226}Ra, in Bq/kg BW, of 39,000, 116,000, or 329,000. Injected ^{226}Ra solutions also contained ^{210}Po, ^{210}Pb, and ^{210}Bi	At lowest dose of 39 kBq/kg BW, kidney was normal, death after 2032 days. At intermediate dose, death in 1210 days; at high dose, death in 581 days. Tubular degeneration and necrosis of kidney at 116 and 329 kBq	7
Beagles, age 7 years, single injection of ^{226}Ra (no contaminating ^{210}Po, ^{210}Pb, or ^{210}Bi) at 45,000 Bq/kg BW, or 122,000 Bq ^{210}Po/kg BW	No kidney damage with ^{226}Ra, but kidney damage with ^{210}Po	7
Beagles, age 7 years given single injection of 1,629,000 Bq ^{226}Ra/kg BW, equivalent to 1.89 Gy (from ^{210}Po contaminants), or 4,831,000 Bq ^{226}Ra/kg BW = 5.15 Gy from ^{210}Po	At low dose, all dead after 516 days; at high dose, all dead in 266 days. Death was from renal failure	7
Guinea pig, *Cavia* sp.; chronically irradiated daily during 8-h exposure. Daily dose, in Gy		
0.000	Mean survival time of 1372 days	8
0.001	50% dead in 1457 days	8
0.011	50% dead in 1224 days	8
0.022	50% dead in 978 days; anemia	8
0.044	50% dead in 653 days; anemia	8
0.088	50% dead in 187 days; anemia	8
Monkey, *Cebus apella*; 1 Gy, whole body, single acute exposure	Leucocyte reduction in 6 days; blood chemistry normal after 90 days	9
Chinese hamster, *Cricetus* sp.; ovary cells, single acute dose ranging between 0.005 and 0.06 Gy	Increased frequency of sister chromatid exchange at 0.005 Gy; increased numbers of chromosomal aberrations at >0.02 Gy; no significant increase in cell death	10
Syrian hamster, *Cricetus* sp.; 0.12–2 Gy, single acute exposure	Genes modifying cytoskeletal development adversely affected at all doses within 3 h by both high LET (neutrons) and low LET (gamma rays, X-rays) radiations	11
Human, *Homo sapiens*		
Developing forebrain, 0.18–0.55 Gy (estimated dose to prenatally exposed Japanese atomic bomb survivors)	Seizures in childhood; reduced school performance at least through age 11 years; some cases of severe mental retardation by age 17 years	12
Fetus, 1 Sv, 8–15 weeks of gestation	40% probability of severe mental retardation; IQ score lowered 30 points	13
Sperm chromosomes, 0.23, 0.45, 0.91, or 1.82 Gy, single acute exposure	Chromosomal aberrations increased linearly from 6.1% at 0.23 Gy to 62% at 1.82 Gy	14
Thyroid, single acute exposure		
0.065 Gy	Minimum dose for induction of thyroid carcinoma	15
3–5 Gy	5% increase in thyroid malignancies 20 years after exposure, with tumors appearing 4–5 years after exposure	16

Table 32.27 (continued) Radiation Effects on Selected Mammals

Species, Dose, and Other Variables	Effect	Reference[a]
7–10 Gy	Linear dose relation to thyroid cancer, and pathology in adjacent parathyroid and salivary glands	16
Whole body		
Single brief exposure		
0.05–0.11 Sv	Doubles rate of cancers	17
0.15 Sv	Temporary sterility, males	13
0.18–0.29 Sv	Doubles rate of pregnancy complications	17
0.5–2.0 Sv	Opacity of lens; depression of hematopoiesis	13
0.68–1.10 Sv	Doubles rate of F1 generation mortality	17
1 Sv, adults	1% probability of hereditary effects; 4% probability of fatal cancer in occupational workers	13
2.5–6.0 Sv	Sterility, females	13
3.5–6.0 Sv	Permanent sterility, males	13
5.0 Sv	Cataracts	13
<1 Gy	Survival almost certain	18
1–2 Gy	Survival probable	18
1–2 Gy	About 5% mortality in several months from infection and hemorrhage	19
2–5 Gy	Survival possible	18
2–7.5 Gy	Hematopoietic syndrome characterized by bone marrow damage, anemia, lowered immune response, hemorrhage, and sometimes death	20
3–5 Gy	Death in 30–60 days, bone marrow damage	13
5–10 Gy	100% adversely affected within weeks with bone marrow abnormalities; about 45% mortality	19
5–15 Gy	Death in 10–20 days; GI tract and lung damage	13
5–20 Gy	Survival improbable	18
7.5–30 Gy	Gastrointestinal damage: nausea, vomiting, anorexia, diarrhea, lethargy, weight loss, dehydration, exhaustion, and death	20
10–15 Gy	All adversely affected with intestinal problems within 30 min; 95% dead in 2 weeks from enterocolitis shock	19
>15 Gy	Death in 1–5 days; nervous system damage	13
>50 Gy	All dead in 48 h, usually from cerebral edema	19
Annual dose rate or protracted annual exposure for many years		
>0.1 Sv	Lens opacity	13
>0.15 Sv	Cataracts	13
>0.2 Sv	Sterility, females	13
0.4 Sv	Temporary sterility, males; hematopoiesis depression	13
2.0 Sv	Permanent sterility, males	13
Rhesus monkey, *Macaca mulatta*		
Females, single acute dose of 0.25–6.5 Gy, observed over a 17-year postexposure period	At doses >2 Gy, 53% developed endometriosis (abnormal uterine growth) vs. 26% in controls; irradiated monkeys weighed 43% less than controls, 35% were anorexic, 89% had abnormal uterine anatomy, and histopathology in most tissues exceed 50% frequency	21
Exposed to single brief whole-body proton irradiation (protons in the energy range encountered by astronauts) ranging between 0.25 and 12 Gy and observed for 24 years until death	Dose-dependent life shortening of at least 40 months at doses >4.5 Gy; mean life shortening was 200–500 monkey days per Gy (equivalent to 500–1250 human days). Brain cancer first observed in 8 Gy group after 13 months. Monkeys receiving 3–8 Gy had a	22

Table 32.27 (continued) Radiation Effects on Selected Mammals

Species, Dose, and Other Variables	Effect	Reference[a]
	significantly higher proportion of cancer deaths than those receiving 0.25–2.8 Gy. Latent period for cancer in animals receiving 4–8 Gy ranged from 13 months to 20 years	
Mammals		
10 species, 2.8–8.05 Gy, single acute exposure	LD50, 30 days after exposure	23
Various species, bioconcentration factors (BCF) of selected radionuclides		
^{60}Co		
Herbivores	Whole body BCF of 0.3	24
Caribou, *Rangifer tarandus*		
Bone vs. kidney	BCF of 0.5 vs. 0.4	24
Liver vs. muscle	BCF of 0.9 vs. 0.02	24
$^{134+137}Cs$		
Herbivores	Whole body BCF of 0.3–2.0	24
Omnivores	Whole body BCF of 1.2–2.0	24
Carnivores	Whole body BCF of 3.8–7.0	24
^{131}I		
Herbivores	Whole body BCF of 0.05	24
Omnivores	Whole body BCF of 0.2	24
Carnivores	Whole body BCF of 0.1	24
^{90}Sr, caribou, muscle vs. bone	BCF of 0.02 vs. 7.0	24
Various species, biological half-life of selected radionuclides		
^{241}Am		
Bone	27.4 years	25
Gonads	>27.4 years	25
Kidney, liver	11 years	25
Muscle	4 years	25
Serum	5 days	25
^{137}Cs, kidney, liver, and muscle	30–50 days	25
$^{238+239+240}Pu$		
Bone	49 years	25
Gonads	>49 years	25
Kidney, liver	19 years	25
Muscle	5.5 years	25
Serum	5 days	25
Singing vole, *Microtus miurus*; 8.46 Gy, single acute dose	LD50, 30 days after exposure	26
Creeping vole, *Microtus oregoni*; 6.51 Gy, single acute dose	LD50, 30 days after exposure; sensitivity may be associated with low chromosome complement	26
Meadow vole, *Microtus pennsylvanicus*; single brief exposure		
7.04 (6.35–7.98) Gy, irradiated in November	LD50, 30 days after exposure; irradiated voles released into environment	27
7.67 (7.01–8.39) Gy, irradiated in May	LD50, 30 days after exposure; irradiated voles released into environment	27
8.44 (8.17–8.77) Gy	LD50, 30 days after exposure; irradiated voles held in laboratory	7
Pine vole, *Microtus pinetorum*; single brief exposure		
7 Gy	None dead 30 days after exposure; weight normal	28
8.8 Gy, males	LD50, 30 days after exposure; weight loss in survivors	28
10.0 Gy, females	LD50, 30 days postexposure; weight loss in survivors	28

Table 32.27 (continued) Radiation Effects on Selected Mammals

Species, Dose, and Other Variables	Effect	Reference[a]
House mouse, *Mus musculus*; single brief exposure		
7.5–8.8 Gy	LD50, 30 days postexposure	26, 29
7.8, 8.1, 8.3, or 9.8 Gy	LD50, 30 days after exposure; 4 different strains	1
Mouse, *Mus* sp.		
Intraperitoneal injection		
Single injection of 850 Bq ^{227}Ac/kg BW alone or in combination with ^{227}Th at 18,500, 74,000, or 185,000 Bq/kg BW	The highest bone cancer incidence was observed at the highest doses of ^{227}Th. The addition of ^{227}Ac resulted in an additional osteosarcoma incidence only at 18,500 Bq ^{227}Th/kg BW	30
Single injection of ^{241}Am at concentrations — in Bq/kg BW — of 0.02, 0.06, 0.19, 0.37, or 1.2	Survival time decreased from 594 days for controls and 0.02 group to 135 days in the 1.2 group; increased frequency of tumors in bone, liver and lymph	31
Adult males, 84 days old, given 2–64 Bq ^{224}Ra/mouse, either as single injection, or 8 injections at 3.5-day intervals over 4 weeks; observed for 24 months	No difference in single or multiple injection effects. No effect at 16 Bq and lower. At 32 and 64 Bq, reduction in bone growth and osteonecrosis of mandible ("radium jaw")	32
Oocytes given single exposures of 0.1, 0.15, or 0.25 Gy; immature oocytes examined 8–12 weeks later	Controls had 100% survival and zero chromosome aberrations; the 0.1 Gy group had 30% survival and 2% chromosome aberrations; 0.15 Gy group had 17% survival and 6% chromosome aberrations; 0.25 Gy group had 5% survival and 23% chromosomal aberrations	33
Whole body, single brief exposure		
1 Gy	Acute exposure may extend life span	34
1.35 Gy	Doubles mutation rate of spermatogonia	17
7.6 Gy	LD50, 30 days postexposure	35
9.5 Gy	LD100, 30 days postexposure	35
10.0 Gy, adult males, 16–20 weeks old, observed for 7 days	At 90 min after irradiation, locomotor activity was suppressed and remained depressed; at 4 days, body weight decreased; at day 7, offensive aggressive behavior	20
12.5 Gy	LD100, days 3–7 postexposure	35
0.12–2.5 Sv	Doubles rate of heritable translocations in males	17
0.25–2.5 Sv	Doubles rate of congenital malformations in females	17
0.4–1.0 Sv	Doubles frequency of dominant lethal mutations	17
0.5–1.0 Sv	Doubles rate of heritable translocations in females	17
0.8–2.5 Sv	Doubles rate of congenital malformations in males	17
1.5–3.0 Sv	Doubles rate of recessive lethal mutations	17
Chronic exposure, daily dose over 8-h period, in Gy		
0.0	Mean survival time of 703 days	8
0.001	Mean survival of 761 days	8
0.011	Mean survival time of 684 days; 50% weight gain over controls	8
0.022	50% dead in 630 days	8
0.044	50% dead in 591 days	8
0.088	50% dead in 488 days	8
Total yearly dose, chronic exposure, 16 Gy (about 0.044 Gy daily)	Tolerated	34
Domestic ferret, *Mustelus putorius*; 2, 4, or 6 Gy; single brief exposure; adult males	All doses depressed locomotion; vomiting in 22 min at 2 Gy, 13 min at 4 Gy, and 11 min at 6 Gy. Various substituted benzamides reduced vomiting	36

Table 32.27 (continued) Radiation Effects on Selected Mammals

Species, Dose, and Other Variables	Effect	Reference[a]
Marsh rice rat, *Oryzomus palustris*; 5.25 Gy, single acute exposure	LD50, 30 days after exposure; this species was the most sensitive of the 10 rodent species tested	1
Domestic sheep, *Ovis aries*		
Ewes fed hay containing 9000 Bq ^{137}Cs/kg DW for 50–60 days, then 40 days on uncontaminated hay; some diets contained 30 or 60 g of vermiculite daily, or 2 g of ammonium ferricyanoferrate (AFCF) daily	Maximum levels of ^{137}Cs were reached in 10 days in milk and 35–40 days in muscle. Radionuclide transfer to milk and meat was reduced 2.5 times at daily intakes of 30 g vermiculite, and 8 times at 60 g vermiculite or 2 g AFCF	37
Ewes given oral dose of 74,000 Bq ^{137}Cs, observed for 76 days	At 76 days, only 26% of ^{137}Cs remained; tissue concentrations, in Bq ^{137}Cs/kg FW, were: 77,000 in salivary gland; 42,000 in muscle; 24,000–36,000 in pancreas, liver, and kidney; 14,000–17,000 in spleen and lung; and <8000 in other tissues	38
Lambs fed 21 kg of vegetation containing 16,600 Bq of $^{238+239+240}$Pu and 14,400 Bq of ^{241}Am over a 14-day period, followed by 4 days on uncontaminated hay	Of the Pu ingested, 46% was in liver, 30% in bone, 12% in muscle, and 2% in lung; for ^{241}Am, these values were 19% in liver, 15% in bone, 6% in meat, and 0.5% in lungs	39
Lamb given single intravenous injection of 23 Bq ^{238}Pu plus 27 Bq ^{241}Am and held for 11 days	Liver retained up to 44% of the injected ^{238}Pu and 28% of the ^{241}Am; bone had 21% of the ^{238}Pu and 20% of the ^{241}Am	39
Great basin pocket mice, *Perognathus parvus*; 8.56 Gy, single exposure	LD50, 30 days after exposure; hair loss within 7 days	26
White-footed mice, *Peromyscus leucopus*; whole body, single brief exposure		
9.5 Gy, both sexes irradiated	No reproduction	40
9.5 Gy, males only irradiated	91% of pairs successful in producing young	40
9.5 Gy, females only irradiated	40% of pairs reproduced successfully	40
10.7 Gy	LD50, 30 days postexposure; this species was the most radioresistant of 10 species of rodents tested	1
Deer mice, *Peromyscus maniculatus*; 9.19 Gy, single brief exposure	LD50, 30 days after irradiation	26
Old-field mouse, *Peromyscus polionotus*; 11.25 Gy, single exposure	LD50, 30 days after exposure	29
Norway rat, *Rattus norvegicus*; 8.67 Gy, single exposure	LD50, 30 days postexposure	1
Laboratory white rat, *Rattus* spp.		
0.001, 0.01, or 0.1 Gy; single acute whole-body exposure; adult males; killed up to 180 days after exposure	Fertility reduction was zero in the low-dose group, 25% at 0.01 Gy, and 66% at 0.1 Gy. Primary sites of damage were tubuli of the testes and spermatogonia. The high-dose group also had altered serum hormone chemistry after 30 days that persisted for 180 days	41
Pregnant rats given single, brief exposure to 0.25, 0.5, 0.75, or 1 Gy on gestational day 15; fetuses examined 24 h after irradiation	No effect at 0.5 Gy and lower on cerebral mantle of developing brain; however, dose-related increase in pyknotic cells and macrophages in cortical mantle of fetus for all doses	42
Pregnant females exposed on day 15 of gestation to 0.75 Gy; fetuses examined 1–3 months after birth	Rats irradiated *in utero* had impaired gait, slower motor behavior, difficulty in learning motor tasks, reduced growth rate, and reduced thickness of cerebral cortex	43
<1 Gy, whole body, single exposure	No brain pathology	44
2 Gy, whole body, single exposure	Brain pathology	44
Adult males conditioned to avoid electric shock by pressing a lever were subjected to various whole-body acute exposures		
1.5 Gy daily for 5 consecutive days (total 7.5 Gy)	No significant change in performance over 8 weeks	45

Table 32.27 (continued) Radiation Effects on Selected Mammals

Species, Dose, and Other Variables	Effect	Reference[a]
4.5 Gy, single exposure	No effect on performance for at least 6 weeks	45
7.5 Gy, single exposure	Significantly decreased response rate over the first 4 weeks; performance normal during weeks 5–6; 9% reduction in body weight	45
9.5 Gy, single exposure	LD100, 30 days postexposure	45
Single local dose of 5–20 Gy to various salivary glands	Dose-dependent decrease in salivary flow rate and sodium composition of saliva; at 10 Gy and higher, changes were irreversible	46
25 Gy, single whole-body exposure	50% dead 30–60 days postirradiation from fatal stomach damage; significant liver damage in survivors	47
Eastern harvest mouse, *Reithrodontomys humulis*; 9.5 Gy, single exposure	LD50, 30 days after exposure	1
Rodents, single brief exposure		
3.8–13.3 Gy, 13 species	LD50, 30 days postexposure; females more sensitive than males	48
4–8 Gy, 5 species	Dead rodents had histopathology of lymph nodes, thymus, bone marrow, liver, lung, and gonads; male survivors had atrophied testes	26
5.3–10.7 Gy, 10 species	Range of LD50 values (30 days after exposure); survivors showed conjunctivitis, ataxia, diarrhea, passiveness, cessation of feeding, aggressiveness, and graying of pelage	1
Squirrel monkey, *Saimiri* spp.; fetuses, 80–90 days postconception; given 0.1 or 1.0 Gy, single acute exposure; young observed from birth to age 2 years	No effect on behavior at 0.1 Gy. At 1 Gy, emotional stability and vision impaired at age 30 days, learning impaired at 1 year; normal at 2 years	44
Cotton rat, *Sigmodon hispidus*		
Females given 5, 7.5, 9, 10.5, or 12 Gy whole body once a month for 4 months, then released into a 0.4-ha impoundment with unirradiated males and females for 15 days	Survival was 91% at 5 Gy and 25% at 12 Gy; intermediate doses had intermediate survival	49
9.58 Gy, single brief exposure	LD50, 30 days after exposure	1
11.2 Gy, single acute exposure, adult females	LD50, 30 days after exposure	49
11.3 Gy, single exposure, adult females	LD50, 15 days postexposure	49
Eastern chipmunk, *Tamias striatus*		
2–4 Gy, single exposure	Irreversible injury throughout life, but life span was increased	50
Given single exposure of 2 or 4 Gy, then released into environment	Irradiated chipmunks had consistently smaller home ranges and moved shorter distances than did controls	51
Iodine-131, half-time release rate from thyroid; females vs. males		
Summer	2.3 h vs. 263 h	52
Spring	156 h vs. 217 h	52
Autumn	126 h vs. 129 h	52

[a] **1,** Dunaway et al. 1969; **2,** Handl et al. 1990; **3,** Handl and Pfau 1989; **4,** Jaenke and Angleton 1990; **5,** Benjamin et al. 1990; **6,** Lloyd et al. 1991; **7,** Bruenger et al. 1990; **8,** Lorenz et al. 1954; **9,** Egami et al. 1991; **10,** Nagasawa et al. 1990; **11,** Woloschak et al. 1990a; **12,** Mole 1990; **13,** ICRP 1991a; **14,** Kamiguchi et al. 1990; **15,** Kim et al. 1990; **16,** Refetoff 1990; **17,** Saknaranarayanan 1991c; **18,** McLean 1973; **19,** United Nations Scientific Committee on the Effects of Atomic Radiation (UNSCEAR) 1988; **20,** Maier and Landauer 1990; **21,** Fanton and Golden 1991; **22,** Wood 1991; **23,** Hobbs and McClellan 1986; **24,** Kitchings et al. 1976; **25,** Gilbert et al. 1989; **26,** O'Farrell 1969; **27,** Iverson and Turner 1976; **28,** Dunaway et al. 1971; **29,** Golley and Gentry 1969; **30,** Muller et al. 1990; **31,** Schoeters et al. 1991; **32,** Robins 1990; **33,** Straume et al. 1991; **34,** Rose 1992; **35,** Cronkite et al. 1955; **36,** King and Landauer 1990; **37,** Daburon et al. 1991; **38,** Vandecasteele et al. 1989; **39,** Ham et al. 1989; **40,** Di Gregorio et al. 1971; **41,** Canfi et al. 1990; **42,** Norton and Kimler 1990; **43,** Norton et al. 1991; **44,** Mole 1990; **45,** Mele et al. 1990; **46,** Vissink et al. 1990; **47,** Geraci et al. 1991; **48,** Whicker and Schultz 1982b; **49,** Pelton and Provost 1969; **50,** Thompson et al. 1990; **51,** Snyder et al. 1976; **52,** Kodrich and Tryon 1971.

32.7.6.1 Survival

Survival time is inversely related to dose in whole-body, acute exposures to ionizing radiation (Figure 32.9). In general, hematopoietic organs are most sensitive, and the gastrointestinal tract and central nervous system are next most sensitive (UNSCEAR 1988). Body weight is an important modifier, and heavier mammals are usually most sensitive to radiation (Figure 32.10). Feral rodent populations are at risk from ionizing radiation through the reduction in numbers from direct kill and indirectly from the radiation-caused diminution of reproduction (Di Gregorio et al. 1971).

Low doses of ionizing radiation are beneficial to many species of mammals. Effects of radiation hormesis include increased survival and longevity, lowered sterility, increased fecundity, and accelerated wound healing (Luckey 1980). Low doses of gamma irradiation cause irreversible injury to the eastern chipmunk (*Tamias striatus*), although the life-span was significantly longer (Thompson et al. 1990). Acquired radioresistance after exposure to a low dose of ionizing radiation has been described in rats, mice, and yeast (Yonezawa et al. 1990). In mice, for example, low doses of X-irradiation (not higher than 0.15 Gy) enhanced 30-day survival if given 2 months prior to a dose of 7.5 Gy. The low-dose exposure seems to stimulate the recovery of blood-forming stem cells after the second irradiation and favors a decrease in the incidence of bone-marrow death. The exact mechanisms of radiation hormesis are unknown because effects are not related to and not predictable from the high-dose exposure (Yonezawa et al. 1990).

Irradiated small mammals released into the environment had a lower survival rate than laboratory populations, suggesting that the extrapolation from laboratory results may overestimate the radioresistance of free-ranging voles and other small animals because of the general level of stress in the population (Iverson and Turner 1976). The opposite was observed in eastern chipmunks given high sublethal doses of X-rays. Chipmunks had an overall reduction in mobility when they were released

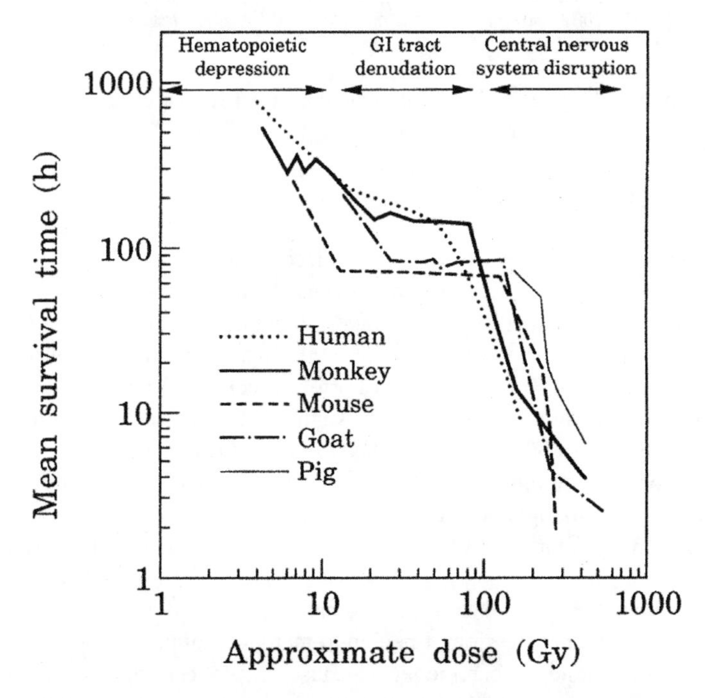

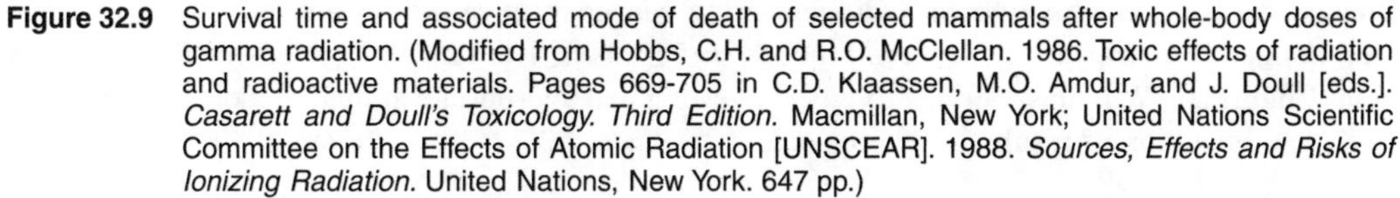
Figure 32.9 Survival time and associated mode of death of selected mammals after whole-body doses of gamma radiation. (Modified from Hobbs, C.H. and R.O. McClellan. 1986. Toxic effects of radiation and radioactive materials. Pages 669-705 in C.D. Klaassen, M.O. Amdur, and J. Doull [eds.]. *Casarett and Doull's Toxicology. Third Edition.* Macmillan, New York; United Nations Scientific Committee on the Effects of Atomic Radiation [UNSCEAR]. 1988. *Sources, Effects and Risks of Ionizing Radiation.* United Nations, New York. 647 pp.)

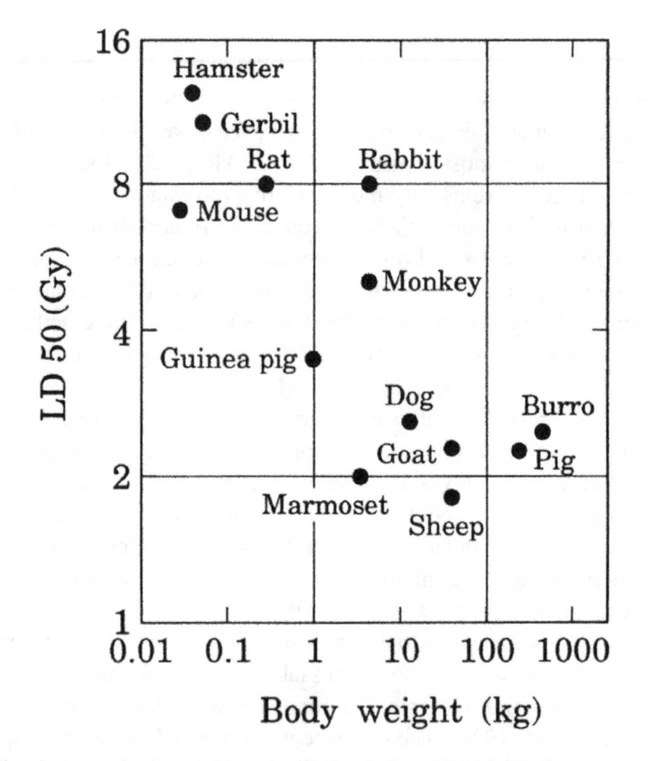

Figure 32.10 Relation between body weight and radiation-induced LD50 (30 days postexposure) for selected mammals. (Modified from United Nations Scientific Committee on the Effects of Atomic Radiation [UNSCEAR]. 1988. *Sources, Effects and Risks of Ionizing Radiation.* United Nations, New York. 647 pp.)

into the environment and a higher survival rate than controls (Snyder et al. 1976), possibly because of increased predation on the more mobile controls.

32.7.6.2 Carcinogenicity

The risk of the induction of cancer is a recognized somatic effect of low doses of ionizing radiation, as judged by epidemiological studies of Japanese survivors of U.S. nuclear bombs (Coggle and Williams 1990) and of Marshall Islanders, underground miners, and radium watch dial workers (Bowden et al. 1990). However, Yoshimoto et al. (1990), in a study on the occurrence of malignant tumors in Japanese children <10 years old and born between 1946 and 1982 to survivors of the atomic bombings in 1945, found no statistically significant increase in malignant tumors in the children of parents exposed to >0.01 Sv whole-body radiation (mean gonadal exposure of 0.43 Sv) at the time of the atomic bombings when compared to a suitable control group. Nutritional status is important when treating malignant tumors. Unlike tumors of nonanemic individuals, tumors in anemic mice and humans frequently do not respond satisfactorily to radiotherapy (McCormack et al. 1990).

Ionizing radiation induces basal cell carcinomas in skin and is active in the initiation of malignant tumors and in the progression of benign to malignant tumors (Bowden et al. 1990). Skin has been widely used in studies of carcinogens because of its accessibility and the visibility of its tumors. All data on experimental radiogenic skin cancer in mice are on a relatively narrow and well-defined response curve. However, mouse skin is about 100 times more sensitive than human skin (Coggle and Williams 1990), strongly suggesting that appropriate animal models are necessary in the extrapolation of results to other species.

Thyroidal cancer in dogs and sheep has been induced with repeated administrations of ^{131}I, although single injections of ^{131}I failed to induce thyroid cancer in adult animals except in some strains of laboratory rodents (Walinder 1990). Humans with a prior history of ^{131}I and other radiation exposure in childhood are at a significantly higher risk of thyroid carcinogenesis, and females are at higher risk than males. The minimum latent period in humans is about 4 years, and neoplastic lesions may develop as late as 40 years after irradiation (Kim et al. 1990). Historically, the human thyroid received radiation from irradiation of the scalp for epilation (up to 0.5 Gy), thymus (up to 5 Gy), tonsils and adenoids (8 Gy), and facial acne (15 Gy). Higher doses of external irradiation (up to 50 Gy) were used from 1920 to 1940 for the treatment of hyperthyroidism in adults and are still used for the treatment of cervical malignancies in people of all ages (Refetoff 1990).

Radium-induced bone malignancies after exposure to ^{226}Ra are similar in beagles and humans, and the tibia in dogs is especially sensitive (Lloyd et al. 1991). "Radium jaw" has been described in humans as a late effect of accidental ingestion or therapeutic administration of long-lived radium isotopes, such as ^{226}Ra and ^{228}Ra, and is characterized by bone tumors, spontaneous fractures, and osteosclerosis (Robins 1990). However, the short-lived ^{224}Ra (Tb 1/2 of 3.6 days) produces similar effects in mice, suggesting that the events that trigger radium-induced bone disorders occur within days of incorporation, although the consequence is a late effect (Robins 1990).

Aerosol exposures of mice, rats, dogs, and hamsters to radon and its decay products resulted in lifetime shortening, pulmonary emphysema, pulmonary fibrosis, and respiratory tract carcinoma. Damage to the skin and kidney was also reported, but the lung seems to be the primary affected organ (Cross 1990). Small mammals and birds that live in burrows containing radon-rich soils (9900 Bq/m^3 of ^{222}Rn) are expected to have an additional 17 lung cancers per 1000 animals than animals not similarly exposed. However, tumors have not been widely reported in these species (Macdonald and Laverock 1998). Radon and ^{222}Rn daughters have caused problems in miners who work underground in uranium mines. These miners had an excessive incidence of disease of the respiratory system, including lung cancer. The problem is related to the emanation of radon into the mines and the decay of the radon, the short-lived radioactive daughters (^{216}Po, ^{214}Pb, ^{214}Bi, ^{214}Po) that attach to dust particles, eventually resulting in alpha radiation exposure of the respiratory airways (Hobbs and McClellan 1986). A similar pattern was evident in rats exposed to ^{239}Pu. Rats exposed to $^{239}PuO_2$ aerosol of about 3700 Bq/lungs and examined 8 to 18 months after exposure had a very high frequency (as much as 80%) of malignant pulmonary neoplasms; genetic mutations were evident in 46% of the radiation-induced tumors (Stegelmeier et al. 1991).

The incidence of ovarian tumors in mice, guinea pigs, and rabbits increased after 3 years of chronic irradiation at doses as low as 1.1 mGy daily (Lorenz et al. 1954). Unlike other tumors, the induction of ovarian tumors depended on a minimum total dose and seemed to be independent of a daily dose (Lorenz et al. 1954). Radiation-induced neoplastic transformation of hamster cells may be associated initially with changes in expression of the genes modifying cytoskeletal elements (Woloschak et al. 1990b).

32.7.6.3 Mutagenicity

In general, ionizing radiation has produced mutations in every plant and animal species studied. Some genetic risks are associated with exposures, but the risk of inducing a dominant genetic disease is quite small because radiation-induced mutations are primarily recessive and usually lethal (Sankaranarayanan 1991c). Residents living near the site of extensive mining and milling of uranium operations in Texas have an increased frequency of chromosomal aberrations and a reduced DNA repair capacity; uranium-238 was much higher in these areas when compared to reference sites, possibly as a result of leaching into the groundwater (McConnell et al. 1998). The genetic doubling dose of radiation is the amount of acute or chronic radiation that doubles the naturally occurring spontaneous mutation rate each generation. For mice, the estimated genetic doubling dose equivalent is 1.35 Sv from acute exposures and 4.0 Sv from chronic exposures to radiation (Neel and Lewis 1990). For protection from radiation, the estimated genetic risks to humans have largely been based on data from mice (Straume et al. 1991). Studies of children of Japanese survivors of nuclear bomb explosions showed that the genetic doubling dose equivalent of acute gonadal radiation is about 2.0 Sv (1.69 to 2.23); from chronic radiation, this value is about 4.0 Sv (Neel and Lewis 1990). Based on results of the study of Japanese survivors of the nuclear explosions, Yoshimoto et al. (1990) and Sankaranarayanan (1991c) concluded that there was no increase in the spontaneous mutation rate after parents were exposed. The high doubling dose of about 4 Sv estimated from these data is another way of stating that, relative to the assumed spontaneous rates, the rate of induction of mutations leading to the measured effects is too small (Sankaranarayanan 1991c). The transmission of radiation-induced genetic effects to offspring has not yet been demonstrated in any human population (Straume et al. 1991).

Specific point mutations were identified in ^{239}Pu-induced preneoplastic lesions and malignant neoplasms in the lungs of rats (Stegelmeier et al. 1991). Mice exposed to a single whole-body dose of 3 Gy produced a radiation-induced mutation that simultaneously generated distinct alleles of the limb deformity and agouti (grizzled fur color) loci, two developmentally important — but not adjoining — regions on a single chromosome. This phenomenon was probably associated with DNA breaks caused by inversion of a segment in another chromosome (Woychik et al. 1990). The plasma membrane in immature oocytes of mice is the hypersensitive lethal target in producing radiation-induced genetic damage (Straume et al. 1991).

32.7.6.4 Organ and Tissue Damage

In the abdomen, the kidneys are one of the most sensitive organs to serious or fatal radiation-induced damage (Jaenke and Angleton 1990). The relatively high incidence of kidney disease among mature beagles injected with ^{226}Ra and its accompanying ^{210}Bi and ^{210}Po resulted from alpha irradiation of the kidneys by the substantial amount of ^{210}Po that was in the injected solution (Bruenger et al. 1990). Hepatic injury induced by ionizing radiation can be a life-threatening complication. The main responses of the liver to acute radiation exposure include enlargement, dilation of blood vessels, fluid accumulation, and histopathology (Geraci et al. 1991). Damaging effects of ionizing radiation on the fetal cerebral cortex has been recognized for many years (Norton and Kimler 1990). The deleterious effects of ionizing radiation on the developing brain are prolonged and progressive. Doses <2 Gy of gamma radiation are harmful to the developing brain; and in humans, mental retardation may occur from doses as low as 0.2 Gy between week 8 and 15 of gestation (Norton et al. 1991).

Irradiated white-footed mice (*Peromyscus leucopus*) frequently had atrophied gonads, degenerating fetuses in the uterus, and greying hair (Di Gregorio et al. 1971). High sublethal doses (7 Gy) of radiation to the pine vole (*Microtus pinetorium*) caused pelage graying, wherein unpigmented hair from damaged follicles replaces molted pigmented hair. Pelage graying may decrease survival from increased predation (Dunaway et al. 1971), although this needs verification.

Human sperm chromosomes retain a high fertilizing ability after a high dose of X-irradiation, although mammalian spermatozoa have little capacity to repair DNA damage induced by radiation (Kamiguchi et al. 1990). Radiation-induced death of lymphoid cells in rats is associated with damage to the cell itself but may also be due to secretions from irradiation-activated natural killer cells that induce pycnosis and interphase death in lymphoid cells (Eidus et al. 1990).

32.7.6.5 Behavior

Numerous behavioral measures have been evaluated for their usefulness in providing a sensitive index of exposure to ionizing radiation. Radiation-related mental retardation is the most likely type of behavioral abnormality in humans; sensitivity peaked between 8 and 15 weeks of conception and at doses >0.4 Gy (UNSCEAR 1988). No specific mechanism for the production of mental retardation has been established, although proposed mechanisms include the loss of cells, migration of neurons, and failure of synaptogenesis (Norton and Kimler 1990). In studies with rats, operant responses decreased (maintained by positive reinforcement such as food or water) at sublethal radiation doses (3.0 to 6.75 Gy) under various schedules of reinforcement (Mele et al. 1990). Disrupted operant responses under shock avoidance at >LD100 levels are reported in pigs and rhesus monkeys (Mele et al. 1990).

32.7.6.6 Absorption and Assimilation

The absorption, bioavailability, and retention of radionuclides in mammals are modified by:

- The age, sex, species, and diet of the organism
- Season of collection
- The chemical form of the radionuclide in tissue and blood
- Residence time in the digestive tract
- Preferential accumulation by selected organs and tissues
- Other variables (Kitchings et al. 1976; Whicker and Schultz 1982a, 1982b; Richmond 1989; Desmet et al. 1991; Harrison 1991).

Assimilation fractions of various elements recommended by the International Commission on Radiological Protection are presented in detail by Whicker and Schultz (1982b).

Many radionuclides preferentially accumulate in certain organs or tissues, but the critical organ is different for different radionuclides: liver for ^{54}Mn, erythrocytes and spleen for ^{55}Fe, liver and kidney for cobalt nuclides, liver and prostate for ^{65}Zn, skeletal muscle for ^{137}Cs, and GI tract for ^{95}Zr (Whicker and Schultz 1982a).The persistence of radionuclides in mammals varies with the chemical form, kinetics, species, and other variables. For example, the time for 50% persistence of selected radionuclides in whole-animal studies ranges from 19 h to 14 days of ^{134}Cs; 4 to 35 days of ^{137}Cs; 5 to 12 h of the short-lived component of ^{60}Co, and 5 to 21 days of the long-lived component; 25 to 593 days of ^{90}Sr; and 4 to 26 days of ^{131}I (Kitchings et al. 1976; Whicker and Schultz 1982b).

Some radionuclides act antagonistically when administered together. The combined incorporation of ^{227}Ac and ^{227}Th at levels tested in mice shows a lower biological effect than the sums of the effects of the components administered singly. The less-than-additive effect is in good agreement with experiments with the incorporation of a mixture of β emitters, in which the effects are also less-than-additive (Muller et al. 1990). Uptake and retention characteristics of essential biological nutrients (i.e., H, C, P, I, K, Ca, Mn, Fe, Co, Zn) are largely controlled by biological processes (Whicker and Schultz 1982a). For example, ^{131}I, regardless of route of administration, is rapidly absorbed into the bloodstream and concentrated in the thyroid. Ionizing radiation associated with high levels of ^{131}I destroys the thyroid, affecting the thyroid hormone production (Hobbs and McClellan 1986).

Alkali metals (K, Rb, Cs) behave similarly and sometimes one is accumulated preferentially when another is deficient. A similar case is made for Sr and Ca (Whicker and Schultz 1982a). The most important alkali metal isotope is ^{137}Cs because of its long physical half-life (30 years) and its abundance as a fission product in fallout from nuclear weapons and in the inventory of a nuclear reactor or a fuel-reprocessing plant. Cesium behaves much like potassium. It is rapidly absorbed into the bloodstream and distributed throughout the active tissues of the body, especially muscle. The β and γ radiation from the decay of ^{137}Cs and its daughter, ^{137}Ba, result in essentially whole-body irradiation that harms bone marrow (Hobbs and McClellan 1986).

Because ^{226}Ra and ^{90}Sr are metabolic analogs of calcium, they are deposited in the skeleton. Both isotopes are associated with bone cancers (Hobbs and McClellan 1986). In pregnant rats, the total amount of ^{226}Ra transferred from the dam to the 8 to 10 fetuses in a litter was low after a single injection and did not exceed 0.3% of the maternal content. The retained whole-body burden in dams was 53% at the first, 48% at the second, and 44% at the third pregnancy, mostly in the skeletal system (Kshirsagar 1990). The rare earths (i.e., ^{144}Ce, ^{152}Eu, ^{140}La, ^{147}Pr, ^{151}Sm) are usually not effectively absorbed from the GI tract, and elimination is rapid (Palumbo 1963). Cerium-144 is one of the more biologically hazardous radionuclides in this group because of its half-life (285 days) and the energetic β emissions from it and its daughter, ^{144}Pr (Hobbs and McClellan 1986).

The greatest uncertainty in dose estimates from the ingestion of long-lived alpha emitters is the values used for their fractional absorption from the GI tract (Harrison 1991). For transuranic elements, the fraction of the ingested material that was assimilated by the whole organism was always <0.01% and usually nearer 0.003% (Whicker and Schultz 1982b). The major hazard of plutonium nuclides to terrestrial organisms comes from inhalation; uptake by plants is low, and further uptake by humans through the gut is low (Noshkin et al. 1971). Americium-241 is an artificial, toxic bone-seeking radionuclide produced through beta decay of ^{241}Pu (Schoeters et al. 1991). Because of its long half-life, its high-energy alpha irradiation, and its accumulation in the liver and skeleton, consideration should be given to ^{241}Am in risk estimates of latent effects, such as induction of liver cancers, bone cancers, and leukemias. In comparison with ^{226}Ra, ^{241}Am is 20 times more effective in reducing life-span in mice and 13 times more effective in the rate of death from bone cancer (Schoeters et al. 1991).

Although radon has long been known as a health hazard to miners in the uranium industry, it was only in the 1980s that radon contamination of buildings was recognized as widely distributed over the Earth. However, years of exposure are required before a health problem develops (Majumdar et al. 1990). Exposure to radon-decay products can be expressed in two different ways: the amount of inhaled decay products (taking into account their potential to emit radiation energy) or the product of the time during which the decay products were inhaled and their concentration in the inhaled air. The potential alpha energy of the inhaled decay products may be expressed in joules (J). The potential alpha energy concentrations in air is expressed in joules per cubic meter; for radon in equilibrium with its decay product, this corresponds to 3700 Bq/m^3 (UNSCEAR 1988).

Rodents dosed with tungsten-185 excreted 80% in 24 h; bone was the major retention site; the half-time persistence ranged from 5.7 days in femurs of mice to 86 days in femurs of rats; some components in the bone of rats persisted with a half-time >3 years (Reed and Martinedes 1971). Niobium-95 is produced directly by nuclear fission and indirectly by decay of ^{95}Zr. Routine discharges of ^{95}Nb from a nuclear fuel reprocessing plant in the United Kingdom in 1970 contributed about 5% of the bone-marrow dose to a 10-year-old child living in the vicinity (Harrison et al. 1990). Gastrointestinal absorption of ^{95}Nb by adult guinea pigs was about 1.1%, and supports the values of 1% absorption in adults and 2% in infants now used to calculate percentage absorption of niobium isotopes by humans (Harrison et al. 1990).

32.8 PROPOSED CRITERIA AND RECOMMENDATIONS

For the protection from radiation, effects of radiation have been characterized as stochastic or nonstochastic. The probability of a stochastic effect — and not its severity — varies as a function of dose in the absence of a threshold (i.e., hereditary effects or carcinogenesis). The probability and severity of nonstochastic effects vary with dose, and a threshold for the dose exists (i.e., cataract of lens, nonmalignant damage to the skin, cell depletion in the bone marrow causing hematological deficiencies, gonadal cell damage leading to impairment of fertility, or pneumotis and pulmonary fibrosis following lung irradiation) (ICRP 1977; Hobbs and McClellan 1986; UNSCEAR 1988). The prevention of nonstochastic effects is achieved by setting dose-equivalent limits at sufficiently low levels so that no threshold dose is reached, not even after exposure for the whole of a lifetime or for the total period of a working life (ICRP 1977, 1991a, 1991b). Guides for the protection from radiation are also predicated on the effective half-life of each isotope, the critical organ, the fraction that reaches the critical organ by ingestion and inhalation, and the maximum tolerable whole-body burdens, as judged by radionuclide concentrations in air, water, and diet (Palumbo 1963).

At present, no radiological criteria or standards have been recommended or established for the protection of fish, wildlife, or other natural resources. All radiological criteria now promulgated or proposed are directed toward the protection of human health. It is generally assumed that humans are comparatively radiosensitive and that guides will probably also protect sensitive natural resources (UNSCEAR 1988; ICRP 1991a, 1991b; NCRP 1991; IAEA 1992; Zach et al. 1993), although this needs verification. Numerous radiological criteria now exist for the protection of human health (Table 32.28). Most authorities agree that some adverse effects to humans are likely under the following conditions:

- >5 mSv whole-body exposure of women during the first 2 months of pregnancy
- >50 mSV whole-body exposure in any single year or >2000 mSv in a lifetime
- An annual inhalation intake by a 60-kg individual — in Bq/kg BW — that exceeds 0.67 ^{232}Th, 3.3 ^{241}Am, 3.3 ^{239}Pu, 16 ^{252}Cf, 33 ^{235}U, 1666 ^{90}Sr, 16,666 ^{60}Co, or 166,666 ^{32}P
- An annual ingestion intake by a 60-kg individual — in Bq/kg BW — that exceeds 3333 ^{129}I, 16,666 ^{125}I, 16,666 ^{131}I, or 66,666 ^{137}Cs
- A total annual intake from all sources — in Bq/kg BW by a 60-kg person — that exceeds 66 ^{210}Pb, 166 ^{210}Po, 333 ^{226}Ra, 666 ^{230}Th, 833 ^{228}Th, or 1333 ^{238}U (see Table 32.28).

Astronauts between 25 and 55 years of age usually receive an average career dose of 2.0 Sv (1.0 to 3.0 Sv in females; 1.5 to 4.0 Sv in males); theoretically, this may cause a life shortening of 2000 to 3000 days (Wood 1991). Other environmental variables also result in life shortening and include cigarette smoking (2250 days), coal mining (1100 days), and being 30% overweight (1300 days); thus, models that assess the harm of a single variable — such as radiation — on life expectancy must incorporate all known data and their interacting effects (Wood 1991).

Environmental dose-response models and animal epidemiological data are most frequently used to assess the risk from ionizing radiation. In its ideal form, a risk assessment should clearly present the rationale for an estimate of risk and should include the recognition of the roles of assumptions, approximations, data, theories, models, and deductions in arriving at an inference and a discussion of the involved uncertainties (Cothern et al. 1990). It now seems clear that current risk assessments of ionizing radiation hazards to all living organisms — not just humans — require additional data and reinterpretation of existing data. Specifically, more effort is needed in the following areas:

1. Measurement of concentrations of naturally occurring radionuclides and natural background doses in the environment as a baseline for studies on radiation effects (Templeton et al. 1971)
2. Refinement of models of radionuclide transfer in food chains to aid in the assessment of radioactive releases from nuclear reactors and other point sources, including possible biomagnification by trophic components and turnover rates by receptor organisms (Kitchings et al. 1976)
3. Continuance of protracted exposure studies to measure carcinogenesis in animal and human cell lines and the role of secondary factors — especially chemical agents — in radiation carcinogenesis (Little 1990)
4. Research on radiation-induced recessive lethal mutations — the predominant type of radiation-induced mutation — and dominant mutation systems (Sankaranarayanan 1991c)
5. Initiation of long-term studies to establish sensitive indicators of radiation stress on individuals and communities, including effects on growth and reproduction (Templeton et al. 1971)
6. Clarification of the role of enzymes and proteins in repair of radiation-damaged cellular DNA and of mechanisms of enzymatic reactions leading to altered nucleotide sequences (Hagen 1990)
7. Reinterpretation of low-level chronic irradiation effects on developing embryos under rigorously controlled conditions (Templeton et al. 1971)
8. Resolution of mathematical shape(s) of radiation dose-response curve(s) (Hobbs and McClellan 1986).

Table 32.28 Recommended Radiological Criteria for the Protection of Human Health

Criterion and Other Variables	Concentration or Dose	Reference[a]
AIR		
United States; radon-222		
Average	<0.0555 Bq (<1.5 pCi)/L or <55 Bq/m^3	1
Acceptable	<0.148 Bq (<4.0 pCi)/L	1, 2
Allowable emission discharge	<0.74 Bq (<20 pCi)/m^2 per sec; should not increase the radon-222 concentration in air at or above any location outside the disposal site by >0.0185 Bq (>0.5 pCi)/L or >18 Bq (>500 pCi)/m^3	3
Unacceptable	>0.185 Bq (>5 pCi)/L	2
ASTRONAUTS		
Age 25–55 years; expected whole-body career dose; females vs. males	1.0–3.0 Sv (100–300 rem) vs. 1.5–4.0 Sv (150–400 rem)	4
Adverse effects expected; lifetime exposure	>2.0 Sv (>200 rem)	4
CANCER RISK AND BIRTH DEFECTS		
Projected 0.04% increase in cancers; 0.01% increase in birth defects	0.11 mSv (0.011 rem) whole-body maximum per year; 0.69 mSv (0.069 rem) whole body over 30 years; or 1.00 mSv (0.1 rem) bone marrow over 30 years	5
Projected 0.18% increase in cancers; 0.07% increase in birth defects	0.51 mSv (0.05 rem) whole-body maximum per year; 3.30 mSv (0.33 rem) whole body over 30 years; or 4.60 mSv (0.46 rem) bone marrow over 30 years	5
Projected 0.92% increase in cancers; 0.38% increase in birth defects	3.03 mSv (0.3 rem) whole-body maximum per year; 19.0 mSv (1.9 rem) whole body over 30 years; or 23.0 mSv (2.3 rem) bone marrow over 30 years	5
Projected 4.4% increase in cancers; 1.8% increase in birth defects	20.1 mSv (2.0 rem) whole-body maximum per year; 91.0 mSv (9.1 rem) whole body over 30 years; or 110.0 mSv (11 rem) bone marrow over 30 years	5
DIET		
All foods; maximum recommended values		
Adults, Italy	600 Bq (16,200 pCi) cesium-134+137/kg fresh weight (FW)	6
Children, Italy	370 Bq (10,000 pCi) cesium-134+137/kg FW	6
Sweden, pre-Chernobyl	300 Bq (8100 pCi) cesium-134+137/kg FW	6
Caribou; muscle; North America	<2260 Bq (<61,000 pCi) cesium-137/kg FW	8
Fish; Great Lakes; muscle	Dose of <0.02 μSv (0.00002 rem)/kg FW fish flesh equivalent to consumers	9
Fish; Sweden	<1500 Bq (<40,500 pCi) cesium-137/kg FW	10
Fraction of ingested dose absorbed; recommended maximum; selected isotopes		
Americium, curium, neptunium, plutonium, thorium	<0.05%	11
Americium, plutonium	<0.1%	12
Californium, and higher mass radionuclides	<0.1%	11
Uranium	<5.0%	11
Milk; maximum values		
Italy	370 Bq (10,000 pCi) cesium-134+137/L	6
Japan	370 Bq (10,000 pCi) cesium-137/L	13
Sweden	300 Bq (8100 pCi) cesium-137/L	14
Meat and fish; Sweden; maximum values	1500 Bq (40,500 pCi) cesium-137/kg FW	14

Table 32.28 (continued) Recommended Radiological Criteria for the Protection of Human Health

Criterion and Other Variables	Concentration or Dose	Reference[a]
Reindeer meat, game, animal meat, fish, berries, mushrooms; Sweden; post-Chernobyl; maximum values	1500 Bq (40,500 pCi) cesium-137/kg FW	15
Sheep, muscle	<1000 Bq (<27,000 pCi) cesium-134+137/kg FW	16
Sheep, muscle	<1000 Bq (<27,000 pCi) cesium-137/kg FW	17
DRINKING WATER		
Natural radioactivity; maximum allowed		
Radium-226+228	0.185 Bq (5 pCi)/L	18
Gross alpha	0.555 Bq (15 pCi)/L	18
Artificial radioactivity; maximum allowed		
Gross beta	1.85 Bq (50 pCi)/L	18
Tritium (Hydrogen-3)	740 Bq (20,000 pCi)/L	18
Strontium-90	0.296 Bq (8 pCi)/L	18
Great Lakes; maximum dose to consumers	10 μSv (0.001 rem)/year	9
GENERAL PUBLIC		
Annual effective dose[b]	<1 mSv (<0.1 rem)	19, 29
Cesium-137; total intake		
Sweden	<50,000 Bq (<1,350,000 pCi)/year, equivalent to <1 mSv (<0.1 rem)	15
North America	<300,666 Bq (<8,100,00 pCi)/year	8
United Kingdom	<400,000 Bq (<10,800,000 pCi)/year, equivalent to <5 mSv (0.5 rem)	20
Maximum permissible dose		
Eye lens	<15 mSv (<1.5 rem)/year	19
Skin	<50 mSv (<5 rem)/year	19
Whole body		
Individual, except students and pregnant women	<5 mSv (<0.5 rem)/year	9, 21–23
Students	<1 mSv (<0.1 rem)/year	21
Pregnant women	<5 mSv (<0.5 rem) during the first 2 months of pregnancy	22
Population dose limits, genetic or somatic	<1.7 mSv (<0.17 rem) yearly average	21
GROUNDWATER		
Maximum allowed		
Radium-226+228	0.185 Bq (5 pCi)/L	3
Alpha emitting radionuclides — including radium-226+228, but excluding radon isotopes	0.555 Bq (15 pCi)/L	3
Total beta and gamma radiation	Total annual whole-body dose equivalent, or dose to any internal organ, <0.04 mSv (<0.004 rem), based on individual consumption of 2 L daily of drinking water from a groundwater source	3
RADIOACTIVE WASTES		
Dose limits from spent nuclear fuel or transuranic radioactive wastes		
Whole body	<0.25 mSv (<0.025 rem)/year	3, 24
Thyroid	<0.75 mSv (<0.075 rem)/year	3, 24
Any other critical organ	<0.25 mSv (<0.025 rem)/year	3, 24
Stored for 10,000 years; maximum cumulative release allowed to the accessible environment per 1000 metric tons of heavy metal during storage		
Americium-241	3.7 trillion (T) Bq (100 TpCi)	3
Americium-243	3.7 TBq (100 TpCi)	3

Table 32.28 (continued) Recommended Radiological Criteria for the Protection of Human Health

Criterion and Other Variables	Concentration or Dose	Reference[a]
Any alpha emitter with physical half-life >20 years	3.7 TBq (100 TpCi)	3
Any non-alpha emitter radionuclide with physical half-life >20 years	37.0 TBq (1000 TpCi)	3
Carbon-14	3.7 TBq (100 TpCi)	3
Cesium-135	37.0 TBq (1000 TpCi)	3
Cesium-137	37.0 TBq (1000 TpCi)	3
Iodine-129	3.7 TBq (100 TpCi)	3
Neptunium-237	3.7 TBq (100 TpCi)	3
Plutonium-238	3.7 TBq (100 TpCi)	3
Plutonium-239	3.7 TBq (100 TpCi)	3
Plutonium-240	3.7 TBq (100 TpCi)	3
Plutonium-242	3.7 TBq (100 TpCi)	3
Radium-226	3.7 TBq (100 TpCi)	3
Strontium-90	37.0 TBq (1000 TpCi)	3
Thorium-230	0.37 TBq (10 TpCi)	3
Thorium-232	0.37 TBq (10 TpCi)	3
Tin-126	37.0 TBq (1000 TpCi)	3
Uranium-233	3.7 TBq (100 TpCi)	3
Uranium-234	3.7 TBq (100 TpCi)	3
Uranium-235	3.7 TBq (100 TpCi)	3
Uranium-236	3.7 TBq (100 TpCi)	3
Uranium-238	3.7 TBq (100 TpCi)	3
Uranium by-product materials; maximum discharge rates allowed into water		
Radium-226+228	0.185 Bq (5 pCi)/L	24
Gross alpha particle activity, excluding radon and uranium isotopes	0.555 Bq (15 pCi)/L	24
Wastes from uranium fuel cycle entering the environment per billion watts/year of electrical energy produced by the fuel cycle; maximum allowed		
Krypton-85	1.85 TBq (50 TpCi)	24
Iodine-129	185 million Bq (5 billion pCi)	24
Plutonium-239 and other alpha emitting transuranics with Tb 1/2 >1 year	2.69 million Bq (72.6 million pCi)	24
OCCUPATIONAL WORKERS		
Annual Limit of Intake[b]		
Inhalation vs. oral		
Americium-241	200 Bq (5400 pCi) vs. 50,000 Bq	25
Californium-252	1000 Bq (27,000 pCi) vs. 200,000 Bq	25
Cesium-137	6 million Bq (162 million pCi) vs. 4 million Bq	25
Cobalt-60	1 million Bq (27 million pCi) vs. 7 million Bq	25
Hydrogen-3	3 billion Bq (81 billion pCi) vs. 3 billion Bq	25
Iodine-125	2 million Bq (54 million pCi) vs. 1 million Bq	25
Iodine-129	300,000 Bq (8,100,000 pCi) vs. 200,000 Bq	25
Iodine-131	2 million Bq (54 million pCi) vs. 1 million Bq	25
Phosphorus-32	10 million Bq (270 million pCi) vs. 20 million Bq	25
Plutonium-239	200 Bq (5400 pCi) vs. 200,000 Bq	25
Polonium-210	20,000 Bq (540,000 pCi) vs. 100,000 Bq	25
Radium-226	20,000 Bq (540,000 pCi) vs. 70,000 Bq	25
Strontium-90	0.1 million Bq (2.7 million pCi) vs. 1.0 million Bq	25
Thorium-232	40 Bq (1000 pCi) vs. 30,000 Bq	25
Uranium-235	2000 Bq (54,000 pCi) vs. 500,000 Bq	25
Total intake from all sources; Canada		
Lead-210	<4000 Bq (<108,000 pCi)	26, 27
Polonium-210	<10,000 Bq (<270,000 pCi)	26, 27

Table 32.28 (continued) Recommended Radiological Criteria for the Protection of Human Health

Criterion and Other Variables	Concentration or Dose	Reference[a]
Radium-226	<20,000 Bq (<540,000 pCi)	26, 27
Thorium-228	<50,000 Bq (<1.35 million pCi)	26, 27
Thorium-230	<40,000 Bq (<1.08 million pCi)	26, 27
Thorium-232	<7000 Bq (<189,000 pCi)	26, 27
Uranium-238	<80,000 Bq (<2.1 million pCi)	26, 27
Effective dose[b]		
Average annual	20 mSv (2 rem), not to exceed 50 mSv (5 rem)	28
5-year maximum	<100 mSv (<10 rem), not to exceed 50 mSv (5 rem) in any year	28
Maximum permissible dose		
Whole body	50 mSv (5 rem) in any one year	21, 22, 29
Long-term accumulation to age N years	(N–18) × 50 mSv (5 rem)	21
Skin	150 mSv (15 rem) in any one year	21
Hands	750 mSv (75 rem) in any one year; not to exceed 250 mSv (25 rem) in 3 months	21
Forearms	300 mSv (30 rem) in any one year; not to exceed 100 mSv (10 rem) in 3 months	21
Skin and hands	500 mSv (50 rem) annually	19, 28
Other organs	150 mSv (15 rem) in any one year; not to exceed 50 mSv (5 rem) in 3 months	21
Pregnant women	5 mSv (0.5 rem) in gestation period	21
Eye lens	150 mSv (15 rem) annually	19, 28
SOIL		
Radium-226; maximum allowed	<185 Bq (<5000 pCi)/kg over background in top 15 cm; <555 Bq (<15,000 pCi)/kg in soils at depth >15 cm	3
Total gamma; maximum allowed	<0.2 μSv (<0.00002 rem)/h over background	3

[a] **1,** Gangopadhyay and Majumdar 1990; **2,** Oge and Dickson 1990; **3,** United States Code of Federal Regulations (USCFR) 1990; **4,** Wood 1991; **5,** Bair et al. 1979; **6,** Battiston et al. 1991; **7,** Andersson et al. 1990; **8,** Allaye-Chan et al. 1990; **9,** Joshi 1991; **10,** Hakanson and Andersson 1992; **11,** Harrison 1991; **12,** Gilbert et al. 1989; **13,** Aii et al. 1990; **14,** Johanson et al. 1989; **15,** Johanson 1990; **16,** Moss et al. 1989; **17,** Crout et al. 1991; **18,** Rose et al. 1990; **19,** International Commission on Radiological Protection (ICRP) 1991a; **20,** Lowe and Horrill 1991; **21,** Hobbs and McClellan 1986; **22,** ICRP 1977; **23,** Gray et al. 1989; **24,** USCFR 1991; **25,** Kiefer 1990; **26,** Clulow et al. 1991; **27,** Clulow et al. 1992; **28,** ICRP 1991b; **29,** National Council on Radiation Protection and Measurements (NCRP) 1991.

[b] The Annual Limit of Intake (ALI) for any radionuclide is obtained by dividing the annual average effective dose limit (20 mSv) by the committed effective dose (E) resulting from the intake of 1 Bq of that radionuclide. ALI data for individual radionuclides are given in ICRP (1991b).

32.9 SUMMARY

This chapter is a selective review and synthesis of the voluminous technical literature on radiation and radionuclides in the environment, notably on fish, wildlife, invertebrates, and other natural resources. The subtopics include the physical and biological properties of the electromagnetic spectrum and of charged particles; radiation sources and uses; concentrations of radionuclides in field collections of abiotic materials and living organisms; lethal and sublethal effects, including effects on survival, growth, reproduction, behavior, metabolism, carcinogenicity, and mutagenicity; a synopsis of two case histories involving massive releases of radionuclides into the biosphere (military weapons tests at the Pacific Proving Grounds, and the Chernobyl nuclear reactor accident); current radiological criteria proposed for the protection of human health and natural resources; and recommendations for additional research. A glossary is included.

Nuclear explosions and nuclear power production are the major sources of human radioactivity in the environment. Other sources include radionuclide use in medicine, industry, agriculture, education, and production; transport and disposal from these activities present opportunities for wastes to enter the environment. Dispersion of radioactive materials is governed by a variety of biogeochemical factors, including winds, water currents, and biological vectors. Living organisms normally receive most of their external exposure to radiation from naturally occurring electromagnetic waves and their internal exposure from naturally occurring radionuclides such as potassium-40. Radiation exposure doses from natural sources of radiation are significantly modified by altitude, amount and type of radionuclides in the immediate vicinity, and route of exposure.

Radionuclide concentrations in representative field collections of biota tend to be elevated in the vicinity of nuclear fuel reprocessing, nuclear power production, and nuclear waste facilities; in locations receiving radioactive fallout from nuclear accidents and atmospheric nuclear tests; and near sites of repeated nuclear detonations. Radionuclide concentrations in field collections of living organisms vary significantly with organism age, size, sex, tissue, diet, and metabolism; season of collection; proximity to point source; and other biological, chemical, and physical variables. To date, no extinction of any animal population has been linked to high background concentrations of radioactivity.

The accident at the Chernobyl, Ukraine, nuclear reactor on April 26, 1986, contaminated much of the northern hemisphere, especially Europe, by releasing large amounts of radiocesium-137 and other radionuclides into the environment. In the immediate vicinity of Chernobyl; at least 30 people died, more than 115,000 others were evacuated, and the consumption of locally produced milk and other foods was banned because of radiocontamination. The most sensitive local ecosystems were the soil fauna and pine forest communities. Elsewhere, fallout from Chernobyl measurably contaminated freshwater, marine, and terrestrial ecosystems, including flesh and milk of domestic livestock. Reindeer (*Rangifer tarandus*) calves in Norway showed an increasing frequency of chromosomal aberrations that seemed to correlate with cesium-137 tissue concentrations; tissue concentrations, in turn, were related to cesium-137 in lichens, an efficient absorber of airborne particles containing radiocesium and the main food source of reindeer during winter. A pattern similar to that of reindeer was documented in moose (*Alces*) in Scandinavia.

A dose- and dose-rate dependent radiation effect on growth, survival, organ development, mutagenicity, fatal neoplasms, and other parameters exists for almost all organisms tested under laboratory conditions. Some discoveries suggest that low acute exposures of ionizing radiation may extend the life-span of certain species, although adverse genetic effects may occur under these conditions. In living organisms, the sensitivity to radiation is governed by ontogeny and phylogeny. Thus, rapidly dividing cells, characteristic of embryos and fetuses, are most radiosensitive and evolutionarily advanced organisms such as mammals are more radiosensitive than primitive organisms. Between species within each taxonomic grouping are large variations in sensitivity to acute and chronic exposures of ionizing radiation and in ability to retain selected radionuclides; these processes are modified by numerous biological and abiotic variables.

Radiosensitive terrestrial plants are adversely affected at single exposures of 0.5 to 1.0 Gy and at chronic daily exposures of 0.2 to 0.65 Gy. Terrestrial insects are comparatively resistant to ionizing radiation; some species show growth stimulation and development at acute doses of 2 Gy — a demonstrably harmful dose for many species of vertebrates. Among aquatic organisms, the developing eggs and young of freshwater fish are among the most sensitive tested organisms; death was observed at acute doses of 0.3 to 0.6 Gy and adverse effects on physiology and metabolism at chronic daily exposure rates of 0.01 Gy. The ability of aquatic organisms to concentrate radionuclides from the medium varies substantially with ecosystem, trophic level, radionuclide, proximity to radiation point source, and many other biological, chemical, and physical modifiers. In amphibians, radiation adversely affects limb regeneration, alters DNA metabolism, causes sterility, and increases the frequency of chromosomal aberrations. Mortality patterns in some species of amphibians begin to stabilize about 200 days after exposure to a single acute dose of ionizing radiation and cannot be evaluated satisfactorily in the typical 30-day postexposure period. In birds, adverse effects on growth were noted at chronic daily exposures as low as 0.9 to 1.0 Gy, and on survival and metabolism at single exposures to 2.1 Gy. Genotoxic effects were associated with whole-body loadings of 2520 becquerels (Bq) of cesium-137/kg in mallards (*Anas platyrhynchos*). The radionuclide retention in birds was modified by sex, season, and reproductive state. In mammals, embryos and fetuses of sensitive species were adversely affected at acute doses of 0.011 to 0.022 Gy. Humans exposed as fetuses to 0.18 to 0.55 Gy scored significantly lower on tests of intelligence.

No radiological criteria now exist for the protection of fish, wildlife, or other sensitive natural resources. All current guides for protection from radiation target human health and are predicated on the assumption that protection of comparatively radiosensitive humans confers a high degree of protection to other life forms. Most authorities agree that significant harmful effects to humans occur under the following conditions:

- Exposure of the whole body of women during the first 2 months of pregnancy to >5 millisieverts (mSv)
- Exposure of the whole body to >50 mSv in any single year or to >2000 mSv in a lifetime
- Annual inhalation intake by a 60-kg individual, in Bq/kg body weight (BW), of more than 0.7 of thorium-232, 3.3 of americium-241, 3.3 of plutonium-239, 16 of californium-252, 33 of uranium-235, 1670 of strontium-90, 16,670 of cobalt-60, or 166,670 of phosphorus-32
- Annual ingestion intake by a 60-kg individual, in Bq/kg BW, of more than 3330 of iodine-129, 16,670 of iodine-125, 16,670 of iodine-131, or 66,670 of cesium-137
- Total annual intake, in Bq/kg BW, from all sources by a 60-kg person exceeds 66 of lead-210, 166 of polonium-210, 333 of radium-226, 670 of thorium-230, 830 of thorium-228, or 1330 of uranium-238.

Current risk assessments of ionizing radiation hazards to living organisms require additional data and reinterpretation of existing data. Specifically, more effort seems needed in eight areas:

1. Establishing a baseline for studies on radiation through measurement of naturally occurring radionuclides and natural background radiation doses
2. Refining radionuclide food-chain transfer models
3. Measuring the role of chemical agents in radiation-induced carcinogenesis
4. Accelerating research on radiation-induced lethal mutations
5. Initiating long-term studies to establish sensitive indicators of radiation stress on individuals and ecosystems
6. Clarifying the role of enzymes and proteins in repair of radiation-damaged cellular DNA
7. Reinterpreting embryotoxic effects of low level chronic irradiation
8. Resolving the mathematical shapes of radiation dose-response curves.

32.10 LITERATURE CITED

Aarkrog, A. 1990. Environmental radiation and radiation releases. *Inter. Jour. Radiation Biol.* 57:619-631.

Abrahamson, S. 1990. Risk estimates: past, present, and future. *Health Physics* 59:99-102.

Ahman, B., S. Forberg, and G. Ahman. 1990a. Zeolite and bentonite as caesium binders in reindeer feed. *Rangifer,* Spec. Issue No. 3:73-82.

Ahman, G., B. Ahman, and A. Rydberg. 1990b. Consequences of the Chernobyl accident for reindeer husbandry in Sweden. *Rangifer,* Spec. Issue No. 3:83-88.

Aii, T., S. Kume, S. Takahashi, M. Kurihara, and T. Mitsuhashi, 1990. The effect of the radionuclides from Chernobyl on iodine-131 and cesium-137 contents in milk and pastures in south-western Japan. *Japan. Soc. Zootech. Sci.* 61:47-53.

Allaye-Chan, A.C., R.G. White, D.F. Holleman, and D.E. Russell. 1990. Seasonal concentrations of cesium-137 in rumen content, skeletal muscles and feces of caribou from the porcupine herd: lichen ingestion rates and implications for human consumption. *Rangifer,* Spec. Issue No. 3:17-24.

Al-Izzi, M.A.J., S.K. Al-Maliky, and N.F. Jabbo. 1990. Effect of gamma rays on males of *Ectomylelois ceratoniae* Zeller (Lepidoptera: Pyralidae) irradiated as pupae or adults. *Ann. Soc. Entomologique France* 26:65-69.

Ananthaswamy, H.N. and W.E. Pierceall. 1990. Molecular mechanisms of ultraviolet radiation carcinogenesis. *Photochem. Photobiol.* 52:1119-1136.

Anderson, D.E. and M. Badzioch. 1991. Association between solar radiation and ocular squamous cell carcinoma in cattle. *Amer. Jour. Veterin. Res.* 52:784-788.

Anderson, L.E. 1990. Biological effects of extremely low-frequency and 60 Hz fields. Pages 196-235 in O.P. Ghandi (ed.). *Biological Effects and Medical Applications of Electromagnetic Energy.* Prentice Hall, Englewood Cliffs, NJ.

Anderson, S.H., G.J. Dodson, and R.I. Van Hook. 1976. Comparative retention of ^{60}Co, ^{109}Cd and ^{137}Cs following acute and chronic feeding in bobwhite quail. Pages 321-324 in C.E. Cushing (ed.). *Radioecology and Energy Resources. Proceedings of the Fourth National Symposium on Radioecology,* 12–14 May 1975, Oregon State University, Corvallis, OR. Ecol. Soc. Amer., Spec. Publ. No. 1.

Anderson, S.L., F.L. Harrison, G. Chan, and D.H. Moore II. 1990. Comparison of cellular and whole-animal bioassays for estimation of radiation effects in the polychaete worm *Neanthes arenaceodentata* (Polychaeta). *Arch. Environ. Contam. Toxicol.* 19:164-174.

Anderson, S.S., F.R. Livens, and D.L. Singleton. 1990. Radionuclides in grey seals. *Mar. Pollut. Bull.* 21:343-345.

Andersson, I., B. Teglof, and K. Elwinger. 1990. Transfer of ^{137}Cs from grain to eggs and meat of laying hens and meat of broiler chickens, and the effect of feeding bentonite. *Swedish Jour. Agricul. Res.* 20:35-42.

Ankley, G.T., J.E. Tietge, D.L. DeFoe, K.M. Jensen, G.W. Holcombe, E.L. Durhan, and S.A. Diamond. 1998. Effects of ultraviolet light and methoprene on survival and development of *Rana pipiens*. *Environ. Toxicol. Chem.* 17:2530-2542.

Anspaugh, L.R., R.J. Catlin, and M. Goldman. 1988. The global impact of the Chernobyl reactor accident. *Science* 242:1513-1519.

Assimakopoulos, P.A., K.G. Ioannides, and A.A. Pakou. 1989. The propagation of the Chernobyl ^{131}I impulse through the air-grass-animal-milk pathway in northwestern Greece. *Sci. Total Environ.* 85:295-305.

Baeza, A., M. del Rio, C. Miro, A. Moreno, E. Navarro, J.M. Paniagua, and M.A. Peris. 1991. Radiocesium and radiostrontium levels in song-thrushes (*Turdus philomelos*) captured in two regions of Spain. *Jour. Environ. Radioactiv.* 13:13-23.

Bagshaw, C. and I.L. Brisbin, Jr. 1984. Long-term declines in radiocesium of two sympatric snake populations. *Jour. Appl. Ecol.* 21:407-413.

Bai, D.H. 1969. Radioactive contamination of the atmosphere, soils, and agricultural products in Korea, 1961–1966. Pages 587-596 in D.J. Nelson and F.C. Evans (eds.). *Symposium on Radioecology. Proceedings of the Second National Symposium.* Available as CONF-670503 from the Clearinghouse for Federal Scientific and Technical Information, National Bureau of Standards, Springfield, VA 22151.

Bair, W.J., J.W. Healy, and B.W. Wachholz. 1979. *The Enewetak Atoll Today.* U.S. Dept. Energy, Washington, D.C. 25 pp.

Baker, R.J., M.J. Hamilton, R.A. Van Den Bussche, L.E. Wiggins, D.W. Sugg, M.H. Smith, M.D. Lomakin, S.P. Gaschak, E.G. Bundova, G.A. Rudenskaya, and R.K. Chesser. 1996. Small mammals from the most radioactive sites near the Chernobyl nuclear power plant. *Jour. Mammal.* 77:155-170.

Barendsen, G.W. 1990. Mechanisms of cell reproductive death and shapes of radiation dose-survival curves of mammalian cells. *Inter. Jour. Radiation Biol.* 57:885-896.

Bastian, R.K. and W.B. Jackson. 1976. ^{137}Cs and ^{60}Co in a terrestrial community at Enewetak Atoll. Pages 314-320 in C.E. Cushing (ed.). *Radioecology and Energy Resources. Proceedings of the Fourth National Symposium on Radioecology,* 12–14 May 1975, Oregon State Univ., Corvallis, OR. Ecol. Soc. Amer., Spec. Publ. No. 1.

Battiston, G.A., S. Degetto, R. Gerbasi, G. Sbrignadello, R. Parigi-Bini, G. Xiccato, and M. Cinetto. 1991. Transfer of Chernobyl fallout radionuclides feed to growing rabbits: cesium-137 balance. *Sci. Total Environ.* 105:1-12.

Becker, C.D. 1990. *Aquatic Bioenvironmental Studies: The Hanford Experience 1944–84. Studies in Environmental Science 39.* Elsevier Science Publishers, New York. 306 pp.

Beach, H. 1990. Coping with the Chernobyl disaster: a comparison of social effects in two reindeer-herding areas. *Rangifer,* Spec. Iss. No. 3:25-34.

Belli, M., A. Drigo, S. Menegon, A. Menin, P. Nazzi, U. Sansone, and M. Toppano. 1989. Transfer of Chernobyl fall-out caesium radioisotopes in the cow food chain. *Sci. Total Environ.* 85:169-177.

Benjamin, S.A., G.M. Angleton, A.C. Lee, W.J. Saunders, and J.S. Williams. 1990. Health effects from whole-body irradiation during development in beagles: studies at Colorado State University. *Radiation Res.* 124:366-368.

Beresford, N.A. 1989. The transfer of Ag-110m to sheep tissues. *Sci. Total Environ.* 85:81-90.

Billen, D. 1990. Spontaneous DNA damage and its significance for the "negligible dose" controversy in radiation protection. *Radiation Res.* 124:242-245.

Blaustein, A.R., P.D. Hoffman, D.G. Hokit, J.M. Kiesecker, S.C. Walls, and J.B. Hays. 1994. UV repair and resistance to solar UV-B in amphibian eggs: a link to population declines? *Proc. Natl. Acad. Sci. U.S.A.* 91:1791-1795.

Blaustein, A.R., P.D. Hoffman, J.M. Kiesecker, and J.B. Hays. 1996. DNA repair activity and resistance to solar UV-B radiation in eggs of the red-legged frog. *Conserv. Biol.* 10:1398-1402.

Boag, T.S., G.I. Johnson, M. Izard, C. Murray, and K.C. Fitzsimmons. 1990. Physiological responses of mangoes cv. Kensington Pride to gamma irradiation treatment as affected by fruit maturity and ripeness. *Ann. Appl. Biol.* 116:177-187.

Bonham, K. and A.D. Welander. 1963. Increase in radioresistance of fish to lethal doses with advancing embryonic development. Pages 353-358 in V. Schultz and A.W. Klement, Jr. (eds.). *Radioecology.* Reinhold, New York.

Book, S.A. 1969. Fallout Cesium-137 Accumulation in Two Subpopulations of Black-Tailed Deer (*Odocoileus hemionus columbianus*). M.A. thesis, Univ. California, Berkeley. 61 pp.

Bothmer, S.V., K.J. Johanson, and R. Bergstrom. 1990. Cesium-137 in moose diet; considerations on intake and accumulation. *Sci. Total Environ.* 91:87-96.

Bowden, G.T., D. Jaffe, and K. Andrews. 1990. Biological and molecular aspects of radiation carcinogenesis in mouse skin. *Radiation Res.* 121:235-241.

Bowen, V.T., J.S. Olsen, C.L. Osterberg, and J. Ravera. 1971. Ecological interactions of marine radioactivity. Pages 200-222 in National Academy of Sciences. *Radioactivity in the Marine Environment.* Natl. Acad. Sci., Panel on Radioactivity in the Marine Environment. Washington, D.C.

Brisbin, I.L., Jr. 1991. Avian radioecology. Pages 69-140 in D.M. Power (ed.). *Current Ornithology, Vol. 8.* Plenum Press, New York.

Brisbin, I.L., Jr. D.D. Breshears, K.L. Brown, M. Ladd, M.H. Smith, M.W. Smith, and A.L. Towns. 1989. Relationships between levels of radiocaesium in components of terrestrial and aquatic food webs of a contaminated streambed and floodplain community. *Jour. Appl. Ecol.* 26:173-182.

Brisbin, I.L., Jr. M.A. Staton, J.E. Pinder, III, and R.A. Geiger. 1974. Radiocesium concentrations of snakes from contaminated and non-contaminated habitats of the AEC Savannah River Plant. *Copeia* 1974 (2):501-506.

Brittain, J.E., A. Storruste, and E. Larsen. 1991. Radiocesium in brown trout (*Salmo trutta*) from a subalpine lake ecosystem after the Chernobyl reactor accident. *Jour. Environ. Radioactivity* 14:181-191.

Brown, B. 1966. Long-Term Radiation Damage Evaluation of Life-Span Studies. Memoran. RM-5083-TAB. Rand Corp., Santa Monica, CA. 66 pp.

Bruenger, F.W., R.D. Lloyd, G.N. Taylor, S.C. Miller, and C.W. Mays. 1990. Kidney disease in beagles injected with polonium-210. *Radiation Res.* 122:241-251.

Bruggeman, D.J., J.A. Bantle, and C. Goad. 1998. Linking teratogenesis, growth, and DNA photodamage to artificial ultraviolet B radiation in *Xenopus laevis* larvae. *Environ. Toxicol. Chem.* 17:2114-2121.

Cadwell, L.L., R.G. Schreckhise, and R.E. Fitzner. 1979. Cesium-137 in coots (*Fulica americana*) on Hanford waste ponds: contribution to population dose and offsite transport estimates. Pages 485-491 in *Low-Level Radioactive Waste Management: Proceedings of Health Physics Society. Twelfth Midyear Topical Symposium.* February 11–15, 1979, Williamsburg, VA.

Calamosca, M., P. Pagano, F. Trent, L. Zaghini, G. Gentile, G. Tarroni, and L. Morandi. 1990. A modelistic approach to evaluate the factors affecting the ^{137}Cs transfer from mother to fetus in cattle. *Deutsche Tier. Wochen.* 97:452-456.

Canfi, A., R. Chayoth, S. Weill, and E. Bedrak. 1990. The reproductive system of male rats exposed to very low doses of ionizing radiation. 1. Hormonal profile of animals exposed after sexual maturity. *Andrologia* 22:129-136.

Carlson, L. and B. Erlandsson. 1991. Seasonal variation of radionuclides in *Fucus vesiculosus* L. from the Oresund, southern Sweden. *Environ. Pollut.* 73:53-70.

Cecil, L.D. and T.F. Gesell. 1992. Sampling and analysis for radon-222 dissolved in ground water and surface water. *Environ. Monitor. Assess.* 20:55-66.

Chakravarti, D. and R. Eisler. 1961. Strontium-90 and gross beta activity in the fat and nonfat fractions of the liver of the coconut crab (*Birgus latro*) collected at Rongelap Atoll during March 1958. *Pacific Sci.* 15:155-159.

Clark, M.J. and F.B. Smith. 1988. Wet and dry deposition of Chernobyl releases. *Nature* 332:245-249.

Cleary, S.F. 1990. Cellular effects of radiofrequency electromagnetic fields. Pages 339-356 in O.P. Ghandi (ed.). *Biological Effects and Medical Applications of Electromagnetic Energy.* Prentice Hall, Englewood Cliffs, NJ.

Clooth, G. and D.C. Aumann. 1990. Environmental transfer parameters and radiological impact of the Chernobyl fallout in and around Bonn (FRG). *Jour. Environ. Radioactivity* 12:97-119.

Clulow, F.V., T.P. Lim, N.K. Dave, and R. Avadhanula. 1992. Radium-226 levels and concentration ratios between water, vegetation, and tissues of ruffed grouse (*Bonasa umbellus*) from a watershed with uranium tailings near Elliot Lake, Canada. *Environ. Pollut.* 77:39-50.

Clulow, F.V., M.A. Mirka, N.K. Dave, and T.P. Lim. 1991. ^{226}Ra and other radionuclides in water, vegetation, and tissues of beavers (*Castor canadensis*) from a watershed containing U tailings near Elliot Lake, Canada. *Environ. Pollut.* 69:277-310.

Colwell, S.V., R.A. Kennamer, and I.L. Brisbin, Jr. 1996. Radiocesium patterns in wood duck eggs and nesting females in a contaminated reservoir. *Jour. Wildl. Manage.* 60:186-194.

Coggle, J.E. and J.P. Williams. 1990. Experimental studies of radiation carcinogenesis in the skin: a review. *Inter. Jour. Radiation Biol.* 57:797-808.

Conkic, L., M. Ivo, S. Lulic, K. Kosatic, J. Simor, P. Vancsura, J. Slivka, and I. Bikit. 1990. The impact of the Chernobyl accident on the radioactivity of the River Danube. *Water Sci. Technol.* 22:195-202.

Cothern, C.R., D.J. Crawford-Brown, and M.E. Wrenn. 1990. Application of environmental dose-response models to epidemiology and animal data for the effects of ionizing radiation. *Environ. Inter.* 16:127-140.

Coughtrey, P.J., J.A. Kirton, and N.G. Mitchell. 1989. Caesium transfer and cycling in upland pastures. *Sci. Total Environ.* 85:149-158.

Coulton, L.A. and A.T. Barker. 1991. The effect of low-frequency pulsed magnetic fields on chick embryonic growth. *Physics Medic. Biol.* 36:369-381.

Crete, M., M.A. Lefebvre, M.B. Cooper, H. Marshall, J.L. Benedetti, P.E. Carriere, and R. Nault. 1990. Contaminants in caribou tissues from northern Quebec. *Rangifer,* Spec. Iss. No. 3:289.

Cristaldi, M., E. D'Arcangelo, L.A. Ieradi, D. Mascanzoni, T. Mattei, and I.V.A. Castelli. 1990. ^{137}Cs determination and mutagenicity tests in wild *Mus musculus domesticus* before and after the Chernobyl accident. *Environ. Pollut.* 64:1-9.

Cristaldi, M., L.A. Ieradi, D. Mascanzoni, and T. Mattei. 1991. Environmental impact of the Chernobyl accident: mutagenesis in bank voles from Sweden. *Inter. Jour. Radiation Biol.* 59:31-40.

Cronkite, E.P., V.P. Bond, W.H. Chapman, and R.H. Lee. 1955. Biological effect of atomic bomb gamma radiation. *Science* 122:148-150.

Cross, F.T. 1990. Health effects and risks of radon exposure. Pages 223-237 in S.K. Majumdar, R.F. Schmalz, and E.W. Miller (eds.). *Environmental Radon: Occurrence, Control and Health Hazards.* Pennsyl. Acad. Science, Easton, PA.

Crout, N.M.J., N.A. Beresford, and B.J. Howard. 1991. The radioecological consequences for lowland pastures used to fatten upland sheep contaminated with radiocaesium. *Sci. Total Environ.* 103:73-87.

Crowley, M., P.I. Mitchell, J. O'Grady, J. Vires, J.A. Sanchez-Cabeza, A. Vidal-Quadras, and T.P. Ryan. 1990. Radiocaesium and plutonium concentrations in *Mytilus edulis* (L.) and potential dose implications for Irish critical groups. *Ocean Shoreline Manage.* 13:149-161.

Cummings, S.L., J.H. Jenkins, T.T. Fendley, L. Bankert, P.H. Bedrosian, and C.R. Porter. 1971. Cesium-137 in white-tailed deer as related to vegetation and soils of the southeastern United States. Pages 123-128 in D.J. Nelson (ed.). *Radionuclides in Ecosystems. Proceedings of the Third National Symposium on Radioecology,* May 10–12, 1971, Oak Ridge, TN. Vol. 1. Available from the National Technical Information Service, Springfield, VA 22151.

Curnow, R.D. 1971. Assimilation and retention of zinc-65 by the mallard duck (*Anas platyrhynchos*). Pages 381-384 in D.J. Nelson (ed.). *Radionuclides in Ecosystems. Proceedings of the Third National Symposium on Radioecology,* May 10–12, 1971, Oak Ridge, TN. Vol. 1. Available from the National Technical Information Service, Springfield, VA 22151.

Curtis, E.J.C., D.S. Popplewell, and G.J. Ham. 1991. Radioactivity in environmental samples taken in the Sellafield, Ravenglass and Morecambe Bay areas of west Cumbria. *Sci. Total Environ.* 105:211-231.

Cushing, C.E. (ed.). 1976. *Radioecology and Energy Resources. Proceedings of the Fourth National Symposium on Radioecology,* 12–14 May 1975, Oregon State Univ., Corvallis, OR. Ecol. Soc. Amer., Spec. Publ. No. 1. 401 pp.

Daburon, F., Y. Archimbaud, J. Cousi, G. Fayart, D. Hoffschir, I. Chevallereau, and H. Le Creff. 1991. Radiocaesium transfer to ewes fed contaminated hay after the Chernobyl accident: effect of vermiculite and AFCF (ammonium ferricyanoferrate) as countermeasures. *Jour. Environ. Radioactivity* 14:73-84.

Daburon, F., G. Fayart, and Y. Tricaud. 1989. Caesium and iodine metabolism in lactating cows under chronic administration. *Sci. Total Environ.* 85:253-261.

Dahlman, R.C. and P.V. Voris. 1976. Cycling of ^{137}Cs in soil and vegetation of a flood plain 30 years after initial contamination. Pages 291-298 in C.E. Cushing (ed.). *Radioecology and Energy Resources. Proceedings of the Fourth National Symposium on Radioecology,* 12–14 May 1975, Oregon State Univ., Corvallis, OR. Ecol. Soc. Amer., Spec. Publ. No. 1.

Davidson, S.F., S.K. Brantley, and S.K. Das. 1991. The effects of ultraviolet radiation on wound healing. *Brit. Jour. Plastic Surgery* 44:210-213.

Desmet, G.M., L.R. Van Loon, and B.J. Howard. 1991. Chemical speciation and bioavailability of elements in the environment and their relevance to radioecology. *Sci. Total Environ.* 100:105-124.

Di Gregorio, D., P.B. Dunaway, and G.E. Cosgrave. 1971. Effect of acute gamma irradiation on the reproduction of *Peromyscus leucopus.* Pages 1076-1081 in D.J. Nelson (ed.). Radionuclides in Ecosystems. *Proceedings of the Third National Symposium on Radioecology,* May 10–12, 1971, Oak Ridge, TN. Vol. 2. Available from the National Technical Information Service, Springfield, VA 22151.

Donaldson, L.R. and K. Bonham. 1970. Effects of chronic exposure of chinook salmon eggs and alevins to gamma irradiation. *Trans. Amer. Fish. Soc.* 99:112-119.

Donaldson, L.R. and R.F. Foster. 1957. Effects of radiation on aquatic organisms. Pages 96-102 in National Academy of Sciences. The Effects of Atomic Radiation on Oceanography and Fisheries. Publ. No. 551. Natl. Acad. Sci.-Natl. Res. Coun., Washington, D.C.

Douka, C.E. and A.C. Xenoulis. 1991. Radioactive isotope uptake in a grass-legume association. *Environ. Pollut.* 73:11-23.

Dunaway, P.B., L.L. Lewis, J.D. Story, J.A. Payne, and J.M. Inglis. 1969. Radiation effects in the soricidae, cricetidae, and muridae. Pages 173-184 in D.J. Nelson and F.C. Evans (eds.). *Symposium on Radioecology. Proceedings of the Second National Symposium.* Available as CONF-670503 from the Clearinghouse for Federal Scientific and Technical Information, National Bureau of Standards, Springfield, VA 22151.

Dunaway, P.B., J.D. Story, and J.T. Kitchings. 1971. Radiation effects and radionuclide excretion in a natural population of pine voles. Pages 1055-1064 in D.J. Nelson (ed.). *Proceedings of the Third National Symposium on Radioecology,* May 10–12, 1971, Oak Ridge, TN. Vol. 2. Available from the National Technical Information Service, Springfield, VA 22151.

Egami, M.I., C. Segreto, J. Kerbauy, and Y. Juliano. 1991. Effects of whole-body X-irradiation on the peripheral blood of primate *Cebus apella. Braz. Jour. Med. Biol. Res.* 24:271-274.

Eidus, L.K., Y.N. Korystov, O.R. Dobrovinskaja, and V.V. Shaposhnikova. 1990. The mechanism of radiation-induced interphase death of lymphoid cells: a new hypothesis. *Radiation Res.* 123:17-21.

Eikelmann, I.M.H., K. Bye, and H.D. Sletten. 1990. Seasonal variation of cesium 134 and cesium 137 in semidomestic reindeer in Norway after the Chernobyl accident. *Rangifer,* Spec. Iss. No. 3:35-38.

Eisler, R. 1961. Effects of visible radiation on salmonoid embryos and larvae. *Growth* 25:281-346.

Eisler, R. 1994. Radiation Hazards to Fish, Wildlife, and Invertebrates: A Synoptic Review. U.S. Natl. Biol. Serv. Biol. Rep. 26. 124 pp.

Eisler, R. 1995. Ecological and toxicological aspects of the partial meltdown of the Chernobyl nuclear plant reactor. Pages 549-564 in D.J. Hoffman, B.A. Rattner, G.A. Burton, Jr., and A.J. Cairns, Jr. (eds.). *Handbook of Ecotoxicology.* Lewis Publ., Boca Raton, FL.

Ekker, M., B.M. Jenssen, and K. Zahlsen. 1990. Radiocesium in Norwegian carnivores following the Chernobyl fallout. Page 275 in S. Myrberget (ed.). *Volume 1: Population Dynamics. Trans. XIXth IUGB Congress.* Inter. Union Game Biol. September 1989. Trondheim, Norway.

Emery, R.M., D.C. Klopfer, T.R. Garland, and W.C. Weimer. 1976. The ecological behavior of plutonium and americium in a freshwater pond. Pages 74-85 in C.E. Cushing (ed.). *Radioecology and Energy Resources. Proceedings of the Fourth National Symposium on Radioecology,* 12–14 May 1975, Oregon State Univ., Corvallis, OR. Ecol. Soc. Amer., Spec. Publ. No. 1.

Engel, D.W. and E.M. Davis. 1976. The effects of gamma radiation on the survival and growth of brine shrimp, *Artemia salina.* Pages 376-380 in C.E. Cushing (ed.). *Radioecology and Energy Resources. Proceedings of the Fourth National Symposium on Radioecology,* 12–14 May 1975, Oregon State Univ., Corvallis, OR. Ecol. Soc. Amer., Spec. Publ. No. 1.

Eriksson, O. 1990. ^{137}Cs in reindeer forage plants 1986-1988. *Rangifer,* Spec. Iss. No. 3:11-14.

Evans, H.H. 1990. Ionizing radiation mutagenesis in mammalian cells. Pages 211-219 in *UCLA Symposia Colloquium. Ionizing Radiation Damage to DNA: Molecular Aspects.* Wiley-Liss, New York.

Fanton, J.W. and J.G. Golden. 1991. Radiation-induced endometriosis in *Macaca mulatta. Radiation Res.* 126:141-146.

Farris, G.C., F.W. Whicker, and A.H. Dahl. 1969. Strontium-90 levels in mule deer and forage plants. Pages 602-608 in D.V. Nelson and F.C. Evans (eds.). *Symposium on Radioecology. Proceedings of the Second National Symposium.* Available as CONF-670503 from The Clearinghouse for Federal Scientific and Technical Information, Natl. Bur. Standards, Springfield, VA 22151.

Folsom, T.R. and J.H. Harley. 1957. Comparisons of some natural radiations received by selected organisms. Pages 28-33 in National Academy of Sciences. The Effects of Atomic Radiation on Oceanography and Fisheries. Publ. No. 551, Natl. Acad. Sci.-Natl. Res. Coun., Washington, D.C.

Folsom, T.R., D.R. Young, V.F. Hodge, and R. Grismore. 1971. Variations of ^{54}Mn, ^{60}Co, ^{65}Zn, ^{110m}Ag, and ^{108m}Ag in tunas. Pages 721-730 in D.J. Nelson (ed.). *Radionuclides in Ecosystems. Proceedings of the Third National Symposium on Radioecology.* May 10–12, 1971, Oak Ridge, TN. Vol. 2. Available from Natl. Tech. Inform. Serv., Springfield, VA 22151.

Fowler, S.W., P. Buat-Menard, Y. Yokoyama, S. Ballestra, E. Holm, and H. Van Nguyen. 1987. Rapid removal of Chernobyl fallout from Mediterranean surface waters by biological activity. *Nature* 329:56-58.

Fry, R.J.M. 1991. Radiation carcinogenesis in the whole-body system. *Radiation Res.* 126:157-161.

Gangopadhyay, A. and S.K. Majumdar. 1990. Radon and health consequences: an overview. Pages 209-237 in S.K. Majumdar, R.F. Schmalz, and E.W. Miller (eds.). *Environmental Radon: Occurrence, Control and Health Hazards.* Pennsylvania Acad. Sci., Easton, PA.

Garaj-Vrhovac, V., D. Horvat, and Z. Koren. 1990. The effect of microwave radiation on the cell genome. *Mutat. Res.* 243:87-93.

Garnier, J. and J.P. Baudin. 1990. Retention of ingested ^{110m}Ag by a freshwater fish, *Salmo trutta* L. *Water Air Soil Pollut.* 50:409-421.

Garnier, J., J.P. Baudin, and L. Folquier. 1990. Accumulation from water and depuration of ^{110m}Ag by a freshwater fish, *Salmo trutta* L. *Water Res.* 24:1407-1414.

Gaubin, Y., M. Depoux, B. Pianezzi, G. Gasset, C. Heilmann, and H. Planel. 1990. Investigations on the effects of cosmic rays on *Artemia* cysts and tobacco seeds; results of Exobloc II experiment, flown aboard Biocosmos 1887. *Inter. Jour. Radiation Applic. Instrument.* 17:133-144.

George, L.S., C.E. Dallas, I.L. Brisbin, Jr., and D.L. Evans. 1991. Flow cytometric DNA analysis of ducks accumulating ^{137}Cs on a reactor reservoir. *Ecotoxicol. Environ. Safety* 21:337-347.

Geraci, J.P., M.S. Mariano, and K.L. Jackson. 1991. Hepatic radiation injury in the rat. *Radiation Res.* 124:65-72.

Ghandi, O.P. 1990. Electromagnetic energy absorption in humans and animals. Pages 174-194 in O.P. Ghandi (ed.). *Biological Effects and Medical Applications of Electromagnetic Energy.* Prentice Hall, Englewood Cliffs, NJ.

Giannetti, M., A. Trux, B. Giannetti, and Z. Zubrzycki. 1990. *Xenopus laevis* South African clawed toad — a potential indicator for radioactive contamination in ecological systems? Pages 279-285 in *First International Symposium on Biology and Physiology of Amphibians.* Karlsruhe, Federal Republic of Germany. 31 August–3 September, 1988. Gustav Fischer, Stuttgart.

Gilbert, R.O., D.W. Engel, and L.R. Anspaugh. 1989. Transfer of aged $^{239+240}Pu$, ^{238}Pu, ^{241}Am, and ^{137}Cs to cattle grazing a contaminated arid environment. *Sci. Total Environ.* 85:53-62.

Glasstone, S. 1958. *Sourcebook on Atomic Energy. Second Edition.* Van Nostrand, Princeton, NJ. 641 pp.

Godoy, J.M., J.R.D. Guimaraes, J.C.A. Pereira, and M.A. Pieres do Rio. 1991. Cesium-137 in the Goiania waterways during and after the radiological accident. *Health Phys.* 60:99-103.

Golley, F.B. and J.B. Gentry. 1969. Response of rodents to acute gamma radiation under field conditions. Pages 166-172 in D.J. Nelson and F.C. Evans (eds.). *Symposium on Radioecology. Proceedings of the Second National Symposium.* Available as CONF-670503 from The Clearinghouse for Federal Scientific and Technical Information, Natl. Bur. Standards, Springfield, VA 22151.

Gorbman, A. and M.S. James. 1963. An exploratory study of radiation damage in the thyroids of coral reef fishes from the Eniwetok Atoll. Pages 385-399 in V. Schultz and A.W. Klement, Jr. (eds.). *Radioecology.* Reinhold, New York.

Grant, K.P. and L.E. Licht. 1995. Effects of ultraviolet radiation on life-history stages of anurans from Ontario, Canada. *Canad. Jour. Zool.* 73:2292-2301.

Gray, R.H., R.E. Jaquish, P.J. Mitchell, and W.H. Rickard. 1989. Environmental monitoring at Hanford, Washington, USA: a brief site history and summary of recent results. *Environ. Manage.* 13:563-572.

Grzybowska, D. 1989. Concentration of ^{137}Cs and ^{90}Sr in marine fish from the southern Baltic Sea. *Acta Hydrobiol.* 31:139-147.

Guary, J.C. and S.W. Fowler. 1990. Experimental study of the transfer of transuranium nuclides in marine decapod crustaceans. *Mar. Ecol. Prog. Ser.* 60:253-270.

Gupta, M.L. and P. Umadevi. 1990. Response of reptilian liver to external gamma irradiation. *Radiobiol. Radiother.* 31:285-288.

Gustafson, P.F. 1969. Cesium-137 in freshwater fish during 1954-1965. Pages 249-257 in D.J. Nelson and F.C. Evans (eds.). *Symposium on Radioecology. Proceedings of the Second National Symposium.* Available as CONF-670503 from The Clearinghouse for Federal Scientific and Technical Information, Natl. Bur. Standards, Springfield, VA 22151.

Hader, D.P. and M. Hader. 1991. Effects of solar and artificial U.V. radiation on motility and pigmentation in the marine *Cryptomonas maculata. Environ. Exper. Botany* 31:33-41.

Hader, D.P., S.M. Liu, M. Hader, and W. Ullrich. 1990. Photoorientation, motility and pigmentation in a freshwater *Peridinium* affected by ultraviolet radiation. *Gen. Physiol. Biophys.* 9:361-371.

Hagen, U. 1990. Molecular radiation biology: future aspects. *Radiation Environ. Biophys.* 29:315-322.

Hakanson, L. and T. Andersson. 1992. Remedial measures against radioactive caesium in Swedish lake fish after Chernobyl. *Aquat. Sci.* 54:141-164.

Halford, D.K., O.D. Markham, and G.C. White. 1983. Biological elimination rates of radioisotopes by mallards contaminated at a liquid radioactive waste disposal area. *Health Phys.* 45:745-756.

Ham, G.J., J.D. Harrison, D.S. Popplewell, and E.J.C. Curtis. 1989. The distribution of ^{137}Cs, plutonium and americium in sheep. *Sci. Total Environ.* 85:235-244.

Handl, J. and A. Pfau. 1989. Long-term transfer of I-129 into the food chain. *Sci. Total Environ.* 85:245-252.

Handl, J., A. Pfau, and F.W. Huth. 1990. Measurements of ^{129}I into the food chain. *Health Phys.* 58:609-618.

Hanson, L.H. 1990. Sterilizing effects of cobalt-60 and cesium-137 radiation on male sea lampreys. *North Amer. Jour. Fish. Manage.* 10:352-361.

Hanson, W.C. 1976. Studies of transuranic elements in Arctic ecosystems. Pages 28-39 in C.E. Cushing (ed.). *Radioecology and Energy Resources. Proceedings of the Fourth National Symposium on Radioecology.* 12–14 May 1975, Oregon State Univ., Corvallis, OR. Ecol. Soc. Amer., Spec. Publ. No. 1.

Harrison, F.L., K.M. Wong, and R.E. Heft. 1976. Role of solubles and particulates in radionuclide accumulation in the oyster *Crassostrea gigas* in the discharge canal of a nuclear power plant. Pages 9-20 in C.E. Cushing (ed.). *Radioecology and Energy Resources. Proceedings of the Fourth National Symposium on Radioecology,* 12–14 May 1975, Oregon State Univ, Corvallis, OR. Ecol. Soc. Amer., Spec. Publ. No. 1.

Harrison, J.D. 1991. The gastrointestinal absorption of the actinide elements. *Sci. Total Environ.* 100:43-60.

Harrison, J.D., J.W. Haines, and D.S. Popplewell. 1990. The gastrointestinal absorption and retention of niobium in adult and newborn guinea pigs. *Inter. Jour Radiation Biol.* 58:177-186.

Hatfield, T.W., D.M. Skauen, and J.S. Rankin, Jr. 1963. Gross beta radioactivity in marine organisms. Pages 141-143 in V. Schultz and A.W. Klement, Jr. (eds.). *Radioecology.* Reinhold, New York.

Havlik, B. and E. Robertson. 1971. Radium uptake by freshwater algae. Pages 372-380 in D.J. Nelson (ed.). *Radionuclides in Ecosystems. Proceedings of the Third National Symposium on Radioecology.* May 10–12, 1971, Oak Ridge, TN. Vol. 1. Available from Natl. Tech. Infor. Serv., Springfield, VA 22151.

Hetherington, J.A., D.F. Jefferies, N.T. Mitchell, R.J. Pentreath, and D.S. Woodhead. 1976. Environmental and public health consequences of the controlled disposal of transuranic elements to the marine environment. Pages 139-154 in *Transuranium Nuclides in the Environment.* IAEA-SM-199/11, Inter. Atom. Ener. Agen., Vienna.

Hinton, T.G. and D.E. Scott. 1990. Radioecological techniques for herpetology, with an emphasis on freshwater turtles. Pages 267-287 in J.W. Gibbons (ed.). *Life History and Ecology of the Slider Turtle.* Smithsonian Instit. Press, Washington, D.C.

Hobbs, C.H. and R.O. McClellan. 1986. Toxic effects of radiation and radioactive materials. Pages 669-705 in C.D. Klaassen, M.O. Amdur, and J. Doull (eds.). *Casarett and Doull's Toxicology. Third Edition.* Macmillan, New York.

Holcomb, C.M., C.G. Jackson, Jr., M.M. Jackson, and S. Kleinbergs. 1971. Occurrence of radionuclides in the exoskeleton of turtles. Pages 385-389 in D.J. Nelson (ed.). *Radionuclides in Ecosystems. Proceedings of the Third National Symposium on Radioecology.* May 10–12, 1971, Oak Ridge, TN. Vol. 1. Available from Natl. Tech. Infor. Serv., Springfield, VA 22151.

Holleman, D.F., R.G. White, and A.C. Allaye-Chan. 1990. Modelling of radiocesium transfer in the lichen-reindeer/caribou-wolf food chain. *Rangifer,* Spec. Iss. No. 3:39-42.

Holtzman, R.B. 1969. Concentrations of the naturally occurring radionuclides ^{226}Ra, ^{210}Pb and ^{210}Po in aquatic fauna. Pages 535-546 in D.J. Nelson and F.C. Evans (eds.). *Symposium on Radioecology. Proceedings of the Second National Symposium.* Available as CONF-670503 from The Clearinghouse for Federal Scientific and Technical Information, Natl. Bur. Standards, Springfield, VA 22151.

Hopewell, J.W. 1990. The skin: its structure and response to ionizing radiation. *Inter. Jour. Radiation Biol.* 57:751-774.

Hove, K., O. Pederson, T.H. Garmo, H.S. Hansen, and H. Staaland. 1990a. Fungi: a major source of radiocesium contamination of grazing ruminants in Norway. *Health Phys.* 59:189-192.

Hove, K., H. Staaland, O. Pedersen, and H.D. Sletten. 1990b. Effect of Prussian blue (ammonium-iron-hexacyanoferrate) in reducing the accumulation of radiocesium in reindeer. *Rangifer,* Spec. Iss. No. 3:43.

Imanaka, T. and H. Koide. 1990. Radiocesium concentration in milk after the Chernobyl accident in Japan. *Jour. Radioanal. Nucl. Chem.* 145:151-158.

International Atomic Energy Agency (IAEA). 1976. Effects of Ionizing Radiation on Aquatic Organisms and Ecosystems. Technical Reports Series No. 172. IAEA, Vienna, Austria. 131 pp.

International Atomic Energy Agency (IAEA). 1992. Effects of Ionizing Radiation on Plants and Animals at Levels Implied by Current Radiation Protection Standards. Technical Reports Series No. 332. IAEA, Vienna, Austria. 74 pp.

International Commission on Radiological Protection (ICRP). 1977. Recommendations of the international commission on radiological protection. ICRP Publ. 26. *Ann. ICRP* 1(3):1-53.

International Commission on Radiological Protection (ICRP). 1991a. 1990 recommendations of the international commission on radiological protection. ICRP Publ. 60. *Ann. ICRP* 21(1-3):1-201.

International Commission on Radiological Protection (ICRP). 1991b. Annual limits on intake of radionuclides by workers based on the 1990 recommendations. ICRP Publ. 61. *Ann. ICRP* 21(4):1-41.

Ioannides, K.G. and A.A. Pakou. 1991. Radioiodine retention in ovine thyroids in northwestern Greece following the reactor accident at Chernobyl. *Health Phys.* 60:517-521.

Iverson, S.L. and B.N. Turnea. 1976. Effects of acute irradiation on survival of captive and free-ranging meadow voles. Pages 359-362 in C.E. Cushing (ed.). *Radioecology and Energy Resources. Proceedings of the Fourth National Symposium on Radioecology,* 12–14 May 1975, Oregon State Univ., Corvallis, OR. Ecol. Soc. Amer., Spec. Publ. No. 1.

Jaenke, R.S. and G.M. Angleton. 1990. Perinatal radiation-induced damage in the beagle. *Radiation Res.* 122:58-65.

Jandl, J., H. Prochazka, and D. Luks. 1991. The biological half-life of ^{137}Cs in snails. *Jour. Radioanal. Nucl. Chem.* 153:327-332.

Jenkins, J.H. and T.T. Fendley. 1971. Radionuclide biomagnification in coastal-plain deer. Pages 116-122 in D.J. Nelson (ed.). *Radionuclides in Ecosystems. Proceedings of the Third National Symposium on Radioecology.* May 10–12, 1971, Oak Ridge, TN. Vol. 1. Available from Natl. Tech. Infor. Serv., Springfield, VA 22151.

Jenkins, J.H., J.R. Monroe, and F.B. Golley. 1969. Comparison of fallout ^{137}Cs accumulation and excretion in certain southeastern mammals. Pages 623-626 in D.J. Nelson and F.C. Evans (eds.). *Symposium on Radioecology. Proceedings of the Second National Symposium.* Available as CONF-670503 from The Clearinghouse for Federal Scientific and Technical Information, Natl. Bur. Standards, Springfield, VA 22151.

Johanson, K.J. 1990. The consequences in Sweden of the Chernobyl accident. *Rangifer,* Spec. Iss. No. 3:9-10.

Johanson, K.J., G. Karlen, and J. Bertilsson. 1989. The transfer of radiocesium from pasture to milk. *Sci. Total Environ.* 85:73-80.

Jones, B. 1990. Radiation effects in reindeer. *Rangifer,* Spec. Iss. No. 3:15-16.

Jones, B.E.V., O. Eriksson, and M. Nordkvist. 1989. Radiocesium uptake in reindeer on natural pasture. *Sci. Total Environ.* 85:207-212.

Jordan, C.F. 1969. Recovery of a tropical rain forest after gamma irradiation. Pages 88-109 in D.J. Nelson and F.C. Evans (eds.). *Symposium on Radioecology. Proceedings of the Second National Symposium.* Available as CONF-670503 from The Clearinghouse for Federal Scientific and Technical Information, Natl. Bur. Standards, Springfield, VA 22151.

Joseph, A.B., P.F. Gustafson, I.R. Russell, E.A. Schuert, H.L. Volchok, and A. Tamplin. 1971. Sources of radioactivity and their characteristics. Pages 6-41 in *National Academy of Sciences. Radioactivity in the Marine Environment.* Natl. Acad. Sci., Panel on Radioactivity in the Marine Environment, Washington, D.C.

Joshi, S.R. 1991. Radioactivity in the Great Lakes. *Sci. Total Environ.* 100:61-104.

Kahn, B. 1971. Radionuclides in the environment at nuclear power stations. Pages 30-36 in D.J. Nelson (ed.). *Radionuclides in Ecosystems. Proceedings of the Third National Symposium on Radioecology.* May 10–12, 1971, Oak Ridge, TN. Vol. 1. Available from Natl. Tech. Infor. Serv., Springfield, VA 22151.

Kalas, J.A., S. Bretten, I. Byrkjedal, and O. Njastad. 1994. Radiocesium (^{137}Cs) from the Chernobyl reactor in Eurasian woodcock and earthworms in Norway. *Jour. Wildl. Manage.* 58:141-147.

Kamiguchi, Y., H. Tateno, and K. Mikamo. 1990. Types of structural chromosome aberrations and their incidences in human spermatozoa X-irradiated *in vitro*. *Mutat. Res.* 228:133-140.

Karu, T.I. 1990. Effects of visible radiation on cultured cells. *Photochem. Photobiol.* 52:1089-1098.

Kaye, S.V. and P.B. Dunaway. 1963. Estimation of dose rate and equilibrium state from bioaccumulation of radionuclides by mammals. Pages 107-111 in V. Schultz and A.W. Klement, Jr. eds. *Radioecology.* Reinhold, New York.

Kedhi, M. 1990. Aerosol, milk and wheat flour radioactivity in Albania caused by the Chernobyl accident. *Jour. Radioanal. Nucl. Chem.* 146:115-124.

Kershaw, P.J. and D.S. Woodhead (eds.). 1991. *Radionuclides in the Study of Marine Processes.* Elsevier, New York. 393 pp.

Kiefer, J. 1990. *Biological Radiation Effects.* Springer-Verlag, Berlin. 444 pp.

Kim, J.H., L.R. Mandell, and R. Leeper. 1990. Radiation effects on the thyroid gland. *Adv. Radiation Biol.* 14:119-156.

King, G.L. and M.R. Landauer. 1990. Effects of zacopride and BMY25801 (batanopride) on radiation-induced emesis and locomotion behavior in the ferret. *Jour. Pharmacol. Exper. Therapeu.* 253:1026-1033.

Kitchings, T., D. DiGregorio, and P.V. Voris. 1976. A review of the ecological parameters of radionuclide turnover in vertebrate food chains. Pages 304-313 in C.E. Cushing, editor. *Radioecology and Energy Resources. Proceedings of the Fourth National Symposium on Radioecology,* 12–14 May 1975, Oregon State Univ., Corvallis, OR. Ecol. Soc. Amer., Spec. Publ. No. 1.

Kligman, L.H. and A.M. Kligman. 1990. Ultraviolet radiation-induced skin aging. *Skinscreens Develop. Eval. Regul. Asp.* 10:55-72.

Kliment, V. 1991. Contamination of pork by caesium radioisotopes. *Jour. Environ. Radioactiv.* 13:117-124.

Kodrich, W.R. and C.A. Tryon. 1971. Effects of season on thyroid release of iodine-131 in free-ranging eastern chipmunks (*Tamias striatus*). Pages 260-264 in D.J. Nelson (ed.). *Radionuclides in Ecosystems. Proceedings of the Third National Symposium on Radioecology.* May 10–12, 1971, Oak Ridge, TN. Vol. 1. Available from Natl. Tech. Infor. Serv., Springfield, VA 22151.

Korhonen, R. 1990. Modeling transfer of ^{137}Cs fallout in a large Finish watercourse. *Health Phys.* 59:443-454.

Kotelevtsev, S.V., F.F. Nagdaliev, and G.A. Skryabin. 1996. The influence of radionuclides on ion transport and its hormonal regulation in membranes of fish erythrocytes. *Jour. Ichthyol.* 36:109-113.

Kryshev, I. 1995. Radioactive contamination of aquatic ecosystems following the Chernobyl accident. *Jour. Environ. Radioactiv.* 27:207-219.

Kshirsagar, S.G. 1990. The metabolism of radium-226 during pregnancy in the rat. *Radiation Res.* 122:294-300.

Kudo, A., Y. Mahara, T. Kauri, and D.C. Santry. 1991. Fate of plutonium released from the Nagasaki A-bomb, Japan. Pages 291-300 in P. Grau, W. Gujer, K.J. Ives, D. Jenkins, H. Kroiss, A.C. Di Pinto, M. Suzuki, D.F. Torien, A. Milburn, E.J. Izod, and P.T. Nagle (eds.). *Water Pollution Research and Control, Kyoto, 1990. Part 1. Proceedings of the Fifteenth Biennial Conference of the International Association on Water Pollution Research and Control.* Kyoto, Japan. 29 July–3 August 1990. Pergamon Press, Oxford, England.

Lamb, T., J.W. Bickman, J.W. Gibbons, M.J. Smolen, and S. McDowell. 1991. Genetic damage in a population of slider turtles (*Trachemys scripta*) inhabiting a radioactive reservoir. *Arch. Environ. Contam. Toxicol.* 20:138-142.

Landesman, R.H. and W.S. Douglas. 1990. Abnormal limb regeneration in adult newts exposed to a pulsed electromagnetic field. *Teratology* 42:137-145.

League of Women Voters Education Fund (LWV) 1985. *The Nuclear Waste Primer.* Nick Lyons Books, New York. 90 pp.

Lesser, M.P. and J.M. Shick. 1990. Effects of visible and ultraviolet radiation on the ultrastructure of zooxanthellae (*Symbiodinium* sp.) in culture and *in situ. Cell Tiss. Res.* 261:501-508.

Lett, J.T. 1990. Damage to DNA and chromatin structure from ionizing radiations, and the radiation sensitivities of mammalian cells. *Prog. Nucleic Acid Res. Molec. Biol.* 39:305-352.

Levy, C.K., K.A. Youngstrom, and C.J. Maletskos. 1976. Whole-body gamma spectroscopic assessment of environmental radionuclides in recapturable wild birds. Pages 113-122 in Cushing, C.E. (ed.). *Radioecology and Energy Resources. Proceedings of the Fourth National Symposium on Radioecology,* 12–14 May 1975, Oregon State Univ., Corvallis, OR. Ecol. Soc. Amer., Spec. Publ. No. 1.

Liepa, V. and V. Balodis. 1994. Monitoring of bird breeding near a powerful radar station. *The Ring* 16:100.

Little, J.B. 1990. Low-dose radiation effects: interactions and synergism. *Health Phys.* 59:49-55.

Lloyd, R.D., F.W. Bruenger, S.C. Miller, W. Angus, G.N. Taylor, W.S.S. Jee, and E. Polig. 1991. Distribution of radium-induced bone cancers in beagles and comparison with humans. *Health Phys.* 60:435-438.

Lorenz, E., L.O. Jacobson, E.W. Heston, M. Shimkin, A.B. Eschenbrenner, M.K. Deringer, J. Doniger, and R. Schweisthal. 1954. Effects of long-continued total-body gamma irradiation on mice, guinea pigs, and rabbits. III. Effects on life span, weight, blood picture, and carcinogenesis and the role of the intensity of radiation. Pages 24-148 in R.E. Zirkle (ed.). *Biological Effects of X and Gamma Radiation.* McGraw-Hill, New York.

Lowe, V.P.W. 1991. Radionuclides and the birds at Ravenglass. *Environ. Pollut.* 70:1-26.

Lowe, V.P.W. and A.D. Horrill. 1991. Caesium concentration factors in wild herbivores and the fox (*Vulpes vulpes* L). *Environ. Pollut.* 70:93-107.

Lowman, F.G., T.R. Rice, and F.A. Richards. 1971. Accumulation and redistribution of radionuclides by marine organisms. Pages 161-199 in *National Academy of Sciences. Radioactivity in the Marine Environment.* Natl. Acad. Sci., panel on radioactivity in the marine environment. Washington, D.C.

Lucke-Huhle, C., B. Gloss, and P. Herrlich. 1990. Radiation-induced gene amplification in rodent and human cells. *Acta Biol. Hungarica* 41:159-171.

Luckey, T.D. 1980. *Hormesis with Ionizing Radiation.* CRC Press, Boca Raton, FL. 222 pp.

Macdonald, C.R. and M.J. Laverock. 1998. Radiation exposure and dose to small mammals in radon-rich soils. *Arch. Environ. Contam. Toxicol.* 35:109-120.

Maier, M.A. and M.R. Landauer. 1990. Onset of behavioral effects in mice exposed to 10 Gy ^{60}Co radiation. *Aviation Space Environ. Med.* 61:893-898.

Majundar, S.K., R.F. Schmalz, and E.W. Miller (eds.). 1990. *Environmental Radon: Occurrence, Control, and Health Hazards.* Pennsylvania Acad. Sci., Easton, PA. 436 pp.

Makeyeva, A.P., N.G. Yemel'yanova, N.V. Belova, and I.N. Ryabov. 1995. Radiobiological analysis of silver carp, *Hypophthalmichthys molitrix*, from the cooling pond of the Chernobyl nuclear power plant since the time of the accident. 2. Development of the reproductive system in the first generation of offspring. *Jour. Ichthyol.* 35:40-64.

Malhotra, N., N. Rani, K. Rana, and R.K. Malhotra. 1990. Radiation induced blood pathology in chick-erythrocytes and related parameters. *Exper. Pathol.* 38:241-248.

Markham, O.D., T.E. Hakonson, and J.S. Morton. 1983. Iodine-129 in mule deer thyroids in the Rocky Mountain west. *Health Phys.* 45:31-37.

Mascanzoni, D., S.V. Bothmer, T. Mattei, and M. Cristaldi. 1990. Small mammals as biological indicators of radioactive contamination of the environment. *Sci. Total Environ.* 99:61-66.

Mathiesen, S.D., L.M. Nordoy, and A.S. Blix. 1990. Elimination of radiocesium in contaminated adult Norwegian reindeer. *Rangifer,* Spec. Iss. No. 3:49.

McConnell, M.A., V.M.S. Ramanujam, N.W. Alcock, G.J. Gabehart, and W.W. Au. 1998. Distribution of uranium-238 in environmental samples from a residential area impacted by mining and milling activities. *Environ. Toxicol. Chem.* 17:841-850.

McCormack, M., A.H.W. Nias, and E. Smith. 1990. Chronic anemia, hyperbaric oxygen and tumour radiosensitivity. *Brit. Jour. Radiol.* 63:752-759.

McCormick, J.F. 1969. Effects of ionizing radiation on a pine forest. Pages 78-87 in D.J. Nelson and F.C. Evans (eds.). *Symposium on Radioecology. Proceedings of the Second National Symposium.* Available as CONF-670503 from The Clearinghouse for Federal Scientific and Technical Information, Natl. Bur. Standards, Springfield, VA 22151.

McInnes, D.O. and T.T.Y. Wong. 1990. Mediterranean fruitfly: interference of oviposition by radiation-sterilized females in field cages. *Entomol. Exper. Appl.* 56:125-130.

McKay, W.A. and N.J. Pattenden. 1990. The transfer of radionuclides from sea to land via the air: a review. *Jour. Environ. Radioactiv.* 12:49-77.

McLean, A.S. 1973. Early adverse effects of radiation. *Brit. Med. Jour.* 29:69-73.

Mele, P.C., C.G. Franz, and J.R. Harrison. 1990. Effects of ionizing radiation on fixed-ratio escape performance in rats. *Neurotoxicol. Teratol.* 12:367-373.

Miettinen, J.K. 1969. Enrichment of radioactivity by Arctic ecosystems in Finnish Lapland. Pages 23-31 in D.J. Nelson and F.C. Evans (eds.). *Symposium on Radioecology. Proceedings of the Second National Symposium.* Available as CONF-670503 from The Clearinghouse for Federal Scientific and Technical Information, Natl. Bur. Standards, Springfield, VA 22151.

Millard, J.B. and F.W. Whicker. 1990. Radionuclide uptake and growth of barn swallows nesting by radioactive leaching ponds. *Health Phys.* 58:429-439.

Mo, T. and F.G. Lowman. 1976. Laboratory experiments on the transfer of plutonium from marine sediments to seawater and to marine organisms. Pages 86-95 in C.E. Cushing (ed.). *Radioecology and Energy Resources. Proceedings of the Fourth National Symposium on Radioecology,* 12–14 May 1975, Oregon State Univ., Corvallis, OR. Ecol. Soc. Amer., Spec. Publ. No. 1.

Mole, R.H. 1990. The effect of prenatal radiation exposure on the developing human brain. *Inter. Jour. Radiation Biol.* 57:647-663.

Monte, L. 1990. Evaluation of the environmental transfer parameters for ^{137}Cs using the contamination produced by the Chernobyl accident at a site in central Italy. *Jour. Environ. Radioactiv.* 12:13-22.

Morse, J.W. and G.R. Choppin. 1991. The chemistry of transuranic elements in natural waters. *Rev. Aquat. Sci.* 4:1-22.

Moss, B.W., E.F. Unsworth, C.H. McMurray, J. Pearce, and D.J. Kilpatrick. 1989. Studies on the uptake, partition and retention of ionic and fallout radiocaesium by suckling and weaned lambs. *Sci. Total Environ.* 85:91-106.

Mufson, S. 1992. G-7 eyes aid for ex-Soviet nuclear plants. *Washington Post* (newspaper), Washington, D.C. 22 May 1992:F1-F3.

Muller, W.A., A.B. Murray, U. Linzner, and A. Luz. 1990. Osteosarcoma risk after simultaneous incorporation of the long-lived radionuclide ^{227}Ac and the short-lived radionuclide ^{227}Th. *Radiation Res.* 121:14-20.

Nadzhafova, R.S., N.S. Bulatova, A.I. Kozlovskii, and I.N. Ryabov. 1994. Identification of a structural chromosomal rearrangement in the karyotype of a root vole from Chernobyl. *Russian Jour. Genet.* 30:318-322

Nagasawa, H., J.B. Little, W.C. Inkret, S. Carpenter, K. Thompson, M.R. Raju, D.J. Chen, and G.F. Strniste. 1990. Cytogenetic effects of extremely low doses of plutonium-238 alpha particle irradiation in CHO K-1 cells. *Mutat. Res.* 244:233-238.

National Academy of Sciences (NAS). 1957. The Effects of Atomic Radiation on Oceanography and Fisheries. Publ. No. 551. NAS — Natl. Res. Coun., Washington, D.C. 137 pp.

National Academy of Sciences (NAS). 1971. Radioactivity in the Marine Environment. NAS, Panel on Radioactivity in the Marine Environment. Washington, D.C. 272 pp.

National Council on Radiation Protection and Measurements (NCRP). 1991. Effects of Ionizing Radiation on Aquatic Organisms. NCRP Rep. No. 109. 115 pp. Available from NCRP Publications, 7910 Woodmont Ave., Bethesda, MD 29814.

Neel, J.V. and S.E. Lewis. 1990. The comparative radiation genetics of humans and mice. *Annu. Rev. Genet.* 24:327-362.

Neel, J.W. and K.H. Larson. 1963. Biological availability of strontium-90 to small native animals in fallout patterns from the Nevada test site. Pages 45-49 in V. Schultz and A.W. Klement, Jr. (eds.). *Radioecology.* Reinhold, New York.

Nelin, P. 1995. Radiocaesium uptake in moose in relation to home range and habitat composition. *Jour. Environ. Radioactiv.* 26:189-203.

Nelson, D.J. 1963. The strontium and calcium relationships in Clinch and Tennessee River mollusks. Pages 203-211 in V. Schultz and A.W. Klement, Jr. (eds.). *Radioecology.* Reinhold, New York.

Nelson, D.J. (ed.). 1971. *Radionuclides in ecosystems. Proceedings of the Third National Symposium on Radioecology.* May 10–12, 1971, Oak Ridge, TN. Vol 1:1-678; Vol. 2:679-1268. Available from the Natl. Tech. Infor. Serv., Springfield, VA 22151, as CONF 710501-P1 and CONF 710501-P2.

Nelson, D.J. and F.C. Evans (eds.). 1969. *Symposium on Radioecology. Proceedings of the Second National Symposium.* 774 pp. Available as CONF-670503 from The Clearinghouse for Federal Scientific and Technical Information, Natl. Bur. Standards, Springfield VA 22151.

Nelson, W.C. and F.W. Whicker 1969. Cesium-137 in some Colorado game fish, 1965-66. Pages 258-265 in Nelson, D.J. and F.C. Evans (eds.). *Symposium on Radioecology. Proceedings of the Second National Symposium.* Available as CONF-670503 from The Clearinghouse for Federal Scientific and Technical Information, Natl. Bur. Standards, Springfield, VA 22151.

Newman, M.C. and I.L. Brisbin, Jr. 1990. Variation of ^{137}Cs levels between sexes, body sizes and collection localities of mosquitofish, *Gambusia holbrooki* (Girard 1859), inhabiting a reactor cooling reservoir. *Jour. Environ. Radioactiv.* 12:131-144.

Norton, S. and B.F. Kimler. 1990. Early effects of low doses of ionizing radiation on the fetal cerebral cortex in rats. *Radiation Res.* 124:235-241.

Norton, S., B.F. Kimler, and P.J. Mullenix. 1991. Progressive behavioral changes in rats after exposure to low levels of ionizing radiation *in utero*. *Neurotoxicol. Teratol.* 13:181-188.

Noshkin, V.E., V.T. Bowen, K.M. Wong, and J.C. Barke. 1971. Plutonium in North Atlantic ocean organisms: ecological relationships. Pages 681-688 in D.J. Nelson (ed.). *Radionuclides in Ecosystems. Proceedings of the Third National Symposium on Radioecology.* May 10–12, 1971, Oak Ridge, TN. Vol. 2. Available from the Natl. Tech. Infor. Serv., Springfield, VA 22151.

Nozaki, Y. 1991. The systematics and kinetics of U/Th decay series nuclides in ocean water. *Rev. Aquat. Sci.* 4:75-105

O'Connor, M.E. 1990. Teratogenesis: nonionizing electromagnetic fields. Pages 357-372 in O.P. Ghandi (ed.). *Biological Effects and Medical Applications of Electromagnetic Energy.* Prentice Hall, Englewood Cliffs, NJ.

O'Farrell, T.P. 1969. Effects of acute ionizing radiation in selected Pacific northwest rodents. Pages 157-165 in D.J. Nelson and F.C. Evans (eds.). *Symposium on Radioecology. Proceedings of the Second National Symposium.* Available as CONF-670503 from The Clearinghouse for Federal Scientific and Technical Information, Natl. Bur. Standards, Springfield, VA 22151.

O'Farrell, T.P. and R.O. Gilbert. 1975. Transport of radioactive materials by jackrabbits on the Hanford Reservation. *Health Phys.* 29:9-15.

Oge, M. and M. Dickson. 1990. The Environmental Protection Agency's radon action program. Pages 331-341 in S.K. Majumdar, R.F. Schmalz, and E.W. Miller (eds.). *Environmental Radon: Occurrence, Control, and Health Hazards.* Pennsylvania Acad. Sci., Easton, PA.

Ophel, I. L, and J.M. Judd. 1976. Strontium and calcium accumulation in fish as affected by food composition. Pages 221-225 in C.E. Cushing (ed.). *Radioecology and Energy Resources. Proceedings of the Fourth National Symposium on Radioecology,* 12–14 May 1975, Oregon State Univ., Corvallis, OR. Ecol. Soc. Amer., Spec. Publ. No. 1.

Osburn, W.S., Jr. 1963. Influence of four Rocky Mountain regional environments on pea plants grown from irradiated seeds. Pages 319-324 in V. Schultz and A.W. Klement, Jr. (eds.). *Radioecology.* Reinhold, New York.

Othman, I. 1990. The impact of the Chernobyl accident on Syria. *Jour. Radiolog. Protect.* 10:103-108.

Palo, R.T., P. Nelin, T. Nylen, and G. Wickman. 1991. Radiocesium levels in Swedish moose in relation to deposition, diet, and age. *Jour. Environ. Qual.* 20:690-695.

Palumbo, R.F. 1963. Factors controlling the distribution of the rare earths in the environment and in living organisms. Pages 533-537 in V. Schultz and A.W. Klement, Jr. (eds.). *Radioecology.* Reinhold, New York.

Patt, H.M. and M.N. Swift. 1948. Influence of temperature on the response of frogs to X irradiation. *Amer. Jour. Physiol.* 155:388-393.

Pearce, J., C.H. McMurray, E.F. Unsworth, B.W. Moss, F.J. Gordon, and D.J. Kilpatrick. 1989. Studies of the transfer of dietary radiocaesium from silage to milk in dairy cows. *Sci. Total Environ.* 85:267-278.

Pelton, M.R. and E.E. Provost. 1969. Effects of radiation on survival of wild cotton rats (*Sigmodon hispidus*) in enclosed areas of natural habitat. Pages 39-45 in D.J. Nelson and F.C. Evans (eds.). *Symposium on Radioecology. Proceedings of the Second National Symposium.* Available as CONF-670503 from The Clearinghouse for Federal Scientific and Technical Information, Natl. Bur. Standards, Springfield, VA 22151.

Pentreath, R.J., D.S. Woodhead, and D.F. Jefferies. 1971. Radioecology of the plaice (*Pleuronectes platessa* L.) in the northeast Irish Sea. Pages 731-737 in D.J. Nelson (ed.). *Radionuclides in Ecosystems. Proceedings of the Third National Symposium on Radioecology.* May 10–12, 1971, Oak Ridge, TN. Vol. 2. Available from Natl. Tech. Infor. Serv., Springfield, VA 22151.

Plummer, G.L., T.M. Pullen, Jr., and E.E. Provost. 1969. Cesium-137 and a population of Georgia white-tailed deer. Pages 609-615 in D.J. Nelson and F.C. Evans (eds.). *Proceedings of the Second National Symposium.* Available as CONF-670503 from The Clearinghouse for Federal Scientific and Technical Information, Natl. Bur. Standards, Springfield, VA 22151.

Poinsot-Balaguer, N., R. Castel, and E. Tabone. 1991. Impact of chronic gamma irradiation on a Mediterranean forest ecosystem in Cadarache (France). *Jour. Environ. Radioactiv.* 14:23-36.

Polikarpov, G.G. (ed.). 1973. Artificial and Natural Radionuclides in Marine Life. Acad. Sci. Ukrainian SSR, A.O. Kovalevskii Inst. Biol. Southern Seas. 115 pp. Available from the U.S. Dept. Commerce, Natl. Tech. Infor. Serv., Springfield, VA 22151.

Pomerantseva, M.D., L.K. Ramaiya, and A.V. Chekhovich. 1996. Genetic consequences of the Chernobyl disaster for house mice (*Mus musculus*). *Russian Jour. Genet.* 32:264-268.

Poston, T.P., D.C. Klopfer, and M.A. Simmons. 1990. Short-term bioconcentration studies of Np in freshwater biota. *Health Phys.* 59:869-877.

Prokopchak, M.J., D.P. Spangenberg, and J. Shaeffer. 1990. The effects of X irradiation on the metamorphosis and budding of *Aurelia aurita*. *Radiation Res.* 124:34-42.

Raabe, O.G., L.S. Rosenblatt, and R.A. Schlenker. 1990. Interspecies scaling of risk for radiation-induced bone cancer. *Inter. Jour. Radiation Biol.* 57:1047-1061.

Reed, J.R. and B.A. Martinedes. 1971. Uptake and retention of tungsten-181 by crayfish (*Cambarus longulus longirostris* Ort.). Pages 390-393 in D.J. Nelson (ed.). *Radionuclides in Ecosystems. Proceedings of the Third National Symposium on Radioecology.* May 10–12, 1971. Available from Natl. Tech. Infor. Serv., Springfield, VA 22151.

Reed, J.R. and D.J. Nelson. 1969. Radiostrontium uptake in blood and flesh in bluegills (*Lepomis macrochirus*). Pages 226-233 in D.J. Nelson and F.C. Evans (eds.). *Symposium on Radioecology. Proceedings of the Second National Symposium.* Available as CONF-370503 from The Clearinghouse for Federal Scientific and Technical Information, Natl. Bur. Standards, Springfield, VA 22151.

Refetoff, S. 1990. A review of risks of external irradiation of the thyroid gland. Pages 101-103 in *Iodine Prophylaxis Following Nuclear Accidents: Proceedings of the Joint WHO/CEC Workshop.* July 1988. Pergamon Press, New York.

Renfro, W.C. and G. Benayoun. 1976. Sediment-worm interaction: transfer of ^{65}Zn from marine silt by the polychaete *Nereis diversicolor.* Pages 250-255 in C.E. Cushing (ed.). *Radioecology and Energy Resources. Proceedings of the Fourth National Symposium on Radioecology,* 12–14 May 1975, Oregon State Univ., Corvallis, OR. Ecol. Soc. Amer., Spec. Publ. No. 1.

Renn, C.E. 1957. Physical and chemical properties of wastes produced by atomic power industry. Pages 26-27 in *National Academy of Sciences. The Effects of Atomic Radiation on Oceanography and Fisheries.* Publ. No. 551, NAS-Natl. Res. Coun., Washington, D.C.

Richmond, C.R. 1989. Transfer of radionuclides to animals — an historical perspective of work done in the United States. *Sci. Total Environ.* 85:1-15.

Rickard, W.H. and K.R. Price. 1990. Strontium-90 in Canada goose eggshells and reed Canary grass from the Columbia River, Washington. *Environ. Monitor. Assess.* 14:71-76.

Rissanen, K. and T. Rahola. 1989. Cs-137 concentration in reindeer and its fodder plants. *Sci. Total Environ.* 85:199-206.

Rissanen, K. and T. Rahola. 1990. Radiocesium in lichens and reindeer after the Chernobyl accident. *Rangifer,* Spec. Iss. No. 3:55-61.

Robins, M.W. 1990. ^{224}Ra-induced osteopenia in male CBA mice. *Calcif. Tissue Inter.* 46:94-100.

Roed, K.H., I.M.H. Eikelmann, M. Jacobsen, and O. Pedersen. 1991. Chromosome aberrations in Norwegian reindeer calves exposed to fallout from the Chernobyl accident. *Hereditas* 115:201-206

Romney, E.M., W.A. Rhoads, A. Wallace, and R.A. Wood. 1971. Persistence of radionuclides in soil, plants, and small mammals in areas contaminated with radioactive fallout. Pages 170-176 in D.J. Nelson (ed.). *Radionuclides in Ecosystems. Proceedings of the Third National Symposium on Radioecology.* May 10–12, 1971, Oak Ridge, TN. Vol. 1. Available from The Natl. Tech. Infor. Serv., Springfield, VA 22151.

Romney, E.M., A. Wallace, and J.D. Childress. 1971a. Revegetation problems following nuclear testing activities at the Nevada test site. Pages 1015-1022 in D.J. Nelson (ed.). *Radionuclides in Ecosystems. Proceedings of the Third National Symposium on Radioecology.* May 10–12, 1971, Oak Ridge, TN. Vol. 1. Available from The Natl. Tech. Infor. Serv., Springfield, VA 22151.

Ronai, Z.A., M.E. Lambert, and I.B. Weinstein. 1990. Inducible cellular responses to ultraviolet light irradiation and other mediators of DNA damage in mammalian cells. *Cell Biol. Toxicol.* 6:105-126.

Rose, A.W., W.A. Jester, and B.C. Ford. 1990. Radioactive elements in Pennsylvania waters. Pages 91-109 in S.K. Majumdar, R.F. Schmalz, and E.W. Miller (eds.). *Environmental Radon: Occurrence, Control, and Health Hazards.* Pennsylvania Acad. Sci., Easton, PA.

Rose, K.S.B. 1992. Lower limits of radiosensitivity in organisms, excluding man. *Jour. Environ. Radioactiv.* 15:113-133.

Sankaranarayanan, K. 1991a. Ionizing radiation and genetic risks II. Nature of radiation-induced mutations in experimental mammalian *in vivo* systems. *Mutat. Res.* 258:51-73.

Sankaranarayanan, K. 1991b. Ionizing radiation and genetic risks III. Nature of spontaneous and radiation-induced mutations in mammalian *in vitro* systems and mechanisms of induction of mutations by radiation. *Mutat. Res.* 258:75-97.

Sankaranarayanan, K. 1991c. Ionizing radiation and genetic risks IV. Current methods, estimates of risk of Mendelian disease, human data and lessons from biochemical and molecular studies of mutations. *Mutat. Res.* 258:99-122.

Sawada, M., J.C. Carlson, and H.E. Enesco. 1990. The effects of UV radiation and antioxidants on life span and lipid peroxidation in the rotifer *Asplanchna brightwelli*. *Arch. Gerontol. Geriatr.* 10:27-36.

Sawidis, T. 1988. Uptake of radionuclides by plants after the Chernobyl accident. *Environ. Pollut.* 50:317-324.

Schell, W.R. 1976. Concentrations and physical-chemical states of ^{55}Fe in Bikini Atoll lagoon. Pages 271-276 in C.E. Cushing (ed.). *Radioecology and Energy Resources. Proceedings of the Fourth National Symposium on Radioecology,* 12–14 May 1975, Oregon State Univ., Corvallis, OR. Ecol. Soc. Amer., Spec. Publ. No. 1.

Schoeters, G.E.R., J.R. Maisin, and O.L.V. Vanderborght. 1991. Toxicity of ^{241}Am in male C57BL mice: relative risk versus ^{226}Ra. *Radiation Res.* 126:198-205.

Schreckhise, R.G. and F.W. Whicker. 1976. A model for predicting strontium-90 levels in mule deer. Pages 148-156 in C.E. Cushing (ed.). *Radioecology and Energy Resources. Proceedings of the Fourth National Symposium on Radioecology,* 12–14 May 1975, Oregon State Univ., Corvallis, OR. Ecol. Soc. Amer., Spec. Publ. No. 1.

Schultz, V. and A.W. Klement, Jr. (eds.). 1963. *Radioecology.* Reinhold, New York. 746 pages.

Schultz, V. and W.M. Longhurst. 1963. Accumulation of strontium-90 in yearling Columbian black-tailed deer, 1950-1960. Pages 73-76 in V. Schultz, and A.W. Klement, Jr. (eds.). *Radioecology.* Reinhold, New York.

Severa, J. and J. Bar. 1991. *Handbook of Radioactive Contamination and Decontamination.* Studies in Environmental Science 47. Elsevier, New York. 363 pp.

Shadley, J.D., J.L. Whitlock, J. Rotmensch, R.W. Atcher, J. Tang, and J.L. Schwartz. 1991. The effects of radon daughter α-particle irradiation in K1 and xrs-5 CHO cell lines. *Mutat. Res.* 248:73-83.

Shick, J.M., M.P. Lesser, and W.R. Stochaj. 1991. Ultraviolet radiation and photooxidative stress in zooxanthellate anthozoa: the sea anemone *Phyllodiscus semoni* and the octocoral *Clavularia* sp. *Symbiosis* 10:145-173.

Shields, L.M. and P.V. Wells. 1963. Recovery of vegetation on atomic target areas at the Nevada test site. Pages 307-310 in V. Schultz and A.W. Klement, Jr. (eds.). *Radioecology.* Reinhold, New York.

Shima, A. and A. Shimada. 1991. Development of a possible nonmammalian test system for radiation-induced germ-cell mutagenesis using a fish, the Japanese medaka (*Oryzias latipes*). *Proc. Natl. Acad. Sci. U.S.A.* 88:2545-2549.

Shure, D.J. and M.R. Gottschalk. 1976. Cesium-137 dynamics within a reactor effluent stream in South Carolina. Pages 234-241 in C.E. Cushing (ed.). *Radioecology and Energy Resources. Proceedings of the Fourth National Symposium on Radioecology,* 12–14 May 1975, Oregon State Univ., Corvallis, OR. Ecol. Soc. Amer., Spec. Publ. No. 1.

Sibley, T.H. and J.S. Stohr. 1990. Accumulation of Am-241 and Cm-244 from water and sediments by *Hyalella* sp. and *Tubifex* spp. *Bull. Environ. Contam. Toxicol.* 44:165-172.

Sicard, R.E. and M.F. Lombard. 1990. Putative immunological influence upon amphibian forelimb regeneration. II. Effects of X-irradiation on regeneration and autograft rejection. *Biol. Bull.* 178:21-24.

Sivinski, J. and B. Smittle. 1990. Effects of gamma radiation on the development of the Caribbean fruit fly (*Anastrepha suspensa*) and the subsequent development of its parasite *Diachasmimorpha longicaudata. Entomolog. Exper. Applic.* 55:295-297.

Skogland, T. and I. Espelien. 1990. The biological effects of radiocesium contamination of wild reindeer in Norway following the Chernobyl accident. Pages 276-279 in S. Myrberget (ed.). Vol. 1: Populations Dynamics. *Trans. XIXth IUGB Congr. Inter. Union Game Biol.* September 1989, Trondheim, Norway.

Slovic, P., J.H. Flynn, and M. Layman. 1991. Perceived risk, trust, and the politics of nuclear waste. *Science* 254:1603-1607.

Smedile, E. and G. Queirazza. 1976. Uptake of ^{60}Co and ^{137}Cs in different components of a river ecosystem connected with discharges of a nuclear power station. Pages 1-8 in C.E. Cushing (ed.). *Radioecology and Energy Resources. Proceedings of the Fourth National Symposium on Radioecology,* 12–14 May 1975, Oregon State Univ., Corvallis, OR. Ecol. Soc. Amer., Spec. Publ. No. 1.

Smith, F.B. and M.J. Clark. 1986. Radionuclide deposition from the Chernobyl cloud. *Nature* 322:690-691.

Snyder, D.P., C.A. Tryon, and D.L. Graybill. 1976. Effect of gamma radiation on range parameters in the eastern chipmunk, *Tamias striatus*. Pages 354-358 in C.E. Cushing (ed.). *Radioecology and Energy Resources. Proceedings of the Fourth National Symposium on Radioecology,* 12–14 May 1975, Oregon State Univ., Corvallis, OR. Ecol. Soc. Amer., Spec. Publ. No. 1.

Sokolov, V.E., D.A. Krivolutzky, I.N. Ryabov, A.E. Taskaev, and V.A. Shevchenko. 1990. Bioindication of biological after-effects of the Chernobyl atomic power station accident in 1986-1987. *Biol. Inter.* 18:6-11.

Spezzano, P. and R. Giacomelli. 1991. Transport of ^{131}I and ^{137}Cs from air to cow's milk produced in northwestern Italian farms following the Chernobyl accident. *Jour. Environ. Radioactiv.* 13:235-250.

Staaland, H., K. Hove, and O. Pedersen. 1990. Transport and recycling of radiocesium in the alimentary tract of reindeer. *Rangifer,* Spec. Iss. No. 3:63-72.

Stanek, Z., M. Penaz, and E. Wohlgemuth. 1990. Accumulation and kinetics of ^{90}Sr in fishes and other components of an artificial aquatic system. *Folia Zool.* 39:375-383.

Steele, A.K. 1990. Derived concentration factors for caesium-137 in edible species of North Sea fish. *Mar. Pollut. Bull.* 21:591-594.

Stegelmeier, B.L., N.A. Gillett, A.H. Rebar, and G. Kelly. 1991. The molecular progression of plutonium-239-induced rat lung carcinogenesis: Ki-*ras* expression and activation. *Molecul. Carcinogen.* 4:43-51.

Steger, W. and C.J. Goodnight. 1976. The influence of environmental factors on the accumulation and retention of cesium-137 by the worm *Lininodrilus hoffmeisteri* (Oligochaeta). Pages 242-249 in C.E. Cushing (ed.). *Radioecology and Energy Resources. Proceedings of the Fourth National Symposium on Radioecology,* 12–14 May 1975, Oregon State Univ., Corvallis, OR. Ecol. Soc. Amer., Spec. Publ. No. 1.

Straume, T., T.C. Kwan, L.S. Goldstein, and R.L. Dobson. 1991. Measurement of neutron-induced genetic damage in mouse immature oocytes. *Mutat. Res.* 248:123-133.

Su, R.Z., M. Kost, and J. Kiefer. 1990. The effect of krypton ions on *Artemia* cysts. *Radiation Environ. Biophys.* 29:103-107.

Sugg, D.W., R.K. Chesser, J.A. Brooks, and B.T. Grasman. 1995. The association of DNA damage to concentrations of mercury and radiocesium in largemouth bass. *Environ. Toxicol. Chem.* 14:661-668.

Sweigert, S.E. and D. Carroll. 1990. Repair and recombination of X-irradiated plasmids in *Xenopus laevis* oocytes. *Molec. Cell. Biol.* 10:5849-5856.

Talmage, S. and L. Meyers-Schone. 1995. Nuclear and thermal. Pages 469-491 in D.J. Hoffman, B.A. Rattner, G.A. Burton, Jr., and J. Cairns, Jr. (eds.). *Handbook of Ecotoxicology.* Lewis Publishers, Boca Raton, FL.

Templeton, W.L., R.E. Nakatani, and E.E. Held. 1971. Radiation effects. Pages 223-229 in National Academy of Sciences. *Radioactivity in the Marine Environment.* NAS, Panel on Radioactivity in the Marine Environment, Washington, D.C.

Thacker, J. 1990. Molecular nature of ionizing radiation-induced mutations of native and introduced genes in mammalian cells. Pages 221-230 in *Ionizing Radiation Damage to DNA: Molecular Aspects. Proceedings of a Radiation Research Society — UCLA Symposia Colloquium.* Lake Tahoe, CA, January 16-21, 1990. Wiley-Liss, New York.

Theodrakis, C.W., J.W. Bickham, T. Elbl, L.R. Shugart, and R.K. Chesser. 1998. Genetics of radionuclide-contaminated mosquitofish populations and homology between *Gambusia affinis* and *G. holbrooki. Environ. Toxicol. Chem.* 17:1992-1998.

Thompson, G.A., J. Smithers, and H. Boxenbaum. 1990. Biphasic mortality response of chipmunks in the wild to single doses of ionizing radiation: toxicity and longevity hormesis. *Drug Metabol. Rev.* 22:269-289.

Tonelli, D., E. Gattavecchia, S. Ghini, C. Porrini, G. Celli, and A.M. Mercuri. 1990. Honey bees and their products as indicators of environmental radioactive pollution. *Jour. Radioanal. Nuclear Chem.* 141:427-436.

Trnovec, T., Z. Kallay, and S. Bezek. 1990. Effects of ionizing radiation on the blood brain barrier permeability to pharmacologically active substances. *Inter. Jour. Radiation Oncol. Biol. Physics* 19:1581-1587.

Turner, F.B., P. Licht, J.D. Thrasher, P.A. Medica, and J.R. Lannom, Jr. 1971. Radiation-induced sterility in natural populations of lizards (*Crotaphytus wislizenii* and *Cnemidophorus tigris*). Pages 1131-1143 in D.J. Nelson (ed.). *Radionuclides in Ecosystems. Proceedings of the Third National Symposium on Radioecology.* May 10–12, 1971, Oak Ridge, TN. Vol. 1. Available from Natl. Tech. Infor. Serv., Springfield, VA 22151.

Ulrikson, G.U. 1971. Radiation effects on serum proteins, hematocrits, electrophoretic patterns and protein components in the bluegill (*Lepomis macrochirus*). Pages 1100-1105 in D.J. Nelson (ed.). *Radionuclides in Ecosystems. Proceedings of the Third National Symposium on Radioecology.* May 10–12, 1971, Oak Ridge, TN. Vol. 1. Available from Natl. Tech. Infor. Serv., Springfield, VA 22151.

United Nations Scientific Committee on the Effects of Atomic Radiation (UNSCEAR). 1988. *Sources, Effects and Risks of Ionizing Radiation.* United Nations, New York. 647 pp.

United States Code of Federal Regulations (USCFR). 1990. Subchapter F — Radiation Protection Programs. 40:6-23.

U.S. Code of Federal Regulations (USCFR). 1991. Subchapter F — Radiation Protection Programs. 40:6-23.

U.S. Environmental Protection Agency (USEPA). 1975. Environmental Radiation Data, Report 2. EPA, Off. Radiation Progr., Washington, D.C. 51 pp.

U.S. Environmental Protection Agency (USEPA). 1977. Environmental Radiation Data, Report 11. EPA, Off. Radiation Progr., Washington, D.C. 37 pp.

U.S. Environmental Protection Agency (USEPA). 1979. Environmental Radiation Data, Report 15, January 1979. EPA, Off. Radiation Progr., Washington, D.C. 57 pp.

U.S. Environmental Protection Agency (USEPA) 1982a. Environmental Radiation Data, Report 28, October–December 1981. EPA Rep. 520/1-83-002. 98 pp.

U.S. Environmental Protection Agency (USEPA). 1982b. Environmental Radiation Data, Report 30, April–June 1982. EPA Rep. 520/5-6-83-006. 90 pp.

U.S. Environmental Protection Agency (USEPA). 1989a. Environmental Radiation Data, Report 55, July–September 1988. EPA Rep. 520/5-89-011. 37 pp.

U.S. Environmental Protection Agency (USEPA). 1989b. Environmental Radiation Data, Report 56, October–December 1988. EPA Rep. 520/5-89-013. 32 pp.

U.S. Environmental Protection Agency (USEPA). 1989c. Environmental Radiation Data, Report 57, January–March 1989. EPA Rep. 520/5-89-021. 41 pp.

U.S. Environmental Protection Agency (USEPA). 1989d. Environmental Radiation Data, Report 58, April–June 1989. EPA Rep. 520/5-89-034. 35 pp.

U.S. Environmental Protection Agency (USEPA). 1990a. Environmental Radiation Data, Report 59, July–September 1989. EPA Rep. 520/5-90-003. 46 pp.

U.S. Environmental Protection Agency (USEPA). 1990b. Environmental Radiation Data, Report 60, October–December 1989. EPA Rep. 520/5-90-018. 33 pp.

U.S. Environmental Protection Agency (USEPA). 1990c. Environmental Radiation Data, Report 61, January–March 1990. EPA Rep. 520/5-90-031. 32 pp.

U.S. Environmental Protection Agency (USEPA). 1990d. Environmental Radiation Data, Report 62, April–June 1990. EPA Rep. 520/5-91-013. 41 pp.

U.S. Environmental Protection Agency (USEPA). 1991. Environmental Radiation Data, Report 63, July–September 1991. EPA Rep. 520/5-91-025. 34 pp.

van de Mortel, T.F. and W.A. Buttemer. 1996. Are *Litoria aurea* eggs more sensitive to ultraviolet-B radiation than eggs of sympatric *L. peronii* or *L. dentata*? *Austral. Zool.* 30:150-157.

Van Den Hoek, J. 1989. European research on the transfer of radionuclides to animals — a historical perspective. *Sci. Total Environ.* 85:17-27.

Vandecasteele, C.M., M. Van Hees, J.P. Culot, and J. Vankerkom. 1989. Radiocaesium metabolism in pregnant ewes and their progeny. *Sci. Total Environ.* 85:213-223.

Veskovic, M.M. and G. Djuric. 1990. Game as a bioindicator of radiocontamination. *Acta Veterin. (Beograd)* 40:229-234.

Vissink, A., E.J. 's-Gravenmade, E.E. Ligeon, and A.W.T. Konings. 1990. A functional and chemical study of radiation effects on rat parotid and submandibular/sublingual glands. *Radiation Res.* 124:259-265.

Voors, P.I. and A.W. Van Weers. 1989. Transfer of Chernobyl ^{134}Cs and ^{137}Cs in cows from silage to milk. *Sci. Total Environ.* 85:179-188.

Voors, P.I. and A.W. Van Weers. 1991. Transfer of Chernobyl radiocaesium (^{134}Cs and ^{137}Cs) from grass silage to milk in dairy cows. *Jour. Environ. Radioactiv.* 13:125-140.

Vreman, K., T.D.B. Van Der Struijs, J. Van Den Hoek, D.L.M. Berende, and P.W. Goedhart. 1989. Transfer of Cs-137 from grass and wilted grass silage to milk of dairy cows. *Sci. Total Environ.* 85:139-147.

Walinder, G. 1990. A review of risks of internal irradiation of the thyroid gland — in utero, neonatally, adults. Pages 105-110 in *Iodine Prophylaxis Following Nuclear Accidents. Proceedings of the Joint WHO/CEC Workshop.* July 1988. Pergamon Press, New York.

Weast, R.C. (ed.). 1985. *CRC Handbook of Chemistry and Physics.* CRC Press, Boca Raton, FL. Pages B233-B454, F65-F108.

Welander, A.D. 1969. Distribution of radionuclides in the environment of Eniwetok and Bikini Atolls, August 1964. Pages 346-354 in D.J. Nelson and F.C. Evans (eds.). *Symposium on Radioecology. Proceedings of the Second National Symposium.* Available as CONF-370503 from The Clearinghouse for Federal Scientific and Technical Information, Natl. Bur. Standards, Springfield, VA 22151.

Wetherbee, D.K. 1966. Gamma Irradiation of Birds' Eggs and the Radiosensitivity of Birds. Mass. Agricul. Exper. Stat., Bull. 561, Univ. Massachusetts, Amherst, MA. 103 pp.

Whicker, F.W., J.E. Pinder III, J.W. Bowling, J.J. Alberts, and I.L. Brisbin, Jr. 1990. Distribution of long-lived radionuclides in an abandoned reactor cooling reservoir. *Ecol. Monogr.* 60:471-496.

Whicker, F.W. and V. Schultz. 1982a. *Radioecology: Nuclear Energy and the Environment.* Vol. I. CRC Press LLC, Boca Raton, FL. 212 pp.

Whicker, F.W. and V. Schultz. 1982b. *Radioecology: Nuclear Energy and the Environment.* Vol. II. CRC Press LLC, Boca Raton, FL. 228 pp.

Whitehead, N.E., S. Ballestra, E. Holm, and L. Huynh-Ngoc. 1988a. Chernobyl radionuclides in shellfish. *Jour. Environ. Radioactiv.* 7:107-121.

Whitehead, N.E., S. Ballestra, E. Holm, and A. Walton. 1988b. Air radionuclide patterns observed at Monaco from the Chernobyl accident. *Jour. Environ. Radioactiv.* 7:249-264.

Wiklund, K., L.E. Holm, and G. Eklund. 1990. Cancer risks in Swedish Lapps who breed reindeer. *Amer. Jour. Epidemiol.* 132:1078-1082.

Willard, W.K. 1963. Relative sensitivity of nestlings of wild passerine birds to gamma radiation. Pages 345-349 in V. Schultz and A.W. Klement, Jr. (eds.). *Radioecology.* Reinhold, New York.

Willis, D.L. and W.L. Lappenbusch. 1976. The radiosensitivity of the rough-skinned newt (*Taricha granulosa*). Pages 363-375 in C.E. Cushing (ed.). *Radioecology and Energy Resources. Proceedings of the Fourth National Symposium on Radioecology,* 12–14 May 1975, Oregon State Univ., Corvallis, OR. Ecol. Soc. Amer., Spec. Publ. No. 1.

Willis, D.L. and B.B. Valett. 1971. Metabolism of iodine-131 in two amphibian species (*Taricha granulosa* and *Rana pipiens*). Pages 394-400 in D.J. Nelson (ed.). *Radionuclides in Ecosystems. Proceedings of the Third National Symposium on Radioecology.* May 10–12, 1971, Oak Ridge, TN. Vol. 1. Available from Natl. Tech. Infor. Serv., Springfield, VA 22151.

Witherspoon, J.P. 1969. Radiosensitivity of forest tree species to acute fast neutron radiation. Pages 120-126 in D.J. Nelson and F.C. Evans (eds.). *Symposium on Radioecology. Proceedings of the Second National Symposium.* Available as CONF-370503 from The Clearinghouse for Federal Scientific and Technical Information, Natl. Bur. Standards, Springfield, VA 22151.

Wolfe, D.A. and C.D. Jennings. 1971. Iron-55 and ruthenium-103 and -106 in the brackish-water clam *Rangia cuneata*. Pages 783-790 in D.J. Nelson (ed.). *Radionuclides in Ecosystems. Proceedings of the Third National Symposium on Radioecology.* May 10–12, 1971, Oak Ridge, TN. Vol. 1. Available from Natl. Tech. Infor. Serv., Springfield, VA 22151.

Wolfe, D.A. and C.L. Schelske. 1969. Accumulation of fallout radioisotopes by bivalve molluscs from the lower Trent and Neuse Rivers. Pages 493-504 in D.J. Nelson and F.C. Evans (eds.). *Symposium on Radioecology. Proceedings of the Second National Symposium.* Available as CONF-670503 from The Clearinghouse for Federal Scientific and Technical Information, Natl. Bur. Standards, Springfield, VA 22151.

Woloschak, G.E., C.M. Chang-Liu, and P. Shearin-Jones. 1990a. Regulation of protein kinase C by ionizing radiation. *Cancer Res.* 50:3963-3967.

Woloschak, G.E., P. Shearin-Jones, and C.M. Chang-Liu. 1990b. Effects of ionizing radiation on expression of genes encoding cytoskeletal elements: kinetics and dose effects. *Molec. Carcinogen.* 3:374-378.

Wood, D.H. 1991. Long-term mortality and cancer risk in irradiated rhesus monkeys. *Radiation Res.* 126:132-140.

Woychik, R.P., W.M. Generoso, L.B. Russell, K.T. Cain, N.L.A. Cacheiro, S.J. Bultman, P.B. Selby, M.E. Dickinson, B.L.M. Hogan, and J.C. Rutledge. 1990. Molecular and structural rearrangement in mouse chromosome 2 causing mutations at the limb deformity and agouti loci. *Proc. Natl. Acad. Sci. U.S.A.* 87:2588-2592.

Wrenn, M.E., J.W. Lentsch, M. Eisenbud, G.J. Lauer, and G.P. Howells. 1971. Radiocesium distribution in water, sediment, and biota in the Hudson River estuary from 1964 through 1970. Pages 334-343 in D.J. Nelson (ed.). *Radionuclides in Ecosystems. Proceedings of the Third National Symposium on Radioecology.* May 10–12, 1971, Oak Ridge, TN. Vol. 1. Available from Natl. Tech. Infor. Serv., Springfield, VA 22151.

Yasuhira, S., H. Mitani, and A. Shima. 1991. Enhancement of photorepair of ultraviolet-damage by preillumination with fluorescent light in cultured fish cells. *Photochem. Photobiol.* 53:211-215.

Yathom, S., R. Padova, S. Tal, and I. Ross. 1990. Effects of gamma radiation on the immature stages of *Liriomyza trifolii. Photoparasitica* 18:117-124.

Yonezawa, M., A. Takeda, and J. Misonoh. 1990. Acquired radioresistance after low dose X-irradiation in mice. *Jour. Radiation Res.* 31:256-262.

Yoshimoto, Y., J.V. Neel, W.J. Schull, H. Kato, M. Soda, R. Eto, and K. Mabuchi. 1990. Malignant tumors during the first 2 decades of life in the offspring of atomic bomb survivors. *Amer. Jour. Human Genet.* 46:1041-1052.

Yusifov, N.I., A.M. Kuzin, F.A. Agaev, and S.G. Alieva. 1990. The effect of low level ionizing radiation on embryogenesis of silkworm, *Bombyx mori* L. *Radiation Environ. Biophys.* 29:323-327.

Zach, R., J.L. Hawkins, and S.C. Sheppard. 1993. Effects of ionizing radiation on breeding swallows at current radiation protection standards. *Environ. Toxicol. Chem.* 12:779-786.

Zakaria, A.H. 1991. Effect of low doses of gamma irradiation prior to egg incubation on hatchability and body weight of broiler chickens. *Brit. Poult. Sci.* 32:103-107.

Zavitkovski, J. and T.D. Rudolph. 1971. Predicted effects of chronic gamma irradiation of northern forest communities. Pages 1007-1014 in D.J. Nelson (ed.). *Radionuclides in Ecosystems. Proceedings of the Third National Symposium on Radioecology.* May 10–12, 1971, Oak Ridge, TN. Vol. 1. Available from Natl. Tech. Infor. Serv., Springfield, VA 22151.

32.11 GLOSSARY*

Actinides: Elements of atomic numbers 89 to 103 (Ac, Th, Pa, U, Np, Pu, Am, Cm, Bk, Cf, Es, Fm, Md, No, Lw).

Activity: The activity of a radioactive material is the number of nuclear disintegrations per unit time. Up to 1977, the accepted unit of activity was the curie (Ci), equivalent to 37 billion disintegrations/s, a number that approximated the activity of 1 g radium-226. The present unit of activity is the becquerel (Bq), equivalent to 1 disintegration/s.

Alpha (α) particles: An α particle is composed of two protons and two neutrons, with a charge of +2; essentially, it is a helium nucleus without orbital electrons. Alpha particles usually originate from the nuclear decay of radionuclides of atomic number >82, and are detected in samples containing U, Th, or Ra. Alpha particles react strongly with matter and consequently produce large numbers of ions per unit length of their paths. As a result, they are not very penetrating and will traverse only a few centimeters of air. Alpha particles are unable to penetrate clothing or the outer layer of skin; however, when internally deposited, α particles are often more damaging than most other types of radiations because comparatively large amounts of energy are transferred within a very small volume of tissue. Alpha particle absorption involves ionization and orbital electron excitation. Ionization occurs whenever the α particle is sufficiently near an electron to pull it from its orbit. The α particle also loses kinetic energy by exciting orbital electrons with interactions that are insufficient to cause ionization.

Atom: The smallest part of an element that has all the properties of that element. An atom consists of one or more protons and neutrons (in the nucleus) and one or more electrons.

Atomic number: The number of electrons outside the nucleus of a neutral (nonionized) atom and the number of protons in the nucleus.

Becquerel (Bq): The presently accepted unit of activity is the becquerel, equivalent to 1 disintegration/s. About 0.037 Bq = 1 picocurie.

Beta (β) particles: Beta particles are electrons that are spontaneously ejected from the nuclei of radioactive atoms during the decay process. They may either be positively or negatively charged. A positively charged beta (β^+), called a *positron*, is less frequently encountered than its negative counterpart, the *negatron* (β^-). The *neutrino*, a small particle, accompanies beta emission. The neutrino has very little mass and is electrically neutral; however, neutrinos conduct a variable part of the energy of transformation and account for the variability in kinetic energies of beta particles emitted from a given radionuclide. Positrons (β^+) are emitted by many of the naturally and artificially produced radionuclides; they are considerably more penetrating than α particles, but less penetrating than X-rays and γ rays. Beta particles interact with other electrons as well as nuclei in the travel medium. The ultimate fate of a beta particle depends on its charge. Negatrons, after their kinetic energy is spent, combine with a positively charged ion or become free electrons. Positrons also dissipate kinetic energy through ionization and excitation; the collision of positrons and electrons causes annihilation and release of energy equal to the sums of their particle masses.

Breeder reactor: A nuclear chain reactor in which transmutation produces a greater number of fissionable atoms than the number of consumed parent atoms.

Cosmic rays: Highly penetrating radiations that originate in outer space.

Curie (Ci): The Ci is equal to that quantity of radioactive material producing 37 billion nuclear transformations/s. One millicurie (mCi) = 0.001 Ci; 1 microcurie (μCi) = 1 millionth of a Ci; 1 picocurie (pCi) = 1 millionth of a millionth Ci = 0.037 disintegrations/s. About 27 pCi = 1 becquerel (Bq).

Decay: Diminution of a radioactive substance because of nuclear emission of α or β particles or of γ rays.

Decay product: A nuclide resulting from the radioactive disintegration of a radionuclide and found as the result of successive transformations in a radioactive series. A decay product may be either radioactive or stable.

Effective dose equivalent: The weighted sum, in sieverts (Sv), of the radiation dose equivalents in the most radiosensitive organs and tissues, including gonads, active bone marrow, bone surface cells, and the lung.

Electron: An electron is a negatively charged particle with a diameter of 10^{-12} cm. Every atom consists of one nucleus and one or more electrons. Cathode rays and negatrons are electrons.

Electron-volt (eV): Energy acquired by any charged particle that carries unit electronic charge when it falls through a potential difference of 1 volt. One eV = 1.602×10^{-19} joule.

Fission: The splitting of an atomic nucleus into two fragments that usually releases neutrons and γ rays. Fission may occur spontaneously or may be induced by capture of bombarding particles. Primary fission products usually decay by β particle emission to radioactive daughter products. The chain reaction that may result in controlled burning of nuclear fuel or in an uncontrolled nuclear weapons explosion results from the release of 2 or 3 neutrons/fission. Neutrons cause additional fissile nuclei in the vicinity to fission, producing still more neutrons, in turn producing still more fissions. The speed of the chain reaction is governed by the density and geometry of fissile nuclei and of materials that slow or capture the neutrons. In nuclear reactors, neutron-absorbing rods are inserted to various depths into the reactor core. A nuclear explosion is not physically possible in a reactor because of fuel density, geometry, and other factors.

Fusion: A nuclear reaction in which smaller atomic nuclei or particles combine to form larger ones with the release of energy from mass transformation.

Gamma (γ) rays: Gamma rays have electromagnetic wave energy that is similar to but higher than the energy of X-rays. Gamma rays are highly penetrating, being able to traverse several cm of lead. See **Photons**.

Genetically significant dose (GSD): A radiation dose that, if received by every member of the population, would produce the same total genetic injury to the population as the actual doses received by the various individuals.

Gray (Gy): 1 Gy = 1 joule/kg = 100 rad.

Half-life: The average time in which half the atoms in a sample of a radioactive element decay.

Hertz (Hz): A measure of frequency equal to 1 cycle/s.

Indirectly ionizing particles: Uncharged particles such as neutrons or photons that directly liberate ionizing particles or initiate nuclear transformations.

Ion: An atomic particle, atom, or chemical radical with either a negative or positive electric charge.

Ionization: The process by which neutral atoms become either positively or negatively electrically charged by the loss or gain of electrons.

Isomer: One of two or more radionuclides having the same mass number and the same atomic number, but with different energies and radioactive properties for measurable durations.

Isotope: One of several radionuclides of the same element (i.e., with the same number of protons in their nuclei) with different numbers of neutrons and different energy contents. A single element may have many isotopes. Uranium, for example, may appear naturally as ^{234}U (142 neutrons), ^{235}U (143 neutrons), or ^{238}U (146 neutrons); however, each uranium isotope has 92 protons.

Joule (J): 1 J = 10^7 ergs.

Latent period: Period of seeming inactivity between time of exposure of tissue to an acute radiation dose and the onset of the final stage of radiation sickness.

Linear energy transfer (LET): A function of the capacity of the radiation to produce ionization. LET is the rate at which charged particles transfer their energies to the atoms in a medium and a function of the energy and velocity of the charged particle. See **Radiation dose**.

Linear hypothesis: The assumption that any radiation causes biological damage in direct proportion of dose to effect.

Mass number: The total number of neutrons and protons in the nucleus of the element, and equal to the sum of the atomic number and the number of neutrons.

Meson: Particles of mass that are intermediate between the masses of the electron and proton.

Neutrinos: Neutrinos and antineutrinos are formed whenever a positron particle is created in a radioactive decay; they are highly penetrating.

Neutrons: Neutrons are electrically neutral particles that consist of an electron and a proton and are not affected by the electrostatic forces of the atom's nucleus or orbital electrons. Because they have no charge, neutrons readily penetrate the atom and may cause a nuclear transformation. Neutrons are produced in the atmosphere by cosmic ray interactions and combine with nitrogen and other gases to form carbon-14, tritium, and other radionuclides. A free neutron has a lifetime of about 19 minutes, after which it spontaneously decays to a proton, a β particle, and a neutrino. A high-energy neutron that encounters biological material is apt to collide with a proton with sufficient force to dislodge the proton from the molecule. The recoil proton may then have sufficient energy to cause secondary damage through ionization and excitation of atoms and molecules along its path.

Nucleus: The dense central core of the atom in which most of the mass and all of the positive charge is concentrated. The charge on the nucleus distinguishes one element from another.

Photons: X-rays and gamma (γ) rays, collectively termed photons, are electromagnetic waves with shorter wavelengths than other members of the electromagnetic spectrum such as visible radiation, infrared radiation, and radiowaves. X- and γ photons have identical properties, behavior, and effects. Gamma rays originate from atomic nuclei, but X-rays arise from the electron shells. All photons travel at the speed of light, but energy is inversely proportional to wavelength. The energy of a photon directly influences its ability to penetrate matter. Many types of nuclear transformations are accompanied by γ ray emission. For example, α and β decay of many radionuclides is frequently accompanied by γ photons. When a parent radionuclide decays to a daughter nuclide, the nucleus of the daughter frequently contains excess energy and is unstable; stability is usually achieved through release of one or more γ photons, a process called *isometric transition*. The daughter nucleus decays from one energy state to another without a change in atomic number or weight.

The most probable fate of a photon with an energy higher than the binding energy of an encountered electron is photoelectric absorption, in which the photon transfers its energy to the electron and photon existence ends. As with ionization from any process, secondary radiations initiated by the photoelectron produce additional excitation of orbital electrons.

Planck's constant (h): A universal constant of nature that relates the energy of a photon of radiation to the frequency of the emitting oscillator. Its numerical value is about 6.626×10^{-27} ergs/s.

Positron: A positively charged particle of mass equal to an electron. Positrons are created either by the radioactive decay of unstable nuclei or by collision with photons.

Proton: A positively charged subatomic particle with a mass of 1.67252×10^{-24} g that is slightly less than the mass of a neutron but about 1836 times greater than the mass of an electron. Protons are identical to hydrogen nuclei; their charge and mass make them potent ionizers.

Radiation: The emission and propagation of energy through space or through a material medium in the form of waves. The term also includes subatomic particles, such as α, β, and cosmic rays and electromagnetic radiation.

Radiation absorbed dose (rad): Radiation-induced damage to biological tissue results from the absorption of energy in or around the tissue. The amount of energy absorbed in a given volume of tissue is related to the types and numbers of radiations and the interactions between radiations and tissue atoms and molecules. The fundamental unit of the radiation absorbed dose is the rad; 1 rad = 100 erg (absorbed)/g material. In the latest nomenclature, 100 rad = 1 gray (Gy).

Radiation dose: The term "radiation dose" can mean several things, including absorbed dose, dose equivalent, or effective dose equivalent. The absorbed dose of radiation is the imparted energy per unit mass of the irradiated material. Until 1977, the rad was the unit of absorbed dose, wherein 1 rad = 0.01 Joule/kg. The present unit of absorbed dose is the gray (Gy), equivalent to 1 Joule/kg. Thus, 1 rad = 0.01 Joule/kg = 0.01 Gy. Different types of radiation have different Relative Biological Effectiveness (RBE). The RBE of one type of radiation in relation to a reference type of radiation (usually X or γ) is the inverse ratio of the absorbed doses of the two radiations needed to cause the same degree of the biological effect for which the RBE is given. Regulatory agencies have recommended certain values of RBE for radiation protection, and absorbed doses of various radiations are multiplied by these values to arrive at radioprotective doses. The unit of this weighted absorbed dose is the roentgen equivalent man (rem). The dose equivalent is the product of the absorbed dose and a quality factor (Q), and its unit is the rem. The quality factor is a function of the capacity to produce ionization, expressed as the linear energy transfer (LET). A Q value is assigned to each type of radiation: 1 to X-rays, γ rays, and β particles; 10 to fast neutrons; and 20 to α particles and heavy particles. The new unit of the effective dose equivalent is the sievert (Sv), replacing rem, where 1 Sv = 100 rem. In addition to absorbed dose and dose equivalent, there is also the exposure. Exposure is the total electrical charge of ions of one sign produced in air by electrons liberated by X- or gamma rays per unit mass of irradiated air. The unit of exposure is Coulomb/kg, but the old unit, the roentgen (R) is still in use. One roentgen = 2.58×10^{-4} Coulomb/kg.

Radioactivity: The process of spontaneous disintegration by a parent radionuclide, which releases one or more radiations and forms a daughter nuclide. When half the radioactivity remains, that time interval is designated the half-life (Tb 1/2). The Tb 1/2 value gives some insight into the behavior of a radionuclide and into its potential hazards.

Radionuclide: An atom that is distinguished by its nucleus composition (number of protons, number of neutrons, energy content), atomic number, mass number, and atomic mass.

Relative biological effectiveness (RBE): The biological effectiveness of any type of ionizing radiation in producing a specific damage (i.e., leukemia, anemia, carcinogenicity). See **Radiation dose**.

Roentgen (R): 1 R = 2.58×10^{-4} Coulombs/kg air = production by X- or γ rays of one electrostatic unit of charge per cm^3 of dry air at 0°C and 760 mmHg = 0.87 rad in air.

Roentgen equivalent man (rem): The amount of ionizing radiation of any type that produces the same damage to humans as 1 roentgen of radiation. One rem = 1 roentgen equivalent physical (rep)/relative biological effectiveness (RBE). In the latest nomenclature, 100 rem = 1 Sievert (Sv).

Roentgen equivalent physical (rep): One rep is equivalent to the amount of ionizing radiation of any type that results in the absorption of energy of 93 ergs/g, and is approximately equal to 1 roentgen of X-radiation in soft tissue.

Shell: Extranuclear electrons are arranged in orbits at various distances from the nucleus in a series of concentric spheres called shells. In order of increasing distance from the nucleus, the shells are designated the K, L, M, N, O, P, and Q shells; the number of electrons that each shell can contain is limited.

Sievert (Sv): New unit of dose equivalent. One Sv = 100 rem = 1 J/kg. See **Radiation dose**.

Specific activity: The ratio between activity (in number of disintegrations/min) and the mass (in grams) of material giving rise to the activity. Biological hazards of radionuclides are directly related to their specific activity and are expressed in Bq/kg mass.

Threshold hypothesis: A radiation-dose-consequence hypothesis that holds that biological radiation effects will occur only above some minimum dose.

Transmutation: A nuclear change that produces a new element from an old one.

Transuranic elements: Elements of atomic number >92. All are radioactive and produced artificially; all are members of the actinide group.

X-rays: See **Photons**.

*From Whicker and Schultz 1982a; League of Women Voters (LWV) 1985; Weast 1985; Hobbs and McClellan 1986; United Nations Scientific Committee on the Effects of Atomic Radiation (UNSCEAR) 1988; U.S. Code of Federal Regulations (USCFR) 1990; Eisler 1994.

CHAPTER 33

Cumulative Index to Chemicals and Species

33.1 INTRODUCTION

Indices are presented of biologically active compounds or substances, and common and scientific names of all living species listed in the three-volume *Handbook of Chemical Risk Assessment* series.

33.2 INDEX TO CHEMICALS

All chemicals, chemical trade names, and other substances with known biological properties listed in the *Handbook of Chemical Risk Assessment* series — a total of about 1600 — are presented in Table 33.1.

33.3 INDEX TO SPECIES

Taxonomic nomenclatures for plants and animals are under constant revision. In the *Handbook of Chemical Risk Assessment* series, the author elected to conform as much as possible to the systems and spellings used by Scott and Wasser (1980) for plants, Swain and Swain (1948) for insects, Turgeon et al. (1988) for aquatic molluscs, Williams et al. (1989) for decapod crustaceans, Pratt (1935) and Hyman (1940, 1951a, 1951b, 1955) for miscellaneous invertebrates, Robins et al. (1991) for fishes, Ditmars (1966) for reptiles, Edwards (1974) and Howard and Moore (1991) for birds, and Nowak and Paradiso (1983) for mammals. Individual species are arranged alphabetically by scientific and common names (Table 33.2). In total, about 2300 species of animals and plants were cited, of which only 23 (1.0%) were listed in at least 20 chapters. The most widely cited species include:

- One species of plant (corn, *Zea mays*)
- Two species of invertebrates (freshwater crustacean, *Daphnia magna*; American oyster, *Crassostrea virginica*)
- Seven species of teleosts (channel catfish, *Ictalurus punctatus*; bluegill, *Lepomis macrochirus*; coho salmon, *Oncorhynchus kisutch*; rainbow trout, *Oncorhynchus mykiss*; fathead minnow, *Pimephales promelas*; brook trout, *Salvelinus fontinalis*; lake trout, *Salvelinus namaycush*)

- Three species of birds (mallard, *Anas platyrhynchos*; domestic chicken, *Gallus* sp.; Japanese quail, *Coturnix japonica*)
- Ten species of mammals (cow, *Bos* spp.; domestic dog, *Canis familiaris*; guinea pig, *Cavia* spp.; domestic cat, *Felis domesticus*; human, *Homo sapiens*; hamster, *Cricetus* spp.; domestic mouse, *Mus* spp.; domestic sheep, *Ovis aries*; laboratory white rat, *Rattus* spp.; domestic pig, *Sus* spp.).

It is probable that these species are not representative of unusually sensitive or endangered species, but they can be considered appropriate sentinel organisms for many species of free-living wildlife.

33.4 SUMMARY

An index is provided to the common and scientific names of approximately 2300 biological species listed in the *Handbook of Chemical Risk Assessment* series. A similar index is shown for the approximately 1600 chemicals, chemical trade names, and other substances with known biological properties.

33.5 LITERATURE CITED

Ditmars, R.L. 1966. *Reptiles of the World.* Macmillan, New York. 321 pp. + 89 plates.

Edwards, E.P. 1974. *A Coded List of Birds of the World.* Ernest P. Edwards, Sweet Briar, VA. 174 pp.

Howard, R. and A. Moore. 1991. *A Complete Checklist of the Birds of the World. Second Edition.* Academic Press, New York. 622 pp.

Hyman, L.H. 1940. *The Invertebrates. Volume 1. Protozoa Through Ctenophora.* McGraw-Hill, New York. 726 pp.

Hyman, L.H. 1951a. *The Invertebrates. Volume 2. Platyhelminthes and Rhyncocoela. The Acoelomate Bilateria.* McGraw-Hill, New York. 550 pp.

Hyman, L.H. 1951b. *The Invertebrates. Volume 3. Acanthocephala, Aschelminthes, and Entoprocta. The Pseudocoelomate Bilateria.* McGraw-Hill, New York. 572 pp.

Hyman, L.H. 1955. *The Invertebrates. Volume 4. Echinodermata. The Coelomate Bilateria.* McGraw-Hill, New York. 763 pp.

Nowak, R.M. and J.L. Paradiso. 1983. *Walker's Mammals of the World. Volumes I and II.* Johns Hopkins Univ. Press, Baltimore. 1362 pp.

Pratt, H.S. 1935. *A Manual of the Common Invertebrate Animals.* P. Blakiston's Son and Company, Philadelphia. 854 pp.

Robins, C.R., R.M. Bailey, C.E. Bond, J.R. Brooker, E.A. Lachner, R.N. Lea, and W.B. Scott. 1991. *Common and Scientific Names of Fishes from the United States and Canada. Fifth Edition.* Amer. Fish. Soc. Spec. Publ. 20. 183 pp.

Scott, T.G. and C.H. Wasser. 1980. Checklist of North American Plants for Wildlife Biologists. The Wildlife Society, 7101 Wisconsin Ave. NW, Washington, D.C. 20014. 58 pp.

Swain, R.B. and S.N. Swain. 1948. *The Insect Guide.* Doubleday, New York. 261 pp.

Turgeon, D.D., A.E. Bogan, E.V. Coan, W.K. Emerson, W.G. Lyons, W.L. Pratt, C.F.E. Roper, A. Scheltema, F.G. Thompson, and J.D. Williams. 1988. *Common and Scientific Names of Aquatic Invertebrates from the United States and Canada: Mollusks.* Amer. Fish. Soc. Spec. Publ.16. 227 pp. + 12 plates.

Williams, A.B., L.G. Abele, D.L. Felder, H.H. Hobbs, Jr., R.B. Manning, P.A. McLaughlin, and I.P. Farfante. 1989. *Common and Scientific Names of Aquatic Invertebrates from the United States and Canada: Decapod Crustaceans.* Amer. Fish. Soc. Spec. Publ. 17. 77 pp. + 4 plates.

Table 33.1 Chemical and Trade Names of Substances Listed in the *Handbook of Chemical Risk Assessment* Series

1068, 824
1080, 1413–1452
38023, 1069

A

Aatrex, 769
Aatrex 4L, 769
Aatrex 4LC, 769
Aatrex nine-o, 769
Aatrex 80W, 769
AC 38023, 1069
Acenaphthalene, 1366, 1367, 1372, 1384, 1394, 1395
Acenaphthanthracene, 1389
Acenaphthene, 1358, 1369
Acenaphthylene, 1358, 1365
Acetaldehyde, 323
Acetamide, 1425
Acetaminophen, 615
Acetazolamide, 615
Acetone cyanohydrin, 939, 943
Acetonitrile, 939, 943, 1250
Acetylaminofluorene, 1384
N-Acetylcysteine, 47, 745, 1164
N-Acetyl-D-glucosamine, 985, 992
N-Acetylglucosamine, 985, 992
Acid copper chromate, 77
Acraaldehyde, 743
Acraldehyde, 743
Acridine, 1372, 1394, 1395
Acrolein, 739–763
Acryladehyde, 743
Acrylic acid, 741, 742, 744, 746, 754, 755
Acrylic aldehyde, 743
Acrylonitrile, 752, 754, 907, 913, 939, 943
Actinium-227, 1712, 1725, 1792, 1799
Actinium-228, 1712
Actinomycin D, 611
Adiponitrile, 913
AG-500, 96
Agent orange, 1022, 1023, 1053
Agmatrin, 1091
Agracide Maggot Killer, 1460
Alachlor, 1195
Aldehyde dehydrogenase, 748, 760
Aldrin, 864
Alfa-tox, 962
Alkyl isocyanates, 907
Alkylmercury, 333
Allethrin, 1089, 1098
Alltox, 1460
Allyl alcohol, 740, 744, 745
Allyl aldehyde, 743
Allylamine, 755
Allyl formate, 740
Aluminum, 102, 131, 741, 1531, 1569, 1613
Aluminum-borate complex, 1569
Aluminum oxide, 1570
Amdro, 1152
American Cyanamid 38023, 1069
Americium, 1719, 1720, 1782, 1783, 1802, 1803
Americium-241, 1713, 1721, 1723, 1725, 1728, 1733, 1735, 1736, 1738, 1744, 1746, 1747, 1748, 1779, 1783, 1791, 1793, 1800, 1801, 1803, 1804
Americium-243, 1713, 1803
O-Aminoacetophone, 932
α-Aminobutyronitrile, 921
2-Amino-4-chloro-6-ethylamino-1,3,5-triazine, 769, 789
2-Amino-4-chloro-6-isopropylamino-1,3,5-triazine, 789
m-Aminophenol, 743
p-Aminopropiophenone, 912
2-Aminothiazoline-4-carboxylic acid, 909
Amitrol, 1160
Ammonia, 917, 921, 926
Ammonium alginate, 1195
Ammonium borate, 1594
Ammonium chromate, 77
Ammonium dichromate, 66
Ammonium diuranate, 1721
Ammonium hexacyanoferrate, 1758
Ammonium hydroxide, 416
Ammonium molybdate, 169, 174, 1634, 1636
Ammonium paramolybdate, 1636
Ammonium thiocyanate, 913
Amobarbitol, 863
Amygdalin, 903, 915
Amylnitrite, 947
Angeleside, 203
Aniline, 911
Anisoles, 1197, 1204, 1287
Anthanthrene, 1384
Anthracene, 1287, 1348, 1355, 1356, 1358, 1359, 1360, 1365, 1367, 1369, 1372, 1375, 1378, 1384, 1385, 1387, 1394, 1395
Antimony, 1678
Antimony-124, 1711
Antimony-125, 1711, 1725, 1734, 1737, 1750, 1751, 1768
Antimony-127, 1711
Aqualin, 743
Aquilin, 743
Argentite, 500
Argentum, 504
Argentum crede CI 77820, 504
Argon-39, 1710, 1721
Aroclor 1016, 1251, 1256, 1257, 1258, 1282, 1314, 1321, 1322
Aroclor 1221, 1251, 1282, 1322
Aroclor 1232, 1282, 1322
Aroclor 1242, 1239, 1251, 1257, 1282, 1296, 1321, 1322
Aroclor 1246, 1312
Aroclor 1248, 1239, 1281, 1282, 1296, 1309, 1314, 1321, 1322
Aroclor 1254, 1239, 1251, 1252, 1253, 1255, 1258, 1281, 1282, 1283, 1296, 1309, 1312, 1314, 1315, 1316, 1317, 1321, 1322
Aroclor 1260, 1251, 1252, 1253, 1281, 1282, 1296, 1309, 1314
Aroclor 1268, 1281
Arsanilic acid, 1534, 1542, 1544, 1550, 1552, 1553

Table 33.1 (continued) Chemical and Trade Names of Substances Listed in the *Handbook of Chemical Risk Assessment* Series

Arsenates, 1504, 1505, 1508, 1509, 1513, 1516, 1523, 1532, 1534, 1535, 1542, 1544, 1547, 1548, 1549, 1550
Arsenic, 28, 77, 166, 420, 615, 1501–1557, 1674, 1678, 1685
Arsenic acid, 1502, 1504, 1526, 1547, 1549
Arsenic pentoxide, 1505, 1547
Arsenic sulfides, 1502, 1506, 1511
Arsenic trioxide, 1502, 1504, 1505, 1507, 1510, 1511, 1531, 1533, 1534, 1542, 1546, 1547, 1548, 1549, 1550, 1555
Arsenites, 1504, 1505, 1507, 1513, 1516, 1523, 1532, 1533, 1547, 1548, 1550
Arsenobetaine, 1509, 1514–1515, 1516, 1520, 1521, 1523, 1526, 1536, 1549, 1550
Arsenocholine, 1514, 1516, 1523, 1526, 1549
Arsenous acid, 1507, 1526
Arsenous oxide, 1549
Arsines, 1503, 1505, 1555
Arsonous acid, 1507
Arsphenamine, 1503
Asarco, 608
Ascorbic acid, 20, 23, 47, 75, 100, 104, 106, 208, 210, 471, 745, 912, 1390
Aspon, 824
Astatine-215, 1712
Astatine-218, 1712
Astatine-219, 1712
Atranex, 769
Atratol, 769
Atratol 8P, 769
Atratol 80W, 769
Atrazine, 767–790, 1168, 1195
Atrazine 4L, 769
Atrazine 80W, 769
Atred, 769

B

BAL, 481
Barium, 420
Barium-137, 1799
Barium-140, 1711, 1721, 1731, 1742, 1753, 1755, 1765, 1787
Barium sulphonate, 72
Basudin, 962
Bay 70142, 800
Belmark, 1091
Belt, 824
Bentonite, 72, 1164, 1758, 1788
Benz*(a)*anthracene, 1347, 1348, 1355, 1358, 1359, 1360, 1362, 1364, 1365, 1366, 1367, 1369, 1370, 1371, 1372, 1375, 1378, 1383, 1385, 1386, 1387, 1388, 1393, 1395
Benzaldehyde, 915, 940
Benzene, 1024, 1068, 1079, 1134, 1196
Benzo*(a)*fluoranthene, 1395
Benzo*(a)*fluorene, 1369, 1372
Benzo*(a)*perylene, 1355
Benzo*(a)*pyrene, 425, 426, 754, 1247, 1347, 1348, 1349, 1352, 1353, 1356, 1358, 1359, 1360, 1362, 1364, 1365, 1366, 1367, 1368, 1369, 1370, 1372, 1373, 1375–1376, 1377, 1378, 1382, 1383, 1384, 1385, 1386, 1387, 1388, 1393, 1395, 1530, 1572
Benzo*(b)*fluoranthene, 1347, 1356, 1358, 1359, 1362, 1364, 1365, 1366, 1369, 1383, 1387, 1388, 1395
Benzo*(b,k)*fluoranthene, 1386
Benzo*(c)*phenanthrene, 1347
Benzo*(e)*pyrene, 1355, 1359, 1362, 1365, 1369, 1383
Benzo*(g,h,i)*fluoranthene, 1358, 1359, 1364, 1383
Benzo*(g,h,i)*perylene, 1348, 1356, 1358, 1364, 1367, 1369, 1386, 1395
Benzo*(j)*fluoranthene, 1358, 1383, 1395
Benzo*(k)*fluoranthene, 1355, 1356, 1358, 1359, 1364, 1365, 1367, 1383, 1387, 1395
Benzoylphenylureas, 990
Benzyl isothiocyanate, 1371
Berkelium, 1782
Beryllium-7, 1710, 1720
Beryllium-10, 1710
Beta-dimethylcysteamine, 745
Beta-ecdysone, 994
2-(Beta-D-glucopyranosyloxy)-isobutyronitrile, 937
Beta-hydroxypropionaldehyde, 743–744
Beta-propionaldehyde, 745
Bicep, 769
Bicep 4.5L, 769
Bincennite, 914
4,4'-Bipyridyl, 1166
O,N-Bisdesmethylfamphur, 1070, 1078, 1082, 1083
Bismuth, 1614, 1717
Bismuth-207, 1750, 1751
Bismuth-210, 1712, 1789, 1798
Bismuth-211, 1712
Bismuth-212, 1712
Bismuth-214, 1712, 1797
Bismuth-215, 1712
Bismuth subnitrate, 911
Blue copperas, 97
Blue powder, 608
Blue vitriol, 97
Bo-Ana, 1069
Boracite, 1568
Boranes, 1567, 1570, 1571, 1572, 1595
Borates, 1568, 1569, 1571, 1581, 1582, 1594, 1595, 1596, 1597, 1598, 1603
Borax, 1567, 1568, 1570, 1571, 1573, 1584, 1585, 1588, 1589, 1590, 1593, 1594, 1596, 1597, 1598, 1599, 1601
Borax pentahydrate, 1568
Borazines, 1571
Bordeaux mixture, 96
Boric acid, 1567, 1571, 1572, 1581, 1584, 1585, 1586, 1587, 1588, 1589, 1590, 1591, 1592, 1593, 1594, 1595, 1596, 1597, 1598, 1599, 1601, 1603, 1604
Boric oxide, 1568
Boron, 1515, 1567–1605
Boron-10, 1569, 1572, 1573
Boronatrocalcite, 1568
Boron oxide, 1571, 1595, 1598, 1599, 1603
Boron tribromide, 1572, 1603
Boron trichloride, 1570, 1572
Boron trifluoride, 1570, 1572, 1595, 1600, 1603
Brifur, 800
Brodan, 884

Table 33.1 (continued) Chemical and Trade Names of Substances Listed in the *Handbook of Chemical Risk Assessment* Series

Brodifacoum, 1418, 1451
Bromphemothrin, 1098
Butyltins, 555, 563, 564, 570, 571, 587
N-Butyronitrile, 943

C

C. I. No. 77775, 416
Cacodylic acid, 1503, 1504, 1530, 1531, 1532, 1533, 1534, 1543, 1544, 1546, 1547, 1548, 1549, 1550, 1551
Cadmium, 1–34, 64, 102–103, 132, 134, 136, 172, 206, 374, 420, 461, 514, 559, 1508, 1655, 1674, 1685, 1783
Cadmium-109, 27, 1711, 1787
Cadmium-115, 27, 1711
Cadmium chloride, 23, 25, 28, 268
Cadmium cyanide, 906
Cadmium fluoride, 19
Cadmium fluroborate, 19
Cadmium-113m, 1711, 1725
Cadmium-115m, 27, 1711
Cadmium selenide, 1651
Calciferol, 1451
Calcineurin, 477
Calcium, 23, 100, 139, 206, 420, 422, 507, 1099, 1425, 1530, 1531, 1571, 1575, 1582, 1614, 1684, 1749, 1784, 1799
Calcium-45, 1710, 1721
Calcium acetate, 425
Calcium arsenate, 1504, 1506, 1531, 1547, 1548, 1551
Calcium borate, 1573, 1597, 1598, 1603
Calcium carbonate, 28, 31, 151, 155, 156, 157, 158, 159, 176, 207, 533, 540, 688, 926
Calcium chromate, 75
Calcium cyanide, 905, 906, 913, 938
Calcium glutonate, 1424, 1425, 1445, 1450
Calcium molybdate, 1634, 1636
Calcium sodium EDTA, 105, 615
Calcium trisodium diethylenetriamine penta acetic acid, 105
Californium, 1782, 1802
Californium-252, 1713, 1801, 1804
Calmodulin, 422
Camphechlor, 1460
Camphofene Huilex, 1460
Captafol, 812
Captan, 977
Carbaryl, 976
Carbofuran, 799–817
Carbofuran phenol, 803, 815
Carbon, 1717, 1799
Carbon-14, 1315, 1710, 1713, 1719, 1720, 1721, 1727, 1728, 1736, 1737, 1738, 1740, 1804
Carbon dioxide, 917, 920, 921
Carbon tetrachloride, 912, 1134, 1196, 1371
Carbonyl nickel powder, 416, 424
Carbopol, 1427
4-Carboxy-l-methylpyridium ion, 1166, 1167
CAS 52-85-7, 1069
CAS 87-86-5, 1196
CAS 107-02-8, 743
CAS 151-50-8, 906
CAS 333-41-5, 962
CAS 1563-66-2, 800
CAS 1746-01-6, 1024
CAS 1910-42-5, 1163
CAS 1912-24-9, 769
CAS 2385-85-5, 1134
CAS 2921-88-2, 884
CAS 4685-14-17, 1163
CAS 5103-71-9, 824
CAS 5103-74-2, 824
CAS 7440-02-0, 416
CAS 7440-22-4, 504
CAS 7440-50-8, 97
CAS 7440-66-6, 608
CAS 7646-85-7, 608
CAS 7733-02-0, 608
CAS 7758-98-7, 97
CAS 7761-88-8, 504
CAS 8001-35-2, 1460
CAS 35367-38-5, 984
CAS 51630-58-1, 1091
Cassiterite, 557
Catechols, 11, 1025
CD-68, 824
Cekuquat, 1163
Celatom MP-78, 913
Cerium, 1783
Cerium-141, 1711, 1719, 1726, 1736, 1737, 1739, 1753, 1755, 1758, 1765
Cerium-143, 1711, 1721
Cerium-144, 1711, 1719, 1720, 1721, 1725, 1726, 1731, 1734, 1736, 1737, 1739, 1740, 1750, 1751, 1753, 1755, 1757, 1758, 1765, 1768, 1781, 1800
Cerium chloride, 1165
Ceruloplasmin, 99, 101, 105, 139, 169, 172, 1180, 1632, 1633
Cerusite, 203
Cesium, 1760, 1783, 1799
Cesium-134, 1711, 1721, 1736, 1737, 1741, 1742, 1743, 1744, 1746, 1747, 1753, 1755, 1757, 1758, 1760, 1761, 1762, 1763, 1764, 1765, 1766, 1768, 1769, 1787, 1791, 1799, 1802, 1803
Cesium-135, 1711, 1755, 1804
Cesium-137, 1711, 1713, 1719, 1720, 1721, 1722, 1725, 1726, 1727, 1728, 1729, 1730, 1731, 1732, 1733, 1734, 1735, 1736, 1737, 1738, 1739, 1740, 1741, 1742, 1743, 1744, 1745, 1746, 1747, 1748, 1749, 1750, 1752, 1753, 1757, 1758, 1759, 1760, 1761, 1762, 1763, 1764, 1765, 1766, 1767, 1768, 1769, 1773, 1775, 1782, 1783, 1786, 1787, 1788, 1791, 1793, 1799, 1801, 1802, 1803, 1804, 1806
Cesium chloride, 1769
Chalcocite, 94
Chalcopyrite, 94, 203
Chem-penta, 1196
Chemtrol, 1196
Chitin, 984, 985, 993–994
Chitinase, 985
Chitin synthetase, 985, 990
Chlordan, 824
Chlor-dan, 824

Table 33.1 (continued) Chemical and Trade Names of Substances Listed in the *Handbook of Chemical Risk Assessment* Series

Chlordane, 799, 823–872, 1143
Chlordene, 825, 869
3-Chlordene, 864
Chlordene epoxide, 864
Chlorfluzuaron, 990
Chlorinated camphene, 1460
Chlorinated diphenyl ethers, 1197
Chlorindan, 824
Chlor-kil, 824
Chlornitofen, 1023
4-Chloroaniline, 984, 985, 993, 1001, 1005, 1008, 1011, 1013
Chloroanisoles, 1214
Chlorobenzenes, 156, 1024
2-Chlorobiphenyl, 1257
4-Chlorobiphenyl, 1308
4'-Chloro-3,4-biphenyldiol, 1312
4'-Chloro-4-biphenylol, 1312
2-Chlorochlordene, 869
3-Chlorochlordene, 864, 869
Chlorodane, 824
Chlorodene epoxide, 869
2-Chloro-4,6-diamino-1,3,5-triazine, 789
2-Chloro-4-ethylamino-6-isopropylamino-1,3,5-triazine, 769. *See also* Atrazine
Chloroform, 769, 1024
4'-Chloroformanilide, 991
4-Chloro-4'-hydroxybiphenyl, 1308
4'Chloro-3-methoxy-4-biphenylol, 1312
Chlorophen, 1196
Chlorophenols, 1022, 1024, 1197, 1198, 1216
N-[[(4-Chlorophenyl)amino]carbonyl]-2,6-difluoro-3-hydroxybenzamide], 1009
1-(4-Chlorophenyl)-3-(2,6-difluorobenzoyl)urea, 983. *See also* Diflubenzuron
4-Chlorophenyl isocyanate, 984
3-(4-Chlorophenyl) isovaleric acid, 1099
2-(4-Chlorophenyl)-3-methylbutyric acid, 1114
4-Chlorophenylurea, 983, 984, 987, 991, 993, 1001, 1005, 1008, 1011, 1013
Chlorpyrifos, 883–897, 1151
Chlorpyrifos-ethyl, 884
Chlorpyrifos oxon, 885
Chlortox, 824
Cholanthrene, 1347
Cholecalciferol, 1451
Cholesterol (2R)-2-(4-chlorophenol) isovalerate, 1101
Cholestyramine, 1216, 1218
Chromated copper arsenate, 77, 113, 128, 132, 133, 165–166
Chromate ion, 47
Chrome lignosulphonate, 72, 77
Chromic acid, 45
Chromic oxide, 46, 69
Chromic trioxide, 75
Chromium, 45–82, 461, 1257, 1616, 1630, 1685, 1783
Chromium-51, 72, 74, 76, 1710, 1742, 1787
Chromium chloride, 69
Chrysene, 1347, 1356, 1358, 1359, 1360, 1362, 1364, 1365, 1366, 1367, 1369, 1373, 1376, 1383, 1386, 1387, 1394, 1395
CI 77949, 608
CI pigment metal, 608
Cismethrin, 1089, 1098
Citric acid, 96
Citrinin, 47
CL 38023, 1069
Clofibrate, 1164
Clophen A30, 1251
Clophen A40, 1251
Clophen A50, 1251
Clophen A60, 1251
Clor chem T-590, 1460
Cobalt, 26, 420, 912, 1614, 1783, 1799
Cobalt-57, 1710, 1721, 1724, 1736, 1737, 1750, 1751
Cobalt-58, 1710, 1721, 1724, 1735, 1787
Cobalt-60, 1710, 1721, 1724, 1730, 1731, 1734, 1735, 1736, 1737, 1739, 1740, 1742, 1746, 1749, 1750, 1751, 1752, 1768, 1787, 1791, 1801, 1804
Cobalt chloride, 912, 1008
Cobalt cyanide, 906
Cobalt edetate, 951
Cobalt histidine, 912
Colemanite, 1568
Compound 1080, 1413–1452
Co-op, 769
Co-op atra-pril, 769
Copper, 17, 26, 28, 93–181, 374, 509, 514, 533, 559, 1257, 1614, 1615, 1632, 1639, 1673, 1674
Copper-64, 100, 101, 1710
Copper acetoarsenite, 1542, 1543, 1544, 1551
Copper acetyl acetonate, 96
Copper arsenate, 132, 1504
Copper arsenite-acetate, 96
Copper carbonate, 96, 98, 100
Copper chloride, 96, 98, 170
Copper dimethyl dithiocarbamate, 96
Copper glutamate, 170
Copper-glycine complex, 105
Copper hydroxide, 98
Copper hydroxycarbonate, 96
Copper 8-hydroxyquinoline, 106
Copper nitrate, 98
Copper oxide, 96, 109
Copper pentachlorophenate, 96
Copper ricinoleate, 96
Copper rosinate, 96
Copper sulfate, 95, 96, 142, 148, 168, 171, 172, 1616
Copper sulfate pentahydrate, 96
Copper sulfide, 109
Copper-tartaric acid, 96
Copper-zinc superoxide dismutase, 20
Corodane, 824
Coronene, 1355
Cortilan-neu, 824
Coumatetralyl, 1418
Cresylic acid, 166
Crisazine, 769
Crisfuran, 800
Crisquat, 1163
Cristatrina, 769
Cristofuran, 800
Cristoxo, 1460
Cupric acetate, 103
Cupric hydroxide, 98
Cupric oleinate, 97
Cupric oxide, 98
Cupric sulfate, 97
Cupric sulfide, 98

Table 33.1 (continued) Chemical and Trade Names of Substances Listed in the *Handbook of Chemical Risk Assessment* Series

Cuprous arsenite, 1504
Cuprous chloride, 98
Cuprous oxide, 98
Curaterr, 800
Curium, 1782, 1783, 1802
Curium-241, 1713
Curium-242, 1713, 1721, 1735, 1748, 1753
Curium-243, 1713, 1748
Curium-244, 1713, 1723, 1748, 1783
Curium-247, 1713, 1723
Curium-248, 1713, 1723
Curium-250, 1713
Cyanic acid, 907
Cyanide, 98, 903–952
Cyanides, weak-acid dissociable, 906
Beta-Cyanoalanine, 921
Cyanocobalamin, 909, 912
Cyanogen, 905, 907
Cyanogen bromide, 913, 942
Cyanogen chloride, 913, 920, 941
Cyanogenic glycosides, 903–904
Cyanohydric acid, 905
Cyanohydrins, 907, 912, 922
α-Cyano-3-phenoxybenzyl α-(4-chlorophenyl)isovalerate, 1091
α-Cyano-3-phenoxybenzyl 2-(4-chlorophenyl)-3-methylbutyrate, 1091
(*RS*)-α-Cyano-3-phenoxybenzyl-(*RS*)-2-4-chlorophenyl-3-methylbutyrate, 1091
Cyano (3-phenoxyphenyl)methyl-4-chloro-α-(1-methylethyl) benzeneacetate, 1091
Cycasin, 904
Cyclam, 421
Cyclethrin, 1089
Cyclohexanediamine tetraacetic acid, 421
Cycloheximide, 24
Cyclopenta*(cd)*pyrene, 1390
Cyclopentadiene, 824
Cyclophosphamide, 740, 745, 754
Cyflee, 1069
Cyhexatin, 591
Cypremethrin, 1089, 1097–1098
Cysteine, 612, 745
Cytochrome oxidase, 101, 139, 180

D

2,4-D, 1053
D-1221, 800
Dalapon, 1160
Dazzel, 962
DDE, 370, 864, 1287
DDT, 379, 622, 799, 827, 828, 831, 1098, 1137, 1139, 1145, 1149, 1256, 1461, 1472
Decaborane, 1567, 1570, 1572, 1596, 1597, 1598, 1600, 1603
Decachlorobiphenyls, 1245, 1259, 1268, 1271, 1286
Decarboxyfenvalerate, 1121
Dechlorane, 1133, 1146, 1148
Dechlorane 510, 1134
Dechlorane 4070, 1134
Deethylated atrazine, 768, 780
Deethylatrazine, 785
Deethylhydroxyatrazine, 774
Deflubenzon, 984
Dehydroascorbic acid, 912
Dehydrodiazinon, 974
Deisopropylated atrazine, 768, 780
Deisopropylatrazine, 785
Deisopropylhydroxyatrazine, 774
Deltamethrin, 1089, 1097
N-Desmethylfamphur, 1070, 1073, 1082
O-Desmethylfamphur, 1070
Dextrone, 1163
Dextrone X, 1163
Dexuron, 1163
DFP, 1451
Dhurrin, 916
Diagran, 962
Dialkyllead, 208
Dialkylorganotins, 555
Dialkyltins, 584
Dianon, 962
DiaterrFos, 962
Diazajet, 962
Diazatol, 962
Diazepam, 1100, 1104, 1105, 1113
Diazide, 962
Diazinon, 961–978, 1151
Diazinon AG 500, 969
Diazinon 4E, 969
Diazinon 14G, 969, 976
Diazinon 4S, 969
Diazinon 50 W, 969
Diazol, 962
Diazoxon, 963, 969, 970, 972, 974, 975
Dibenz*(a,c)*anthracene, 1387
Dibenz*(a,h)*anthracene, 1358, 1359, 1364, 1367, 1369, 1373, 1386, 1387, 1394
Dibenz*(a,j)*anthracene, 1387
Dibenzo*(a,c)*fluorene, 1347
Dibenzo*(a,e)*pyrene, 1347, 1387
Dibenzo*(a,g)*fluorene, 1347
Dibenzo*(a,h)*fluorene, 1347
Dibenzo*(a,h)*pyrene, 1347, 1388
Dibenzo*(a,i)*pyrene, 1347, 1387
Dibenzo*(a,l)*pyrene, 1347
Dibenzofurans, 1045, 1197, 1198, 1214, 1216, 1217, 1225, 1287, 1316
Diborane, 1567, 1570, 1603
Dibutylmethyltins, 554
Dibutyltin dichloride, 580, 583, 592
Dibutyltin disulfide, 592
Dibutyltins, 563, 564, 567, 570, 574, 584, 587, 593
Dichlorobenzene, 1024
2,2'-Dichlorobiphenyl, 1257
2,6'-Dichlorobiphenyl, 1257
4,4'-Dichloro-3-biphenylol, 1312
Dichlorobiphenyls, 1239, 1241, 1257, 1267
Dichlorobornane, 1460
Dichlorochlordene, 824, 826
1,2-Dichlorochlordene, 826
Dichlorodene, 824
2,8-Dichlorodibenzodioxin, 1049
Dichloromaleic acid, 1199
2,4-Dichlorophenol, 1197
Dichromate ion, 46, 47
Dicobalt ethylenediamine tetraacetic acid, 912

Table 33.1 (continued) Chemical and Trade Names of Substances Listed in the *Handbook of Chemical Risk Assessment* Series

Dicrotophos, 1076
Dicyclohexyltin dichloride, 580
Dieldrin, 833, 1143, 1149, 1461
Diethyldithiocarbamate, 425, 1180
Diethylenetriamine pentaacetic acid, 422
Diethyl 2-isopropyl-6-methylpyrimidin-4-yl phosphate, 963. *See also* Diazoxon
Diethyllead, 263
Diethylmercury, 341
Diethylnitrosamine, 424, 754
Diethyl phosphoric acid, 974
Diethyl phosphorothioic acid, 974
Diethyltin dichloride, 580
Diethyltin diiodide, 551
Diethyltins, 574, 587
O,O-Diethyl-*O*-(3,5,6-trichloro-2-pyridyl) phosphate, 889
O,O-Diethyl-*O*-(3,5,6-trichloro-2-pyridyl) phosphorothioate, 976
Diflubenuron, 984
Diflubenzuron, 983–1014
Difluoroacetone, 1425
2,6-Difluorobenzamide, 983, 984, 985, 987, 991, 1005, 1011, 1013
2,6-Difluorobenzoic acid, 984, 985, 987, 991, 993, 1005, 1008, 1009, 1011, 1013
2,6-Difluorohippuric acid, 1009
2,6-Difluoro-3-hydroxydiflubenzuron, 1009
Dihydrodiol epoxide, 1352
Dihydrodiols, 1352
2,8-Dihydromirex, 1147
3,8-Dihydromirex, 1147
3,4-Dihydroxybenzoic acid, 1025
Dihydroxybiphenyls, 1197, 1313
2,4-dihydroxy-7-methoxy-1,4-[2H]-benzoxazin-3-[4H]-1, 774
4,4'-Dihydroxy 2,'3',5,'6'-tetrachlorobiphenyl, 1312–1313
4,4'-Dihydroxy 3,5,3,'5'-tetrachlorobiphenyl, 1312
2,3-Dimercapto-1-propane sulfonic acid, 615
2,3-Dimercaptopropanol, 321, 481
2,3-Dimercaptopropyl) phthalamidic acid, 1508
2,3-Dimercaptosuccinic acid, 615
Dimethoate, 1151
Dimethoxytetrachlorobenzenes, 1204
4-Dimethylaminophenol, 911, 912, 951
O-[4-1(Dimethylamino)sulfonyl] phenyl phosphorothioic acid *O,O*-dimethyl ester, 1069
Dimethylarsinate, 1507, 1514, 1548
Dimethylarsine, 1506
Dimethylarsinic acid, 1504, 1505, 1506, 1507, 1509, 1516, 1523, 1526, 1527, 1530, 1534, 1535, 1546, 1549, 1550
Dimethylarsinous acid, 1507
Dimethylarsonic acid, 1549
Dimethylbenz*(a)*anthracene, 1347, 1384, 1386
7,12-Dimethylbenz*(a)*anthracene, 1370, 1372, 1383, 1387, 1388, 1390
Dimethylbenz*(a,h)*anthracene, 1387
1,1'-Dimethyl-4,4'-bipyridinium, 1163
1,1'-Dimethyl-4,4'-bipyridinium dichloride, 1163
Dimethyl diselenide, 1650, 1651
Dimethyllead, 263
Dimethyl maleate, 1165, 1180
Dimethylmercury, 319, 321, 341
Dimethylnaphthalenes, 1376, 1378
N-N-Dimethylquinoneimine, 912
Dimethyl selenide, 933, 1508, 1650, 1651
Dimethyl sulfoxide, 769
Dimethyltin chloride, 583
Dimethyltin dichloride, 580
Dimethyltins, 563, 564, 568, 570, 571, 573, 587
Dimethyl (2,2,2-trichloro-1-hydroxyethyl) phosphonate, 976
Dimilin, 984
Dinoseb, 1171
Dioctyltin, 555, 584, 588
Diorganotins, 551, 555, 560, 580, 584
Dioxins, 1021–1056, 1197, 1216, 1217, 1225, 1287
Diphacinone, 1416
Diphenyl ethers, 1214, 1287
5,5-Diphenyl hydantoin, 915
Diphenyltin dichloride, 580
Diphenyltins, 555
Dipropyltin dichloride, 580
Disodium calcium cyclohexanediamine tetraacetate, 615
Disodium ethylenediamine tetraacetic acid, 615
Disodium heptahydrate, 1533–1534
Disodium methanearsonate, 1504, 1532
Disodium methylarsenate, 1543
Disodium methylarsonate, 1530
Disodium octaborate tetrahydrate, 1571
Dithiocarbamates, 105, 618
Dithiomolybdates, 1616
Dizinon, 962
Dodecachlorooctahydro-1,3,4-metheno-2H-cyclobuta(c,d)pentalene, 1133
Dodecyclamine p-chlorophenylarsonate, 1543
Dovip, 1069
Dowchlor, 824
Dowicide 7, 1196
Dowicide EC-7, 1196
Dowicide G, 1196
Dow pentachlorophenol, 1196
DP-2, 1196
D-Penicillamine, 105, 1164
DU 112307, 984
Dual Paraquat, 1163
Duphar BV, 984
Durotox, 1196
Dursban, 883, 976
Dylox, 976
Dyzol, 962
D.z.n., 962

E

Ectrin, 1091
Emanay zinc dust, 608
Endrin, 864
ENT 9932, 824
ENT 19507, 962
ENT 25644, 1069
ENT 25719, 1134
ENT 27164, 800
ENT 28244, 769
ENT 29054, 984

Table 33.1 (continued) Chemical and Trade Names of Substances Listed in the *Handbook of Chemical Risk Assessment* Series

EPTC, 812
Erbon, 1022
Esfenvalerate, 1106
Esgram, 1163
Ethanol, 1450
Ethanolamine, 95, 96
2-Ethoxynaphthalene, 1388
Ethoxyquin, 166
Ethyl acetate, 769
Ethyl chloride, 863
Ethylmercury, 371, 372
N-Ethyl-*N*-hydroxyethyl nitrosamine, 425
N-Ethyl-*N*-nitrosourea, 616
Ethyltins, 555, 587
O-Ethyl trichloropyridyl phosphorothioate, 885
Europium-152, 1711, 1800
Europium-155, 1711, 1737

F

Famaphos, 1069
Famfos, 1069
Famophos, 1069
Famoxon, 1068, 1069, 1070, 1071, 1079, 1084
Famphos, 1069
Famphur, 1067–1085
Fanfos, 1069
Farmco atrazine, 769
Fenbutatin oxide, 591
Fendeet, 1117
Fenkill, 1091
Fenoprop, 1022
Fenpropthrion, 1089
Fenvalerate, 1089–1125
Ferriamicide, 1151
Ferric sulfate, 421, 921, 1510
Ferrihemoglobin cyanide, 912
Ferrimolybdate, 1614
Ferritin, 99
Ferrochelatase, 208
Ferrochrome lignosulfonate, 72
Ferrous chloride, 1151
Ferrous sulfate, 421
Fezadin, 962
Fluoranthene, 1356, 1358, 1359, 1364, 1365, 1366, 1367, 1368, 1369, 1372, 1373, 1376, 1378, 1383, 1384, 1387, 1394, 1395
Fluorene, 1287, 1356, 1358, 1360, 1362, 1365, 1366, 1368, 1373, 1376, 1378, 1384, 1386, 1394, 1395
Fluorides, 1616
Fluoroacetic acid, 1419
Fluorocitrate, 1422, 1425, 1426, 1427
5-Fluorocystine, 220
FMC 10242, 800
Formaldehyde, 755
Formaldoxime, 907
Formamide, 921
Formic acid, 908–909
N-Formylglycine, 1166
FP Tracerite yellow, 914
Francium-223, 1712
Fratol, 1420
Free cyanide, 905, 952
Fulvic acid, 98, 105
Furadan, 800, 807

G

G-24480, 962
G-30027, 769
Galena, 203
Gallium arsenide, 1548
Gamma-mercaptopropionylglycine, 745
Gardentox, 962
Gasparim, 769
Geniphene, 1460
Germanium, 1678
Gesaprim, 769
Gesaprim 500 FW, 769
Giese salt, 1758
Gliftor, 1451
Glucuronic acid, 1352
Glutaraldehyde, 741
Glutathione, 745, 755
Glyceraldehyde, 742, 744
Glyceric acid, 750
Glycerol, 740, 744, 750
Glycerol monoacetate, 1424, 1425, 1450
Glycidaldehyde, 742, 744, 745, 746, 754
Glycidol, 744, 745
Gold-198, 1712
Gramonol, 1163
Gramoxone, 1160, 1163, 1180
Gramuron, 1163
Griffex, 769
Guaiacols, 1197

H

HCS 3260, 824
Heptachlor, 799, 824, 825, 826, 827, 1143
Heptachlor epoxide, 824, 825, 826, 870
Heptachlorinated dibenzodioxins, 1029
Heptachlorinated diphenyl ethers, 1214
2,2',3,4,4,'5,5'-Heptachlorobiphenyl, 1244, 1299
Heptachlorobiphenyls, 1244, 1259, 1267, 1268, 1271, 1286, 1299
1,2,3,4,6,7,8-Heptachlorodibenzodioxin, 1030, 1031, 1032, 1034, 1035, 1036, 1037, 1043, 1045, 1049
Heptachlorodibenzofurans, 1198
Heptachlorodioxins, 1023, 1031, 1032, 1034, 1035, 1036, 1037, 1048, 1198
Heptachlorophenoxyphenols, 1198
Herbaxon, 1163
Hexachlorobenzene, 1134, 1196, 1198, 1203, 1217, 1225, 1287
2,2',3,4,'6,6'-Hexachlorobiphenyl, 1244
2,2',3,4,4,'5'-Hexachlorobiphenyl, 1243, 1299
2,2',3,5,5,'6-Hexachlorobiphenyl, 1244
2,2',4,4,'5,4'-Hexachlorobiphenyl, 1238
2,2',4,4,'5,5'-Hexachlorobiphenyl, 1244, 1299
2,3,3',4,4,'5-Hexachlorobiphenyl, 1244, 1299

Table 33.1 (continued) Chemical and Trade Names of Substances Listed in the *Handbook of Chemical Risk Assessment* Series

2,3,3',4,5,6-Hexachlorobiphenyl, 1244
2,3,4,4',5,5'-Hexachlorobiphenyl, 1244, 1299
2,3,4,4',5,6-Hexachlorobiphenyl, 1244
3,3',4,4,'5,5'-Hexachlorobiphenyl, 1244, 1299
Hexachlorobiphenyls, 1025, 1052, 1243, 1256, 1259, 1267, 1268, 1271, 1286, 1299, 1307
Hexachloro-1,3-butadiene, 47
Hexachlorocyclopentadiene, 824, 1134
1,2,3,4,7,8-Hexachlorodibenzodioxin, 1028, 1029, 1030, 1031, 1032, 1034, 1035, 1036, 1037, 1043, 1045, 1049, 1194
1,2,3,6,7,8-Hexachlorodibenzodioxin, 1029, 1031, 1032, 1034, 1035, 1036, 1037, 1049
1,2,3,7,8,9-Hexachlorodibenzodioxin, 1030, 1031, 1032, 1034, 1035, 1049
Hexachlorodibenzodioxins, 1029, 1031, 1033, 1035, 1197
Hexachlorodibenzofurans, 1197, 1198
Hexachlorodioxins, 1023, 1024, 1031, 1032, 1033, 1034, 1035, 1036, 1037, 1048, 1198
Hexachlorophene, 1022, 1030
Hexafluron, 990
1,2,6-Hexane thiol, 741
Hexavalent chromium, 45, 46, 47, 60, 61, 62–63, 65, 66, 67, 75, 79, 80, 374, 1630
4-Hexylresorcinol, 743
Humic acid, 98, 105, 131, 456, 533, 609, 670, 1506
Hyaluronic acid, 986, 1005
Hydantoin, 915
Hydantonic acid, 915
Hydrochromate ion, 47, 61
Hydrocyanic acid, 904, 925
Hydrogen, 1717, 1799
Hydrogen-3, 1710, 1713, 1719, 1720, 1721, 1725, 1726, 1727, 1728, 1736, 1737, 1738, 1740, 1780, 1803, 1804
Hydrogen cyanide, 905, 907, 909, 913, 914, 915, 916, 917, 922, 925, 930, 933, 934, 935, 938, 941, 942, 943, 948, 951
Hydrogen fluoride, 1420
Hydrogen peroxide, 106, 148, 456, 932, 1163, 1164, 1165
Hydrogen selenide, 1651
Hydroquinone, 742
Hydroxocobalamin, 762, 909, 912
1-Hydroxy, 2,3-epoxy chlordene, 869
3-Hydroxyanthronilic acid, 973
Hydroxyatrazine, 768, 774, 778, 780, 786, 787
4-Hydroxybiphenyl, 1313
3-Hydroxycarbofuran, 801, 803, 808
3-Hydroxycarbofuran-7-phenol, 815
3-Hydroxychlordane, 826
1-Hydroxychlordene, 869
4-Hydroxy 2-chlorobiphenyl, 1312
4-Hydroxy 4'-chlorobiphenyl, 1312
Hydroxydiazinon, 974
20-Hydroxyecdysone, 991
Hydroxyethylenediamine triacetic acid, 422
4-Hydroxyfenvalerate, 1114
N-Hydroxymethyl carbofuran, 802
Hydroxymonomethoxytetrachlorobenzenes, 1204
3-Hydroxynitrosocarbofuran, 815
3-4' Hydroxyphenoxy) benzoic acid, 1116
3-4' Hydroxyphenoxy) benzyl alcohol, 1116
Hydroxy-4-picolinic acid, 1166
3-Hydroxypropylmercapturic acid, 755
4-Hydroxy-3,5,4'-trichlorobiphenyl, 1312

I

Ilsemannite, 1614
Imidazole, 612, 915
2-Imidazolidinone, 915
2-Iminothiazolidene-4-carboxylic acid, 909
Indeno*(1,2,3-cd)*pyrene, 1355, 1358, 1359, 1364, 1365, 1383–1384, 1386, 1387, 1395
Indium, 28
Indoleacetic acid, 1582
Indoleacetonitrile, 916
Iodine, 951, 1799
Iodine-125, 1711, 1801, 1804
Iodine-129, 1711, 1721, 1727, 1733, 1734, 1744, 1763, 1788, 1801, 1804
Iodine-130, 1711, 1721
Iodine-131, 1711, 1720, 1721, 1726, 1727, 1742, 1753, 1755, 1757, 1760, 1761, 1763, 1764, 1765, 1768, 1784, 1787, 1791, 1794, 1797, 1799, 1801, 1804
Iron, 23, 102, 103, 169, 206, 1530, 1531, 1614, 1616, 1783, 1799
Iron-55, 1710, 1721, 1724, 1732, 1736, 1738, 1739, 1750, 1784
Iron-59, 1710, 1721, 1724
Iron oxide, 331, 434, 1502, 1570
Isobornyl thiocyanate, 1585

J

JASAD Merrillite, 608
Jordisite, 1614

K

Kadethrin, 1098
Kanechlor, 1254
Kaolinite, 71, 162, 938
Kayazinon, 962
Kayazol, 962
Kelocyanor, 911
Kepone, 1134
4-Ketobenztriazine, 932
3-Ketocarbofuran, 801, 803
3-Ketocarbofuran phenol, 801, 803, 815
Knox out, 962
Krypton-85, 1710, 1721, 1753, 1804
L-Kynurenine, 973
Kypclor, 824

L

L15, 608
Lactic acid, 750, 937

Table 33.1 (continued) Chemical and Trade Names of Substances Listed in the *Handbook of Chemical Risk Assessment* Series

Laetrile, 904, 914, 915
Lannate, 976
Lanthanum-140, 1711, 1731, 1755, 1757, 1800
Largon, 984
Lasso, 769
Laterite, 413
Lauxtol A, 1196
Lead, 23, 26, 28, 77, 104, 131, 132, 134, 136, 201–290, 374, 420, 456, 461, 559, 1257, 1508, 1614, 1616, 1674, 1685, 1717, 1783
Lead-204, 206
Lead-206, 206
Lead-207, 206
Lead-208, 206
Lead-210, 206, 275, 1712, 1723, 1724, 1725, 1726, 1728, 1736, 1737, 1738, 1739, 1740, 1744, 1789, 1804
Lead-211, 1712
Lead-212, 206, 1712
Lead-214, 1712, 1797
Lead acetate, 204, 209, 211, 260, 269, 272, 276, 277, 278, 279, 280
Lead arsenate, 213, 270, 275, 277, 287, 1502, 1504, 1511, 1531, 1543, 1544, 1548, 1551
Lead carbonate, 207, 211, 278
Lead chloride, 206, 207
Lead chromate, 287
Lead halides, 204
Lead hydroxide, 207
Lead nitrate, 206, 260, 269, 271, 272
Lead oxide, 204, 206
Lead phosphate, 204
Lead subacetate, 269
Lead sulfate, 204, 206, 207
Lead sulfide, 206, 211
Linamurin, 915, 916, 922, 937
Lorsban, 883, 892
Lotaustralin, 916
Lunar caustic fused silver nitrate, 504

M

M 140, 824
M 410, 824
Magnacide H, 741
Magnesium, 104, 139, 420, 421, 456, 651, 1571, 1575, 1614
Magnesium acetate, 425
Magnesium oxide, 331, 1613, 1634
Magnesium sulfate, 863, 1425
Malachite, 94
Malathion, 1451
Malayaite, 557
Malic acid, 1166
Malondialdehyde, 99, 148
Malonitrile, 939, 943
Mandelonitrile, 922
Manganese, 94, 103, 205, 420, 421, 425, 479, 514, 1581, 1616, 1783, 1799
Manganese-54, 1710, 1721, 1724, 1726, 1730, 1731, 1734, 1735, 1736, 1737, 1738, 1740, 1750, 1751, 1799
Manganese-56, 1710
Manganese dioxide, 505
Manganese oxide, 434, 1506
Manganite, 914
Maxolon, 1439
2-Mercaptoethanol, 745
Mercapturic acids, 755
Mercuric acetate, 374
Mercuric ammonium chloride, 356
Mercuric chloride, 47, 318, 374, 375, 383
Mercuric cyanide, 906
Mercuric selenide, 321–322, 1651
Mercuric sulfide, 313, 317
Mercurous chloride, 318
Mercurous ion, 321
Mercury, 23, 28, 313–391, 420, 514, 559, 1654, 1655, 1656, 1673, 1674, 1685
Mercury-203, 1712
Mercury-206, 1712
Methanearsonic acid, 1506, 1547
Methanol, 97, 769, 1024
Methionine, 741, 1616, 1678
Methomyl, 1416
9-Methoxyanthracene, 1373
Methoxychlor, 864
2-Methoxy-3,4-dihydro-2PH-pyran, 755
Methoxyethyl mercury, 369, 372
2-Methoxy naphthalene, 1388
Methylamine, 1166
Methylamine hydrochloride, 1166
Methylanthracene, 1373
9-Methylanthracene, 1375
Methylarsines, 1505, 1506
Methylarsonate, 1511
Methylarsonic acid, 1507, 1523, 1526
Methylarsonous acid, 1507
Methyl bromide, 1151
3-Methylcholanthrene, 1008, 1249, 1347, 1383, 1384, 1385, 1388, 1390
20-Methylcholanthrene, 422, 425
Methylcobalamin, 554
Methylene blue, 911
Methylene dioxyphenyls, 963, 970
Methyl fluoride, 1420
Methyl iodide, 554
Methylmercury, 268, 314, 315, 318, 319, 320, 322, 325, 330, 340, 356, 361, 369, 370, 371, 372, 374, 376, 377, 378, 379, 380, 381, 382, 383, 384, 386, 387, 388, 390, 813, 1674
Methyl methacrylate, 913
Methyl methanesulfonate, 1509
1-Methylnaphthalene, 1360, 1374, 1378
2-Methylnaphthalene, 1356, 1360, 1374, 1378
Methyl parathion, 1073
1-Methylphenanthrene, 137
1-Methyl-4-phenyl-1,2,3,6-tetrahydropyridine (MPTP), 1170
Methyl prednisolone, 1164
4-Methylpyrazole, 1424, 1425, 1450, 1451
Methyltin(s), 554, 563, 564, 570, 571, 587
Methyl viologen, 1163
Metoclopramide, 1439, 1440
Metribuzin, 1172
Micromite, 984
Mirex, 1133–1152, 1256, 1287
Mitomycin C, 421

Table 33.1 (continued) Chemical and Trade Names of Substances Listed in the *Handbook of Chemical Risk Assessment* Series

Molded silver nitrate argenti, 504
Molybdenite, 1614
Molybdenum, 61, 103–104, 169, 1601, 1613–1641, 1783
Molybdenum-99, 1711, 1753, 1765
Molybdenum trioxide, 1614, 1632, 1636, 1637
Molybdic acid, 1630
Monobutyltins, 567, 568, 574
Monochlorobiphenyls, 1239, 1241
Monochlorodihydrochlordene, 826
Monocrotophos, 1076
Monofluoroacetic acid, 1422
8-Monohydro mirex, 1147
9-Monohydro mirex, 1147
10-Monohydro mirex, 1147
Monomethylarsinic acid, 1527, 1530
Monomethylarsonate, 1514
Monomethylarsonic acid, 1505, 1509, 1511, 1535, 1550
Monomethylmercury, 321
Monomethyltins, 568, 573
Monoorganotins, 555, 5560
Monosodium fluoroacetate, 1420
Monosodium fluoroacetic acid, 1413
Monosodium methanearsonate, 1504, 1546, 1547
morphine, 105
Morsodren, 813
Motox, 1460

N

Naphthalene, 1287, 1348, 1355, 1356, 1358, 1367, 1368, 1369, 1376, 1377, 1378, 1383, 1384, 1386, 1387, 1394, 1395
Naphthene, 1386
1-Napthyl *N*-methylcarbamate, 976
Neocidol, 962
Neodymium-147, 1711, 1721
Neptunium, 1782, 1783, 1802
Neptunium-235, 1712, 1783
Neptunium-237, 1713, 1723, 1783, 1804
Neptunium-239, 1713, 1721, 1723, 1738, 1753
Neptunium-241, 1713
Niagra NIA-10242, 800
Nickel, 26, 134, 411–487, 559, 1783
Nickel-56, 417
Nickel-57, 417
Nickel-58, 417
Nickel-59, 417
Nickel-60, 417
Nickel-61, 417
Nickel-62, 417
Nickel-63, 417, 1710, 1736, 1737
Nickel-64, 417
Nickel-65, 417, 1710
Nickel-66, 417
Nickel-67, 417
Nickel acetate, 415, 424, 425, 427, 469, 472, 474, 475, 476
Nickel ammonium sulfate, 415, 424
Nickel antimonide, 424
Nickel arsenide, 424
Nickel bromide, 412
Nickel carbonate, 415, 417, 424, 467, 468, 469, 473
Nickel carbonyl, 412, 415, 416, 417, 418, 419, 422, 423, 424, 467, 471, 473, 474, 475, 476, 480, 484, 485, 1371
Nickel chloride, 415, 418, 421, 422, 425, 426, 427, 463, 464, 465, 467, 469, 470, 471, 472, 473, 474, 475, 476, 483
Nickel chloride hexahydrate, 417, 466, 472
Nickel chromate, 424
Nickel cyanide, 906
Nickel disodium EDTA, 419
Nickel fluoborate, 415, 476
Nickel fluoride, 415, 424
Nickel hexahydrate, 476, 477
Nickel hydroxide, 415, 417, 424
Nickel mesotetraphenylporphine, 427
Nickel monosulfide, 415, 424, 426
Nickel monoxide, 484
Nickel nitrate, 415, 426, 471, 476
Nickelocene, 415, 424, 472, 476
Nickeloplasmin, 416
Nickel oxides, 412, 413, 414, 418, 420, 422, 423, 424, 466, 467, 469, 470, 471, 472, 474, 475, 476, 483
Nickel perchlorate heptahydrate, 472
Nickel selenide, 424
Nickel subsulfide, 411, 412, 414, 415, 420, 422, 424, 425, 426, 467, 471, 474, 475, 480, 483
Nickel sulfamate, 415
Nickel sulfate, 414, 415, 417, 418, 423, 424, 425, 467, 468, 470, 471, 472, 474, 475, 476, 615, 926, 930
Nickel sulfate hexahydrate, 417, 418, 468, 469, 473
Nickel sulfide, 417, 484, 617
Nickel telluride, 424
Nicotinamide, 1591
Nicotinamide adenine dinucleotide phosphate (NADPH), 1163
Nicotine, 1451
Nigrosine, 1415
Niobium-95, 1711, 1725, 1731, 1736, 1737, 1738, 1739, 1740, 1742, 1755, 1800
Nipsan, 962
Niran, 824
Nitric acid silver (I) salt, 504
Nitric acid silver (1+) salt, 504
Nitrilotriacetc acid, 104, 105
Nitrilotriacetic acid, 104, 211
Nitrobenzene, 911
3-Nitro-4-hydroxyphenyl arsonic acid, 1504, 1542, 1544, 1550, 1551, 1552, 1553
4-Nitroquinoline-*N*-oxide, 616
N-Nitrosoatrazine, 787
Nitrosocarbofuran, 814, 815
N-Nitroso-*N*-methylurea, 1631
N-Nitrososarcosine ethyl ester, 1631
N-[4-(5-Nitro-2-furyl)-2 thiazoyl] formamide, 755
Nonachlorobiphenyls, 1245, 1259, 1268, 1271, 1286
Nonachlorobornane, 1464
Nonachlorophenoxyphenols, 1198
Nonachlors, 826, 827, 829, 832
Nucidol, 962

Table 33.1 (continued) Chemical and Trade Names of Substances Listed in the *Handbook of Chemical Risk Assessment* Series

O

OCDD, 1024, 1031, 1032, 1034, 1035, 1036, 1037, 1043, 1045
Octachlor, 824
Octachlor epoxide, 825
Octachlorinated diphenyl ethers, 1214
Octachlorobiphenyls, 1244, 1259, 1271, 1286, 1314
Octachlorobornane, 1464
Octachlorocamphene, 1460
Octachlorocyclopentene, 824
Octachlorodibenzofurans, 1198
Octachlorodioxins, 1023, 1024, 1032, 1040–1041, 1198, 1199
Octachloronaphthalene, 1240, 1245
Octachlorophenoxyphenols, 1198
Octachlorostyrenes, 1256
Octa-klor, 824
Octaterr, 824
Octyltins, 555, 587
OMS 864, 800
OMS 1804, 984
Ontrack WE-1, 1196
Organoarsenicals, 1503, 1504, 1507, 1531, 1545
Organoleads, 206
Organomercurials, 314
Organotins, 551, 553–557, 560–566, 568, 593
Orthoarsenic acid, 1505
Ortho-klor, 824
Ortho paraquat, 1163
Orvar, 1163
Oxalic acid, 1166
Oxazolone, 864
Oxine, 98
Oxychlordane, 824, 825, 826, 827, 830, 832, 870, 871

P

PAHs, 1343–1397
Palladium-109, 1711
Papite, 741
PAPP, 1451
Paracol, 1163
Paramolybdate ion, 1615
Paraoxon, 861
Paraquat, 1159–1186, 1674
Paraquat CL, 1163
Paraquat dichloride, 1163, 1177
Paraquat dimethylsulfate, 1160
Parathion, 370, 1076
Paris green, 96, 1504
PASCO, 608
Pathclear, 1163
PCBs, 833, 1237–1322
PCDDs, 1021, 1023
PCP, 1196
PDD 6040-I, 984
Penchlorol, 1196
Penicillamide, 745
Penicillamine, 105, 481, 615
Penta, 1196
Pentaborane, 1567, 1570, 1596, 1597, 1600, 1603
2,2',4,5,5'-Pentachlorobiphenyl, 1243, 1299
2,3,3',4,4'-Pentachlorobiphenyl, 1243
3,3',4,4,'5-Pentachlorobiphenyl, 1243
Pentachlorobiphenyls, 1242, 1256, 1259, 1267, 1271, 1285, 1286, 1299, 1314
Pentachlorocyclopentadiene, 824
1,2,3,7,8-Pentachlorodibenzodioxin, 1029, 1030, 1031, 1032, 1033, 1034, 1035, 1036, 1037, 1049
1,2,4,7,8-Pentachlorodibenzodioxin, 1031, 1043, 1049
2,3,4,7,8-Pentachlorodibenzodioxin, 1032
Pentachlorodibenzofurans, 1198, 1246
Pentachlorodioxins. *See* 1,2,3,7,8-Pentachlorodibenzodioxin; Pentachlorodibenzodioxins
Pentachlorophenol, 1022, 1023, 1028, 1193–1226
Pentachlorophenol laurate, 1195
Pentacon, 1196
Pentafluorobenzyl bromide, 1420
Penta general weed killer, 1196
N-Pentane, 769
Pentanol, 1196
Pentasol, 1196
Pentlandite, 413
Pentyltins, 555
Penwar, 1196
Permacide, 1196
Permaguard, 1196
Permasem, 1196
Permatox, 1196
Permethrin, 1098
Perylene, 1355, 1356, 1360, 1369, 1376, 1383, 1385
PH 60-40, 984
Phenacide, 1460
Phenanthrene, 1356, 1358, 1359, 1360, 1362, 1364, 1365, 1366, 1367, 1368, 1369, 1374, 1376, 1377, 1383, 1384, 1387, 1388, 1394
1,10-Phenanthroline, 1164
Phenatox, 1460
Phenethyl isothiocyanate, 1371
Phenobarbital, 105, 869, 1164, 1216, 1248, 1249
Phenoloxidase, 985
Phenothrin, 1098
3-Phenoxybenzaldehyde, 1095, 1114
3-Phenoxybenzoic acid, 1099, 1114
3-Phenoxybenzoyl cyanide, 1095
3-Phenoxybenzoyl glycine, 1116
3-Phenoxybenzyl alcohol, 1090, 1091
3-Phenoxybenzyl cyanide, 1095
3-Phenoxybenzyl methylbutyric acid, 1090, 1095
Phenoxyphenols, 1197, 1198
2-Phenoxyphenols, 1214
Phenvalerate, 1091
Phenylarsonic acids, 1504, 1508, 1553
Phenylmercury, 319, 320, 341, 371, 372, 375
Phenyltins, 586
Phenytoin, 105
Phosphates, 2, 1506, 1509, 1530, 1536, 1594
Phosphomolybdic acid, 1572
Phosphorothioic acid, 884
Phosphorothioic acid, *O*-[4 (dimethylamino)sulfonyl], phenyl *O,O*-dimethyl ester, 1067, 1069
Phosphorothioic acid *O,O*-diethyl *O*-(6-methyl-2-(1-methylethyl)-4-pyrimidinyl ester, 962, 977
Phosphorus, 623, 1531, 1535, 1571, 1616, 1799

Table 33.1 (continued) Chemical and Trade Names of Substances Listed in the *Handbook of Chemical Risk Assessment* Series

Phosphorus-32, 1710, 1721, 1780, 1801, 1804
Photochlordane, 857
Photomirex, 1139, 1147
4-Picolinic acid, 1166
Pillarfuran, 800
Pillarquat, 1163
Pillarxone, 1163
Pindone, 1451
Piperonyl butoxide, 1100, 1371
Planar PCBs, 1250
Platinum, 906
Plutonium, 1719, 1720, 1782, 1783, 1802
Plutonium-237, 1779
Plutonium-238, 1713, 1720, 1725, 1726, 1727, 1728, 1734, 1735, 1736, 1737, 1738, 1743, 1744, 1748, 1753, 1768, 1779, 1781, 1791, 1793, 1804
Plutonium-239, 1713, 1720, 1721, 1723, 1725, 1726, 1727, 1728, 1729, 1733, 1734, 1735, 1736, 1737, 1738, 1739, 1743, 1744, 1746, 1747, 1748, 1750, 1768, 1784, 1791, 1793, 1798, 1801, 1804
Plutonium-240, 1713, 1726, 1728, 1729, 1733, 1734, 1735, 1736, 1737, 1738, 1743, 1744, 1746, 1747, 1748, 1753, 1768, 1784, 1791, 1793, 1804
Plutonium-241, 1713, 1748, 1753, 1800
Plutonium-242, 1713, 1723, 1804
Plutonium-244, 1713, 1723
Plutonium dioxide, 1722, 1797
PNAs, 1344
Polonium, 1717
Polonium-210, 1712, 1723, 1726, 1728, 1729, 1733, 1736, 1737, 1738, 1739, 1740, 1744, 1789, 1798, 1801, 1804
Polonium-211, 1712
Polonium-212, 1712
Polonium-214, 1712
Polonium-215, 1712
Polonium-216, 1712, 1797
Polonium-218, 1712, 1797
Polyacrylonitrile, 942, 944
Polychlorinated biphenyls, 831, 864, 1022, 1028, 1237–1322. *See also* PCBs
Polychlorinated dibenzofurans, 1214
Polychlorocamphene, 1460
Polycyclic aromatic hydrocarbons, 1343–1397
Polycyclic organic matter, 1344
Polymethacrylates, 1160
Polynuclear aromatic hydrocarbons, 1344
POM, 1344
Potassium, 102, 139, 623, 1105, 1582, 1717, 1758, 1799
Potassium-40, 1710, 1713, 1717, 1725, 1726, 1728, 1729, 1730, 1731, 1735, 1736, 1737, 1738, 1740, 1742, 1744, 1747, 1763, 1764, 1765, 1768
Potassium-42, 1710, 1721
Potassium arsenite, 1504
Potassium borate, 1594
Potassium chloride, 914
Potassium cyanide, 906, 922, 923, 928, 932, 934, 939, 941, 942, 943, 945
Potassium dichromate, 47, 66, 75
Potassium monofluoroacetate, 1413
Potassium pentachlorophenate, 1196
Potassium sorbate, 1418
Powellite, 1614
PP 148, 1163
PP 910, 1163
Pralidoxime chloride, 1068
Praseodymium-143, 1711
Praseodymium-144, 1711
Praseodymium-147, 1711, 1737, 1800
Preeglone, 1163
Priltox, 1196
Primatol A, 769
Profenofos, 991, 1100
Promethium-147, 1711
1,3-Propanediol, 750
Propenal, 743
2-Propenal, 743
Propionitrile, 907, 943
Propranolol, 1100
Propylene, 741
Propyltins, 555
Protactinium-231, 1712, 1724
Protactinium-234, 1712
Protactinium-234m, 1712
Prussian blue, 1758
Prussic acid, 9905
Pydrin, 1091
Pyrene, 1355, 1356, 1358, 1359, 1362, 1364, 1365, 1366, 1367, 1368, 1369, 1372, 1376, 1383, 1384, 1386, 1388
2-Pyridinealdoxime methochloride, 973
Pyridine nucleotides, 1571
Pyridoxal 5-phosphate, 912
Pyrinex, 884
Pyrite, 203, 1502, 1510

Q

Quinolinic acid, 973

R

Radium, 1717
Radium-223, 1712
Radium-224, 1712, 1797
Radium-226, 1712, 1725, 1726, 1728, 1732, 1733, 1734, 1737, 1739, 1740, 1741, 1744, 1778, 1789, 1797, 1798, 1799, 1801, 1803, 1804, 1805
Radium-228, 1712, 1725, 1728, 1797, 1803, 1804
Radon, 1713, 1717
Radon-218, 1712
Radon-219, 1712
Radon-220, 1712
Radon-222, 1712, 1714, 1726, 1729, 1797, 1802
Ramrod, 769
Red squill, 1413
Resmethrin, 1089, 1098
Resorcinol, 743, 1025, 1215
Rhodanese, 908
Rhodium-106, 1710

Table 33.1 (continued) Chemical and Trade Names of Substances Listed in the *Handbook of Chemical Risk Assessment* Series

Roman vitriol, 97
Ronnel, 1022
Rotenone, 1459
Rubidium, 1799
Rubidium-86, 1710
Rubidium-87, 1710, 1713, 1717, 1726, 1728, 1736, 1738, 1740
Ruthenium, 1783
Ruthenium-103, 1711, 1721, 1726, 1731, 1736, 1737, 1739, 1753, 1755, 1762, 1763, 1764, 1765
Ruthenium-106, 1710, 1719, 1720, 1721, 1725, 1726, 1734, 1736, 1737, 1738, 1739, 1740, 1750, 1751, 1753, 1755, 1757, 1762, 1763, 1765, 1768, 1784

S

S-5602, 1091
Salzburg vitriol, 97
Samarium-143, 1711
Samarium-151, 1711, 1800
Sanmarton, 1091
Santobrite, 1196
Santophen, 1196
Sarolex, 962
Scandium, 48
SD 43775, 1091
Selenates, 1650, 1651, 1658, 1659, 1673, 1675, 1676, 1677, 1678, 1683, 1684
Selenites, 1650, 1651, 1655, 1658, 1659, 1673, 1675, 1676, 1677, 1678, 1683, 1684, 1685
Selenium, 23, 170, 321, 334, 335, 370, 389, 509, 615, 616, 933, 945, 1177, 1503, 1508, 1515, 1591, 1649–1692
Selenium-74, 1650
Selenium-75, 1652, 1655, 1656, 1684, 1710, 1742, 1787
Selenium-76, 1650
Selenium-77, 1650
Selenium-78, 1650
Selenium-80, 1650
Selenium-82, 1650
Selenium dioxide, 1650, 1652
Selenium sulfide, 1651
Selenocystathionine, 1672
Selenocysteine, 321, 1508, 1592, 1650, 1651, 1672, 1680
Selenocystine, 1672, 1673
Selenomethionine, 378, 1508, 1592, 1650, 1672, 1673, 1675, 1677, 1680, 1681, 1682, 1685, 1686
Selenopurine, 1675
Selocide, 1655
Shell atrazine herbicide, 769
Shell silver, 504
Silber, 504
Silflake, 504
Silicon, 131, 646
Silpowder, 504
Silver, 104, 134, 499–543, 509, 1673, 1674, 1685, 1783
Silver-110, 1711
Silver-111, 1711
Silver-113, 1711
Silver acetate, 507
Silver acetylide, 504
Silver albuminate, 507, 535
Silver arsenate, 538
Silver atom, 504
Silver carbonate, 506
Silver chloride, 505, 507, 508, 522, 532
Silver colloidal, 504
Silver cyanide, 535, 538, 906, 925
Silver fulminate, 535
Silver iodide, 501, 502
Silver 108m, 1711
Silver-108m, 1740
Silver-110m, 501, 504, 507, 508, 509, 527, 538, 1711, 1736, 1739, 1740, 1760, 1768, 1769, 1781, 1784
Silver nitrate, 499, 502, 503, 504, 507, 535, 536, 538
Silver (1+) nitrate, 504
Silver oxide, 535
Silver phosphate, 508
Silver selenide, 1651
Silver sulfate, 506
Silver sulfide, 505, 508, 521, 523, 533
Silver thiosulfate, 502, 504, 533
Silvex, 1022, 1053, 1054
Silvisar-510, 1543, 1549
Sinituho, 1196
S-Methyl-*N*-((methylcarbamoyl)oxy)-thioacetimidate, 976
Sodium, 138, 1105
Sodium-22, 1710
Sodium-24, 1710, 1721, 1742
Sodium acetate, 279, 1424, 1425, 1450
Sodium alpha ketoglutarate, 1425
Sodium arsanilate, 1547, 1552
Sodium arsenate, 1502, 1504, 1529, 1531, 1543, 1547, 1548, 1549, 1550, 1551, 1553
Sodium arsenite, 741, 1504, 1508, 1530, 1531, 1532, 1536, 1542, 1543, 1545, 1546, 1547, 1548, 1549, 1550, 1551
Sodium bicarbonate, 1420
Sodium bisulfite, 761
Sodium borate, 1569, 1571, 1589, 1594, 1597, 1598, 1599 1600, 1603, 1604
Sodium borohydride, 1595
Sodium cacodylate, 1531, 1543, 1548, 1549, 1551
Sodium chlorite, 103
Sodium chromate, 66, 77
Sodium chromate tetrahydrate, 69
Sodium citrate, 211
Sodium cyanide, 905, 906, 912, 913, 914, 917, 923, 924, 926, 928, 930, 933, 934, 935, 939, 941, 942, 943, 947, 1416
Sodium dichromate, 45, 66
Sodium diethyldithiocarbamate, 425
Sodium ferrocyanide, 914
Sodium fluoacetate, 1420
Sodium fluoride, 907, 1427
Sodium fluoroacetate, 1420
Sodium hexacyanoferrate, 914
Sodium hydroxide, 390
Sodium molybdate, 1626, 1629, 1631
Sodium monofluoroacetate, 1413–1452
Sodium nitrite, 103, 911, 936, 951

Table 33.1 (continued) Chemical and Trade Names of Substances Listed in the *Handbook of Chemical Risk Assessment* Series

Sodium nitroprusside, 914, 933
Sodium pentachlorophenate, 77, 1193, 1196, 1201
Sodium phosphate, 1501
Sodium pyruvate, 911, 1591
Sodium selenite, 321, 1680
Sodium succinate, 1424, 1445, 1450
Sodium sulfate, 169, 174, 1250
Sodium tetraborate, 1568, 1593, 1598, 1603
Sodium thiocyanate, 912, 913
Sodium thiosulfate, 911, 936, 951, 1508
Spectracide, 962
Spermidine, 740, 1182
Spermine, 740
Sphalerite, 203
Stalinon, 551
Stannite, 557
Stannous fluoride, 557
Strobane-T, 1460
Strontium, 1783
Strontium-85, 1710, 1719, 1721, 1738
Strontium-89, 1710, 1719, 1721, 1726, 1727, 1753
Strontium-90, 1710, 1713, 1719, 1720, 1721, 1722, 1725, 1726, 1727, 1728, 1729, 1731, 1732, 1733, 1734, 1735, 1737, 1738, 1741, 1742, 1745, 1746, 1747, 1748, 1749, 1750, 1753, 1755, 1773, 1780, 1781, 1782, 1783, 1784, 1791, 1799, 1801, 1803, 1804
Strychnine, 1413
Strychnine sulfate, 1678
Succinic acid, 1166
Succinonitrile, 907, 939, 943
Sulfates, 1536, 1625, 1673, 1675
Sulfhydryl groups, 223, 320, 321, 507, 508, 556, 744, 754, 763, 1507
Sulfotep, 962
Sulfur, 174, 389, 623, 905, 936, 1089, 1614, 1632, 1634, 1690
Sulfur-35, 1710
Sulfur dioxide, 755, 1651
Sumicidin, 1091
Sumifly, 1091
Sumipower, 1092
Sumitox, 1091
Superoxide dismutase, 20, 139, 456, 610, 1164, 1168, 1172, 1180
Sweep, 1163
Synklor, 824
Synthetic 3956, 1460

T

Tartrazine, 116
Tat-chlor 4, 824
Technetium-99, 1711
Technetium-99m, 1711
Teflubenzuron, 990
Tellurium, 321, 389
Tellurium-129, 1711
Tellurium-132, 1711, 1753
Tellurium-127m, 1711
Tellurium-129m, 1711
Tenaklene, 1163
Ten-eighty, 1413–1452
Term-l-trol, 1196
Tetraalkyllead, 204, 208, 218, 283
Tetraalkyltins, 584
1,4,8,11-Tetraazacyclotetradecane, 421
Tetrabutyltins, 563, 579, 583, 586, 590
Tetrachlorobenzoquinones, 1215
2,2',4,4'-Tetrachlorobiphenyl, 1242
2,2',4,5'-Tetrachlorobiphenyl, 1242, 1299
2,2',5,5'-Tetrachlorobiphenyl, 1242, 1299
3,4,4',5-Tetrachlorobiphenyl, 1242
2,3,6,7-Tetrachlorobiphenylene, 1025
Tetrachlorobiphenyls, 1241, 1259, 1267, 1271, 1285, 1286, 1293, 1299, 1314
Tetrachlorobornane, 1460
Tetrachlorocatechols, 1199, 1215
1,2,3,7-Tetrachlorodibenzodioxin, 1045
1,2,3,8-Tetrachlorodibenzodioxin, 1024
1,3,6,8-Tetrachlorodibenzodioxin, 1023, 1024, 1041, 1043, 1045
1,3,7,9-Tetrachlorodibenzodioxin, 1023
2,3,7,8-Tetrachlorodibenzodioxins. *See* 2,3,7,8-Tetrachlorodibenzo-para-dioxin
2,3,7,8-Tetrachlorodibenzofuran. *See* Dioxins
2,3,7,8-Tetrachlorodibenzo-para-dioxin, 105, 1021, 1025, 1029, 1030, 1031, 1032, 1033, 1034, 1035, 1036, 1037, 1038, 1039, 1040, 1041, 1042, 1043, 1044, 1045, 1046, 1047, 1049, 1051, 1054, 1245, 1246, 1256
Tetrachlorodihydroxyl benzenes, 1199
Tetrachlorodiols, 1199
Tetrachlorodioxins, 1198
Tetrachlorohydroquinones, 1199, 1215, 1216, 1220, 1222
2,3,4,5-Tetrachlorophenol, 1198, 1215
2,3,4,6-Tetrachlorophenol, 1023, 1197, 1215
Tetrachlorophenols, 1198
Tetraethyl dithiopyrophosphate, 962
Tetraethyllead, 204, 206, 207, 254, 262, 263, 268, 269, 277, 279
Tetraethyltins, 574, 583, 590, 592
Tetrahydro-5,5-dimethyl-2(1H)-pyrimidine, 1152
Tetrahydrotetrols, 1352
Tetrahydrotriols, 1352
Tetramethrin, 1089, 1098
Tetramethylarsonium, 1516, 1523, 1526
Tetramethylarsonium iodide, 1549
Tetramethyl lead, 204, 206, 218, 254, 262, 263, 274, 279
Tetramethyltins, 554, 563, 568, 574, 586
Tetraorganotins, 555, 557, 560, 584, 586
Tetraphenyltins, 583
Tetrapropyltins, 586
Tetrathiomolybdates, 1616
Tetrodotoxin, 1099
TH 6040, 984
Thallium, 1413, 1674, 1713
Thallium-206, 1712
Thallium-207, 1712
Thallium-208, 1712
Thallium-210, 1712
Thermosaline, 174
Thiamin, 210, 211
Thiocyanates, 904, 907, 920, 1099
Thiomolybdates, 103, 1616

Table 33.1 (continued) Chemical and Trade Names of Substances Listed in the *Handbook of Chemical Risk Assessment* Series

Thorium, 1717, 1802
Thorium-227, 1712, 1792, 1799
Thorium-228, 1712, 1725, 1728, 1733, 1737, 1801, 1805
Thorium-230, 1712, 1728, 1801, 1804, 1805
Thorium-231, 1712
Thorium-232, 1712, 1713, 1715, 1721, 1726, 1727, 1728, 1733, 1739, 1801, 1804, 1805
Thorium-234, 1712
Thymidine, 1771
Tin, 551–595
Tin-112, 552
Tin-114, 552
Tin-116, 552
Tin-117, 552
Tin-118, 552
Tin-119, 552
Tin-120, 552
Tin-122, 552
Tin-123, 1711
Tin-124, 552
Tin-126, 1711, 1804
Titanium, 48, 1783
Toluene, 1287, 1385
Topiclor, 824
Totacol, 1163
Toxafeen, 1460
Toxakil, 1460
Toxaphene, 1095, 1459–1477
Toxer total, 1163
Toxichlor, 824
Toxon 63, 1460
Trialkyllead, 208, 254, 266, 283
Trialkyllead chlorides, 254
Trialkyltins, 552, 584, 585
Tributylmethyltins, 554
Tributyltin chloride, 374, 573, 583, 589
Tributyltin fluoride, 586
Tributyltin oxide, 580, 583
Tributyltins, 374, 551, 552, 555, 556, 564–565, 567, 568, 569, 570, 580, 581, 583, 585, 586, 589, 592, 594, 1247
Trichlorobenzenes, 1022, 1054
Trichlorobenzoquinones, 1199, 1215
2,2',5-Trichlorobiphenyl, 1241
2,4,5'-Trichlorobiphenyl, 1241, 1299
Trichlorobiphenyls, 1239, 1241, 1259, 1271, 1285, 1286
Trichlorobornane, 1460
2,3,7-Trichlorodibenzodioxin, 1049
Trichlorodiols, 1199
Trichlorohydroquinones, 1199, 1215
2,4,5-Trichlorophenol, 1022
2,4,6-Trichlorophenol, 1023, 1197
Trichlorophenols, 1028, 1038, 1048, 1198
2,4,5-Trichlorophenoxyacetic acid (2,4,5-T), 1022, 1053
3,5,6-Trichloro-2-pyridinol, 884
Trichloropyrophos, 884
Tricyclohexyltin bromide, 580
Tricyclohexyltin chloride, 583
Tricyclohexyltin hydroxide, 584
Tricyclohexyltins, 557, 590, 593
Triethanolamine cacodylate, 1543, 1549
Triethylenetetramine, 421
Triethyllead, 254, 262, 263, 272, 279
Triethyltin acetate, 551
Triethyltin chloride, 583
Triethyltin hydroxide, 580
Triethyltins, 557, 573, 574, 584, 585, 588, 591, 592, 593
Trihexyltins, 590
Trimethylarsine, 1506, 1523, 1536
Trimethylarsinoxide, 1549, 1551
Trimethyllead, 262, 263, 273, 279
Trimethylnaphthalenes, 1374
Trimethylselenomium chloride, 1508
Trimethyltin, 555, 556, 563, 564, 570, 573–574, 584, 585
Trimethyltin chloride, 583
Trimethyltin hydroxide, 580
Trimethyltins, 556, 557, 568, 570, 571, 573, 583, 584, 588, 591
Trimethyltin sulfate, 584
Trioctyltins, 590
Triorganotins, 551, 554, 555, 560, 580, 584, 592, 594
Tripentyltins, 573, 579
Triphenyltin acetate, 590
Triphenyltin chloride, 583
Triphenyltin hydroxide, 580
Triphenyltins, 555, 556, 569, 573, 579, 582, 583, 586, 589, 593
Tripropyltin chloride, 583
Tripropyltin oxide, 580
Tripropyltins, 556, 573, 574, 589
Trithiomolybdates, 1616
Trivalent chromium, 45, 46, 47, 60, 61, 63–64, 65, 66, 67, 79, 80
Trivalent copper, 98
Tungsten, 1614, 1616, 1678
Tungsten-181, 1711, 1784
Tungsten-185, 1712
Tungsten-187, 1712
Tyrosinase, 101

U

Uranium, 77, 1613, 1614, 1713, 1717, 1721, 1802
Uranium-232, 1713, 1779, 1781
Uranium-233, 1712, 1804
Uranium-234, 1712, 1726, 1727, 1728, 1737, 1739, 1740, 1804
Uranium-235, 1712, 1713, 1715, 1721, 1726, 1727, 1728, 1737, 1739, 1740, 1801, 1804
Uranium-236, 1712, 1804
Uranium-238, 1712, 1713, 1714, 1715, 1717, 1721, 1726, 1727, 1728, 1737, 1738, 1740, 1801, 1804, 1805
Uranium dioxide, 1721
Uranium hexafluoride, 1721, 1722
Uridine diphospo *N*-acetyl glucosamine, 985, 1005

V

Vanadium, 61, 616, 1614, 1616
Vapotone, 1460

Table 33.1 (continued) Chemical and Trade Names of Substances Listed in the *Handbook of Chemical Risk Assessment* Series

Vectal, 769
Vectal SC, 769
Velsicol 1068, 824
Vigilante, 984
Vinyl acetate, 754

W

Warbex, 1067, 1069
Warfarin, 1418
Weed-beads, 1196
Weedol, 1163
Weedone, 1196
WL 43775, 1091
Wolframate, 1630
Wulfenite, 1614

X

Xanthines, 1287
Xanthones, 1287
Xenon-131, 1711, 1721
Xenon-133, 1711, 1721, 1753
Xenon-135, 1711
Xylene, 1134

Y

Yaltox, 800
Yttrium-90, 1711, 1781
Yttrium-91, 1711, 1721, 1781

Z

Zinc, 17, 23, 26, 28, 64, 104, 134, 136, 138, 166, 169, 374, 456, 514, 559, 614, 1257, 1616, 1673, 1674, 1684, 1685, 1783, 1799
Zinc-65, 649, 669, 675, 1710, 1721, 1724, 1731, 1735, 1736, 1738, 1739, 1740, 1742, 1746, 1780, 1784, 1787, 1799
Zinc acetate, 615, 617
Zinc aquo ion, 608, 686, 695
Zinc bicarbonate, 609
Zinc borate, 1594
Zinc carbonate, 609
Zinc chloride, 608, 609, 616, 617, 677, 686, 691
Zinc chromate, 77
Zinc cyanide, 906, 920
Zinc humic acid, 609, 670
Zinc hydroxide, 609
Zinc oxide, 617, 680, 691
Zinc phosphide, 676, 679
Zinc sulfate, 174, 608, 609, 617, 679, 684
Zinc sulfide, 608, 610
Zirconium, 1713, 1783
Zirconium-95, 1711, 1719, 1720, 1721, 1724, 1725, 1731, 1735, 1736, 1737, 1738, 1739, 1740, 1742, 1753, 1755, 1765, 1799, 1800

Table 33.2 Common and Scientific Names of Plants and Animals Listed in the *Handbook of Chemical Risk Assessment series* (References are provided from common names to scientific names)

A

Abalone
  Black, *Haliotis cracherodii,* 147
  Red,*Haliotis rufescens,* 5, 52, 147, 256, 442, 515, 659
  Various, *Haliotis tuberculata,* 442
Abarenicola pacifica, 1377
Abelmoschus esculentus, 810, 1095, 1096, 1626, 1638
Abies pindrow, 142, 653
Abies spp., 1005, 1006
Acacia georginae, 1413, 1426
Acacia spp., 1426
Acanthocepthalan, *Paratenuisentis* sp., 229
Acanthogobius flavimanus, 837
Acartia clausi, 63, 64, 149, 377, 927, 1540, 1677
Acartia spp., 672
Acartia tonsa, 131, 142, 149, 368, 527, 577, 660, 778, 784, 997, 1468, 1677
Accipiter sp., 333
Accipiter cooperii, 4, 13, 845, 1149
Accipiter fasciatus, 1432
Accipiter gentilis, 346, 1289
Accipiter nisus, 333, 346, 349, 379, 1289
Accipiter striatus, 232, 346, 833, 844, 845, 1289
Acer macrophyllum, 1583
Acer rubrum, 110, 128, 143, 438, 653, 685, 815
Acheta domesticus, 613, 614
Acheta pennsylvanicus, 855, 1102, 1103
Acholla multispinosa, 990
Achyla sp., 456
Acipenser nudiventris, 574, 591
Acipenser oxyrhynchus, 1739
Acipenser stellatus, 574
Acmaea digitalis, 226
Acris crepitans, 119, 435, 445, 783
Acris sp., 367
Acrobeloides sp., 991
Acrocheilus alutaceus, 1738
Actinomycete, *Nocardiopsis* sp., 826
Actinomycetes sp., 924
Adamussium colbecki, 112
Adelotus brevis, 1174
Adinia xenica, 1141
Aechmophorus occidentalis, 119, 346, 348, 844, 1524, 1659
Aedes aegypti, 995, 1041, 1072
Aedes albopictus, 994, 995
Aedes melanimon, 986, 1004
Aedes nigromaculis, 986, 996, 1004, 1107
Aedes sp., 1107
Aegolius funereus, 1289
Aeolanthus bioformifolius, 110, 132
Aeolanthus sp., 132
Aepyceros melampus, 123, 136
Aeromonas hydrophila, 156
Aeromonas sp., 1651
Agapornis roseicollis, 676
Agaricus bisporus, 989
Agaricus campestris, 520
Agelaius phoeniceus, 269, 370, 800, 806, 861, 862, 890, 964, 977, 1006, 1036, 1076, 1077, 1096, 1135, 1289, 1466
Agkistrodon piscivorus, 1466
Agosia chrysogaster, 663
Agriolimax reticulatus, 976
Agrobacter sp., 67
Agropyron cristatum, 1427
Agropyron smithii, 1667
Agrostis capillaris, 143
Agrostis orthogonia, 894
Agrostis stolonifera, 15
Agrostis tenuis, 1168, 1516
Agrotis ipsilon, 1733
Aix sponsa, 12, 20, 332, 347, 1027, 1034, 1055, 1741, 1787
Ajaia ajaja, 350
Albacore, *Thunnus alalunga,* 1740
Albatross, 121
  Black-footed, *Diomedea nigripes,* 1239
  Laysan, *Diomedea immutabilis,* 203, 220, 238, 1239
Alburnus alburnus, 579, 1110
Alcaligenes sp., 453
Alca torda, 1289, 1308
Alces alces, 13, 59, 61, 380, 437, 449, 1649, 1669, 1756, 1759, 1766, 1806
Alces alces gigas, 1623, 1633
Alcyonia alcyonium, 628
Aldrichetta forsteri, 63, 64
Alectoris chukar, 371, 890, 1543
Alectoris graeca, 1434
Alewife, *Alosa pseudoharengus,* 55, 1148, 1202, 1288, 1320, 1517
Alfalfa, *Medicago sativa,* 137, 337, 456, 483, 746, 1095, 1161, 1168, 1427, 1516, 1530, 1584, 1621, 1763
Algae
  Amphidinium sp., 656
  Amphora coffeaeformis, 582
  Anabaena sp., 748, 777, 1302
  Anacystis sp., 457, 458, 522, 655, 1586, 1675
  Ankistrodesmus sp., 893
  Asterionella formosa, 21
  Asteromonas gracillis, 573
  Bellerochea polymorpha, 1587
  Bladder wrack, *Fucus vesiculosus,* 5, 52, 224, 440, 656, 1519, 1534, 1730, 1735
  Blidingia minima, 224
  Blue-green
    Anabaena flos-aquae, 96
    Aphanizomenon flos-aquae, 96
    Microcystis aeruginosa, 96, 257, 927
  Brown
    Ascophyllum nodosum, 5, 52, 111, 440, 655, 656
    Egregia laevigata, 256
    Pelvetia canaliculata, 111
  Chara sp., 747, 1207
  Chlamydomonas, 71, 857
  Chlorella sp., 144, 145, 524, 781, 927, 1304
  Chlorococcum sp., 1171
  Chroomonas sp., 1511

Table 33.2 (continued) Common and Scientific Names of Plants and Animals Listed in the *Handbook of Chemical Risk Assessment series* (References are provided from common names to scientific names)

Cladophora sp., 224, 747, 1202
Coelastrium cambricum, 1778
Cricosphaera, 656
Cyclotella, 779, 995
Cylindrotheca fusiformis, 1585
Cymodocea sp., 440
Diatoma sp., 655
Dunaliella, 145, 262, 670, 748, 1171, 1207, 1535
Enteromorpha sp., 96, 165, 568, 1207
Epithemia sp., 781
Exuviella baltica, 857
Glenodinium halli, 656, 689
Gymnodinium splendens, 656
Haslea ostrearia, 165
Hymenomonas carterae, 1535
Isochrysis galbana, 582, 656, 1107, 1171
Microspora sp., 655
Mougeotia, 96, 655, 781
Nannochloris oculata, 780, 973
Nitella, 1628
Nitzschia, 145, 582, 656, 1107, 1778
Nostoc muscorum, 775, 810, 1628
Ochromonas danica, 145
Oedogonium sp., 67, 96, 781
Olisthodiscus lutens, 70
Peridinium gatunense, 1770
Phormidium inundatum, 524
Platymonas subcordiformis, 255
Plectonema boryanum, 994, 995, 1002
Polygonum sp., 1170
Procentrum micans, 656
Prorocentrum mariae-lehouriae, 524
Red
Champia parvula, 1540
Plumaria elegans, 1540
Rhizosolenia sp., 656, 689
Scenedesmus spp., 524, 1302
Schroederella, 655, 657, 689
Scripsiella faeroense, 145, 375
Selenastrum capricornutum, 256, 257, 657, 811, 995, 1195, 1207, 1302, 1372, 1536, 1783
Skeletonema costatum, 22, 165, 261, 522, 524, 573, 574, 577, 579, 657, 689, 995, 1107, 1207, 1540, 1587
Spirogyra, 96, 747, 1161
Symiodinium sp., 1770
Synedra sp., 655, 1302
Tabellaria sp., 655
Tetraedron sp., 893
Tetraselmis chui, 1535
Thalassiosira sp., 523, 993, 995
Tolypothrix sp., 781
Ulothrix sp., 655
Ulva sp., 5, 165, 225
Zygnema sp., 451
Alligator, *Alligator mississippiensis*, 345, 634, 1286, 1523
Alligator mississipiensis, 345, 634, 1286, 1523
Allium cepa, 438
Allium sp., 426, 453
Allolobophora caliginosa, 1039, 1205, 1766
Allolobophora spp., 441
Allorchestes compressa, 149, 614, 660, 1628
Allysum spp., 434, 439, 456
Almond, *Prunus amygdalus*, 918
Alopex lagopus, 124, 1416, 1441, 1760
Alosa fallax, 1466
Alosa pseudoharengus, 55, 1148, 1202, 1288, 1320, 1517
Alpaca, *Lama pacos*, 137
Alphitobius diaperinus, 1071, 1073
Alphitobius sp., 108
Amantia muscaria, 1516
Amblema sp., 112
Ambloplites rupestris, 116, 340, 443, 461, 1528
Amblyomma americanum, 1079
Amblyomma maculatum, 1079
Ambystoma gracile, 17, 22, 25
Ambystoma jeffersonianum, 135, 261, 1586, 1588
Ambystoma maculatum, 131, 1373, 1586, 1588
Ambystoma mexicanus, 577, 1213
Ambystoma opacum, 160, 264, 368, 463, 668, 1538
Ambystoma sp., 1385
Ambystoma tigrinum, 1355
Ameiurus melas, 1267, 1279, 1468
Ameiurus nebulosus, 444, 452, 1354
Ameiva exsul, 365
Amia calva, 313, 632, 750, 1038, 1465
Amnicola sp., 63
Ampelisca abdita, 263, 522, 527, 1027
Amphidinium carteri, 656
Amphioxus, *Branchiostoma caribaeum*, 578, 1109
Amphipods
Allorchestes, 149, 614, 660, 1628
Ampelisca abdita, 263, 522, 527, 1027
Corophium, 150, 1540
Crangonyx, 1209
Elasmopus, 1374
Gammarus, 62, 64, 142, 459, 778, 1136, 1629, 1783
Hyalella azteca, 142, 151, 460, 528, 533, 856, 858, 887, 998, 1003, 1378, 1676
Leptocheirus plumulosus, 17
Orchestia, 115, 134, 578, 629
Parhalella, 152
Pontoporeia, 22, 836, 1304, 1378
Rhepoxynius, 1627
Talitrus, 629
Talorchestia, 629
Themisto, 116, 629
Amphora coffeaeformis, 582
Amrasca biguttula biguttula, 1102
Amygdalus cummunis, 904
Anabaena flos-aquae, 96
Anabaena inequalis, 458
Anabaena oryzae, 775
Anabaena oscillaroides, 1627, 1628
Anabaena spp., 748, 777, 1302
Anabaena variabilis, 1675
Anabas scandens, 69
Anabas sp., 62
Anabas testudineus, 62, 64, 260, 461, 804
Anacardium occidentale, 1626
Anacystis marina, 522
Anacystis nidulans, 457, 458, 655, 1586, 1675
Anadara granosa, 112
Anadara trapezium, 133
Ananas comusus, 767
Anarhichas minor, 229
Anas acuta, 202, 242, 267, 799, 1434

Table 33.2 (continued) Common and Scientific Names of Plants and Animals Listed in the *Handbook of Chemical Risk Assessment series* (References are provided from common names to scientific names)

Anas americana, 799, 961, 1434
Anas carolinensis, 799, 1787
Anas clypeata, 862, 1787
Anas cyanoptera, 1579, 1665
Anas discors, 347, 624, 634, 862, 1290, 1787
Anas fulvigula, 242
Anas platyrhynchos, 17, 120, 136, 167, 209, 232, 234, 241, 267, 283, 348, 371, 378, 384, 427, 436, 445, 464, 465, 480, 482, 583, 584, 625, 634, 676, 677, 689, 751, 752, 786, 806, 808, 833, 845, 862, 890, 891, 893, 932, 933, 961, 968, 1004, 1006, 1047, 1074, 1076, 1135, 1136, 1142, 1167, 1175, 1214, 1290, 1308, 1386, 1430, 1434, 1469, 1470, 1508, 1524, 1543, 1553, 1579, 1592, 1665, 1666, 1680, 1689, 1742, 1786, 1807
Anas rubripes, 23, 57, 74, 120, 136, 202, 219, 232, 242, 379, 445, 833, 845, 961, 1076, 1471, 1524
Anas spp., 120, 166, 167, 268, 676, 861, 961, 977, 1524, 1666, 1685
Anas strepera, 219, 446, 1524, 1579, 1665
Anas superciliosa, 1431, 1434
Anastrepha ludens, 1585
Anastrepha suspensa, 1776, 1777
Anchovetta, *Cetengraulis mysticetus,* 1579
Anchovy
  Adriatic, *Engraulis encrasicolus,* 117
  Northern, *Engraulis mordax,* 623
Ancylus fluviatilis, 657, 782
Andropogon scoparius, 1667
Anemone
  Plumose, *Metridium senile,* 628
  Sea, *Anemonia viridis, Anthopleura xanthogrammica,* 146, 161, 1778
Anemonia viridis, 146, 161
Anguilla anguilla, 11, 20, 229, 256, 260, 888, 964, 965, 970, 1028, 1031, 1212, 1370, 1379
Anguilla rostrata, 461, 1030, 1148, 1283, 1739
Anguilla sp., 1278
Anguilla vulgaris, 377
Anguillicola sp., 229
Anhinga, *Anhinga anhinga,* 1149
Anhinga anhinga, 1149
Ankistrodesmus falcatus, 55, 574, 578, 1536, 1778
Ankistrodesmus sp., 893
Annelids. *See Nereis; Tubifex*
Anocentor nitens, 1072, 1079
Anodonta anatina, 441, 574, 578, 1207
Anodonta cygnea, 102, 146, 613, 994, 997
Anodonta grandis, 112, 133, 146
Anodonta nuttalliana, 669
Anodonta piscinalis, 112, 129
Anomalocera sp., 672
Anopheles quadrimaculatus, 1428
Anopheles stephensi, 1103
Anser albifrons, 232
Anser anser, 202, 219, 1308, 1743
Anser spp., 120, 961, 968, 977
Ant
  Big-head, *Formica* sp., 1527
  Fire, *Solenopsis invicta,* 855, 1102, 1104, 1133, 1152
  Formica sp., 1527
  Harvester, *Pogonomyrmex* spp., 1427
  Liometopum occidentale, 1432
  Veromessor andrei, 1432
Antechinus
  Brown, *Antechinus stuartii,* 1441
  Dusky, *Antechinus swainsonii,* 1441
Antechinus stuartii, 1441
Antechinus swainsonii, 1441
Anthocidaris crassispina, 73, 263, 663, 1587
Anthonomus grandis, 983, 991, 1013, 1504
Anthopleura xanthogrammica, 1778
Antilocapra americana, 59, 1673, 1733
Antimora rostrata, 340, 514, 1145
Apeltes quadracus, 1678
Apheloria corrugata, 918
Apheloria kleinpeteri, 918
Apheloria spp., 922
Aphids
  Various, *Lipaphis erysimi, Macrosiphium gei,* 1102
Aphinizomenon flos-aquae, 96
Apis cerama indica, 1103
Apis mellifera, 111, 226, 621, 627, 961, 990, 991, 1012, 1013, 1102, 1103, 1169, 1428, 1532, 1533, 1585
Apis sp., 226, 809, 894, 1123, 1764
Apium graveolans, 434, 438
Apium sp., 1093
Aplexa hypnorium, 749
Aplodinotus grunniens, 340, 839
Aplodontia rufa, 1029
Apodemus flavicollis, 644
Apodemus sylvaticus, 124, 242, 248, 275, 355, 1515
Aporrectodea calignosa, 141, 144
Aporrectodea rosea, 128
Aporrectodea tuberculata, 653
Apple, *Malus,* 918, 989, 1092
Apricot, *Prunus armenaica,* 904
Apteryx spp., 1433
Aquila audax, 1431, 1432, 1434
Aquila chrysaetos, 220, 233, 347, 637, 679, 905, 1290, 1432, 1434
Aquila heliaca adalberti, 1288, 1290
Aquipecten irradians, 26
Ara ararauna, 676
Arabidopsis thaliana, 1755
Arachis hypogea, 810, 1427
Arachis sp., 1093
Aramus guarauna, 1463
Araneus umbricatus, 217
Arbacia lixula, 226, 529
Arbacia punctulata, 460
Arbutus menziesii, 1583
Arctica islandica, 112, 441
Arctocephalus pusillus, 355
Arcularia gibbosula, 339
Ardea cinerea, 235, 314, 349, 1290
Ardea herodias, 334, 347, 348, 350, 354, 845, 1028, 1035, 1290, 1297, 1319
Ardea herodias occidentalis, 347, 350
Ardea sp., 1524
Arenicola cristata, 578, 579, 582, 1136
Arenicola marina, 153, 163, 631, 1737
Argiope aurantia, 112
Argopecten irradians, 146, 161, 368, 525, 658
Arianta arbustorum, 112
Arion ater, 627, 653, 654

Table 33.2 (continued) Common and Scientific Names of Plants and Animals Listed in the *Handbook of Chemical Risk Assessment series* (References are provided from common names to scientific names)

Arion hortensis, 627
Arion subfuscus, 627
Arius felis, 838, 1384, 1465
Arius graffei, 1369
Arius sp., 364
Artemia franciscana, 105
Artemia nauplii, 528
Artemia salina, 997, 1535, 1778
Artemia sp., 660, 1707
Artemisia tridentata, 56, 1577
Arthrobacter sp., 67, 1199
Arundinaria spp., 918
Arvicola terrestris, 1160
Ascophyllum nodosum, 5, 52, 111, 440, 655, 656
Asellus aquaticus, 21, 258
Asellus communis, 671, 927
Asellus meridianus, 257, 1171
Asellus racovitai, 1209
Asellus sp., 338, 1171
Ash, mountain, *Sorbus aucuparia,* 12, 1767
Asio flammeus, 1291
Asparagus, *Asparagus officinale,* 1667
Asparagus officinale, 1667
Aspen
  Largetooth, *Populus grandidentata,* 1734
  Trembling, *Populus tremuloides,* 436, 1734
  Various, *Populus* sp., 12
Aspen sp., *Populus,* 12
Aspergillus clavatus, 456
Aspergillus flavus, 456
Aspergillus fumigatus, 1421
Asplanchna brightwelli, 1770
Asplanchna sp., 1004
Ass, domestic, *Equus asinus,* 243
Astacus astacus, 115
Astacus fluviatilis, 377, 1209
Aster, *Aster, Astragalus,* 1652, 1655, 1667
Aster, tansy, *Machaeranthera* spp., 1652
Aster caerulescens, 1667
Aster commutatus, 1667
Asterias forbesi, 63, 460
Asterias rubens, 443, 663, 1354, 1628
Asterionella formosa, 21
Aster mul, 1667
Aster multiflora, 1667
Aster occidentalis, 1667
Asteroid, *Echinus esculentus,* 7
Asteromonas gracilis, 573
Aster spp., 1652, 1655
Astragalus argillosus, 1667
Astragalus beathii, 1667
Astragalus bisulcatus, 1667
Astragalus confertiflorus, 1667
Astragalus crotulariae, 1667
Astragalus racemosus, 1655
Astragalus spp., 1652, 1655, 1667
Astralium rogosum, 70
Astyanax bimaculatus, 1205
Ateles geoffroyi, 1441
Athene cunicularia, 816, 905, 1432
Atherinasoma microstoma, 63
Atherinops affinis, 153, 164, 1110
Atherix sp., 1107, 1468
Atriplex spp., 1652, 1667
Aulosira fertissima, 775
Aurelia aurita, 1778
Australorbis glabratus, 578, 579, 749, 1208
Australorbis sp., 525
Austropotamobius pallipes, 671, 672
Avena sativa, 253, 434, 456, 521, 774, 775, 1168, 1667
Avocet, American, *Recurvirostra americana,* 1515, 1580, 1657, 1666
Aythya affinis, 8, 235, 242
Aythya americana, 120, 242, 446
Aythya collaris, 122, 235, 446
Aythya ferina, 122, 202, 219
Aythya fuligula, 219
Aythya marila, 348, 446, 518, 1291, 1666
Aythya spp., 1524
Aythya valisineria, 4, 57, 120, 136, 202, 221, 233, 234, 284, 446, 634, 845, 1368, 1463
Azospirillum sp., 885
Azotobacter sp., 885
Azotobacter vinelandii, 989

B

Babesia caballi, 1072
Baboon, *Papio anubis,* 278, 758, 867, 912, 1417
Bacillus cereus, 453
Bacillus spp., 1651
Bacillus subtilis, 67, 1216
Bacteria
  Actinomycetes sp., 924
  Aeromonas sp., 651
  Alcaligenes sp., 453
  Arthrobacter sp., 67, 1199
  Aulosira fertissima, 775
  Azobacter vinelandii, 989
  Bacillus sp., 1651
  Clostridium posterianum, 453
  Escherichia coli, 425, 534, 657, 746, 1039, 1165, 1169, 1216
  Flavobacterium sp., 1199, 1651
  Klebsiella pneumoniae, 754, 1546
  Methanobacterium sp., 453
  Oscillatoria sp., 453
  Pseudomonas spp., 657, 803, 963, 1200, 1421, 1651
  Renibacterium salmoninarum, 1674
  Salmonella sp., 28, 617, 754, 759, 1473
  Streptococcus, 524, 754
  Tolypothrix sp., 781
Badger, *Taxidea taxus,* 905, 1440, 1448
Baetis pygmaeus, 995
Baetis thermicus, 100, 115, 162
Balaena mysticetus, 1660
Balaenoptera acutorostrata, 1685
Balaenoptera physalis, 336, 355, 1525
Balaenoptera rostrata, 1660
Balanus amphitrite, 578, 629
Balanus balanoides, 629, 671
Balanus eburneus, 749, 995, 997
Balanus sp., 72, 623
Balistoides viridescens, 633
Bamboo
  Arundinaria spp., 918

Table 33.2 (continued) Common and Scientific Names of Plants and Animals Listed in the *Handbook of Chemical Risk Assessment series* (References are provided from common names to scientific names)

Bambusa spp., 903
Dendrocalamus spp., 918
Bambusa spp., 903
Bandicoot
Golden, *Isoodon auratus barrowensis,* 1443
Gunn's, *Perameles gunni,* 1446
Long-nosed, *Perameles nasuta,* 1446
Northern brown, *Isoodon macrourus,* 1443
Southern brown, *Isoodon obesulus,* 1443
Barb
Rosy, *Barbus conchonius,* 260
Thai silver, *Puntius gonionotus,* 1173
Barbus conchonius, 260
Barilus vagra, 966
Barley, *Hordeum vulgare,* 652, 653, 775, 1168, 1530, 1583, 1621, 1761
Barnacle, *Balanus, Elminius, Semibalanus,* 614
Barnardius zonarius, 1434
Barn-owl, common, *Tyto alba,* 24, 863, 1074, 1432
Bass
Barred sand, *Paralabrax nebulifer,* 1282
Black sea, *Centropristis striata,* 1522
European sea, *Dicentrarchus labrax,* 164, 1379
Kelp, *Paralabrax clathratus,* 118, 444
Largemouth, *Micropterus salmoides,* 55, 56, 313, 314, 340, 342, 367, 444, 451, 462, 530, 533, 632, 750, 838, 860, 928, 1038, 1205, 1210, 1221, 1280, 1382, 1464, 1468, 1471, 1517, 1519, 1589, 1602, 1662, 1663, 1729, 1739, 1782
Micropterus, 747
Rock, *Ambloplites rupestris,* 116, 340, 443, 461, 1528
Sea, *Lateolabrax japonicus,* 841
Smallmouth, *Micropterus dolomieui,* 230, 260, 444, 1026, 1031, 1173, 1518
Striped, *Morone saxatilis,* 7, 18, 22, 62, 155, 332, 342, 376, 462, 518, 575, 665, 767, 799, 841, 860, 886, 888, 1021, 1268, 1284, 1303, 1468, 1517, 1523, 1589, 1659, 1663, 1677, 1678, 1739
White, *Morone chrysops,* 839
Bat
Big brown, *Eptesicus fuscus,* 243, 639, 823
Common pipistrelle, *Pipistrellus pipestrellus,* 1217, 1219, 1299
Dutch pond, *Myotis dasycneme,* 1202
Eastern big-eared, *Plecotus phyllotis,* 990
Fruit, *Pteropus* sp., 247
Gray, *Myotis grisescens,* 641, 834, 853
Greater horseshoe, *Rhinolophus ferrumequinum,* 1299
Indiana, *Myotis sodalis,* 990
Little brown, *Myotis lucifugus,* 222, 245, 834, 943
Mexican free-tailed, *Tadarida brasiliensis,* 15, 249, 1526
Schreiber's, *Miniopterus schreibersi,* 1299
Southeastern, *Myotis austroriparius,* 641
Bean
Buck, *Menyanthes trifoliata,* 1767
Castor, *Ricinus communis,* 746, 922
Green, *Phaseolus vulgaris,* 252, 922, 1039, 1584
Jack, *Canavalia* sp., 453, 918
Lima, *Phaseolus lunatus,* 903, 989
Phaseolus sp., 521, 746, 812, 918, 1093
Vicia faba, 141, 426, 1776
Bears
Brown, *Ursus arctos,* 59, 361, 383
Polar, *Ursus maritimus,* 103, 123, 135, 336, 361, 363, 520, 642, 834, 1264, 1298, 1301, 1464, 1671
Beaver
Common, *Castor canadensis,* 336, 356, 438, 572, 1669, 1733, 1744
European, *Castor fiber,* 13
Mountain, *Aplodontia rufa,* 1029
Beaver(s), 4
Becium homblei, 132
Bee
Alfalfa leaf cutter, *Megachile rotundata,* 1103, 1104
Alkali, *Nomia melanderi,* 1104
Bumble, *Bombus* spp., 112
Cuckoo bumble, *Psithyrus bohemicus,* 112
Honey, *Apis mellifera,* 111, 226, 621, 627, 961, 990, 991, 1012, 1013, 1102, 1103, 1169, 1428, 1532, 1533, 1585
Beet, *Beta vulgaris,* 434, 1531, 1577, 1583, 1626
Beetle
aquatic
Hydrophilus triangularis, 996
Thermonectes basillaris, 996
Tropisternus lateralis, 996
Burying, *Nicrophorus tomentosus,* 228
Colorado potato, *Leptinotarsa decemlineata,* 1504
Convergent lady, *Hippodamia convergens,* 991
Cotesia melanoscela, 991
Dendroides sp., 644
Diachasmimorpha longicaudata, 1777
Ground, *Calosoma* sp., 1096
Japanese, *Popillia japonica,* 837
Leaf, *Paropsis atomaria,* 922
Perga dorsalis, 1428
Sawtoothed grain, *Oryzaephilus surinamensis,* 883
Tiger, *Megacephala virginica,* 922
Yellow mealworm, *Tenebrio molitor,* 991
Bellerochea polymorpha, 1587
Bemisia tabaci, 1102
Berardius bairdii, 1271
Beroe cucumis, 1579
Berry
Bil, *Vaccinium myrtillus,* 1759, 1767
Bog whortle, *Vaccinium uliginosum,* 1767
Cow, *Vaccinium vitis-idaea,* 1767
Beta vulgaris, 434, 1531, 1577, 1583, 1626
Beta vulgaris cicla, 224, 568
Bettong
Brush-tailed, *Bettongia penicillata,* 1444
Burrowing, *Bettongia leseur,* 1441
Bettongia leseur, 1441
Bettongia penicillata, 1444
Betula nana, 1765
Betula papyrifera, 438
Betula sp., 1762, 1767
Bilberry, *Vaccinium* sp., 1762
Bilby, greater, *Macrotus lagotis,* 1444
Biomphalaria alexandrina, 1170
Biomphalaria glabrata, 146, 574, 575, 579, 658
Biomphalaria spp., 560, 578, 750
Biomphalaria sudanica, 579

Table 33.2 (continued) Common and Scientific Names of Plants and Animals Listed in the *Handbook of Chemical Risk Assessment series* (References are provided from common names to scientific names)

Birch
  Dwarf, *Betula nana,* 1765
  Paper, *Betula papyrifera,* 438
  Various, *Betula* sp., 1762, 1767
Birgus latro, 1749
Bison
  American, *Bison bison,* 124, 835, 1669
  European, *Bison bonasus,* 124
Bison bison, 124, 835, 1669
Bison bonasus, 124
Blaberus giganteus, 1776
Blackbird
  Brewer's, *Euphagus cyanocephalus,* 1432, 1435
  Common, *Turdus merula,* 241, 638, 895, 1668
  Red-winged, *Agelaius phoeniceus,* 269, 370, 800, 806, 861, 862, 890, 964, 977, 1006, 1036, 1076, 1077, 1096, 1135, 1289, 1466
Blackfish
  Common, *Gadopsis marmoratus,* 341
  Largescale, *Girella punctata,* 324
  Sacramento, *Orthodon microlepidotus,* 55, 331, 1517
Blackfly, *Simulium* sp., 228
Blacktail, *Diplodus sargus,* 340
Blarina brevicauda, 13, 242, 248, 250, 336, 365, 645, 807, 1788
Blattella germanica, 883, 991, 1169, 1584, 1585
Bleak, *Alburnus alburnus,* 579, 1110
Blepharisma undulans, 1102
Blidingia minima, 224
Blissus leucopterus listus, 895
Bloater, *Coregonus hoyi,* 117, 1266, 1283, 1514
Blueberry
  Common, *Vaccinium pallidum,* 225
  Lowbush, *Vaccinium angustifolium,* 128, 440, 1516
Bluebird, eastern, *Sialia sialis,* 846, 1048, 1786, 1788
Bluefish, *Pomatomus saltatrix,* 8, 332, 435
Bluegill, *Lepomis macrochirus,* 19, 28, 55, 56, 62, 65, 68, 155, 229, 258, 340, 451, 462, 529, 532, 578, 664, 747, 750, 783, 805, 841, 856, 859, 888, 893, 924, 928, 964, 966, 970, 972, 993, 1000, 1110, 1136, 1141, 1173, 1205, 1211, 1372, 1375, 1377, 1380, 1395, 1428, 1464, 1468, 1517, 1518, 1534, 1538, 1553, 1578, 1589, 1602, 1629, 1656, 1662, 1675, 1677, 1779, 1784
Bluestem, little, *Andropogon scoparius,* 1667
Boar, wild, *Sus scrofa,* 59, 127, 245, 968, 969, 975, 978, 1414, 1417, 1448, 1747, 1756
Bobcat, *Lynx rufus,* 357, 914, 1144, 1440, 1745
Bobwhite, common, *Colinus virginianus,* 265, 269, 371, 786, 806, 861, 862, 890, 891, 968, 978, 1047, 1114, 1136, 1142, 1167, 1176, 1309, 1430, 1435, 1469, 1542, 1543, 1742, 1787
Boleophthalmus boddaerti, 923, 928
Boleophthalmus dussumieri, 62, 69, 1683
Boll weevil, *Anthonomus grandis,* 983, 991, 1013, 1504
Bolti, *Tilapia zilli,* 668
Bombina variegata, 119
Bombus spp., 112
Bombyx mori, 1776, 1777
Bonasa umbellus, 120, 235, 436, 446, 518, 571, 1732, 1741
Bonytail (fish), *Gila elegans,* 1588, 1676
Booby
  Red-footed, *Sula sula,* 1660
  Various, *Sula,* 624
Boophilus annulatus, 1079
Boophilus microlopus, 1079
Boreogadus saida, 1264, 1465
Bos bovis, 13, 223, 243, 572, 851, 865, 984, 1013, 1525, 1547
Bos indicus, 1078, 1085
Bosmina longirostris, 150, 1003, 1536
Bosmina longispina, 1766
Bosmina sp., 1003
Bos spp., 124, 170, 177, 234, 276, 285, 454, 469, 483, 484, 638, 649, 681, 690, 753, 756, 787, 851, 892, 905, 940, 1008, 1009, 1079, 1084, 1116, 1117, 1179, 1216, 1368, 1438, 1441, 1469, 1473, 1547, 1595, 1624, 1634, 1639, 1744, 1788
Bos taurus, 1078, 1217, 1414
Bothromestoma sp., 1004
Bothrops jararaca, 615
Bouteloua dactyloides, 1667
Bovicola crassipes, 1081
Bovicola limbatus, 1010, 1081
Bowdleria punctata, 1433
Bowfin, *Amia calva,* 313, 632, 750, 1038, 1465
Brachidontes exustus. see Ischadium recurvum
Brachionus calyciflorus, 146, 965, 972
Brachydanio rerio, 154, 259, 367, 375, 461, 613, 674, 784, 965, 966, 1042, 1173, 1304, 1588, 1675
Brachythecium rivulare, 223
Brachythecium salebrosum, 434
Branchiostoma caribaeum, 578, 1109
Brant, Atlantic, *Branta bernicla hrota,* 221
Brant, *Branta bernicla,* 895, 961
Branta bernicla, 895, 961
Branta bernicla hrota, 221
Branta canadensis, 219, 220, 235, 269, 800, 890, 895, 961, 1076, 1418
Branta canadensis leucoparlia, 1416
Branta canadensis maxima, 624, 679
Branta canadensis minima, 846
Branta canadensis moffitti, 1732, 1741
Brassica campestri, 521
Brassica juncea, 774, 775
Brassica oleracea botrytis, 1095, 1097
Brassica oleracea capitata, 434, 440, 989, 1427
Brassica oleracea italica, 1583
Brassica rapa, 521
Brassica spp., 989, 1093, 1626
Bream, Red sea, *Pagrus major,* 583, 636
Brevoortia patronus, 642
Brevoortia tyrannus, 578, 623
Briareum abestium, 835
Brittle star, *Ophioderma brevispina,* 576
Broadleaf, *Griselinia littoralis,* 1418
Broccoli, *Brassica oleracea italica,* 1583
Bromus inermis, 1626
Bromus japonicum, 130, 644
Bromus spp., 224
Broomweed, *Gutierezia* spp., *Haplopappus* spp., 1652, 1667
Bruguiera caryophylloides, 568
Bryconamericus iheringii, 1172
Bryoria fuscescens, 1767

Table 33.2 (continued) Common and Scientific Names of Plants and Animals Listed in the *Handbook of Chemical Risk Assessment series* (References are provided from common names to scientific names)

Bryozoans, *Bugula, Victorella, Watersipora,* 517, 527, 660
Bubalus sp., 124
Bubo bubo, 333, 348, 1291
Bubo sp., 333
Bubo virginianus, 834, 846, 1071, 1076, 1085, 1150, 1431, 1434
Bubulcus ibis, 348, 365, 1149, 1668
Buccinum striatissimum, 1514
Buccinum undatum, 113, 441
Bucephala clangula, 220, 352, 446, 846, 1308
Bucephala islandica, 1623
Budworm, western spruce, *Choristoneura occidentalis,* 1532, 1533
Buffalo (fish)
  Bigmouth, *Ictiobus cyprinellus,* 839
  Smallmouth, *Ictiobus bubalus,* 839, 1466
Buffalo (mammal)
  Indian, *Bubalus* sp., 124
  Water, *Bubalus* sp., 124
Bufo americanus, 160, 165, 261, 436, 784, 1586, 1588
Bufo arenarum, 17, 260, 264, 420, 613, 647, 860
Bufo bufo, 119, 345, 860
Bufo fowleri, 160, 464, 1096, 1588
Bufo marinus, 119, 135
Bufo regularis, 464
Bufo sp., 109, 676
Bufo terrestris, 1025, 1523, 1665
Bufo valliceps, 1786
Bufo vulgaris, 1588
Bufo woodhousei fowleri, 1174
Bugula neritina, 660
Bulinus globosus, 146
Bulinus sp., 560, 750
Bulinus truncatas, 1170
Bullfinch, *Pyrrhula,* 636
Bupalus piniarius, 112, 132
Bursaphelencchus xylophilus, 976
Busycon canaliculatum, 113, 147, 162
Buteo buteo, 235, 637, 809, 895, 1291
Buteo jamaicensis, 269, 370, 800, 802, 1075
Buteo lagopus, 1291, 1431, 1434
Buteo lineatus, 800, 834
Buteo regalis, 1431, 1435
Buteo sp., 333, 1524
Butorides striatus, 240, 1149, 1464
Butorides virescens, 1665
Butorides virescens virescens, 1096
Butterfly, large white, *Pieris brassicae,* 963, 992, 1427
Buzzard
  Common, *Buteo buteo,* 235, 637, 809, 895, 1291
  Honey, *Pernis apivorus,* 202

C

Cabbage
  Brassica oleracea capitata, 434, 440, 989, 1427
  Chinese, *Brassica campestri,* 521
Cabezon, *Scorpaenichthys marmoratus,* 654, 667
Cabomba spp., 927
Cacatua galerita, 1431, 1435
Cacatua roseicapilla, 1431, 1435
Caddisfly, *Triaenodus tardus,* 811
Caenorhabditis elegans, 141, 143, 146, 525
Cajanus sp., 918
Calamospiza melanocorys, 976
Calamus bajonado, 117
Calanus marshallae, 1627, 1628
Calcarius *mccownii,* 1432
Calcarius ornatus, 976
Calidris alba, 57, 121
Calidris alpina, 8, 846, 1660
Calidris canutus, 57, 235
Calidris pusilus, 57
Callaeas cinera, 1433
Callibaetis skokianus, 1209
Callibaetis sp., 996
Callinectes sapidus, 54, 63, 365, 520, 611, 642, 767, 857, 893, 995, 998, 1021, 1038, 1136, 1468, 1521, 1527, 1671, 1677, 1778
Callinectes similis, 150
Callipepla californica, 861, 862, 1431, 1435, 1469, 1542, 1543
Callithrix jacchus, 588, 961, 1049, 1055, 1314, 1546
Callorhinus ursinus, 9, 335
Calluna vulgaris, 1516, 1767
Caloenas nicobarica, Nicobar pigeon, 634, 676
Calonectris diomedea, 634
Calosoma sp., 1096
Cambarus bartoni, 102, 150, 435
Cambarus latimanus, 21
Cambarus longulus longirostris, 1784
Cambarus sp., 227
Camel, Bactrian, *Camelus bactrianus,* 124, 139
Camelus bactrianus, 124, 139
Campanularia flexuosa, 146, 577
Campeloma decisum, 147
Canary, *Serinus canarius,* 935, 1215
Canavalia sp., 453, 918
Cancer irroratus, 6, 53, 71–72, 517, 1364
Cancer magister, 72, 263, 368, 804, 857, 1374, 1521, 1540, 1579, 1677
Canis aureus, 1417
Canis familiaris, 66, 124, 137, 243, 276, 373, 469, 534, 536, 590, 639, 681, 690, 756, 787, 806, 851, 865, 870, 892, 904, 940, 964, 969, 975, 1010, 1014, 1050, 1116, 1118, 1179, 1217, 1299, 1442, 1473, 1503, 1547, 1596, 1745, 1789
Canis familiaris dingo, 1414, 1442
Canis familiaris familiaris, 1418
Canis latrans, 13, 59, 124, 802, 905, 940, 1414, 1442, 1451, 1724, 1745
Canis lupus, 1033, 1416, 1756, 1759
Canis mesomelas, 1417
Canis sp., 484, 808, 1135, 1136
Canvasback, *Aythya valisineria,* 4, 57, 120, 136, 202, 221, 233, 234, 284, 446, 634, 845, 1368, 1463
Capitella capitata, 63, 368, 662
Capra hircus, 124, 454, 469, 890, 1414, 1469
Capra sp., 124, 276, 454, 469, 639, 649, 691, 865, 905, 911, 1010, 1033, 1081, 1104, 1179, 1427, 1442, 1547, 1761
Capreolus capreolus, 24, 125, 244, 355, 895, 1525, 1756, 1769
Capreolus sp., 1747, 1767
Capricornus crispus, 336, 356
Caracara, crested, *Polyborus plancus,* 362

Table 33.2 (continued) Common and Scientific Names of Plants and Animals Listed in the *Handbook of Chemical Risk Assessment series* (References are provided from common names to scientific names)

Carassius auratus, 62, 69, 154, 165, 260, 377, 457, 461, 576, 579, 583, 750, 783, 805, 856, 859, 966, 1136, 1141, 1206, 1212, 1468, 1471, 1538, 1552, 1588, 1676, 1738, 1770
Carassius auratus gibelio, 78
Carassius carassius, 1755
Carassius carassius grandoculis, 578
Carcharhinus longimanus, 117, 443, 1522
Carcharhinus spp., 344
Carcharodon carcharius, 837
Carcinus maenas, 99, 150, 442, 578, 671, 672, 1628, 1779
Carcinus mediterraneus, 134
Cardinal, *Richmondena cardinalis,* 1096
Carduelis chloris, 636, 1433
Caretta caretta, 345, 1145, 1285, 1286
Carex spp., 1767
Caribou, *Rangifer tarandus,* 4, 32, 126, 380, 437, 450, 451, 538, 1072, 1082, 1085, 1758, 1761, 1762, 1791, 1806
Carp
  Common, *Cyprinus carpio,* 21, 55, 69, 154, 229, 259, 340, 457, 461, 529, 574, 612, 784, 837, 838, 859, 966, 970, 1026, 1031, 1042, 1094, 1110, 1165, 1167, 1172, 1210, 1221, 1316, 1343, 1353, 1379, 1465, 1468, 1517, 1578, 1662, 1676, 1739, 1779, 1781
  Cyprinus carpo communis, 461
  Indian, *Saccobranchus fossilis,* 69, 159, 164, 805, 860
  Prussian, *Carassius auratus gibelio,* 78
  Round Crucian, *Carassius carassius grandoculis,* 578
  Silver, *Hypophthalmichthys moltrix,* 1756
Carpiodes carpio, 839
Carpocapsa pomonella, 1504
Carpodacus mexicanus, 806
Cashew, *Anacardium occidentale,* 1626
Casmerodius albus, 350, 1665
Cassava, *Manihot esculenta,* 903, 918, 950, 951
Cassia spp., 224, 253
Cassiope sp., 451
Castanea sp., 1757
Castilleja spp., 1652, 1667
Castor canadensis, 336, 356, 438, 572, 1669, 1733, 1744
Castor fiber, 13
Cat
  Bob, *Lynx rufus,* 357, 914, 1144, 1440, 1745
  Domestic, *Felis domesticus,* 66, 277, 324, 356, 373, 381, 389, 470, 590, 682, 691, 756, 806, 852, 1117, 1118, 1180, 1218, 1443, 1469, 1548, 1635, 1640
  Eastern native, *Dasyurus viverrinus,* 1424, 1438, 1443
  Feral, *Felis cattus,* 1417, 1443
  Northern native, *Dasyurus hallucatus,* 1439, 1443
  Tiger, *Dasyurus maculatus,* 1421, 1438
Caterpillar
  Erannis defoliara, 1288
  Gypsy, *Porthetria dispar,* 228, 435, 441, 644
  Operophtera brumata, 1288
  Plutella xylostella, 1093, 1103
  Saltmarsh, *Estigmene acrea, Estigmene* sp., 812, 1002
  Spodoptera, 922, 990, 992, 1093
  Tent
    Eastern, *Malacosoma americanum,* 216, 228
    Malacosoma disstria, 988
  Tortrix viridana, 1288
Catfish
  African, *Mystus vittatus,* 805, 811, 1111
  African sharp-tooth, *Clarias gariepinus,* 117, 128, 154, 632
  Air-breathing, *Clarias batrachus, Clarias lazera,* 260, 367, 369, 375, 614, 664, 811
  Blue, *Ictalurus furcatus,* 643, 839, 1267, 1463, 1465
  Channel, *Ictalurus punctatus,* 20, 62, 65, 155, 259, 342, 367, 377, 462, 578, 614, 615, 675, 688, 750, 778, 784, 805, 839, 859, 885, 888, 1000, 1002, 1030, 1042, 1106, 1110, 1136, 1173, 1205, 1212, 1267, 1303, 1305, 1465, 1468, 1469, 1471, 1472, 1518, 1538, 1589, 1663, 1676
  Clarias sp., 154
  Cnidoglanis macrocephalus, 1536
  Flathead, *Pylodictis olivaris,* 839, 1267
  Indian, *Heteropneustes fossilis,* 154, 164, 859
  Mystus vittatus, 805, 811, 1111
  Sea, *Arius felis,* 838, 1384, 1465
  Spotted, *Pseudoplatysoma coruscans,* 362
Catharacta skua, 348, 354, 1659
Catharacta spp., 333
Catharcta maccormicki, 361
Cathartes aura, 234, 348, 446, 625, 634, 933, 1431, 1435
Catoptrophorus semipalmatus, 120, 136, 634, 1524, 1579, 1666
Catostomus columbianus, 1738
Catostomus commersoni, 65, 117, 127, 154, 165, 259, 443, 624, 632, 750, 1026, 1030, 1031, 1042, 1212, 1279, 1353, 1363, 1379, 1462, 1517, 1676
Catostomus latipinnis, 1629
Catostomus macrocheilus, 1738
Cattail, *Typha latifolia,* 1171, 1577
Cattle, *Bos* spp., 124, 170, 177, 234, 276, 285, 454, 469, 483, 484, 649, 681, 690, 753, 756, 787, 851, 892, 905, 940, 1008, 1009, 1079, 1084, 1116, 1117, 1179, 1216, 1368, 1438, 1441, 1469, 1473, 1547, 1595, 1624, 1634, 1639, 1744, 1788
Cauliflower, *Brassica oleracea botrytis,* 1095, 1097
Cavia cobaya, 276, 806, 968
Cavia porcellus, 890, 1217
Cavia sp., 19, 66, 210, 470, 535, 536, 588, 589, 590, 615, 650, 682, 690, 753, 756, 808, 909, 1049, 1050, 1100, 1177, 1180, 1438, 1442, 1469, 1473, 1547, 1596, 1634, 1789
Cebus apella, 1789
Cedar, eastern red, *Juniperus virginiana,* 1168
Celery, *Apium graveolans,* 434, 438
Centipede, *Lithobius,* 25
Centropristis striata, 1522
Cephanomyia sp., 1084
Cephenomyia trompe, 1072, 1083
Cepphus sp., 1463, 1466
Cerastoderma edule, 133
Ceratitis capitata, 1776, 1777
Ceratophyllum demersum, 1516

Table 33.2 (continued) Common and Scientific Names of Plants and Animals Listed in the *Handbook of Chemical Risk Assessment series* (References are provided from common names to scientific names)

Ceratophyllum sp., 747, 749, 1041, 1656
Ceratopteris richardii, 1168
Ceriodaphnia affinis, 1676
Ceriodaphnia dubia, 150, 459, 480, 887, 965, 1222
Ceriodaphnia lacustris, 1107
Ceriodaphnia reticulata, 660, 688, 1207
Ceriodaphnia sp, 1003
Ceriodaphnia sp., 1003
Cervus canadensis, 59
Cervus elaphus, 125, 243, 245, 356, 1669, 1673
Cervus elephus, 625, 639, 1418, 1427, 1525, 1769
Cervus sp., 13, 936
Cetengraulis mysticetus, 1579
Chachalaca, gray-headed, *Ortalis cinereiceps,* 676
Chaenocephalus aceratus, 117, 443, 517
Chaetoceros gracilis, 582
Chaetognath, *Sagitta elegans,* 1579
Champia parvula, 1540
Channa punctatus, 19, 69, 154, 376, 663, 811, 966
Chaoborus astictopus, 1003
Chaoborus sp., 1136
Char, Arctic, *Salvelinus alpinus,* 73, 218, 230, 1213, 1284, 1465
Charadrius vociferus, 1075
Chara sp., 747, 1207
Chard, Swiss, *Beta vulgaris cicla,* 224, 568
Chelonia mydas, 1145
Chelon labrosus, 73, 1541
Chelydra serpentina, 231, 1031, 1285, 1286, 1741
Chen caerulescens caerulescens, 236, 284
Chenonetta jubatta, 1435
Chenopodium album, 775
Cherry
Black, *Prunus serotina,* 216, 225
Red, *Prunus avium,* 337
Chestnut, *Castanea* sp., 1757
Chickadee
Black-capped, *Parus atricapillus,* 1005
Mountain, *Parus gambeli,* 1006
Chicken
Domestic, *Gallus gallus, Gallus* sp., 20, 61, 66, 142, 167, 209, 210, 267, 270, 427, 446, 454, 464, 465, 482, 534, 571, 584, 625, 635, 646, 676, 677, 679, 689, 752, 785, 786, 806, 812, 847, 861, 862, 890, 891, 892, 961, 968, 977, 1005, 1006, 1048, 1055, 1076, 1115, 1137, 1142, 1167, 1176, 1214, 1292, 1302, 1307, 1310, 1322, 1436, 1503, 1543, 1623, 1786, 1787
Prairie, *Tympanuchus cupido,* 371
Chilomonas paramecium, 68
Chinocetes bairdii, 1521
Chipmunk
Common, *Eutamias townsendii,* 244
Eastern, *Tamias striatus,* 1221, 1794, 1795
Chiracanthium mildei, 1102
Chironomid
Chironomus sp., 153, 778, 993, 1468
Glyptotendipes, 996
Goeldichironomus holoprasinus, 996
Labrundinia pilosella, 778
Psectocladius sp., 1172
Tanypus grodhausi, 990
Chironomidae, 26
Chironomus decorus, 990, 996, 1003, 1587
Chironomus ninevah, 153
Chironomus plumosus, 996, 1469, 1471
Chironomus riparius, 25, 150, 257, 782, 811, 1197, 1208, 1375, 1380
Chironomus sp., 153, 778, 993, 1468
Chironomus tentans, 68, 258, 527, 782, 857, 887, 1107
Chiselmouth, *Acrocheilus alutaceus,* 1738
Chlamydomonas bullosa, 145
Chlamydomonas reinhardtii, 67, 145, 257, 1107, 1108, 1109, 1655
Chlamydomonas spp., 71, 857
Chlamydotis undulata maqueenii, 236
Chlamys ferrei nipponensis, 323
Chlamys operculis, 113, 441
Chlamys varia, 525
Chlorella emersonii, 811
Chlorella kessleri, 778
Chlorella pyrenoidosa, 67, 777, 811, 972, 1207, 1302, 1587
Chlorella sp., 144, 145, 524, 781, 927, 1304
Chlorella vulgaris, 26, 27, 67, 573, 775, 1364, 1535, 1628, 1778
Chloris gayana, 1583
Chlorococcum sp., 1171
Chlorophyte
Hydrodictyon sp., 747
Oedogonium sp., 67, 96, 781
Choerodon azurio, 324
Chondrus crispus, 1517, 1519
Chorioptes bovis, 1104
Choristoneura occidentalis, 1532, 1533
Chroomonas sp., 1511
Chrysanthemum cinariaefolium, 1089
Chrysemys scripta, 1145
Chrysopa carnea, 340
Chrysopa oculata, 990
Chrysoperia carnea, 991
Chrysophrys auratus, 340, 1659
Chrysophrys sp., 991
Chukar, *Alectoris chukar, A. graeca,* 371, 890, 1543
Chydorus sphaericus, 150
Cicada, 17-year, *Magicicada* spp., 112, 132
Cicer arietinum, 774
Cichlasoma bimaculatum, 1205
Cichlasoma cyanoguttatum, 663
Cichlasoma sp., 856
Cichlid
Texas, *Cichlasoma cyanoguttatum,* 663
Various, *Cichlasoma,* 856
Ciconia ciconia, 638, 1291
Cimex lectularius, 1082
Cinclus cinclus, 1289, 1291
Cipangopaludina japonica, 1107
Cipangopaludina malleata, 971
Circus aeruginosus, 236
Circus approximans, 1433
Circus cyaneus, 800, 1431, 1435
Citellus spp., 1442
Citharichthys stigmaeus, 63, 74, 578
Citrus limonia osbeck, 1583
Citrus sinensis, 140
Citrus tachibana, 337
Claassenia sabulosa, 887, 1468
Claassenia sp., 1468
Cladina spp., 1767

Table 33.2 (continued) Common and Scientific Names of Plants and Animals Listed in the *Handbook of Chemical Risk Assessment series* (References are provided from common names to scientific names)

Cladoceran
- *Bosmina,* 1003
- *Ceriodaphnia,* 1003
- *Chydorus sphaericus,* 150
- *Daphnia* (*See Daphnia*)
- *Moina irrasa,* 152
- *Simocephalus,* 1468

Cladonia alpestris, 1734
Cladonia uncialis, 653
Cladophora glomerata, 748, 927, 1781
Cladophora sp., 224, 747, 1202
Clam
- Asiatic, *Corbicula fluminea,* 147, 162, 525, 540, 658, 835, 1662
- Blood, *Anadara granosa,* 112
- *Elliptio compalanata,* 750, 1026
- False quahog, *Pitar morrhuanus,* 53, 516, 629
- Giant, *Tridacna derasa, Tridacna maxima,* 149, 161, 1366, 1514
- *Glebula rotundata,* 812, 813
- Hardshell or Quahaug, *Mercenaria mercenaria,* 52, 70, 148, 368, 442, 458, 526, 654, 659, 1378, 1468, 1520
- *Hormomya mutabilis,* 323
- *Lamellidens marginalis,* 147, 458, 858
- *Margaretifera margaretifera,* 1629
- *Meretrix lusoria,* 1528
- *Paphia undulata,* 115
- *Potamocorbula amurensis,* 516, 527
- Pullet-shell, *Venerupis pallustra,* 1629
- *Rangia cuneata,* 71, 805, 812, 1375, 1376, 1377, 1737
- Razor, *Ensis minor,* 1207
- *Scrobicularia plana,* 6, 442, 451, 507, 516, 527
- Short-necked, *Tapes phillippinarum,* 836, 1208
- Softshell, *Mya arenaria,* 70, 71, 142, 148, 222, 262, 339, 368, 457, 459, 515, 526, 655, 659, 669, 1201, 1365
- Southern quahaug, *Mercenaria campechiensis,* 669
- *Strophites rugosus,* 782
- Surf, *Spisula solidissima,* 527, 660
- *Tapes* sp., 1264
- *Tellina tenuis,* 165
- *Villorita cyprinoides,* 149

Clangula hyemalis, 1291
Clarias batrachus, 367, 369, 375, 811
Clarias gariepinus, 117, 128, 154, 632
Clarias lazera, 260, 367, 614, 664
Clarias sp., 154
Clethrionomys glareolus, 23, 125, 136, 243, 276, 437, 615, 639, 644, 691, 1515, 1756, 1767
Clethrionomys rufocanus, 136, 1671
Clethrionomys rutilis, 437, 853, 1150, 1300
Clibanarius olivaceous, 660
Clinocardium nuttali, 805
Clistoronia magnifica, 459, 996
Cloeon dipterum, 1172
Clostridium posterianum, 453
Clover
- *Lotus* sp., 1619
- *Melilotus* sp., 1616
- *Trifolium* sp., 137

Clupea harengus, 632, 664, 674, 689, 837, 841, 1031, 1038
Clupea harengus harengus, 154, 570, 837, 1464
Clupea harengus pallasi, 154, 570
Cnemidophorus sexlineatus, 1025
Cnemidophorus tigris, 1786
Cnesterodon decemmaculatus, 1173
Cnidoglanis macrocephalus, 1536
Coccinella septempunctata, 129
Coccyzus americus, 23
Cochliomyia hominivorax, 1776
Cockatoo, sulphur-crested, *Cacatua galerita,* 1431, 1435
Cockles
- *Anadara,* 112, 113
- *Cerastoderma,* 133
- *Clinocardium,* 805

Cockroach
- American, *Periplaneta americana,* 992, 1072, 1073, 1102, 1104, 1585
- German, *Blattella germanica,* 883, 991, 1169, 1584, 1585
- Various, *Blaberus giganteus,* 1776

Cod
- Arctic, *Boreogadus saida,* 1264, 1465
- Atlantic, *Gadus morhua,* 8, 118, 341, 444, 514, 517, 571, 841, 1038, 1043, 1263, 1266, 1285, 1464, 1466, 1522, 1766, 1782
- Hump rock, *Notothenia gibberifrons,* 118, 444, 518
- Tom, *Microgadus tomcod,* 1383

Coelastrium cambricum, 1778
Coenobita sp., 1747
Coffea arabacia, 438
Coffee, *Coffea arabacia,* 438
Coleomegilla sp., 991
Coleoptera, 226
Colinus virginianus, 265, 269, 371, 786, 806, 861, 862, 890, 891, 968, 978, 1047, 1114, 1136, 1142, 1167, 1176, 1309, 1430, 1435, 1469, 1542, 1543, 1742, 1787
Colisa fasciata, 461
Collard, *Brassica* spp., 989, 1093, 1626
Colluricincla harmonica, 1435
Collybia butyracea, 1516
Collybia maculata, 1516
Colpoda cucullus, 1102
Colpoda steini, 141
Colpophyllia amaranthus, 835
Colpophyllia natans, 835
Columba livia, 20, 50, 57, 221, 236, 269, 271, 371, 379, 806, 890, 933, 1308, 1435
Columba palumbus, 895
Columba sp., 584, 912
Comandra spp., 1652
Comephorus dybowski, 837, 1465
Comephorus sp., 230
Comptonia peregrina, 438
Compylium polyanum, 438
Condor
- Andean, *Vultur gryphus,* 202, 935
- California, *Gymnogyps californianus,* 202, 234, 238, 446, 625, 635, 905, 934, 1415

Contarina medicaginis, 1103
Contopus virens, 1005
Coontail, *Ceratophyllum demersum,* 1516
Coot

Table 33.2 (continued) Common and Scientific Names of Plants and Animals Listed in the *Handbook of Chemical Risk Assessment series* (References are provided from common names to scientific names)

American, *Fulica americana,* 446, 1579, 1666, 1685, 1732, 1742
Red-knobbed, *Fulica cristata,* 57, 129, 136
Copepods
Acartia sp., 63, 64, 142, 660, 672
Anomalocera sp., 672
Calanus marshallae, 1627, 1628
Cyclops sp., 459, 1003, 1012
Diaptomus, 1003, 1172
Eucyclops, 1172
Eudiaptomus padanus, 459, 661
Eurytemora affinis, 17, 575, 784, 998, 1374, 1541
Mesocyclops thermocyclopoides, 998
Nitroca spinipes, 460, 578, 579, 1109
Pseudodiaptomus coronatus, 152
Temora sp., 576, 672
Tisbe, 63, 152, 615, 662, 689
Tropocyclops prasinus mexicanus, 662
Coptotermes formosanus, 1626
Coptotermes heimi, 991
Coptotermes spp., 142
Coragyps atratus, 362, 933, 1435
Coral
Alcyonia alcyonium, 628
Briareum, 835
Colpophyllia, 835
Diploria, 835
Gorgonia, 835
Montastrea annularis, 228, 835
Porites, 835
Pseudopterogorgia, 835
Siderastrea, 835
Corbicula fluminea, 147, 162, 525, 542, 658, 835, 1662
Corbicula manilensis, 812
Corbicula sp., 515, 1520
Coregonus artedi, 1042
Coregonus autumnalis migratorius, 837, 1465
Coregonus clupeaformis, 117, 443, 571, 837
Coregonus fera, 782
Coregonus hoyi, 117, 1266, 1283, 1514
Coregonus spp., 217, 229
Corixid, *Sigara,* 1172
Cormorant
Blue-eyed, *Phalacrocorax atriceps,* 120, 446, 518
Double-crested, *Phalacrocorax auritus,* 54, 1026, 1028, 1035, 1047, 1287, 1288, 1294, 1311
Various, *Phalacrocorax* spp., 1150
White-necked, *Phalacrocorax carbo,* 11, 353, 850, 1027, 1028, 1036, 1294, 1524
Corn, *Zea mays,* 57, 132, 136, 252, 440, 521, 568, 653, 746, 767, 775, 801, 831, 903, 989, 1039, 1093, 1168, 1531, 1584, 1626, 1639, 1668
Corophium volutator, 150, 1540
Cortinarius spp., 337
Corvus bennetti, 1431, 1432, 1435
Corvus brachyrhynchos, 800, 890, 1076
Corvus corax, 234, 446, 625, 634
Corvus corone, 895, 1743
Corvus coronoides, 1432, 1435
Corvus mellori, 1435
Corvus monedula, 370
Corvus orru, 1432
Corvus spp., 905
Costia sp., 96
Cotesia melanoscela, 991
Cottocomophorus sp., 230
Cotton, *Gossypium hirsutum,* 746, 810, 989, 1092, 1095
Cottonmouth, *Agkistrodon piscivorus,* 1466
Cottontail
Desert, *Sylvilagus audubonii,* 1439, 1669
Eastern, *Sylvilagus floridanus,* 15, 130, 250, 800, 1745
Cottus bairdi, 529, 782
Cottus cognatus, 341
Coturnix coturnix, 267, 890, 968, 1135, 1542, 1787
Coturnix coturnix cournix, 369
Coturnix coturnix japonica, 1135, 1136, 1470
Coturnix japonica, 20, 269, 371, 379, 464, 465, 583, 584, 616, 677, 689, 786, 806, 807, 808, 861, 862, 890, 891, 934, 967, 977, 1074, 1076, 1114, 1115, 1174, 1176, 1214, 1222, 1307, 1309, 1435
Coturnix risoria, 464, 465, 891
Cow, *Bos bovis,* 13, 223, 243, 572, 851, 865, 984, 1013, 1525, 1547
Cowbird, brown-headed, *Molothrus ater,* 370, 806, 861, 862, 1076, 1096, 1542, 1544
Cowpea, *Vigna* sp., 918, 1532
Coyote, *Canis latrans,* 13, 59, 124, 802, 905, 940, 1414, 1442, 1451, 1724, 1745
Crab
Alaskan king, *Paralithodes camtschatica,* 623, 1521
Alaskan snow, *Chionocetes bairdii,* 1521
Blue, *Callinectes sapidus,* 54, 63, 365, 520, 611, 642, 767, 857, 893, 995, 998, 1021, 1038, 1136, 1468, 1521, 1527, 1671, 1677, 1778
Carcinus mediterraneus, 134
Coconut, *Birgus latro,* 1749
Dorippe granulata, 115
Drift-line, *Sesarma cinereum,* 782, 1468
Dungeness, *Cancer magister,* 72, 263, 368, 804, 857, 1374, 1521, 1540, 1579, 1677
Eriocheir japonicus, 1029, 1264
Fiddler, *Uca pugilator,* 22, 377, 577, 582, 613, 671, 785, 994, 1000
Green, *Carcinus maenas,* 99, 150, 442, 578, 671, 672, 1628, 1779
Hermit, *Clibanarius olivaceous, Coenobita* sp., *Eupagurus bernhardus, Pagurus,* 630, 660, 1171, 1629, 1749
Horseshoe, *Limulus polyphemus,* 152, 374, 573, 647, 661, 997
Lesser blue, *Callinectes similis,* 150
Mud, *Rithropanopeus harrissii, Neopanope texana,* 578, 614, 662, 689, 785, 999, 1468
Neptunus pelagicus, 323
Paragrapsus quadridentatus, 152
Podophthalmus vigil, 72
Portunus pelagicus, 671
Rock, *Cancer irroratus,* 6, 53, 71–72, 517
Soldier, *Mictyris longicarpus,* 162
Stone, *Menippe mercenaria,* 998, 1468, 1521
Crane
Lesser sandhill, *Grus canadensis canadensis,* 1469
Mississippi sandhill, *Grus canadensis pulla,* 238
Sandhill, *Grus canadensis,* 202, 890, 1076, 1580

Table 33.2 (continued) Common and Scientific Names of Plants and Animals Listed in the *Handbook of Chemical Risk Assessment series* (References are provided from common names to scientific names)

Whooping, *Grus americana,* 202, 238, 1580
Crangon allmanni, 442
Crangon crangon, 17, 150, 262, 578, 1521
Crangon septemspinosa, 857, 1107
Crangonyx pseudogracilis, 1209
Crappie
Black, *Pomoxis nigromaculatus,* 632, 930, 1002, 1004, 1280
White, *Pomoxis annularis,* 11, 1001, 1212, 1280
Crassostrea commercialis, 5, 628
Crassostrea gigas, 5, 16, 70, 96, 113, 129, 147, 323, 364, 442, 523, 525, 567, 568, 575, 623, 658, 689, 1208, 1528, 1541, 1586, 1677, 1731
Crassostrea madrasensis, 128
Crassostrea spp., 927, 1264, 1520
Crassostrea virginica, 25, 26, 64, 70, 71, 113, 133, 147, 148, 217, 227, 262, 368, 442, 458, 514, 515, 520, 523, 526, 557, 567, 569, 577, 622, 628, 642, 655, 658, 749, 767, 784, 831, 835, 857, 1107, 1171, 1207, 1221, 1264, 1281, 1365, 1375, 1377, 1385, 1468, 1471, 1472, 1520, 1535, 1541, 1629, 1657, 1735
Crataegus spp., 226
Crayfish, 115, 338, 451
Astacus, 115, 377, 1209
Austropotamobius pallipes, 671, 672
Cambarus spp., 227
Orconectes limosus, 367
Orconectes sp., 782
Procambarus spp., 1106, 1141, 1464
Red, *Procambarus clarki,* 152, 163, 836, 887, 971, 1099, 1106, 1109, 1171, 1537, 1656
Rusty, *Orconectes rusticus,* 163, 1109
Crepidula fornicata, 368, 375, 442, 526
Cricetomys gambianus, 941
Cricetus spp., 28, 61, 470, 588, 754, 892, 1008, 1049, 1050, 1118, 1180, 1217, 1442, 1507, 1509, 1547, 1789
Cricket, *Acheta,* Gryllidae, 613, 614, 855, 1102, 1103
Cricosphaera carterae, 656
Cricotopus spp., 996
Cristigera sp., 263, 657
Croaker, Atlantic, *Micropogonias undulatus,* 74, 263, 444, 643
Crocodile, American, *Crocodylus acutus,* 119, 345, 833, 844, 1145, 1523
Crocodylus acutus, 119, 345, 833, 844, 1145, 1523
Crocothemis erythryaea, 887
Crotaphytus wislizenii, 1785, 1786
Crow
American, *Corvus brachyrhynchos,* 800, 890, 1076
Australian, *Corvus orru,* 1432
Carrion, *Corvus corone,* 895, 1743
Little, *Corvus bennetti,* 1431, 1432, 1435
Cryptococcus terrens, 457
Cryptotis parva, 448
Ctenocephalides felis, 992
Ctenocephalides spp., 892
Ctenochaetus strigosus, 633
Ctenophore
Beroe cucumis, 1579
Pleurobrachia pileus, 146
Cuckoo, Yellow-billed, *Coccyzus americus,* 23
Cucumber, *Cucumis sativus,* 140
Cucumis sativus, 140
Cucumis sp., 1093
Cucurbita spp., 746, 1093
Culex fatigans, 1676
Culex pipiens, 996
Culex pipiens pipiens, 1108
Culex pipiens quinquefasciatus, 992, 996, 1000, 1206, 1375
Culex quinquefasciatus, 1108
Culex spp., 1108
Culex tarsalis, 1004
Cunner, *Tautogolabrus adspersus,* 522, 531
Cunninghamella blakesleeana, 456
Cunninghamella elegans, 1370
Curlew
Eurasian, *Numenius arquata,* 1743
Long-billed, *Numenius americanus,* 834, 849
Currawong, pied, *Strepera graculina,* 1432, 1436
Cuterebra spp., 1079, 1081
Cuttlefish, *Sepia officinalis,* 115, 149, 515
Cyanea capillata, 111
Cyanocitta cristata, 846, 1742, 1787
Cyclops abyssorum prealpinus, 459
Cyclops spp., 459, 1003, 1012
Cyclopterus lumpus, 443
Cyclotella cryptica, 995
Cyclotella meneghiniana, 779
Cydonia sp., 1577
Cygnus atratus, 221
Cygnus buccinator, 202, 220, 237, 284, 635
Cygnus columbianus, 219, 221, 237
Cygnus cygnus, 237
Cygnus olor, 136, 202, 220, 237, 284
Cylindrotheca fusiformis, 1585
Cymodocea sp., 440
Cynodon dactylon, 1532, 1621
Cynodon plectostachyus, 918, 936
Cynomus ludovicianus, 905, 1440, 1442
Cynoscion nebulosus, 117, 445, 1465
Cynthia claudicans, 517
Cypericercus sp., 1002
Cypicerus sp., 998
Cypridopsis sp., 998, 1003
Cyprinion macrostomus, 452
Cyprinodon macularis, 110
Cyprinodon variegatus, 17, 529, 555 576, 583, 785, 804, 805, 811, 859, 887, 888, 966, 971, 972, 1110, 1212, 1374, 1468, 1469, 1472, 1629, 1678
Cyprinotus sp., 1002
Cyprinus carpio, 21, 55, 69, 154, 229, 259, 340, 457, 461, 529, 574, 612, 784, 837, 838, 859, 966, 970, 1026, 1031, 1042, 1094, 1110, 1165, 1167, 1172, 1210, 1221, 1343, 1353, 1379, 1465, 1468, 1517, 1578, 1662, 1676, 1739, 1779, 1781
Cyprinus carpio communis, 461
Cypselurus cyanopterus, 1266
Cyrinofus sp., 1003
Cystophora cristata, 639, 1363

Table 33.2 (continued) Common and Scientific Names of Plants and Animals Listed in the *Handbook of Chemical Risk Assessment series* (References are provided from common names to scientific names)

D

Dab, *Limanda limanda,* 7, 118, 573, 841, 1267, 1285, 1466, 1541, 1589
Dace
  Longfin, *Agosia chrysogaster,* 663
  Pearl, *Semotilus margarita,* 159
  Speckled, *Rhinichthys osculus,* 531
Dacelo novaguineae, 1435
Dacryodes excelsa, 1775
Dactylus glomerata, 1168
Damalina bovis, 1080
Dama sp., 1747
Danio rerio. See Zebrafish
Daphnia ambigua, 150
Daphnia carinata, 150
Daphnia galeata mendotae, 21, 660, 661, 688, 1108
Daphnia hyalina, 258, 459, 1171
Daphnia laevis, 1003
Daphnia longispina, 579, 887
Daphnia magna, 17, 18, 21, 27, 62, 63, 65, 68, 142, 150, 162, 208, 209, 256, 257, 367, 368, 459, 528, 573, 576, 577, 579, 580, 613, 646, 661, 688, 749, 782, 811, 858, 886, 887, 927, 965, 972, 998, 1041, 1108, 1172, 1206, 1208, 1303, 1375, 1377, 1428, 1468, 1469, 1470, 1471, 1536, 1587, 1676, 1783
Daphnia parvula, 150
Daphnia pulex, 21, 150, 151, 646, 661, 857, 858, 887, 892, 927, 965, 1108, 1171, 1372, 1373, 1375, 1376, 1377, 1468, 1537, 1676, 1779
Daphnia pulicaria, 151, 459, 782
Daphnia sp., 142, 162, 528, 778, 812, 1136
Dasyuroides byrnei, 1442
Dasyurus hallucatus, 1439, 1443
Dasyurus maculatus, 1421, 1438
Dasyurus viverrinus, 1424, 1438, 1443
Daucus carota, 252
Deer, 4
  Black-tailed, *Odoicoileus hemionus columbianus,* 1673, 1732
  Capreolus sp., 1747, 1767
  Fallow, *Dama* spp., 1747
  Mule, *Odocoileus hemionus,* 222, 234, 286, 450, 641, 1624, 1633, 1640, 1641, 1669, 1673, 1745
  Red, *Cervus elephus,* 625, 639, 1418, 1427, 1525, 1769
  Roe, *Capreolus capreolus,* 24, 125, 244, 355, 895, 1525, 1756, 1769
  White-tailed, *Odocoileus virginianus,* 14, 126, 130, 136, 223, 245, 250, 336, 358, 450, 572, 625, 641, 1418, 1427, 1526, 1544, 1545, 1549, 1633, 1670, 1733, 1745
Delphinapterus leucas, 25, 244, 362, 374, 851, 1032, 1263, 1271, 1304, 1363, 1467, 1660
Delphinus delphis, 54
Dendraster excentricus, 663, 689
Dendrocalamus spp., 918
Dendrocygna bicolor, 371, 805, 806, 1469
Dendrodrilus rubidus, 621, 627
Dendroica spp., 1005
Dendroica townsendi, 1006
Dendroides sp., 644
Dermacentor andersoni, 1072
Dermacentor variabilis, 1079
Dermochelys coriaca, 250, 634, 1368
Deroceras caruane, 627
Deroceras reticulatum, 621, 627
Deschampia flexuosa, 110, 128, 132, 434, 439, 1767
Diachasmimorpha longicaudata, 1777
Diaptomus oregonensis, 1108
Diaptomus sp., 1003, 1172
Diatoma sp., 655
Dicentrarchus labrax, 164, 1379
Diceros bicornis, 1673
Dichapetalum braunii, 1426
Dichapetalum cymosum, 1413, 1426
Dichapetalum spp., 1413
Dichapetalum toxicarium, 1413
Dickcissel, *Spiza americana,* 1094, 1096
Dicranum scoparium, 337
Didelphis marsupialis, 1145
Didelphis virginiana, 823, 1440
Digitaria sanguinalis, 1504
Dingo, *Canis familiaris dingo,* 1414, 1442
Diomedea immutabilis, 139, 203, 220, 238
Diomedea nigripes, 1239
Diplodus sargus, 340
Diploria clivosa, 835
Diploria strigosa, 835
Dipodomys heermanni, 1439
Dipodomys heermanni morroensis, 1450
Dipodomys spp., 1417, 1443, 1449
Dipper, *Cinclus cinclus,* 1289, 1291
Dog
  Black-tailed prairie, *Cynomus ludovicianus,* 905, 1440, 1442
  Domestic, *Canis familiaris,* 66, 124, 137, 243, 276, 373, 469, 536, 590, 639, 681, 690, 756, 787, 806, 851, 865, 870, 892, 904, 940, 964, 969, 975, 1010, 1014, 1050, 1116, 1118, 1179, 1217, 1299, 1442, 1473, 1503, 1547, 1596, 1745, 1789
  Wild, *Canis familiaris familiaris,* 1418
Dogfish
  Lesser spotted, *Scyliorhinus caniculus,* 344, 634
  Smooth, *Mustelus canis,* 53, 518
  Spiny, *Squalus acanthias,* 119, 421, 463, 571, 1466, 1523
Dolphin
  Bottle-nosed, *Tursiops truncatus,* 54, 123, 249, 275, 642, 1661
  Common, *Delphinus delphis,* 54
  Hump-backed, *Sousa chinensis,* 1620, 1625
  La Plata river, *Pontoporia blainvellei,* 123
  Pacific white-sided, *Lagenorhynchus obliquidens,* 1272
  Stenella attenuata, 360
  Striped, *Stenella coeruleoalba,* 123, 335, 336, 360, 642, 1263, 1276, 1661
  Tursiops gephyreus, 123
  White-beaked, *Lagenorhynchus albirostris,* 640, 852, 1467
  White-sided, *Lagenorhynchus acutus,* 54
Donax cuneatus, 165
Donax incarnatus, 147
Donax venustus, 339
Donkey, *Equus asinus,* 243
Dorippe granulata, 115

Table 33.2 (continued) Common and Scientific Names of Plants and Animals Listed in the *Handbook of Chemical Risk Assessment series* (References are provided from common names to scientific names)

Dorosoma cepedianum, 642, 778, 1379, 1463, 1465
Dorosoma petenense, 642, 1464, 1663
Dorosoma spp., 838
Dove
- Laughing, *Streptopelea senegalensis,* 1437
- Mourning, *Zenaida macroura,* 202, 242, 273, 355, 1437, 1576
- Ringed, *Streptopelia* sp., 20
- Ringed turtle-, *Streptopelia risoria,* 272, 890, 1047, 1311
- Ring-necked, *Streptopelia capicola,* 1137
- Rock, *Columba livia,* 20, 50, 57, 221, 236, 269, 271, 371, 379, 806, 890, 933, 1308, 1435

Dowitcher, long-billed, *Limnodromus scolopaceus,* 1660
Dragon, bearded, *Pogona barbatus,* 1430
Dreissena polymorpha, 55, 114, 134, 147, 227, 526, 569, 628, 1278, 1303, 1304, 1365
Drill, oyster, *Ocenebra erinacea,* 622, 629
Dromaius novaehollandiae, 1431
Drosophila melanogaster, 9, 141, 382, 426, 746, 763, 1707
Drosophila sp., 745, 1199, 1755
Drum
- Freshwater, *Aplodinotus grunniens,* 340, 839
- Red, *Sciaenops ocellatus,* 96, 118, 159, 445, 838

Duck
- American black, *Anas rubripes,* 23, 57, 74, 120, 136, 202, 219, 232, 242, 379, 445, 833, 845, 961, 1076, 1471, 1524
- American wood, *Aix sponsa,* 12, 20, 332, 347, 1027, 1034, 1055, 1741, 1787
- *Anas* spp., 120, 166, 167, 268, 676, 861, 961, 977, 1524, 1666, 1685
- Canvasback, *Aythya valisineria,* 4, 57, 120, 136, 202, 221, 233, 234, 284, 446, 634, 845, 1368, 1463
- Fulvous whistling, *Dendrocygna bicolor,* 371, 805, 806, 1469
- Long-tailed, *Clangula hyemalis,* 1291
- Maned, *Chenonetta jubatta,* 1435
- Mottled, *Anas fulvigula,* 242
- Pacific black, *Anas superciliosa,* 1431, 1434
- Ring-necked, *Aythya collaris,* 122, 235, 446
- Ruddy, *Oxyura jamaicensis,* 348
- *Tadorna tadorna,* 1731
- Tufted, *Aythya fuligula,* 219

Duckweed, *Lemna media, Lemna minor,* 67, 96, 441, 524, 578, 579, 811, 1041, 1170, 1207, 1222, 1587
Dugesia dorotocephala, 368, 375, 858, 995
Dugesia lugubris, 1208
Dunaliella bioculata, 748
Dunaliella marina, 1535
Dunaliella salina, 145
Dunaliella tertiolecta, 145, 262, 670, 1171, 1207
Dunlin, *Calidris alpina,* 8, 846, 1660
Dunnart
- Fat-tailed, *Sminthopsis crassicaudata,* 1448
- Stripe-faced, *Sminthopsis macroura,* 1448

Dunnock, *Prunella modularis,* 1075
Dysdera crocata, 653

E

Eagle
- Australian little, *Hieraetus morphnoides,* 1432
- Bald, *Haliaeetus leucocephalus,* 120, 202, 238, 270, 284, 334, 351, 802, 833, 847, 905, 1075, 1149, 1288, 1466
- Booted, *Hieraetus pennatus,* 638, 1292
- Golden, *Aquila chrysaetos,* 220, 233, 347, 637, 679, 905, 1290, 1432, 1434
- Imperial, *Aquila heliaca adalberti,* 1288, 1290
- Wedge-tailed, *Aquila audax,* 1431, 1432, 1434
- White-tailed sea-, *Haliaeetus albicilla,* 379, 635, 1288, 1292

Earthworms, 12, 132, 138, 1531
Echinometra lucunter, 443
Echinometra mathaei, 153
Echinus esculentus, 7
Echium sp., 105
Ecklonia radiata, 1514
Ectomyelois ceratoniae, 1777
Edwardsiella ictaluri, 615
Eel
- American, *Anguilla rostrata,* 461, 1030, 1148, 1283, 1739
- Electric, *Electrophorus electricus,* 1170
- European, *Anguilla anguilla,* 11, 20, 229, 256, 260, 888, 964, 965, 970, 1028, 1031, 1212, 1370, 1379

Eelpout, glacial, *Lycodes frigidus,* 1465
Eels, *Anguilla,* 1278
Eggplant, *Solanum melongena,* 1093
Egregia laevigata, 256
Egret
- Cattle, *Bubulcus ibis,* 348, 365, 1149, 1668
- Great, *Casmerodius albus,* 350, 1665
- Little, *Egretta garzetta,* 635
- Snowy, *Egretta thula,* 354, 1297

Egretta garzetta, 635
Egretta thula, 354, 1297
Egretta tricolor, 354
Eichhornia crassipes, 337, 923, 927
Eichhornia sp., 747, 1748
Eider
- Common, *Somateria mollissima,* 9, 58, 121, 122, 233, 235, 514, 519, 1038, 1150, 1311, 1385
- Spectacled. *Somateria fischeri,* 202, 233

Eisenia andrei, 138, 141
Eisenia fetida, 141, 143
Eisenia fetida andrei, 141, 1204
Eisenia foetida, 57, 129, 373, 382, 456, 653, 809, 1416
Eisenia rosea, 227
Eisenoides carolinensis, 227
Elaphe obsoleta, 1741
Elasmopus pectenicrus, 1374
Elderberry, *Sambucus* spp., 905
Electrophorus electricus, 1170
Eledone cirrhosa, 114, 515
Elephant, Indian, *Elephas maximus,* 639
Elephas maximus, 639
Elk
- *Cervus canadensis,* 59
- *Cervus elaphus,* 125, 243, 245, 356, 1669, 1673
- *Cervus* sp., 13, 936

Table 33.2 (continued) Common and Scientific Names of Plants and Animals Listed in the *Handbook of Chemical Risk Assessment series* (References are provided from common names to scientific names)

Elliptio complanata, 750, 1026
Elm, *Ulmus americana,* 111, 568
Elminius modestus, 517, 614, 671
Elodea canadensis, 96, 255, 524, 748, 779, 812, 1170
Elodea sp., 741, 747, 812, 927, 1041, 1577
Elsholtzia spp., 132
Emblema temporalis, 1431, 1435
Emu, *Dromaius novaehollandiae,* 1431
Engraulis encrasicolus, 117
Engraulis mordax, 623
Engraulis sp., 1779
Enhydra lutris, 10, 215
Ensis minor, 1207
Enteromorpha linza, 224
Enteromorpha sp., 96, 165, 568, 1207
Enteropmorpha intestinalis, 128, 749
Entosiphon sulcatum, 257
Eopsetta grigorjewi, 1522
Epeorus latifolium, 662, 672, 688
Ephemerella grandis, 528
Ephemerella sp., 18, 26, 1108
Ephemerella walkeri, 750
Ephydatia fluviatilis, 646, 657, 688
Epicoccum nigrum, 774
Epilobium angustifolium, 1767
Epithemia sp., 781
Epitrimerus pyri, 1102
Eptesicus fuscus, 243, 639, 823
Equis asinus X E. *caballus,* 243, 1443
Equisetum sp., 1762
Equus asinus, 243, 1443
Equus caballus, 125, 137, 223, 243, 277, 448, 639, 682, 691, 905, 1427, 1443, 1548, 1624
Equus sp., 170, 1635
Erannis defoliara, 1288
Eremophila alpestris, 679, 892, 894, 976, 1096, 1432, 1469
Erethizon dorsatum, 13, 125, 1443
Erignathus barbatus, 249, 1660, 1743
Eriocheir japonicus, 1029, 1264
Erithacus rubecula, 1075
Erolia spp., 800
Erpobdella octoculata, 631, 662, 672
Escherichia coli, 425, 534, 657, 746, 1039, 1165, 1169, 1216
Eschrichtius robustus, 122
Esox lucius, 65, 117, 154, 260, 332, 340, 444, 457, 571, 832, 838, 1042, 1202, 1284, 1375, 1378, 1382, 1517, 1654, 1676, 1739, 1758, 1762
Esox masquinongy, 444
Esox niger, 444, 1027
Esox sp., 444, 457
Estigmene acrea, 812
Estigmene sp., 812, 1002
Etroplus maculatus, 461
Eucalyptus cladocalyx, 936
Eucalyptus sp., 936
Eucalyptus viminalis, 936
Eucyclops agilis, 21
Eucyclops sp., 1172
Eudiaptomus padanus, 459, 661
Euglena, *Euglena gracilis,* 144, 145, 646, 655, 993, 1628, 1778
Euglena gracilis, 144, 145, 646, 655, 993, 1628, 1778
Euglena sp., 627
Eulimnadia spp., 998
Eumetopias jubata, 626, 639
Eumetopias jubatus, 1263, 1272
Eupagurus bernhardus, 630, 1629
Euphagus cyanocephalus, 1432, 1435
Euphausia pacifica, 1627, 1629
Euphausia superba, 115, 630, 835
Euphausiid, *Euphausia, Meganyctiphanes,* 115, 630, 835, 1627, 1629, 1658, 1684
Euplotes vannus, 145
Eurotium sp., 1169
Eurycea bislineata, 160
Eurytemora affinis, 17, 575, 784, 998, 1374, 1541
Eutamias townsendii, 244
Euthynnus pelamis, 444
Exuviella baltica, 857

F

Fabrea salina, 524
Falco bevigora, 1432
Falco cenchroides, 1432
Falco columbarius, 1150, 1291
Falco mexicanus, 269, 846
Falcon
  Brown, *Falco bevigora,* 1432
  Peregrine, *Falco peregrinus,* 237, 350, 638, 833, 846, 1150, 1287, 1291
  Prairie, *Falco mexicanus,* 269, 846
  Swedish gyr, *Falco rusticolus,* 333, 350, 1291
Falco naumanni, 120
Falco peregrinus, 237, 350, 638, 833, 846, 1150, 1287, 1291
Falco rusticolus, 333, 350, 1291
Falco sp., 333
Falco sparverius, 209, 270, 284, 806, 934, 1115, 1137, 1142, 1167, 1176, 1288, 1307, 1309, 1681, 1685
Falco tinnunculus, 349, 370, 1291
Fasciola hepatica, 1171
Felis cattus, 1417, 1443
Felis concolor coryi, 336, 356, 385
Felis domesticus, 66, 277, 324, 356, 373, 381, 389, 470, 590, 682, 691, 756, 806, 852, 1117, 1118, 1180, 1218, 1443, 1469, 1548, 1635, 1640
Felis lynx, 1760
Fern
  Sweet, *Comptonia peregrina,* 438
Fernbird, *Bowdleria punctata,* 1433
Ferret
  Black-footed, *Mustela nigripes,* 1415
  Domestic, *Mustelus putorius,* 101, 1445, 1792
  European, *Mustela putorius furo,* 245, 683, 693
Fescue, *Festuca arundinacea,* 56, 1168
Festuca arundinacea, 56, 1168
Festuca rubra, 224, 1168
Festuca sp., 439
Finch
  Bull, *Pyrrhula,* 636
  Green, *Carduelis chloris,* 636, 1433
  House, *Carpodacus mexicanus,* 806
  Zebra, *Peophila guttata,* 271, 370, 371, 1436

Table 33.2 (continued) Common and Scientific Names of Plants and Animals Listed in the *Handbook of Chemical Risk Assessment series* (References are provided from common names to scientific names)

Fir
  Abies pindrow, 142, 653
  Douglas, *Pseudotsuga menziesii,* 143, 1005, 1006
  True, *Abies* spp., 1005, 1006
Firetail, red-browed, *Emblema temporalis,* 1431, 1435
Fish
  Atlantic guitar, *Rhinobatis lentiginosus,* 118
  Benthic, *Trematomus bernacchii,* 844
  Bigmouth buffalo, *Ictiobus cyprinellus,* 839
  Cyprinion macrostomus, 452
  Cyprinofirm, *Labeo rohita,* 664, 1111
  Goat, *Mullus, Upeneus,* 633
  Goat, *Upeneus* spp., 633
  Harlequin, *Rasbora heteromorpha,* 579, 751
  Lump, *Cyclopterus lumpus,* 443
  Margined flying, *Cypselurus cyanopterus,* 1266
  Monk, *Squatina squatina,* 634
  Parrot, *Scarus* spp., 633
  Rabbit, *Siganus* spp., 633
  Smallmouth buffalo, *Ictiobus bubalus,* 839, 1466
  Star, *Asterias rubens,* 443, 663, 1354, 1628
  Stolothrissa sp., 117
  Surgeon, *Ctenochaetus* spp., 633
  Tile, *Lopholatilus chamaeleonticeps,* 435, 1267, 1285
Flagfish, *Jordanella floridae,* 529, 664, 688, 966, 972, 1210, 1536, 1538, 1676
Flamingo
  Caribbean, *Phoenicopterus ruber ruber,* 240
  Greater, *Phoenicopterus ruber,* 240, 625, 637, 1666
  Lesser, *Phoeniconaias minor,* 11, 1524
  Lesser, *Phoenicopterus minor,* 351
  Phoenicopterus ruber roseus, 121
Flavobacterium sp., 1199, 1651
Flax, *Linum* spp., 903
Fleas, *Ctenocephalides* spp., Siphonaptera, 892
Florida coerulea, 1096
Flounder
  Baltic or European, *Platichthys flesus,* 7, 158, 230, 343, 344, 1269, 1285, 1377, 1523
  Limanda sp., 7, 1515
  Paralichthys sp., 100, 158
  Pleuronectes, 533
  Pleuronectes sp., 533
  Roundnose, *Eopsetta grigorjewi,* 1522
  Southern, *Paralichthys lethostigma,* 118, 445, 643
  Starry, *Platichthys stellatus,* 571
  Starry, *Pleuronectes stellatus,* 768
  Summer, *Paralichthys dentatus,* 142, 158, 531, 1678
  Windowpane, *Scophthalmus aquosus,* 518, 1367, 1523
  Winter, *Pleuronectes americanus,* 20, 22, 118, 159, 444, 514, 518, 531, 571, 832, 842, 1040, 1269, 1282, 1367, 1378, 1381, 1541, 1678
  Witch, *Glyptocephalus cynoglossus,* 1548
  Yellowtail, *Pleuronectes ferruginea* (formerly *Limanda limanda),* 7, 118, 444, 573, 841, 1267, 1285, 1541, 1589
Fluke, liver, *Fasciola hepatica,* 1171
Fly
  Black, *Simulium vittatum,* 995, 996
  Bot, *Hypoderma* spp., 1067, 1068, 1071, 1084
  Caddis
    Clistoronia magnifica, 459, 996
    Hydropsyche bettani, 995
    Philarctus quaeris, 1209
  Caribbean fruit, *Anastrepha suspensa,* 1776, 1777
  Crane, *Playtcentropus radiatus, Tipula,* 1003
  Damsel, *Ischnura,* 26
  Dragon
    Crocothemis erythryaea, 887
    Orthemis sp., 996
    Pantala sp., 996
    Pseudagrion spp., 887
  Face, *Musca autumnalis,* 988, 1116
  Fruit, *Anastrepha, Drosophila,* 745, 1199
  Horn, *Haematobia irritans,* Muscidae, 892, 988, 1072, 1079, 1116
  House, *Musca domestica, Musca* sp., 809, 812, 969, 990, 1072, 1079, 1103, 1104, 1116, 1585
  Leafmining, *Liriomyza trifolii,* 1777
  May
    Baetis, 100, 115, 162, 995
    Callibaetis sp., 996
    Cloeon dipterum, 1172
    Epeorus latifolium, 662, 672, 688
    Ephemerella sp., 18, 26, 1108
    Hexagenia sp., 338, 1265, 1280, 1375, 1662
    Isonychia bicolor, 528, 532
    Leptophlebia sp., 995
    Siphlonurus lacustris, 1766
    Stenonema sp., 528
  Mediterranean fruit, *Ceratitis capitata,* 1776, 1777
  Reindeer nostril, *Cephenomyia trompe,* 1072, 1083
  Reindeer warble, *Oedemagena tarandi,* 1072, 1083
  Rhagionid, *Atherix* sp., 1107, 1468
  Rodent bot, *Cuterebra* spp., 1079
  Screw-worm, *Cochliomyia hominivorax,* 1776
  Sheep blow, *Lucilia cuprina,* 1011
  Snipe, *Atherix* sp., 1107, 1468
  Stable, *Stomoxys calcitrans,* 990, 1116
  Stone
    Claassenia sp., 1468
    Leuctra sp., 528
    Paragnetina media, 995
    Pteronarcella badia, 886, 887
    Pteronarcys sp., 1468
    Skwala sp., 997
  Warble, *Hypoderma* spp., 1067, 1068, 1071, 1084
  White, *Bemisia tabaci,* 1102
Flycatcher
  Ash-throated, *Myiarchus cinerascens,* 1432
  Great crested, *Myiarchus crinitus,* 1005, 1786, 1788
Flyingfish, margined, *Cypselurus cyanopterus,* 1266
Folsomia candida, 653, 1169
Fontinalis antipyretica, 927
Fontinalis dalecarlica, 25
Fontinalis sp., 26
Fontinalis squamosa, 627
Formica sp., 1527
Fox
  Arctic, *Alopex lagopus,* 124, 1416, 1441, 1760
  Desert kit, *Vulpes macrotis arsipus,* 1449
  Gray, *Urocyon cinereoargentatus,* 1448, 1745
  Red, *Vulpes vulpes, Vulpes fulva,* 127, 336, 360, 380, 823, 1298, 1301, 1417, 1438, 1449, 1745, 1756, 1769
  San Joaquin kit, *Vulpes macrotis mutica,* 1450

Table 33.2 (continued) Common and Scientific Names of Plants and Animals Listed in the *Handbook of Chemical Risk Assessment series* (References are provided from common names to scientific names)

Vulpes sp., 572, 914, 1144, 1417, 1527
Foxtail
Giant, *Setaria faberii,* 644
Setaria sp., 130
Fragaria vesca, 802
Fratercula arctica, 54, 354, 847
Frog
Adelotus brevis, 1174
Bull, *Rana catesbeiana,* 11, 55, 218, 231, 261, 264, 345, 778, 886, 892, 966, 1047, 1372, 1373, 1429, 1523, 1665, 1670
Columbia spotted, *Rana luteiventris,* 18, 668
Cricket, *Acris* sp., 367
European, *Rana temporaria,* 119, 346, 579, 1170
Gray tree, *Hyla versicolor,* 119, 435, 445
Green, *Rana clamitans,* 119, 130, 218, 250, 261, 264, 445, 1096
Green tree, *Hyla cinerea,* 1096
Leap, *Rana dalmutina,* 668
Leopard
Rana pipiens, Rana sphenocephala, 160, 264, 346, 367, 375, 532, 784, 1105, 1113, 1174, 1306, 1373, 1385, 1429, 1468, 1590, 1770, 1785
Rana sphenocephala, 1468
Limnodynastes peroni, 1174
Litoria aurea, 860, 1770
Northern cricket, *Acris crepitans,* 119, 435, 445, 783
Northern leopard, *Rana pipiens pipiens,* 1113
Pseudis paradoxa, 1205
Rana spp., 231, 507, 676, 1524
River, *Rana heckscheri,* 367
Scinax nasica, 1174
South African clawed, *Xenopus laevis,* 232, 427, 611, 668, 751, 1213, 1373, 1385, 1429, 1586, 1590, 1676, 1682, 1770, 1786
Southern gray tree, *Hyla chrysoscelis,* 160
Southern leopard, *Rana utricularia,* 264, 1096
Spotted grass, *Limnodynastes tasmaniensis,* 1429
Tree, *Hyla* sp., 367
Water, *Rana ridibunda,* 18, 25
Western chorus, *Pseudacris triseriata,* 1174
Wood, *Rana sylvatica,* 783, 1586, 1588
Fucus disticus, 224, 643
Fucus serratus, 656, 689
Fucus spp., 1362, 1519
Fucus vesiculosus, 5, 52, 224, 440, 656, 1519, 1534, 1730, 1735
Fulica americana, 446, 862, 1579, 1666, 1685, 1732, 1742
Fulica cristata, 57, 129, 136
Fulmar
Southern, *Fulmarus glacialoides,* 1292
Fulmarus glacialoides, 1292
Fundulus diaphanus, 462
Fundulus grandis, 838
Fundulus heteroclitus, 26, 27, 28, 63, 118, 154, 263, 369, 374, 375, 377, 462, 529, 578, 655, 664, 674, 893, 1000, 1040, 1074, 1110, 1205, 1213, 1283, 1304, 1739
Fundulus kansae, 1683
Fundulus majalis, 838
Fundulus similis, 750, 888, 1173, 1212, 1377, 1472
Fundulus sp., 55
Fungus
Achyla sp., 456
Aspergillus, 456, 1421
Cortinarius spp., 337
Cunninghamella, 13, 456
Epicoccum nigrum, 774
Eurotium sp., 1169
Fusarium spp., 774, 1169
Helminthosporium sp., 1169
Ophiolobus sp., 1169
Penicillium spp., 1421, 1422
Phomopsis leptostromiformes, 615
Rhizoctonia solani, 774, 921
Rhizopus sp., 1169
Saprolegnia sp., 456
Sclerotium rolfsii, 774
Thielaviopsis basicola, 915
Trichoderma viride, 774
Fusarium oxysporum, 1421
Fusarium solani, 921
Fusarium spp., 774, 1169

G

Gadopsis marmoratus, 341
Gadus morhua, 8, 118, 341, 444, 514, 517, 571, 841, 1038, 1043, 1263, 1266, 1285, 1464, 1466, 1522, 1766, 1782
Gadwall, *Anas strepera,* 219, 446, 1524, 1579, 1665
Galah, *Cacatua roseicapilla,* 1431, 1435
Galeorhinus galeus, 435
Galleria melonella, 809
Gallinula chloropus, 365
Gallinule, purple, *Porphyrula martinica,* 1463
Gallus domestica, 138, 371, 931, 934, 1385, 1593, 1685
Gallus gallus, 1543
Gallus spp., 20, 61, 66, 142, 167, 210, 267, 270, 427, 446, 454, 464, 465, 482, 534, 584, 625, 635, 646, 676, 677, 679, 689, 752, 785, 786, 812, 861, 862, 890, 891, 892, 1005, 1006, 1048, 1055, 1076, 1115, 1137, 1142, 1167, 1176, 1214, 1292, 1302, 1307, 1310, 1322, 1436, 1503, 1544, 1623, 1786, 1787
Gambusia affinis, 26, 55, 340, 367, 377, 529, 664, 674, 750, 805, 856, 886, 888, 1000, 1002, 1042, 1096, 1110, 1139, 1173, 1206, 1374, 1380, 1382, 1465, 1517, 1578, 1589, 1656, 1664, 1676
Gambusia holbrooki, 1731, 1739
Gambusia sp., 250, 369, 893, 1777
Gammarus duebeni, 369, 661, 671
Gammarus fasciatus, 782, 858, 965, 1172, 1468
Gammarus lacustris, 886, 887 1766
Gammarus oceanicus, 577
Gammarus pseudolimnaeus, 142, 151, 257, 367, 528, 782, 927, 998, 1108, 1209, 1304, 1373, 1469, 1533, 1537
Gammarus pulex, 151, 613, 887, 923, 927
Gammarus spp., 64, 459, 778, 1136, 1629, 1783
Gannet, northern, *Morus bassanus, Sula bassanus,* 354, 834, 851, 1297
Gar

Table 33.2 (continued) Common and Scientific Names of Plants and Animals Listed in the *Handbook of Chemical Risk Assessment series* (References are provided from common names to scientific names)

Lepisosteus sp., 313
Longnose, *Lepistoseus osseus,* 924
Spotted, *Lepisosteus oculatus,* 643, 1464, 1466
Garra rufa, 452
Gasterosteus aculeatus, 18, 260, 341, 529, 576, 614, 675, 859, 888, 1468
Gastrolobium grandiflorum, 1426
Gastrolobium spp., 1413, 1426, 1429
Gastrophryne carolinensis, 160, 367, 464, 480, 668, 675, 688, 1533, 1538
Gavia immer, 202, 238, 333, 334, 349, 379, 446, 1666
Geese, white-fronted, *Anser albifrons,* 232
Geissosis prainosa, 439
Geocoris punctipes, 991
Geomys breviceps, 1443
Geomys bursarius, 938
Geomys floridanus, 1443
Geomys personatus, 1437
Geranium, *Geranium* spp., 746
Geranium spp., 746
Gerbil, *Gerbillus* sp., *Meriones,* 588
Gerbillus sp., 588
Geukensia demissa, 1170, 1172, 1283
Gifblaar, *Dichapetalum cymosium,* 1413
Gila bicolor, 571
Gila elegans, 1588, 1676
Gillia altilis, 1208
Ginger, *Zingiber officinale,* 1735
Girella punctata, 324
Glebula rotundata, 812, 813
Glenodinium halli, 656, 689
Globicephala macrorhynchus, 9, 1660
Globicephala melaena, 122, 640, 852, 1467
Globicephala melas, 9, 25
Glossiphonia complanata, 782
Glycera dibranchiata, 909
Glyceria borealis, 1004
Glycine max, 12, 140, 253, 421, 453, 456, 521, 767, 774, 775, 909, 989, 1093, 1168, 1427, 1530, 1532, 1583
Glyptocephalus cynoglossus, 1548
Glyptotendipes paripes, 996
Gnat, *Chaoborus* spp., 1136
Gnatcatcher, blue-gray, *Polioptila caerulea,* 1005
Gnathonemus sp, 928
Gnathopodon caerulescens, 583, 889
Goat, domestic, *Capra hircus, Capra* sp., 124, 276, 454, 469, 639, 649, 691, 865, 890, 905, 911, 1010, 1033, 1081, 1179, 1414, 1427, 1442, 1469, 1547, 1761
Goatfish, *Mullus, Upeneus,* 633
Goby
Edible, *Boleophthalmus dussumieri,* 62, 69, 1683
Lizard, *Rhinogobius flumineus,* 842
Various, *Acanthogobius flavimanus,* 837
Godwit, bar-tailed, *Limosa lapponica lapponica,* 1743
Goeldichironomus holoprasinus, 996
Goldeneye
Barrows, *Bucephala islandica,* 1623
Common, *Bucephala clangula,* 220, 352, 446, 846, 1308
Goldfish, *Carassius auratus,* 62, 69, 154, 165, 260, 377, 457, 461, 576, 579, 583, 750, 783, 805, 856, 859, 966, 1136, 1141, 1206, 1212, 1468, 1471, 1538, 1552, 1588, 1676, 1738, 1770
Gonatopsis borealis, 835
Goose
Aleutian Canada, *Branta canadensis leucoparlia,* 1416
Cackling Canada, *Branta canadensis minima,* 846
Canada, *Branta canadensis, Branta canadensis moffitti,* 219, 220, 235, 269, 895, 961, 1076, 1418, 1732, 1741
Giant Canada, *Branta canadensis maxima,* 624, 679
Greylag, *Anser anser,* 202, 219, 1308, 1743
Lesser snow, *Chen caerulescens caerulescens,* 236, 284
Various, *Anser* spp., *Branta,* 961, 968, 977
Goosefish, *Lophius piscatorius,* 74
Gopher
Breviceps pocket, *Geomys breviceps,* 1443
Texas pocket, *Geomys personatus,* 1437
Tuza pocket, *Geomys floridanus,* 1443
Various, *Geomys, Thomomys* sp., 961
Gopherus polyphemus, 1741
Gorgonia flabellum, 835
Gorilla, *Gorilla gorilla gorilla,* 244
Gorilla gorilla gorilla, 244, 640
Goshawk
Brown, *Accipiter fasciatus,* 1432
Various, *Accipiter gentilis,* 346, 1289
Gossypium hirsutum, 746, 810, 989, 1092, 1095
Gossypium sp., 812
Gourami
Blue, *Trichogaster* sp., 369
Giant, *Colisa fasciata,* 461
Grackle
Common, *Quiscalus quiscula,* 370, 806, 807, 846, 861, 890, 1076
Great-tailed, *Quiscalus mexicanus,* 1466
Grallina cyanoleuca, 1432, 1436
Gram, *Cicer arietinum,* 774
Grape, *Vitis* sp., 1093, 1668
Grass
Arrow, *Triglochin* spp., 905
Bahia, *Paspalum notatum,* 1143
Barley, *Hordeum glaucum,* 1168
Blue, *Poa annua,* 1504
Brome, *Bromus inermis, Bromus* spp., 224, 1626
Buffalo, *Bouteloua dactyloides,* 1667
Colonial bent-, *Agrostis tenuis,* 1168, 1516
Common Bermuda, *Cynodon dactylon,* 1532, 1621
Copper-tolerant, *Agrostis capillaris,* 143
Crab, *Digitaria sanguinalis,* 1504
Creeping bent, *Agrostis tenuis,* 15, 1168, 1516
Crested wheat, *Agropyron cristatum,* 1427
Eel, *Zostera* spp., 111
Hair, *Deschampia flexuosa,* 110, 128, 132, 434, 439, 1767
Johnson, *Sorghum halepense,* 812, 905
Kentucky blue, *Poa pratensis,* 1168, 1531
Marsh, *Spartina* sp., 52, 435, 452
Orchard, *Dactylus glomerata,* 1168
Perenniel rye, *Lolium perenne,* 521, 895, 1168, 1621, 1757, 1763
Red fescue, *Festuca rubra,* 224, 1168
Redhead, *Potamogeton perfoliatus,* 776, 779
Reed canary, *Phalaris arundinacea,* 1732, 1734
Rhodes, *Chloris gayana,* 1583

Table 33.2 (continued) Common and Scientific Names of Plants and Animals Listed in the *Handbook of Chemical Risk Assessment series* (References are provided from common names to scientific names)

Saltmarsh, *Spartina alterniflora,* 330, 441, 777, 780, 782, 1534
Sea, *Heterozostera tasmanica,* 144
Shoal, *Halodule wrightii,* 54, 780, 1527, 1671
Star, *Cynodon plectostachyus,* 918, 936
Sudan, *Sorghum sudanense, Sorghum almum,* 905
Turtle, *Thalassia testudinum,* 451
Western wheat, *Agropyron smithii,* 1667
Widgeon, *Ruppia maritima,* 1517, 1578, 1591
Grasshopper
Oxya velox, 106
Senegalese, *Oedaleus senegalensis,* 894
Grayling, Arctic, *Thymallus arcticus,* 160, 463, 531, 619, 667
Grebe
Eared, *Podiceps nigricollis,* 1666, 1667
Great crested, *Podiceps cristata,* 354
Pied-billed, *Podilymbus podiceps,* 1466
Western, *Aechmophorus occidentalis,* 119, 346, 348, 1524, 1659
Greenfinch, *Carduelis chloris,* 636, 1433
Grindelia spp., 1652
Grindelia squarrosa, 1667
Griselinia littoralis, 1418
Groundnut, *Arachis hypogea,* 810, 1427
Grouse
Black, *Tetrao tetrix,* 1768
Red, *Lagopus lagopus,* 1768
Ruffed, *Bonasa umbellus,* 120, 235, 436, 446, 518, 571, 1732, 1741
Sharp-tailed, *Tympanuchus phasianellus,* 1469
Grunion, California, *Leuresthes tenuis,* 886, 888, 889, 1111, 1375
Grus americana, 202, 238, 1580
Grus canadensis, 202, 890, 1076, 1580
Grus canadensis canadensis, 1469
Grus canadensis pulla, 238
Gryllidae, 613, 614, 855, 1102, 1103
Gudgeon, topmouth, *Pseudorasbora parva,* 971
Guillemot
Brunnich's, *Uria lomvia,* 834, 851, 1661
Cepphus sp., 1463, 1466
Common, *Uria aalge,* 136, 1027, 1297, 1660
Gull
Audouin's, *Larus audouinii,* 1289, 1293
Bonaparte's, *Larus philadelphia,* 352
California, *Larus californicus,* 352, 799
Common black-headed, *Larus ridibundus,* L. *r. ridibundus,* 350, 1311, 1731, 1743, 1768, 1788
Franklin's, *Larus pipixcan,* 4, 9, 58, 121, 239, 352, 1466, 1659
Glaucous, *Larus hyperboreus,* 636
Glaucous-winged, *Larus glaucescens,* 849, 1524
Great black-backed, *Larus marinus,* 849, 1150, 1743
Grey, *Larus modestus,* 121
Herring, *Larus argentatus,* 54, 58, 239, 271, 285, 333, 350, 351, 446, 451, 848, 1021, 1035, 1146, 1148, 1252, 1288, 1292, 1311, 1319, 1466, 1659, 1733
Kelp, *Larus dominicanus,* 121
Laughing, *Larus atricilla,* 8, 1466
Lesser black-blacked, *Larus fuscus,* 58, 120, 446, 518
Ring-billed, *Larus delawarensis,* 1048, 1466
Yellow-legged herring, *Larus cachinnans,* 1289, 1293
Gulo gulo, 1760
Gum
Manna, *Eucalyptus viminalis,* 936
Sugar, *Eucalyptus cladocalyx,* 939
Gumweed, *Grindelia* spp., *Grindelia squarrosa,* 1652, 1667
Guppy, *Poecilia reticulata,* 18, 21, 63, 159, 256, 377, 421, 463, 531, 578, 667, 674, 812, 888, 965, 966, 1040, 1045, 1073, 1173, 1213, 1302, 1306, 1375, 1382, 1468
Gutierezia spp., 1652, 1667
Gymnodinium splendens, 656
Gymnogyps californianus, 202, 234, 238, 446, 625, 635, 905, 934, 1415
Gymnorhina tibicen, 1420, 1432, 1436
Gyps fulvus, 238

H

Haddock, *Melanogrammus aeglefinus,* 368, 841, 1466, 1678, 1782
Haemaphysalis longicornis, 1072
Haematobia irritans, 892, 988, 1072, 1079, 1116
Haematobia sp., 1084
Haematopinus eurysternus, 1071, 1080
Haematopinus spp., 1071, 1084
Haematopus ostralegus, 121, 1659, 1731, 1743, 1768
Hagfish, Atlantic, *Myxine glutinosa,* 1383
Hake
Argentinian, *Merluccius merluccius hubbsi,* 1268
Blue, *Antimora rostrata,* 340, 514, 1145
Pacific, *Merluccius productus,* 633
Red, *Urophycus chuss,* 1368
Halfbeak, *Hemiramphus marginatus,* 633
Haliaeetus albicilla, 379, 635, 1288, 1292
Haliaeetus leucocephalus, 120, 202, 238, 270, 284, 334, 351, 802, 833, 847, 905, 1075, 1149, 1288, 1466
Haliaeetus sp., 333
Haliastur sphenurus, 1432
Halibut
Atlantic, *Hippoglossus hippoglossus,* 341, 571
Greenland, *Reinhardtius hippoglossoides,* 341
Halichoerus grypus, 122, 335, 358, 450, 640, 852, 1032, 1263–1264, 1272, 1467, 1661, 1733, 1744
Halichondria sp., 441
Haliotis cracherodii, 147
Haliotis rufescens, 5, 52, 147, 256, 442, 515, 659
Haliotis tuberculata, 442
Halobates spp., 442
Halocynthia roretzi, 443
Halodule wrightii, 54, 780, 1527, 1671
Hamster
Chinese or Syrian, *Cricetus* spp., 28, 61, 470, 588, 754, 892, 1008, 1049, 1050, 1118, 1180, 1217, 1442, 1507, 1509, 1547, 1789
Golden, *Cricetus* spp., 28, 61, 470, 588, 754, 892, 1008, 1049, 1050, 1118, 1180, 1217, 1442, 1507, 1509, 1547, 1789

Table 33.2 (continued) Common and Scientific Names of Plants and Animals Listed in the *Handbook of Chemical Risk Assessment series* (References are provided from common names to scientific names)

Syrian golden, *Mesocricetus auratus,* 75, 757
Haplopappus spp., 1667
Hardy head, small-mouthed, *Atherinasoma microstoma,* 63
Hare
Blue, *Lepus timidus,* 14, 1671, 1769
Brown, *Lepus capensis,* 895, 1769
European, *Lepus europaeus,* 14, 244, 271, 558, 590, 640, 1178, 1756
Mountain, Lepus timidus, 14, 1671, 1769
Snowshoe, *Lepus americanus,* 449, 853, 1150, 1300, 1733
Harrier
Marsh, *Circus aeruginosus,* 236
Northern, *Circus cyaneus,* 800, 1431, 1435
Haslea ostrearia, 165
Hawk
Australian harrier, *Circus approximans,* 1433
Cooper's, *Accipiter cooperii,* 4, 13, 845, 1149
Falco, 333
Ferruginous, *Buteo regalis,* 1431, 1435
Pigeon, *Falco columbarius,* 1150, 1291
Red-shouldered, *Buteo lineatus,* 800, 834
Red-tailed, *Buteo jamaicensis,* 269, 370, 800, 802, 1075
Rough-legged, *Buteo lagopus,* 1291, 1431, 1434
Sharp-shinned, *Accipiter striatus,* 232, 346, 833, 844, 845, 1289
Hawthorn, *Crataegus* spp., 226
Heather, Scotch, *Calluna vulgaris,* 1516, 1767
Hediste diversicolor, 153
Helianthus annuus, 1583
Heliotropium sp., 105
Helisoma campanulata, 1537
Helisoma sp., 747, 1041
Helisoma trivolvis, 887, 1108
Helix aspersa, 654, 812
Helix pomatia, 253, 1782
Helix spp., 253
Helminthosporium sp., 1169
Helobdella stagnalis, 782
Hemicentrotus sp., 73
Hemichromus bimaculatus, 579
Hemifusus spp., 1520, 1552
Hemiramphus marginatus, 633
Hen
Guinea, *Gallus* sp., 61, 66, 142, 210, 270, 427, 446, 454, 464, 465, 482, 534, 584, 625, 635, 646, 676, 677, 679, 689, 752, 785, 786, 812, 847, 861, 862, 890, 891, 892, 1005, 1006, 1048, 1055, 1076, 1115, 1137, 1142, 1167, 1176, 1214, 1292, 1302, 1307, 1310, 1322, 1436, 1503, 1623, 1786, 1787
Moor, *Gallinula chloropus,* 365
Hermione hystrix, 72
Heron
Black-crowned night-, *Nycticorax nycticorax,* 58, 636, 1035, 1150, 1289, 1294, 1319, 1665
Gray, *Ardea cinerea,* 235, 314, 349
Great blue, *Ardea herodias,* 334, 347, 348, 350, 354, 845, 1028, 1035, 1290, 1297, 1319
Great white, *Ardea herodias occidentalis,* 347, 350
Green-backed, *Butorides virescens virescens*
Green-backer (also known as Green, or Striated), *Butorides striatus,* 240, 1096, 1149, 1464
Little blue, *Florida coerulea,* 1096
Little green, *Butorides virescens,* 1665
Trocolored, *Egretta tricolor,* 354
Yellow-crowned night-, *Nycticorax violaceus,* 849, 1464
Herring
Atlantic, *Clupea harengus harengus,* 154, 570, 837, 1464
Baltic, *Clupea harengus,* 632, 664, 674, 689, 837, 841, 1031, 1038
Lake, *Coregonus artedi,* 1042
Pacific, *Clupea harengus pallasi,* 154, 570
Heteropneustes fossilis, 154, 164, 859
Heterotermes indicola, 992
Heterotermes spp., 142
Heterozostera tasmanica, 144
Hexagenia bilineata, 1264, 1374
Hexagenia limbata, 1264
Hexagenia sp., 338, 1265, 1280, 1375, 1662
Hieraetus morphnoides, 1432
Hieraetus pennatus, 638, 1292
Himantopus mexicanus, 1657, 1666
Hippodamia convergens, 991
Hippoglossoides elassodon, 1282
Hippoglossoides platessoides, 341, 583
Hippoglossus hippoglossus, 341, 571
Hirundo neoxena, 1433
Hirundo nigricans, 1433
Hirundo rustica, 221, 239, 1657, 1668, 1731, 1742
Hog, *Sus* spp., 97, 127, 173, 280, 373, 450, 454, 483, 652, 685, 694, 869, 891, 905, 946, 1011, 1014, 1221, 1448, 1505, 1551, 1637, 1761
Hogchoker, *Trinectes maculatus,* 1141
Holmesimysis costata, 151
Holothuria forskalii, 53
Holothuria mexicana, 451
Holothurian, sea cucumber, *Holothuria mexicana, Stichopus,* 451
Homalium spp., 434, 439
Homalothecium sericium, 1763
Homarus americanus, 7, 26, 151, 263, 339, 443, 517, 528, 569, 577, 630, 661, 671, 835, 1108, 1362, 1365, 1370, 1515, 1521
Homarus gammarus, 671
Homarus vulgaris, 116
Homo sapiens, 13, 59, 125, 170, 244, 325, 356, 373, 381, 448, 470, 519, 536, 572, 588, 640, 682, 691, 756, 808, 852, 865, 918, 941, 1033, 1180, 1203, 1218, 1299, 1469, 1473, 1525, 1548, 1580, 1596, 1624, 1635, 1669, 1745, 1763, 1789
Honeybee, *Apis mellifera, Apis* spp., 111, 226, 621, 627, 809, 894, 961, 990, 991, 1012, 1013, 1102, 1103, 1123, 1169, 1428, 1532, 1533, 1585, 1764
Hoopoe, *Upupa epops,* 625, 638
Hoplias sp., 1747
Hoplosternum littorale, 1205
Hordeum glaucum, 1168
Hordeum vulgare, 652, 653, 775, 1168, 1530, 1583, 1621, 1761
Hormomya mutabilis, 323
Hornworm, tobacco, *Manduca sexta,* 975

Table 33.2 (continued) Common and Scientific Names of Plants and Animals Listed in the *Handbook of Chemical Risk Assessment series* (References are provided from common names to scientific names)

Horse, *Equus caballus,* 125, 137, 223, 243, 277, 448, 639, 682, 691, 1427, 1443, 1548, 1624
Horsetails, *Equisetum* sp., 1762
Human, *Homo sapiens,* 13, 59, 126, 170, 244, 325, 356, 373, 381, 448, 470, 519, 572, 588, 640, 682, 691, 756, 852, 865, 918, 941, 1033, 1180, 1203, 1218, 1299, 1469, 1473, 1525, 1548, 1580, 1596, 1624, 1635, 1669, 1745, 1763, 1789
Hyacinth, water, *Eichhornia crassipes,* 337, 923, 927
Hyalella azteca, 142, 151, 460, 528, 533, 856, 858, 887, 998, 1003, 1378, 1676
Hyalella sp., 1783
Hybanthus spp., 434, 439
Hydra, *Hydra oligactis, Hydra* sp., 995, 1136, 1676
Hydra littoralis, 1207
Hydra oligactis, 995
Hydra sp., 1136, 1676
Hydrilla verticillata, 166, 255
Hydrodictyon spp., 747
Hydroid, *Campanularia flexuosa,* 146, 577
Hydromys chrysogaster, 1443
Hydrophilus triangularis, 996
Hydropsyche bettani, 995
Hydrurga leptonyx, 122, 355, 519
Hyla chrysoscelis, 160
Hyla cinerea, 1096
Hyla crucifer, 436
Hyla sp., 367
Hyla versicolor, 119, 435, 445
Hylocichla mustelina, 1005
Hylocomium splendens, 1531
Hymenomonas carterae, 1535
Hypnum cupressiforme, 110, 223, 439, 1364, 1621
Hypoderma bovis, 1079
Hypoderma lineatum, 1079
Hypoderma spp., 1067, 1068, 1071, 1084
Hypogymnia physodes, 337
Hypophthalmichthys molitrix, 1756
Hystrix indica, 1417

I

Ibis, white-faced, *Plegadis chihi,* 353, 1659, 1667
Icefish, blackfin, *Chaenocephalus aceratus,* 117, 443, 517
Ichthyopthirius sp., 96
Ictalurus furcatus, 643, 839, 1267, 1463, 1465
Ictalurus melas. see Ameiurus melas
Ictalurus nebulosa, 838, 1000, 1002, 1027, 1030, 1038, 1202, 1353, 1367, 1370, 1380, 1383
Ictalurus nebulosus, 118, 154, 164, 632, 928, 1000, 1027, 1383
Ictalurus punctatus, 20, 62, 65, 155, 259, 342, 367, 377, 462, 578, 614, 615, 675, 688, 750, 778, 784, 805, 839, 859, 885, 888, 1000, 1002, 1030, 1042, 1106, 1110, 1136, 1173, 1205, 1212, 1267, 1303, 1305, 1465, 1468, 1469, 1471, 1472, 1518, 1538, 1589, 1663, 1676
Ictiobus bubalus, 839, 1466
Ictiobus cyprinellus, 839
Illex illecebrosus argentinus, 1265
Ilyanassa obsoleta, 114
Impala, *Aepyceros melampus,* 123, 136
Indoplanorbis exustus, 971, 1044
Ipomoea batatas, 903, 1734
Ipomoea sp., 746, 749, 1092
Ischadium exustus, 520
Ischadium recurvum, 642
Ischnura sp., 26
Isochrysis galbana, 582, 656, 1107, 1171
Isonychia bicolor, 528, 532
Isoodon auratus barrowensis, 1443
Isoodon macrourus, 1443
Isoodon obesulus, 1443
Isoperla sp., 338
Isopod, *Asellus* sp., 338, 1171
Isurus oxyrinchus, 633, 1523

J

Jacana, wattled, *Jacana jacana,* 1214
Jacana jacana, 1214
Jackal
Asiatic, *Canis aureus,* 1417
Black-backed, *Canis mesomelas,* 1417
Jackdaw, *Corvus monedula,* 370
Jackrabbit, black-tailed, *Lepus californicus,* 1444, 1724, 1733, 1745
Jararaca, *Bothrops jararaca,* 615
Jay, blue, *Cyanocitta cristata,* 846, 1742, 1787
Jewelfish, *Hemichromus bimaculatus,* 579
Jird, tristram, *Meriones tristrami,* 1445
Jordanella floridae, 529, 664, 688, 966, 972, 1210, 1536, 1538, 1676
Juga plicifera, 458, 997
Juncus roemerianus, 777, 781
Juniperus virginiana, 1168
Jussiaea sp., 747

K

Kakapo, *Strigops habroptilus,* 1433
Kangaroo
Eastern gray, *Macropus giganteus,* 1444
Red, *Macropus rufus,* 1444
Western gray, *Macropus fuliginosus,* 1444
Kelp
Brown, *Ecklonia radiata,* 154
Giant, *Macrocystis pyrifera,* 70, 458, 1207
Kestrel
American, *Falco sparverius,* 209, 270, 284, 806, 934, 1115, 1137, 1142, 1167, 1176, 1307, 1309, 1681, 1685
Australian, *Falco cenchroides,* 1432
Lesser, *Falco naumanni,* 120
Various, *Falco tinnunculus,* 349, 370, 1291
Killdeer, *Charadrius vociferus,* 1075
Killifish
Banded, *Fundulus diaphanus,* 462
Diamond, *Adinia xenica,* 1141
Freshwater, *Fundulus kansae,* 1683
Longnose, *Fundulus similis,* 750, 888, 1173, 1212, 1377, 1472

Table 33.2 (continued) Common and Scientific Names of Plants and Animals Listed in the *Handbook of Chemical Risk Assessment series* (References are provided from common names to scientific names)

Kingfish, yellowtail, *Seriola grandis,* 344, 388
Kinglet, golden-crowned, *Regulus satrapa,* 1006
Kinosternon sabrubrum, 1741
Kite
  Black, *Milvus migrans,* 638, 809, 1294, 1432, 1436
  Black-eared, *Milvus migrans lineatus,* 352
  Snail, *Rostrhamus sociabilis,* 166, 1214, 1215
  Whistling, *Haliastur sphenurus,* 1432
Kittiwake, black-legged, *Rissa tridactyla,* 9, 333, 354, 624, 637, 1660
Kiwi, *Apteryx* spp., 1432
Klebsiella pneumoniae, 754, 1546
Knifejaw, striped, *Oplegnathus fasciatus,* 965, 966
Knot, red, *Calidris canutus,* 57, 235
Knotted wrack, *Ascophyllum nodosum,* 5, 52, 111, 440, 655, 656
Kogia breviceps, 122
Kokako, *Callaeas cinera,* 1433
Konosirus punctatis, 841
Kookaburra, laughing, *Dacelo novaeguineae,* 1435
Kowari, *Dasyuroides byrnei,* 1442
Krill, *Euphausia superba,* 115, 630, 835
Krobia, *Cichlasoma bimaculatum,* 1205
Kwi kwi, *Hoplosternum littorale,* 1205

L

Labeo rohita, 664, 1111
Labrundinia pilosella, 778
Laccophilis spp., 996
Lacewing
  Common green, *Chrysopa carnea,* 340
  Various, *Chrysopa oculata,* 990
Lactarius spp., 1766
Lactuca sativa, 225, 439, 440, 521, 921, 1361, 1364, 1427, 1626, 1638
Lactuca spp., 1093
Lagenorhynchus acutus, 54
Lagenorhynchus albirostris, 640, 852, 1467
Lagenorhynchus obliquidens, 1272
Lagodon rhomboides, 54, 859, 1136, 1211, 1468, 1472, 1527, 1671, 1678, 1779
Lagopus lagopus, 12, 103, 120, 271, 333, 446, 1671, 1768
Lagopus mutus, 333
Lagostrophus fasciatus, 1444
Lama pacos, 137
Lamellidens marginalis, 147, 458, 858
Laminaria digitata, 656, 1519
Laminaria hyperborea, 656, 1519
Lamprey, Petromyzontidae, *Petromyzon marinus,* 842, 860, 1213, 1780
Lanistes carinatus, 887
Lanius ludovicianus, 848, 961, 1149, 1292
Laomedea loveni, 22
Larch, *Larix* sp., 1527
Larix sp., 1527
Lark
  Australian magpie-, *Grallina cyanoleuca,* 1432, 1436
  Eastern meadow, *Sturnella magna,* 1096
  Horned, *Eremophila alpestris,* 679, 892, 894, 976, 1096, 1432, 1469
  Southern meadow, *Sturnella magna argutula,* 1026
  Western meadow, *Sturnella neglecta,* 976, 1432
Larus argentatus, 54, 58, 239, 271, 285, 333, 350, 351, 446, 451, 848, 1021, 1035, 1146, 1148, 1252, 1288, 1292, 1311, 1319, 1466, 1659, 1733
Larus atricilla, 8, 1466
Larus audouinii, 1289, 1293
Larus cachinnans, 1289, 1293
Larus californicus, 352, 799
Larus delawarensis, 1048, 1466
Larus dominicanus, 121
Larus fuscus, 58, 120, 446, 518
Larus glaucescens, 849, 1524
Larus hyperboreus, 636
Larus marinus, 849, 1150, 1743
Larus modestus, 121
Larus novaehollandiae, 1369
Larus philadelphia, 352
Larus pipixcan, 4, 9, 58, 121, 239, 352, 1466, 1659
Larus ridibundus, 350, 1311, 1731, 1743
Larus ridibundus ridibundus, 1743, 1768, 1788
Larus spp., 121
Lasallia papulosa, 627, 653
Lasiorhinus latifrons, 1439, 1444
Laspeyresia pomonella, 1102
Lateolabrax *japonicus,* 841
Laurel, cherry, *Prunus laurocerasus,* 904
Laurus nobilis, 1364
Ledum sp., 337
Leech
  Erpobdella, 631, 662, 672
  Glossiphonia, 782
  Helobdella, 782
Leiopotherapon unicolor, 664, 674
Leiostomus xanthurus, 155, 462, 643, 654, 664, 785, 1363, 1471
Lemming, *Lemmus* sp., 136
Lemmus sp., 136, 437
Lemna gibba, 1372
Lemna media, 578, 579
Lemna minor, 67, 96, 441, 524, 811, 1041, 1170, 1207, 1222, 1587
Lemna obscura, 255
Lemna sp., 144, 781
Lemon, *Citrus limonia osbeck,* 1583
Lens esculenta, 774
Lentil, *Lens esculenta,* 774
Leopomis humilis, 129, 632
Lepidium latifolium, 318
Lepisosteus oculatus, 643, 1464, 1466
Lepisosteus osseus, 924, 928
Lepisosteus sp., 313
Lepomis auritis, 1782
Lepomis cyanellus, 155, 165, 805, 893, 924, 1173, 1468, 1518, 1536, 1664, 1675
Lepomis gibbosus, 56, 342, 462
Lepomis gulosus, 632
Lepomis macrochirus, 19, 28, 55, 56, 62, 65, 68, 155, 229, 258, 340, 451, 462, 529, 532, 578, 664, 674, 747, 750, 783, 805, 841, 856, 859, 888, 893, 924, 928, 964, 966, 970, 972, 993, 1000, 1073, 1110, 1136, 1141, 1173, 1205, 1211, 1372, 1375, 1377, 1380, 1395, 1428, 1464,

Table 33.2 (continued) Common and Scientific Names of Plants and Animals Listed in the *Handbook of Chemical Risk Assessment series* (References are provided from common names to scientific names)

1468, 1517, 1518, 1534, 1538, 1553, 1578, 1589, 1602, 1629, 1656, 1662, 1675, 1677, 1779, 1784
Lepomis megalotis, 217, 230
Lepomis microlophus, 56, 451, 632, 1468
Lepomis punctatus, 1026
Lepomis sp., 1380
Leptinotarsa decemlineata, 1504
Leptocheirus plumulosus, 17
Leptonychotes weddelli, 103, 122, 355, 362, 519, 853
Leptophlebia sp., 995
Lepus americanus, 449, 853, 1150, 1300, 1733
Lepus californicus, 1444, 1724, 1733, 1745
Lepus capensis, 895, 1769
Lepus europaeus, 14, 244, 271, 558, 590, 640, 1178, 1756
Lepus sp., 277, 589, 590, 963, 969, 1135, 1180, 1416, 1546, 1747
Lepus timidus, 14, 1671, 1769
Lerista puctorittata, 860
Lethrinus spp., 633
Lettuce
  Common, *Lactuca sativa, Lactuca* spp., 225, 439, 440, 521, 921, 1093, 1361, 1364, 1427, 1626, 1638
  Sea, *Ulva lactuca,* 657, 1735
Leuciscus hakonensis, 1004
Leuciscus idus melanotus, 574, 579
Leucospius delineatus, 27
Leuctra sp., 528
Leuresthes tenuis, 886, 888, 889, 1111, 1375
Lichens
  Bryoria fuscescens, 1767
  Cetraria nivalis, 451
  Cladina spp., 1767
  Cladonia, 653, 1734
  Compylium polyanum, 438
  Hypogymnia physodes, 337
  Lasallia papulosa, 627, 653
  Lecanora conizaeoides, 217
  Parimelia baltimorensis, 110, 216, 223
  Ramalina fraxinea, 1763
  Umbilicaria sp., 440
Lily
  Water, *Nuphar luteum,* 10
  Yellow pond, *Nuphar* sp., 441, 1577
Limanda limanda, 7, 118, 573, 841, 1267, 1285, 1466, 1541, 1589
Limanda sp., 7, 1515
Limnodrilus hoffmeisteri, 27, 562, 805, 1783
Limnodrilus sp., 134, 1378
Limnodromus scolopaceus, 1660
Limnodynastes peroni, 1174
Limnodynastes tasmaniensis, 1429
Limosa lapponica lapponica, 1743
Limpet
  Acmaea digitalis, 226
  Common, *Patella vulgata,* 149, 442, 516
  Littorina littorea, 5, 52, 507, 515, 659, 1520
  Patella caerulea, 567, 570
  Slipper, *Crepidula fornicata,* 368, 375, 442, 526
Limpkin, *Aramus guarauna,* 1463
Limulus polyphemus, 152, 374, 573, 647, 661, 997
Linognathus sp., 1084
Linognathus vituli, 1071, 1080
Linseed, *Linum* sp., 903
Linum sp., 903
Liometopum occidentale, 1432
Lipaphis erysimi, 1102
Liriodendron tulipifera, 1003
Liriomyza trifolii, 1777
Lissorhopterus oryzophilus, 1004
Lithobius forficatus, 25
Litoria aurea, 1770
Litoria caerulea, 860
Litoria dentata, 1770
Litoria peronii, 860
Littorina irroratea, 669
Littorina littorea, 5, 52, 507, 515, 659, 1520
Lizard
  Ameiva exsul, 365
  Blotched blue-tongued, *Tiliqua nigrolutea,* 1430
  Leopard, *Crotaphytus wislizenii,* 1785, 1786
  Shingle-back, *Tiliqua rugosa,* 1423, 1429, 1430
  Side-blotched, *Uta stansburiana,* 1786
  Spiny tailed, *Uromastix hardwickii,* 1785
  Western whiptail, *Cnemidophorus tigris,* 1786
Loach
  Misgurnis fossilis, 574
  Stone, *Noemacheilus barbatulis,* 255, 619, 665
  Stone, *Noemacheilus barbatulus,* 255, 619, 665
Lobodon carcinophagus, 122, 355, 520
Lobster
  American, *Homarus americanus,* 7, 26, 151, 263, 339, 443, 517, 528, 569, 577, 630, 661, 671, 835, 1108, 1362, 1365, 1370, 1515, 1521
  Caribbean spiny, *Panulirus argus,* 443
  Spiny, *Nephrops norvegicus,* 152, 339, 836
  Spiny, *Panulirus,* 7
  Western rock, *Panulirus cygnus,* 1514
Locoweed, *Astragalus* spp., 1652
Locust, desert, *Schistoma gregaria,* 894
Loligo forbesi, 25
Loligo vulgaris, 1520
Lolium perenne, 521, 895, 1168, 1621, 1757, 1763
Longspur
  Chestnut-collared, *Calcarius ornatus,* 976
  McCown's, *Calcarius mccownii,* 1432
Loon
  Common, *Gavia immer,* 202, 238, 333, 334, 349, 379, 446, 1666
Lophius piscatorius, 74
Lophodytes cucullatus, 352
Lopholatilus chamaeleonticeps, 435, 1267, 1285
Lophortyx gambeli, 1436
Lota sp., 1031
Lotus sp., 1619
Louse
  Angora goat biting, *Bovicola limbatus,* 1010, 1081
  Cattle, *Haematopinus* spp., 1071, 1084
  Cattle-biting, *Damalina bovis,* 1080
  Hairy goat, *Bovicola crassipes,* 1081
  Long-nosed cattle, *Linognathus vituli,* 1071
  Short-nosed cattle, *Haematopinus eurysternus,* 1071, 1080
  Wood, *Oniscus asellus, Porcellio scaber,* 628
Lovebird, peach-faced, *Agapornis roseicollis,* 676
Lucilia cuprina, 1011
Lugworm, *Arenicola cristata,* 578, 579, 582, 1136
Luidia clathrata, 517

Table 33.2 (continued) Common and Scientific Names of Plants and Animals Listed in the *Handbook of Chemical Risk Assessment series* (References are provided from common names to scientific names)

Lumbriculus variegatus, 153, 163, 460, 533, 1372
Lumbricus herculeus, 809
Lumbricus rubellus, 112, 128, 130, 141, 143–144, 621, 627, 644, 1039, 1204, 1766
Lumbricus terrestris, 141, 227, 809, 1169, 1204, 1205, 1532, 1533
Lupinus spp., 169
Lutianus fulviflamma, 633
Lutianus griseus, 1376
Lutjanus griseus, 1376
Lutra canadensis, 336, 357, 373, 449, 853, 1144–1145, 1670, 1689
Lutra lutra, 336, 357, 385, 625, 1298, 1300, 1301, 1756
Lutra sp., 1298
Lycastis ouanaryensis, 116
Lycium andersonii, 1577
Lycodes frigidus, 1465
Lycopersicon esculentum, 110, 521, 746, 1093, 1097
Lycosa sp., 1004
Lycosidae, 112
Lymantria dispar, 140, 142, 983, 992, 1013, 1585
Lymnaea acuminata, 1208
Lymnaea calliaudi, 1170
Lymnaea luteola, 659, 1208
Lymnaea palustris, 257
Lymnaea peregra, 256
Lymnaea sp., 578, 747, 1171
Lymnaea stagnalis, 21, 574, 1208
Lynx, *Felis lynx,* 1760
Lynx rufus, 357, 914, 1144, 1440, 1745
Lysmata caudata, 1684
Lysmata seticaudata, 1534, 1535, 1658
Lytechinus pictus, 460, 647
Lytechinus variegatus, 363, 451

M

Macaca fascicularis, 278, 381, 588, 865, 1686
Macaca fuscata, 1180
Macaca iris, 278
Macaca mulatta, 278, 373, 381, 640, 650, 692, 1050, 1051, 1142, 1215, 1218, 1302, 1314, 1322, 1444, 1790
Macaca nigra, Celebes ape, 683
Macaca sp., 381, 808, 866, 940, 942, 1181, 1548, 1597
Macaw, blue- and yellow, *Ara ararauna,* 676
Machaeranthera spp., 1652
Machrobrachium sp., 152, 452
Mackerel
  Atlantic, *Scomber scomber,* 634
  King, *Scomberomorus cavalla,* 634
  Spanish, *Scomberomorus maculatus,* 119
Macoma balthica, 114, 133, 148, 266, 513, 515, 523, 629, 654, 1520, 1541
Macoma liliana, 148
Macoma nasuta, 515, 805, 1303
Macrobrachium hendersodyanum, 671
Macrobrachium lamarrei, 62
Macrobrachium rosenbergerii, 152, 1134
Macrobrachium rude, 152
Macrocystis pyrifera, 70, 458, 1207
Macronectes giganteus, 120, 446, 518
Macrophytes
  aquatic
    Cabomba spp., 927
    Chara sp., 747, 1207
    Elodea sp., 741, 747, 812, 927, 1041, 1577
    Lemna sp., 144, 781
    Myriophyllum sp., 111, 780, 927, 1577, 1656
    Polygonum sp., 1170
    Potamogeton spp., 96, 111, 225, 451, 741, 747, 780, 1517, 1577
    Ruppia spp., 780, 927
    Spirodella oligorrhiza, 1171
    Typha latifolia, 1171, 1577
    Zannichellia sp., 747, 780
    Zostera spp., 111
Macropus agilis, 1444
Macropus eugenii, 1444
Macropus fuliginosus, 1444
Macropus giganteus, 1444
Macropus irma, 1444
Macropus rufogriseus, 1417, 1418, 1444
Macropus rufus, 1444
Macrotus lagotis, 1444
Madrone, *Arbutus menziesii,* 1583
Magicicada spp., 112, 132
Magpie
  Australian, *Gymnorhina tibicen,* 1420, 1432, 1436
  Black-billed, *Pica pica,* 370, 895, 914, 1067, 1085, 1436
  Yellow-billed, *Pica nuttalli,* 1433
Mahoe, *Melicytus ramiflorus,* 1427
Makaira indica, 7, 118, 342, 1659
Makaira nigricans, 118, 342, 365, 1659
Malacosoma americanum, 216, 228
Malacosoma disstria, 988
Mallard, *Anas platyrhynchos,* 17, 120, 136, 167, 209, 232, 234, 241, 267, 283, 348, 371, 378, 384, 427, 436, 445, 464, 465, 480, 482, 583, 584, 625, 634, 676, 677, 689, 751, 752, 786, 806, 808, 833, 845, 862, 890, 891, 893, 932, 933, 961, 968, 1004, 1006, 1047, 1074, 1076, 1135, 1136, 1142, 1167, 1175, 1214, 1290, 1308, 1386, 1430, 1434, 1469, 1470, 1508, 1524, 1543, 1553, 1579, 1592, 1665, 1666, 1680, 1689, 1742, 1786
Malus malus, 904
Malus spp., 918, 989, 1092
Malus sylvestris, 905, 911, 1516
Mamestra brassicae, 768, 775, 992
Manatee, *Trichechus manatus,* 109, 123
Manduca sexta, 975
Mangifera indica, 1775
Mango, *Mangifera indica,* 1775
Manihot esculenta, 903, 918, 951
Maple
  Bigleaf, *Acer macrophyllum,* 1583
  Red, *Acer rubrum,* 110, 128, 143, 438, 653, 685, 815
Margaretifera margaretifera, 1629
Marlin
  Black, *Makaira indica,* 7, 118, 342, 1659
  Blue, *Makaira nigricans,* 118, 342, 365, 1659
Marmoset, *Callithrix jacchus,* 588, 961, 1049, 1055, 1314, 1546

Table 33.2 (continued) Common and Scientific Names of Plants and Animals Listed in the *Handbook of Chemical Risk Assessment series* (References are provided from common names to scientific names)

Marmota monax, 126, 572, 1670
Marphysa sanguinea, 517
Marten, *Martes martes,* 380, 1444
Martes martes, 380, 1444
Martin
  Purple, *Progne subis subis,* 1096
  Tree, *Hirundo nigricans,* 1433
Mayfly, *Hexagenia* sp., 338, 1265, 1280, 1375, 1662
Mealworm, lesser, *Alphitobius diaperinus,* 1071, 1073
Medaka, *Oryzias latipes,* 22, 811, 971, 1042, 1044, 1173, 1205, 1206, 1211, 1381, 1777, 1780
Medicago lupulina, 1621
Medicago sativa, 137, 337, 456, 483, 746, 1095, 1161, 1168, 1427, 1516, 1530, 1584, 1621, 1763
Medicago sp., 812
Megacephala virginica, 922
Megachile rotundata, 1103, 1104
Megachile sp., 1123
Megalops atlantica, 365
Meganyctiphanes norvegica, 115, 630, 1658, 1684
Melanitta perspicillata, 121, 136, 348, 518, 1666
Melanogrammus aeglefinus, 368, 841, 1466, 1678, 1782
Melanoides sp., 452
Melanophis spp., 1542
Melanotaenia fluviatilis, 1111
Melanperes formicivorus, 1432
Meleagris gallopavo, 60, 121, 138, 168, 202, 534, 612, 625, 636, 678, 889, 891, 905, 964, 968, 977, 1175, 1177, 1308, 1311, 1436, 1463, 1503, 1544, 1553
Melicytus ramiflorus, 1427
Melilotus sp., 1616
Melomys burtoni, 1444
Melospiza melodia, 1005
Menhaden, Atlantic, *Brevoortia tyrannus,* 578, 623
Menidia beryllina, 888, 889, 1111, 1200, 1222
Menidia menidia, 63, 155, 462, 529, 888, 889, 1111, 1678
Menidia peninsula, 155, 462, 665, 888, 889, 1111
Menippe mercenaria, 998, 1468, 1521
Menyanthes trifoliata, 1767
Mephitis mephitis, 823, 1440, 1444
Mephitis sp., 1145
Mercenaria campechiensis, 669
Mercenaria mercenaria, 52, 70, 148, 368, 442, 458, 526, 654, 659, 1378, 1468, 1520
Meretrix lusoria, 1528
Merganser
  Common, *Mergus merganser,* 352, 1623, 1666
  Hooded, *Lophodytes cucullatus,* 352
  Red-breasted, *Mergus serrator,* 636, 849, 932, 1287, 1293, 1666, 1743
Mergus merganser, 352, 1623, 1666
Mergus serrator, 636, 849, 932, 1287, 1293, 1666, 1743
Meriones hurrianae, 866
Meriones tristrami, 1445
Meriones unguiculatus, 1509
Merlangius merlangius, 344, 841, 1523, 1782
Merlin, *Falco columbarius,* 1150, 1291
Merluccius merluccius hubbsi, 1268
Merluccius productus, 633
Merluccius sp., 1466
Mesocricetus auratus, 74, 75, 757
Mesocyclops thermocyclopoides, 998
Mesotoma sp., 1004
Metamysidopsis elongata, 577
Metapenaeus ensis, 152
Methanobacterium sp., 453
Metridium senile, 628
Mice
  Deer, *Peromyscus maniculatus,* 15, 130, 246, 247, 248, 641, 1095, 1096, 1388, 1528, 1668, 1793
  Field, *Microtus arvalis,* 1028, 1033, 1179
  White-footed, *Peromyscus leucopus,* 246, 248, 250, 365, 590, 645, 807, 814, 944, 962, 975, 978, 1095, 1440, 1445, 1793, 1798
Microcystis aeruginosa, 96, 257, 927
Microhyla ornata, 367
Micropogonias undulatus, 74, 263, 444, 643
Micropterus dolomieui, 230, 260, 444, 1026, 1031, 1173, 1518
Micropterus ochrooaster. see Microtus ochrogaster
Micropterus salmoides, 55, 56, 313, 314, 340, 342, 367, 444, 451, 462, 530, 533, 632, 750, 838, 860, 928, 1038, 1205, 1210, 1221, 1280, 1382, 1464, 1468, 1471, 1517, 1519, 1589, 1602, 1662, 1663, 1729, 1739, 1782
Micropterus sp., 747
Microspora sp., 655
Microstomus pacificus, 444
Microtermes obesi, 855
Microtermes spp., 142
Microtus agrestis, 15, 248, 643, 1179, 1515, 1670
Microtus arvalis, 1028, 1033, 1179
Microtus californicus, 247, 1668, 1669
Microtus guentheri, 1445
Microtus haydeni, 1445
Microtus miurus, 1791
Microtus ochrogaster, 245, 1135, 1136, 1181
Microtus oeconomus, 1756
Microtus oregoni, 1791
Microtus pennsylvanicus, 14, 130, 248, 449, 1096, 1445, 1528, 1791
Microtus pinetorum, 270, 1791, 1798
Microtus spp., 136, 248, 1451
Mictyris longicarpus, 162
Midge
  Chaoborus sp., 1136
  Chironomus riparius, 25, 150, 257, 782, 811, 1197, 1208, 1375
  Chironomus sp., 153, 778, 993, 1468
  Cricotopus spp., 996
  Flower, *Contarina medicaginis,* 1103
  Procladius sp., 1003
  Tanytarsus sp., 1003
Midshipman, plainfin, *Porichthys notatus,* 930
Milkvetch, *Astragalus* spp., 1652
Millet, *Panicum millaceum,* 903
Millipede, *Apheloria* spp., 922
Milo, *Sorghum* spp., 746, 918
Milvus migrans, 638, 809, 1294, 1432, 1436
Milvus migrans lineatus, 352
Milvus milvus, 809
Mimosa pudica, 363
Mimus polyglottos, 1542

Table 33.2 (continued) Common and Scientific Names of Plants and Animals Listed in the *Handbook of Chemical Risk Assessment series* (References are provided from common names to scientific names)

Miniopterus schreibersi, 1299
Mink, 362
Mink, *Mustela vison,* 139, 171, 245, 336, 357, 373, 381, 449, 650, 693, 1050, 1144, 1150, 1298, 1300, 1302, 1312, 1316, 1322, 1421, 1446, 1689
Minnow
  Cyprinid, *Phoxinus, Puntius,* 17
  Eastern mud, *Umbra pygmaea,* 56, 1384
  Fathead, *Pimephales promelas,* 63, 64, 65, 69, 158, 260, 375, 463, 531, 532, 579, 666, 674, 751, 784, 805, 860, 888, 894, 924, 930, 966, 970, 972, 1001, 1042, 1045, 1112, 1136, 1140, 1211, 1269, 1306, 1372, 1374, 1375, 1395, 1468, 1469, 1471, 1472, 1536, 1539, 1630, 1677, 1682, 1781
  Leuciscus, 574, 579
  Mud, *Umbra limi,* 970
  Poeciliopsis spp., 1373, 1382, 1384
  Sheepshead, *Cyprinodon variegatus,* 17, 529, 555, 576, 583, 785, 804, 805, 811, 859, 887, 888, 966, 971, 972, 1110, 1212, 1374, 1468, 1469, 1472, 1629, 1678
Mint plant
  *Aeolanthus,*sp, 132
  Copper, *Aeolanthus biformifolius,* 110, 132
  Elsholtzia spp., 132
Misgurnis fossilis, 574
Mite
  Chorioptes bovis, 1104
  European red, *Panonychus ulmi,* 591
  McDaniel spider, *Tetranychus mcdanieli,* 591
  Northern fowl, *Ornithonyssus sylvilarum,* 1071, 1076
  Oribatid, *Platynothrus peltifer,* 138, 142
  Pear rust, *Epitrimerus pyri,* 1102
  Two-spotted spider, *Tetranychus urticae,* 591, 990, 1102, 1169
  Typhlodromus sp., 1169
Mizuhopecten yessoensis, 148
Mnesampla privata, 1428
Mockingbird, northern, *Mimus polyglottos,* 1542
Mohoua ochrocephala, 1433
Moina irrasa, 152
Mole, *Talpa europaea,* 59, 450, 1620, 1625, 1670
Molly, shortfin, *Poecilia mexicana,* 1173
Molothrus ater, 370, 806, 861, 862, 1076, 1096, 1542, 1544
Monitor
  Gould's, *Varanus gouldi,* 1430
  Lace, *Varanus varius,* 1430
Monkey
  Black-handed spider, *Ateles geoffroyi,* 1441
  Cebus apella, 1789
  Cynomolgus, *Macaca* spp., 381, 808, 940, 942, 1181, 1548, 1597
  Japanese, *Macaca fuscata,* 1180
  Marmoset, cotton top, *Callithrix jacchus,* 588, 961, 1049, 1055, 1314, 1546
  Rhesus, *Macaca mulatta,* 278, 373, 381, 640, 650, 692, 1050, 1142, 1215, 1218, 1302, 1314, 1322, 1444, 1790
  Squirrel, *Saimiri sciurea, Saimiri* spp., 761, 1794
Monodon monoceros, 245, 1467
Montastrea annularis, 228, 835
Moose, *Alces alces,* 13, 59, 61, 380, 437, 449, 1669, 1756, 1759, 1766
Morethia boulengeri, 860
Morning glory, *Ipomoea* sp., 746
Morone americana, 134, 462, 1026, 1739
Morone chrysops, 839
Morone saxatilis, 7, 18, 22, 62, 155, 332, 342, 376, 462, 518, 665, 767, 799, 841, 860, 886, 888, 1021, 1268, 1284, 1303, 1468, 1517, 1523, 1589, 1659, 1663, 1677, 1678, 1739
Morus bassanus, 1297
Mosquito
  Aedes, 1107
  Anopheles, 1103, 1148
  Culex, 1108
  Psorophora, 1004, 1109
Mosquitofish
  Gambusia affinis, 26, 55, 340, 367, 377, 529, 664, 674, 750, 805, 856, 886, 888, 1002, 1042, 1096, 1110, 1139, 1173, 1206, 1374, 1380, 1382, 1465, 1517, 1578, 1589, 1656, 1664, 1676
  Gambusia holbrooki, 1731, 1739
Moss
  Brachythecium, 223, 434
  Fontinalis dalecarlica, 25
  Fontinalis sp., 26
  Homalothecium sericium, 1763
  Hylocomium splendens, 1531
  Hypnum cupressiforme, 110, 223, 439, 1364, 1621
  Irish, *Chondrus crispus,* 1517, 1519
  Pleurozium schreberi, 440
  Rhacomitrium sp., 451
  Rhyncostegium riparioides, 145, 927
  Tamenthypnum sp., 451
Moth
  Cabbage, *Mamestra brassicae,* 768, 775, 992
  Codling, *Carpocapsa pomonella, Laspeyresia pomonella,* 1102, 1504
  Diamondback, *Plutella xylostella,* 1093, 1103
  Ectomyelois ceratoniae, 1777
  Greater wax, *Galleria melonella,* 809
  Gypsy
    Lymantria dispar, 140, 142, 983, 992, 1013, 1585
    Porthetria dispar, 228, 435, 441, 644
  Pine, *Bupalus* spp., 112
  Pineapple gummosis (*see* Lepidoptera)
  pine noctuid, *Panolis flammea,* 112, 132
  Zygaenid, *Zygaena filipendulae, Z. trifolii,* 918, 922
Mougeotia sp., 96, 655, 781
Mouse
  Beach, or Oldfield, *Peromyscus polionotus,* 806, 807, 808, 1026, 1136, 1142, 1793
  Deer
    Peromyscus maniculatus, 15, 130, 246, 247, 248, 641, 1095, 1096, 1388, 1528, 1668, 1793
    Peromyscus spp., 1439, 1446
  Domestic, *Mus* spp., 23, 66, 126, 170, 277, 286, 471, 537, 587, 588, 589, 590, 613, 650, 683, 692, 757, 787, 863, 866, 939, 942, 1010, 1049, 1050, 1078, 1081, 1085, 1118, 1135, 1136, 1181, 1216, 1219, 1315, 1387, 1418, 1438, 1445, 1469, 1597, 1635, 1686, 1792
  Eastern harvest, *Reithrodontomys humulis,* 1794

Table 33.2 (continued) Common and Scientific Names of Plants and Animals Listed in the *Handbook of Chemical Risk Assessment series* (References are provided from common names to scientific names)

Field, *Microtus arvalis,* 1028, 1033, 1179
Great Basin pocket, *Perognathus parvus,* 1793
House, *Mus musculus,* 129, 247, 270, 449, 768, 806, 942, 964, 969, 975, 978, 1095, 1181, 1388, 1445, 1668, 1669, 1756, 1792
Little pocket, *Perognathus longimembris,* 1439
Long-tailed, *Pseudomys higginsi,* 1447
Meadow, *Microtus haydeni,* 1445
Mitchell's hopping, *Notomys mitchelli,* 1446
Plains, *Pseudomys australis,* 1447
Pocket, *Perognathus inornatus,* 1446
Salt marsh harvest, *Reithrodontomys raviventris,* 1450
Sandy inland, *Pseudomys hermannsburgensis,* 1447
Spinifex hopping, *Notomys alexis,* 1446
Western chestnut, *Pseudomys nanus,* 1447
Western harvest, *Reithrodontomys megalotis,* 1439
Western jumping, *Zapus princeps,* 60
White-footed, *Peromyscus leucopus,* 246, 248, 250, 365, 590, 645, 807, 814, 944, 962, 975, 978, 1095, 1440, 1445, 1793, 1798
Wood, *Apodemus sylvaticus,* 124, 242, 248, 275, 355, 1515
Yellow-necked field, *Apodemus flavicollis,* 644
Mouth brooder, southern, *Pseudocrenilabrus philander,* 118
Moxostoma duquesnei, 230
Mudalia potosensis, 1374
Mudpuppy, *Necturus maculosus,* 612, 909, 1785
Mugil cephalus, 860, 888, 1111, 1141, 1173, 1213, 1268, 1281, 1369, 1465, 1523
Mugil curema, 643
Mule, *Equus asinus* X *Equus caballus,* 243, 1443
Mulinia lateralis, 1207
Mullet
Gray, *Chelon labrosus,* 73, 1541
Striped, *Mugil cephalus,* 860, 888, 1111, 1141, 1173, 1213, 1268, 1281, 1369, 1465, 1523
Various, *Rhinomugil corsula,* 1213
White, *Mugil curema,* 643
Yellow-eye, *Aldrichetta forsteri,* 63, 64
Mullus barbatus, 118, 567, 1269
Mummichog, *Fundulus heteroclitus,* 26, 27, 28, 63, 118, 154, 263, 369, 374, 375, 377, 462, 529, 578, 655, 664, 674, 893, 1000, 1040, 1074, 1110, 1205, 1213, 1283, 1304, 1739
Murex trunculus, 364
Murre
Common, *Uria aagle,* 136, 1027
Thick-billed, *Uria lomvia,* 834, 851, 1660
Murrel, *Channa punctatus,* 19, 69, 154, 376, 663, 811, 966
Murrelet, marbled, *Synthliboramphus antiguis,* 1524
Musca autumnalis, 988, 1009, 1116
Musca domestica, 809, 969, 990, 1009, 1079, 1103, 1104, 1116, 1585
Musca sp., 812, 1072
Mus domesticus, 245
Mushroom, *Agaricus, Amantia muscaria, Collybia, Lactarius,* 1516
Muskellunge, *Esox masquinongy,* 444
Muskox, *Ovibus moschatus,* 4, 32, 126
Muskrat, *Ondatra zibethicus,* 126, 136, 246, 336, 358, 572, 1733, 1746
Mus musculus, 129, 247, 270, 449, 768, 806, 942, 964, 969, 975, 978, 1095, 1181, 1388, 1445, 1668, 1669, 1756, 1792
Mus musculus domesticus, 449, 1759, 1764
Mussel
Amblema sp., 112
Anodonta, 102, 146, 441, 574, 578, 613, 994, 997, 1072, 1079
Brown, *Perna indica,* 149
California, *Mytilus californianus,* 336, 339, 514, 516, 669, 1384
Common, *Mytilus edulis,* 6, 52, 71, 114, 134, 148, 161, 228, 262, 339, 368, 442, 459, 516, 523, 526, 569, 575, 582, 615, 622, 623, 629, 643, 655, 659, 669, 670, 742, 750, 805, 916, 928, 1038, 1201, 1202, 1265, 1281, 1362, 1366, 1378, 1520, 1541, 1622, 1627, 1629, 1658, 1684, 1731, 1736
Duck, *Anodonta anatina,* A. *nuttalliana,* 441, 574, 578, 1207
Green-lipped, *Perna viridis,* 115, 149, 570, 629, 660
Hooked, *Ischadium recurvum* (formerly *Brachidontes exustus),* 642
Ischadium exustus, 520
Lagoon, *Mytela strigata,* 114
Lake, *Anodonta piscinalis,* 112, 129
Mytilopsis sallei, 19, 22
Mytilus sp., 114, 134, 161, 748, 1264, 1366
Ribbed, *Geukensia demissa,* 1170, 1172, 1283
Villosa iris, 149
Zebra, *Dreissena polymorpha,* 55, 114, 134, 147, 227, 526, 569, 628, 1278, 1303, 1304, 1365
Mus spp., 23, 66, 126, 170, 277, 286, 471, 537, 587, 588, 589, 590, 613, 650, 683, 692, 757, 787, 863, 866, 939, 942, 1010, 1049, 1050, 1078, 1081, 1085, 1118, 1135, 1136, 1181, 1216, 1219, 1315, 1387, 1418, 1438, 1445, 1469, 1597, 1635, 1686, 1792
Mustard, *Brassica juncea,* 774, 775
Mustela erminea, 1298, 1301
Mustela nigripes, 1415
Mustela nivalis, 24, 1298, 1301
Mustela putorius, 380, 939, 1298, 1300
Mustela putorius furo, 245, 683, 693
Mustela vison, 139, 171, 245, 336, 357, 373, 382, 449, 650, 693, 1050, 1144, 1150, 1298, 1300, 1302, 1312, 1316, 1322, 1421, 1446
Mustelus antarcticus, 1514
Mustelus canis, 53, 518
Mustelus manazo, 1515
Mustelus putorius, 101, 1445, 1792
Mya arenaria, 70, 71, 142, 148, 222, 262, 339, 368, 457, 459, 515, 526, 655, 659, 669, 1201, 1365
Mya sp., 25
Mycteria americana, 332, 352, 1668
Myiarchus cinerascens, 1432
Myiarchus crinitus, 1005, 1786, 1788
Myiarchus sp., 1771
Myocastor coypus, 1446
Myotis austroriparius, 641
Myotis dasycneme, 1202
Myotis grisescens, 641, 834, 853
Myotis lucifugus, 222, 245, 834, 943

Table 33.2 (continued) Common and Scientific Names of Plants and Animals Listed in the *Handbook of Chemical Risk Assessment series* (References are provided from common names to scientific names)

Myotis sodalis, 990
Myotis spp., 14, 245
Myoxocephalus quadricornis, 842
Myoxocephalus scorpius, 8
Myoxocephalus sp., 1464
Myriophyllum sp., 111, 780, 927, 1577, 1656
Myriophyllum spicatum, 67, 96, 776, 780, 1171
Mysidaceans, *Mysidopsis, Mysis relicta, Praunus flexuosus,* 623, 661, 672, 836, 1303, 1304
Mysidopsis bahia, 65, 152, 368, 375, 460, 528, 578, 661, 784, 886, 887, 928, 966, 972, 998, 1106, 1109, 1208, 1468, 1471, 1541, 1629, 1677
Mysidopsis bigelowi, 460
Mysidopsis formosa, 460
Mysidopsis spp., 22
Mysis relicta, 623, 672, 836, 1303, 1304
Mystus gulio, 633
Mystus vittatus, 805, 811, 1111
Mytela strigata, 114
Mytilopsis sallei, 19, 22
Mytilus californianus, 336, 339, 514, 516, 669, 1384
Mytilus edulis, 6, 52, 71, 114, 134, 148, 161, 228, 262, 339, 368, 442, 459, 516, 523, 526, 569, 575, 582, 615, 622, 623, 629, 643, 655, 659, 669, 670, 742, 750, 805, 916, 928, 1038, 1201, 1202, 1265, 1281, 1362, 1366, 1378, 1520, 1541, 1622, 1627, 1629, 1658, 1684, 1731, 1736
Mytilus edulis aoteanus, 516, 1622
Mytilus edulis planulatus, 6
Mytilus galloprovincialis, 128, 148, 364, 527, 1369, 1534, 1658, 1675, 1684, 1765
Mytilus smarangdium, 114
Mytilus sp., 114, 134, 161, 748, 1264, 1366
Myxine glutinosa, 1383

N

Nabis sp., 991
Nais sp., 63
Najas spp., 747, 749
Najus guadalupensis, 255
Nannochloris oculata, 780, 973
Nannochloris sp., 582
Narwhal, *Monodon monoceros,* 245, 1467
Nassarius obsoletus, 165, 459, 515, 527, 578, 581, 660, 1541
Nassarius sp., 442
Natrix sp., 1665, 1783
Navicula pelliculosa, 458
Navicula sp., 255, 655
Neanthes arenaceodentata, 63, 64, 65, 73, 262, 662, 673, 1209, 1373, 1376, 1779
Neanthes virens, 63, 64, 65, 73, 262, 662, 673
Necturus maculosus, 612, 909, 1785
Necturus sp., 1785
Nematode
  Acrobeloides sp., 991
  Anguillicola sp., 229
  Bursaphelencchus xylophilus, 976
  Caenorhabditis elegans, 141, 143, 146, 525
  Panagrellus redivivus, 976, 992
  Parafilaria bovicola, 1508
Nematolosa comi, 1369
Neopanope texana, 785
Neophocaena phocaenoides, 1272
Neotoma albigula, 1446
Neotoma intermedia, 1446
Neotoma lepida, 1439
Nephrops norvegicus, 152, 339, 836
Neptunus pelagicus, 323
Nereis diversicolor, 7, 53, 122, 153, 163, 443, 451, 460, 517, 623, 631, 655, 662, 673, 1737, 1784
Nereis virens, 63, 856, 858, 1303, 1378
Neritina sp., 578
Nerodia sipedon, 231, 844
Nerodia spp., 1466, 1665, 1783
Nerodia taxispilota, 1741
Newt
  California, *Taricha torosa,* 833, 844
  Eastern, *Notophthalmus viridescens,* 1771, 1785
  Rough-skinned, *Taricha granulosa,* 1784, 1785
  Triturus, 668, 886, 1385
Nicotiana sp., 813
Nicotiana tabacum, 12, 56, 78, 337, 453, 802, 1093, 1168, 1667, 1774, 1776
Nicrophorus tomentosus, 228
Nitella flexilis, 1628
Nitroca spinipes, 460, 578, 579, 1109
Nitzschia angularis, 1107
Nitzschia closterium, 656, 1778
Nitzschia liebethrutti, 582
Nitzschia longissima, 656
Nitzschia palea, 145
Nocardia spp., 1421
Nocardiopsis sp., 826
Noemacheilus barbatulus, 255, 619, 665
Noemacheilus sp., 665
Nomia melanderi, 1104
Nostoc muscorum, 775, 810, 1628
Notemigonus crysoleucas, 62, 68, 632, 1096, 1303, 1740
Notiomystis cincta, 1433
Notomys alexis, 1446
Notomys mitchelli, 1446
Notophthalmus viridescens, 1771, 1785
Notopterus notopterus, 367, 376
Notornis mantelli, 1433
Notothenia coriiceps neglecta, 1367
Notothenia gibberifrons, 118, 444, 518
Notropis cornutus, 1213
Notropis hudsonius, 1030, 1147, 1538
Notropis venustus, 1464
Nucella lapillus, 574, 575, 581
Nucellus lapilus, 570
Numenius americanus, 834, 849
Numenius arquata, 1743
Numenius phaeopus, 121
Nuphar luteum, 10
Nuphar sp., 441, 1577
Nuthatch, white-breasted. *Sitta carolensis,* 1432
Nutria, *Myocastor coypus,* 1446
Nycticorax nycticorax, 58, 624, 636, 1035, 1150, 1289, 1294, 1319, 1665
Nycticorax violaceus, 849, 1464

Table 33.2 (continued) Common and Scientific Names of Plants and Animals Listed in the *Handbook of Chemical Risk Assessment series* (References are provided from common names to scientific names)

O

Oak
  Red, *Quercus rubra,* 440, 653, 1005
  Various, *Quercus* spp., 621, 685, 1005, 1006, 1288
Oat, *Avena sativa,* 253, 434, 456, 521, 774, 775, 1168, 1667
Oceanodroma furcata, 1150
Oceanodroma leucorhoa, 54
Ocenebra erinacea, 622, 629
Ochotona sp., 914
Ochrogaster lunifer, 1428
Ochromonas danica, 145
Octochaetus pattoni, 373, 383
Octopus
  Eledone cirrhosa, 114, 515
  Paroctopus sp., 1515
  Polypus bimaculatus, 1579
Odobenus rosmarus divergens, 9, 10, 25, 853
Odocoileus hemionus, 222, 234, 286, 450, 641, 1624, 1633, 1640, 1641, 1669, 1673, 1745
Odocoileus hemionus hemionus, 370, 373, 1446, 1469, 1549, 1732
Odocoileus sp., 126, 936
Odocoileus virginianus, 14, 126, 130, 136, 223, 245, 250, 336, 358, 450, 572, 625, 641, 1418, 1427, 1526, 1544, 1545, 1549, 1633, 1670, 1733, 1745
Odoicoileus hemionus columbianus, 1673, 1732
Odontotermes obesus, 855
Odontotermes spp., 142
Odontotermes transvaalensis, 441
Oedaleus senegalensis, 894
Oedemagena tarandi, 1072, 1083
Oedemogena sp., 1084
Oedogonium cardiacum, 812, 1000, 1041, 1206, 1375, 1675
Oedogonium sp., 67, 96, 781
Oikomonas termo, 1102
Okra, *Abelmoschus esculentus,* 810, 1095, 1096, 1626, 1638
Olea spp., 164
Oligocottus maculosus, 530, 533
Olisthodiscus lutens, 70
Olive, *Olea* spp., 1364
Ommastrephes bartrami, 5, 115
Omul (fish), *Coregonus autumnalis migratorius,* 837, 1465
Onchiurus apuanicus, 774
Oncopeltes fasciatus, 1072, 1073
Oncorhynchus clarki, 155, 520, 665, 842, 860, 888, 928, 966, 1001, 1738
Oncorhynchus gorbuscha, 567, 1374, 1541
Oncorhynchus keta, 530, 567, 842, 1539
Oncorhynchus kisutch, 18, 68, 155, 254, 462, 530, 665, 805, 860, 929, 1001, 1030, 1040, 1044, 1136, 1148, 1202, 1211, 1212, 1280, 1374, 1381, 1468, 1519, 1589, 1627, 1629, 1656, 1664, 1677, 1739, 1779
Oncorhynchus mykiss, 11, 18, 19, 21, 26, 56, 62, 64, 65, 69, 100, 156, 176, 258, 367, 375, 384, 444, 462, 480, 482, 506, 530, 533, 576, 579, 611, 647, 665, 688, 747, 750, 768, 783, 805, 832–833, 860, 888, 905, 929, 964, 965, 1001, 1026, 1030, 1040, 1043, 1054, 1073, 1100, 1111, 1112, 1137, 1173, 1202, 1209, 1221, 1269, 1305, 1353, 1373, 1375, 1380, 1428, 1461, 1468, 1519, 1533, 1536, 1539, 1581, 1586, 1589, 1602, 1620, 1623, 1629, 1641, 1677, 1738, 1779
Oncorhynchus nerka, 666, 1210, 1221, 1579, 1586, 1590, 1620, 1738
Oncorhynchus spp., 62, 1030, 1147
Oncorhynchus tshawytscha, 18, 68, 158, 531, 567, 666, 675, 751, 860, 929, 1030, 1148, 1212, 1269, 1282, 1284, 1302, 1322, 1539, 1590, 1630, 1664, 1674, 1677, 1770, 1780
Ondatra zibethicus, 126, 136, 246, 336, 358, 572, 1733, 1746
Onion, *Allium* sp., 426, 453
Oniscus asellus, 628
Oonopsis spp., 1652
Operophtera brumata, 1288
Ophiocephalus punctatus, 158
Ophioderma brevispina, 576
Ophiolobus sp., 1169
Ophryotrocha diadema, 1207
Oplegnathus fasciatus, 965, 966
Oporornis tolmiei, 1006
Opossum, *Didelphis virginiana,* 823, 1440
Opsanus beta, 887, 888, 889
Opsanus tau, 581, 1112, 1381
Orange
  Mandarin, *Citrus tachibana,* 337
  Sweet, *Citrus sinensis,* 140
Orchestia gammarellus, 115, 134, 629
Orchestia mediterranea, 629
Orchestia traskiana, 578
Orcinus orca, 1272
Orconectes immunis, 887
Orconectes limosus, 367
Orconectes nais, 129, 255, 858
Orconectes rusticus, 163, 1109
Orconectes sp., 782
Orconectes virilis, 338, 630, 661, 672, 750
Oreochromis aureus, 887, 888
Oreochromis hornorum, 1173
Oreochromis mossambicus, 21, 102
Oreochromis niloticus, 27, 158, 260, 463, 886, 888, 1381
Orfe, golden, *Leuciscus idus melanotus,* 574, 579
Orius sp., 991
Ornithonyssus sylvilarum, 1071, 1076
Ortalis cinereiceps, 676
Orthemis sp., 996
Orthodon microlepidotus, 55, 331, 1517
Orthosia gothica, 768
Oryctolagus cuniculus, 788, 890, 895, 937, 1014, 1219, 1414, 1417, 1446
Oryctolagus sp., 66, 126, 171, 421, 450, 471, 534, 537, 683, 757, 867, 912, 943, 1010, 1050, 1072, 1078, 1082, 1120, 1181, 1438, 1550, 1597, 1636, 1760, 1764, 1769
Orygia pseudotsuga, 1005
Oryzaephilus surinamensis, 883
Oryza sativa, 225, 337, 382, 439, 453, 1204, 1530, 1532, 1667
Oryzias latipes, 22, 811, 971, 1042, 1044, 1173, 1205, 1206, 1211, 1381, 1777, 1780
Oryzomus palustris, 1793

Table 33.2 (continued) Common and Scientific Names of Plants and Animals Listed in the *Handbook of Chemical Risk Assessment series* (References are provided from common names to scientific names)

Oscillatoria sp., 453
Osmerus mordax, 341, 571, 1030, 1148, 1202
Osprey, *Pandion haliaetus,* 58, 249, 348, 353, 625, 636, 833, 849, 1036, 1294, 1463, 1515, 1524
Ostracods
  Cypericercus sp., 1002
  Cypicerus sp., 998
  Cypridopsis sp., 998, 1003
  Cyprinotus sp., 1002
  Cyrinofus sp., 1003
Ostrea edulis, 551, 570, 575, 629
Ostrea equestris, 516
Ostrea sinuata, 516
Ostrea sp., 1264
Otter
  European, *Lutra lutra,* 336, 357, 385, 625, 1298, 1300, 1301, 1756
  Lutra spp., 1298
  River, *Lutra canadensis,* 336, 357, 373, 449, 853, 1144–1145, 1670, 1689
  Sea, *Enhydra lutris,* 10, 215
Otus asio, 806, 837, 846, 934, 1682
Ovenbird, *Seiurus aurocapillus,* 1005
Ovibus moschatus, 4, 32, 126
Ovis aries, 93, 126, 171, 177, 234, 246, 277, 358, 449, 454, 572, 641, 650, 684, 693, 758, 774, 787, 802, 867, 905, 944, 969, 975, 984, 1011, 1014, 1078, 1082, 1120, 1182, 1219, 1414, 1446, 1469, 1550, 1746, 1764, 1793
Ovis canadensis, 59, 1649, 1673
Ovis canadensis californiana, 1669
Ovis sp., 93, 126, 171, 234, 336, 590, 641, 1635, 1640
Owl
  Burrowing, *Athene cunicularia,* 816, 905, 1432
  Common barn, *Tyto alba,* 24, 863, 1074, 1432
  Eagle, *Bubo bubo,* 333, 348, 1291
  Great horned, *Bubo virginianus,* 834, 846, 1071, 1076, 1085, 1150, 1431, 1434
  Laughing, *Sceloglaux albifacies,* 1433
  Screech, *Otus asio,* 806, 837, 846, 934, 1682
  Short-eared, *Asio flammeus,* 1291
  Tawny, *Strix aluco,* 1297
  Tengmalm's, *Aegolius funereus,* 1289
Oxya velox, 106
Oxylobium spp., 1429
Oxyura jamaicensis, 348
Oyster
  American, *Crassostrea virginica,* 25, 26, 64, 70, 71, 113, 133, 147, 148, 217, 227, 262, 368, 442, 458, 514, 515, 520, 523, 526, 557, 567, 569, 577, 622, 628, 642, 655, 658, 749, 767, 784, 831, 835, 857, 1107, 1171, 1207, 1221, 1264, 1281, 1365, 1375, 1377, 1385, 1468, 1471, 1472, 1520, 1535, 1541, 1629, 1657, 1735
  Crassostrea, Ostrea, 927, 1264, 1520
  European flat, *Ostrea edulis,* 551, 570, 575, 629
  Ostrea, 1264
  Pacific, *Crassostrea gigas,* 5, 16, 70, 96, 113, 129, 147, 323, 364, 442, 523, 525, 567, 568, 575, 623, 658, 689, 1208, 1528, 1541, 1586, 1677, 1731
  Rock, *Saccostrea cucullata,* 115, 612, 629
  Saccostrea sp., 516
  Sydney rock, *Crassostrea commercialis,* 5, 628
Oystercatcher, Eurasian, *Haematopus ostralegus,* 121, 1659, 1731, 1743, 1768

P

Pachymetopan grande, 74
Pacifastacus leniusculus, 115, 1629
Pacifastacus sp., 338
Paddlefish, *Polyodon spathula,* 839, 1269, 1284
Pademelon, *Thylogale billardierii,* 1417, 1448
Pagodroma nivea, 361
Pagophilus groenlandica, 373, 381, 1032
Pagrus major, 583, 636, 1005
Pagurus longicarpus, 17, 460
Pagurus sp., 1171
Paintbrush, Indian, *Castilleja* spp., 1652, 1667
Palaemon elegans, 661, 671, 672
Palaemonetes kadiakensis, 1468
Palaemonetes pugio, 7, 27, 72, 228, 528, 532, 578, 582, 631, 785, 887, 999, 1106, 1109, 1208, 1283, 1303, 1373, 1468, 1534
Palaemonetes sp., 54, 365, 1527, 1671
Palaemonetes varians, 672
Palaemonetes vulgaris, 17, 1074, 1141
Palaemon macrodactylus, 858, 1468
Palaemon serratus, 613, 623
Palaemon spp., 671
Palicourea marcgravii, 1413, 1426
Palm, Iraqi date, *Phoenix* sp., 895
Panagrellus redivivus, 976, 992
Pandalus borealis, 1514, 1521
Pandalus jordant, 115
Pandalus montagui, 7, 228, 443, 613, 631, 671, 672
Pandalus spp., 1521
Pandion haliaetus, 58, 249, 348, 353, 625, 636, 833, 849, 1036, 1294, 1463, 1515, 1524
Panicum millaceum, 903
Panolis flammea, 112, 132
Panonychus ulmi, 591
Pantala hymenea, 26
Pantala sp., 996
Panther, Florida, *Felis concolor coryi,* 336, 385
Pan troglodytes, 275
Panulirus argus, 443
Panulirus cygnus, 1514
Panulirus interruptus, 7
Panulirus japonicus, 1099
Papaver orientale, 110
Paphia undulata, 115
Papio anubis, 278, 758, 867, 912, 1417
Paracentrotus lividus, 153, 228, 1208
Paracottus sp., 230
Parafilaria bovicola, 1508
Paragnetina media, 995
Paragrapsus quadridentatus, 152
Paralabrax clathratus, 118, 444
Paralabrax nebulifer, 1282
Paralichthys dentatus, 142, 158, 531, 1678
Paralichthys lethostigma, 118, 445, 643
Paralichthys sp., 100, 158
Paralithodes camtschatica, 623, 1521
Paramecium caudatum, 1384

Table 33.2 (continued) Common and Scientific Names of Plants and Animals Listed in the *Handbook of Chemical Risk Assessment series* (References are provided from common names to scientific names)

Paramecium multimicronucleatum, 657
Paramecium spp., 425
Paranais sp., 1041
Paratenuisentis sp., 229
Paratya australiensis, 105, 152, 782
Paratya sp., 105
Parauronema acutum, 1385
Pardosa sp., 1004
Parhalella natalensis, 152
Parimelia baltimorensis, 110, 216, 223
Paroctopus sp., 1515
Parophrys vetulus. see Pleuronectes vetulus
Paropsis atomaria, 922
Parrot
  Port Lincoln, *Barnardius zonarius,* 1434
  Red-rumped, *Psephotus haematonotus,* 1436
Partridge, gray, *Perdix perdix,* 371, 377, 1469, 1544
Parus atricapillus, 1005
Parus bicolor, 1005
Parus gambeli, 1006
Parus major, 636, 1288
Parus montanus, 636
Parus palustris, 636
Parus sp., 1771
Paspalum notatum, 1143
Passer domesticus, 235, 239, 371, 508, 806, 807, 890, 967, 1436, 1668
Passerina amoena, 1006
Passerina cyanea, 1005
Passer montanus, 235, 1005
Patella caerulea, 567, 570
Patella vulgata, 149, 442, 516
Pea, *Pisum sativum,* 774, 1165, 1530, 1532, 1621, 1774
Peach, *Prunus persica,* 904
Peanut, *Arachis hypogea,* 810, 1427
Pear, *Pyrus communis,* 904, 1584, 1667
Pear scylla, *Psylla pyricola,* 1102
Pearsonia metallifera, 439
Pecten alba, 1515
Pecten jacobeus, 115
Pecten maximus, 6, 567
Pecten novae-zelandiae, 1622
Peewee, eastern wood-, *Contopus virens,* 1005
Pelecanus conspicillatus, 1369
Pelecanus occidentalis, 9, 58, 121, 240, 353, 636, 849, 1463, 1515, 1524, 1659
Pelecanus onocrotalus, 11, 351, 1524
Pelican
  Brown, *Pelecanus occidentalis,* 9, 58, 121, 240, 353, 636, 849, 1463, 1515, 1524, 1659
  White, *Pelecanus onocrotalus,* 11, 351, 1524
Pelodera sp., 992
Pelvetia canaliculata, 111
Penaeopsis joyneri, 971
Penaeus aztecus, 6, 54, 116, 443, 642, 750, 784, 858, 886, 966, 1374, 1377, 1521, 1527, 1671, 1677
Penaeus brasiliensis, 631
Penaeus duorarum, 785, 858, 1106, 1109, 1376, 1377, 1468, 1629
Penaeus indicus, 165, 368
Penaeus latisulcatus, 1515
Penaeus setiferus, 377, 642, 894, 1521
Penaeus spp., 364
Penaeus vannamei, 1206
Penguin, 121
  Adelie, *Pygoscelis adeliae,* 8, 120, 347, 362, 446, 518, 850, 1295
  Chinstrap, *Pygoscelis antarctica,* 120, 347, 446, 518
  Gentoo, *Pygoscelis papua,* 120, 267, 347, 446, 518, 1295
Penicillium spp., 1421, 1422
Peophila guttata, 271, 370, 371, 1436
Pepper, *Piper* spp., 1093, 1516
Perameles gunni, 1446
Perameles nasuta, 1446
Perca flavescens, 118, 331, 340, 343, 452, 571, 632, 633, 805, 930, 1001, 1027, 1030, 1045, 1202 1280, 1284, 1378, 1468, 1519, 1654, 1677, 1739, 1762
Perca fluviatilis, 129, 230, 343
Perch
  Climbing
    Anabas scandens, 69
    Anabas testudineus, 62, 64, 260, 461, 804
  Spangled, *Leipotherapon unicolor,* 664, 674
  Various, *Perca fluviatilis,* 129, 230, 343
  White, *Morone americana,* 134, 462, 1026, 1739
  Yellow, *Perca flavescens,* 118, 331, 340, 343, 452, 571, 632, 633, 805, 930, 1001, 1027, 1030, 1045, 1202, 1280, 1284, 1378, 1468, 1519, 1654, 1677, 1739, 1762
Perdix perdix, 371, 377, 1469, 1544
Perga dorsalis, 1428
Peridinium gatunense, 1770
Periplaneta americana, 992, 1072, 1073, 1102, 1104, 1585
Periwinkle, *Littorina littorea,* 5, 52, 507, 515, 659, 1520
Perna canaliculus, 516
Perna indica, 149
Perna viridis, 115, 149, 570, 629, 660
Pernis apivorus, 202
Perognathus inornatus, 1446
Perognathus longimembris, 1439
Perognathus parvus, 1793
Peromyscus leucopus, 246, 248, 250, 365, 590, 645, 807, 814, 944, 962, 975, 978, 1095, 1440, 1445, 1793, 1798
Peromyscus maniculatus, 15, 130, 246, 247, 248, 641, 1095, 1096, 1388, 1528, 1668, 1793
Peromyscus polionotus, 806, 807, 808, 1026, 1136, 1142, 1793
Peromyscus spp., 1439, 1446
Petrel, 121, 361
  Grey-faced, *Pterodroma macroptera,* 1418
  Southern giant, *Macronectes giganteus,* 120, 446, 518
  Storm, *Oceanodroma furcata,* 1150
Petrogale penicillata, 1420, 1433
Petromyzon marinus, 842, 860, 1213, 1780
Petromyzontidae, 1363
Phaeodactylum tricornutum, 22, 255, 261, 458, 656, 1171
Phalacrocorax atriceps, 120, 446, 518
Phalacrocorax auritus, 54, 1026, 1028, 1035, 1047, 1287, 1288, 1294, 1311
Phalacrocorax carbo, 11, 353, 850, 1027, 1028, 1036, 1294, 1524

Table 33.2 (continued) Common and Scientific Names of Plants and Animals Listed in the *Handbook of Chemical Risk Assessment series* (References are provided from common names to scientific names)

Phalacrocorax carbo sinensis, 1295
Phalacrocorax sp., 1150
Phalaris arundinacea, 1732, 1734
Phalcrocorax aristotelis, 1287
Phaseolus aureus, 1204
Phaseolus lunatus, 903, 989
Phaseolus sp., 521, 746, 812, 918, 1093
Phaseolus vulgaris, 252, 922, 1039, 1584
Phasianus colchicus, 13, 209, 240, 353, 371, 384, 679, 786, 806, 808, 861, 862, 890, 891, 895, 961, 968, 977, 1047, 1076, 1135, 1136, 1177, 1215, 1311, 1436, 1469, 1470, 1472, 1544, 1668
Pheasant, ring-necked, *Phasianus colchicus,* 13, 209, 240, 353, 371, 384, 679, 786, 806, 808, 861, 862, 890, 891, 895, 961, 968, 977, 1047, 1076, 1135, 1136, 1177, 1215, 1311, 1436, 1469, 1470, 1544, 1668
Philarctus quaeris, 1209
Philesturnus carunculatus, 1433
Philodena acuticornis, 62, 458, 657
Philohela minor. see Scolopax minor
Phleum pratense, 434, 774
Phoca groenlandica, 54, 335, 358, 1032
Phoca hispida, 25, 245, 250, 336, 358, 380, 450, 625, 641, 834, 1032, 1263, 1264, 1273, 1464, 1467, 1660, 1661
Phoca hispida ladogensis, 249
Phoca hispida saimensis, 853, 1661
Phoca largha, 1744
Phoca sibirica, 10, 25, 123, 335, 359, 641, 853, 1262, 1273, 1467
Phoca spp., 362
Phoca vitulina, 123, 222, 358, 450, 572, 853, 1032, 1274, 1319, 1368, 1467, 1526, 1579, 1661
Phoca vitulina richardi, 359
Phocoena phocoena, 123, 336, 359, 641, 693, 1032, 1262, 1263, 1273, 1275, 1319, 1363, 1368, 1467, 1661
Phocoena sinus, 450
Phocoenoides dalli, 641, 854, 1276
Phoeniconaias minor, 11, 1524
Phoenicopterus minor, 351
Phoenicopterus ruber, 240, 625, 637, 1666
Phoenicopterus ruber roseus, 121
Phoenicopterus ruber ruber, 240
Phoenix sp., 895
Phomopsis leptostromiformes, 615
Phormidium inundatum, 524
Phoxinus lagowski, 1004
Phoxinus phoxinus, 17, 577, 581, 582, 666, 1539, 1766
Phragmites australis, 129
Phragmites sp., 55, 1169
Phyllophaga sp., 837
Physa gyrina, 146, 1208
Physa heterostropha, 62, 660, 928, 1778
Physa integra, 149
Physa sp., 142, 750, 997, 1003, 1041, 1206, 1376, 1382, 1676
Physeter macrocephalus, 336, 359, 450, 909, 1744
Pica nuttalli, 1433
Pica pica, 370, 895, 914, 1067, 1085, 1436
Pica sp., 905, 1433
Picea abies, 110, 225, 382
Pickerel
  Chain, *Esox niger,* 444, 1027
  Various, *Esox,* 444, 457
Pieris brassicae, 963, 992, 1427
Pig
  Domestic, *Sus* spp., 97, 127, 173, 280, 373, 450, 454, 483, 652, 685, 694, 869, 891, 905, 946, 1011, 1014, 1221, 1448, 1505, 1551, 1637, 1761
  Feral, *Sus scrofa,* 59, 127, 245, 968, 969, 975, 978, 1414, 1417, 1448, 1747, 1756
  Guinea, *Cavia* spp., 19, 66, 210, 470, 536, 588, 589, 590, 593, 615, 650, 682, 753, 756 808, 909, 1049, 1050, 1177, 1180, 1438, 1442, 1469, 1473, 1547, 1596, 1634, 1789
Pigeon
  Domestic, *Columba livia,* 20, 50, 57, 221, 236, 269, 271, 371, 379, 806, 890, 933, 1308, 1435
  Nicobar, *Caloenas nicobarica,* 634, 676
Pika, *Ochotona princeps, Ochotona* sp., 914
Pike
  Northern, *Esox lucius,* 65, 117, 154, 260, 332, 340, 444, 457, 571, 832, 838, 1042, 1202, 1284, 1375, 1378, 1382, 1517, 1654, 1676, 1739, 1758, 1762
  Various, *Hoplias* sp., 1747
Pimephales promelas, 63, 64, 65, 69, 158, 260, 375, 463 531, 532, 579, 666, 674, 751, 784, 805, 860, 888, 894, 924, 930, 966, 970, 972, 1001, 1042, 1045, 1112, 1136, 1140, 1211, 1269, 1306, 1372, 1374, 1375, 1395, 1468, 1469, 1471, 1472, 1536, 1539, 1630, 1677, 1682, 1781
Pine
  Digger, *Pinus sabiniana,* 1584
  Lodgepole, *Pinus contorta,* 143
  Longleaf, *Pinus palustris,* 1776
  Maritime, *Pinus pinaster,* 140
  Mugho, *Pinus* sp., 140, 740, 813
  Scotch, *Pinus silvestris,* 1516, 1532, 1755, 1767
  Shortleaf, *Pinus echinata,* 225
  Slash, *Pinus elliottii,* 1776
  Southern, *Pinus* sp., 140, 740, 813
  Stone, *Pinus pinea,* 140
  Sugar, *Pinus lambertiana,* 1776
  White, *Pinus strobus,* 143
Pineapple, *Ananas comusus,* 767
Pinfish, *Lagodon rhomboides,* 54, 85, 1136, 1211, 1468, 1472, 1527, 1671, 1678, 1779
Pinna nobilis, 70, 442
Pintail, northern, *Anas acuta,* 202, 242, 267, 799, 1434
Pinus contorta, 143
Pinus echinata, 225
Pinus elliottii, 1776
Pinus lambertiana, 1776
Pinus nigra, 364
Pinus palustris, 1776
Pinus pinaster, 140
Pinus pinea, 140
Pinus sabiniana, 1584
Pinus silvestrus, 1516, 1532, 1755, 1767
Pinus sp., 140, 740, 813
Pinus strobus, 143
Piper spp., 1093, 1516

Table 33.2 (continued) Common and Scientific Names of Plants and Animals Listed in the *Handbook of Chemical Risk Assessment series* (References are provided from common names to scientific names)

Pipilo erythrophthalmus, 1005
Pipilo spp., 1432
Pipistrellus pipestrellus, 1217, 1219, 1299
Piranga olivacea, 1005
Piranha
  Black River, *Pygocentrus nattereri,* 362
  Various, *Seerasalmus* sp., 1748
Pisaster brevispinus, 116
Pistia sp., 747
Pisum sativum, 774, 1165, 1530, 1532, 1621, 1774
Pisum sp., 426, 1093
Pitar morrhuana. see Pitar morrhuanus
Pitar morrhuanus, 53, 516, 629
Pituophis catenifer, 1430
Pituophis melanoleucas, 1670
Pitymys pinetorium. see Microtus pinetorum
Placopecten magellanicus, 6, 442, 516, 1520
Plaice
  American, *Hippoglossoides platessoides,* 341, 583
  Platichthys flesus, 7, 158, 230, 343, 344, 1269, 1285, 1377, 1523
  Pleuronectes platessa, 165, 263, 341, 343, 533, 611, 624, 674, 841, 1212, 1523, 1729, 1740, 1782
Planchonella oxyedra, 439
Planorbis corneus, 1781
Plant
  Copper-tolerant, *Becium homblei,* 132
  Floating, *Eichornia* sp., *Jussiaea* sp., *Pistia* sp., 747
  Heliotrope, *Echium* sp., *Heliotropium* sp., *Senecio,* 105
  Lupinosis , *Lupinus* spp., 169
  Marine flowering, *Posidonia oceanica,* 337
  Nickel hyperaccumlator
    Allysum spp., 434, 439, 456
    Geissosis prainosa, 439
    Homalium spp., 434, 439
    Hybanthus spp., 434, 439
    Pearsonia metallifera, 439
    Planchonella oxyedra, 439
    Psychotria douarrei, 439
    Sebertia acuminata, 439
  Perennial, *Rubus* spp., 130
  poisonous (fluoroacetate-containing)
    Acacia spp., 1426
    Dichapetalum, 1413
    Gastrolobium spp., 1413, 1426, 1429
    Oxylobium spp., 1429
  Vascular, *Cassiope* sp., 451
Platalea leucorodia, 1295
Platichthys stellatus, 571
Platichythys flesus, 7, 158, 230, 343, 344, 1269, 1285, 1523
Platycentropus radiatus, 1002
Platycephalus bassensis, 1367
Platycercus elegans, 1433
Platyhypnidium riparoides, 25
Platyhypnidium sp., 26
Platymonas subcordiformis, 255
Platynothrus peltifer, 138, 142
Plecostomus commersonii, 1173
Plecotus phyllotis, 990
Plectonema boryanum, 994, 995, 1002
Plegadis chihi, 353, 1659, 1667
Plethodon cinereus, 130, 250
Pleurobrachia pileus, 146
Pleurodeles walti, 1381
Pleuronectes americanus, 20, 22, 118, 159, 444, 514, 518, 531, 571, 832, 842, 1040, 1269, 1282, 1367, 1378, 1381, 1541, 1678
Pleuronectes ferruginea, 444
Pleuronectes flesus, 377, 1766
Pleuronectes platessa, 165, 263, 341, 343, 533, 611, 624, 674, 841, 1212, 1523, 1729, 1740, 1782
Pleuronectes sp., 533
Pleuronectes stellatus, Starry flounder, 768
Pleuronectes vetulus, 768, 1279, 1378, 1381, 1523
Pleurozium schreberi, 440
Plover, black-bellied, *Pluvialis squatorola,* 1660
Plum, *Prunus* spp., 904
Plumaria elegans, 1540
Plutella xylostella, 1093, 1103
Pluvialis squatorola, 1660
Poa annua, 1504
Poa pratensis, 1168, 1531
Poa spp., 130, 644
Pochard, *Aythya ferina,* 122, 202, 219
Podiceps cristata, 354
Podiceps grisigena, 1149
Podiceps nigricollis, 1666, 1667
Podilymbus podiceps, 1466
Podoclavella moluccensis, 53
Podophthalmus vigil, 72
Poecilia mexicana, 1173
Poecilia reticulata, 18, 21, 63, 159, 256, 377, 421, 463, 531, 578, 667, 674, 812, 888, 965, 966, 1040, 1045, 1073, 1173, 1213, 1302, 1306, 1375, 1382, 1468
Poeciliopsis spp., 1373, 1382, 1384
Pogona barbatus, 1430
Pogonomyrmex spp., 1427
Polecat, *Mustela putorius furo,* 245, 683, 693
Polioptila caerulea, 1005
Pollack, walleye, *Theragra chalcogrammus,* 843, 1465
Polyborus plancus, 362
Polycarpa peduculata, 644
Polychaetes
  Abarenicola pacifica, 1377
  Capitella capitata, 63, 368, 662
  Glycera dibranchiata, 909
  Lycastis ouanaryensis, 116
  Marphysa sanguinea, 517
  Neanthes arenaceodentata, 63, 64, 65, 73, 262, 662, 673, 1209, 1373, 1376, 1779
  Ophryotrocha diadema, 1207
  Sabella pavonina, 529
Polygonum sp., 1170
Polyodon spathula, 839, 1269, 1284
Polypus bimaculatus, 1579
Pomacea glauca, 1215
Pomacea lineata, 1215
Pomacea paludosa, 129, 149, 166
Pomacea spp., 1205
Pomatomus saltatrix, 8, 332, 435
Pomoxis annularis, 11, 1001, 1212, 1280
Pomoxis nigromaculatus, 632, 930, 1002, 1004, 1280
Pompano, Florida, *Trachinotus carolinus,* 160
Pontoporeia affinis, 22, 26
Pontoporeia hoyi, 836, 1304, 1378

Table 33.2 (continued) Common and Scientific Names of Plants and Animals Listed in the *Handbook of Chemical Risk Assessment series* (References are provided from common names to scientific names)

Pontoporia blainvillei, 123
Popillia japonica, 837
Poplar, tulip, *Liriodendron tulipifera,* 1003
Poppy, *Papaver orientale,* 110
Populus grandidentata, 1734
Populus sp., 12
Populus tremula, 436, 1767
Populus tremuloides, 1734
Porcellio scaber, 174, 177, 217, 253, 621, 628, 652, 654
Porcellio sp., 1668
Porcupine
  Indian crested, *Hystrix indica,* 1417
  North American, *Erethizon dorsatum,* 13, 125, 1443
Porgy
  Black, *Sparus macrocephalus,* 324
  Jolthead, *Calamus bajonado,* 117
  Pachymetopan grande, 74
Porgy, *Pachymetopan grande,* 74
Porichthys notatus, 930
Porites asteroides, 835
Porphyra sp., 1735
Porphyra umbilicalis, 441
Porphyrula martinica, 1463
Porpoise
  Dall's, *Phocoenoides dalli,* 641, 854, 1276
  Finless, *Neophocaena phocaenoides,* 1272
  Harbor, *Phocoena phocoena,* 123, 336, 359, 641, 693, 1032, 1262, 1263, 1273, 1275, 1319, 1363, 1368, 1467, 1661
  Various, *Phocoena sinus,* 450
Porthetria dispar, 228, 435, 441, 644
Portunus pelagicus, 671
Porzana carolina, 240
Posidonia oceanica, 337
Possum, brush-tailed, *Trichosurus vulpecula,* 938, 1414, 1417, 1420, 1428, 1448
Posthodiplostomum minimum, 146
Potamocorbula amurensis, 516, 527
Potamogeton carinatus, 749
Potamogeton crispus, 2, 749
Potamogeton foliosus, 255
Potamogeton pectinatus, 129, 781, 811
Potamogeton perfoliatus, 776, 779
Potamogeton pusillus, 1170
Potamogeton richardsoni, 10
Potamogeton spp., 96, 111, 225, 451, 741, 747, 780, 1517, 1577
Potamogeton tricarinatus, 749
Potato
  Common, *Solanum tuberosum,* 225, 1204, 1667
  Sweet, *Ipomoea batatas,* 903, 1734
Praunus flexuosus, 661
Prawn
  Deep sea, *Pandalus borealis,* 1514, 1521
  Freshwater, *Macrobrachium* spp., 152, 452
  Pandalus, 7
  Penaeus sp., 364
  White, *Penaeus indicus,* 165, 368
Prionace glauca, 435, 633, 1515
Pristina sp., 21
Procambarus acutus, 672, 805
Procambarus blandingi, 1135
Procambarus clarki, 152, 163, 836, 887, 971, 1099, 1106, 1109, 1171, 1537, 1656
Procambarus spp., 1106, 1141, 1464
Procentrum micans, 656
Procladius sp., 1003
Procyon lotor, 126, 247, 286, 336, 359, 450, 835, 854, 1438, 1447, 1467, 1670
Progne subis subis, 1096
Pronghorn, *Antilocapra americana,* 59, 1673, 1733
Prorocentrum mariae-lehouriae, 524
Prosopium cylindraceum, 343
Prosopium williamsoni, 914, 1026, 1738
Prosthemadera novae-seelandiae, 1433
Prototheca zopfi, 927
Protozoans
  Blepharisma undulans, 1102
  Chilomonas paramecium, 68
  Colpoda, 141, 1102
  Costia sp., 96
  Cristigera sp., 263, 657
  Entosiphon sulcatum, 257
  Fabrea salina, 524
  Ichthyopthirius sp., 96
  Oikomonas termo, 1102
  Paramecium spp., 425
  Spirostomum ambiguum, 525
  Tetrahymena pyriformis, 145, 1683
  Trichodina sp., 96
  Uronema sp., 257
  Vorticella convallaria, 657
Prunella modularis, 1075
Prunus amygdalus, 918
Prunus armenaica, 904
Prunus avium, 337
Prunus laurocerasus, 904
Prunus persica, 904
Prunus serotina, 216, 225
Prunus spp., 904, 905
Psectocladius sp., 1172
Psephotus haematonotus, 1436
Psettichthys melanostictus, 1376, 1377
Pseudacris triseriata, 1174
Pseudagrion spp., 887
Pseudemys floridana peninsularis, 1741
Pseudemys scripta, 1741
Pseudis paradoxa, 1205
Pseudocrenilabrus philander, 118
Pseudodiaptomus coronatus, 152
Pseudomonas cepacia, 1421
Pseudomonas fluorescens, 915
Pseudomonas spp., 657, 803, 963, 1200, 1421, 1651
Pseudomys australis, 1447
Pseudomys hermannsburgensis, 1447
Pseudomys higginsi, 1447
Pseudomys nanus, 1447
Pseudoplatysoma coruscans, 362
Pseudopleuronectes americanus. see Pleuronectes americanus
Pseudopterogorgia acerosa, 835
Pseudorasbora parva, 971
Pseudotsuga menziesii, 143, 1005, 1006
Psithyrus bohemicus, 112
Psorophora columbiae, 1004, 1109
Psychotria douarrei, 439
Psylla pyricola, 1102
Ptarmigan
  Common, *Lagopus mutus,* 333

Table 33.2 (continued) Common and Scientific Names of Plants and Animals Listed in the *Handbook of Chemical Risk Assessment series* (References are provided from common names to scientific names)

Willow, *Lagopus lagopus,* 12, 103, 120, 271, 333, 446, 1671, 1768
Pterodroma macroptera, 1418
Pteronarcella badia, 886, 887
Pteronarcys californica, 162, 528, 859, 887, 965, 1172, 1537
Pteronarcys dorsata, 1109, 1538
Pteronarcys sp., 1468
Pteropus sp., 247
Ptychocheilus lucius, 1588, 1676, 1677
Ptychocheilus oregonensis, 332
Puccinia graminis, 740
Puffer, northern, *Sphoeroides maculatus,* 1073
Puffin
Atlantic, *Fratercula arctica,* 54, 354, 847
Puffinus pacificus, 1659
Puffinus sp., 121
Pumpkinseed, *Lepomis gibbosus,* 56, 342, 462
Pungitius pungitius, Ninespine stickleback, 888
Puntius conchonius, 260
Puntius gonionotus, 1173
Pupfish, desert, *Cyprinodon macularis,* 1110
Pusa hispida, 1274, 1276
Pygocentrus nattereri, 362
Pygoscelis adeliae, 8, 120, 347, 362, 446, 518, 850, 1295
Pygoscelis antarctica, 120, 347, 446, 518
Pygoscelis papua, 120, 267, 347, 446, 518, 1295
Pylodictis olivaris, 839, 1267
Pyrethrum flower, *Chrysanthemum cinariaefolium,* 1089
Pyrrhula, Bullfinch, 636
Pyrus communis, 904, 1584, 1667
Pyrus spp., 918, 1092
Pytiscidae, 26

Q

Quahog or quahaug
False, *Pitar morrhuanus,* 53, 516, 629
Northern, *Mercenaria mercenaria,* 52, 70, 148, 442, 458, 526, 654, 659, 1378, 1468, 1520
Ocean, *Arctica islandica,* 112, 441
Southern, *Mercenaria campechiensis,* 669
Quail
California, *Callipepla californica,* 861, 862, 1431, 1435, 1469, 1542, 1543
Common, *Coturnix coturnix,* 267, 890, 968, 1135, 1542, 1787
Coturnix, *Coturnix risoria,* 464, 891
Japanese, *Coturnix japonica,* 20, 269, 371, 379, 464, 465, 583, 584, 616, 677, 689, 786, 806, 807, 808, 861, 862, 890, 891, 934, 967, 977, 1074, 1076, 1114, 1115, 1174, 1176, 1214, 1222, 1307, 1309, 1435
Quelea, *Quelea quelea,* 806
Quelea quelea, 806
Quercus prinus, 1005
Quercus rubra, 440, 653, 1005
Quercus spp., 621, 685, 1005, 1006, 1288
Quercus velutina, 1005
Quiscalus mexicanus, 1466
Quiscalus quiscula, 370, 806, 807, 846, 861, 890, 1076
Quokka, *Setonix brachyurus,* 1448
Quoll
Eastern native, *Dasyurus viverrinus,* 1424, 1438, 1443
Tiger, *Dasyurus maculatus,* 1421, 1438

R

Rabbit
European, *Oryctolagus cuniculus,* 788, 890, 895, 937, 1014, 1219, 1414, 1417, 1446
Lepus, 590
Oryctolagus sp., 66, 126, 171, 421, 450, 471, 534, 537, 683, 757, 867, 912, 943, 1010, 1050, 1072, 1078, 1082, 1120, 1181, 1438, 1550, 1597, 1636, 1760, 1764, 1769
Sylvilagus sp., 1550
Rabbitfish, *Siganus oramin,* 633
Raccoon, *Procyon lotor,* 126, 247, 286, 336, 359, 450, 835, 854, 1438, 1447, 1467, 1670
Racerunner, six-lined, *Cnemidophorus sexlineatus,* 1025
Radish, *Raphanus sativus, Raphanus* spp., 25, 250, 434, 989
Ragwort, tansy, *Senecio jacobaea,* 1632
Rail
Clapper, *Rallus longirostris,* 850, 1463
Sora, *Porzana carolina,* 240
Rainbow trout, *Oncorhynchus mykiss,* 11, 18, 19, 21, 26, 56, 62, 64, 65, 69, 100, 156, 258, 367, 375, 384, 444, 462, 480, 482, 506, 530, 533, 576, 579, 611, 647, 665, 688, 747, 750, 768, 783, 805, 832–833, 860, 888, 905, 929, 964, 965, 1001, 1026, 1030, 1040, 1043, 1054, 1073, 1100, 1111, 1112, 1137, 1173, 1202, 1209, 1221, 1269, 1305, 1353, 1375, 1380, 1428, 1461, 1468, 1519, 1533, 1536, 1539, 1581, 1586, 1589, 1602, 1620, 1623, 1629, 1641, 1677, 1738, 1779
Raja clavata, 533
Raja erinacea, 321, 1040
Raja radiata, 341
Raja sp., 314, 1514, 1523
Rallus longirostris, 850, 1150, 1463
Ramalina fraxinea, 1763
Rana catesbeiana, 11, 55, 218, 231, 261, 264, 345, 778, 886, 892, 966, 1047, 1372, 1373, 1429, 1523, 1665, 1670
Rana clamitans, 119, 130, 218, 250, 261, 264, 445, 1096
Rana cyanophlyctis, 367
Rana dalmutina, 668
Rana esculenta, 1099
Rana heckscheri, 367
Rana japonica, 745
Rana luteiventris, 18, 668
Rana pipiens, 160, 264, 346, 367, 375, 532, 784, 1105, 1113, 1174, 1306, 1373, 1385, 1429, 1590, 1770, 1785
Rana pipiens pipiens, 1113
Rana ridibunda, 18, 25

Table 33.2 (continued) Common and Scientific Names of Plants and Animals Listed in the *Handbook of Chemical Risk Assessment series* (References are provided from common names to scientific names)

Rana sphenocephala, 1468
Rana spp., 231, 507, 676, 1524
Rana sylvatica, 783, 1586, 1588
Rana temporaria, 119, 346, 579, 1170
Rana utricularia, 264, 1096
Rangia cuneata, 71, 805, 812, 1375, 1376, 1377, 1737
Rangifer tarandus, 4, 32, 126, 380, 437, 450, 451, 538, 1072, 1082, 1085, 1758, 1761, 1762, 1791
Rangifer tarandus fennica, 15, 247
Rangifer tarandus granti, 1759, 1761
Raphanus sativus, 25, 250
Raphanus spp., 434, 746, 989
Rasbora daniconus neilgeriensis, 1211
Rasbora heteromorpha, 579, 751
Raspberry
  Various, *Rubus,* 130
Rat
  African giant, *Cricetomys gambianus,* 941
  Alexandrine, *Rattus alexandricus,* 1447
  Black, *Rattus rattus,* 964, 968, 975, 978, 1441, 1447, 1749, 1751
  Bush, *Rattus fuscipes,* 1447
  Canefield, *Rattus sordidus,* 1447
  Cotton, *Sigmodon hispidus,* 59, 78, 280, 1095, 1448, 1551, 1733, 1745, 1794
  Desert wood, *Neotoma lepida,* 1439
  Grassland melomys, *Melomys burtoni,* 1444
  Heermann's kangaroo, *Dipodomys heermanni,* 1439
  Kangaroo, *Dipodomys* spp., 1417, 1443, 1449
  Laboratory white, *Rattus* spp., 19, 23, 66, 127, 169, 172, 177, 208, 278, 373, 450, 454, 473, 483, 484, 535, 537, 587, 588, 589, 590, 613, 642, 651, 684, 693, 758, 774, 787, 806, 808, 812, 867, 870, 944, 984, 1011, 1014, 1055, 1068, 1078, 1083, 1085, 1114, 1120, 1135, 1136, 1142, 1182, 1215, 1220, 1302, 1308, 1313, 1317, 1322, 1387, 1388, 1426, 1441, 1469, 1474, 1476, 1546, 1550, 1624, 1636, 1640, 1793
  Marsh rice, *Oryzomus palustris,* 1793
  Morro Bay kangaroo, *Dipodomys heermanni morroensis,* 1450
  Norway, *Rattus norvegicus,* 247, 382, 768, 890, 1363, 1418, 1441, 1447, 1670, 1793
  Roof (black), *Rattus rattus,* 964, 968, 975, 978, 1441, 1447, 1749, 1751
  Swamp, *Rattus lutreolus,* 1447
  Thick-tailed, *Zyzomys argurus,* 1449
  Tunney's, *Rattus tunneyi,* 1447
  Water-, *Hydromys chrysogaster,* 1443
  White-throated wood, *Neotoma albigula,* 1446
  Wood, *Neotoma intermedia,* 1446
Ratsbane, *Dichapetalum toxicarium,* 1413
Rattus alexandricus, 1447
Rattus fuscipes, 1447
Rattus lutreolus, 1447
Rattus norvegicus, 247, 382, 768, 890, 1363, 1418, 1441, 1447, 1670, 1793
Rattus rattus, 964, 968, 975, 978, 1441, 1447, 1749, 1751
Rattus sordidus, 1447
Rattus spp., 19, 23, 66, 127, 172, 177, 208, 278, 373, 382, 450, 454, 473, 483, 484, 535, 537, 587, 588, 589, 590, 613, 642, 651, 684, 693, 758, 774, 787, 806, 808, 812, 867, 870, 944, 984, 1011, 1014, 1055, 1068, 1078, 1083, 1085, 1114, 1120, 1135, 1136, 1142, 1182, 1215, 1220, 1302, 1308, 1313, 1317, 1322, 1387, 1388, 1426, 1441, 1469, 1474, 1476, 1546, 1550, 1624, 1636, 1640, 1793
Rattus tunneyi, 1447
Raven
  Australian, *Corvus coronoides,* 1432, 1435
  Common, *Corvus corax,* 234, 446, 625
  Little, *Corvus mellori,* 1435
Ray
  Electric, *Torpedo* sp., 1099
  Raja radiata, 341
  Thornback, *Raja clavata,* 533
  Various, *Raja* sp., 341, 1514, 1523
Razorbill, *Alca torda,* 1289, 1308
Recurvirostra americana, 1515, 1580, 1657, 1666
Redhead, *Aythya americana,* 120, 242, 446
Redhorse, black, *Moxostoma duquesnei,* 230
Redshank
  Common, *Tringa totanus,* 122
  Icelandic, *Tringa totanus robusta,* 1525
Regulus satrapa, 1006
Reindeer, *Rangifer tarandus,* 4, 32, 126, 380, 437, 450, 451, 1072, 1085, 1761, 1762, 1791
Reinhardtius hippoglossoides, 341
Reithrodontomys humulis, 1794
Reithrodontomys megalotis, 1439
Reithrodontomys raviventris, 1450
Renibacterium salmoninarum, 1674
Reticulitermes flavipes, 1626
Rhacomitrium sp., 451
Rhepoxynius abronius, 1627
Rheumobates spp., 442
Rhinichthys osculus, 531
Rhinobatis lentiginosus, 118
Rhinoceros, black, *Diceros bicornis,* 1673
Rhinogobius flumineus, 842
Rhinolophus ferrumequinum, 1299
Rhinomugil corsula, 1213
Rhizobium sp., 885
Rhizoctonia solani, 774, 921
Rhizophora mangle, 451
Rhizopus sp., 1169
Rhizosolenia sp., 656, 689
Rhodotorula sp., 457
Rhyncostegium riparioides, 145, 927
Rice
  Domestic, *Oryza sativa,* 225, 337, 382, 439, 453, 1204, 1530, 1532, 1667
  Wild, *Zizania aquatica,* 255
Richmondena cardinalis, 1096
Ricinus communis, 746, 922
Rissa tridactyla, 9, 333, 354, 624, 637, 1660
Rithropanopeus harrisii, 578, 614, 662, 689, 999, 1171, 1468
Rivulus, mangrove (fish), *Rivulus marmoratus,* 159
Rivulus marmoratus, 159
Roach (fish), *Rutilus rutilus,* 229, 1762
Robin

Table 33.2 (continued) Common and Scientific Names of Plants and Animals Listed in the *Handbook of Chemical Risk Assessment series* (References are provided from common names to scientific names)

American, *Turdus migratorius,* 13, 234, 241, 810, 846, 1542, 1623
European, *Erithacus rubecula,* 1075
Rosella, crimson, *Platycercus elegans,* 1433
Rostrhamus sociabilis, 166, 1214, 1215
Rotifer, *Asplanchna, Brachionus,* 1004
Rubus frondosus, 644
Rubus sp, 130
Rudd, *Scardinus erythrophthalmus,* 69
Ruppia maritima, 1517, 1578, 1591
Ruppia spp., 780, 927
Rutilus rutilus, 229, 1762
Rye, perennial, *Lolium perenne,* 521, 895, 1168, 1621, 1757, 1763
Rye, *Secale cereale,* 57, 1667
Rynchops niger, 355, 850, 1659, 1660

S

Sabella pavonina, 529
Saccharomyces cerevisiae, 746, 864, 1008, 1216
Saccharomyces spp., 1204
Saccharum officinarum, 746, 767, 1097, 1167
Saccharum sp., 1093
Saccobranchus fossilis, 69, 159, 164, 805, 860
Saccostrea commercialis, 115
Saccostrea cucullata, 115, 612, 629
Saccostrea sp., 516
Saddleback, *Philesturnus carunculatus,* 1433
Sagebrush, big, *Artemisia tridentata,* 56, 1577
Sagitta elegans, 1579
Saguinus fuscicollis, 961
Saimiri sciurea, 761
Saimiri spp., 1794
Salamander
Eastern red-backed, *Plethodon cinereus,* 130, 250
Jefferson, *Ambystoma jeffersonianum,* 135, 261, 1586, 1588
Marbled, *Ambystoma opacum,* 160, 264, 368, 463, 668, 1538
Necturus, 1785
Northwestern, *Ambystoma gracile,* 17, 22, 25
Spotted, *Ambystoma maculatum,* 131, 1373, 1586, 1588
Tiger, *Ambystoma tigrinum,* 1355
Two-lined, *Eurycea bislineata,* 160
Salix spp., 4, 1765, 1767
Salmo clarki. see Oncorhynchus clarki
Salmo gairdneri. see Oncorhynchus mykiss
Salmon
Atlantic, *Salmo salar,* 21, 135, 159, 567, 571, 607, 624, 633, 667, 674, 688, 842, 914, 930, 970, 1001, 1113, 1139, 1141, 1212, 1270, 1375, 1519, 1649, 1781
Chinook, *Oncorhynchus tshawytscha,* 18, 68, 158, 531, 567, 666, 675, 751, 860, 929, 1030, 1148, 1212, 1269, 1282, 1284, 1302, 1322, 1539, 1590, 1630, 1664, 1674, 1677, 1770, 1780
Chum, *Oncorhynchus keta,* 530, 567, 842, 1539
Coho, *Oncorhynchus kisutch,* 18, 68, 155, 254, 462, 530, 665, 805, 860, 929, 1001, 1030, 1040, 1044, 1136, 1148, 1202, 1211, 1212, 1280, 1374, 1381, 1468, 1519, 1589, 1627, 1629, 1656, 1664, 1677, 1739, 1779
Kokanee (*see* Sockeye)
Pink, *Oncorhynchus gorbuscha,* 567, 1374, 1541
Sockeye, *Oncorhynchus nerka,* 666, 1210, 1221, 1579, 1586, 1590, 1620, 1738
Salmonella pullorum-gallinarum, 534
Salmonella sp., 28, 617, 754, 759, 1473
Salmonella typhimurium, 67, 746, 815, 864, 1008, 1039, 1216, 1370, 1385, 1515, 1572
Salmo salar, 21, 135, 159, 567, 571, 607, 624, 633, 667, 674, 688, 914, 930, 970, 1001, 1113, 1139, 1141, 1212, 1270, 1375, 1519, 1649, 1781
Salmo spp., 343
Salmo trutta, 11, 21, 102, 159, 259, 531, 667, 673, 688, 751, 805, 860, 914, 931, 1026, 1173, 1202, 1270, 1363, 1372, 1519, 1738, 1758, 1766, 1781, 1784
Salmo trutta fario, 1031, 1270
Salpa fusiformes, 1579
Saltbush, *Atriplex* spp., 1652, 1667
Salvelinus alpinus, 73, 218, 230, 1213, 1284, 1465
Salvelinus fontinalis, 18, 19, 20, 21, 65, 159, 259, 343, 367, 375, 384, 463, 634, 667, 783, 860, 931, 966, 970, 971, 972, 1001, 1040, 1046, 1139, 1141, 1213, 1471, 1472, 1540, 1628, 1677, 1738
Salvelinus namaycush, 56, 65, 159, 259, 343, 445, 463, 571, 578, 805, 842, 856, 888, 966, 1026, 1027, 1029, 1030, 1040, 1046, 1054, 1147, 1202, 1270, 1282, 1462, 1519, 1577, 1682, 1738
Sambucus spp., 905
Sanddab, speckled, *Citharichthys stigmaeus,* 63, 74, 578
Sand dollar, *Dendraster excentricus,* 663, 689
Sanderling, *Calidris alba,* 57, 121
Sandpiper
Calidris pusilus, 57
Erolia spp., 800
Saprolegnia parasitica, 929
Saprolegnia sp., 456
Sarcophilus harrisii, 1424, 1448
Sarcorhampus papa, 202
Sargassum fluvitans, 1519
Sargassum sp., 441, 452, 1519
Sauger, *Stizostedion canadense,* 11, 839
Scallop
Antarctic, *Adamussium colbecki,* 112
Bay, *Argopecten irradians,* 146, 161, 368, 525, 658
Mizuhopecten yessoensis, 148
Pacific, *Chlamys ferrei nipponensis,* 323
Sea, *Placopecten magellanicus,* 6, 442, 516, 1520
Scaphirhynchus platyrynchus, 832, 839
Scardinus erythrophthalmus, 69
Scarus gyttatus, 633
Scaup
Greater, *Aythya marila,* 348, 446, 518, 1291, 1666
Lesser, *Aythya affinis,* 8, 235, 242
Sceloglaux albifascies;, 1433
Scenedesmus acutiformis, 145, 458
Scenedesmus costatum, 1207
Scenedesmus obliquus, 1533, 1536, 1778

Table 33.2 (continued) Common and Scientific Names of Plants and Animals Listed in the *Handbook of Chemical Risk Assessment series* (References are provided from common names to scientific names)

Scenedesmus quadricauda, 573, 656, 777, 856, 857, 927
Scenedesmus spp., 524, 1302
Scenedesmus subspicatus, 145
Schistoma gregaria, 894
Schistosoma mansoni, 750
Schroederella schroederi, 655, 657, 689
Sciaenops ocellatus, 96, 118, 159, 445, 838
Scinax nasica, 1174
Sciurus carolinensis, 15, 127, 360, 450, 854
Sciurus hudsonicus. see Tamiasciurus hudsonicus
Sclerotium rolfsii, 774
Scolopax minor, 1055, 1149
Scolopax rusticola, 1766, 1768
Scomberomorus cavalla, 634
Scomberomorus maculatus, 119
Scomber scombrus, 634
Scophthalmus aquosus, 518, 1367, 1523
Scophthalmus maeoticus, 1781
Scophthalmus maximus, 675
Scorpaena guttata, 20
Scorpaena porcus, 1781
Scorpaenichthys marmoratus, 654, 667
Scoter, surf, *Melanitta perspicillata,* 121, 136, 348, 518, 1666
Screech-owl, eastern, *Otus asio,* 806, 837, 846, 934, 1682
Scripsiella faeroense, 145, 375
Scrobicularia plana, 6, 442, 451, 507, 516, 527
Scud, *Gammarus, Hyalella,* 528
Scud, *Gammarus pseudolimnaeus,* 142, 151, 257, 367, 528, 782, 927, 998, 1108, 1209, 1304, 1373, 1469, 1533, 1537
Sculpin
  Comephorus dybowski, 837, 1465
  Fourhorn, *Myoxocephalus quadricornis,* 842
  Mottled, *Cottus bairdi,* 529, 782
  Shorthorn, *Myoxocephalus scorpius,* 8
  Slimy, *Cottus cognatus,* 341
  Tidepool, *Oligocottus maculosus,* 530, 533
Scup, *Stenotomus chrysops,* 445, 581, 1303, 1307
Scyliorhinus caniculus, 344, 634
Scyliorhinus sp., 667
Scylla serrata, 1368
Sea-eagle, white-tailed, *Haliaeetus albicilla,* 379, 635, 1288, 1292
Seal
  Australian fur, *Arctocephalus pusillus,* 355
  Baikal, *Phoca sibirica,* 10, 25, 123, 335, 359, 641, 853 1262, 1273, 1467
  Bearded, *Erignathus barbatus,* 249, 1660, 1743
  Crabeater, *Lobodon carcinophagus,* 122, 355, 520
  Gray, *Halichoerus grypus,* 122, 335, 358, 450, 640, 852, 1032 1263–1264, 1272, 1467, 1661, 1733, 1744
  Harbor, *Phoca vitulina, Pusa vitulina,* 123, 222, 358, 359, 450, 572, 853, 1032, 1274, 1319, 1467, 1526, 1579, 1661
  Harp, *Pagophilus groenlandica, Phoca groenlandica,* 54, 335, 358, 373, 382, 389, 1032
  Hooded, *Cystophora cristata,* 639, 1363
  Leopard, *Hydrurga leptonyx,* 122, 355, 519
  Northern fur, *Callorhinus ursinus,* 9, 335
  Ringed, 362
    Phoca hispida, 25, 245, 250, 336, 358, 380, 450, 625, 641, 834, 1032, 1263, 1264, 1273, 1464, 1467, 1660, 1661
    Phoca hispida ladogensis, 249
    Pusa hispida, 1274, 1276
  Saimaa ringed, *Phoca hispida saimensis,* 853, 1661
  Spotted, *Phoca largha,* 1744
  Weddell, *Leptonychotes weddell,* 103, 122, 355, 362
Sea lion
  California, *Zalophus californianus,* 10, 222, 335, 361, 514, 520, 1661, 1679
  Northern, *Eumetopias jubata,* 627
  Stellar, *Eumetopias jubatus,* 1263, 1272
Seaskaters (oceanic insects), *Halobates* spp., *Rheumobates* spp., 442
Sea star (Asteroidea), *Pisaster brevispinus,* 116
Seatrout, spotted, *Cynoscion nebulosus,* 117, 445, 1465
Sea urchin
  Anthocidaris crassispina, 73, 263, 663, 1587
  Arbacia lixula, 226, 529
  Hemicentrotus sp., 73
  Lytechinus sp., 363, 451
  Paracentrotus lividus, 153, 228, 1208
  Strongylocentrotus, 153
  Tripneustes esculentus, 443
Sebertia acuminata, 439
Secale cereale, 57, 1667
Sedges, *Carex* spp., 1767
Seerasalmus sp., 1748
Seiurus aurocapillus, 1005
Selenastrum capricornutum, 256, 257, 657, 811, 995, 1197, 1207, 1302, 1372, 1536, 1783
Semibalanus balanoides, 152
Semotilus atromaculatus, 1528
Semotilus margarita, 159
Senecio jacobaea, 1632
Senecio sp., 105
Senecio vulgaris, 768
Sepia officinalis, 115, 149, 515
Sepioteuthis australis, 1515
Sergestes lucens, 1521
Serinus canarius, 935, 1215
Seriola grandis, 344, 388
Seriola quinqueradiata, 965
Serow, *Capricornus crispus,* 336, 356
Sesarma cinereum, 782, 1468
Sesarma heamatocheir, 64
Sesarma reticulatum, 999
Setaria faberii, 644
Setaria sp., 130
Setonix brachyurus, 1448
Shad
  Dorosoma spp., 838
  Gizzard, *Dorosoma cepedianum,* 642, 778, 1379, 1463, 1465
  Konosirus punctatis, 841
  Threadfin, *Dorosoma petenense,* 642, 1464, 1663
  Twaite, *Alosa fallax,* 1466
Shag
  Common, *Phalacrocorax aristotelis,* 1287
  Imperial, *Phalacrocorax atriceps,* 120, 446, 518

Table 33.2 (continued) Common and Scientific Names of Plants and Animals Listed in the *Handbook of Chemical Risk Assessment series* (References are provided from common names to scientific names)

Shark
  Blue, *Prionace glauca,* 435, 633, 1515
  Shortfin mako, *Isurus oxyrinchus,* 633, 1523
  Tope, *Galeorhinus galeus,* 435
  Various
    Carcharhinus spp, 344
    Mustelus, 53, 101, 518, 1445, 1514, 1515, 1792
    Sphyrna spp., 344
  White, *Carcharodon carcharius,* 837
  Whitetip, *Carcharhinus longimanus,* 117, 443, 1522
Shearwater
  Wedge-tailed, *Puffinus pacificus,* 1659
Sheep
  Bighorn, *Ovis canadensis,* 59, 1649, 1673
  California bighorn, *Ovis canadensis californiana,* 1669
  Domestic, *Ovis aries, Ovis* sp., 93, 126, 171, 177, 234, 245, 277, 336, 358, 449, 454, 572, 590, 641, 650, 684, 693, 757, 774, 787, 802 867, 905, 944, 969, 975, 984, 1011, 1014, 1078, 1082, 1120, 1182, 1219, 1414, 1446, 1469, 1550, 1635, 1640, 1746, 1764, 1793
Shelduck, *Tadorna tadorna,* 1731
Shell
  Ivory, *Buccinum striatissimum,* 1514
  Pen, *Pinna nobilis,* 70, 442
  Spindle, *Hemifusus* spp., 1520, 1552
Shiner
  Blacktail, *Notropis venustus,* 1464
  Common, *Notropis cornutus,* 1213
  Golden, *Notemigonus crysoleucas,* 62, 68, 632, 1096, 1303, 1740
  Spottail, *Notropis hudsonius,* 1030, 1147, 1538
  Willow, *Gnathopodon caerulescens,* 583, 889
Shoveler, northern, *Anas clypeata,* 862, 1787
Shrew
  Common, *Sorex araneus,* 15, 32, 173, 248, 436, 450, 643, 1515, 1671, 1747, 1767
  Least, *Cryptotis parva,* 448
  Long-tailed, *Sorex minutus,* 450
  Masked, *Sorex cinereus,* 450
  Short-tailed, *Blarina brevicauda,* 13, 242, 248, 250, 336, 365, 645, 807, 1788
Shrike, loggerhead, *Lanius ludovicianus,* 848, 961, 1149, 1292
Shrimp
  Aesop, *Pandalus montagui,* 7, 228, 443, 613, 631, 671, 672
  Brine, *Artemia salina, Artemia* sp., 660, 997, 1535, 1707, 1778
  Brown, *Penaeus aztecus,* 6, 54, 116, 443, 642, 750, 784, 858, 886, 966, 1374, 1377, 1521, 1527, 1627, 1677
  Clam, *Eulimnadia* spp., 998
  Freshwater, *Paratya australiensis,* 105, 152, 782
  Glass, *Palaemonetes kadiakensis,* 1468
  Grass
    Palaemonetes pugio, 7, 27, 72, 228, 528, 532, 578, 582, 631, 785, 887, 999, 1106, 1109, 1208, 1283, 1303, 1373, 1468, 1534
    Palaemonetes vulgaris, 17, 1074, 1141
  Korean, *Palaemon macrodactylus,* 858, 1468
  Lysmata, 1534, 1535, 1658, 1684
  Mysid, *Mysidopsis* spp., 22, 65
  Palaemonetes *vulgaris,* 17, 1074, 1141
  Pandalus jordant, 115
  Penaeopsis joyneri, 971
  Pink, *Penaeus duorarum,* 785, 858, 1106, 1109, 1376, 1377, 1468, 1629
  Sand
    Crangon allmanni, 442
    Crangon crangon, 17, 150, 262, 578, 1521
    Crangon *septemspinosa,* 857, 1107
  Sergestes lucens, 1521
  Tadpole, *Triops longicaudatus,* 1000
  White, *Penaeus setiferus,* 377, 642, 894, 1521
Sialia sialis, 846, 1048, 1786, 1788
Sialia sp., 1171
Sicklepod, *Cassia* spp., 224, 253
Siderastrea siderea, 835
Siganus oramin, 633
Sigara sp., 1172
Sigmodon hispidus, 59, 78, 280, 1095, 1448, 1551, 1733, 1745, 1794
Silkworm, *Bombyx mori,* 1776, 1777
Sillago bassensis, 1536
Silverside
  Atlantic, *Menidia menidia,* 63, 155, 462, 529, 888, 889, 1111, 1678
  Inland, *Menidia beryllina,* 888, 889, 1111, 1200, 1222
  Tidewater, *Menidia peninsula,* 155, 462, 665, 888, 889, 1111
Simocephalus serrulatus, 18, 859, 965, 1171, 1538
Simocephalus sp., 1468
Simocephalus vetulus, 887, 1208
Simulium sp., 228
Simulium vittatum, 995, 996
Siphamia cephalotes, 644
Siphlonurus lacustris, 1766
Siphlonurus sp., 893
Siphonaptera, 892
Sitophilus granarius, 923
Sitta carolinensis, 1432
Skate, little, *Raja erinacea,* 321, 1040
Skeletonema costatum, 22, 165, 261, 522, 524, 573, 574, 577, 579, 657, 689, 995, 1107, 1207, 1540, 1587
Skeletonema sp., 993
Skimmer, black, *Rynchops niger,* 355, 850, 1659, 1660
Skink, *Lerista puctorittata, Morethia boulengeri, Tiliqua,* 860
Skipper, mud, *Boleophthalmus dussumieri,* 62, 69, 1683
Skua
  Great, *Catharacta skua,* 348, 354, 1659
  Various, *Catharcta maccormicki,* 361
Skunk, striped, *Mephitis mephitis,* 823, 1440, 1444
Skwala sp., 997
Slippersnail, common Atlantic, *Crepidula fornicata,* 368, 375, 442, 526
Slug
  Grey field, *Deroceras reticulatum,* 621, 627
  Various, *Agriolimax reticulatus, Arion* spp., *Deroceras,* 976
Smelt
  Rainbow, *Osmerus mordax,* 341, 571, 1030, 1148, 1202
  Top, *Atherinops affinis,* 153, 164, 1110

Table 33.2 (continued) Common and Scientific Names of Plants and Animals Listed in the *Handbook of Chemical Risk Assessment series* (References are provided from common names to scientific names)

Sminthopsis crassicaudata, 1448
Sminthopsis macroura, 1448
Snail
  Amnicola sp., 63
  Ancylus fluviatilis, 657, 782
  Aplexa hypnorium, 749
  Apple, *Pomacea paludosa,* 129, 149, 166
  Arcularia gibbosula, 339
  Australorbis sp., 525
  Biomphalaria spp., 560, 578, 750
  Bulinus sp., 560, 750
  Campeloma decisum, 147
  Gillia altilis, 1208
  Helisoma sp., 747, 1041
  Helix spp., 253
  Juga plicifera, 458, 997
  Land, *Helix aspersa, Arianta arbustorum,* 112, 654, 812
  Lanistes carinatus, 887
  Lymnaea palustris, 257
  Lymnaea peregra, 256
  Melanoides sp., 452
  Mud
    Ilyanassa obsoleta, 114
    Nassarius obsoletus, 165, 459, 515, 527, 578, 581, 660, 1541
    Nassarius sp., 442
  Mudalia potosensis, 1374
  Neritina sp., 578
  Physa sp., 142, 750, 997, 1003, 1041, 1206, 1376, 1382, 1676
  Planorbis corneus, 1781
  Pond, *Cipangopaludina malleata,* 971
  Ram's horn, *Helisoma trivolvis,* 887, 1108
  Red, *Indoplanorbis exustus,* 971, 1044
  Thiara tuberculata, 149
Snake
  Cottonmouth, *Agkistrodon piscivorus,* 1466
  Elaphe obsoleta, 1741
  Garter, *Thamnophis* sp., 1094, 1385
  Gopher, *Pituophis catenifer, Pituophis melanoleucas,* 1430, 1670
  Northern water, *Nerodia sipedon,* 231, 844
  Water, *Natrix* sp., *Nerodia* sp., 1466, 1665, 1783
  Western ribbon, *Thamnophis proximos,* 1096
Snakehead, green, *Ophiocephalus punctatus,* 158
Snapper
  Lutianus fulviflamma, 633
  Mangrove, *Lutjanus griseus,* 1376
Solanum melongena, 1093
Solanum sp., 1092
Solanum tuberosum, 225, 1204, 1667
Sole
  Dover, *Microstomus pacificus,* 444
  English, *Pleuronectes vetulus* (formerly *Parophrys vetulus),* 768, 1279, 1378, 1423
  Flathead, *Hippoglossoides elassodon,* 1282
  Limanda sp., 7, 1515
  Sand, *Psettichthys melanostictus,* 1376, 1377
  Solea solea, 578
  Trinectes maculatus, 1141
Solea solea, 578
Solenopsis invicta, 855, 1102, 1104, 1133, 1152
Solenopsis richteri, 1133
Solenopsis sp., 1143, 1626
Somateria fischeri, 202, 233
Somateria mollissima, 9, 58, 121, 122, 233, 235, 514, 519, 1038, 1150, 1311, 1385
Sorbus aucuparia, 12, 1767
Sorex araneus, 15, 32, 173, 248, 436, 450, 643, 1515, 1671, 1747, 1767
Sorex cinereus, 450
Sorex minutus, 450
Sorex sp., 127, 1179
Sorghum, *Sorghum halepense, Sorghum* spp., *Sorghum vulgare,* 746, 767, 812, 905, 918
Sorghum almum, 905
Sorghum halepense, 812, 905
Sorghum spp., 746, 918
Sorghum vulgare, 767
Sousa chinensis, 1620, 1625
Sowbug, *Asellus* sp., 338, 1171
Soybean, *Glycine max,* 12, 140, 253, 421, 453, 456, 521, 767, 774, 909, 989, 1093, 1168, 1427, 1530, 1532, 1583
Sparrow
  Chipping, *Spizella passerina,* 27, 1005
  House, *Passer domesticus,* 235, 239, 371, 508, 806, 807, 890, 967, 1436, 1668
  Song, *Melospiza melodia,* 1005
  Tree, *Passer montanus,* 235, 1005
Sparrowhawk, European, *Accipiter nisus,* 333, 346, 349, 379, 1289
Spartina alterniflora, 330, 441, 777, 780, 782, 1534
Spartina spp., 52, 435, 452
Sparus aurata, 1382
Sparus macrocephalus, 324
Spermophilus beecheyi, 1431, 1448
Spermophilus spp., 1442
Spermophilus variegatus, 127, 1670
Sphagnum, *Sphagnum* sp., 337, 1768
Sphagnum spp., 337, 1768
Sphoeroides maculatus, 1073
Sphyrna spp., 344
Spiders
  Araneus umbricatus, 217
  Argiope aurantia, 112
  Chiracanthium mildei, 1102
  Dysdera crocata, 653
  Lycosa sp., 1004
  Pardosa sp., 1004
Spiderwort, *Aribidopsis thaliana,* 1755
Spilogale sp., 1145
Spilosoma spp., 1428
Spinach, *Spinacia oleracea,* 440, 521, 1095, 1097
Spinacia oleracea, 440, 521, 1095, 1097
Spirodela oligorhiza, 1428
Spirodella oligorrhiza, 1171
Spirogyra sp., 96, 747, 1161
Spirorbis lamellora, 663
Spirostomum ambiguum, 525
Spisula solidissima, 527, 660
Spiza americana, 1094, 1096
Spizella passerina, 27, 1005
Spodoptera eridania, 922
Spodoptera littoralis, 990, 992
Spodoptera litura, 1093
Sponge
  Various, *Halichondria* sp., 441

Table 33.2 (continued) Common and Scientific Names of Plants and Animals Listed in the *Handbook of Chemical Risk Assessment series* (References are provided from common names to scientific names)

Various, *Halichondria* sp., *Spongilla* sp., 441, 607
Spongilla sp., 441, 607
Spoonbill
Roseate, *Ajaia ajaja,* 350
White, *Platalea leucorodias*, 1295
Spot, *Leiostomus xanthurus,* 155, 462, 643, 654, 664, 785, 1363, 1471
Springtails, *Folsomia candida, Tullbergia granulata,* 653, 1169
Squalus acanthias, 119, 421, 463, 571, 1466, 1523
Squash, *Cucurbita* spp., 746, 1093
Squatina squatina, 634
Squawfish
Colorado, *Ptychocheilus lucius,* 1588, 1676, 1677
Northern, *Ptychocheilus oregonensis,* 332
Squids, 25, 1515
Squilla sp., 325
Squirrel
California ground, *Spermophilus beecheyi,* 1431, 1448
Gray, *Sciurus carolinensis,* 15, 127, 360, 450, 854
Ground, *Citellus* spp., *Spermophilus* spp., 1442
Red, *Tamiasciurus hudsonicus,* 15, 450
Rock, *Spermophilus variegatus,* 127, 1670
Srieba, *Astyanax bimaculatus,* 1205
Stanleya spp., 1652
Starfish. *See Asterias; Luidia clathrata*
Starling, European, *Sturnus vulgaris,* 13, 221, 241, 272, 370, 379, 800, 806, 833, 846, 850, 861, 862, 890, 932, 935, 968, 1074, 1077, 1135, 1386, 1431, 1525, 1788
Steelhead trout, *Oncorhynchus mykiss,* 530
Stelgidopteryx serripennis, 241
Stenella attenuata, 360
Stenella coeruleoalba, 123, 335, 336, 360, 642, 1273, 1276, 1661
Stenonema sp., 528
Stenotomus chrysops, 445, 581, 1303, 1307
Sterna caspia, 354, 1147, 1288, 1296
Sterna dougallii, 58
Sterna forsteri, 355, 850, 1026, 1027, 1035, 1036, 1289, 1296, 1319, 1466, 1660
Sterna fuscata, 1660
Sterna hirundo, 9, 221, 241, 272, 350, 354, 379, 384, 436, 1036, 1297, 1311, 1363, 1660
Sterna maxima, 1660
Sterna nilotica, 850
Sterna paradisaea, 1288, 1297
Sterna paridisaea, 1288
Sternotherus minor, 1741
Sternotherus odoratus, 1741
Stichopus japonicus, 1515
Stichopus tremulus, 673
Stickleback
Fourspine, *Apeltes quadracus,* 1678
Ninespine, *Pungitius pungitius,* 888
Threespine, *Gasterosteus aculeatus,* 18, 260, 341, 529, 576, 675, 859, 888, 1468
Stilt, black-necked, *Himantopus mexicanus,* 1657, 1666
Stitchbird, *Notiomystis cincta,* 1433
Stizostedion canadense, 11, 839
Stizostedion vitreum vitreum, 11, 19, 65, 159, 344, 747, 839, 1284
Stoat, *Mustela erminea,* 1298, 1301
Stolothrissa sp., 117
Stomoxys calcitrans, 990, 1116
Stonefly, *Isoperla* sp., 338
Stork
White, *Ciconia ciconia,* 638, 1291
Wood, *Mycteria americana,* 332, 352, 1668
Strawberry, *Fragaria vesca,* 802
Strepera graculina, 1432, 1436
Streptococcus faecalis, 524
Streptococcus zooepidemicus, 754
Streptopelea senegalensis, 1437
Streptopelia capicola, 1137
Streptopelia risoria, 272, 890, 1047, 1311
Streptopelia sp., 20
Strigops habroptilus, 1433
Strix aluco, 1297
Strongylocentrotus franciscanus, 663
Strongylocentrotus nudus, 153
Strongylocentrotus purpuratus, 460, 647, 663, 689, 1738
Strophites rugosus, 782
Sturgeon
Atlantic, *Acipenser oxyrhynchus,* 1739
Sevyuga, *Acipenser stellatus,* 574
Sheep, *Acipenser nudiventris,* 574
Shovelnose, *Scaphirhynchus platyrynchus,* 832, 839
Sturnella magna, 1096
Sturnella magna argutula, 1026
Sturnella neglecta, 976, 1432
Sturnus vulgaris, 13, 221, 241, 272, 370, 379, 800, 806, 833, 846, 850, 861, 862, 890, 932, 935, 968, 1074, 1076, 1135, 1386, 1431, 1525, 1788
Stylodrilus sp., 1378
Sucker
Bridgelip, *Catostomus columbianus,* 1738
Flannelmouth, *Catostomus latipinnis,* 1629
Largescale, *Catostomus macrocheilus,* 1738
Razorback, *Xyrauchen texanus,* 1589, 1649, 1664, 1676
River carp, *Carpiodes carpio,* 839
White, *Catostomus commersoni,* 65, 117, 154, 165, 259, 443, 624, 632, 750, 1026, 1030, 1031, 1042, 1212, 1279, 1353, 1363, 1379, 1462, 1517, 1676
Sudex, *Sorghum bicolor* X *Sorghum sudanense,* 143
Sugarcane, *Saccharum officinarum,* 746, 767, 1097, 1167
Sula bassanus, 354, 834, 851
Sula sp., 624
Sula sula, 1660
Sunfish
Green, *Lepomis cyanellus,* 155, 165, 805, 893, 924, 1173, 1468, 1518, 1536, 1664, 1675
Longear, *Lepomis megalotis,* 217, 230
Orangespot, *Lepomis humilis,* 129, 632
Redbreast, *Lepomis auritis,* 1782
Redear, *Lepomis microlophus,* 56, 451, 632, 1468
Spotted, *Lepomis punctatus,* 1026
Sunflower, *Helianthus annuus,* 1583
Sus scrofa, 59, 127, 245, 968, 969, 975, 978, 1414, 1417, 1448, 1747, 1756
Sus scrofa scrofa, 360

Table 33.2 (continued) Common and Scientific Names of Plants and Animals Listed in the *Handbook of Chemical Risk Assessment series* (References are provided from common names to scientific names)

Sus spp., 97, 127, 173, 280, 373, 450, 454, 483, 652, 685, 694, 869, 891, 905, 946, 1011, 1014, 1221, 1448, 1505, 1551, 1637, 1761
Swallow
  Barn, *Hirundo rustica,* 221, 239, 1657, 1668, 1731, 1742
  Northern rough-winged, *Stelgidopteryx serripennis,* 241
  Tree, *Tachycineta bicolor,* 58, 122, 136, 354, 436, 851, 1037, 1288, 1297, 1788
  Welcome, *Hirundo neoxena,* 1433
Swan
  Black, *Cygnus atratus,* 221
  Mute, *Cygnus olor,* 136, 202, 220, 237, 284
  Trumpeter, *Cygnus buccinator,* 202, 220, 237, 284, 635
  Tundra, *Cygnus columbianus,* 219, 221, 237
  Whooper, *Cygnus cygnus,* 237
Swine, *Sus* spp., 97, 127, 173, 280, 373, 450, 652, 685, 869, 891, 1011, 1221, 1448, 1505, 1551, 1637, 1761
Swordfish, *Xiphias gladius,* 345, 1659
Sylvilagus audubonii, 1439, 1669
Sylvilagus floridanus, 15, 130, 250, 800, 1745
Sylvilagus sp., 1550
Symiodinium sp, 1770
Synechococcus sp., 582
Synechocystis aquatilis, 573
Synedra sp., 655, 1302
Synthlioboramphus antiguis, 1524

T

Tabellaria sp., 655
Tachycineta bicolor, 58, 122, 136, 354, 436, 851, 1038, 1288, 1297, 1788
Tadarida brasiliensis, 15, 249, 1526
Tadorna tadorna, 1731
Tadpoles, 1000
Takake, *Notornis mantelli,* 1433
Talitrus saltator, 629
Talorchestia deshayesii, 629
Talpa europea, 59, 450, 1620, 1625, 1670
Tamarin, *Saguinus fuscicollis,* 961
Tamenthypnum sp., 451
Tamiasciurus hudsonicus, 15, 450
Tamias striatus, 1221, 1794, 1795
Tanager, scarlet, *Piranga olivacea,* 1005
Tanypus grodhausi, 990
Tanytarsus dissimilis, 18, 153, 257, 529, 688, 750, 997, 1676
Tanytarsus sp., 1003
Tapes decussatus, 377
Tapes japonica, 515
Tapes philippinarum, 8, 136, 1208
Tapes sp., 1264
Taphius sp., 525
Taraxacum sp., 1577
Taricha granulosa, 1784, 1785
Taricha torosa, 833, 844
Tarpon, *Megalops atlantica,* 365
Tasmanian devil, *Sarcophilus harrisii,* 1424, 1448
Tautog, *Tautoga onitis,* 435
Tautoga onitis, 435
Tautogolabrus adspersus, 522, 531
Taxidea taxus, 905, 1440, 1448
Tea, Labrador, *Ledum* sp., 337
Teal
  Blue-winged, *Anas discors,* 347, 624, 634, 1290, 1787
  Cinnamon, *Anas cyanoptera,* 1579, 1665
  Green-winged, *Anas carolinensis,* 799, 1787
Tellina tenuis, 165
Temora longicarpus, 576
Temora sp., 672
Tench, *Tinca tinca,* 100
Tenebrio molitor, 991
Termite
  Coptotermes spp., 142
  Heterotermes spp., 142
  Microtermes spp., 142
  Odontotermes spp., 142
  Reticulitermes flavipes, 1626
  Rhodesian, *Trinervitermes dispar,* 441
Tern
  Arctic, *Sterna paradisaea,* 1288, 1297
  Caspian, *Sterna caspia, Hydroprogne caspia,* 354, 1147, 1288, 1296
  Common, *Sterna hirundo,* 9, 221, 241, 272, 350, 354, 379, 384, 436, 1036, 1297, 1311, 1363, 1660
  Forster's, *Sterna forsteri,* 355, 850, 1026, 1027, 1035, 1036, 1289, 1296, 1319, 1466, 1660
  Gull-billed, *Gelochelidon nilotica, Sterna nilotica,* 850
  Roseate, *Sterna dougallii,* 58
  Royal, *Sterna maxima,* 1660
  Sooty, *Sterna fuscata,* 1660
Terrapene carolina, 231, 1145, 1741
Tetraedron sp., 893
Tetrahymena pyriformis, 145, 1683
Tetranychus mcdanieli, 591
Tetranychus urticae, 591, 990, 1102, 1169
Tetrao tetrix, 1768
Tetraselmis chui, 1535
Thais lapillus, 6
Thais sp., 166, 582
Thalassia testudinum, 451
Thalassiosira aestivalis, 1540
Thalassiosira nordenskioldi, 995
Thalassiosira pseudonana, 145, 574, 582, 657, 670, 1107, 1207
Thalassiosira rotula, 457, 458
Thalassiosira sp., 523, 993, 995
Thalassiosira weissflogii, 613, 995
Thamnophis proximos, 1096
Thamnophis sirtalis, 833, 844
Thamnophis sp., 1094, 1385
Themisto spp., 116, 629
Theragra chalcogrammus, 843, 1465
Thermonectes basillaris, 996
Thiara tuberculata, 149
Thielaviopsis basicola, 915
Thomomys sp., 961
Thorn, box, *Lycium andersonii,* 1577
Thrush
  Grey shrike, *Colluricincla harmonica,* 1435

Table 33.2 (continued) Common and Scientific Names of Plants and Animals Listed in the *Handbook of Chemical Risk Assessment series* (References are provided from common names to scientific names)

New Zealand, *Turnagra capensis,* 1433
Song, *Turdus philomelos,* 1759, 1766
Wood, *Hylocichla mustelina,* 1005
Thunnus alalunga, 1740
Thunnus albacares, 345, 365, 1579, 1740
Thunnus obesus, 345
Thunnus thynnus, 119, 741
Thylogale billardierii, 1417, 1448
Thymallus arcticus, 160, 463, 531, 619, 667
Tick
American dog, *Dermacentor variabilis,* 1079
Cattle, *Haemaphysalis longicornis,* 1072
Gulf Coast, *Amblyomma maculatum,* 1079
Lone star, *Amblyomma americanum,* 1079
Rocky mountain wood, *Dermacentor andersoni,* 1072
Tropical horse, *Anocentor nitens,* 1072, 1079
Tigriopus californicus, 1000
Tigriopus japonicus, 152
Tilapia
Mozambique, *Tilapia* (also *Oreochromis) mossambica,* 27, 102, 217, 365
Nile, *Oreochromus niloticus,* 27, 158, 260
Tilapia mossambica, 27, 102, 134, 217, 365, 578, 784, 1113
Tilapia nilotica, 457, 463, 966
Tilapia sparrmanii, 68, 668, 784
Tilapia zilli, 668
Tiliqua nigrolutea, 1430
Tiliqua rugosa, 1423, 1429, 1430
Timothy, *Phleum pratense,* 434, 774
Tinca Tinca, 100
Tipula abdominalis, 1003
Tipula sp., 895, 1468, 1668
Tisbe furcata, 152
Tisbe holothuriae, 63, 615, 662, 689
Tit
Great, *Parus major,* 636, 1288
Marsh, *Parus palustris,* 636
Willow, *Parus montanus,* 636
Titmouse, tufted, *Parus bicolor,* 1005
Toad
American, *Bufo americanus,* 160, 165, 261, 436, 784, 1586, 1588
Bombina variegata, 119
Bufo sp., 109, 676
Egyptian, *Bufo regularis,* 464
European, *Bufo bufo,* 119, 345, 860
Fowler's, *Bufo fowleri, Bufo woodhousei fowleri,* 160, 464, 1096, 1174, 1588
Giant, *Bufo marinus,* 119, 135
Gulf coast, *Bufo valliceps,* 1786
Narrow-mouthed, *Gastrophryne carolinensis,* 160, 367, 464, 480, 668, 675, 688, 1533, 1538
Southern, *Bufo terrestris,* 1025, 1523, 1665
Toadfish
Gulf, *Opsanus beta,* 887, 888, 889
Oyster, *Opsanus tau,* 581, 1112, 1381
Toadflax, bastard, *Comandra* spp., 1652
Tobacco, *Nicotiana tabacum,* 12, 56, 78, 337, 453, 802, 1093, 1168, 1667, 1774, 1776
Tolypothrix sp., 781
Tolypothrix tenuis, 775
Tomato, *Lycopersicon esculentum,* 110, 521, 746, 1093, 1097
Tomcod, *Microgadus tomcod,* 1383
Torpedo sp., 1099
Tortoise, gopher, *Gopherus polyphemus,* 1741
Tortrix viridana, 1288
Torulopsis glabrata, 457
Towhee
Rufous-sided, *Pipilo erythrophthalmus,* 1005
Various, *Pipilo* spp., 1432
Trachemys scripta, 25, 261, 1741, 1785, 1786
Trachinotus carolinus, 160
Trachymys floridana hoyi, 1741
Tree, tropical rainforest, *Dacryodes excelsa,* 1775
Trematomus bernacchii, 844
Triaenodus tardus, 811
Trichechus manatus, 109, 123
Trichoderma viride, 774
Trichodina sp., 96
Trichogaster pectoralis, 811
Trichogramma brassicae sp., 885
Trichosurus vulpecula, 938, 1414, 1417, 1420, 1428, 1448
Tridacna derasa, 149
Tridacna maxima, 1366, 1514
Trifolium pratense, 1168
Trifolium repens, 1621
Trifolium sp., 137
Triglochin spp., 905
Trinectes maculatus, 1141
Trinervitermes dispar, 441
Tringa totanus, 122
Tringa totanus robusta, 1525
Triops longicaudatus, 1000
Tripneustes esculentus, 443
Triticum aestivum, 12, 48, 440, 465, 740, 767, 775, 1168, 1621, 1668, 1776
Triticum spp., 989, 1761
Triticum vulgare, 568
Triturus cristatus, 668, 1385
Triturus vulgaris, 886
Troglodytes aedon, 1788
Tropisternus lateralis, 996
Tropocyclops prasinus mexicanus, 662
Trout
Brook, *Salvelinus fontinalis,* 18, 19, 20, 21, 65, 159, 259, 343, 367, 375, 384, 463, 634, 667, 783, 860, 931, 966, 970, 971, 972, 1001, 1040, 1046, 1139, 1141, 1213, 1471, 1472, 1540, 1628, 1677, 1738
Brown, *Salmo trutta,* 11, 21, 102, 159, 259, 531, 667, 673, 688, 751, 805, 860, 914, 931, 1026, 1173, 1202, 1270, 1363, 1372, 1519, 1738, 1758, 1766, 1781, 1784
Cutthroat, *Oncorhynchus clarki* (formerly *Salmo clarki),* 155, 520, 665, 842, 860, 888, 928, 966, 1001, 1738
Lake, *Salvelinus namaycush,* 56, 65, 159, 259, 343, 444, 463, 571, 578, 805, 843, 856, 888, 966, 1026, 1027, 1029, 1030, 1040, 1046, 1054, 1147, 1202, 1270, 1282, 1462, 1519, 1577, 1682, 1738
Rainbow, *Oncorhynchus mykiss* (formerly *Salmo gairdneri),* 11, 18, 19, 21, 26, 56, 62, 64, 65, 69, 100, 156, 258, 367, 375, 384, 444, 462, 480, 506, 533, 576, 579, 611, 647, 665, 688, 747, 750, 768, 783, 805, 832–833, 860, 888,

Table 33.2 (continued) Common and Scientific Names of Plants and Animals Listed in the *Handbook of Chemical Risk Assessment series* (References are provided from common names to scientific names)

905, 929, 964, 965, 1001, 1026, 1030, 1040, 1043, 1054, 1073, 1100, 1111, 1112, 1137, 1173, 1202, 1209, 1221, 1269, 1305, 1353, 1373, 1375, 1380, 1428, 1461, 1468, 1519, 1533, 1536, 1539, 1581, 1586, 1589, 1602, 1620, 1623, 1629, 1641, 1677, 1738, 1779
Spotted sea, *Cynoscion nebulosus,* 117, 445, 1465
Steelhead, *Oncorhynchus mykiss,* 530
Trumpeter, six-lined, *Siphamia cephalotes,* 644
Tubifex costatus, 1541
Tubifex sp., 68, 1783
Tubifex tubifex, 116, 460, 562, 578, 1208
Tui, *Prosthemadera novae-seelandiae,* 1433
Tui chub, *Gila bicolor,* 571
Tullbergia granulata, 1169
Tuna
Big-eye, *Thunnus obesus,* 345
Bluefin, *Thunnus thynnus,* 119, 741
Skipjack, *Euthynnus pelamis,* 444
Yellowfin, *Thunnus albacares,* 345, 365, 1579, 1740
Tunicates
Ciona intestinalis, 116
Cynthia claudicans, 517
Halocynthia roretzi, 443
Podoclavella moluccensis, 53
Polycarpa peduculata, 644
Salpa fusiformes, 1579
Turbellarians, *Bothromestoma* sp., *Mesotoma* sp., 1004
Turbot, *Scopthalmus maximus,* S. *maeoticus maeoticus,* 675
Turdus merula, 241, 638, 895, 1668
Turdus migratorius, 13, 234, 241, 810, 846, 1542, 1623
Turdus philomelos, 1759, 1766
Turkey, *Meleagris gallopavo,* 60, 121, 138, 168, 202, 534, 612, 625, 636, 678, 889, 891, 905, 964, 968, 977, 1175, 1177, 1308, 1311, 1436, 1463, 1503, 1544, 1553
Turnagra capensis, 1433
Turnip, *Brassica rapa,* 521
Tursiops gephyreus, 123
Tursiops truncatus, 54, 123, 249, 275, 642, 1661
Turtle
Atlantic green, *Chelonia mydas,* 1145
Common box, *Terrapene carolina,* 231, 1145, 1741
Common mud, *Kinosternon sabrubrum,* 1741
Common musk, *Sternotherus odoratus,* 1741
Leatherback, *Dermochelys coriaca,* 250, 634, 1368
Loggerhead, *Caretta caretta,* 345, 1145, 1285, 1286
Loggerhead musk, *Sternotherus minor,* 1741
Missouri slider, *Trachymys floridana hoyi,* 1741
Peninsula cooter, *Pseudemys floridana peninsularis,* 1741
Pond slider, *Pseudemys scripta,* 1741
Red-eared, *Trachemys scripta,* 25, 261, 1741, 1785, 1786
Slider
Chrysemys scripta, 1145
Trachemys scripta, 25, 261, 1785, 1786
Snapping, *Chelydra serpentia,* 231, 1031, 1285, 1286, 1741
Turtle-dove, ringed, *Streptolia risoria,* 272, 890
Tuskfish, scarbreast, *Choerodon azurio,* 324
Tympanuchus cupido, 371
Tympanuchus phasianellus, 1469
Typha latifolia, 1171, 1577
Typha sp., 1169
Typhlodromus sp., 1169
Tyto alba, 24, 863, 1074, 1432

U

Uca minax, 812
Uca pugilator, 22, 377, 577, 582, 613, 671, 785, 994, 1000
Uca pugnax, 785
Uca sp., 165, 893
Uca tangeri, 928
Ulmus americana, 111, 568
Ulothrix sp., 655
Ulva lactuca, 657, 1735
Ulva rigida, 5
Ulva sp., 5, 165, 225
Umbilicaria sp., 440
Umbra limi, 970
Umbra pygmaea, 56, 1384
Upeneus moluccensis, 567
Upeneus tragula, 633
Upupa epops, 625, 638
Uria aalge, 136, 1027, 1297, 1660
Uria lomvia, 834, 851, 1660
Urocyon cinereoargentatus, 1448, 1745
Urocyon sp., 1144
Uromastix hardwickii, 1785
Uronema marinum, 368
Uronema nigricans, 377
Uronema sp., 257
Urophycis chuss, 119
Urophycus chuss, 1368
Ursus arctos, 59, 361, 383
Ursus maritimus, 103, 123, 135, 336, 361, 363, 520, 642, 834, 1264, 1298, 1301, 1464, 1671
Ursus spp., 914, 1448
Uta sp., 1785
Uta stansburiana, 1786

V

Vaccinium angustifolium, 128, 440, 1516
Vaccinium myrtillus, 1759, 1767
Vaccinium pallidum, 225
Vaccinium sp., 1762
Vaccinium uliginosum, 1767
Vaccinium vitis-idaea, 1767
Vallisneria americana, 4, 776, 779
Vallisneria gigantia, 749
Vallisneria spiralis, 749
Vallisneria spp., 742
Vaquita (porpoise), *Phocoena sinus,* 450
Varanus gouldi, 1430
Varanus varius, 1430
Vasum turbinellus, 582
Venerupis pallustra, 1629
Veromessor andrei, 1432

Table 33.2 (continued) Common and Scientific Names of Plants and Animals Listed in the *Handbook of Chemical Risk Assessment series* (References are provided from common names to scientific names)

Vespula germanica, 1428
Vespula vulgaris, 1428
Vicia faba, 141, 426, 1776
Victorella sp., 517, 527, 528
Vigna sp., 918, 1532
Villorita cyprinoides, 149
Villosa iris, 149
Vireo
  Red-eyed, *Vireo olivaceous,* 1005
  Warbling, *Vireo gilvus,* 1006
Vireo gilvus, 1006
Vireo olivaceous, 1005
Vitis sp., 1093, 1668
Viviparus ater, 210, 258
Viviparus bengalensis, 1209
Vole
  Arvicola terrestris, 1160
  Bank, *Clethrionomys glareolus,* 23, 125, 136, 243, 276, 437, 615, 639, 644, 691, 1515, 1756, 1767
  Brown-backed, *Clethrionomys rufocanus,* 136, 1671
  California, *Microtus californicus,* 247, 1668, 1669
  Clethrionomys rutilis, 437, 853, 1150, 1300
  Creeping, *Microtus oregoni,* 1791
  Grey-sided, *Clethrionomus rufocanus,* 136, 1671
  Lemmus sp., 136
  Levant, *Microtus guentheri,* 1445
  Meadow, *Microtus pennsylvanicus,* 14, 130, 248, 448, 1096, 1445, 1528, 1791
  Root, *Microtus oeconomus,* 1756
  Short-tailed field, *Microtus agrestis,* 15, 248, 643, 1179, 1515, 1670
  Singing, *Microtus miurus,* 1791
Vombatus ursinus, 1439, 1449
Vorticella convallaria, 657
Vulpes fulva, 336, 1745
Vulpes macrotis arsipus, 1449
Vulpes macrotis mutica, 1450
Vulpes sp., 572, 914, 1144, 1417, 1527
Vulpes vulpes, 127, 380, 823, 1298, 1301, 1414, 1438, 1449, 1756, 1769
Vulture
  Black, *Coragyps atratus,* 362, 933, 1435
  Griffon, *Gyps fulvus,* 238
  King, *Sarcorhampus papa,* 202
  Turkey, *Cathartes aura,* 234, 348, 446, 625, 634, 933, 1431, 1435
Vultur gryphus, 202, 935

W

Wallaby
  Agile, *Macropus agilis,* 1444
  Banded hare-, *Lagostrophus fasciatus,* 1444
  Bennett's (also known as red-necked), *Macropus rufogriseus,* 1417, 1418, 1444
  Rock, *Petrogale penicillata,* 1420, 1433
  Tammar, *Macropus eugenii,* 1444
  Western brush, *Macropus irma,* 1444
Walleye, *Stizostedion vitreum vitreum,* 11, 19, 65, 159, 344, 747, 839
Walrus, *Odobenus rosmarus divergens,* 9, 10, 25, 853
Warbler
  MaGillivray's, *Oporornis tolmiei,* 1006
  Townsend's, *Dendroica townsendi,* 1006
  Various, *Dendroica* spp., 1005
Warmouth, *Lepomis gibbosus,* 56, 342, 462
Wasp
  Common, *Vespula vulgaris,* 1428
  German, *Vespula germanica,* 1428
Watermilfoil, Eurasian, *Myriophyllum spicatum,* 67, 96, 776, 780, 1171
Watersipora cucullata, 660
Weasel, *Mustela nivalis,* 24, 1298, 1301
Weed
  Chenopodium album, 775
  Cladophora sp., 224, 747, 1202
  Duck, *Lemna, Spirodella oligorrhiza,* 1171
  Fire, *Epilobium angustifolium,* 1767
  Hydrilla aquatic, *Hydrilla verticillata,* 166, 255
  Loco, *Astragalus,* 1652
  Pond, 96, 111
    Ceratophyllum sp., 1656
    Najus guadalupensis, 255
    Navicula sp., 255, 655
    Potamogeton, 96, 111, 1517
  Rat, *Palicourea marcgravii,* 1413, 1426
  Ribbon, *Vallisneria spiralis,* 749
  Sargassum, *Sargassum fluvitans,* 1519
  Sea, *Chondrus crispus, Fucus, Porphyra* sp., *Sargassum, Ulva,* 441, 1517, 1519, 1735
  Water, *Elodea,* 741, 812
Weevil
  Cotton boll, *Anthonomus grandis,* 983, 991, 1013, 1504
  Granary, *Sitophilus granarius,* 923
  Rice water, *Lissorhopterus oryzophilus,* 1004
Whale
  Beluga, *Delphinapterus leucas,* 25, 362, 851, 1032, 1263, 1271, 1304, 1363, 1467, 1660
  Bowhead, *Balaena mysticetus,* 1660
  Cuvier's goosebeaked, *Ziphius cavirostris,* 1277
  Fin, *Balaenoptera physalis,* 336, 355, 1525
  Giant bottlenosed, *Berardius bairdii,* 1271
  Gray, *Eschrichtius robustus,* 122
  Killer, *Orcinus orca,* 1272
  Minke, *Balaenoptera rostrata,* 1660
  Narwhal, *Monodon monoceros,* 245, 1467
  Pilot, *Globicephala macrorhynchus,* 9, 1660
  Pilot, *Globicephala melas,* 9, 25
  Pygmy sperm, *Kogia breviceps,* 122
  Sperm, *Physeter catodon, Physeter macrocephalus,* 336, 359, 450, 909, 1744
  Various, *Ziphius* sp., 1264
  White, *Delphinapterus leucas,* 25, 244, 851, 1032, 1263, 1271, 1304, 1467, 1660
Wheat
  Triticum aestivum, 12, 48, 440, 465, 767, 775, 1168, 1621, 1668, 1776
  Triticum spp, 989, 1761
  Triticum vulgare, 568
Whelk
  Channeled, *Busycon canaliculatum,* 113, 147, 162
  Common dog, *Nucella lapillus,* 574, 575, 582
  Kelletia kelletia, Thais lapillus, 6
  Thais lapillus, 6
  Waved, *Buccinum undatum,* 113, 441

Table 33.2 (continued) Common and Scientific Names of Plants and Animals Listed in the *Handbook of Chemical Risk Assessment series* (References are provided from common names to scientific names)

Whelks, 582
Whimbrel, *Numenius phaeopus,* 121
Whistling-duck, fulvous, *Dendrocygna bicolor,* 371, 805, 806, 1469
Whitefish
  Lake, *Coregonus clupeaformis,* 117, 443, 571, 837
  Mountain, *Prosopium williamsoni,* 914, 1026, 1738
  Round, *Prosopium cylindraceum,* 343
  Various, *Coregonus* spp., 217, 229
Whiting, *Merlangius merlangius, Sillago bassensis,* 344, 841, 1523, 1536, 1782
Wigeon, American, *Anas americana,* 799, 961, 1434
Wildcelery, *Vallisneria americana,* 4, 776, 779
Willet, *Catoptrophorus semipalmatus,* 120, 136, 634, 1524, 1579, 1666
Willow, *Salix* spp., 4, 1765, 1767
Wolf, *Canis lupus,* 1033, 1416, 1756, 1759
Wolffish, spotted, *Anarhichas minor,* 229
Wolverine, *Gulo gulo,* 1760
Wombat
  Common, *Vombatus ursinus,* 1439, 1449
  Southern hairy-nosed, *Lasiorhinus latifrons,* 1439, 1444
Woodchuck, *Marmota monax,* 126, 572, 1670
Woodcock
  American, *Scolopax minor,* 1055, 1149
  Eurasian, *Scolopax rusticola,* 1766, 1768
Woodlice, *Porcellio scaber, Oniscus asellus,* 177, 217, 253, 621, 628, 652, 654
Woodpeckers
  Acorn, *Melanperes formicivorus,* 1432
Worm
  Black cut, *Agrotis ipsilon,* 1733
  Cotton leaf, *Spodoptera littoralis,* 990, 992
  Earth, *Allolobophora, Eisenia, Lumbricus, Octochaetus,* 441, 1039
  Lesser meal, *Alphitobius diaperinus,* 1071, 1073
  Limnodrilus sp., 134, 1378
  Lug, *Arenicola cristata,* 578, 579, 582, 1136
  Manure, *Eisenia foetida,* 57, 129, 382, 456, 653, 809, 1416
  Marine, *Nereis, Nephtys,* 7
  Nais sp., 63
  Neanthes arenaceodentata, 63, 64, 65, 73, 262, 662, 673, 1209, 1373, 1376, 1779
  Oligochaete, *Lumbriculus variegatus,* 153, 163, 460, 533, 1372
  Paranais sp., 1041
  Rag, *Hediste diversicolor,* 153
  Sand, *Nereis diversicolor,* 7, 53, 122, 153, 163, 443, 451, 460, 517, 623, 631, 655, 662, 673, 1737, 1784
  Silk, *Bombyx mori,* 1776, 1777
  Southern army, *Spodoptera eridanea,* 922
  Spirorbis lamellora, 663
  Stylodrilus sp., 1378
  Tiger, *Eisenia foetida,* 57, 129, 382, 456, 653, 809, 1416
  Tobacco horn, *Manduca sexta,* 975
  Tubifex sp., 68, 1783
Wren
  Bush, *Xenicus longipus,* 1418, 1433
  House, *Troglodytes aedon,* 1788
  Rock, *Xenicus gilviventris,* 1433

X

Xenicus gilviventris, 1433
Xenicus longipes, 1418, 1433
Xenopus laevis, 232, 427, 611, 668, 751, 1213, 1373, 1385, 1429, 1586, 1590, 1676, 1682, 1770, 1786
Xenopus sp., 1385
Xiphias gladius, 119, 345, 1659
Xyrauchen texanus, 1589, 1649, 1664, 1676

Y

Yeast, 1532, 1678
  Cryptococcus terrens, 457
  Rhodotorula sp., 457
  Saccharomyces spp., 1204
  Torulopsis glabrata, 457
Yellowhead, *Mohoua ochrocephala,* 1433
Yellowtail, *Seriola quinqueradiata* (also *Seriola lalandei),* 965

Z

Zalophus californianus, 10, 222, 335, 361, 514, 520, 1661, 1679
Zannichellia palustris, 781
Zannichellia sp., 747, 780
Zapus princeps, 60
Zea mays, 57, 132, 138, 252, 440, 521, 568, 653, 746, 767, 775, 801, 831, 903, 989, 1039, 1093, 1168, 1531, 1584, 1626, 1639, 1668
Zebrafish, *Brachydanio rerio, Danio rerio,* 154, 259, 367, 375, 461, 613, 674, 784, 965, 966, 1042, 1173, 1304, 1588, 1675
Zenaida (also *Zenaidura) macroura,* 202, 242, 273, 355, 1437, 1576
Zingiber officinale, 1735
Ziphius cavirostris, 1277
Ziphius sp., 1264
Zizania aquatica, 255
Zostera marina, 622, 627, 655, 781
Zostera spp., 111
Zygaena filipendulae, 918
Zygaena trifolii, 918, 922
Zygnema sp., 451
Zylorhiza, 1652
Zyzomys argurus, 1449

Index

A

Abiotic materials
  arsenic concentrations, 1510–1515
  boron concentrations, 1573–1576
  molybdenum concentrations, 1617–1619
  radionuclide concentrations, 1725–1729
  selenium concentrations, 1653–1657
Adriatic Sea, 1520
Aerosol, selenium concentrations, 1692
Africa
  Central, 1754
  East, 1754
  North, 1754
  West, 1754
Agricultural drainage water, 1574
Aiken, South Carolina, 1741, 1785
Air
  arsenic concentrations, 1511, 1531, 1551, 1555
  boron concentrations, 1574, 1603
  molybdenum concentrations, 1619, 1640
  radionuclide concentrations, 1717, 1802
  selenium concentrations, 1654, 1690
Alabama, 1613, 1746
Alaska
  arsenic concentrations, 1527
  molybdenum concentrations, 1622, 1623
  molybdenum reserves, 1614
  radionuclide concentrations, 1734, 1743, 1744, 1761
  selenium concentrations, 1660
Albania, 1761
Algae
  radiation effects, 1778
  radionuclide concentrations, 1730, 1735
  selenium concentrations, 1657, 1662
    toxic, 1675
Alkali disease, 1679
Amazon River, 1653
Amphibian(s)
  arsenic concentrations, 1523–1524
  boron toxicity, 1588, 1590–1591
  radiation effects, 1784–1786
  selenium concentrations, 1665
    toxic, 1676
Annelids
  radiation effects, 1779 (*See also* Invertebrate(s))
Antarctic Ocean, 1660
Apalachicola River, 1663, 1665
Aquatic invertebrate(s)
  molybdenum concentrations, 1622
  radionuclide concentrations, 1735–1738
Aquatic organism(s)
  arsenic concentrations
    toxic, 1533–1542, 1553
  boron concentrations, 1577–1579
    toxic, 1585–1591, 1602
  Chernobyl fallout, 1757–1758
  molybdenum toxicity, 1626–1630
  radiation effects, 1777–1784
  selenium toxicity, 1675–1678, 1682–1685
Aquatic plant(s)
  arsenic concentrations, 1517, 1519–1520
  molybdenum concentrations, 1622
    toxic, 1628
  radionuclide concentrations, 1735
Aransas/Wood Buffalo National Park, 1580
Argentina, 1653
Arizona, 1614
Arkansas, 1746
Arsenic, 1501–1557
  in abiotic material, 1510–1515
    air, 1511, 1531, 1551, 1555
    detergents, synthetic, 1513
    drinking water, 1511, 1529, 1554
    dust, 1511
    fossil fuels, 1511
    groundwater, 1512
    lake water, 1512
    rain, 1512
    river water, 1512
    rock, 1512
    seawater, 1512–1513
    sediment, 1513
    snow, 1513
    soil, 1513
    soil pore waters, 1513
  antagonism, 1508
  carcinogenicity, 1529
  dietary, 1507
    limits, 1554
  epidemiological studies, 1501
  field collections, 1509–1528
    abiotic material, 1510–1515

amphibian, 1523–1524
aquatic plant, 1517, 1519–1520
bird, 1524–1525
crustacean, 1521
fish, 1517–1519, 1522–1523
integrated studies, 1527–1528
mammal, 1525–1527
mollusc, 1520
reptile, 1523
terrestrial plant, 1516–1517
global production, 1502
lethal effects, 1528–1551 (*See also* Arsenic, toxicity)
physical properties, 1505
recommendations, 1551–1555
sources, 1502
sublethal effects, 1528–1551 (*See also* Arsenic, toxicity)
teratogenicity, 1530
tolerance, 1509
toxicity, 1501, 1528–1551
antagonism, 1508
aquatic organism, 1533–1542, 1553
bird, 1542–1544, 1553
human, 1553–1554
invertebrate, 1531–1532, 1533, 1536–1538, 1540–1541
mechanism, 1507
plant, 1530–1531, 1532, 1536, 1540
protection against, 1553–1551
terrestrial invertebrate, 1531–1532, 1533
terrestrial plant, 1530–1531, 1532
uses, 1503–1505
agricultural, 1504
therapeutic, 1503
Asia
East, 1754
South, 1754
Southeast, 1754
Southwest, 1754
Astronaut, radiation effects, 1801, 1802
Aswan, Egypt, 1668
Atlantic Ocean, 1512, 1722, 1744
Atomic number, 1714–1715
Australia
arsenic concentrations, 1512
molybdenum production, 1614
molybdenum toxicity, 1613
selenium concentrations, 1653, 1659
selenium toxicity, 1679
Austria, 1516, 1753

B

Baja California, Mexico, 1576, 1579
Bakersfield, California, 1511
Baltic Sea, 1621, 1757
Bangladesh, 1501, 1546
Belews Lake, 1649
Belgium
radioactive waste disposal, 1722
radionuclide concentrations
cattle, 1733
drinking water, 1726
milk, 1727
soil, 1728
terrestrial mammal, 1744
terrestrial plant, 1734
Beta radiation, 1719, 1731
Bikini Atoll, 1707, 1749, 1750
Bird(s)
arsenic concentrations, 1524–1525
toxic, 1542–1544, 1553
boron concentrations, 1579–1580
toxic, 1591–1593
molybdenum concentrations, 1623
toxic, 1630–1631, 1639
radiation effects, 1786–1788
radionuclide concentrations, 1731–1732, 1741–1743
selenium concentrations, 1659–1660, 1665–1667, 1668
toxic, 1679–1682, 1685–1686
Boron, 1567–1605
in abiotic material, 1573–1576
agricultural drainage water, 1574
air, 1574, 1603
coal-fired power plant, 1574
drinking water, 1574, 1603, 1604
groundwater, 1574
mine drainage water, 1574
rain, 1574
river water, 1575
seawater, 1575
sediment, 1575
sewage water, 1575
soil, 1575
surface water, 1575–1576
thermal springs, 1576
well water, 1576
chemical properties, 1570–1571
dietary, 1571, 1603
effects, 1581–1600 (*See also* Boron, toxicity)
field collections, 1573–1581
abiotic material, 1573–1576
aquatic organism, 1577–1579
bird, 1579–1580
mammal, 1580–1581
terrestrial plant, 1577
interactions, 1571
physical properties, 1568
production, 1568
recommendations, 1600–1604
sources, 1568, 1569
toxicity, 1581–1600
amphibian, 1588, 1590–1591
aquatic organism, 1585–1591, 1602
bird, 1591–1593
fish, 1588–1590
invertebrate, 1581–1584, 1587–1588
mammal, 1593–1600
plant, 1581–1584, 1586–1587
protection against, 1602–1604
terrestrial invertebrate, 1581–1584, 1584–1585

        terrestrial plant, 1581–1584
    uses, 1568–1569
Boron, California, 1568
Brazil, 1747
British Columbia, 1575, 1579, 1614, 1616
Brown's Lake, 1510
Bulgaria, 1753

C

Cairo, Egypt, 1668
California
    arsenic concentrations, 1512
        bird, 1524
        marine mollusc, 1520
        pond, 1515
    boron concentrations, 1569
        bird, 1579
        drinking water, 1574
        fresh water organism, 1578
    molybdenum concentrations
        groundwater, 1618
        sediment, 1618
        surface water, 1618
    molybdenum mining site, 1614
    radionuclide concentrations, 1732
        fish, 1740
        mammal, 1745
        terrestrial plant, 1734
    selenium concentrations
        fish, 1655, 1663
        integrated studies, 1654
        mammal, 1670, 1673
            aquatic, 1679
        mollusc, 1657
        pond, 1691
Camden, New Jersey, 1655
Canada
    arsenic concentrations, 1512, 1513
        terrestrial plant, 1517
    arsenic toxicity, protection against, 1554
    boron concentrations, 1575
        bird, 1580
    molybdenum production, 1614
        aquatic plant, 1622
    molybdenum toxicity, 1613
    radionuclide concentrations
        atmospheric, 1720
        bird, 1732
        Chernobyl fallout, 1753
        lakewater, 1727
        mammal, 1761
    selenium concentrations, 1653
        bird, 1666
    selenium use, 1651
Cap de la Hague reprocessing plant, 1722
Cape May, 1660
Carcinogenicity, 1529, 1796–1797, 1802
Caribbean, 1754
Carson Lake, 1667
Carson Valley, Nevada, 1621
Castle Lake, California, 1628
Central America, 1754
Central Valley, California, 1580
Charlotte Harbor, Florida, 1535
Chernobyl, Ukraine, 1752, 1753, 1806
Chernobyl accident, 1707, 1752–1769
    air plume behavior, 1754
    effects, 1755–1769
        aquatic organism, 1757–1758
        local, 1755–1756
        mammal, 1758–1761
        nonlocal, 1756–1769
        plant, 1757
        soil, 1757
        in specific geographic areas, 1761–1769
        wildlife, 1758–1760
    fission products, 1753
    human dose equivalent commitment, 1754
Chesapeake Bay, 1510, 1515
Chile, 1501, 1529, 1574, 1614
China
    arsenic concentrations, 1516
    boron deposits, 1568
    molybdenum concentrations, 1614
    nuclear weapons testing, 1707, 1725
    radionuclide concentrations, 1753
    selenium concentrations, 1666
        toxic, 1649
Cibola Lake, 1656, 1663
Cigarette smoke, radiation, 1723
Clark Fork River, 1510, 1519
Climax, Colorado, 1614, 1615
Coal-fired power plant, boron concentrations, 1574
Coelenterates
    radiation effects, 1778
    selenium toxicity, 1676
Colombia, 1649
Colorado
    molybdenum concentrations
        soil, 1619
        surface water, 1618, 1651
        toxic, 1613
    molybdenum production, 1614, 1628
    nuclear fuel reprocessing plants, 1733
    radionuclide concentrations
        fish, 1729, 1730, 1738
        mammal, 1730, 1745
        terrestrial plant, 1734
Colorado River, 1654
Colorado River Valley, 1656, 1663, 1671
Columbia River
    neptunium-239 concentrations, 1738
    strontium-90 concentrations, 1732, 1734, 1741
    zinc-65 concentrations, 1787
Connecticut, 1735
Cook Inlet, 1527
Corpus Christi, Texas, 1515, 1525, 1657
Cosmic rays, 1709
Costa Rica, 1668

Crustacean(s)
arsenic concentrations, 1521
radiation effects, 1778–1779
radionuclide concentration factors, 1782
radionuclide concentrations, 1736
selenium concentrations, 1658
toxic, 1676, 1677
Cumbrian Coast, United Kingdom, 1735, 1760
Czechoslovakia
arsenic concentrations, 1511, 1531
limits, 1555
Chernobyl fallout, 1753
radionuclide concentrations, 1760, 1761
integrated studies, 1781

D

Danube River, 1757, 1762
Dead Sea, 1654
Delaware, 1668
Delaware River, 1655
Denmark, 1732, 1745
Denver, Colorado, 1655
Detergents, synthetic, 1513
Drainage water, 1574
boron concentrations, 1574
Drinking water
arsenic concentrations, 1511, 1529, 1554
molybdenum concentrations, 1618, 1639, 1640
radionuclide concentrations, 1726, 1803
selenium concentrations, 1653, 1690, 1691
Dust, 1511

E

Echinoderm, selenium concentrations, 1658
Elbe River, 1512
Electromagnetic spectrum, 1709
Elliot Lake, 1734, 1741
England
arsenic concentrations, 1513
terrestrial plant, 1517
toxic, 1501
boron concentrations, 1575
molybdenum toxicity, 1613
nuclear fuel reprocessing plant, 1733, 1736, 1737
radionuclide concentrations, 1733
aquatic invertebrate, 1736, 1737
fish, 1740
mammal, 1733, 1746, 1747
Eniwetok Atoll, 1749, 1750–1752, 1751
Europe, 1567, 1744
arsenic toxicity, 1501
Central, 1754
Northern, 1754
Southeastern, 1754
West, 1754

F

Fairbanks, Alaska, 1511
Finfeather Lake, 1512
Finland
arsenic concentrations, 1526
molybdenum concentrations, 1625
radionuclide concentrations, 1762
Chernobyl fallout, 1753, 1762
terrestrial plant, 1734
selenium concentrations, 1654
Fish(es)
arsenic concentrations, 1517–1519, 1522–1523
boron toxicity, 1588–1590
molybdenum concentrations, 1622–1623
toxic, 1629–1630
radiation effects, 1779–1781
radionuclide concentration factors, 1782
radionuclide concentrations, 1729–1730, 1738–1740
selenium concentrations, 1658–1659, 1662–1665
toxic, 1676–1677, 1678
Florida
arsenic concentrations, 1512, 1521
bird, 1524
crab, 1521
boron concentrations, 1575
molybdenum deficiency, 1626
radionuclide concentrations
cesium, 1734, 1745
strontium, 1731
selenium concentrations
bird, 1668
fish, 1663
mammal, 1670
mollusc, 1662
reptile, 1665
selenium deficiency, 1649
Florida Everglades, 1649
Food, radionuclide concentrations, 1717
Fossil fuels
arsenic concentrations, 1511
molybdenum concentrations, 1619
selenium concentrations, 1653
Four Corners Power Plant, 1574
France
boron concentrations, 1575
nuclear weapons testing, 1707, 1719
radioactive waste disposal, 1722
radionuclide concentrations, 1763
terrestrial plant, 1775
selenium concentrations, 1653
sodium selenite use, 1651
Freshwater
boron concentrations, 1576, 1578
radionuclide concentrations, 1726–1727
Fresno County, California, 1524

G

Gamma rays, 1709, 1729

Georgia
radionuclide concentrations, 1731
terrestrial mammal, 1745, 1746
terrestrial plant, 1734
Germany
arsenic concentrations
air, 1511
integrated studies, 1527
lake water, 1512
boron concentrations, 1575
Chernobyl fallout, 1753, 1757, 1763
molybdenum concentrations, 1621
radioactive waste disposal, 1722
radionuclide concentrations, 1744, 1753, 1757, 1763
selenium concentrations, 1653
Ghana, 1511
Great Britain, 1729
Great Lakes
arsenic concentrations, 1517
radioactive fallout, 1720
radionuclide concentrations, 1720, 1727, 1748
limits, 1802
selenium concentrations, 1653, 1654
Greece
boron concentrations, 1574, 1576
Chernobyl fallout, 1761
molybdenum concentrations, 1620, 1753
radionuclide concentrations, 1761, 1763–1764
Green River, 1588, 1649, 1664
Greenland, 1671, 1729, 1734, 1754
Groundwater
arsenic concentrations, 1512
boron concentrations, 1574
molybdenum concentrations, 1618
radionuclide concentrations, 1727, 1803
selenium concentrations, 1653

H

Hanford, Washington
radionuclide concentrations
aquatic plant, 1735
bird, 1742
fish, 1738
sediment, 1728
soil, 1728
from water-cooled reactor, 1787
Hawaii, 1655, 1740
Hiroshima, Japan, 1707
Hong Kong
arsenic concentrations
crustacean, 1521, 1522
fish, 1522
mollusc, 1520
molybdenum concentrations, 1620, 1625
Hudson River
radionuclide concentrations, 1725, 1729
aquatic plant, 1735
fish, 1739, 1740
sediment, 1728
mollusc, 1736
plankton, 1737
Hudson River Estuary, 1728, 1729, 1740
Human
arsenic toxicity, 1553–1554
Chernobyl fallout, 1754
molybdenum toxicity, 1635, 1640
radiation exposure, 1718–1723
annual whole-body, 1717, 1732, 1802
radon exposure, 1732
selenium concentrations, 1669
Humboldt Bay, 1731
Humboldt River, 1576
Hungary, 1613, 1753, 1762

I

Idaho, 1614, 1733, 1787
India
arsenic toxicity, 1501
boron concentrations, 1575, 1576
Chernobyl fallout, 1753
nuclear weapons testing, 1707
Infrared waves, 1709
Insect. *See also* Invertebrate(s); Terrestrial invertebrate(s)
selenium toxicity, 1676
Invertebrate(s). *See also* Terrestrial invertebrate(s)
arsenic toxicity, 1531–1532, 1533, 1536–1538, 1540–1541
boron toxicity, 1581–1584, 1587–1588
molybdenum toxicity, 1628–1629
radiation effects, 1776–1777
selenium concentrations, 1668
Iowa, 1619
Irapuato, Mexico, 1649
Ireland, 1594, 1613, 1736
Irish Sea, 1523, 1729, 1748
Israel, 1569, 1653, 1654, 1753
Italy
boron concentrations, 1575
radionuclide concentrations, 1729
atmospheric deposition, 1761
Chernobyl fallout, 1753, 1757, 1761, 1764
fish, 1739
limits, 1802
plankton, 1737
terrestrial mammal, 1744
selenium concentrations, 1653

J

Japan
arsenic concentrations, 1511
lake water, 1512
limits, 1554
soil, 1513
toxic, 1501
boron concentrations

rain water, 1575
seaweed, 1578
water, 1576
molybdenum production, 1614
molybdenum toxicity, 1613
radionuclide concentrations, 1802
Chernobyl fallout, 1753, 1761, 1764–1765
toxicity, 1798
selenium concentrations, 1653, 1654
limits, 1690

K

Kelly Lake, 1510
Kentucky, 1746
Kenya, 1524, 1577, 1673
Kern National Wildlife Refuge, 1517
Kesterson National Wildlife Refuge
boron concentrations, 1573
freshwater organism, 1578
plant and animal, 1576
water, 1574
selenium concentrations
animal, 1669, 1671
bird, 1666
integrated studies, 1670–1671
pond water, 1657
Korea, 1527, 1614, 1725
Kuwait, 1753

L

Laguna Madre, 1527, 1671
Lake Chatauqua, 1512
Lake Erie, 1518, 1519, 1664
Lake George, 1653
Lake Michigan, 1513, 1514
Lake Nakaru, 1524
Lake Ohakuri, 1512
Lake Ontario, 1739
Lake Paijanne, 1758, 1762
Lake Superior, 1512
Lake Washington, 1510
Lake water
arsenic concentrations, 1512
radionuclide concentrations, 1727
selenium concentrations, 1654
Linear energy transfer, 1716
Loch Lomond, 1510
London, England, 1761
Long Island, 1659
Louisiana, 1745, 1746

M

Macrophyte(s)
radionuclide concentrations, 1735
selenium concentrations, 1657, 1667–1668
Mammal(s)
arsenic concentrations, 1525–1527
boron concentrations, 1580–1581
toxic, 1593–1600
Chernobyl fallout, 1758–1761
molybdenum concentrations, 1623–1625
toxic, 1631–1637, 1639–1640
radionuclide concentrations, 1730, 1732–1733, 1743–1747, 1745
selenium concentrations, 1660–1662, 1668–1670
toxic, 1678–1679, 1686
Marine mammals. *See also* Mammal(s)
radionuclide concentrations, 1743–1744
Maryland, 1521, 1746
Massachusetts, 1660, 1742
Medicine Bow, Wyoming, 1649
Mendocino County, California, 1745
Merced County, California, 1524
Mexicali Valley, 1579
Mexico, 1511, 1577, 1614, 1652
Michigan, 1512, 1517, 1619, 1654
Microwaves, 1709
Milk, radionuclide concentrations, 1726, 1727
Mine drainage water, boron concentrations, 1574
Minnesota, 1511, 1689
Mississippi, 1521, 1575, 1745, 1746
Mississippi River
arsenic concentrations, 1513
fish, 1517, 1518
selenium concentrations, 1653
fish, 1655, 1663
Missouri, 1613
Mollusc(s)
arsenic concentrations, 1520
radiation effects, 1778
radionuclide concentration factors, 1782
radionuclide concentrations, 1736
selenium concentrations, 1657–1658, 1662
toxic, 1676, 1677, 1686
Molybdenum, 1613–1641
in abiotic material, 1617–1619
air, 1619, 1640
drinking water, 1618, 1639, 1640
fossil fuel, 1619
groundwater, 1618
seawater, 1618
sediment, 1618
sewage sludge, 1619
soil, 1618–1619
surface water, 1618
chemical properties, 1615
effects, 1625–1637 (*See also* Molybdenum, toxicity)
field collections, 1617–1625
abiotic material, 1617–1619
aquatic invertebrate, 1622
aquatic plant, 1622
bird, 1623
fish, 1622–1623
mammal, 1623–1625

plant, 1621–1622
terrestrial plant, 1621–1622
interactions, 1615–1616
mining sites, 1614
recommendations, 1637–1640
toxicity, 1613–1614, 1625–1637
aquatic organism, 1626–1630
aquatic plant, 1628
bird, 1630–1631, 1639
fish, 1629–1630
human, 1635, 1640
invertebrate, 1628–1629
mammal, 1631–1637, 1639–1640
terrestrial invertebrate, 1626, 1639
terrestrial plant, 1626, 1638–1639
uses, 1614
world production, 1614
Monaco, 1753, 1765
Mongolia, 1501
Montana, 1616

N

Nagasaki, Japan, 1723, 1733, 1735, 1748
Nebraska, 1580, 1653, 1654
Netherlands
arsenic concentrations
bird, 1525
crustacean, 1521
fish, 1518, 1522
molybdenum toxicity, 1613
radioactive waste disposal, 1722
radionuclide concentrations, 1744
Chernobyl fallout, 1761, 1765
Neuse River, 1731, 1737
Nevada
arsenic concentrations, 1511
boron concentrations, 1576
boron deposits, 1568
molybdenum mining sites, 1614
nuclear weapons testing, 1733, 1774
radionuclide concentrations, 1744, 1745
selenium concentrations, 1655
New Jersey, 1660, 1732
New Mexico
arsenic concentrations, 1526
molybdenum concentrations, 1618
molybdenum mining site, 1614
nuclear weapons testing, 1707, 1733
New York
arsenic concentrations, 1504
fish, 1518, 1519
mammal, 1526
toxic, 1545
radon concentrations, human residences, 1732
selenium concentrations, 1662
bird, 1668
New Zealand, 1516, 1525, 1613, 1679
North America, 1754, 1802, 1803
North Carolina, 1682, 1737, 1745, 1746
North Dakota, 1613
North Sea, 1748
Northwest Territories, 1510
Norway
boron concentrations, 1575
molybdenum concentrations, 1621
molybdenum production, 1614
radionuclide concentrations
Chernobyl fallout, 1753, 1757, 1758, 1765–1766, 1806
selenium concentrations, 1660
integrated studies, 1671
Nuclear power station, 1731
Nuclear reactor, 1719, 1786
fission products, 1722
water-cooled, 1787
Nuclear weapons testing
China, 1707, 1725
France, 1707, 1719
India, 1707
Nevada, 1733, 1774
New Mexico, 1707, 1733
United States, 1707

O

Oak Ridge, Tennessee, 1733, 1746
Ohio River, 1653
Oklahoma, 1526
Ontario, Canada, 1528
Oregon, 1568

P

Pacific Ocean, 1512, 1618, 1722, 1724
Pacific Proving Grounds, 1749–1752
radiation dispersion
algae, 1750
bird, 1750
fish, 1750
mammal, 1751
plankton, 1751
plant, 1750
seawater, 1750
sediment, 1750, 1751
soil, 1751
terrestrial invertebrate, 1750
Pakistan, 1707
Pennsylvania, 1586, 1613, 1614, 1732
Plankton, 1751
radionuclide concentrations, 1730–1731, 1737
Plant(s)
arsenic toxicity, 1530–1531, 1532, 1536, 1540
boron toxicity, 1581–1584, 1586–1587
Chernobyl fallout, 1757
molybdenum concentrations, 1621–1622
Poland, 1753, 1766

Pond water
  arsenic concentrations, 1515
  radionuclide concentrations, 1728–1729
  selenium concentrations, 1657, 1691
Precipitation, radionuclide concentrations, 1727–1728
Protozoan, radiation effects, 1778
Puerto Rico, 1668, 1775
Puget Sound, 1504, 1510, 1659

Q

Quebec, 1759

R

Radiation, 1707–1807
  annual whole-body, human, 1717, 1732, 1805
  anthropogenic, 1718–1723
  beta, 1719, 1731
  cigarette smoke, 1723
  dispersion, 1723–1725, 1749–1769 (*See also* Chernobyl accident; Pacific Proving Grounds)
  effects, 1788–1794 (*See also* Chernobyl accident; Pacific Proving Grounds)
    algae, 1778
    amphibian, 1784–1786
    annelid, 1779
    aquatic organism, 1777–1784
    astronaut, 1801, 1802
    bird, 1786–1788
    body organs, 1798
    carcinogenic, 1796–1797, 1802
    coelenterate, 1778
    crustacean, 1778–1779
    fish, 1779–1781
    integrated studies, 1781
    invertebrate, 1776–1777
    mollusc, 1778
    mutagenic, 1797–1798
    protection against, 1800–1805
    protozoan, 1778
    psychological, 1798–1799
    reptile, 1785, 1786
    survival time, 1795–1796
    teratogenic, 1802
    terrestrial plant, 1774–1776
  gamma, 1719
  human exposure, 1718–1723
  ionizing, 1771–1794
  lethal dose range, 1772
  measurement, 1716
  natural, 1717
  nonionizing, 1769–1771
  physical properties, 1708–1716
  ultraviolet, 1770
Radio waves, 1709
Radioactive decay, 1713
Radioactive waste, 1803–1804
  disposal, 1722
Radionuclides, 1710–1715
  in abiotic material, 1725–1729
    air, 1717, 1802
    drinking water, 1726, 1803
    food, 1717
    freshwater, 1726–1727
    groundwater, 1727, 1803
    lake water, 1727
    milk, 1727
    pond, 1728–1729
    precipitation, 1727–1728
    rock, 1726
    sea water, 1728
    sediment, 1728
    soil, 1717, 1725, 1728, 1805
    water, 1717
  absorption, 1799–1800
  atomic number, 1714–1715
  beta radiation, 1719
  biological transport, 1724
  concentration factors, 1782, 1783
  decay, 1713
  field collections, 1725–1749
    abiotic material, 1725–1729
    algae, 1730, 1735
    aquatic invertebrate, 1735–1738
    aquatic plant, 1735
    bird, 1731–1732, 1741–1743
    crustacean, 1736
    fish, 1729–1730, 1738–1740
    integrated studies, 1747–1749
    macrophyte, 1735
    mammal, 1732–1733, 1743–1747
    marine mammal, 1743–1744
    mollusc, 1736
    plankton, 1730–1731, 1737
    reptile, 1741
    terrestrial mammal, 1744–1747
    terrestrial plant, 1734–1735
  formation, 1713
  gamma radiation, 1719
  groups, 1713
  occupational exposure, 1804–1805
  transuranic, 1723
Rain
  arsenic concentrations, 1512
  boron concentrations, 1574, 1575
Red Lakes, 1738
Redwood Creek, 1657
Reptile(s)
  arsenic concentrations, 1523
  radiation effects, 1785, 1786
  radionuclide concentrations, 1741
  selenium concentrations, 1665
Rhode Island, 1512, 1535
Rhone River, 1784
River Beaulieu, 1506
River water
  arsenic concentrations, 1512

boron concentrations, 1575
selenium concentrations, 1653
Rock
arsenic concentrations, 1512
radionuclide concentrations, 1726
Romania, 1511, 1621, 1753
Runit Island, 1751
Russia, 1515, 1524, 1568, 1690

S

Salt water
arsenic concentrations, 1512–1513
boron concentrations, 1575
molybdenum concentrations, 1618
radionuclide concentrations, 1728
selenium concentrations, 1654
San Diego, California, 1740
San Diego Bay, 1579
San Francisco, California, 1663
San Francisco Bay, 1657, 1666
San Joaquin River
arsenic concentrations, 1517
boron concentrations, 1576
fish, 1578
molybdenum concentrations, 1620
selenium concentrations, 1655
San Joaquin Valley, California
boron concentrations, 1574
fish, 1590
irrigation systems, 1578
selenium concentrations, 1656, 1657
bird, 1649, 1685
fish, 1663
Sargasso Sea, 1535
Scandinavia, 1575, 1806
Searles Lake, 1512
Seattle, Washington, 1512
Seawater. *See* Salt water
Sediment(s)
arsenic concentrations, 1513
boron concentrations, 1575
molybdenum concentrations, 1618
Pacific Proving Ground radiation dispersion, 1750, 1751
radionuclide concentrations, 1728
Selenium, 1649–1692
in abiotic material, 1653–1657
aerosol, 1692
air, 1654, 1690
drinking water, 1653, 1690, 1691
fossil fuel, 1653
groundwater, 1653
lake water, 1654
river water, 1653
seawater, 1654
soil, 1653
deficiency, 1672–1674
protection against, 1687–1688
field collections, 1653–1672
abiotic material, 1653–1655
algae, 1657, 1662
amphibian, 1665
bird, 1659–1660, 1665–1667, 1668
crustacean, 1658
echinoderm, 1658
fish, 1658–1659, 1662–1665
human, 1669
integrated studies, 1654, 1670–1672
invertebrate, 1668
macrophyte, 1657, 1667–1668
mammal, 1660–1662, 1668–1670
mollusc, 1657–1658, 1662
reptile, 1665
lethal effects, 1675–1686 (*See also* Selenium, toxicity)
protective action, 1674
sublethal effects, 1682–1686 (*See also* Selenium, toxicity)
toxicity, 1675–1686
algae, 1675
amphibian, 1676
aquatic organism, 1675–1678, 1682–1685
bird, 1679–1682, 1685–1686
coelenterate, 1676
crustacean, 1676, 1677
fish, 1676–1677, 1678
insect, 1676
mammal, 1678–1679, 1686
mollusc, 1676, 1677
protection against, 1688–1690
terrestrial invertebrate, 1685
Sellafield, United Kingdom, 1722, 1723, 1733, 1752
Sellafield reprocessing plant, 1722, 1723
Serbia, 1747
Serpent River, 1733
Sewage sludge, molybdenum concentrations, 1619
Sewage water, boron concentrations, 1575
Simav River, 1573
Slovakia, 1525
Snow, arsenic concentrations, 1513
Soil
arsenic concentrations, 1513
boron concentrations, 1575
Chernobyl fallout, 1757
molybdenum concentrations, 1618–1619
radionuclide concentrations, 1717, 1725, 1728, 1805
selenium concentrations, 1653
South America, 1501, 1754
South Carolina
arsenic concentrations, 1515
amphibian, 1523
bird, 1524
nuclear reactor, 1786
radionuclide concentrations
aquatic plant, 1735
fish, 1739
integrated studies, 1748
mammal, 1745, 1746
selenium concentrations, 1665
South Dakota, 1613, 1649, 1667, 1669
Soviet Union, former, 1640, 1653, 1729, 1754

Spain, 1525, 1759, 1766
Stewart Lake, 1667
Sudbury, Ontario, 1652, 1654, 1692
Surface water
  boron concentrations, 1575–1576
  molybdenum concentrations, 1618
Sweden
  arsenic production, 1502
    fish, 1517
  boron concentrations, 1575
  molybdenum concentrations, 1621
  radionuclide concentrations, 1758, 1766–1767
    dietary intake, 1759
    limits, 1802, 1803
  selenium concentrations, 1654
Switzerland, 1618, 1753
Syria, 1757, 1768

T

Tacoma, Washington, 1502, 1504, 1511, 1513
Taiwan
  arsenic concentrations, 1515
    integrated studies, 1528
    mammal, 1526
  arsenic toxicity, 1501, 1529
Tamar, England, 1510
Tejo, Portugal, 1510
Tennessee
  arsenic concentrations, 1526
  radionuclide concentrations, 1728
    mammal, 1746, 1761
    terrestrial plant, 1735
Tennessee River, 1735
Teratogenicity, 1530, 1802
Terrestrial invertebrate(s)
  arsenic toxicity, 1531–1532, 1533
  boron toxicity, 1581–1584, 1584–1585
  molybdenum toxicity, 1626, 1639
  selenium toxicity, 1685
Terrestrial mammals. *See also* Mammal(s)
  radionuclide concentrations, 1744–1747
Terrestrial plant(s)
  arsenic concentrations, 1516–1517
  boron concentrations, 1577
  boron toxicity, 1581–1584
  molybdenum concentrations, 1621–1622
  molybdenum toxicity, 1626, 1638–1639
  radiation effects, 1774–1776
  radionuclide concentrations, 1734–1735
Texas
  arsenic concentrations
    air, 1511
    bird, 1524
    fish, 1518
  molybdenum toxicity, 1613
  selenium concentrations, 1655
    bird, 1660, 1666, 1668
    integrated studies, 1671
Thailand, 1501
Thermal springs, boron concentrations, 1576
Three Mile Island, Pennsylvania, 1752
Three Mile Island accident, 1752
Tijuana Slough National Wildlife Refuge, 1579
Tomales Bay, 1657
Turkey
  boron concentrations, 1573
    freshwater, 1576
  boron deposits, 1568
  boron mine drainage waters, 1574
  radionuclide concentrations, Chernobyl fallout, 1753

U

U. S. S. R., former, 1511, 1614, 1640, 1753
Ultraviolet waves, 1709
United Kingdom
  arsenic concentrations, 1506
    mammal, 1526
    salt water, 1513
  boron concentrations, 1575
  molybdenum concentrations, limits, 1640
  radioactive waste disposal, 1722
  radionuclide concentrations, 1768–1769
    bird, 1742
    Chernobyl fallout, 1761
    dietary limit, 1760
    nuclear fuel reprocessing plant, 1800
    Sellafield facility, 1725
    soil and vegetation, 1757
United States
  annual whole-body radiation doses to humans, 1719
  arsenic concentrations, 1504
    bird, 1525
    fish, 1518, 1519
    limits, 1554
    mollusc, 1520
    soil, 1513
  arsenic production, 1502
  boron concentrations
    daily intake, 1576–1577
    groundwater, 1574
    rain, 1575
    surface water, 1575
  boron production, 1568
  molybdenum concentrations
    air, 1619
    daily intake, 1633
    drinking water, 1618
    groundwater, 1618
    limits, 1640
    sediment, 1618
    soil, 1618
  molybdenum production, 1614
  nuclear weapons testing, 1707, 1733, 1774
  radionuclide concentrations
    abiotic material, 1725, 1726, 1727
    Chernobyl fallout, 1753

drinking water, 1726
groundwater, 1726
limits, air, 1802
Western, 1576
mammal, 1745
milk, 1726
precipitation, 1726
selenium concentrations, 1653
air, 1690
fish, 1655
human, 1669
lake waters, 1654
limits, 1690
toxic, 1652
uranium production, 1721
Utah, 1614, 1655

V

Vancouver Island, British Columbia, 1524
Victoria, Australia, 1618
Virginia, 1745, 1746
Volta Wildlife Area, 1666

W

Wailoa River, 1510
Wales, 1501, 1760
Washington, 1649
Water, radionuclide concentrations, 1717
Well water, boron concentrations, 1576
West Virginia, 1746
Wisconsin
arsenic concentrations, 1512
fish, 1517, 1518, 1519
radionuclide concentrations, 1775
Wyoming, 1616, 1664, 1669, 1733

X

X-rays, 1709

Y

Yugoslavia, 1520, 1753, 1762
Yukon Territories, 1761